D0117416

INTRODUCTORY
algebra
A REAL-WORLD APPROACH

THIRD EDITION

Ignacio Bello
Hillsborough Community College
Tampa, Florida

McGraw-Hill
Higher Education

Boston   Burr Ridge, IL   Dubuque, IA   New York   San Francisco   St. Louis
Bangkok   Bogotá   Caracas   Kuala Lumpur   Lisbon   London   Madrid   Mexico City
Milan   Montreal   New Delhi   Santiago   Seoul   Singapore   Sydney   Taipei   Toronto

McGraw-Hill
Higher Education

INTRODUCTORY ALGEBRA: A REAL-WORLD APPROACH, THIRD EDITION

Published by McGraw-Hill, a business unit of The McGraw-Hill Companies, Inc., 1221 Avenue of the Americas, New York, NY 10020. Copyright © 2009 by The McGraw-Hill Companies, Inc. All rights reserved. Previous edition © 2006. No part of this publication may be reproduced or distributed in any form or by any means, or stored in a database or retrieval system, without the prior written consent of The McGraw-Hill Companies, Inc., including, but not limited to, in any network or other electronic storage or transmission, or broadcast for distance learning.

Some ancillaries, including electronic and print components, may not be available to customers outside the United States.

This book is printed on acid-free paper.

1 2 3 4 5 6 7 8 9 0 QPV/QPV 0 9 8

ISBN 978–0–07–353343–8
MHID 0–07–353343–2

ISBN 978–0–07–319969–6 (Annotated Instructor's Edition)
MHID 0–07–319969–9

Editorial Director: *Stewart K. Mattson*
Senior Sponsoring Editor: *David Millage*
Developmental Editor: *Michelle Driscoll*
Marketing Manager: *Victoria Anderson*
Senior Project Manager: *Vicki Krug*
Senior Production Supervisor: *Sherry L. Kane*
Senior Media Project Manager: *Sandra M. Schnee*
Senior Designer: *David W. Hash*
Cover/Interior Designer: *Asylum Studios*
(USE) Cover Image: *©Royalty-Free/CORBIS*
Senior Photo Research Coordinator: *Lori Hancock*
Photo Research: *Connie Mueller*
Supplement Coordinator: *Melissa M. Leick*
Compositor: *ICC Macmillan Inc.*
Typeface: *10/12 Times Roman*
Printer: *Quebecor World Versailles, KY*

The credits section for this book begins on page C-1 and is considered an extension of the copyright page.

www.mhhe.com

Ignacio Bello

attended the University of South Florida (USF), where he earned a B.A. and M.A. in Mathematics. He began teaching at USF in 1967, and in 1971 he became a member of the Faculty at Hillsborough Community College (HCC) and Coordinator of the Math and Sciences Department. Professor Bello instituted the USF/HCC remedial program, a program that started with 17 students taking Intermediate Algebra and grew to more than 800 students with courses covering Developmental English, Reading, and Mathematics. Aside from the present series of books (*Basic College Mathematics, Introductory Algebra,* and *Intermediate Algebra*), Professor Bello is the author of more than 40 textbooks including *Topics in Contemporary Mathematics*, *College Algebra, Algebra and Trigonometry,* and *Business Mathematics*. Many of these textbooks have been translated into Spanish. With Professor Fran Hopf, Bello started the Algebra Hotline, the only live, college-level television help program in Florida. Professor Bello is featured in three television programs on the award-winning Education Channel. He has helped create and develop the USF Mathematics Department Website (http://mathcenter.usf.edu), which serves as support for the Finite Math, College Algebra, Intermediate Algebra, and Introductory Algebra, and CLAST classes at USF. You can see Professor Bello's presentations and streaming videos at this website, as well as at http://www.ibello.com. Professor Bello is a member of the MAA and AMATYC and has given many presentations regarding the teaching of mathematics at the local, state, and national levels.

▶ Contents

Chapter
two

2 ▶ Equations, Problem Solving, and Inequalities

Chapter
three

3 ▶ Graphs of Linear Equations, Inequalities, and Applications

▶ Contents

⊙ Contents

⊙ Contents

⯈ From the Author

The Inspiration for My Teaching

I was born in Havana, Cuba, and I encountered some of the same challenges in mathematics that many of my current students face, all while attempting to overcome a language barrier. In high school, I failed my freshman math course, which at the time was a complex language for me. However, perseverance being one of my traits, I scored 100% on the final exam the second time around. After working in various jobs (roofer, sheetrock installer, and dock worker), I finished high school and received a college academic scholarship. I enrolled in calculus and made a "C". Never one to be discouraged, I became a math major and learned to excel in the courses that had previously frustrated me. While a graduate student at the University of South Florida (USF), I taught at a technical school, Tampa Technical Institute, a decision that contributed to my resolve to teach math and make it come alive for my students the way brilliant instructors such as Jack Britton, Donald Rose, and Frank Cleaver had done for me. My math instructors instilled in me the motivation to be successful. I have learned a great deal about the way in which students learn and how the proper guidance through the developmental mathematics curriculum leads to student success. I believe I have accomplished a strong level of guidance in my textbook series to further explain the language of mathematics carefully to students to help them to reach success as well.

A Lively Approach to Reach Today's Students

Teaching math at the University of South Florida was a great new career for me, but I was disappointed by the materials I had to use. A rather imposing, mathematically correct but boring book was in vogue. Students hated it, professors hated it, and administrators hated it. I took the challenge to write a better book, a book that was not only mathematically correct, but **student-oriented** with **interesting applications**—many suggested by the students themselves—and even, dare we say, entertaining! That book's approach and philosophy proved an instant success and was a precursor to my current series.

Students fondly called my class "The Bello Comedy Hour," but they worked hard, and they performed well. Because my students always ranked among the highest on the common final exam at USF, I knew I had found a way to motivate them through **common-sense language** and humorous, **realistic math applications.** I also wanted to show students they could overcome the same obstacles I had in math and become successful, too. If math has been a subject that some of your students have never felt comfortable with, then they're not alone! I wrote this book with the **math-anxious** student in mind, so they'll find my tone is jovial, my explanations are patient, and instead of making math seem mysterious, I make it down-to-earth and easily digestible. For example, after I've explained the different methods for simplifying fractions, I speak directly to readers: "Which way should you simplify fractions? The way you understand!" Once students realize that math is within their grasp and not a foreign language, they'll be surprised at how much more confident they feel.

A Real-World Approach: Applications, Student Motivation, and Problem Solving

What is a "real-world approach"? I found that most textbooks put forth "real-world" applications that meant nothing to the real world of my students. How many of my students would really need to calculate the speed of a bullet (unless they are in its way) or cared to know when two trains traveling in different directions would pass by each other (disaster will certainly occur if they are on the same track)? For my students, both traditional and nontraditional, the real world consists of questions such as, "How do I find the best cell phone plan?" and "How will I pay my tuition and fees if they increase by x%?" That is why I introduce mathematical concepts through everyday applications with **real data** and give homework using similar, well-grounded situations (see the Getting Started application that introduces every section's topic and the word problems in every exercise section). Putting math in a real-world context has helped me overcome one of the problems we all face as math educators: **student motivation.** Seeing math in the real world makes students perk up in a math class in a way I have never seen before, and realism has proven to be the best motivator I've ever used. In addition, the real-world approach has enabled me to enhance students' **problem-solving skills** because they are far more likely to tackle a real-world problem that matters to them than one that seems contrived.

Diverse Students and Multiple Learning Styles

We know we live in a pluralistic society, so how do you write one textbook for everyone? The answer is to build a flexible set of teaching tools that instructors and students can adapt to their own situations. Are any of your students members of a **cultural minority?** So am I! Did they learn **English as a second language?** So did I! You'll find my book speaks directly to them in a way that no other book ever has, and fuzzy explanations in other books will be clear and comprehensible in mine.

Do your students all have the same **learning style?** Of course not! That's why I wrote a book that will help students learn mathematics regardless of their personal learning style. **Visual learners** will benefit from the text's clean page layout, careful use of color highlighting, "*Web Its,*" and the video lectures on the text's website. **Auditory learners** will profit from the audio *e-Professor lectures* on the text's website, and both **auditory** and **social learners** will be aided by the *Collaborative Learning* projects. **Applied** and **pragmatic learners** will find a bonanza of features geared to help them: *Pretests* can be found in MathZone providing practice problems by every example, Practice Tests are at the end of each chapter, and *Mastery Tests,* at the end of every exercise section, to name just a few. **Spatial learners** will find the chapter *Summary* is designed especially for them, while **creative learners** will find the *Research Questions* to be a natural fit. Finally, **conceptual learners** will feel at home with features like "*The Human Side of Algebra*" and the "*Write On*" exercises. Every student who is accustomed to opening a math book and feeling like they've run into a brick wall will find in my books that a number of doors are standing open and inviting them inside.

Listening to Student and Instructor Concerns

McGraw-Hill has given me a wonderful resource for making my textbook more responsive to the immediate concerns of students and faculty. In addition to sending my manuscript out for review by instructors at many different colleges, several times a year McGraw-Hill holds symposia and focus groups with math instructors where the emphasis is *not* on selling products but instead on the **publisher listening** to the needs of faculty and their students. These encounters have provided me with a wealth of ideas on how to improve my chapter organization, make the page layout of my books more readable, and fine-tune exercises in every chapter so that students and faculty will feel comfortable using my book because it incorporates their specific suggestions and anticipates their needs.

R-I-S-E to Success in Math

Why are some students more successful in math than others? Often it is because they know how to manage their time and have a plan for action. Students can use models similar to the following tables to make a weekly schedule of your time (classes, study, work, personal, etc.) and a semester calendar indicating major course events like tests, papers, and so on. Then, try to do as many of the suggestions on the **"R-I-S-E"** list as possible. (Larger, printable versions of these tables can be found in MathZone at www.mhhe.com/bello.)

Weekly Time Schedule							
Time	**S**	**M**	**T**	**W**	**R**	**F**	**S**
8:00							
9:00							
10:00							
11:00							
12:00							
1:00							
2:00							
3:00							
4:00							
5:00							
6:00							
7:00							
8:00							
9:00							
10:00							
11:00							

Semester Calendar					
Wk	**M**	**T**	**W**	**R**	**F**
1					
2					
3					
4					
5					
6					
7					
8					
9					
10					
11					
12					
13					
14					
15					
16					

R—Read and/or view the material before and after each class. This includes the textbook, the videos that come with the book, and any special material given to you by your instructor.

I—Interact and/or practice using the CD that comes with the book or the Web exercises suggested in the sections, or seeking tutoring from your school.

S—Study and/or discuss your homework and class notes with a study partner/group, with your instructor, or on a discussion board if available.

E—Evaluate your progress by checking the odd homework questions with the answer key in the back of the book, using the mastery questions in each section of the book as a selftest, and using the Chapter Reviews and Chapter Practice Tests as practice before taking the actual test.

As the items on this list become part of your regular study habits, you will be ready to **"R–I–S–E"** to success in math.

⊳ Preface

⊳ Improvements in the Third Edition

Based on the valuable feedback of numerous reviewers and users over the years, the following improvements were made to the third edition of *Introductory Algebra*.

Organizational Changes

- Chapter 3 has been revised to include:
  Section 3.4: The Slope of a Line: Parallel and Perpendicular lines
  Section 3.5: Graphing Lines Using Points and Slopes
  Section 3.6: Applications of Equation of Lines
  Section 3.7: Graphing Inequalities in Two Variables
- Chapter 6 has been revised to include Direct and Inverse Variation

Pedagogical Changes

- *Real-World Applications*—many examples, applications, and real-data problems have been added or updated to keep the book's content current.

- *Web Its*—now found in the margin of the Exercises and on MathZone (www.mhhe.com/bello) to encourage students to visit math sites while they're Web surfing and discover the many informative and creative sites that are dedicated to stimulating better education in math.

- *Calculator Corners*—found before the exercise sets, these have been updated with recent information and keystrokes relevant to currently popular calculators.

- *Concept Checkers*—have been added to the end-of-section exercises to help students reinforce key terms and equations.

- *Pretests*—can be found in MathZone (www.mhhe.com/bello) providing practice problems for every example. These Pretest results can be compared to the Practice Tests at the end of the chapter to evaluate and analyze student success.

- *The RSTUV approach* to problem solving has been expanded and used throughout this edition as a response to positive comments from both students and users of the previous edition.

- *Translate It*—boxes appear periodically before word problem exercises to help students translate phrases into equations, reinforcing the RSTUV method.

- *Skill Checker*—now appears at the end of the exercises sets, making sure the students have the necessary skills for the next section.

- *Practice Tests (Diagnostics)*—at the end of every chapter give students immediate feedback and guidance on which section, examples, and pages to review.

▶ Acknowledgments

I would like to thank the following people associated with the third edition:
David Dietz, who provided the necessary incentives and encouragement for creating this series with the cooperation of Bill Barter; Christien Shangraw, my first developmental editor who worked many hours getting reviewers and gathering responses into concise and usable reports; Randy Welch, who continued and expanded the Christien tradition into a well-honed editing engine with many features, including humor, organization, and very hard work; Liz Haefele, my former editor and publisher, who was encouraging, always on the lookout for new markets; Lori Hancock and her many helpers (LouAnn, Emily, David, and Connie Mueller), who always get the picture; Dr. Tom Porter, of Photos at Your Place, who improved on the pictures I provided; Vicki Krug, one of the most exacting persons at McGraw-Hill, who will always give you the time of day and then solve the problem; Hal Whipple, for his help in preparing the answers manuscript; Cindy Trimble, for the accuracy of the text; Jeff Huettman, one of the best 100 producers in the United States, who learned Spanish in anticipation of this project; Marie Bova, for her detective work in tracking down permission rights; and to Professor Nancy Mills, for her expert advice on how my books address multiple learning styles. David Millage, senior sponsoring editor; Michelle Driscoll and Lisa Collette, developmental editors; Torie Anderson, marketing manager; and especially Pat Steele, our very able copy editor. Finally, thanks to our attack secretary, Beverly DeVine, who still managed to send all materials back to the publisher on time. To everyone, my many thanks.

I would also like to extend my gratitude to the following reviewers of the Bello series for their many helpful suggestions and insights. They helped me write better textbooks:

Tony Akhlaghi, *Bellevue Community College*

Theresa Allen, *University of Idaho*

John Anderson, *San Jacinto College–South Campus*

Farouck Assaad, *Clark Atlanta University*

Keith A. Austin, *Devry University–Arlington*

Sohrab Bakhtyari, *St. Petersburg College–Clearwater*

Russell E. Banks, *Guilford Technical Community College*

Fatemah Bicksler, *Delgado Community College*

Ann Brackebusch, *Olympic College*

Donald Bridgewater, *Broward Community College*

Gail G. Burkett, *Palm Beach Community College*

Linda Burton, *Miami Dade Community College*

G. Robert Carlson, *Victor Valley College*

Judy Carlson, *Indiana University–Purdue University Indianapolis*

Gail O. Carter, *St. Petersburg College*

Randall Crist, *Creighton University*

Mark Czerniak, *Moraine Valley Community College*

Joseph De Blassio, *Community College of Allegheny County*

Parsla Dineen, *University of Nebraska–Omaha*

Sue Duff, *Guilford Technical Community College*

▶ Acknowledgments

Lynda Fish, *St. Louis Community College–Forest Park*

Donna Foster, *Piedmont Technical College*

Dot French, *Community College of Philadelphia*

Deborah D. Fries, *Wor-Wic Community College*

Jeanne H. Gagliano, *Delgado Community College*

Mary Ellen Gallegos, *Santa Fe Community College*

Debbie Garrison, *Valencia Community College*

Paul Gogniat, *Community College of Allegheny County*

Donald K. Gooden, *Northern Virginia Community College–Woodbridge*

Edna Greenwood, *Tarrant County College, Northwest Campus*

Ken Harrelson, *Oklahoma City Community College*

Joseph Lloyd Harris, *Gulf Coast Community College*

Marelise Hartley, *Madison Area Technical College*

Tony Hartman, *Texarkana College*

J. D. Herdlick, *St. Louis Community College at Meramec*

Susan Hitchcock, *Palm Beach Community College*

Patricia Carey Horacek, *Pensacola Junior College*

Peter Intarapanich, *Southern Connecticut State University*

Judy Ann Jones, *Madison Area Technical College*

Cheryl Kane, *University of Nebraska–Lincoln*

Linda Kass, *Bergen Community College*

Joe Kemble, *Lamar University*

Joanne Kendall, *Blinn College–Brenham*

Bernadette Kocyba, *J. S. Reynolds Community College*

Jeff A. Koleno, *Lorain County Community College*

Shawn Krest, *Genesee Community College*

Theodore Lai, *Hudson County Community College*

Marcia Lambert, *Pitt Community College*

Marie Agnes Langston, *Palm Beach Community College*

Betty Larson, *South Dakota State University*

Kathryn Lavelle, *Westchester Community College*

Angela Lawrenz, *Blinn College–Bryan*

Richard Leedy, *Polk Community College*

Judith L. Maggiore, *Holyoke Community College*

Timothy Magnavita, *Bucks Community College*

Tsun-Zee Mai, *University of Alabama*

Harold Mardones, *Community College of Denver*

Lois Martin, *Massasoit Community College*

Gary McCracken, *Shelton State Community College*

Sandra Luna McCune, *Stephen F. Austin State University*

Tania McNutt, *Community College of Aurora*

Barbara Miller, *Lexington Community College*

Danielle Morgan, *San Jacinto College–South Campus*

Benjamin Nicholson, *Montgomery College*

Janet Orwick, *Dutchess Community College*

Nora Othman, *Community College of Allegheny County*

Joanne Peeples, *El Paso Community College*

Faith Peters, *Miami Dade College–Wolfson*

Jane Pinnow, *University of Wisconsin–Parkside*

Janice F. Rech, *University of Nebraska–Omaha*

Libbie Reeves, *Mitchell Community College*

Tian Ren, *Queensborough Community College*

Sylvester Roebuck, *Olive-Harvey College*

Karen Roothaan, *Harold Washington College*

Don Rose, *College of the Sequoias*

Pascal Roubides, *Miami Dade College–Wolfson*

Juan Saavedra, *Albuquerque Technical Vocational Institute*

Judith Salmon, *Fitchburg State College*

Mansour Samimi, *Winston-Salem State University*

Susan Santolucito, *Delgado Community College*

Jorge Sarmiento, *County College of Morris*

Ellen Sawyer, *College of DuPage*

Kenneth Takvorian, *Mount Wachusett Community College*

Sara Taylor, *Dutchess Community College*

Sharon Testone, *Onondaga Community College*

Tommy Thompson, *Cedar Valley College*

Pricilla Ann Wake, *San Jacinto College*

Robert E. White, *Allan Hancock College*

▶ Features and Supplements

Motivation for a Diverse Student Audience

A number of features exist in every chapter to motivate students' interest in the topic and thereby increase their performance in the course:

❯ *The Human Side of Algebra*

To personalize the subject of mathematics, the origins of numerical notation, concepts, and methods are introduced through the lives of real people solving ordinary problems.

The Human Side of Algebra

In the "Golden Age" of Greek mathematics, 300–200 B.C., three mathematicians "stood head and shoulders above all the others of the time." One of them was Apollonius of Perga in Southern Asia Minor. Around 262–190 B.C., Apollonius developed a method of "tetrads" for expressing large numbers, using an equivalent of exponents of the single myriad (10,000). It was not until about the year 250 that the *Arithmetica* of Diophantus advanced the idea of exponents by denoting the square of the unknown as Δ^Y, the first two letters of the word *dunamis,* meaning "power." Similarly, K^Y represented the cube of the unknown quantity. It was not until 1360 that Nicole Oresme of France gave rules equivalent to the product and power rules of exponents that we study in this chapter. Finally, around 1484, a manuscript written by the French mathematician Nicholas Chuquet contained the *denominacion* (or power) of the unknown quantity, so that our algebraic expressions $3x$, $7x^2$, and $10x^3$ were written as $.3.$ and $.7.^2$ and $.10.^3$. What about zero and negative exponents? $8x^0$ became $.8.^0$ and $8x^{-2}$ was written as $.8.^{2.m}$, meaning ".8. *seconds moins,*" or 8 to the negative two power. Some things do change!

❯ *Getting Started*

Each topic is introduced in a setting familiar to students' daily lives, making the subject personally relevant and more easily understood.

▶ Getting Started
Don't Forget the Tip!

Jasmine is a server at CDB restaurant. Aside from her tips, she gets \$2.88/hour. In 1 hour, she earns \$2.88; in 2 hr, she earns \$5.76; in 3 hr, she earns \$8.64, and so on. We can form the set of ordered pairs (1, 2.88), (2, 5.76), (3, 8.64) using the number of hours she works as the first coordinate and the amount she earns as the second coordinate. Note that the ratio of second coordinates to first coordinates is the same number:

$$\frac{2.88}{1} = 2.88, \quad \frac{5.76}{2} = 2.88, \quad \frac{8.64}{3} = 2.88,$$

and so on.

❯ Web It

Appearing in the margin of the section exercises, this URL refers students to the abundance of resources available on the Web that can show them fun, alternative explanations and demonstrations of important topics.

❯ Exercises **7.4**

MathZone

Boost your grad
> Practice
  Problems
> NetTutor

Web IT *go to* **mhhe.com/bello** *for more lessons*

⟨**A**⟩ Solving Coin and Money Problems
⟨**B**⟩ Solving General Problems

In Problems 1–6, solve the money problems.

1. Mida has $2.25 in nickels and dimes. She has four times as many dimes as nickels. How many dimes and how many nickels does she have?

2. Dora has $5.50 in nickels and quarters. Sh quarters as she has nickels. How many o have?

3. Mongo has 20 coins consisting of nickels and dimes. If the nickels were dimes and the dimes were nickels, he would have 50¢ more than he now has. How many nickels and how many dimes does he have?

4. Desi has 10 coins consisting of pennies a enough, if the nickels were pennies and the she would have the same amount of money many pennies and nickels does she have?

5. Don had $26 in his pocket. If he had only $1 bills and $5 bills, and he had a total of 10 bills, how many of each of the bills did he have?

6. A person went to the bank to deposit $30 $10 and $20 bills, 25 bills in all. How n person have?

In Problems 7–14, find the solution.

7. The sum of two numbers is 102. Their difference is 16. What are the numbers?

8. The difference between two numbers is What are the numbers?

❯ Write On

Writing exercises give students the opportunity to express mathematical concepts and procedures in their own words, thereby internalizing what they have learned.

❯ ❯ ❯ **Write On**

71. In the expression "$\frac{1}{2}$ of x," what operation does the word *of* signify?

72. Most people believe that the word *and* always means addition.
　a. In the expression "the sum of x and y," does "and" signify the operation of addition? Explain.
　b. In the expression "the product of 2 and three more than a number," does "and" signify the operation of addition? Explain.

73. Explain the difference between "x divided by y" and "x divided into y."

74. Explain the difference between "a less than b" and "a less b."

❯ Collaborative Learning

Concluding the chapter are exercises for collaborative learning that promote teamwork by students on interesting and enjoyable exploration projects.

❯ **Collaborative Learning**

How fast can you go?

How fast can you obtain information to solve a problem? Form three groups: library, the Web, and bookstore (where you can look at books, papers, and so on for free). Each group is going to research car prices. Select a car model that has been on the market for at least 5 years. Each of the groups should find:

1. The new car value and the value of a 3-year-old car of the same model
2. The estimated depreciation rate for the car
3. The estimated value of the car in 3 years
4. A graph comparing age and value of the car for the next 5 years
5. An equation of the form $C = P(1 - r)^n$ or $C = rn + b$, where n is the number of years after purchase and r is the depreciation rate

Which group finished first? Share the procedure used to obtain your information so the most efficient research method can be established.

⊙ Guided Tour

❯ Research Questions

Research questions provide students with additional opportunities to explore interesting areas of math, where they may find the questions can lead to surprising results.

Research Questions

1. In the *Human Side of Algebra* at the beginning of this chapter, we mentioned the Hindu numeration system. The Egyptians and Babylonians also developed numeration systems. Write a report about each of these numeration systems, detailing the symbols used for the digits 1–9, the base used, and the manner in which fractions were written.

2. Write a report on the life and works of Mohammed al-Khowarizmi, with special emphasis on the books he wrote.

3. We have now studied the four fundamental operations. But do you know where the symbols used to indicate these operations originated?

   a. Write a report about Johann Widmann's *Mercantile Arithmetic* (1489), indicating which symbols of operation were found in the book for the first time and the manner in which they were used.

   b. Introduced in 1557, the original equals sign used longer lines to indicate equality. Why were the two lines used to denote equality, what was the name of the person who introduced the symbol, and in what book did the

Abundant Practice and Problem Solving

Bello offers students many opportunities and different skill paths for developing their problem-solving skills.

❯ Pretest

An optional Pretest can be found in MathZone at www. mhhe.com/bello and is especially helpful for students taking the course as a review who may remember some concepts but not others. The answer grid is also found online and gives students the page number, section, and example to study in case they missed a question. The results of the Pretest can be compared with those of the Practice Test at the end of the chapter to evaluate progress and student success.

❯ Pretest Chapter 1

1. Write in symbols:
   a. The sum of m and n
   b. $7m$ plus $3n$ minus 4
   c. 4 times m
   d. $\frac{1}{9}$ of m

2. Write in symbols:
   a. The quotient of m and 3
   b. The quotient of 3 and n
   c. The sum of m and n, divided by the difference of m and n

3. For $m = 6$ and $n = 3$, evaluate:
   a. $m + n$
   b. $m - n$
   c. $2m - 3n$
   d. $\frac{2m + n}{n}$

4. Find the additive inverse (opposite) of:
   a. -8
   b. $\frac{2}{5}$
   c. $0.666\ldots$

5. Find the absolute value:
   a. $\left|-\frac{1}{3}\right|$
   b. $|12|$
   c. $-|0.82|$

6. Consider the set $\{-9, \frac{5}{3}, \sqrt{2}, 0, 8.4, 0.333\ldots, -3\frac{1}{2}, 8, 0.123\ldots\}$. List the numbers in the set that are:
   a. Natural numbers
   b. Whole numbers
   c. Integers
   d. Rational numbers

❯ Answers to Pretest Chapter 1

Answer			If You Missed	Review		
			Question	Section	Examples	Page
1. a. $m + n$	**b.** $7m + 3n - 4$		1	1.1	1, 2	37
c. $4m$	**d.** $\frac{1}{9}m$					
2. a. $\frac{m}{3}$	**b.** $\frac{3}{n}$	**c.** $\frac{m + n}{m - n}$	2	1.1	3	38
3. a. 9	**b.** 3		3	1.1	4	38
c. 3	**d.** 5					
4. a. 8	**b.** $-\frac{2}{5}$	**c.** $-0.666\ldots$	4	1.2	1, 2	43–44
5. a. $\frac{1}{3}$	**b.** 12	**c.** -0.82	5	1.2	3, 4	45
6. a. 8	**b.** 0, 8	**c.** $-9, 0, 8$	6	1.2	5	46–47

⟩ *Paired Examples/ Problems*

Examples are placed adjacent to similar problems intended for students to obtain immediate reinforcement of the skill they have just observed. These are especially effective for students who learn by doing and who benefit from frequent practice of important methods. Answers to the problems appear at the bottom of the page.

EXAMPLE 4 Evaluating algebraic expressions

Evaluate the given expressions by substituting 10 for x and 5 for y.

a. $x + y$ **b.** $x - y$ **c.** $4y$ **d.** $\frac{x}{y}$ **e.** $3x - 2y$

SOLUTION

a. Substitute 10 for x and 5 for y in $x + y$.
We obtain: $x + y = 10 + 5 = 15$.
The number 15 is called the **value** of $x + y$.

b. $x - y = 10 - 5 = 5$ **c.** $4y = 4(5) = 20$

d. $\frac{x}{y} = \frac{10}{5} = 2$ **e.** $3x - 2y = 3(10) - 2(5)$
$= 30 - 10$
$= 20$

PROBLEM 4

Evaluate the expressions by substituting 22 for a and 3 for b.

a. $a + b$ **b.** $2a - b$

c. $5b$ **d.** $\frac{2a}{b}$

e. $2a - 3b$

⟩ *RSTUV Method*

The easy-to-remember **"RSTUV"** method gives students a reliable and helpful tool in demystifying word problems so that they can more readily translate them into equations they can recognize and solve.

- **R**ead the problem and decide what is being asked.
- **S**elect a letter or □ to represent this unknown.
- **T**ranslate the problem into an equation.
- **U**se the rules you have studied to solve the resulting equation.
- **V**erify the answer.

RSTUV Method for Solving Word Problems
1. **R**ead the problem carefully and decide what is asked for (the unknown).
2. **S**elect a variable to represent this unknown.
3. **T**hink of a plan to help you write an equation.
4. **U**se algebra to solve the resulting equation.
5. **V**erify the answer.

TRANSLATE THIS

1. The history of the formulas for calculating ideal body weight W began in **1871** when Dr. P. P. Broca (a French surgeon) created this formula known as Broca's index. The ideal weight W (in pounds) for a woman h inches tall is 100 pounds for the first 5 feet and 5 pounds for each additional inch over 60.

2. The ideal weight W (in pounds) for men h inches tall is 110 pounds for the first 5 feet and 5 pounds for each additional inch over 60.

3. In 1974, Dr. B. J. Devine suggested a formula for the weight W in kilograms (kg) of men h inches tall: 50 plus 2.3 kilograms per inch over 5 feet (60 inches).

4. For women h inches tall, the formula for W is 45.5 plus 2.3 kilograms per inch over 5 feet. By the way, a kilogram (kg) is about 2.2 pounds.

5. In 1983, Dr. J. D. Robinson published a modification of the formula. For men h inches tall, the weight W should be 52 kilograms and 1.9 kilograms for each inch over 60.

In Problems 1–10 **TRANSLATE** the sentence and match the correct translation with one of the equations A–O.

A. $W = 50 + 2.3h - 60$
B. $W = 49 + 1.7(h - 60)$
C. $LBW = B + M + O$
D. $W = 100h + 5(h - 60)$
E. $W = 110 + 5(h - 60)$
F. $LBW = 0.32810C + 0.33929W - 29.5336$
G. $LBW = 0.32810W + 0.33929C - 29.5336$
H. $W = 100 + 5(h - 60)$
I. $LBW = 0.29569W + 0.41813C - 43.2933$
J. $W = 50h + 2.3(h - 60)$
K. $W = 110h + 5(h - 60)$
L. $W = 56.2 + 1.41(h - 60)$
M. $W = 50 + 2.3(h - 60)$
N. $W = 52 + 1.9(h - 60)$
O. $W = 45.5 + 2.3(h - 60)$

6. The Robinson formula W for women h inches tall is 49 kilograms and 1.7 kilograms for each inch over 5 feet.

7. A minor modification of Robinson formula is Miller's formula which defines the weight W for a man h inches tall as 56.2 kilograms added to 1.41 kilograms for each inch over 5 feet

8. There are formulas that suggest your lean body weight (**LBW**) is the sum of the weight of your bones (**B**), muscles (**M**), and organs (**O**). Basically the sum of everything other than fat in your body.

9. For men over the age of 16, C centimeters tall and with weight W kilograms, the lean body weight (**LBW**) is the product of W and 0.32810, plus the product of C and 0.33929, minus 29.5336.

10. For women over the age of 30, C centimeters tall and weighting W kilograms the lean body weight (**LBW**) is the product of 0.29569 and W, plus the product of 0.41813 and C, minus 43.2933

⟩ *Translate This*

These boxes appear periodically before word-problem exercises to help students translate phrases into equations, reinforcing the RSTUV method.

⟩ *Exercises*

A wealth of exercises for each section are organized according to the learning objectives for that section, giving students a reference to study if they need extra help.

MathZone

Boost your grade at mathzone.com!
› Practice Problems
› NetTutor
› Self-Tests
› e-Professors
› Videos

⟩ Exercises 1.6

⟨ **A** ⟩ **Identifying the Associative, Commutative, and Distributive Properties**   In Problems 1–10, name the property illustrated in each statement.

1. $9 + 8 = 8 + 9$ **2.** $b \cdot a = a \cdot b$ **3.** $4 \cdot 3 = 3 \cdot 4$

4. $(a + 4) + b = a + (4 + b)$ **5.** $3(x + 6) = 3x + 3 \cdot 6$ **6.** $8(2 + x) = 8 \cdot 2 + 8x$

7. $a \cdot (b \cdot c) = a \cdot (c \cdot b)$ **8.** $a \cdot (b \cdot c) = (a \cdot b) \cdot c$ **9.** $a + (b + 3) = (a + b) + 3$

10. $(a + 3) + b = (3 + a) + b$

⊙ Guided Tour

⟩ Applications

Students will enjoy the exceptionally creative applications in most sections that bring math alive and demonstrate that it can even be performed with a sense of humor.

⟩ ⟩ ⟩ **Applications**

79. *Distance traveled* Tyrone drove from San Luis Obispo to Los Angeles via Santa Barbara (101 + 96) miles. He then drove 62 more miles to San Bernardino. Write an expression that models the situation and then find the distance from San Luis Obispo to San Bernardino.

80. *Distance traveled* Maria started from San Luis Obispo but first stopped at Santa Barbara, 101 miles. She then drove to San Bernardino via Los Angeles, (96 + 62) miles. Write an expression that models the situation and then find the driving distance for Maria.

81. *Distance traveled* Referring to Problems 79 and 80, which property tells us that the driving distances for Tyrone and Maria are the same?

82. *Price of a car* The price P of a car is its base price (B) plus destination charges D, that is, $P = B + D$. Tran bought a Nissan in Smyrna, Tennessee, and there was no destination charge.

   a. What is D?

   b. Fill in the blank in the equation $P = B +$ _____

   c. What property tells you that the equation in part **b** is correct?

83. *Area* The area of a rectangle is found by multiplying its length L times its width W.

$$W = (b + c)$$

a. The length of the entire rectangle is a and its width is $(b + c)$. What is the area A of the rectangle?

84. *Area* The length of the entire rectangle is a and its width is b.

⟩ Using Your Knowledge

Optional, extended applications give students an opportunity to practice what they've learned in a multistep problem requiring reasoning skills in addition to numerical operations.

⟩ ⟩ ⟩ **Using Your Knowledge**

Tweedledee and Tweedledum Have you ever read *Alice in Wonderland?* Do you know who the author is? It's Lewis Carroll, of course. Although better known as the author of *Alice in Wonderland,* Lewis Carroll was also an accomplished mathematician and logician. Certain parts of his second book, *Through the Looking Glass,* reflect his interest in mathematics. In this book, one of the characters, Tweedledee, is talking to Tweedledum. Here is the conversation.

   Tweedledee: The sum of your weight and twice mine is 361 pounds.

   Tweedledum: Contrariwise, the sum of your weight and twice mine is 360 pounds.

41. If Tweedledee weighs x pounds and Tweedledum weighs y pounds, find their weights using the ideas of this section.

Study Aids to Make Math Accessible

Because some students confront math anxiety as soon as they sign up for the course, the Bello system provides many study aids to make their learning easier.

⟩ Objectives

The objectives for each section not only identify the specific tasks students should be able to perform, they organize the section itself with letters corresponding to each section heading, making it easy to follow.

3.4   The Slope of a Line: Parallel and Perpendicular Lines

⊙ **Objectives**

A ⟩ Find the slope of a line given two points.

B ⟩ Find the slope of a line given the equation of the line.

C ⟩ Determine whether two lines are parallel, perpendicular, or neither.

D ⟩ Solve applications involving slope.

⊙ **To Succeed, Review How To . . .**

1. Add, subtract, multiply, and divide signed numbers (pp. 52–56, 60–64).
2. Solve an equation for a specified variable (pp. 136–140).

⊙ **Getting Started**

Online Services Are Sloping Upward

Can you tell from the graph when the number of subscribers to commercial online services soared? After year 1. The annual increase in subscribers the first year was

$$\frac{\text{Difference in subscribers}}{\text{Difference in years}} = \frac{3.5 - 3.0}{1 - 0} = \frac{0.5}{1} = 0.5 \text{ (million per year)}$$

The annual increase in subscribers in years 1–5 was

$$\frac{13.5 - 3.5}{5 - 1} = \frac{10}{4} = 2.5 \text{ (million per year)}$$

⟩ Reviews

Every section begins with "To succeed, review how to . . . ," which directs students to specific pages to study key topics they need to understand to successfully begin that section.

❯ *Concept Checker*

This feature has been added to each end-of-section exercises to help students reinforce key terms and concepts.

❯ ❯ ❯ **Concept Checker**

Fill in the blank(s) with the correct word(s), phrase, or mathematical statement.

103. If m and n are **positive integers**, $x^m \cdot x^n =$ _____

104. When **multiplying** numbers with the **same (like) signs**, the **product** is _____.

105. When **multiplying** numbers with **different (unlike) signs**, the **product** is _____.

an integer	x^{m+n}
0	**negative**

❯ ❯ ❯ **Mastery Test**

In Problems 80–87, translate into an algebraic expression.

80. The product of 3 and xy

81. The difference of $2x$ and y

82. The quotient of $3x$ and $2y$

83. The sum of $7x$ and $4y$

84. The difference of b and c divided by the sum of b and c

85. Evaluate the expression $2x + y - z$ for $x = 3$, $y = 4$, and $z = 5$.

❯ *Mastery Tests*

Brief tests in every section give students a quick checkup to make sure they're ready to go on to the next topic.

❯ *Skill Checkers*

These brief exercises help students keep their math skills well honed in preparation for the next section.

❯ ❯ ❯ **Skill Checker**

In Problems 88–91, add the numbers.

88. $20 + (-20)$

89. $-3.8 + 3.8$

90. $-\frac{2}{7} + \frac{2}{7}$

91. $1\frac{2}{3} + \left(-1\frac{2}{3}\right)$

Calculator Corner

You can follow the steps in Example 7 with a calculator. Let's do it for the number 4. Here are the keystrokes:

4 `+` 2 `ENTER` `×` 3 `−` 6 `ENTER` `÷` 3 `ENTER`

The answer is the original number, 4. Your window should look like this:

```
4+2
                 6
Ans*3−6
                12
Ans/3
                 4
```

If you have a calculator, try it for -2. The keystrokes are identical and your final answer should be -2.

❯ *Calculator Corner*

When appropriate, optional calculator exercises are included to show students how they can explore concepts through calculators and verify their manual exercises with the aid of technology.

❯ **Summary** Chapter **1**

Section	Item	Meaning	Example
1.1A	Arithmetic expressions	Expressions containing numbers and operation signs	$3 + 4 - 8$, $9 \cdot 4 \div 6$ are arithmetic expressions.
1.1A	Algebraic expressions	Expressions containing numbers, operation signs, and variables	$3x - 4y + 3z$, $2x \div y + 9z$, and $7x - 9y \div 3z$ are algebraic expressions.

❯ *Summary*

An easy-to-read grid summarizes the essential chapter information by section, providing an item, its meaning, and an example to help students connect concepts with their concrete occurrences.

⊙ Guided Tour

❭ Review Exercises

Chapter review exercises are coded by section number and give students extra reinforcement and practice to boost their confidence.

❭ Review Exercises **Chapter 2**

(If you need help with these exercises, look in the section indicated in brackets.)

1. ⟨**2.1A**⟩ *Determine whether the given number satisfies the equation.*
   a. $5; 7 = 14 - x$ **b.** $4; 13 = 17 - x$
   c. $-2; 8 = 6 - x$

2. ⟨**2.1B**⟩ *Solve the given equation.*
   a. $x - \frac{1}{3} = \frac{1}{3}$ **b.** $x - \frac{5}{7} = \frac{2}{7}$
   c. $x - \frac{5}{9} = \frac{1}{9}$

3. ⟨**2.1B**⟩ *Solve the given equation.*
   a. $-3x + \frac{5}{9} + 4x - \frac{2}{9} = \frac{5}{9}$
   b. $-2x + \frac{4}{7} + 3x - \frac{2}{7} = \frac{6}{7}$
   c. $-4x + \frac{5}{6} + 5x - \frac{1}{6} = \frac{5}{6}$

4. ⟨**2.1C**⟩ *Solve the given equation.*
   a. $3 = 4(x - 1) + 2 - 3x$
   b. $4 = 5(x - 1) + 9 - 4x$
   c. $5 = 6(x - 1) + 8 - 5x$

❭ Practice Test with Answers

The chapter Practice Test offers students a non-threatening way to review the material and determine whether they are ready to take a test given by their instructor. The answers to the Practice Test give students immediate feedback on their performance, and the answer grid gives them specific guidance on which section, example, and pages to review for any answers they may have missed.

❭ Practice Test **Chapter 2**

(Answers on page 200)

Visit www.mhhe.com/bello to view helpful videos that provide step-by-step solutions to several of the problems below.

1. Does the number 3 satisfy the equation $6 = 9 - x$?

2. Solve $x - \frac{2}{7} = \frac{3}{7}$.

3. Solve $-2x + \frac{7}{8} + 3x - \frac{5}{8} = \frac{5}{8}$.

4. Solve $2 = 3(x - 1) + 5 - 2x$.

5. Solve $2 + 5(x + 1) = 8 + 5x$.

6. Solve $-3 - 2(x - 1) = -1 - 2x$.

❭ Answers to Practice Test **Chapter 2**

| Answer | If You Missed | | Review | |
	Question	Section	Examples	Page
1. Yes	1	2.1	1	107
2. $x = \frac{5}{7}$	2	2.1	2	108
3. $x = \frac{3}{8}$	3	2.1	3	109–110
4. $x = 0$	4	2.1	4, 5	111–112
5. No solution	5	2.1	6	113
6. All real numbers	6	2.1	7	113
7. $x = -6$	7	2.2	1, 2, 3	119–122

❭ Cumulative Review

The Cumulative Review covers material from the present chapter and any of the chapters prior to it and can be used for extra homework or for student review to improve their retention of important skills and concepts.

❭ Cumulative Review **Chapters 1–2**

1. Find the additive inverse (opposite) of -7.

2. Find: $\left| -9\frac{9}{10} \right|$

3. Find: $-\frac{2}{7} + \left(-\frac{2}{9} \right)$

4. Find: $-0.7 - (-8.9)$

5. Find: $(-2.4)(3.6)$

6. Find: $-(2^4)$

7. Find: $-\frac{7}{8} \div \left(-\frac{5}{24} \right)$

8. Evaluate $y \div 5 \cdot x - z$ for $x = 6$, $y = 60$, $z = 3$.

9. Which property is illustrated by the following statement?

   $9 \cdot (8 \cdot 5) = 9 \cdot (5 \cdot 8)$

10. Multiply: $6(5x + 7)$

Supplements for Instructors

Annotated Instructor's Edition

This version of the student text contains **answers** to all odd- and even-numbered exercises in addition to helpful **teaching tips.** The answers are printed on the same page as the exercises themselves so that there is no need to consult a separate appendix or answer key.

Computerized Test Bank (CTB) Online

Available through MathZone, this **computerized test bank,** utilizes Brownstone Diploma® algorithm-based testing software to quickly create customized exams. This user-friendly program enables instructors to search for questions by topic, format, or difficulty level; to edit existing questions or to add new ones; and to scramble questions and answer keys for multiple versions of the same test. Hundreds of text-specific open-ended and multiple-choice questions are included in the question bank. Sample chapter tests and final exams in Microsoft Word® and PDF formats are also provided.

Instructor's Solutions Manual

Available on MathZone, the Instructor's Solutions Manual provides comprehensive, **worked-out solutions** to all exercises in the text. The methods used to solve the problems in the manual are the same as those used to solve the examples in the textbook.

 MathZone www.mathzone.com

McGraw-Hill's MathZone is a complete online tutorial and homework management system for mathematics and statistics, designed for greater ease of use than any other system available. Instructors have the flexibility to create and share courses and assignments with colleagues, adjunct faculty, and teaching assistants with only a few clicks of the mouse. All algorithmic exercises, online tutoring, and a variety of video and animations are directly tied to text-specific materials. Completely customizable, MathZone suits individual instructor and student needs. Exercises can be easily edited, multimedia is assignable, importing additional content is easy, and instructors can even control the level of help available to students while doing their homework. Students have the added benefit of full access to the study tools to individually improve their success without having to be part of a MathZone course. MathZone allows for automatic grading and reporting of easy-to assign algorithmically generated homework, quizzes and tests. Grades are readily accessible through a fully integrated grade book that can be exported in one click to Microsoft Excel, WebCT, or BlackBoard.

 MathZone Offers.

- Practice exercises, based on the text's end-of-section material, generated in an unlimited number of variations, for as much practice as needed to master a particular topic.
- Subtitled videos demonstrating text-specific exercises and reinforcing important concepts within a given topic.
- Net Tutor™ integrating online whiteboard technology with live personalized tutoring via the Internet.

- Assessment capabilities, powered through ALEKS, which provide students and instructors with the diagnostics to offer a detailed knowledge base through advanced reporting and remediation tools.

- Faculty with the ability to create and share courses and assignments with colleagues and adjuncts, or to build a course from one of the provided course libraries.

- An Assignment Builder that provides the ability to select algorithmically generated exercises from any McGraw-Hill math textbook, edit content, as well as assign a variety of MathZone material including an ALEKS Assessment.

- Accessibility from multiple operating systems and internet browsers.

 ALEKS® (www.aleks.com)

ALEKS (**A**ssessment and **LE**arning in **K**nowledge **S**paces) is a dynamic online learning system for mathematics education, available over the Web 24/7. ALEKS assesses students, accurately determines their knowledge, and then guides them to the material that they are most ready to learn. With a variety of reports, Textbook Integration Plus, quizzes, and homework assignment capabilities, ALEKS offers flexibility and ease of use for instructors.

- ALEKS uses artificial intelligence to determine exactly what each student knows and is ready to learn. ALEKS remediates student gaps and provides highly efficient learning and improved learning outcomes.

- ALEKS is a comprehensive curriculum that aligns with syllabi or specified textbooks. Used in conjunction with a McGraw-Hill texts, students also receive links to text-specific videos, multimedia tutorials, and textbook pages.

- Textbook Integration Plus allows ALEKS to be automatically aligned with syllabi or specified McGraw-Hill textbooks with instructor chosen dates, chapter goals, homework, and quizzes.

- ALEKS with AI-2 gives instructors increased control over the scope and sequence of student learning. Students using ALEKS demonstrate a steadily increasing mastery of the content of the course.

- ALEKS offers a dynamic classroom management system that enables instructors to monitor and direct student progress toward mastery of course objectives.

Supplements for Students

Student's Solutions Manual

This supplement contains complete worked-out solutions to all odd-numbered exercises and all odd- and even-numbered problems in the Review Exercises and Cumulative Reviews in the textbook. The methods used to solve the problems in the manual are the same as those used to solve the examples in the textbook. This tool can be an invaluable aid to students who want to check their work and improve their grades by comparing their own solutions to those found in the manual and finding specific areas where they can do better.

www.mathzone.com

McGraw-Hill's MathZone is a complete online tutorial and homework management system for mathematics and statistics, designed for greater ease of use than any other system available. All algorithmic exercises, online tutoring, and a variety of video and animations are directly tied to text-specific materials.

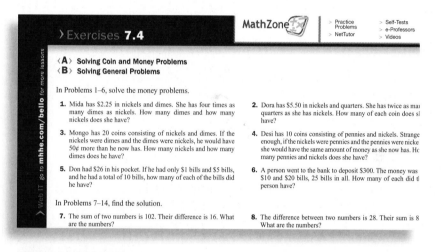

ALEKS® (www.aleks.com)

ALEKS (**A**ssessment and **LE**arning in **K**nowledge **S**paces) is a dynamic online learning system for mathematics education, available over the Web 24/7. ALEKS assesses students, accurately determines their knowledge, and then guides them to the material that they are most ready to learn. With a variety of reports, Textbook Integration Plus, quizzes, and homework assignment capabilities, ALEKS offers flexibility and ease of use for instructors.

Bello Video Series

The video series is available on DVD and VHS tape and features the authors introducing topics and working through selected odd-numbered exercises from the text, explaining how to complete them step by step. The DVDs are **closed-captioned** for the hearing impaired and also **subtitled in Spanish.**

Math for the Anxious: Building Basic Skills, by Rosanne Proga

Math for the Anxious: Building Basic Skills is written to provide a practical approach to the problem of math anxiety. By combining strategies for success with a pain-free introduction to basic math content, students will overcome their anxiety and find greater success in their math courses.

▶ Applications Index

▶ Applications Index

Section

Chapter

R

▷ Prealgebra Review

R.1   Fractions: Building and Reducing

Objectives

A > Write an integer as a fraction.

B > Find a fraction equivalent to a given one, but with a specified denominator.

C > Reduce fractions to lowest terms.

To Succeed, Review How To . . .

Add, subtract, multiply, and divide natural numbers.

Getting Started

Algebra and Arithmetic

The symbols on the sundial and the Roman clock have something in common: they both use numerals to name the numbers from 1 to 12. Algebra and arithmetic also have something in common: they use the same numbers and the same rules.

In arithmetic you learned about the counting numbers. The numbers used for counting are the **natural numbers:**

$$1, 2, 3, 4, 5, \text{ and so on}$$

These numbers are also used in algebra. We use the **whole numbers**

$$0, 1, 2, 3, 4, \text{ and so on}$$

as well. Later on, you probably learned about the **integers.** The integers include the **positive integers,**

$$+1, +2, +3, +4, +5, \text{ and so on} \quad \text{Read "positive one, positive two," and so on.}$$

the **negative integers,**

$$-1, -2, -3, -4, -5, \text{ and so on} \quad \text{Read "negative one, negative two," and so on.}$$

and the number **0,** which is neither positive nor negative. Thus, the integers are

$$\ldots, -2, -1, 0, 1, 2, \ldots$$

where the dots (. . .) indicate that the enumeration continues without end. Note that $+1 = 1$, $+2 = 2$, $+3 = 3$, and so on. Thus, the positive integers are the natural numbers.

A ⟩ Writing Integers as Fractions

All the numbers discussed can be written as **common fractions** of the form:

$$\text{Fraction bar} \longrightarrow \frac{a}{b} \begin{matrix} \longleftarrow \text{Numerator} \\ \longleftarrow \text{Denominator} \end{matrix}$$

When a and b are integers and b is not 0, this ratio is called a **rational number.** For example, $\frac{1}{3}$, $\frac{-5}{2}$, and $\frac{0}{7}$ are rational numbers. In fact, all natural numbers, all whole numbers, and all integers are also rational numbers.

When the numerator a of a fraction is smaller than the denominator b, the fraction $\frac{a}{b}$ is a **proper fraction.** Otherwise, the fraction is **improper.** Improper fractions are often written as **mixed numbers.** Thus, $\frac{9}{5}$ may be written as $1\frac{4}{5}$, and $\frac{13}{4}$ may be written as $3\frac{1}{4}$.

Of course, any integer can be written as a fraction by writing it with a denominator of 1. For example,

$$4 = \frac{4}{1}, \quad 8 = \frac{8}{1}, \quad 0 = \frac{0}{1}, \quad \text{and} \quad -3 = \frac{-3}{1}$$

EXAMPLE 1 **Writing integers as fractions**

Write the given numbers as fractions with a denominator of 1.

a. 10

b. -15

SOLUTION

a. $10 = \frac{10}{1}$

b. $-15 = \frac{-15}{1}$

PROBLEM 1

Write the given number as a fraction with a denominator of 1.

a. 18 **b.** -24

The rational numbers we have discussed are part of a larger set of numbers, the set of *real numbers.* The **real numbers** include the *rational numbers* and the *irrational numbers*. The **irrational numbers** are numbers that cannot be written as the ratio of two integers. For example, $\sqrt{2}$, π, $-\sqrt[3]{10}$, and $\frac{\sqrt{3}}{2}$ are irrational numbers. Thus, each real number is either rational or irrational. We shall say more about the irrational numbers in Chapter 1.

B ⟩ Equivalent Fractions

In Example 1(a) we wrote 10 as $\frac{10}{1}$. Can you find other ways of writing 10 as a fraction? Here are some:

$$10 = \frac{10}{1} = \frac{10 \times 2}{1 \times 2} = \frac{20}{2}$$

$$10 = \frac{10}{1} = \frac{10 \times 3}{1 \times 3} = \frac{30}{3}$$

and

$$10 = \frac{10}{1} = \frac{10 \times 4}{1 \times 4} = \frac{40}{4}$$

Note that $\frac{2}{2} = 1, \frac{3}{3} = 1$, and $\frac{4}{4} = 1$. As you can see, the fraction $\frac{10}{1}$ is **equivalent to** (has the same value as) many other fractions. We can always obtain other fractions equivalent to any given fraction by *multiplying* the numerator and denominator of the original fraction by the *same* nonzero number, a process called **building up** the fraction. This is the same as multiplying the fraction by 1, where 1 is written as $\frac{2}{2}, \frac{3}{3}, \frac{4}{4}$, and so on. For example,

$$\frac{3}{5} = \frac{3 \times 2}{5 \times 2} = \frac{6}{10}$$

$$\frac{3}{5} = \frac{3 \times 3}{5 \times 3} = \frac{9}{15}$$

and

$$\frac{3}{5} = \frac{3 \times 4}{5 \times 4} = \frac{12}{20}$$

Answers to PROBLEMS

1. a. $\frac{18}{1}$ b. $\frac{-24}{1}$

EXAMPLE 2   Finding equivalent fractions

Find a fraction equivalent to $\frac{3}{5}$ with a denominator of 20.

SOLUTION   We must solve the problem

$$\frac{3}{5} = \frac{?}{20}$$

Note that the denominator, 5, was multiplied by 4 to get 20. So, we must also multiply the numerator, 3, by 4.

Thus, the equivalent fraction is $\frac{12}{20}$.

PROBLEM 2

Find a fraction equivalent to $\frac{4}{7}$ with a denominator of 21.

Here is a slightly different problem. Can we find a fraction equivalent to $\frac{15}{20}$ with a denominator of 4? We do this in Example 3.

EXAMPLE 3   Finding equivalent fractions

Find a fraction equivalent to $\frac{15}{20}$ with a denominator of 4.

SOLUTION   We proceed as before.

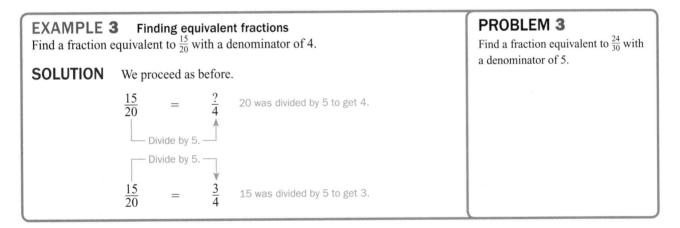

PROBLEM 3

Find a fraction equivalent to $\frac{24}{30}$ with a denominator of 5.

We can summarize our work with equivalent fractions in the following procedure.

PROCEDURE

Obtaining Equivalent Fractions

To obtain an **equivalent fraction,** *multiply* or *divide* both numerator and denominator of the fraction by the *same* nonzero number.

C ⟩ Reducing Fractions to Lowest Terms

The preceding rule can be used to *reduce* (simplify) fractions to lowest terms. A fraction is *reduced* to lowest terms (simplified) when there is no number (except 1) that will divide the numerator and the denominator exactly. The procedure is as follows.

Answers to PROBLEMS

2. $\frac{12}{21}$ 3. $\frac{4}{5}$

PROCEDURE

Reducing Fractions to Lowest Terms

To **reduce** a fraction to *lowest* terms, divide the numerator and denominator by the *largest* natural number that will divide them exactly.

"Divide them exactly" means that both remainders are zero.

To reduce $\frac{12}{30}$ to lowest terms, we divide the numerator and denominator by 6, the *largest* natural number that divides 12 and 30 exactly. [6 is sometimes called the *greatest common divisor* (GCD) of 12 and 30.] Thus,

$$\frac{12}{30} = \frac{12 \div 6}{30 \div 6} = \frac{2}{5}$$

This reduction is sometimes shown like this:

$$\frac{\overset{2}{\cancel{12}}}{\underset{5}{\cancel{30}}} = \frac{2}{5}$$

EXAMPLE 4   Reducing fractions

Reduce to lowest terms:

a. $\frac{15}{20}$ **b.** $\frac{30}{45}$ **c.** $\frac{60}{48}$

SOLUTION

a. The *largest* natural number exactly dividing 15 and 20 is 5. Thus,

$$\frac{15}{20} = \frac{15 \div 5}{20 \div 5} = \frac{3}{4}$$

b. The *largest* natural number exactly dividing 30 and 45 is 15. Hence,

$$\frac{30}{45} = \frac{30 \div 15}{45 \div 15} = \frac{2}{3}$$

c. The *largest* natural number dividing 60 and 48 is 12. Therefore,

$$\frac{60}{48} = \frac{60 \div 12}{48 \div 12} = \frac{5}{4}$$

PROBLEM 4

Reduce to lowest terms:

a. $\frac{30}{50}$ **b.** $\frac{45}{60}$ **c.** $\frac{84}{72}$

What if you are unable to see at once that the *largest* natural number dividing the numerator and denominator in, say, $\frac{30}{45}$, is 15? No problem; it just takes a little longer. Suppose you notice that 30 and 45 are both divisible by 5. You then write

$$\frac{30}{45} = \frac{30 \div 5}{45 \div 5} = \frac{6}{9}$$

Now you can see that 6 and 9 are both divisible by 3. Thus,

$$\frac{6}{9} = \frac{6 \div 3}{9 \div 3} = \frac{2}{3}$$

which is the same answer we got in Example 4(b). The whole procedure can be written as

$$\frac{\overset{2}{\cancel{\overset{6}{\cancel{30}}}}}{\underset{3}{\cancel{\underset{9}{\cancel{45}}}}} = \frac{2}{3}$$

We can also reduce $\frac{15}{20}$ using *prime factorization* by writing

$$\frac{15}{20} = \frac{3 \cdot 5}{2 \cdot 2 \cdot 5} = \frac{3}{4}$$

Answers to PROBLEMS

4. a. $\frac{3}{5}$ b. $\frac{3}{4}$ c. $\frac{7}{6}$

(The dot, ·, indicates multiplication.) Note that 15 is written as the *product* of 3 and 5, and 20 is written as the product 2 · 2 · 5. A **product** is the answer to a multiplication problem. When 15 is written as the product 3 · 5, 3 and 5 are the **factors** of 15. As a matter of fact, 3 and 5 are **prime numbers,** so 3 · 5 is the prime factorization of 15; similarly, 2 · 2 · 5 is the prime factorization of 20, because 2 and 5 are primes. In general,

> A natural number is prime if it has *exactly two different factors*, itself and 1.

The first few primes are

$$2, 3, 5, 7, 11, 13, 17, 19, 23, 29, 31, \quad \text{and so on.}$$

A natural number that is *not* prime is called a **composite number.** Thus, the numbers missing in the previous list, 4, 6, 8, 9, 10, 12, 14, 15, 16, and so on, are **composite.** Note that 1 is considered neither prime (because it does not have *two different* factors) nor composite.

When using prime factorization, we can keep better track of the factors by using a **factor tree.** For example, to reduce $\frac{30}{45}$ using prime factorization, we make two trees to factor 30 and 45 as shown:

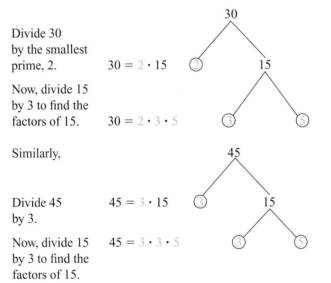

Divide 30 by the smallest prime, 2. $30 = 2 \cdot 15$

Now, divide 15 by 3 to find the factors of 15. $30 = 2 \cdot 3 \cdot 5$

Similarly,

Divide 45 by 3. $45 = 3 \cdot 15$

Now, divide 15 by 3 to find the factors of 15. $45 = 3 \cdot 3 \cdot 5$

Thus,

$$\frac{30}{45} = \frac{2 \cdot 3 \cdot 5}{3 \cdot 3 \cdot 5} = \frac{2}{3} \quad \text{as before.}$$

Caution: When reducing a fraction to lowest terms, it is easier to look for the *largest common factor* in the numerator and denominator, but if no largest factor is obvious, *any common factor* can be used and the simplification can be done in stages. For example,

$$\frac{200}{250} = \frac{20 \cdot 10}{25 \cdot 10} = \frac{4 \cdot 5 \cdot 10}{5 \cdot 5 \cdot 10} = \frac{4}{5}$$

Which way should you simplify fractions? The way you understand! Make sure you follow the instructor's preferences regarding the procedure used.

❯Exercises R.1

❮ A ❯ Writing Integers as Fractions   In Problems 1–8, write the given number as a fraction with a denominator of 1.

1. 28   **2.** 93   **3.** −42   **4.** −86

5. 0   **6.** 1   **7.** −1   **8.** −17

❮ B ❯ Equivalent Fractions   In Problems 9–30, find the missing number that makes the fractions equivalent.

9. $\frac{1}{8} = \frac{?}{24}$ **10.** $\frac{7}{1} = \frac{?}{2}$ **11.** $\frac{7}{1} = \frac{?}{6}$ **12.** $\frac{5}{6} = \frac{?}{48}$ **13.** $\frac{5}{3} = \frac{?}{15}$

14. $\frac{9}{8} = \frac{?}{32}$ **15.** $\frac{7}{11} = \frac{?}{33}$ **16.** $\frac{11}{7} = \frac{?}{35}$ **17.** $\frac{1}{8} = \frac{4}{?}$ **18.** $\frac{3}{5} = \frac{27}{?}$

19. $\frac{5}{6} = \frac{5}{?}$ **20.** $\frac{9}{10} = \frac{9}{?}$ **21.** $\frac{8}{7} = \frac{16}{?}$ **22.** $\frac{9}{5} = \frac{36}{?}$ **23.** $\frac{6}{5} = \frac{36}{?}$

24. $\frac{5}{3} = \frac{45}{?}$ **25.** $\frac{21}{56} = \frac{?}{8}$ **26.** $\frac{12}{18} = \frac{?}{3}$ **27.** $\frac{36}{180} = \frac{?}{5}$ **28.** $\frac{8}{24} = \frac{4}{?}$

29. $\frac{18}{12} = \frac{3}{?}$ **30.** $\frac{56}{49} = \frac{8}{?}$

❮ C ❯ Reducing Fractions to Lowest Terms   In Problems 31–40, reduce the fraction to lowest terms by writing the numerator and denominator as products of primes.

31. $\frac{15}{12}$ **32.** $\frac{30}{28}$ **33.** $\frac{13}{52}$ **34.** $\frac{27}{54}$ **35.** $\frac{56}{24}$

36. $\frac{56}{21}$ **37.** $\frac{22}{33}$ **38.** $\frac{26}{39}$ **39.** $\frac{100}{25}$ **40.** $\frac{21}{3}$

❯❯❯ Applications

41. *AOL ad*   If you take advantage of the AOL offer and get 1000 hours free for 45 days:

a. How many hours will you get per day? (Assume you use the same number of hours each day and do not simplify your answer.)

All-New AOL 9.0
1000 Hours Free!
for 45 days

b. Write the answer to part **a** in lowest terms.

c. Write the answer to part **a** as a mixed number.

d. If you used AOL 24 hours a day, for how many days would the 1000 free hours last?

42. *Census data   Are you "poor"?*   In 1959, the U.S. Census reported that about 40 million of the 180 million people living in the United States were "poor." In the year 2000, about 30 million out of 275 million were "poor" (income less than $8794).

a. What reduced fraction of the people was poor in the year 1959?

b. What reduced fraction of the people was poor in the year 2000?

43. *High school completion rate*   In a recent year, about 41 out of 100 persons in the United States had completed 4 years of high school or more. What fraction of the people is that?

44. *High school completion rates*   In a recent year, about 84 out of 100 Caucasians, 76 out of 100 African-Americans, and 56 out of 100 Hispanics had completed 4 years of high school or more. What reduced fractions of the Caucasians, African-Americans, and Hispanics had completed 4 years of high school or more?

45. *Pizza consumption* The pizza shown here consists of six pieces. Alejandro ate $\frac{1}{3}$ of the pizza.

a. Write $\frac{1}{3}$ with a denominator of 6.

b. How many pieces did Alejandro eat?

c. If Cindy ate two pieces of pizza, what fraction of the pizza did she eat?

d. Who ate more pizza, Alejandro or Cindy?

Fuel gauge Problems 46–50 refer to the photo of the fuel gauge. What fraction of the tank is full if the needle:

46. Points to the line midway between E and $\frac{1}{2}$ full?

47. Points to the first line to the right of empty (E)?

48. Is midway between empty and $\frac{1}{8}$ of a tank?

49. Is one line past the $\frac{1}{2}$ mark?

50. Is one line before the $\frac{1}{2}$ mark?

〉〉〉 *Using Your Knowledge*

Interpreting Fractions During the 1953–1954 basketball season, the NBA (National Basketball Association) had a problem with the game: it was boring. Danny Biasone, the owner of the Syracuse Nationals, thought that limiting the time a team could have the ball should encourage more shots. Danny figured out that in a fast-paced game, each team should take 60 shots during the 48 minutes the game lasted (4 quarters of 12 minutes each). He then looked at the fraction $\frac{\text{Seconds}}{\text{Shots}}$.

51. a. How many seconds does the game last?

b. How many total shots are to be taken in the game?

c. What is the reduced fraction $\frac{\text{Seconds}}{\text{Shots}}$?

Now you know where the $\frac{\text{Seconds}}{\text{Shots}}$ clock came from!

You have learned how to work with fractions. Now you can use your knowledge to interpret fractions from diagrams. As you see from the diagram (circle graph), "CNN Headline News" devotes the first quarter of every hour $\left(\frac{15}{60} = \frac{1}{4}\right)$ to National and International News.

52. What total fraction of the hour is devoted to: Dollars & Sense?

53. What total fraction of the hour is devoted to: Sports?

54. What total fraction of the hour is devoted to: Local News or People & Places?

55. Which feature uses the most time, and what fraction of an hour does it use?

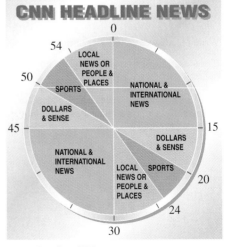

Source: Data from CNN.

As you can see from the illustration, "Bay News 9" devotes $8 + 6 = 14$ minutes of the hour (60 minutes) to News. This represents $\frac{14}{60} = \frac{7}{30}$ of the hour.

56. What fraction of the hour is devoted to Traffic?

57. What fraction of the hour is devoted to Weather?

58. What fraction of the hour is devoted to News & Beyond the Bay?

59. Which features use the most and least time, and what fraction of the hour does each one use?

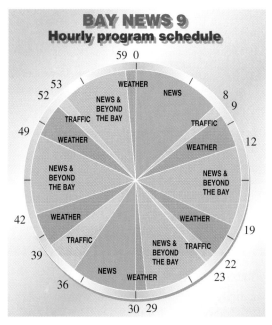

Source: Data from Bay News 9.

> ⟩ ⟩ ⟩   *Write On*

60. Write the procedure you use to reduce a fraction to lowest terms.

62. Write a procedure to determine whether two fractions are equivalent.

61. What are the advantages and disadvantages of writing the numerator and denominator of a fraction as a product of primes before reducing the fraction?

R.2   Operations with Fractions and Mixed Numbers

▶ Objective

A ⟩ Add, subtract, multiply, and divide fractions.

▶ To Succeed, Review How To . . .

1. Reduce fractions (pp. 4–6).

2. Write a number as a product of primes (p. 6).

▶ Getting Started
A Sweet Problem

Each of the cups contains $\frac{1}{4}$ cup of sugar. How much sugar do the cups contain altogether?

To find the answer, we can multiply 3 by $\frac{1}{4}$; that is, we can find $3 \cdot \frac{1}{4}$.

A ⟩ Adding, Subtracting, Multiplying, and Dividing Fractions

How do we perform the multiplication $3 \cdot \frac{1}{4}$ required to determine the total amount of sugar in the three cups? Note that

$$3 \cdot \frac{1}{4} = \frac{3}{1} \cdot \frac{1}{4} = \frac{3 \cdot 1}{1 \cdot 4} = \frac{3}{4}$$

Similarly,

$$\frac{4}{9} \cdot \frac{2}{5} = \frac{4 \cdot 2}{9 \cdot 5} = \frac{8}{45}$$

Here is the general rule for multiplying fractions.

⊞ ◈ ▦ **Calculator Corner**

Multiplying Fractions

To multiply 3 by $\frac{1}{4}$ using a calculator with an $\boxed{x/y}$ key, enter **3** ✕ **1** $\boxed{x/y}$ **4** **ENTER** and you get $\frac{3}{4}$. If your calculator has a $\boxed{/}$ or a $\boxed{\div}$ key, the strokes will be **3** ✕ **1** $\boxed{\div}$ **4** **ENTER** and you will get the same answer.

RULE

Multiplying Fractions

To multiply fractions, multiply the numerators, multiply the denominators, and then simplify. In symbols,

$$\frac{a}{b} \cdot \frac{c}{d} = \frac{a \cdot c}{b \cdot d} \quad b \cdot d \neq 0$$

Note: To avoid repetition, from now on we will assume that the denominators are *not* 0.

EXAMPLE 1   Multiplying fractions

Multiply: $\frac{9}{5} \cdot \frac{3}{4}$

SOLUTION   We use our rule for multiplying fractions:

$$\frac{9}{5} \cdot \frac{3}{4} = \frac{9 \cdot 3}{5 \cdot 4} = \frac{27}{20}$$

PROBLEM 1

Multiply: $\frac{9}{7} \cdot \frac{3}{5}$

When multiplying fractions, we can save time if we divide out common factors *before* we multiply. For example,

$$\frac{2}{5} \cdot \frac{5}{7} = \frac{2 \cdot 5}{5 \cdot 7} = \frac{10}{35} = \frac{2}{7}$$

We can save time by writing

$$\frac{2}{\cancel{5}} \cdot \frac{\cancel{5}}{7} = \frac{2 \cdot 1}{1 \cdot 7} = \frac{2}{7} \qquad \text{This can be done because } \tfrac{5}{5} = 1.$$

CAUTION

Only factors that are common to *both* numerator *and* denominator can be divided out.

EXAMPLE 2   Multiplying fractions with common factors

Multiply:

a. $\dfrac{3}{7} \cdot \dfrac{7}{8}$

b. $\dfrac{5}{8} \cdot \dfrac{4}{15}$

SOLUTION

a. $\dfrac{3}{\cancel{7}} \cdot \dfrac{\overset{1}{\cancel{7}}}{8} = \dfrac{3 \cdot 1}{1 \cdot 8} = \dfrac{3}{8}$

b. $\dfrac{5}{\underset{2}{\cancel{8}}} \cdot \dfrac{\overset{1}{\cancel{4}}}{\underset{3}{\cancel{15}}} = \dfrac{1 \cdot 1}{2 \cdot 3} = \dfrac{1}{6}$

PROBLEM 2

Multiply:

a. $\dfrac{4}{9} \cdot \dfrac{9}{11}$ **b.** $\dfrac{5}{14} \cdot \dfrac{7}{20}$

If we wish to multiply a fraction by a **mixed number,** such as $3\frac{1}{4}$, we must convert the mixed number to a fraction first. The number $3\frac{1}{4}$ (read "3 and $\frac{1}{4}$") means $3 + \frac{1}{4} = \frac{12}{4} + \frac{1}{4} = \frac{13}{4}$. (This addition will be clearer to you after studying the addition of fractions.) For now, we can shorten the procedure by using the following diagram.

$$3\frac{1}{4} \;=\; \frac{13}{4}$$

Same denominator

Work clockwise. *First* multiply the denominator 4 by the whole number part 3; add the numerator. This is the new numerator. Use the same denominator.

EXAMPLE 3   Multiplying fractions and mixed numbers

Multiply: $5\frac{1}{3} \cdot \dfrac{9}{16}$

SOLUTION   We first convert the mixed number to a fraction:

$$5\frac{1}{3} = \frac{3 \cdot 5 + 1}{3} = \frac{16}{3}$$

Thus,

$$5\frac{1}{3} \cdot \frac{9}{16} = \frac{\overset{1}{\cancel{16}}}{\underset{1}{\cancel{3}}} \cdot \frac{\overset{3}{\cancel{9}}}{\underset{1}{\cancel{16}}} = \frac{1 \cdot 3}{1 \cdot 1} = 3$$

PROBLEM 3

Multiply: $2\frac{3}{4} \cdot \dfrac{8}{11}$

If we wish to divide one number by another nonzero number, we can indicate the division by a fraction. Thus, to divide 2 by 5 we can write

Multiply.

$$2 \div 5 = \frac{2}{5} = 2 \cdot \frac{1}{5}$$

The divisor 5 is inverted.

Note that to divide 2 by 5 we multiplied 2 by $\frac{1}{5}$, where the fraction $\frac{1}{5}$ was obtained by *inverting* $\frac{5}{1}$ to obtain $\frac{1}{5}$. (In mathematics, $\frac{5}{1}$ and $\frac{1}{5}$ are called **reciprocals.**) *Note:* Only the 5 (the divisor) is replaced by its reciprocal, $\frac{1}{5}$.

Now let's try the problem $5 \div \frac{5}{7}$. If we do it like the preceding example, we write

Multiply.

$$5 \div \frac{5}{7} = 5 \cdot \frac{7}{5} = \frac{\cancel{5}}{1} \cdot \frac{7}{\cancel{5}} = 7$$

Invert.

In general, to divide $\frac{a}{b}$ by $\frac{c}{d}$, we multiply $\frac{a}{b}$ by the *reciprocal* of $\frac{c}{d}$ —that is, $\frac{d}{c}$. Here is the rule.

Answers to PROBLEMS

2. a. $\dfrac{4}{11}$ **b.** $\dfrac{1}{8}$ **3.** 2

RULE

Dividing Fractions

To divide fractions, multiply the first fraction by the reciprocal of the second fraction and then simplify. In symbols,

$$\frac{a}{b} \div \frac{c}{d} = \frac{a}{b} \cdot \frac{d}{c}$$

EXAMPLE 4   Dividing fractions

Divide:

a. $\frac{3}{5} \div \frac{2}{7}$

b. $\frac{4}{9} \div 5$

SOLUTION

a. $\frac{3}{5} \div \frac{2}{7} = \frac{3}{5} \cdot \frac{7}{2} = \frac{21}{10}$

b. $\frac{4}{9} \div 5 = \frac{4}{9} \cdot \frac{1}{5} = \frac{4}{45}$

PROBLEM 4

Divide:

a. $\frac{3}{4} \div \frac{5}{7}$ b. $\frac{3}{7} \div 5$

As in the case of multiplication, if the problem involves mixed numbers, we change them to fractions first, like this:

$$2\frac{1}{4} \div \frac{3}{5} = \frac{9}{4} \div \frac{3}{5} = \frac{9}{4} \cdot \frac{\overset{3}{5}}{\underset{1}{3}} = \frac{15}{4}$$

Change. — Invert.

EXAMPLE 5   Dividing fractions and mixed numbers

Divide:

a. $3\frac{1}{4} \div \frac{7}{8}$

b. $\frac{11}{12} \div 7\frac{1}{3}$

SOLUTION

a. $3\frac{1}{4} \div \frac{7}{8} = \frac{13}{4} \div \frac{7}{8} = \frac{13}{\underset{1}{4}} \cdot \frac{\overset{2}{8}}{7} = \frac{26}{7}$

b. $\frac{11}{12} \div 7\frac{1}{3} = \frac{11}{12} \div \frac{22}{3} = \frac{\overset{1}{11}}{\underset{4}{12}} \cdot \frac{\overset{1}{3}}{\underset{2}{22}} = \frac{1}{8}$

PROBLEM 5

Divide:

a. $2\frac{1}{5} \div \frac{7}{10}$ b. $\frac{11}{15} \div 7\frac{1}{3}$

Now we are ready to *add* fractions.

The photo shows that 1 quarter plus 2 quarters equals 3 quarters. In symbols,

$$\frac{1}{4} + \frac{2}{4} = \frac{1+2}{4} = \frac{3}{4}$$

In general, to add fractions with the *same* denominator, we add the numerators and keep the denominator.

RULE

Adding Fractions with the Same Denominator

To add fractions with the same denominators, add the numerators and keep the same denominators. In symbols,

$$\frac{a}{b} + \frac{c}{b} = \frac{a+c}{b}$$

Answers to PROBLEMS

4. a. $\frac{21}{20}$ b. $\frac{3}{35}$

5. a. $\frac{22}{7}$ b. $\frac{1}{10}$

Thus,

$$\frac{1}{5} + \frac{2}{5} = \frac{1+2}{5} = \frac{3}{5}$$

and

$$\frac{3}{8} + \frac{1}{8} = \frac{3+1}{8} = \frac{4}{8} = \frac{1}{2}$$

> **NOTE**
>
> In the last addition we reduced $\frac{4}{8}$ to $\frac{1}{2}$ by dividing the numerator and denominator by 4. When a result involves fractions, we always **reduce** to lowest terms.

Now suppose you wish to add $\frac{5}{12}$ and $\frac{1}{18}$. Since these two fractions do not have the same denominators, the rule does not work. To add them, you must learn how to write $\frac{5}{12}$ and $\frac{1}{18}$ as equivalent fractions with the same denominator. To keep things simple, you should also try to make this denominator as small as possible; that is, you should first find the **least common denominator (LCD)** of the fractions.

The LCD of two fractions is the **least** (*smallest*) **common multiple (LCM)** of their denominators. Thus, to find the LCD of $\frac{5}{12}$ and $\frac{1}{18}$, we find the LCM of 12 and 18. There are several ways of doing this. One way is to select the larger number (18) and find its multiples. The first multiple of 18 is $2 \cdot 18 = 36$, and 12 divides into 36. Thus, 36 is the LCD of 12 and 18.

Unfortunately, this method is not practical in algebra. A more convenient method consists of writing the denominators 12 and 18 in completely *factored* form. We can start by writing 12 as $2 \cdot 6$. In turn, $6 = 2 \cdot 3$; thus,

$$12 = 2 \cdot 2 \cdot 3$$

Similarly,

$$18 = 2 \cdot 9, \quad \text{or} \quad 2 \cdot 3 \cdot 3$$

Now we can see that the smallest number that is a multiple of 12 and 18 must have at least two 2's (there are two 2's in 12) and two 3's (there are two 3's in 18). Thus, the LCD of $\frac{5}{12}$ and $\frac{1}{18}$ is

$$\underbrace{2 \cdot 2}_{\text{Two 2's}} \cdot \underbrace{3 \cdot 3}_{\text{Two 3's}} = 36$$

Fortunately in arithmetic, there is an even shorter way of finding the LCM of 12 and 18. Note that what we want to do is find the *common* factors in 12 and 18, so we can divide 12 and 18 by the smallest divisor common to both numbers (2) and then the next (3), and so on. If we multiply these divisors by the final quotient, the result is the LCM. Here is the shortened version:

Divide by 2. $2)\ \underline{12 \quad 18}$
Divide by 3. $3)\ \underline{\ 6 \quad\ 9}$
                   $\quad\ 2 \quad\ 3 \quad\longrightarrow$ Multiply $2 \cdot 3 \cdot 2 \cdot 3$.

The LCM is $2 \cdot 3 \cdot 2 \cdot 3 = 36$.

In general, we use the following procedure to find the LCD of two fractions.

> **PROCEDURE**
>
> **Finding the LCD of Two Fractions**
>
> **1.** Write the denominators in a horizontal row and divide each number by a *divisor* common to both numbers.
> **2.** Continue the process until the resulting quotients have no common divisor (except 1).
> **3.** The product of the *divisors* and the *final quotients* is the LCD.

Now that we know that the LCD of $\frac{5}{12}$ and $\frac{1}{18}$ is 36, we can add the two fractions by writing each as an equivalent fraction with a denominator of 36. We do this by multiplying numerators and denominators of $\frac{5}{12}$ by 3 and of $\frac{1}{18}$ by 2. Thus, we get

$$\frac{5}{12} = \frac{5 \cdot 3}{12 \cdot 3} = \frac{15}{36} \longrightarrow \frac{15}{36}$$

$$\frac{1}{18} = \frac{1 \cdot 2}{18 \cdot 2} = \frac{2}{36} \longrightarrow + \frac{2}{36}$$

$$\frac{17}{36}$$

We can also write the results as

$$\frac{5}{12} + \frac{1}{18} = \frac{15}{36} + \frac{2}{36} = \frac{15 + 2}{36} = \frac{17}{36}$$

EXAMPLE 6   Adding fractions with different denominators

Add: $\frac{1}{20} + 1\frac{1}{18}$

SOLUTION   We first find the LCM of 20 and 18.

Divide by 2. $2) \underline{20 \quad 18}$
          $10 \longrightarrow 9 \longrightarrow$ Multiply $2 \cdot 10 \cdot 9$.

Since 10 and 9 have no common divisor, the LCM is

$$2 \cdot 10 \cdot 9 = 180$$

(You could also find the LCM by writing $20 = 2 \cdot 2 \cdot 5$ and $18 = 2 \cdot 3 \cdot 3$; the LCM is $2 \cdot 2 \cdot 3 \cdot 3 \cdot 5 = 180$.)

Now, write the fractions with denominators of 180 and add.

$$\frac{1}{20} = \frac{1 \cdot 9}{20 \cdot 9} = \frac{9}{180} \longrightarrow \frac{9}{180}$$

$$1\frac{1}{18} = \frac{19}{18} = \frac{19 \cdot 10}{18 \cdot 10} = \frac{190}{180} \longrightarrow + \frac{190}{180}$$

$$\frac{199}{180}$$

If you prefer, you can write the procedure like this:

$$\frac{1}{20} + 1\frac{1}{18} = \frac{9}{180} + \frac{190}{180} = \frac{199}{180}, \quad \text{or} \quad 1\frac{19}{180}$$

PROBLEM 6

Add: $\frac{3}{20} + 1\frac{1}{16}$

Can we use the same procedure to add three or more fractions? Almost; but we need to know how to find the LCD for three or more fractions. The procedure is very similar to that used for finding the LCM of two numbers. If we write 15, 21, and 28 in factored form,

$$15 = 3 \cdot 5$$

$$21 = 3 \cdot 7$$

$$28 = 2 \cdot 2 \cdot 7$$

we see that the LCM must contain at least 2 · 2, 3, 5, and 7. (Note that we select each factor the *greatest* number of times it appears in any factorization.) Thus, the LCM of 15, 21, and 28 is $2 \cdot 2 \cdot 3 \cdot 5 \cdot 7 = 420$.

We can also use the following shortened procedure to find the LCM of 15, 21, and 28.

PROCEDURE

Finding the LCD of Three or More Fractions

1. Write the denominators in a horizontal row and divide the numbers by a divisor common to two or more of the numbers. If any of the other numbers is not divisible by this divisor, *circle* the number and carry it to the next line.

$$\text{Divide by } 3. \quad 3\overline{)15 \quad 21 \quad ㉘}$$

2. Repeat step 1 with the quotients and carrydowns until *no two numbers* have a common divisor (except 1).

$$\text{Divide by } 7. \quad 7\overline{)⑤ \quad 7 \quad 28}$$
$$5 \quad 1 \quad 4$$

$$\text{Multiply } 3 \cdot 7 \cdot 5 \cdot 1 \cdot 4.$$

3. The LCD is the *product* of the divisors from the preceding steps and the numbers in the final row.

$$\text{The LCD is } 3 \cdot 7 \cdot 5 \cdot 1 \cdot 4 = 420.$$

Thus, to add

$$\frac{1}{16} + \frac{3}{10} + \frac{1}{28}$$

we first find the LCD of 16, 10, and 28. We can use either method.

Method 1. Find the LCD by writing

$$16 = 2 \cdot 2 \cdot 2 \cdot 2$$
$$10 = 2 \cdot 5$$
$$28 = 2 \cdot 2 \cdot 7$$

and selecting each factor the *greatest* number of times it appears in any factorization. Since 2 appears four times and 5 and 7 appear once, the LCD is

$$\underbrace{2 \cdot 2 \cdot 2 \cdot 2}_{\text{four times}} \cdot \underbrace{5 \cdot 7}_{\text{once}} = 560$$

Method 2. Use the three-step procedure

Step 1. Divide by 2. $2\overline{)16 \quad 10 \quad 28}$

Step 2. Divide by 2. $2\overline{)8 \quad ⑤ \quad 14}$
$$4 \quad 5 \quad 7$$
Multiply $2 \cdot 2 \cdot 4 \cdot 5 \cdot 7.$

Step 3. The LCD is
$$2 \cdot 2 \cdot 4 \cdot 5 \cdot 7 = 560.$$

We then write $\frac{1}{16}, \frac{3}{10},$ and $\frac{1}{28}$ with denominators of 560.

$$\frac{1}{16} = \frac{1 \cdot 35}{16 \cdot 35} = \frac{35}{560} \longrightarrow \qquad \frac{35}{560}$$

$$\frac{3}{10} = \frac{3 \cdot 56}{10 \cdot 56} = \frac{168}{560} \longrightarrow \quad + \frac{168}{560}$$

$$\frac{1}{28} = \frac{1 \cdot 20}{28 \cdot 20} = \frac{20}{560} \longrightarrow \quad + \frac{20}{560}$$
$$\frac{223}{560}$$

Or, if you prefer,

$$\frac{1}{16} + \frac{3}{10} + \frac{1}{28} = \frac{35}{560} + \frac{168}{560} + \frac{20}{560} = \frac{223}{560}$$

NOTE

Method 1 for finding the LCD is preferred because it is more easily generalized to algebraic fractions.

Now that you know how to add fractions, *subtraction* is no problem. All the rules we have mentioned still apply! For example, we subtract fractions with the same denominator as shown.

$$\frac{5}{8} - \frac{2}{8} = \frac{5-2}{8} = \frac{3}{8}$$

$$\frac{7}{9} - \frac{1}{9} = \frac{7-1}{9} = \frac{6}{9} = \frac{2}{3}$$

The next example shows how to subtract fractions involving *different* denominators.

EXAMPLE 7   Subtracting fractions with different denominators

Subtract: $\frac{7}{12} - \frac{1}{18}$

SOLUTION   We first find the LCD of the fractions.

Method 1.   Write

$$12 = 2 \cdot 2 \cdot 3$$
$$18 = 2 \cdot 3 \cdot 3$$

The LCD is $2 \cdot 2 \cdot 3 \cdot 3 = 36$.

Method 2.

Step 1. $2\,\underline{)\,12\quad 18}$

Step 2. $3\,\underline{)\,\ 6\quad\ \ 9}$
                         $\quad\ \ 2\qquad 3$

Step 3.   The LCD is
                  $2 \cdot 3 \cdot 2 \cdot 3 = 36$.

We then write each fraction with 36 as the denominator.

$$\frac{7}{12} = \frac{7 \cdot 3}{12 \cdot 3} = \frac{21}{36} \quad \text{and} \quad \frac{1}{18} = \frac{1 \cdot 2}{18 \cdot 2} = \frac{2}{36}$$

Thus,

$$\frac{7}{12} - \frac{1}{18} = \frac{21}{36} - \frac{2}{36} = \frac{21-2}{36} = \frac{19}{36}$$

PROBLEM 7

Subtract: $\frac{7}{18} - \frac{1}{12}$

The rules for adding fractions also apply to subtraction. If mixed numbers are involved, we can use horizontal or vertical subtraction, as illustrated in Example 8.

EXAMPLE 8   Subtracting mixed numbers

Subtract: $3\frac{1}{6} - 2\frac{5}{8}$

SOLUTION   *Horizontal subtraction.* We first find the LCM of 6 and 8, which is 24. Then we convert the mixed numbers to improper fractions.

$$3\frac{1}{6} = \frac{19}{6} \quad \text{and} \quad 2\frac{5}{8} = \frac{21}{8}$$

Thus,

$$3\frac{1}{6} - 2\frac{5}{8} = \frac{19}{6} - \frac{21}{8}$$

$$= \frac{19 \cdot 4}{6 \cdot 4} - \frac{21 \cdot 3}{8 \cdot 3}$$

$$= \frac{76}{24} - \frac{63}{24}$$

$$= \frac{13}{24}$$

Vertical subtraction. Some students prefer to subtract mixed numbers by setting the problem in a column, as shown:

$$3\frac{1}{6}$$

$$\underline{-2\frac{5}{8}}$$

PROBLEM 8

Subtract: $4\frac{1}{8} - 2\frac{5}{6}$

Answers to PROBLEMS

7. $\frac{11}{36}$ **8.** $\frac{31}{24}$ or $1\frac{7}{24}$

The fractional part is subtracted first, followed by the whole number part. Unfortunately, $\frac{5}{8}$ cannot be subtracted from $\frac{1}{6}$ at this time, so we rename $1 = \frac{6}{6}$ and rewrite the problem as

$$2\frac{7}{6} \qquad 3\frac{1}{6} = 2 + \frac{6}{6} + \frac{1}{6} = 2\frac{7}{6}$$

$$-2\frac{5}{8}$$

Since the LCM is 24, we rewrite $\frac{7}{6}$ and $\frac{5}{8}$ with 24 as the denominator.

$$2\frac{7 \cdot 4}{6 \cdot 4} = 2\frac{28}{24}$$

$$-2\frac{5 \cdot 3}{8 \cdot 3} = 2\frac{15}{24}$$

$$\frac{13}{24}$$

The complete procedure can be written as follows:

$$
\begin{array}{ccccc}
3\frac{1}{6} & & 2\frac{7}{6} & & 2\frac{28}{24} \\
& \longrightarrow & & \longrightarrow & \\
-2\frac{5}{8} & & -2\frac{5}{8} & & -2\frac{15}{24} \\
\hline
& & & & \frac{13}{24}
\end{array}
$$

Of course, the answer is the same as before.

Now we can do a problem involving both addition and subtraction of fractions.

EXAMPLE 9   Adding and subtracting fractions

Perform the indicated operations: $1\frac{1}{10} + \frac{5}{21} - \frac{3}{28}$

SOLUTION   We first write $1\frac{1}{10}$ as $\frac{11}{10}$ and then find the LCM of 10, 21, and 28. (Use either method.)

Method 1.

$$10 = 2 \cdot 5$$
$$21 = 3 \cdot 7$$
$$28 = 2 \cdot 2 \cdot 7$$

The LCD is $2 \cdot 2 \cdot 3 \cdot 5 \cdot 7 = 420$.

Method 2.

Step 1.   2) 10 ㉑ 28

Step 2.   7) ⑤ 21 14
                    5   3   2

Step 3.   The LCD is
              $2 \cdot 7 \cdot 5 \cdot 3 \cdot 2 = 420$.

Now

$$1\frac{1}{10} = \frac{11}{10} = \frac{11 \cdot 42}{10 \cdot 42} = \frac{462}{420} \qquad\longrightarrow\qquad \frac{462}{420}$$

$$\frac{5}{21} = \frac{5 \cdot 20}{21 \cdot 20} = \frac{100}{420} \qquad\longrightarrow\qquad + \frac{100}{420}$$

$$\frac{3}{28} = \frac{3 \cdot 15}{28 \cdot 15} = \frac{45}{420} \qquad\longrightarrow\qquad - \frac{45}{420}$$

$$\frac{517}{420}$$

Horizontally, we have

$$1\frac{1}{10} + \frac{5}{21} - \frac{3}{28} = \frac{462}{420} + \frac{100}{420} - \frac{45}{420}$$

$$= \frac{462 + 100 - 45}{420}$$

$$= \frac{517}{420} \quad \text{or} \quad 1\frac{97}{420}$$

PROBLEM 9

Perform the indicated operations:

$1\frac{1}{8} + \frac{7}{21} - \frac{3}{28}$

Answers to PROBLEMS

9. $\frac{227}{168}$ or $1\frac{59}{168}$

EXAMPLE 10   Application: Operations with fractions

How many total cups of water and oat bran should be mixed to prepare 3 servings? (Refer to the instructions in the illustration.)

SOLUTION   We need $3\frac{3}{4}$ cups of water and $1\frac{1}{2}$ cups of oat bran,

a total of $3\frac{3}{4} + 1\frac{1}{2}$ cups.

Write $3\frac{3}{4}$ and $1\frac{1}{2}$ as improper fractions. $\dfrac{15}{4} + \dfrac{3}{2}$

The LCD is 4. Rewrite as $\dfrac{15}{4} + \dfrac{6}{4}$

Add the fractions. $\dfrac{21}{4}$

Rewrite as a mixed number. $5\frac{1}{4}$

Note that this time it is more appropriate to write the answer as a *mixed number*. In general, it is acceptable to write the result of an operation involving fractions as an *improper fraction*.

PROBLEM 10

How many total cups of water and oat bran should be mixed to prepare 1 serving?

Hot Cereal

MICROWAVE AND STOVE TOP AMOUNTS

SERVINGS	1	2	3
WATER	1-1/4 cups	2-1/2 cups	3-3/4 cups
OAT BRAN	1/2 cup	1 cup	1-1/2 cups
SALT (optional)	dash	dash	1/8 tsp

MathZone

Boost your grade at mathzone.com!
> Practice Problems   > Self-Tests
> NetTutor   > e-Professors
> Videos

❭ Exercises R.2

❬ **A** ❭   **Adding, Subtracting, Multiplying, and Dividing Fractions**   In Problems 1–50, perform the indicated operations and reduce your answers to lowest terms.

1. $\dfrac{2}{3} \cdot \dfrac{7}{3}$

2. $\dfrac{3}{4} \cdot \dfrac{7}{8}$

3. $\dfrac{6}{5} \cdot \dfrac{7}{6}$

4. $\dfrac{2}{5} \cdot \dfrac{5}{3}$

5. $\dfrac{7}{3} \cdot \dfrac{6}{7}$

6. $\dfrac{5}{6} \cdot \dfrac{3}{5}$

7. $7 \cdot \dfrac{8}{7}$

8. $10 \cdot 1\frac{1}{5}$

9. $2\frac{3}{5} \cdot 2\frac{1}{7}$

10. $2\frac{1}{3} \cdot 4\frac{1}{2}$

11. $7 \div \dfrac{3}{5}$

12. $5 \div \dfrac{2}{3}$

13. $\dfrac{3}{5} \div \dfrac{9}{10}$

14. $\dfrac{2}{3} \div \dfrac{6}{7}$

15. $\dfrac{9}{10} \div \dfrac{3}{5}$

16. $\dfrac{3}{4} \div \dfrac{3}{4}$

17. $1\frac{1}{5} \div \dfrac{3}{8}$

18. $3\frac{3}{4} \div 3$

19. $2\frac{1}{2} \div 6\frac{1}{4}$

20. $3\frac{1}{8} \div 1\frac{1}{3}$

21. $\dfrac{1}{5} + \dfrac{2}{5}$

22. $\dfrac{1}{3} + \dfrac{1}{3}$

23. $\dfrac{3}{8} + \dfrac{5}{8}$

24. $\dfrac{2}{9} + \dfrac{4}{9}$

25. $\dfrac{7}{8} + \dfrac{3}{4}$

26. $\dfrac{1}{2} + \dfrac{1}{6}$

27. $\dfrac{5}{6} + \dfrac{3}{10}$

28. $3 + \dfrac{2}{5}$

29. $2\frac{1}{3} + 1\frac{1}{2}$

30. $1\frac{3}{4} + 2\frac{1}{6}$

31. $\dfrac{1}{5} + 2 + \dfrac{9}{10}$

32. $\dfrac{1}{6} + \dfrac{1}{8} + \dfrac{1}{3}$

33. $3\frac{1}{2} + 1\frac{1}{7} + 2\frac{1}{4}$

34. $1\frac{1}{3} + 2\frac{1}{4} + 1\frac{1}{5}$

35. $\dfrac{5}{8} - \dfrac{2}{8}$

36. $\dfrac{3}{7} - \dfrac{1}{7}$

37. $\dfrac{1}{3} - \dfrac{1}{6}$

38. $\dfrac{5}{12} - \dfrac{1}{4}$

39. $\dfrac{7}{10} - \dfrac{3}{20}$

40. $\dfrac{5}{20} - \dfrac{7}{40}$

41. $\dfrac{8}{15} - \dfrac{2}{25}$

42. $\dfrac{7}{8} - \dfrac{5}{12}$

43. $2\frac{1}{5} - 1\frac{3}{4}$

44. $4\frac{1}{2} - 2\frac{1}{3}$

45. $3 - 1\frac{3}{4}$

46. $2 - 1\frac{1}{3}$

47. $\dfrac{5}{6} + \dfrac{1}{9} - \dfrac{1}{3}$

48. $\dfrac{3}{4} + \dfrac{1}{12} - \dfrac{1}{6}$

49. $1\frac{1}{3} + 2\frac{1}{3} - 1\frac{1}{5}$

50. $3\frac{1}{2} + 1\frac{1}{7} - 2\frac{1}{4}$

Answers to PROBLEMS

10. $1\frac{3}{4}$

>>> Applications

51. *Weights on Earth and the moon*   The weight of an object on the moon is $\frac{1}{6}$ of its weight on Earth. How much did the Lunar Rover, weighing 450 pounds on Earth, weigh on the moon?

52. *Do you want to meet a millionaire?*   Your best bet is to go to Idaho. In this state, $\frac{1}{38}$ of the population happens to be millionaires! If there are about 912,000 persons in Idaho, about how many millionaires are there in the state?

53. *Union membership*   The Actors Equity has 28,000 members, but only $\frac{1}{5}$ of the membership is working at any given moment. How many working actors is that?

54. *Battery voltage*   The voltage of a standard C battery is $1\frac{1}{2}$ volts (V). What is the total voltage of 6 such batteries?

55. *Excavation*   An earthmover removed $66\frac{1}{2}$ cubic yards of sand in 4 hours. How many cubic yards did it remove per hour?

56. *Bond prices*   A bond is selling for $\$3\frac{1}{2}$. How many can be bought with \$98?

57. *Household incomes*   In a recent survey, it was found that $\frac{9}{50}$ of all American households make more than \$25,000 annually and that $\frac{3}{25}$ make between \$20,000 and \$25,000. What total fraction of American households make more than \$20,000?

58. *Immigration*   Between 1971 and 1977, $\frac{1}{5}$ of the immigrants coming to America were from Europe and $\frac{9}{20}$ came from this hemisphere. What total fraction of the immigrants were in these two groups?

59. *Employment*   U.S. workers work an average of $46\frac{3}{5}$ hours per week, while Canadians work $38\frac{9}{10}$. How many more hours per week do U.S. workers work?

60. *Bone composition*   Human bones are $\frac{1}{4}$ water, $\frac{3}{10}$ living tissue, and the rest minerals. Thus, the fraction of the bone that is minerals is $1 - \frac{1}{4} - \frac{3}{10}$. Find this fraction.

>>> Using Your Knowledge

Hot Dogs and Buns   In this section, we learned that the LCM of two numbers is the *smallest* multiple of the two numbers. Can you ever apply this idea to anything besides adding and subtracting fractions? You bet! Hot dogs and buns. Have you noticed that hot dogs come in packages of 10 but buns come in packages of 8 (or 12)?

61. What is the smallest number of packages of hot dogs (10 to a package) and buns (8 to a package) you must buy so that you have as many hot dogs as you have buns? (*Hint:* Think of multiples.)

62. If buns are sold in packages of 12, what is the smallest number of packages of hot dogs (10 to a package) and buns you must buy so that you have as many hot dogs as you have buns?

>>> Write On

63. Write the procedure you use to find the LCM of two numbers.

64. Write the procedure you use to convert a mixed number to an improper fraction.

65. Write the procedure you use to:
   a. Multiply two fractions.
   b. Divide two fractions.

66. Write the procedure you use to:
   a. Add two fractions with the same denominator.

   b. Add two fractions with different denominators.

   c. Subtract two fractions with the same denominator.

   d. Subtract two fractions with different denominators.

>>> Skill Checker

The *Skill Checker* exercises practice skills previously studied. You will need those skills to succeed in the next section!
From now on, the Skill Checker Exercises will appear at the end of every exercise set.

67. Divide 128 by 16.   **68.** Divide 2100 by 35.   **69.** Reduce: $\frac{20}{100}$ **70.** Reduce: $\frac{75}{100}$

R.3   Decimals and Percents

▶ Objectives

A ⟩ Write a decimal in expanded form.

B ⟩ Write a decimal as a fraction.

C ⟩ Add and subtract decimals.

D ⟩ Multiply and divide decimals.

E ⟩ Write a fraction as a decimal.

F ⟩ Convert decimals to percents and percents to decimals.

G ⟩ Convert fractions to percents and percents to fractions.

H ⟩ Round numbers to a specified number of decimal places.

▶ To Succeed, Review How To . . .

1. Divide one whole number by another one.
2. Reduce fractions (pp. 4–6).

▶ Getting Started
Decimals at the Gas Pump

The price of Unleaded Plus gas is $2.939 (read "two point nine three nine"). This number is called a **decimal.** The word *decimal* means that we count by *tens*—that is, we use *base ten*. The dot in 2.939 is called the *decimal point*. The decimal 2.939 consists of two parts: the whole-number part (the number to the *left* of the decimal point) and the decimal part, 0.939.

$$2.939$$

Whole number

Decimal

A ⟩ Writing Decimals in Expanded Form

The number 11.664 can be written in a diagram, as shown. With the help of this diagram, we can write the number in **expanded form,** like this:

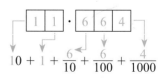

$$10 + 1 + \frac{6}{10} + \frac{6}{100} + \frac{4}{1000}$$

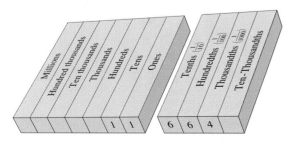

The number 11.664 is read by reading the number to the left of the decimal (*eleven*), using the word *and* for the decimal point, and then reading the number to the right of the decimal point followed by the place value of the *last* digit: *eleven and six hundred sixty-four thousandths.*

EXAMPLE 1   Writing decimals in expanded form

Write in expanded form: 35.216

SOLUTION

$$30 + 5 + \frac{2}{10} + \frac{1}{100} + \frac{6}{1000}$$

PROBLEM 1

Write in expanded form: 46.325

B Writing Decimals as Fractions

Decimals can be converted to fractions. For example,

$$0.3 = \frac{3}{10}$$

$$0.18 = \frac{18}{100} = \frac{9}{50}$$

and

$$0.150 = \frac{150}{1000} = \frac{3}{20}$$

Here is the procedure for changing a decimal to a fraction.

> **PROCEDURE**
>
> **Changing Decimals to Fractions**
> 1. Write the nonzero digits of the number, omitting the decimal point, as the *numerator* of the fraction.
> 2. The *denominator* is a 1 *followed by as many zeros as there are decimal digits in the decimal.*
> 3. Reduce, if possible.

EXAMPLE 2   Changing decimals to fractions

Write as a reduced fraction:

a. 0.035

b. 0.0275

SOLUTION

a. $0.035 = \frac{35}{1000} = \frac{7}{200}$

   3 digits   3 zeros

b. $0.0275 = \frac{275}{10,000} = \frac{11}{400}$

   4 digits   4 zeros

PROBLEM 2

Write as a reduced fraction:

a. 0.045   **b.** 0.0375

If the decimal has a whole-number part, we write all the digits of the number, omitting the decimal point, as the numerator of the fraction. For example, to write 4.23 as a fraction, we write

$$4.23 = \frac{423}{100}$$

2 digits 2 zeros

EXAMPLE 3   Changing decimals with whole-number parts

Write as a reduced fraction:

a. 3.11

b. 5.154

SOLUTION

a. $3.11 = \frac{311}{100}$

b. $5.154 = \frac{5154}{1000} = \frac{2577}{500}$

PROBLEM 3

Write as a reduced fraction:

a. 3.13   **b.** 5.258

Answers to PROBLEMS

1. $40 + 6 + \frac{3}{10} + \frac{2}{100} + \frac{5}{1000}$ **2. a.** $\frac{9}{200}$ **b.** $\frac{3}{80}$ **3. a.** $\frac{313}{100}$ **b.** $\frac{2629}{500}$

C › Adding and Subtracting Decimals

Adding decimals is similar to adding whole numbers, as long as we first line up (align) the decimal points by writing them in the *same* column and then make sure that digits of the same place value are in the same column. We then add the columns from right to left, as usual. For example, to add 4.13 + 5.24, we write

$$
\begin{array}{r}
4.13 \\
+\,5.24 \\
\hline
9.37
\end{array}
$$

Align the decimal points; that is, write them in the same column.

Add hundredths first.
Add tenths next.
Then add units.

The result is 9.37.

EXAMPLE 4   Adding decimals

Add: 5 + 18.15

SOLUTION   Note that 5 = 5. and attach 00 to the 5. so that both addends have the same number of decimal digits.

$$
\begin{array}{r}
5.00 \\
+\,18.15 \\
\hline
23.15
\end{array}
$$

Align the decimal points.

Add hundredths.
Add tenths.
Add units. 5 + 8 = 13. Write 3, carry 1.
Add tens.

The result is 23.15.

PROBLEM 4

Add: 6 + 17.25

Subtraction of decimals is also like subtraction of whole numbers as long as you remember to align the decimal points and attach any needed zeros so that both numbers have the same number of decimal digits. For example, if you earned $231.47 and you made a $52 payment, you have $231.47 − $52. To find how much you have left, we write

$$
\begin{array}{r}
\$\,231.47 \\
-\quad 52.00 \\
\hline
179.47
\end{array}
$$

Align the decimal points.
Attach two zeros {$52 is the same as 52 dollars and no cents}.

The decimal point is in the same column.

Thus, you have $179.47 left. You can check this answer by adding 52 and 179.47 to obtain 231.47.

▤ ◈ ▥   Calculator Corner

If you have a calculator, you do not have to worry about placing the decimal point or inserting place-holder zeros when adding or subtracting decimal numbers. For example, to add 4.13 + 5.24 simply enter [4] [·] [1] [3] [+] [5] [·] [2] [4] and press [ENTER]. The answer 9.37 will appear on the display screen. Similar keystrokes can be used to subtract decimal numbers.

Answers to PROBLEMS

4. 23.25

EXAMPLE 5   Subtracting decimals
Subtract: $383.43 - 17.5$

SOLUTION

$$
\begin{array}{r}
^{7}\,{}^{12}\!\!\downarrow^{14} \\
3\ \not{8}\ \not{3}.4\ 3 \\
-\quad 1\ 7.5\ 0 \\
\hline
3\ 6\ 5.9\ 3
\end{array}
$$

Align the decimal points.
Attach one zero.
Subtract.
The decimal point is in the same column.

PROBLEM 5
Subtract: $273.32 - 14.5$

EXAMPLE 6   Subtracting decimals
Subtract: $347.8 - 182.231$

SOLUTION

$$
\begin{array}{r}
{}^{9} \\
{}^{2}\,{}^{14}\ \ {}^{7}\,{}^{10}\,{}^{10} \\
\not{3}\ \not{4}\ 7.\not{8}\ 0\ 0 \\
-1\ 8\ 2.2\ 3\ 1 \\
\hline
1\ 6\ 5.5\ 6\ 9
\end{array}
$$

Align the decimal points.
Attach two zeros.

Subtract.
The decimal point is in the same column.

PROBLEM 6
Subtract: $458.6 - 193.341$

D › Multiplying and Dividing Decimals

When multiplying decimals, the number of decimal digits in the product is the *sum* of the numbers of decimal digits in the factors. For example, since 0.3 has *one* decimal digit and 0.0007 has *four*, the product

$$0.3 \times 0.0007 = \frac{3}{10} \times \frac{7}{10{,}000} = \frac{21}{100{,}000} = 0.00021$$

$1 + 4 = 5$ digits

has $1 + 4 = 5$ decimal digits.

Here is the procedure used to multiply decimals.

PROCEDURE

Multiplying Decimals

1. *Multiply* the two decimal numbers *as if they were whole numbers.*
2. The *number of decimal digits* in the product is the *sum* of the numbers of decimal digits in the factors.

EXAMPLE 7   Multiplying Decimals
Multiply:

a. 5.102×21.03

b. 5.213×0.0012

SOLUTION

a.
$$
\begin{array}{r}
5.102 \\
\times\ 21.03 \\
\hline
15306 \\
51020\ \ \\
10204\ \ \ \ \\
\hline
107.29506
\end{array}
$$
3 decimal digits
2 decimal digits

$3 + 2 = 5$ decimal digits

b.
$$
\begin{array}{r}
5.213 \\
\times\ 0.0012 \\
\hline
10426 \\
5213\ \ \\
\hline
0.0062556
\end{array}
$$
3 decimal digits
4 decimal digits

$3 + 4 = 7$ decimal digits

Attach two zeros to obtain 7 decimal digits.

PROBLEM 7
Multiply:

a. 6.203×31.03

b. 3.123×0.0015

Answers to PROBLEMS

5. 258.82   **6.** 265.259
7. a. 192.47909   **b.** 0.0046845

Now suppose you want to divide 52 by 6.5. You can write

$$\frac{52}{6.5} \quad \longleftarrow \text{This is called the dividend.}$$
$$\phantom{\frac{52}{6.5}} \quad \longleftarrow \text{This is called the divisor.}$$

If we multiply the numerator and denominator of this fraction by 10, we obtain

$$\frac{52}{6.5} = \frac{52 \times 10}{6.5 \times 10} = \frac{520}{65} = 8$$

Thus $\frac{52}{6.5} = 8$, as can be easily checked, since $52 = 6.5 \times 8$. This problem can be shortened by using the following steps.

Step 1. Write the problem in the usual long division form.

$$6.5\overline{)52}$$

Divisor   Dividend

Step 2. Move the decimal point in the divisor, 6.5, to the right until a whole number is obtained. (This is the same as multiplying 6.5 by 10.)

$$65.\overline{)52}$$

Multiply by 10.

Step 3. Move the decimal point in the dividend the *same* number of places as in step 2. (This is the same as multiplying the dividend 52 by 10.) Attach zeros if necessary.

$$65.\overline{)520}$$

Multiply by 10.

Step 4. Place the decimal point in the answer directly above the new decimal point in the dividend.

$$65\overline{)520.}$$

Step 5. Divide exactly as you would divide whole numbers. The result is 8, as before.

$$\begin{array}{r} 8. \\ 65\overline{)520.} \\ \underline{520} \\ 0 \end{array}$$

Here is another example. Divide

$$\frac{1.28}{1.6}$$

Step 1. Write the problem in the usual long division form.

$$1.6\overline{)1.28}$$

Step 2. Move the decimal point in the divisor to the right until a whole number is obtained.

$$16.\overline{)1.28}$$

Decimal moved

Step 3. Move the decimal point in the dividend the *same* number of places as in step 2.

$$16.\overline{)12.8}$$

Decimal moved

Step 4. Place the decimal point in the answer directly above the new decimal point in the dividend.

$$16\overline{)12.8}$$

Step 5. Divide exactly as you would divide whole numbers.

$$\begin{array}{r} 0.8 \\ 16\overline{)12.8} \\ \underline{12.8} \\ 0 \end{array}$$

Thus,

$$\frac{1.28}{1.6} = 0.8$$

Calculator Corner

Dividing Decimals

To divide decimals using a calculator, recall that the bar in $\frac{1.28}{1.6}$ means to divide. Accordingly, we enter 1.28 ÷ 1.6 ENTER. Do not worry about the decimal point; the answer comes out as 0.8 as before.

EXAMPLE 8   Dividing decimals

Divide: $\frac{2.1}{0.035}$

SOLUTION   We move the decimal point in the divisor (and also in the dividend) three places to the right. To do this, we need to attach two zeros to 2.1.

$$035.\overline{)2100.}$$

Multiply by 1000.

PROBLEM 8

Divide: $\frac{1.8}{0.045}$

Answers to PROBLEMS

8. 40

We next place the decimal point in the answer directly above the one in the dividend and proceed in the usual manner.

$$
\begin{array}{r}
60. \\
35\overline{)2100.} \\
\underline{210} \\
00
\end{array}
$$

The answer is 60; that is,

$$
\frac{2.1}{0.035} = 60
$$

CHECK $0.035 \times 60 = 2.100$

E ⟩ Writing Fractions as Decimals

Since a fraction indicates a division, we can write a fraction as a decimal by dividing. For example, $\frac{3}{4}$ means $3 \div 4$, so

$$
\begin{array}{r}
0.75 \\
4\overline{)3.00} \\
\underline{2\,8} \\
20 \\
\underline{20} \\
0
\end{array}
$$ Note that 3 = 3.00.

Here the division **terminates** (has a 0 remainder). Thus, $\frac{3}{4} = 0.75$, and 0.75 is called a **terminating decimal.**

Since $\frac{2}{3}$ means $2 \div 3$,

$$
\begin{array}{r}
0.666\ldots \\
3\overline{)2.000} \\
\underline{1\,8} \\
20 \\
\underline{18} \\
20 \\
\underline{18} \\
2
\end{array}
$$ The ellipsis (. . .) means that the 6 repeats.

The division continues because the remainder 2 repeats.

Hence, $\frac{2}{3} = 0.666\ldots$. Such decimals (called **repeating decimals**) can be written by placing a bar over the repeating digits; that is, $\frac{2}{3} = 0.\overline{6}$.

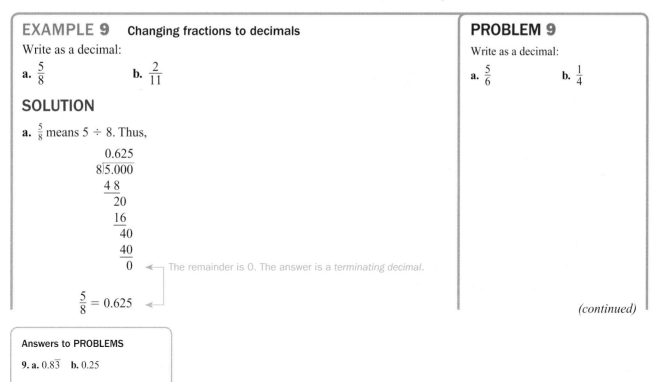

EXAMPLE 9   Changing fractions to decimals

Write as a decimal:

a. $\frac{5}{8}$ **b.** $\frac{2}{11}$

SOLUTION

a. $\frac{5}{8}$ means $5 \div 8$. Thus,

$$
\begin{array}{r}
0.625 \\
8\overline{)5.000} \\
\underline{4\,8} \\
20 \\
\underline{16} \\
40 \\
\underline{40} \\
0
\end{array}
$$ ← The remainder is 0. The answer is a *terminating decimal.*

$$
\frac{5}{8} = 0.625
$$ ←

PROBLEM 9

Write as a decimal:

a. $\frac{5}{6}$ **b.** $\frac{1}{4}$

(continued)

Answers to PROBLEMS

9. a. $0.8\overline{3}$ **b.** 0.25

b. $\frac{2}{11}$ means $2 \div 11$. Thus,

The remainders 9 and 2 alternately repeat. The answer 0.1818 . . . is a *repeating decimal.*

$\frac{2}{11} = 0.\overline{18}$ The bar is written over the digits that repeat.

⊞ ◈ ▦ Calculator Corner

Changing Fractions to Decimals

To change a fraction to a decimal using a calculator, just perform the indicated operations. Thus, to convert $\frac{2}{3}$ to a decimal, enter **2** **÷** **3** **ENTER** and you get the answer 0.666666666 or 0.666666667. Note that you get eight 6's and one 7. How do you know the 6 repeats indefinitely? At this point, you do not. We shall discuss this in the Collaborative Learning section.

F › Converting Decimals and Percents

An important application of decimals is the study of **percents.** The word *percent* means "per one hundred" and is written using the symbol %. Thus,

$$29\% = \frac{29}{100}$$

or "twenty-nine hundredths," that is, 0.29.

Similarly,

$$43\% = \frac{43}{100} = 0.43, \quad 87\% = \frac{87}{100} = 0.87, \quad \text{and} \quad 4.5\% = \frac{4.5}{100} = \frac{45}{1000} = 0.045$$

Note that in each case, the number is divided by 100, which is equivalent to moving the decimal point two places to the *left.*

Here is the general procedure.

PROCEDURE

Converting Percents to Decimals

Move the decimal point in the number *two* places to the *left* and omit the % symbol.

EXAMPLE 10   Converting percents to decimals

Write as a decimal:

a. 98%   **b.** 34.7%   **c.** 7.2%

SOLUTION

a. $98\% = 98. = 0.98$ Recall that a decimal point follows every whole number.

b. $34.7\% = 34.7 = 0.347$

c. $7.2\% = 07.2 = 0.072$ Note that a 0 was inserted so we could move the decimal point two places to the left.

PROBLEM 10

Write as a decimal:

a. 86%   **b.** 48.9%   **c.** 8.3%

Answers to PROBLEMS

10. a. 0.86   **b.** 0.489   **c.** 0.083

As you can see from Example 10, $0.98 = 98\%$ and $0.347 = 34.7\%$. Thus, we can reverse the previous procedure to convert decimals to percents. Here is the way we do it.

> **PROCEDURE**
>
> **Converting Decimals to Percents**
> Move the decimal point in the number *two* places to the *right* and attach the % symbol.

EXAMPLE 11   Converting decimals to percents

Write as a percent:

a. 0.53   **b.** 3.19   **c.** 64.7

SOLUTION

a. $0.53 = 0.53\% = 53\%$

b. $3.19 = 3.19\% = 319\%$

c. $64.7 = 64.70\% = 6470\%$ Note that a 0 was inserted so we could move the decimal point two places to the right.

PROBLEM 11

Write as a percent:

a. 0.49   **b.** 4.17   **c.** 89.2

G ﹥ Converting Fractions and Percents

Since % means *per hundred*,

$$5\% = \frac{5}{100}$$

$$7\% = \frac{7}{100}$$

$$23\% = \frac{23}{100}$$

$$4.7\% = \frac{4.7}{100}$$

$$134\% = \frac{134}{100}$$

Thus, we can convert a number written as a percent to a fraction by using the following procedure.

> **PROCEDURE**
>
> **Converting Percents to Fractions**
> Write the *number* over 100, omit the % sign, and *reduce* the fraction, if possible.

EXAMPLE 12   Converting percents to fractions

Write as a fraction:

a. 49%   **b.** 75%

SOLUTION

a. $49\% = \frac{49}{100}$

b. $75\% = \frac{75}{100} = \frac{3}{4}$

PROBLEM 12

Write as a fraction:

a. 37%   **b.** 25%

Answers to PROBLEMS

11. a. 49%   **b.** 417%   **c.** 8920%

12. a. $\frac{37}{100}$ **b.** $\frac{1}{4}$

How do we write a fraction as a percent? If the fraction has a denominator that is a factor of 100, it's easy. To write $\frac{1}{5}$ as a percent, we first multiply numerator and denominator by a number that will make the denominator 100. Thus,

$$\frac{1}{5} = \frac{1 \times 20}{5 \times 20} = \frac{20}{100} = 20\%$$

Similarly,

$$\frac{3}{4} = \frac{3 \times 25}{4 \times 25} = \frac{75}{100} = 75\%$$

Note that in both cases the denominator of the fraction was a factor of 100.

EXAMPLE 13   **Converting fractions to percents**

Write as a percent: $\frac{4}{5}$

SOLUTION   We multiply by $\frac{20}{20}$ to get 100 in the denominator.

$$\frac{4}{5} = \frac{4 \times 20}{5 \times 20} = \frac{80}{100} = 80\%$$

PROBLEM 13

Write as a percent: $\frac{3}{5}$

In Example 13 the denominator of the fraction, $\frac{4}{5}$ (the 5), was a *factor* of 100. Suppose we wish to change $\frac{1}{6}$ to a percent. The problem here is that 6 is *not* a factor of 100. But don't panic! We can write $\frac{1}{6}$ as a percent by dividing the numerator by the denominator. Thus, we divide 1 by 6, continuing the division until we have two decimal digits.

$$
\begin{array}{r}
0.16 \\
6\overline{)1.00} \\
\underline{6} \\
40 \\
\underline{36} \\
4 \quad \text{Remainder}
\end{array}
$$

The answer is 0.16 with remainder 4; that is,

$$\frac{1}{6} = 0.16\frac{4}{6} = 0.16\frac{2}{3} = \frac{16\frac{2}{3}}{100} = 16\frac{2}{3}\%$$

Similarly, we can write $\frac{2}{3}$ as a percent by dividing 2 by 3, obtaining

$$
\begin{array}{r}
0.66 \\
3\overline{)2.00} \\
\underline{1\,8} \\
20 \\
\underline{18} \\
2 \quad \text{Remainder}
\end{array}
$$

Thus,

$$\frac{2}{3} = 0.66\frac{2}{3} = \frac{66\frac{2}{3}}{100} = 66\frac{2}{3}\%$$

Here is the procedure we use.

PROCEDURE

Converting Fractions to Percents

Divide the *numerator* by the *denominator* (carry the division to two decimal places), convert the resulting decimal to a percent by *moving the decimal point two places to the right,* and attach the % symbol. (Don't forget to include the remainder, if there is one.)

Answers to PROBLEMS

13. 60%

EXAMPLE 14 Converting fractions to percents	**PROBLEM 14**
Write as a percent: $\frac{5}{8}$	Write as a percent: $\frac{3}{8}$

SOLUTION   Dividing 5 by 8, we have

$$\begin{array}{r} 0.62 \\ 8\overline{)5.00} \\ \underline{4\ 8} \\ 20 \\ \underline{16} \\ 4 \end{array}$$

Note that the remainder is $\frac{4}{8} = \frac{1}{2}$.

Thus,

$$\frac{5}{8} = 0.62\frac{1}{2} = 0.62\frac{1}{2}\% = 62\frac{1}{2}\%$$

H ⟩ Rounding Numbers

We can shorten some of the decimals we have considered in this section if we use **rounding**. For example, the numbers 107.29506 and 0.062556 can be rounded to three decimal places—that is, to the nearest *thousandth*. Here is the rule we use.

> **RULE**
>
> **To Round Numbers**
> **Step 1.** **Underline** the digit in the place to which you are rounding.
> **Step 2.** If the first digit to the right of the underlined digit is *5 or more*, add 1 to the underlined digit. Otherwise, do not change the underlined digit.
> **Step 3.** **Drop** all the numbers to the right of the underlined digit (attach 0's to fill in the place values if necessary).

EXAMPLE 15 Rounding decimals	**PROBLEM 15**
Round to three decimal places:	Round to two decimal places:
a. 107.29506 **b.** 0.062556	**a.** $0.\overline{6}$ **b.** $0.\overline{18}$

SOLUTION

a. Step 1. Underline the 5.   1 0 7 . 2 9 5 0 6

    Step 2. The digit to the right   1 0 7 . 2 9 5 0 6
          of 5 is 0, so we do not
          change the 5.

    Step 3. Drop all numbers to   1 0 7 . 2 9 5
          the right of 5.

Thus, when 107.29506 is rounded to three decimal places, that is, to the nearest thousandth, the answer is 107.295.

b. Step 1. Underline the 2.   0 . 0 6 2 5 5 6

    Step 2. The digit to the right   0 . 0 6 2 5 5 6
          of 2 is 5, so we add 1
          to the underlined digit.

    Step 3. Drop all numbers to   0 . 0 6 3
          the right of 3.

Thus, when 0.062556 is rounded to three decimal places, that is, to the nearest thousandth, the answer is 0.063.

Answers to PROBLEMS
14. $37\frac{1}{2}\%$ **15. a.** 0.67   **b.** 0.18

› Exercises **R.3**

‹ **A** › Writing Decimals in Expanded Form   In Problems 1–10, write the given decimal in expanded form.

1. 4.7

2. 3.9

3. 5.62

4. 9.28

5. 16.123

6. 18.845

7. 49.012

8. 93.038

9. 57.104

10. 85.305

‹ **B** › Writing Decimals as Fractions   In Problems 11–20, write the given decimal as a reduced fraction.

11. 0.9

12. 0.7

13. 0.06

14. 0.08

15. 0.12

16. 0.18

17. 0.054

18. 0.062

19. 2.13

20. 3.41

‹ **C** › Adding and Subtracting Decimals   In Problems 21–40, add or subtract as indicated.

21. $648.01 + $341.06

22. $237.49 + $458.72

23. 72.03 + 847.124

24. 13.12 + 108.138

25. 104 + 78.103

26. 184 + 69.572

27. 0.35 + 3.6 + 0.127

28. 5.2 + 0.358 + 21.005

29. 27.2 − 0.35

30. 4.6 − 0.09

31. $19 − $16.62

32. $99 − $0.61

33. 9.43 − 6.406

34. 9.08 − 3.465

35. 8.2 − 1.356

36. 6.3 − 4.901

37. 6.09 + 3.0046

38. 2.01 + 1.3045

39. 4.07 + 8.0035

40. 3.09 + 5.4895

‹ **D** › Multiplying and Dividing Decimals   In Problems 41–60, multiply or divide as indicated.

41. 9.2×0.613

42. 0.514×7.4

43. 8.7×11

44. 78.1×108

45. 7.03×0.0035

46. 8.23×0.025

47. 3.0012×4.3

48. 6.1×2.013

49. 0.0031×0.82

50. 0.51×0.0045

51. $15\overline{)9}$

52. $48\overline{)6}$

53. $5\overline{)32}$

54. $8\overline{)36}$

55. $8.5 \div 0.005$

56. $4.8 \div 0.003$

57. $4 \div 0.05$

58. $18 \div 0.006$

59. $2.76 \div 60$

60. $31.8 \div 30$

‹ **E** › Writing Fractions as Decimals   In Problems 61–76, write the given fraction as a decimal.

61. $\frac{1}{5}$

62. $\frac{1}{2}$

63. $\frac{7}{8}$

64. $\frac{1}{8}$

65. $\frac{3}{16}$

66. $\frac{5}{16}$

67. $\frac{2}{9}$

68. $\frac{4}{9}$

69. $\frac{6}{11}$ **70.** $\frac{7}{11}$ **71.** $\frac{3}{11}$ **72.** $\frac{5}{11}$

73. $\frac{1}{6}$ **74.** $\frac{5}{6}$ **75.** $\frac{10}{9}$ **76.** $\frac{11}{9}$

⟨ **F** ⟩ **Converting Decimals and Percents**   In Problems 77–86, write the given percent as a decimal.

77. 33%   **78.** 52%   **79.** 5%   **80.** 9%

81. 300%   **82.** 500%   **83.** 11.8%   **84.** 89.1%

85. 0.5%   **86.** 0.7%

In Problems 87–96, write the given decimal as a percent.

87. 0.05   **88.** 0.07   **89.** 0.39   **90.** 0.74

91. 0.416   **92.** 0.829   **93.** 0.003   **94.** 0.008

95. 1.00   **96.** 2.1

⟨ **G** ⟩ **Converting Fractions and Percents**   In Problems 97–106, write the given percent as a reduced fraction.

97. 30%   **98.** 40%   **99.** 6%   **100.** 2%

101. 7%   **102.** 19%   **103.** $4\frac{1}{2}$%   **104.** $2\frac{1}{4}$%

105. $1\frac{1}{3}$%   **106.** $5\frac{2}{3}$%

In Problems 107–116, write the given fraction as a percent.

107. $\frac{3}{5}$ **108.** $\frac{4}{25}$ **109.** $\frac{1}{2}$ **110.** $\frac{3}{50}$

111. $\frac{5}{6}$ **112.** $\frac{1}{3}$ **113.** $\frac{4}{8}$ **114.** $\frac{7}{8}$

115. $\frac{4}{3}$ **116.** $\frac{7}{6}$

⟨ **H** ⟩ **Rounding Numbers**   In Problems 117–126, round each number to the specified number of decimal places.

117. 27.6263; 1   **118.** 99.6828; 1   **119.** 26.746706; 4   **120.** 54.037755; 4

121. 35.24986; 3   **122.** 69.24851; 3   **123.** 52.378; 2   **124.** 6.724; 2

125. 74.846008; 5   **126.** 39.948712; 5

⟩ ⟩ ⟩   *Applications*

Source: www.clickz.com

Coupons   Do you clip coupons or do you get them from the Web? Here is a quote from the article Coupons Converge Online "*The survey finds 95.5 percent of respondents clip or print coupons, while 62.2 percent clip as often as once a week. Just over half (51.3 percent) the respondents redeem most of the coupons they save.*"

127. Write 95.5 percent as a decimal.   **128.** Write 95.5 percent as a reduced fraction.

129. Write one-half as a fraction, as a decimal, and as a percent.   **130.** Write 51.3% as a reduced fraction and as a decimal.

Is 51.3 percent just over one half? Explain.

〉〉〉 *Using Your Knowledge*

The Pursuit of Happiness   Psychology Today conducted a survey about happiness. Here are some conclusions from that report.

131. Seven out of ten people said they had been happy over the last six months. What percent of the people is that?

132. Of the people surveyed, 70% expected to be happier in the future than now. What fraction of the people is that?

133. The survey also showed that 0.40 of the people felt lonely.

   a. What percent of the people is that?

   b. What fraction of the people is that?

134. Only 4% of the men were ready to cry. Write this percent as a decimal.

135. Of the people surveyed, 49% were single. Write this percent as:

   a. A fraction.

   b. A decimal.

Do you wonder how they came up with some of these percents? They used their knowledge. You do the same and fill in the spaces in the accompanying table, which refers to the marital status of the 52,000 people surveyed. For example, the first line shows that 25,480 persons out of 52,000 were single. This is

$$\frac{25,480}{52,000} = 49\%$$

Fill in the percents in the last column.

	Marital Status	Number	Percent
	Single	25,480	49%
136.	Married (first time)	15,600	____
137.	Remarried	2600	____
138.	Divorced, separated	5720	____
139.	Widowed	520	____
140.	Cohabiting	2080	____

〉〉〉 *Skill Checker*

141. Find $4(10) - 2(5)$

142. Find $25,000 - 4750 - 2(3050)$

〉Collaborative Learning

Form several groups of four or five students each. The first group will convert the fractions $\frac{1}{2}, \frac{1}{3}, \frac{1}{4}, \frac{1}{5}$, and $\frac{1}{6}$ to decimals. The second group will convert the fractions $\frac{1}{7}, \frac{1}{8}, \frac{1}{9}, \frac{1}{10}$, and $\frac{1}{11}$ to decimals. The next group will convert $\frac{1}{12}, \frac{1}{13}, \frac{1}{14}, \frac{1}{15}$, and $\frac{1}{16}$ to decimals. Ask each group to show the fractions that have terminating decimal representations. You should have $\frac{1}{2}, \frac{1}{4}$, and so on. Now, look at the fractions whose decimal representations are not terminating. You should have $\frac{1}{3}, \frac{1}{6}$, and so on. Can you tell which fractions will have terminating decimal representations? (*Hint:* Look at the denominators of the terminating fractions!)

(Answers on page 34)

Visit www.mhhe.com/bello to view helpful videos that provide step-by-step solutions to several of the problems below.

1. Write -18 as a fraction with a denominator of 1.

2. Find a fraction equivalent to $\frac{3}{7}$ with a denominator of 21.

3. Find a fraction equivalent to $\frac{9}{15}$ with a denominator of 5.

4. Reduce $\frac{27}{54}$ to lowest terms.

5. Multiply: $5\frac{1}{4} \cdot \frac{32}{21}$.

6. Divide: $2\frac{1}{4} \div \frac{3}{8}$.

7. Add: $1\frac{1}{10} + \frac{5}{12}$.

8. Subtract: $4\frac{1}{6} - 1\frac{9}{10}$.

9. Write 68.428 in expanded form.

10. Write 0.045 as a reduced fraction.

11. Write 3.12 as a reduced fraction.

12. Add: $847.18 + 29.365$.

13. Subtract: $447.58 - 27.6$.

14. Multiply: 4.315×0.0013.

15. Divide: $\dfrac{4.2}{0.035}$.

16. **a.** Write $\frac{8}{11}$ as a decimal.

    b. Round the answer to two decimal places.

17. Write 84.8% as a decimal.

18. Write 0.69 as a percent.

19. Write 52% as a reduced fraction.

20. Write $\frac{5}{9}$ as a percent.

> Answers to Practice Test **Chapter R**

Answer	If You Missed	Review		
	Question	Section	Examples	Page
1. $\dfrac{-18}{1}$	1	R.1	1	3
2. $\dfrac{9}{21}$	2	R.1	2	4
3. $\dfrac{3}{5}$	3	R.1	3	4
4. $\dfrac{1}{2}$	4	R.1	4	5
5. 8	5	R.2	1, 2, 3	10–11
6. 6	6	R.2	4, 5	12
7. $\dfrac{91}{60}$	7	R.2	6	14
8. $\dfrac{34}{15}$	8	R.2	7, 8, 9	16–17
9. $60 + 8 + \dfrac{4}{10} + \dfrac{2}{100} + \dfrac{8}{1000}$	9	R.3	1	21
10. $\dfrac{9}{200}$	10	R.3	2	21
11. $\dfrac{78}{25}$	11	R.3	3	21–22
12. 876.545	12	R.3	4	22
13. 419.98	13	R.3	5, 6	23
14. 0.0056095	14	R.3	7	23
15. 120	15	R.3	8	24–25
16. a. $0.\overline{72}$ b. 0.73	16	R.3	9, 15	25–26, 29
17. 0.848	17	R.3	10	26
18. 69%	18	R.3	11	27
19. $\dfrac{13}{25}$	19	R.3	12	27
20. $55\frac{5}{9}\%$	20	R.3	13, 14	28–29

Section

Chapter

1

one

▶ ## Real Numbers and Their Properties

The Human Side of Algebra

The digits 1–9 originated with the Hindus and were passed on to us by the Arabs. Mohammed ibn Musa al-Khowarizmi (ca. A.D. 780–850) wrote two books, one on algebra (*Hisak al-jabr w'almuqabala,* "the science of equations"), from which the name *algebra* was derived, and one dealing with the Hindu numeration system. The oldest dated European manuscript containing the Hindu-Arabic numerals is the *Codex Vigilanus* written in Spain in A.D. 976, which used the nine symbols

Zero, on the other hand, has a history of its own. According to scholars, "At the time of the birth of Christ, the idea of zero as a symbol or number had never occurred to anyone." So, who invented zero? An unknown Hindu who wrote a symbol of his own, a dot he called *sunya,* to indicate a column with no beads on his counting board. The Hindu notation reached Europe thanks to the Arabs, who called it *sifr*. About A.D. 150, the Alexandrian astronomer Ptolemy began using *o* (omicron), the first letter in the Greek word for *nothing,* in the manner of our zero.

1.1   Introduction to Algebra

▶ Objectives

A ▷ Translate words into algebraic expressions.

B ▷ Evaluate algebraic expressions.

▶ To Succeed, Review How To . . .

Add, subtract, multiply, and divide numbers (see Chapter R).

▶ Getting Started

The poster uses the language of algebra to tell you how to be successful. The letters X, Y, and Z are used as *placeholders, unknowns,* or *variables.* The letters t, u, v, w, x, y, and z are frequently used as unknowns, and the letter x is used most often (x, y, and z are used in algebra because they are seldom used in ordinary words).

if...
A = SUCCESS

then...   when...
A = X + Y + Z   X = WORK
                          Y = PLAY
                          Z = LISTEN

A ▷ Translating into Algebraic Expressions

In arithmetic, we express ideas using **arithmetic expressions.** How do we express our ideas in algebra? We use **algebraic expressions,** which contain the variables x, y, z, and so on, of course. For example, $3k + 2$, $2x + y$, and $3x - 2y + 7z$ are *algebraic* expressions. Here is a comparison.

Arithmetic	Algebra
$9 + 51$	$9 + x$
$4.3 - 2$	$4.3 - y$
7×8.4	$7 \times y$ or $7y$
$\dfrac{3}{4}$	$\dfrac{a}{b}$

In algebra, it is better to write $7y$ instead of $7 \times y$ because the multiplication sign $\times$ can be easily confused with the letter x. Besides, look at the confusion that would result if we wrote x multiplied by x as $x \times x$! (You probably know that $x \times x$ is written as x^2.) From now on, we will try to avoid using $\times$ to indicate multiplication.

There are many words that indicate addition or subtraction. We list some of these here for your convenience.

Add (+)	Subtract (−)
Plus, sum, increase, more than	**Minus, difference, decrease, less than**
$a + b$ (read "a plus b") means:	$a - b$ (read "a minus b") means:
1. The sum of a and b	**1.** The difference of a and b
2. a increased by b	**2.** a decreased by b
3. b more than a	**3.** b less than a
4. b added to a	**4.** b subtracted from a
5. The total of a and b	**5.** a take away b

With this in mind, try Example 1.

EXAMPLE 1   Translations involving addition and subtraction

Translate into an algebraic expression.

a. The sum of x and y **b.** x minus y **c.** $7x$ plus $2a$ minus 3

SOLUTION

a. $x + y$ **b.** $x - y$ **c.** $7x + 2a - 3$

PROBLEM 1

Translate into an algebraic expression.

a. The sum of p and q

b. q minus p

c. $3q$ plus $5y$ minus 2

How do we write multiplication problems in algebra? We use the raised dot ($\cdot$) or parentheses ( ). Here are some ways of writing the product of a and b.

A raised dot:	$a \cdot b$
Parentheses:	$a(b)$, $(a)b$, or $(a)(b)$
Juxtaposition (writing a and b next to each other):	ab

In each of these cases, a and b (the variables, or numbers, to be multiplied) are called **factors.** Of course, the last notation must not be used when multiplying specific numbers because then "5 times 8" would be written as "58," which looks like fifty-eight. To the right are some words that indicate a multiplication.

We will use them in Example 2.

Multiply ($\times$ or $\cdot$)
___
times, of, product, multiplied by

EXAMPLE 2   Writing products using juxtaposition

Write the indicated products using juxtaposition.

a. 8 times x **b.** $-x$ times y times z

c. 4 times x times x **d.** $\frac{1}{5}$ of x

e. The product of 3 and x

SOLUTION

a. 8 times x is written as $8x$.   **b.** $-x$ times y times z is written as $-xyz$.

c. 4 times x times x is written as $4xx$.   **d.** $\frac{1}{5}x$

e. $3x$

PROBLEM 2

Write using juxtaposition.

a. -3 times x

b. a times b times c

c. 5 times a times a

d. $\frac{1}{2}$ of a

e. The product of 5 and a

What about division? In arithmetic we use the division ($\div$) sign to indicate division. In algebra we usually use fractions to indicate division. Thus, in arithmetic we write $15 \div 3$ (or $3\overline{)15}$) to indicate the quotient of 15 and 3. However, in algebra usually we write

$$\frac{15}{3}$$

Similarly, "the quotient of x and y" is written as

$$\frac{x}{y}$$

[We avoid writing $\frac{x}{y}$ as x/y because more complicated expressions such as $\frac{x}{y+z}$ then need to be written as $x/(y+z)$.]

Here are some words that indicate a division.

Divide ($\div$ or fraction bar —)
___
Divided by, quotient

We will use them in Example 3.

EXAMPLE 3   **Translations involving division**

Translate into an algebraic expression.

a. The quotient of x and 7

b. The quotient of 7 and x

c. The quotient of $(x + y)$ and z

d. The sum of a and b, divided by the difference of a and b

SOLUTION

a. $\dfrac{x}{7}$ **b.** $\dfrac{7}{x}$ **c.** $\dfrac{x + y}{z}$ **d.** $\dfrac{a + b}{a - b}$

PROBLEM 3

Translate into an algebraic expression.

a. The quotient of a and b

b. The quotient of b and a

c. The quotient of $(x - y)$ and z

d. The difference of x and y, divided by the sum of x and y

B ⟩ Evaluating Algebraic Expressions

As we have seen, algebraic expressions contain variables, operation signs, and numbers. If we substitute a value for one or more of the variables, we say that we are **evaluating the expression.** We shall see how this works in Example 4.

EXAMPLE 4   **Evaluating algebraic expressions**

Evaluate the given expressions by substituting 10 for x and 5 for y.

a. $x + y$ **b.** $x - y$ **c.** $4y$ **d.** $\dfrac{x}{y}$ **e.** $3x - 2y$

SOLUTION

a. Substitute 10 for x and 5 for y in $x + y$.

We obtain: $x + y = 10 + 5 = 15$.

The number 15 is called the **value** of $x + y$.

b. $x - y = 10 - 5 = 5$

c. $4y = 4(5) = 20$

d. $\dfrac{x}{y} = \dfrac{10}{5} = 2$

e. $3x - 2y = 3(10) - 2(5)$
$= 30 - 10$
$= 20$

PROBLEM 4

Evaluate the expressions by substituting 22 for a and 3 for b.

a. $a + b$ **b.** $2a - b$

c. $5b$ **d.** $\dfrac{2a}{b}$

e. $2a - 3b$

The terminology of this section and evaluating expressions are extremely important concepts in everyday life. Examine the federal income tax form excerpt shown in the figure. Can you find the words *subtract* and *multiply?* Suppose we use the variables A, S, and E to represent the adjusted gross income, the standard deduction, and the total *number* of exemptions, respectively. What expression will represent the taxable income, and how can we evaluate it? See Example 5!

Form 1040A (2003)			Page **2**
Tax, credits, and payments	**22**	Enter the amount from line 21 (adjusted gross income).	22 A
	23a	Check if: ☐ **You** were born before January 2, 1939, ☐ Blind ⎱ Total boxes ☐ **Spouse** was born before January 2, 1939, ☐ Blind ⎰ checked ▶ 23a	☐
Standard Deduction for—	**b**	If you are married filing separately and your spouse itemizes deductions, see page 32 and check here ▶ 23b	☐
• People who checked any box on line 23a or 23b **or** who can be	**24**	Enter your **standard deduction** (see left margin).	24 S
	25	Subtract line 24 from line 22. If line 24 is more than line 22, enter -0-.	25
	26	Multiply $3,050 by the total number of exemptions claimed on line 6d.	26 E
	27	Subtract line 26 from line 25. If line 26 is more than line 25, enter -0-. This is your **taxable income.** ▶ 27	

Source: Internal Revenue Service.

Answers to PROBLEMS

3. a. $\dfrac{a}{b}$ **b.** $\dfrac{b}{a}$ **c.** $\dfrac{x - y}{z}$ **d.** $\dfrac{x - y}{x + y}$ **4. a.** 25   **b.** 41   **c.** 15   **d.** $\dfrac{44}{3}$ **e.** 35

EXAMPLE 5   Application: Evaluating expressions

Look at the part of Form 1040A, shown where A represents the adjusted gross income; S, the standard deduction; and E, the total *number* of exemptions.

a. What algebraic expression should be written on line 25?

b. What algebraic expression should be written on line 26?

c. What algebraic expression should be written on line 27?

d. Suppose your adjusted gross income is $25,000, the standard deduction is $4750, and you have six exemptions. What will be the taxable income entered on line 27?

SOLUTION

a. Line 25 directs you to subtract line 24 (S) from line 22 (A). Thus, the expression $A - S$ will go on line 25.

b. Line 26 says to multiply $3050 by the total *number* of exemptions E. Thus, the entry on line 26 is $3050E$.

c. Line 27 says to subtract line 26, which is $3050E$, from line 25, which is $A - S$. The entry will be $A - S - 3050E$.

d. The taxable income is $A - S - 3050E$, where $A = \$25{,}000$, $S = \$4750$, and $E = 6$. Evaluating this expression, we obtain

$$A - S - 3050E = 25{,}000 - 4750 - 3050(6)$$
$$= 25{,}000 - 4750 - 18{,}300$$
$$= \$1950$$

PROBLEM 5

Suppose A, the adjusted gross income, was $20,000.

a. What algebraic expression should be written on line 27?

b. Assuming that you still have six exemptions, what arithmetic expression should be written on line 27? (Do not simplify.)

c. According to the instructions, what number should you enter on line 27?

MathZone +x

❯ Exercises 1.1

⟨ **A** ⟩ **Translating into Algebraic Expressions**   In Problems 1–40, translate into an algebraic expression.

1. The sum of a and c

2. The sum of u and v

3. The sum of $3x$ and y

4. The sum of 8 and x

5. $9x$ plus $17y$

6. $5a$ plus $2b$

7. The difference of $3a$ and $2b$

8. The difference of $6x$ and $3y$

9. $-2x$ less 5

10. $-7y$ less $3x$

11. 7 times a

12. -9 times y

13. $\frac{1}{7}$ of a

14. $\frac{1}{9}$ of y

15. The product of b and d

16. The product of 4 and c

17. xy multiplied by z

18. $-a$ multiplied by bc

19. $-b$ times $(c + d)$

20. $(p + q)$ multiplied by r

21. $(a - b)$ times x

22. $(a + d)$ times $(x - y)$

23. The product of $(x - 3y)$ and $(x + 7y)$

24. The product of $(a - 2b)$ and $(2a - 3b)$

25. $(c - 4d)$ times $(x + y)$

26. x divided by $2y$

27. y divided by $3x$

Answers to PROBLEMS

5. a. $A - S - 3050E$

  b. $20{,}000 - 4750 - 18{,}300$

  c. 0

❯ Web IT   go to **mhhe.com/bello** for more lessons

28. The quotient of $2a$ and b

29. The quotient $2b$ and a

30. The quotient of $2b$ and ac

31. The quotient of a and the sum of x and y

32. The quotient of $(a + b)$ and c

33. The quotient of the difference of a and b, and c

34. The sum of a and b, divided by the difference of x and y

35. The quotient when x is divided into y

36. The quotient when y is divided into x

37. The quotient when the sum of p and q is divided into the difference of p and q

38. The quotient when the difference of $3x$ and y is divided into the sum of x and $3y$

39. The quotient obtained when the sum of x and $2y$ is divided by the difference of x and $2y$

40. The quotient obtained when the difference of x and $3y$ is divided by the sum of x and $3y$

‹ **B** › **Evaluating Algebraic Expressions**   In Problems 41–60, evaluate the expression for the given values.

41. The sum of a and c for $a = 7$ and $c = 9$

42. The sum of u and v for $u = 15$ and $v = 23$

43. $9x$ plus $17y$ for $x = 3$ and $y = 2$

44. $5a$ plus $2b$ for $a = 5$ and $b = 2$

45. The difference of $3a$ and $2b$ for $a = 5$ and $b = 3$

46. The difference of $6x$ and $3y$ for $x = 3$ and $y = 6$

47. $2x$ less 5 for $x = 4$

48. $7y$ less $3x$ for $x = 3$ and $y = 7$

49. 7 times ab for $a = 2$ and $b = 4$

50. 9 times yz for $y = 2$ and $z = 4$

51. The product of b and d for $b = 3$ and $d = 2$

52. The product of 4 and c for $c = 5$

53. xy multiplied by z for $x = 10$, $y = 5$, and $z = 1$

54. a multiplied by bc for $a = 5$, $b = 7$, and $c = 3$

55. The quotient of a and the sum of x and y for $a = 3$, $x = 1$, and $y = 2$

56. The quotient of a plus b divided by c for $a = 10$, $b = 2$, and $c = 3$

57. The quotient when x is divided into y for $x = 2$ and $y = 8$

58. The quotient when $2y$ is divided into x for $x = 3$ and $y = 5$

59. The quotient when x is divided by y for $x = 2$ and $y = 8$

60. The quotient when $2y$ is divided by x for $x = 3$ and $y = 5$

❯ ❯ ❯   *Using Your Knowledge*

The words we have studied are used in many different fields. Perhaps you have seen some of the following material in your classes! Use the knowledge gained in this section to write it in symbols. The word in italics indicates the field from which the material is taken.

61. *Electricity*   The voltage V across any part of a circuit is the product of the current I and the resistance R.

62. *Economics*   The total profit TP equals the total revenue TR minus the total cost TC.

63. *Chemistry*   The total pressure P in a container filled with gases A, B, and C is equal to the sum of the partial pressures P_A, P_B, and P_C.

64. *Psychology*   The intelligence quotient (IQ) for a child is obtained by multiplying his or her mental age M by 100 and dividing the result by his or her chronological age C.

65. *Physics*   The distance D traveled by an object moving at a constant rate R is the product of R and the time T.

66. *Astronomy*   The square of the period P of a planet's orbit equals the product of a constant C and the cube of the planet's distance R from the sun.

67. *Physics*   The energy E of an object equals the product of its mass m and the square of the speed of light c.

68. *Engineering*   The depth h of a gear tooth is found by dividing the difference between the major diameter D and the minor diameter d by 2.

69. *Geometry*   The square of the hypotenuse c of a right triangle equals the sum of the squares of the sides a and b.

70. *Auto mechanics*   The horsepower (hp) of an engine is obtained by multiplying 0.4 by the square of the diameter D of each cylinder and by the number N of cylinders.

⟩⟩⟩ *Write On*

71. In the expression "$\frac{1}{2}$ of x," what operation does the word *of* signify?

72. Most people believe that the word *and* always means addition.
   a. In the expression "the sum of x and y," does "and" signify the operation of addition? Explain.
   b. In the expression "the product of 2 and three more than a number," does "and" signify the operation of addition? Explain.

73. Explain the difference between "x divided by y" and "x divided into y."

74. Explain the difference between "a less than b" and "a less b."

⟩⟩⟩ *Concept Checker*

This feature is found in every exercise set and is designed to check the student's understanding of the concepts covered in the section.

Fill in the blank(s) with the correct word(s), phrase, or mathematical statement.

75. In symbols, the **sum** of a and b can be written as _____.

76. In symbols, the **difference** of a and b can be written as _____.

77. In symbols, the **product** of a and b can be written as _____.

78. In symbols, the **quotient** of a and b can be written as _____.

79. If we **substitute a value** for one or more of the variables in an expression, we say that we are _____ the expression.

solving	**ab or $a \cdot b$**
evaluating	**$b - a$**
$a - b$	**$\frac{a}{b}$ or $a \div b$**
$a + b$	**$\frac{b}{a}$ or $b \div a$**

⟩⟩⟩ *Mastery Test*

In Problems 80–87, translate into an algebraic expression.

80. The product of 3 and xy

81. The difference of $2x$ and y

82. The quotient of $3x$ and $2y$

83. The sum of $7x$ and $4y$

84. The difference of b and c divided by the sum of b and c

85. Evaluate the expression $2x + y - z$ for $x = 3$, $y = 4$, and $z = 5$.

86. Evaluate the expression $\frac{p - q}{3}$ for $p = 9$ and $q = 3$.

87. Evaluate the expression $\frac{2x - 3y}{x + y}$ for $x = 10$ and $y = 5$.

⟩⟩⟩ *Skill Checker*

In Problems 88–91, add the numbers.

88. $20 + (-20)$

89. $-3.8 + 3.8$

90. $-\frac{2}{7} + \frac{2}{7}$

91. $1\frac{2}{3} + \left(-1\frac{2}{3}\right)$

1.2   The Real Numbers

▶ Objectives

A ❭ Find the additive inverse (opposite) of a number.

B ❭ Find the absolute value of a number.

C ❭ Classify numbers as natural, whole, integer, rational, or irrational.

D ❭ Solve applications using real numbers.

▶ To Succeed, Review How To . . .

Recognize a rational number (see Chapter R).

▶ Getting Started
Temperatures and Integers

To study algebra we need to know about numbers. In this section, we examine *sets* of numbers that are related to each other and learn how to classify them. To visualize these sets more easily and to study some of their properties, we represent (graph) them on a number line. For example, the thermometer shown uses *integers* (not fractions) to measure temperature; the integers are . . . , -3, -2, -1, 0, 1, 2, 3, ... You will find that you use integers every day. For example, when you earn $20, you *have* 20 dollars; we write this as $+20$ dollars. When you spend $5, you *no longer* have it; we write this as -5 dollars. The number 20 is a *positive integer,* and the number -5 (read "negative 5") is a *negative integer.* Here are some other quantities that can be represented by positive and negative integers.

A loss of $25	-25	A $25 gain	25
10 ft below sea level	-10	10 ft above sea level	10
15° below zero	-15	15° above zero	15

These quantities are examples of *real numbers.* There are other types of real numbers; we shall learn about them later in this section.

A ❭ Finding Additive Inverses (Opposites)

The temperature 15 degrees below zero is highlighted on the Fahrenheit scale of the thermometer in the *Getting Started*. If we take the scale on this thermometer and turn it sideways so that the positive numbers are on the right, the resulting scale is called a **number line** (see Figure 1.1). Clearly, on a number line the positive integers are to the right of 0, the negative integers are to the left of 0, and 0 itself is neither positive nor negative.

>**Figure 1.1**

Note that the number line in Figure 1.1 is drawn a little over 5 units long on each side; the arrows at each end indicate that the line could be drawn to any desired length. Moreover, for every *positive* integer, there is a corresponding *negative* integer. Thus, for the positive integer 4, we have the negative integer -4. Since 4 and -4 are the same distance from 0 but in opposite directions, 4 and -4 are called **opposites.** Moreover, since $4 + (-4) = 0$, we call 4 and -4 **additive inverses.** Similarly, the additive inverse (opposite) of -3 is 3, and the additive inverse (opposite) of 2 is -2. Note that $-3 + 3 = 0$ and $2 + (-2) = 0$. In general, we have

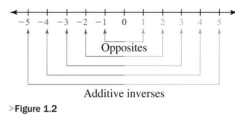

$a + (-a) = (-a) + a = 0$ for any integer a
This means that the sum of an integer a and its additive inverse $-a$ is always 0.

Figure 1.2 shows the relation between the negative and the positive integers.

Opposites

Additive inverses

>Figure 1.2

ADDITIVE INVERSE   The **additive inverse** (opposite) of any number a is $-a$.

You can read $-a$ as "the opposite of a" or "the additive inverse of a." Note that a and $-a$ are additive inverses of each other. Thus, 10 and -10 are additive inverses of each other, and -7 and 7 are additive inverses of each other. You can verify this since $10 + (-10) = 0$ and $-7 + 7 = 0$.

EXAMPLE 1   Finding additive inverses of integers

Find the additive inverse (opposite) of:

a. 5   **b.** -4 **c.** 0

SOLUTION

a. The additive inverse of 5 is -5 (see Figure 1.3).
b. The additive inverse of -4 is $-(-4) = 4$ (see Figure 1.3).
c. The additive inverse of 0 is 0.

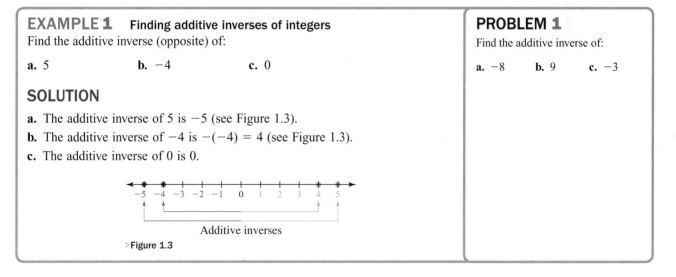

Additive inverses

>Figure 1.3

PROBLEM 1

Find the additive inverse of:

a. -8 **b.** 9   **c.** -3

Note that $-(-4) = 4$ and $-(-8) = 8$. In general, we have

$$-(-a) = a \quad \text{for any number } a$$

As with the integers, every rational number—that is, every fraction written as the ratio of two integers and every decimal—has an *additive inverse* (*opposite*). Here are some rational numbers and their additive inverses.

Rational Number	Additive Inverse (Opposite)
$\dfrac{9}{2}$	$-\dfrac{9}{2}$
$-\dfrac{3}{4}$	$-\left(-\dfrac{3}{4}\right) = \dfrac{3}{4}$
2.9	-2.9
-1.8	$-(-1.8) = 1.8$

EXAMPLE 2   Finding additive inverses of fractions and decimals

Find the additive inverse (opposite) of:

a. $\dfrac{5}{2}$ **b.** -4.8 **c.** $-3\dfrac{1}{3}$ **d.** 1.2

SOLUTION

a. $-\dfrac{5}{2}$ **b.** $-(-4.8) = 4.8$

c. $-\left(-3\dfrac{1}{3}\right) = 3\dfrac{1}{3}$ **d.** -1.2

These rational numbers and their inverses are graphed on the number line in Figure 1.4. Note that to locate $\dfrac{5}{2}$, it is easier to first write $\dfrac{5}{2}$ as the mixed number $2\dfrac{1}{2}$.

>Figure 1.4

PROBLEM 2

Find the additive inverse of:

a. $\dfrac{3}{11}$ **b.** -7.4

c. $-9\dfrac{8}{13}$ **d.** 3.4

B › Finding the Absolute Value of a Number

Let's look at the number line again. What is the distance between 3 and 0? The answer is 3 units. What is the distance between -3 and 0? The answer is *still* 3 units.

The distance between any number n and 0 is called the *absolute value* of the number and is denoted by $|n|$. Thus, $|-3| = 3$ and $3 = |3|$. (See Figure 1.5.)

>Figure 1.5

ABSOLUTE VALUE

The **absolute value** of a number n is its distance from 0 and is denoted by $|n|$.

In general,

$$|n| = \begin{cases} n, & \text{if } n \text{ is positive} \\ -n, & \text{if } n \text{ is negative} \\ 0, & \text{if } n = 0 \end{cases}$$

You can think of the absolute value of a number as the number of units it represents disregarding its sign. For example, suppose Pedro and Tamika leave class together. Pedro walks 2 miles east while Tamika walks 2 miles west. Who walked farther? They walked the same distance! They are both 2 miles from the starting point.

NOTE

Since the absolute value of a number can represent a distance and a distance is **never** negative, the absolute value $|a|$ of a nonzero number a is *always* positive. Because of this, $-|a|$ is *always* negative. Thus, $-|6| = -6$ and $-|-8| = -8$.

Answers to PROBLEMS

2. a. $-\dfrac{3}{11}$ **b.** 7.4

   c. $9\dfrac{8}{13}$ **d.** -3.4

EXAMPLE 3   Finding absolute values of integers

Find the absolute value:

a. $|-8|$ **b.** $|7|$ **c.** $|0|$ **d.** $-|-3|$

SOLUTION

a. $|-8| = 8$ −8 is 8 units from 0.

b. $|7| = 7$ 7 is 7 units from 0.

c. $|0| = 0$ 0 is 0 units from 0.

d. $-|-3| = -3$ −3 is 3 units from 0.

PROBLEM 3

Find the absolute value:

a. $|-5|$ **b.** $|10|$

c. $|2|$ **d.** $-|-5|$

Every fraction and decimal also has an absolute value, which is its distance from zero. Thus, $\left|-\frac{1}{2}\right| = \frac{1}{2}, |3.8| = 3.8$, and $\left|-1\frac{1}{7}\right| = 1\frac{1}{7}$, as shown in Figure 1.6.

>**Figure 1.6**

EXAMPLE 4   Finding absolute values of rational numbers

Find the absolute value of:

a. $\left|-\frac{3}{7}\right|$ **b.** $|2.1|$ **c.** $\left|-2\frac{1}{2}\right|$

d. $|-4.1|$ **e.** $-\left|\frac{1}{4}\right|$

SOLUTION

a. $\left|-\frac{3}{7}\right| = \frac{3}{7}$ **b.** $|2.1| = 2.1$ **c.** $\left|-2\frac{1}{2}\right| = 2\frac{1}{2}$

d. $|-4.1| = 4.1$ **e.** $-\left|\frac{1}{4}\right| = -\frac{1}{4}$

PROBLEM 4

Find the absolute value of:

a. $\left|-\frac{5}{7}\right|$ **b.** $|3.4|$ **c.** $\left|-3\frac{1}{8}\right|$

d. $|-3.8|$ **e.** $-\left|-\frac{1}{5}\right|$

C > Classifying Numbers

The real-number line is a picture (graph) used to represent the *set* of real numbers. A **set** is a collection of objects called the **members** or **elements** of the set. If the elements can be listed, a pair of braces { } encloses the list with individual elements separated by commas. Here are some sets of numbers contained in the real-number line:

The set of **natural** numbers {1, 2, 3, . . .}

The set of **whole** numbers {0, 1, 2, 3, . . .}

The set of **integers** {. . . , −2, −1, 0, 1, 2, 3, . . .}

The three dots (an ellipsis) at the end or beginning of a list of elements indicates that the list continues indefinitely in the same manner.

As you can see, every natural number is a whole number and every whole number is an integer. In turn, every integer is a *rational number,* a number that can be written in the form $\frac{a}{b}$, where a and b are integers and b is not zero. Thus, an integer n can always be written as the rational number $\frac{n}{1} = n$. (The word *rational* comes from the word *ratio,* which indicates a quotient.) Since the fraction $\frac{a}{b}$ can be written as a decimal by dividing the numerator a by the denominator b, to obtain either a terminating (as in $\frac{3}{4} = 0.75$) or a repeating (as in $\frac{1}{3} = 0.\overline{3}$) decimal, all terminating or repeating decimals are also rational numbers. The set of rational numbers is described next in words, since it's impossible to make a list containing all the rational numbers.

Answers to PROBLEMS

3. a. 5   **b.** 10   **c.** 2   **d.** −5

4. a. $\frac{5}{7}$ **b.** 3.4   **c.** $3\frac{1}{8}$

    d. 3.8   **e.** $-\frac{1}{5}$

<table>
<tr><td>**RATIONAL NUMBERS**</td><td>The set of **rational numbers** consists of all numbers that can be written as quotients $\frac{a}{b}$, where a and b are integers and $b \neq 0$.</td></tr>
</table>

The set of *irrational numbers* is the set of all real numbers that are not rational.

<table>
<tr><td>**IRRATIONAL NUMBERS**</td><td>The set of **irrational numbers** consists of all real numbers that *cannot* be written as the quotient of two integers.</td></tr>
</table>

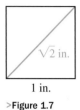

1 in.

>Figure 1.7

For example, if you draw a square 1 inch on a side, the length of its diagonal is $\sqrt{2}$ (read "the square root of 2"), as shown in Figure 1.7. This number cannot be written as a quotient of integers. It is irrational. Since a rational number $\frac{a}{b}$ can be written as a fraction that terminates or repeats, the decimal form of an irrational number never terminates and never repeats. Here are some irrational numbers:

$$\sqrt{2}, \quad -\sqrt{50}, \quad 0.123\ldots, \quad -5.1223334444\ldots, \quad 8.101001000\ldots, \quad \text{and} \quad \pi$$

All the numbers we have mentioned are *real numbers,* and their relationship is shown in Figure 1.8. Of course, you may also represent (graph) these numbers on a number line. Note that a number may belong to more than one category. For example, -15 is an integer, a rational number, and a real number.

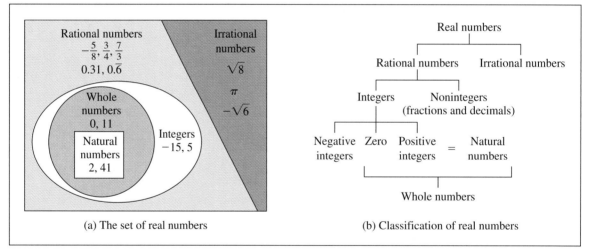

(a) The set of real numbers   (b) Classification of real numbers

>Figure 1.8

EXAMPLE 5   Classifying numbers

Classify as whole, integer, rational, irrational, or real number:

a. -3 **b.** 0 **c.** $\sqrt{5}$

d. 0.3 **e.** $0.101001000\ldots$ **f.** 0.101001000

SOLUTION

a. -3 is an integer, a rational number, and a real number.

b. 0 is a whole number, an integer, a rational number, and a real number.

c. $\sqrt{5}$ is an irrational number and a real number.

d. 0.3 is a terminating decimal, so it is a rational number and a real number.

PROBLEM 5

Classify as whole, integer, rational, irrational, or real number:

a. -9 **b.** 200

c. $\sqrt{7}$ **d.** 0.9

e. $0.010010001\ldots$ **f.** 0.010010001

Answers to PROBLEMS

5. a. Integer, rational, real   **b.** Whole number, integer, rational, real   **c.** Irrational, real   **d.** Rational, real   **e.** Irrational, real   **f.** Rational, real

e. 0.101001000. . . never terminates and never repeats, so it is an irrational number and a real number.

f. 0.101001000 terminates, so it is a rational number and a real number.

D › Applications Involving Real Numbers

EXAMPLE 6   Using real numbers

Use real numbers to write the quantities in the following applications:

a. The price of a product *went down* $\$\frac{3}{4}$.

b. A record low of 128.6 degrees Fahrenheit *below* zero was recorded at Vostok, Antarctica, on July 21, 1983.

c. The oldest mathematical puzzle is contained in the Rhind Papyrus and dates back to 1650 B.C.

SOLUTION

a. $-\dfrac{3}{4}$ **b.** $-128.6°$ **c.** -1650

PROBLEM 6

Use real numbers to write the quantities in the applications:

a. The price of a product *went up* $\$\frac{1}{2}$.

b. A record low of 80 degrees Fahrenheit *below* zero was recorded at Prospect Creek, Alaska, on January 23, 1971.

c. The Moscow Papyrus was copied by an unknown scribe around 1850 B.C.

EXAMPLE 7   Using + or − to indicate height or depth

The following table lists the altitude (the distance above or below sea level or zero altitude) of various locations around the world.

Location	Altitude (feet)
Mount Everest	+29,029
Mount McKinley	+20,320
Sea level	0
Caspian Sea	−92
Marianas Trench (deepest descent)	−35,813

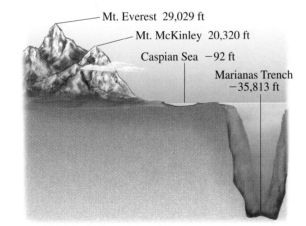

Mt. Everest 29,029 ft
Mt. McKinley 20,320 ft
Caspian Sea −92 ft
Marianas Trench −35,813 ft

Use this information to answer the following questions:

a. How far above sea level is Mount Everest?

b. How far below sea level is the Caspian Sea?

c. How far below sea level was the deepest descent made?

SOLUTION

a. 29,029 feet above sea level   **b.** 92 feet below sea level

c. 35,813 feet below sea level

PROBLEM 7

a. Refer to the table. How far above sea level is Mount McKinley?

b. The Dead Sea, Israel-Jordan, is located at altitude −1349 feet. How far below sea level is that?

c. Some scientists claim that the altitude of the Marianas Trench is really −36,198 feet. How far below sea level is that?

Answers to PROBLEMS

6. a. $+\frac{1}{2}$ **b.** $-80°$ **c.** -1850 **7. a.** 20,320 feet above sea level   **b.** 1349 feet below sea level   **c.** 36,198 feet below sea level

Calculator Corner

Additive Inverse and Absolute Value

The additive inverse and absolute value of a number are so important that graphing calculators have special keys to handle them. To find the additive inverse, press (−). Don't confuse the additive inverse key with the minus sign key. (Operation signs usually have color keys; the (−) key is gray.)

    To find absolute values with a TI-83 Plus calculator, you have to do some math, so press MATH. Next, you have to deal with a special type of number, absolute value, so press ▶ to highlight the NUM menu at the top of the screen. Next press 1, which tells the calculator you want an absolute value; finally, enter the number whose absolute value you want, and close the parentheses. The display window shows how to calculate the additive inverse of -5, the absolute value of 7, and the absolute value of -4.

```
--5
                    5
abs (7)
                    7
abs (-4)
                    4
```

> Exercises **1.2**

Web IT go to **mhhe.com/bello** for more lessons

< **A** > **Finding Additive Inverses (Opposites)**   In Problems 1–18, find the additive inverse (opposite) of the given number.

1. 4

2. 11

3. -49

4. -56

5. $\frac{7}{3}$

6. $-\frac{8}{9}$

7. -6.4

8. -2.3

9. $3\frac{1}{7}$

10. $-4\frac{1}{8}$

11. 0.34

12. 0.85

13. $-0.\overline{5}$

14. $-3.\overline{7}$

15. $\sqrt{7}$

16. $-\sqrt{17}$

17. π

18. $\frac{\pi}{2}$

< **B** > **Finding the Absolute Value of a Number**   In Problems 19–38, find the absolute value.

19. $|-2|$

20. $|-6|$

21. $|48|$

22. $|78|$

23. $|-(-3)|$

24. $|-(-17)|$

25. $\left|-\frac{4}{5}\right|$

26. $\left|-\frac{9}{2}\right|$

27. $|-3.4|$

28. $|-2.1|$

29. $\left|-1\frac{1}{2}\right|$

30. $\left|-3\frac{1}{4}\right|$

31. $-\left|-\frac{3}{4}\right|$

32. $-\left|-\frac{1}{5}\right|$

33. $-|-0.\overline{5}|$

34. $-|-3.\overline{7}|$

35. $-|-\sqrt{3}\,|$

36. $-|-\sqrt{6}\,|$

37. $-|-\pi|$

38. $-\left|-\frac{\pi}{2}\right|$

< **C** > **Classifying Numbers**   In Problems 39–54, classify the given numbers. (See Figure 1.8; some numbers belong in more than one category.)

39. 17

40. -8

41. $-\frac{4}{5}$

42. $-\frac{7}{8}$

43. 0

44. 0.37

45. 3.76

46. $3.\overline{8}$

47. $17.\overline{28}$

48. $\sqrt{10}$

49. $-\sqrt{3}$

50. 0.777...

51. $-0.888...$

52. 0.202002000...

53. 0.202002000

54. $\frac{\pi}{2}$

In Problems 55–62, consider the set $\left\{-5, \frac{1}{5}, 0, 8, \sqrt{11}, 0.\overline{1}, 2.505005000..., 3.666...\right\}$. List the numbers in the given set that are members of the specified set.

55. Natural numbers

56. Whole numbers

57. Positive integers

58. Negative integers

59. Nonnegative integers

60. Irrational

61. Rational numbers

62. Real numbers

In Problems 63–74, determine whether the statement is true or false. If false, give an example that shows it is false.

63. The opposite of any positive number is negative.

64. The opposite of any negative number is positive.

65. The absolute value of any real number is positive.

66. The negative of the absolute value of a number is equal to the absolute value of its negative.

67. The absolute value of a number is equal to the absolute value of its opposite.

68. Every integer is a rational number.

69. Every rational number is an integer.

70. Every terminating decimal is rational.

71. Every nonterminating decimal is rational.

72. Every nonterminating and nonrepeating decimal is irrational.

73. A decimal that never repeats and never terminates is a real number.

74. The decimal representation of a real number never terminates and never repeats.

‹ D › Applications Involving Real Numbers   In Problems 75–88, use real numbers to write the indicated quantities.

75. *Football*   A 20-yard *gain* in a football play.

76. *Football*   A 10-yard *loss* in a football play.

77. *Below sea level*   The Dead Sea is 1312 feet *below* sea level.

78. *Above sea level*   Mount Everest reaches a height of 29,029 feet *above* sea level.

79. *Temperature*   On January 22, 1943, the temperature in Spearfish, South Dakota, rose from 4°F *below* zero to 45°F *above* zero in a period of 2 minutes.

80. *Births and deaths*   Every hour, there are 460 *births* and 250 *deaths* in the United States.

81. *Internet searches*   In a 1-year period the number of U.S. Internet searches *grew* 55%.

82. *Internet searches*   In a 1-year period Yahoo searches *declined* by 0.3%.

83. *Internet searches*   In a 1-year period MSN searches *went down* 3.1%.
Source: www.clickz.com; http://www.clickz.com.

84. *Advertising*   In a recent month, the advertising placements in the automotive industry *declined* 20.91%.

85. *Advertising*   In a recent month, public service ads *increased* by 22.50%.
Source: Nielsen/Net Ratings, 2006; http://www.clickz.com.

86. *Population*   It is projected that between 2006 and 2050 the country that will *lose* the highest percent of people will be Ukraine with a projected 19.6 million *loss*.

87. *Population*   It is projected that between 2006 and 2050, the population of Russia will *dwindle* by about 30.78 million.

88. *Population*   It is projected that between 2006 and 2050 the world population will grow by an amazing 2.54 billion people.
Source: www.msnbc.com; http://www.msnbc.msn.com.

››› *Using Your Knowledge*

Weather   According to *USA Today Weather Almanac,* the coldest city in the United States (based on its average annual temperature in degrees Fahrenheit) is International Falls, Minnesota (see Figure 1.9). Note that the lowest ever temperature there was −46°F.

89. What was the next lowest ever temperature (February) in International Falls?

90. What was the highest record low?

91. What was the record low in December?

92. What was the lowest average low?

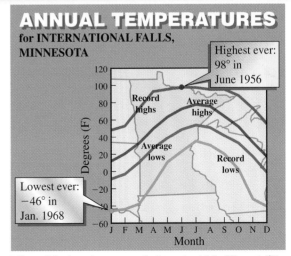

>**Figure 1.9**   Annual temperatures for International Falls, Minnesota (°F)
Source: Data from USA Today Weather Almanac.

〉〉〉 *Write On*

93. What do we mean by the following phrases?

    a. the additive inverse of a number

    b. the absolute value of a number

95. The rational numbers have been defined as the set of numbers that can be written in the form $\frac{a}{b}$, where a and b are integers and b is not 0. Define the rational numbers in terms of their decimal representation.

97. Write a paragraph explaining the set of real numbers as it relates to the natural numbers, the integers, the rational numbers, and the irrational numbers.

94. Explain why every integer is a rational number.

96. Define the set of irrational numbers in terms of their decimal representation.

〉〉〉 *Concept Checker*

Fill in the blank(s) with the correct word(s), phrase, or mathematical statement.

98. The **additive inverse** (opposite) of any number a is _____.

99. The **absolute value** of a number n is its distance from _____.

100. The set of **rational numbers** consists of all numbers that can be written as _____.

101. The set of _____ **numbers** consists of all real numbers that *cannot* **be written** as the quotient of two integers.

natural	**−a**
rational	**−1**
irrational	**n**
0	**numbers**
a	**quotients**

〉〉〉 *Mastery Test*

Find the additive inverse of the number.

102. $0.\overline{7}$

103. $-\sqrt{19}$

104. $\frac{\pi}{3}$

105. $-8\frac{1}{4}$

Find the absolute value.

106. $\left|\sqrt{23}\right|$

107. $\left|-0.\overline{4}\right|$

108. $-\left|\frac{3}{5}\right|$

109. $-\left|-\frac{1}{2}\right|$

Classify the number as a natural, whole, integer, rational, irrational, or real number. (*Hint:* More than one category may apply.)

110. $\sqrt{21}$

111. -2

112. $0.010010001\ldots$

113. $4\frac{1}{3}$

114. $0.333\ldots$

115. 0

116. 39

〉〉〉 *Skill Checker*

Add or subtract:

117. $-7.8 + 7.8$

118. $8.6 - 3.4$

119. $2.3 + 4.1$

120. $\frac{5}{8} - \frac{2}{5}$

121. $\frac{5}{6} + \frac{7}{4}$

1.3   Adding and Subtracting Real Numbers

▶ Objectives

A ⟩ Add two real numbers.

B ⟩ Subtract one real number from another.

C ⟩ Add and subtract several real numbers.

D ⟩ Solve application problems involving real numbers.

▶ To Succeed, Review How To . . .

1. Add and subtract fractions (pp. 12–18).
2. Add and subtract decimals (pp. 22–23).

▶ Getting Started
Signed Numbers and Population Changes

Now that we know what real numbers are, we will use the real-number line to visualize the process used to add and subtract them. For example, what is the U.S. population change per hour? To find out, examine the graph and add the births ($+456$), the deaths (-273), and the new immigrants ($+114$). The result is

$$456 + (-273) + 114 = 570 + (-273) \qquad \text{Add 456 and 114.}$$
$$= +297 \qquad\qquad\qquad \text{Subtract 273 from 570.}$$

Source: Population Reference Bureau (2000, 2001, and 2002 data).

This answer means that the U.S. population is increasing by 297 persons every hour! Note that

1. The addition of 456 and -273 is written as $456 + (-273)$, instead of the confusing $456 + -273$.

2. To add 570 and -273, we *subtracted* 273 from 570 because subtracting 273 from 570 is the *same* as adding 570 and -273. We will use this idea to define subtraction.

If you want to know what the population change is per day or per minute, you must learn how to multiply and divide real numbers, as we will do in Section 1.4.

A ⟩ Adding Real Numbers

The number line we studied in Section 1.2 can help us add real numbers. Here's how we do it.

> **PROCEDURE**
>
> **Adding on a Number Line**
>
> To add $a + b$ on a number line,
>   1. Start at zero and move to a (to the *right* if a is *positive*, to the *left* if a is *negative*).
>   2. **A.** If b is *positive*, move *right* b units.
>   **B.** If b is *negative*, move *left* $|b|$ units.
>   **C.** If b is zero, stay at a.

For example, the sum $2 + 4$, or $(+2) + (+4)$, is found by starting at zero, moving 2 units to the right, followed by 4 more units to the right ending at 6. Thus, $2 + 4 = 6$, as shown in Figure 1.10

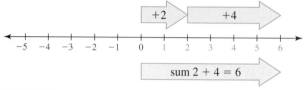

>Figure 1.10

A number along with the sign ($+$ or $-$) indicating a direction on the number line is called a **signed number.**

EXAMPLE 1   Adding integers with different signs
Add: $5 + (-3)$

SOLUTION   Start at zero. Move 5 units to the right and then 3 units to the left. The result is 2. Thus, $5 + (-3) = 2$, as shown in Figure 1.11.

>Figure 1.11

PROBLEM 1

Add: $4 + (-2)$

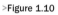

⌨ ⬡ ▦ **Calculator Corner**

Adding Integers:

To enter $-3 + (-2)$ using a scientific calculator, enter **3** [+/−] [**+**] **2** [+/−] [ENTER]. If your calculator has a set of parentheses, then you can enter parentheses around the -3 and -2.

This same procedure can be used to add two negative numbers. However, we have to be careful when writing such problems. For example, to add -3 and -2, we should write

$$-3 + (-2)$$

Why the parentheses? Because writing

$$-3 + -2$$

is confusing. *Never* write two signs together without parentheses.

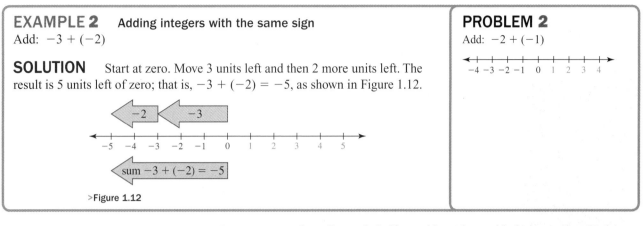

EXAMPLE 2   Adding integers with the same sign

Add: $-3 + (-2)$

SOLUTION   Start at zero. Move 3 units left and then 2 more units left. The result is 5 units left of zero; that is, $-3 + (-2) = -5$, as shown in Figure 1.12.

>Figure 1.12

PROBLEM 2

Add: $-2 + (-1)$

As you can see from Example 2, if we add numbers with the *same* sign (both + or both −), the result is a number with the same sign. Thus,

$$2 + 4 = 6 \quad \text{and} \quad -3 + (-2) = -5$$

If we add numbers with *different* signs, the answer carries the sign of the number with the larger absolute value. Hence, in Example 1,

$$5 + (-3) = 2 \qquad \text{The answer is positive because 5 has a larger absolute value than } -3. \text{ (See Figure 1.11.)}$$

but note that

$$-5 + 3 = -2 \qquad \text{The answer is negative because } -5 \text{ has a larger absolute value than 3.}$$

The following rules summarize this discussion.

> **RULES**
>
> **Adding Signed Numbers**
>
> **1.** With the **same (like)** sign: *Add* their absolute values and give the sum the *common* sign.
>
> **2.** With **different (unlike)** signs: *Subtract* their absolute values and give the difference the sign of the number with the *larger* absolute value.

For example, to add $8 + 5$ or $-8 + (-5)$, we note that the 8 and 5, and -8 and -5 have the same signs. So we add their absolute values and give the sum the common sign. Thus,

$$8 + 5 = 13 \quad \text{and} \quad -8 + (-5) = -13$$

To add $-8 + 5$, we first notice that the numbers have *different* signs. So we subtract their absolute values and give the difference the sign of the number with the larger absolute value. Thus,

Use the sign of the number with the larger absolute value (−). $-8 + 5 = -(8 - 5) = -3$ Subtract the smaller number from the larger one.

Similarly,

$$8 + (-5) = +(8 - 5) = 3$$

Here we have used the sign of the number with the larger absolute value, 8, which is understood to be +.

EXAMPLE 3   Adding integers with different signs

Add:

a. $-14 + 6$ **b.** $14 + (-6)$

SOLUTION

a. $-14 + 6 = -(14 - 6) = -8$ **b.** $14 + (-6) = +(14 - 6) = 8$

PROBLEM 3

Add:

a. $-10 + 4$ **b.** $10 + (-4)$

Answers to PROBLEMS

2. -3 **3. a.** -6 **b.** 6

The addition of rational numbers uses the same rules for signs, as we illustrate in Examples 4 and 5.

EXAMPLE 4   Adding decimals with different signs

Add:

a. $-8.6 + 3.4$ **b.** $6.7 + (-9.8)$ **c.** $-2.3 + (-4.1)$

SOLUTION

a. $-8.6 + 3.4 = -(8.6 - 3.4) = -5.2$

b. $6.7 + (-9.8) = -(9.8 - 6.7) = -3.1$

c. $-2.3 + (-4.1) = -(2.3 + 4.1) = -6.4$

PROBLEM 4

Add:

a. $-7.8 + 2.5$ **b.** $5.4 + (-7.8)$

c. $-3.4 + (-5.1)$

EXAMPLE 5   Adding fractions with different signs

Add:

a. $-\dfrac{3}{7} + \dfrac{5}{7}$ **b.** $\dfrac{2}{5} + \left(-\dfrac{5}{8}\right)$

SOLUTION

a. Note that $\left|\dfrac{5}{7}\right|$ is larger than $\left|-\dfrac{3}{7}\right|$; hence

$$-\frac{3}{7} + \frac{5}{7} = +\left(\frac{5}{7} - \frac{3}{7}\right) = \frac{2}{7}$$

b. As usual, we must first find the LCM of 5 and 8, which is 40. We then write

$$\frac{2}{5} = \frac{16}{40} \quad \text{and} \quad -\frac{5}{8} = -\frac{25}{40}$$

Thus,

$$\frac{2}{5} + \left(-\frac{5}{8}\right) = \frac{16}{40} + \left(-\frac{25}{40}\right) = -\left(\frac{25}{40} - \frac{16}{40}\right) = -\frac{9}{40}$$

PROBLEM 5

Add:

a. $-\dfrac{4}{9} + \dfrac{5}{9}$ **b.** $\dfrac{2}{5} + \left(-\dfrac{3}{4}\right)$

B ⟩ Subtracting Real Numbers

We are now ready to subtract signed numbers. Suppose you use *positive* integers to indicate money *earned* and *negative* integers to indicate money *spent* (expenditures). If you earn $10 and then spend $12, you owe $2. Thus,

$$\underline{10 - 12} = 10 + (-12) = \underline{-2}$$

Earn $10. Spend $12.   Owe $2.

Also,

$$-5 - 10 = -5 + (-10) = -15$$ To take away (subtract) earned money is the same as adding an expenditure.

because if you spend $5 and then spend $10 more, you now owe $15. What about $-10 - (-3)$? We claim that

$$-10 - (-3) = -7$$

because if you spend $10 and then subtract (take away) a $3 expenditure (represented by -3), you save $3; that is,

$$-10 - (-3) = -10 + 3 = -7$$ When you take away (subtract) a $3 expenditure, you save (add) $3.

In general, we have the following definition.

PROCEDURE

Subtraction

To **subtract** a number b, add its inverse $(-b)$. In symbols,

$$a - b = a + (-b)$$

Thus,

To subtract 8, add its inverse.

$$5 - 8 = 5 + (-8) = -3$$
$$7 - 3 = 7 + (-3) = 4$$
$$-4 - 2 = -4 + (-2) = -6$$
$$-6 - (-4) = -6 + 4 = -2$$

EXAMPLE 6   Subtracting integers

Subtract:

a. $17 - 6$ 　　　　　　　　　　　 **b.** $-21 - 4$
c. $-11 - (-5)$ 　　　　　　　　　 **d.** $-4 - (-6)$

SOLUTION　Use the fact that $a - b = a + (-b)$ to rewrite as an addition.

a. $17 - 6 = 17 + (-6) = 11$ 　　　 **b.** $-21 - 4 = -21 + (-4) = -25$
c. $-11 - (-5) = -11 + 5 = -6$ 　　 **d.** $-4 - (-6) = -4 + 6 = 2$

PROBLEM 6

Subtract:

a. $15 - 8$ 　　 **b.** $-23 - 5$
c. $-12 - (-4)$ 　 **d.** $-5 - (-7)$

EXAMPLE 7   Subtracting decimals or fractions

Subtract:

a. $-4.2 - (-3.1)$ 　　　　　　　　 **b.** $-2.5 - (-7.8)$
c. $\dfrac{2}{9} - \left(-\dfrac{4}{9}\right)$ 　　　　　　　　　 **d.** $-\dfrac{5}{6} - \dfrac{7}{4}$

SOLUTION

a. $-4.2 - (-3.1) = -4.2 + 3.1 = -1.1$　　Note that $-(-3.1) = 3.1$.
b. $-2.5 - (-7.8) = -2.5 + 7.8 = 5.3$　　Note that $-(-7.8) = 7.8$.
c. $\dfrac{2}{9} - \left(-\dfrac{4}{9}\right) = \dfrac{2}{9} + \dfrac{4}{9} = \dfrac{6}{9} = \dfrac{2}{3}$　　Note that $-\left(-\dfrac{4}{9}\right) = \dfrac{4}{9}$.

d. The LCD is 12. Now,

$$-\frac{5}{6} = -\frac{10}{12} \quad \text{and} \quad \frac{7}{4} = \frac{21}{12}$$

Thus,

$$-\frac{5}{6} - \frac{7}{4} = -\frac{5}{6} + \left(-\frac{7}{4}\right) = -\frac{10}{12} + \left(-\frac{21}{12}\right) = -\frac{31}{12}$$

PROBLEM 7

Subtract:

a. $3.2 - (-2.1)$ 　**b.** $-3.4 - (-6.9)$
c. $\dfrac{2}{7} - \left(-\dfrac{3}{7}\right)$ 　　 **d.** $-\dfrac{5}{8} - \dfrac{9}{4}$

C › Adding and Subtracting Several Real Numbers

Suppose you wish to find $18 - (-10) + 12 - 10 - 17$. Using the fact that $a - b = a + (-b)$, we write

$$18 - (-10) + 12 - 10 - 17 = 18 + 10 + 12 + (-10) + (-17)$$

$$10 + (-10) = 0$$

$$= 18 + 12 + (-17)$$
$$= 30 + (-17)$$
$$= 13$$

Answers to PROBLEMS

6. a. 7 　**b.** -28 　**c.** -8 　**d.** 2

7. a. 5.3 　**b.** 3.5 　**c.** $\dfrac{5}{7}$ 　**d.** $-\dfrac{23}{8}$

EXAMPLE 8   Adding and subtracting numbers

Find: $12 - (-13) + 10 - 25 - 13$

SOLUTION   First, rewrite as an addition.

$$12 - (-13) + 10 - 25 - 13 = 12 + 13 + 10 + (-25) + (-13)$$

$$13 + (-13) = 0$$

$$= 12 + 10 + (-25)$$
$$= 22 + (-25)$$
$$= -3$$

PROBLEM 8

Find: $14 - (-15) + 10 - 23 - 15$

D > Applications Involving Real Numbers

EXAMPLE 9   Finding temperature differences

The greatest temperature variation in a 24-hour period occurred in Browning, Montana January 23 to 24, 1916. The temperature fell from 44°F to −56°F. How many degrees did the temperature fall?

SOLUTION   We have to find the difference between 44 and −56; that is, we have to find $44 - (-56)$:

$$44 - (-56) = 44 + 56$$
$$= 100$$

Thus, the temperature fell 100°F.

PROBLEM 9

The greatest temperature variation in a 12-hour period occurred in Fairfield, Montana. The temperature fell from 63°F at noon to −21°F at midnight. How many degrees did the temperature fall? (When did this happen? December 24, 1924. They certainly did have a cool Christmas!)

MathZone

> Exercises **1.3**

Web IT go to **mhhe.com/bello** *for more lessons*

< **A** > **Adding Real Numbers**   In Problems 1–40, perform the indicated operations (verify your answer using a number line).

1. $3 + 3$ **2.** $2 + 1$ **3.** $-5 + 1$ **4.** $-4 + 3$

5. $6 + (-5)$ **6.** $5 + (-1)$ **7.** $-2 + (-5)$ **8.** $-3 + (-3)$

9. $3 + (-3)$ **10.** $-4 + 4$ **11.** $-18 + 21$ **12.** $-3 + 5$

13. $19 + (-6)$ **14.** $8 + (-1)$ **15.** $-9 + 11$ **16.** $-8 + 13$

17. $-18 + 9$ **18.** $-17 + 4$ **19.** $-17 + (+5)$ **20.** $-4 + (+8)$

21. $-3.8 + 6.9$ **22.** $-4.5 + 7.8$ **23.** $-7.8 + (3.1)$ **24.** $-6.7 + (2.5)$

25. $3.2 + (-8.6)$ **26.** $4.1 + (-7.9)$ **27.** $-3.4 + (-5.2)$ **28.** $-7.1 + (-2.6)$

29. $-\frac{2}{7} + \frac{5}{7}$ **30.** $-\frac{5}{11} + \frac{7}{11}$ **31.** $-\frac{3}{4} + \frac{1}{4}$ **32.** $-\frac{5}{6} + \frac{1}{6}$

33. $\frac{3}{4} + \left(-\frac{5}{6}\right)$ **34.** $\frac{5}{6} + \left(-\frac{7}{8}\right)$ **35.** $-\frac{1}{6} + \frac{3}{4}$ **36.** $-\frac{1}{8} + \frac{7}{6}$

37. $-\frac{1}{3} + \left(-\frac{2}{7}\right)$ **38.** $-\frac{4}{7} + \left(-\frac{3}{8}\right)$ **39.** $-\frac{5}{6} + \left(-\frac{8}{9}\right)$ **40.** $-\frac{4}{5} + \left(-\frac{7}{8}\right)$

Answers to PROBLEMS

8. 1   **9.** 84°F

⟨ **B** ⟩ **Subtracting Real Numbers**   In Problems 41–60, perform the indicated operations.

41. $-5 - 11$

42. $-4 - 7$

43. $-4 - 16$

44. $-9 - 11$

45. $7 - 13$

46. $8 - 12$

47. $9 - (-7)$

48. $8 - (-4)$

49. $0 - 4$

50. $0 - (-4)$

51. $-3.8 - (-1.2)$

52. $-6.7 - (-4.3)$

53. $-3.5 - (-8.7)$

54. $-6.5 - (-9.9)$

55. $4.5 - 8.2$

56. $3.7 - 7.9$

57. $\frac{3}{7} - \left(-\frac{1}{7}\right)$

58. $\frac{5}{6} - \left(-\frac{1}{6}\right)$

59. $-\frac{5}{4} - \frac{7}{6}$

60. $-\frac{2}{3} - \frac{3}{4}$

⟨ **C** ⟩ **Adding and Subtracting Several Real Numbers**   In Problems 61–66, perform the indicated operations.

61. $8 - (-10) + 5 - 20 - 10$

62. $15 - (-9) + 8 - 2 - 9$

63. $-15 + 12 - 8 - (-15) + 5$

64. $-12 + 14 - 7 - (-12) + 3$

65. $-10 + 9 - 14 - 3 - (-14)$

66. $-7 + 2 - 6 - 8 - (-6)$

⟨ **D** ⟩ **Applications Involving Real Numbers**

67. *Earth temperatures*   The temperature in the center core of the Earth reaches $+5000\,°C$. In the thermosphere (a region in the upper atmosphere), the temperature is $+1500\,°C$. Find the difference in temperature between the center of the Earth and the thermosphere.

68. *Extreme temperatures*   The record high temperature in Calgary, Alberta, is $+99\,°F$. The record low temperature is $-46\,°F$. Find the difference between these extremes.

69. *Stock price fluctuations*   The price of a certain stock at the beginning of the week was $47. Here are the changes in price during the week: $+1, +2, -1, -2, -1$. What is the price of the stock at the end of the week?

70. *Stock price fluctuations*   The price of a stock on Friday was $37. On the following Monday, the price went up $2; on Tuesday, it went down $3; and on Wednesday, it went down another $1. What was the price of the stock then?

71. *Temperature variations*   Here are the temperature changes (in degrees Celsius) by the hour in a certain city:

1 P.M.	$+2$
2 P.M.	$+1$
3 P.M.	-1
4 P.M.	-3

If the temperature was initially 15°C, what was it at 4 P.M.?

How many calories do you eat for lunch? How many do you use during exercise? The table shows the number of calories in some fast foods and the number of calories used in different activities.

Food	Calories (+)	Activity	Calories Burned (−)
Hamburger (McD)	280	Bicycling 6 mph	240 cal./hr
Fries (McD)	210	Bicycling 12 mph	410 cal./hr
Hamburger (BK)	310	Jogging 5 mph	740 cal./hr
Fries (BK)	230	Jogging 7 mph	920 cal./hr
Salad (McD)	20	Swimming 25 yd/min	275 cal./hr
Salad (BK)	25	Swimming 50 yd/min	500 cal./hr.

Calories   In Problems 72–76 find the number of calories gained or lost in each situation.

72. You have a McDonald's (McD) hamburger and bicycle at 6 mph for 1 hour.

73. You have a Burger King (BK) hamburger with fries and you jog at 7 mph for 1 hour.

74. You have a hamburger, fries, and a salad at McDonald's, then bicycle at 12 mph for an hour and jog at 7 mph for another hour.

75. You have a hamburger, fries, and a salad at BK, and then jog at 7 mph for an hour.

76. You eat fries and a salad at BK then go swimming at 25 yards per minute for an hour. What else can you eat so your caloric intake will be 0?

〉〉〉 *Using Your Knowledge*

A Little History   The following chart contains some important historical dates.

Important Historical Dates	
323 B.C.	Alexander the Great died
216 B.C.	Hannibal defeated the Romans
A.D. 476	Fall of the Roman Empire
A.D. 1492	Columbus landed in America
A.D. 1776	The Declaration of Independence signed
A.D. 1939	World War II started
A.D. 1988	Reagan–Gorbachev summit

We can use negative integers to represent years B.C. For example, the year Alexander the Great died can be written as -323, whereas the fall of the Roman Empire occurred in $+476$ (or simply 476). To find the number of years that elapsed between the fall of the Roman Empire and their defeat by Hannibal, we write

$$476 - (-216) = 476 + 216 = 692$$

Fall of the   Hannibal defeats   Years elapsed
Roman Empire   the Romans
(A.D. 476)   (216 B.C.)

Use these ideas to find the number of years elapsed between the following:

77. The fall of the Roman Empire and the death of Alexander the Great

78. Columbus's landing in America and Hannibal's defeat of the Romans

79. Columbus landing in America and the signing of the Declaration of Independence

80. The year of the Reagan–Gorbachev summit and the signing of the Declaration of Independence

81. The start of World War II and the death of Alexander the Great

〉〉〉 **Write On**

82. Explain what the term *signed numbers* means and give examples.

83. State the rule you use to add signed numbers. Explain why the sum is sometimes positive and sometimes negative.

84. State the rule you use to subtract signed numbers. How do you know when the answer is going to be positive? How do you know when the answer is going to be negative?

85. The definition of subtraction is as follows: To subtract a number, add its inverse. Use a number line to explain why this works.

〉〉〉 *Concept Checker*

Fill in the blank(s) with the correct word(s), phrase, or mathematical statement.

86. To **add** signed numbers with the **same** sign, add their absolute value and give the sum the _____ sign.

87. To add numbers with **different** signs, subtract their absolute value and give the difference the sign of the number with the _____ absolute value.

88. The **definition of subtraction** states that $a - b = a +$ _____ .

89. To **subtract** a number b from a number a, add the _____ of b.

reciprocal	smaller
inverse	larger
(b)	common
$(-b)$	different

〉〉〉　*Mastery Test*

Add or subtract

90. $7 - 13$

91. $3.8 - 6.9$

92. $\frac{1}{5} - \frac{3}{4}$

93. $-3.5 - 4.2$

94. $-\frac{2}{3} - \frac{4}{5}$

95. $-5 - 15$

96. $-6 - (-4)$

97. $-3.4 - (-4.6)$

98. $-\frac{3}{4} - \left(-\frac{1}{5}\right)$

99. $3.9 + (-4.2)$

100. $\frac{3}{4} + \left(-\frac{1}{6}\right)$

101. $-3.2 + (-2.5)$

102. $7 - (-11) + 13 - 11 - 15$

103. $\frac{3}{4} - \left(-\frac{2}{5}\right) + \frac{4}{5} - \frac{2}{5} + \frac{5}{4}$

104. The temperature in Verkhoyansk, Siberia, ranges from 98°F to −94°F. What is the difference between these temperatures?

〉〉〉　*Skill Checker*

Perform the indicated operation.

105. $\frac{5}{8} \cdot \frac{16}{25}$

106. $5\frac{1}{4} \cdot 3\frac{1}{8}$

107. $6\frac{1}{6} \cdot 5\frac{7}{10}$

108. $6\frac{1}{4} \cdot \frac{20}{21}$

109. $\frac{18}{11} \div 5\frac{1}{2}$

110. $2\frac{1}{4} \div 1\frac{3}{8}$

1.4　Multiplying and Dividing Real Numbers

▶ Objectives

A 〉 Multiply two real numbers.

B 〉 Evaluate expressions involving exponents.

C 〉 Divide one real number by another.

D 〉 Solve an application involving multiplying and dividing real numbers.

▶ To Succeed, Review How To . . .

1. Multiply and divide whole numbers, decimals, and fractions (Chapter R, pp. 12–18, 22–23).

2. Find the reciprocal of a number (Chapter R, pp. 11–12).

▶ Getting Started
A Stock Market Loss

In Section 1.3, we learned how to add and subtract real numbers. Now we will learn how to multiply and divide them. Pay particular attention to the notation used when multiplying a number by itself and also to the different applications of the multiplication and division of real numbers. For example, suppose you own 4 shares of Eastman Kodak (East Kodak), and the closing price today is *down* $3 (written as −3). Your loss then is

$$4 \cdot (-3) \quad \text{or} \quad 4(-3)$$

How do we multiply positive and negative integers? As you may recall, the result of a multiplication is a *product,* and the numbers being multiplied (4 and −3) are called *factors.* Now you can think of multiplication as *repeated addition.* Thus,

$$4 \cdot (-3) = \underbrace{(-3) + (-3) + (-3) + (-3)}_{\text{four } (-3)\text{'s}} = -12$$

Also note that

$$(-3) \cdot 4 = -12$$

So your stock has gone down $12. As you can see, the product of a *negative* integer and a *positive* integer is negative. What about the product of two negative integers, say $-4 \cdot (-3)$? Look for the pattern in the following table:

The number in this column decreases by 1.

The number in this column increases by 3.

$$4 \cdot (-3) = -12$$
$$3 \cdot (-3) = -9$$
$$2 \cdot (-3) = -6$$
$$1 \cdot (-3) = -3$$
$$0 \cdot (-3) = 0$$
$$-1 \cdot (-3) = 3$$
$$-2 \cdot (-3) = 6$$
$$-3 \cdot (-3) = 9$$
$$-4 \cdot (-3) = 12$$

You can think of $-4 \cdot (-3)$ as subtracting -3 four times; that is,

$$-(-3) - (-3) - (-3) - (-3) = 3 + 3 + 3 + 3 = 12$$

So, when we multiply two integers with *different* (*unlike*) signs, the product is *negative*. If we multiply two integers with the *same* (*like*) signs, the product is *positive*. This idea can be generalized to include the product of any two real numbers, as you will see later.

A ⟩ Multiplying Real Numbers

Here are the rules that we used in the *Getting Started.*

RULES

Multiplying Signed Numbers

1. When two numbers with the *same* (*like*) sign are multiplied, the product is *positive* (+).

2. When two numbers with *different* (*unlike*) signs are multiplied, the product is *negative* (−).

Here are some examples.

$$9 \cdot 4 = 36$$
$$-9 \cdot (-4) = 36$$

9 and 4 have the same sign (+); −9 and −4 have the same sign (−); thus, the product is positive.

$$-9 \cdot 6 = -54$$
$$9 \cdot (-6) = -54$$

−9 and 6 have different signs; 9 and −6 have different signs; thus, the product is negative.

EXAMPLE 1   Finding products of integers

Multiply:

a. $7 \cdot 8$ **b.** $-8 \cdot 6$ **c.** $4 \cdot (-8)$ **d.** $-7 \cdot (-9)$

SOLUTION   **a.** $7 \cdot 8 = 56$

b. $-8 \cdot 6 = -48$

Different signs   Negative product

c. $4 \cdot (-8) = -32$

Different signs   Negative product

d. $-7 \cdot (-9) = 63$

Same sign   Positive product

PROBLEM 1

Multiply:

a. $9 \cdot 6$ **b.** $-7 \cdot 8$

c. $5 \cdot (-4)$ **d.** $-5 \cdot (-6)$

Answers to PROBLEMS

1. a. 54   **b.** −56   **c.** −20   **d.** 30

The multiplication of rational numbers also uses the same rules for signs, as we illustrate in Example 2. Remember that $-3.1(4.2)$ means $-3.1 \cdot 4.2$. Parentheses are one of the ways we indicate multiplication.

EXAMPLE 2   Finding products of decimals and fractions

Multiply:

a. $-3.1(4.2)$ **b.** $-1.2(-3.4)$ **c.** $-\dfrac{3}{4}\left(-\dfrac{5}{2}\right)$ **d.** $\dfrac{5}{6}\left(-\dfrac{4}{7}\right)$

SOLUTION

a. -3.1 and 4.2 have different signs. The product is *negative*. Thus,

$$-3.1(4.2) = -13.02$$

b. -1.2 and -3.4 have the same sign. The product is *positive*. Thus,

$$-1.2(-3.4) = 4.08$$

c. $-\dfrac{3}{4}$ and $-\dfrac{5}{2}$ have the same sign. The product is *positive*. Thus,

$$-\frac{3}{4}\left(-\frac{5}{2}\right) = \frac{15}{8}$$

d. $\dfrac{5}{6}$ and $-\dfrac{4}{7}$ have different signs. The product is *negative*. Thus,

$$\frac{5}{6}\left(-\frac{4}{7}\right) = -\frac{20}{42} = -\frac{10}{21}$$

PROBLEM 2

Multiply:

a. $-4.1 \cdot (3.2)$ **b.** $-1.3(-4.2)$

c. $-\dfrac{3}{5}\left(-\dfrac{7}{10}\right)$ **d.** $\dfrac{5}{9}\left(-\dfrac{6}{7}\right)$

B 〉 Evaluating Expressions Involving Exponents

Sometimes a number is used several times as a factor. For example, we may wish to find or evaluate the following products:

$$3 \cdot 3 \quad \text{or} \quad 4 \cdot 4 \cdot 4 \quad \text{or} \quad 5 \cdot 5 \cdot 5 \cdot 5$$

In the expression $3 \cdot 3$, the 3 is used as a factor twice. In such cases it's easier to use **exponents** to indicate how many times the number is used as a factor. We then write

3^2 (read "3 squared") instead of $3 \cdot 3$

4^3 (read "4 cubed") instead of $4 \cdot 4 \cdot 4$

5^4 (read "5 to the fourth") instead of $5 \cdot 5 \cdot 5 \cdot 5$

The expression 3^2 uses the exponent 2 to indicate how many times the **base** 3 is used as a factor. Similarly, in the expression 5^4, the 5 is the base and the 4 is the exponent. Now,

$$3^2 = \underline{3 \cdot 3} = 9 \qquad \text{3 is used as a factor 2 times.}$$
$$4^3 = \underline{4 \cdot 4 \cdot 4} = 64 \qquad \text{4 is used as a factor 3 times.}$$

and

$$\left(\frac{1}{5}\right)^4 = \underline{\frac{1}{5} \cdot \frac{1}{5} \cdot \frac{1}{5} \cdot \frac{1}{5}} = \frac{1}{625} \qquad \tfrac{1}{5}\text{ is used as a factor 4 times.}$$

What about $(-2)^2$? Using the definition of exponents, we have

$$(-2)^2 = (-2) \cdot (-2) = 4 \qquad \text{−2 and −2 have the same sign; thus, their product is positive.}$$

Moreover, -2^2 means $-(2 \cdot 2)$. To emphasize that the multiplication is to be done *first*, we put parentheses around the $2 \cdot 2$. Thus, the placing of the parentheses in the expression $(-2)^2$ is very important. Clearly, since $(-2)^2 = 4$ and $-2^2 = -(2 \cdot 2) = -4$,

$$\boxed{(-2)^2 \neq -2^2} \qquad \text{"}\neq\text{" means "is not equal to."}$$

EXAMPLE 3   **Evaluating expressions involving exponents**

Evaluate:

a. $(-4)^2$ **b.** -4^2 **c.** $\left(-\dfrac{1}{3}\right)^2$ **d.** $-\left(\dfrac{1}{3}\right)^2$

SOLUTION

a. $(-4)^2 = (-4)(-4) = 16$ Note that the base is -4.

b. $-4^2 = -(4 \cdot 4) = -16$ Here the base is 4.

c. $\left(-\dfrac{1}{3}\right)^2 = \left(-\dfrac{1}{3}\right)\left(-\dfrac{1}{3}\right) = \dfrac{1}{9}$ The base is $-\frac{1}{3}$.

d. $-\left(\dfrac{1}{3}\right)^2 = -\left(\dfrac{1}{3}\right)\left(\dfrac{1}{3}\right) = -\dfrac{1}{9}$ The base is $\frac{1}{3}$.

From parts **a** and **b,** you can see that $(-4)^2 \neq -4^2$.

PROBLEM 3

Evaluate:

a. $(-7)^2$ **b.** -7^2

c. $\left(-\dfrac{1}{4}\right)^2$ **d.** $-\left(\dfrac{1}{4}\right)^2$

EXAMPLE 4   **Evaluating expressions involving exponents**

Evaluate:

a. $(-2)^3$ **b.** -2^3

SOLUTION

a. $(-2)^3 = \underbrace{(-2) \cdot (-2)}_{4} \cdot (-2)$

$ = 4 \cdot (-2)$

$ = -8$

b. $-2^3 = -\underbrace{(2 \cdot 2 \cdot 2)}$

$ = -(8)$

$ = -8$

PROBLEM 4

Evaluate:

a. $(-4)^3$ **b.** -4^3

Note that

$$(-4)^2 = 16$$ Negative number raised to an even power; positive result.

but

$$(-2)^3 = -8$$ Negative number raised to an odd power; negative result.

C › Dividing Real Numbers

What about the rules for division? As you recall, a division problem can always be checked by multiplication. Thus, the division

$$3\overline{)18} \quad \text{or} \quad \dfrac{18}{3} = 6$$
$$\underline{-18}$$
$$0$$

is correct because $18 = 3 \cdot 6$. In general, we have the following definition for division.

DIVISION

$$\dfrac{a}{b} = c \quad \text{means} \quad a = b \cdot c, \quad b \neq 0$$

Note that the operation of division is defined using multiplication. Because of this, the same rules of sign that apply to the multiplication of real numbers also apply to the division of real numbers. Note that the result of the division of two numbers is called the **quotient.**

Answers to PROBLEMS

3. a. 49   **b.** -49 **c.** $\dfrac{1}{16}$ **d.** $-\dfrac{1}{16}$

4. a. -64 **b.** -64

RULES

Dividing with Signed Numbers

1. When dividing two numbers with the *same* (*like*) sign, give the quotient a *positive* ($+$) sign.

2. When dividing two numbers with *different* (*unlike*) signs, give the quotient a *negative* ($-$) sign.

Here are some examples:

$$\frac{24}{6} = 4$$ 24 and 6 have the same sign; the quotient is positive.

$$\frac{-18}{-9} = 2$$ -18 and -9 have the same sign; the quotient is positive.

$$\frac{-32}{4} = -8$$ -32 and 4 have different signs; the quotient is negative.

$$\frac{35}{-7} = -5$$ 35 and -7 have different signs; the quotient is negative.

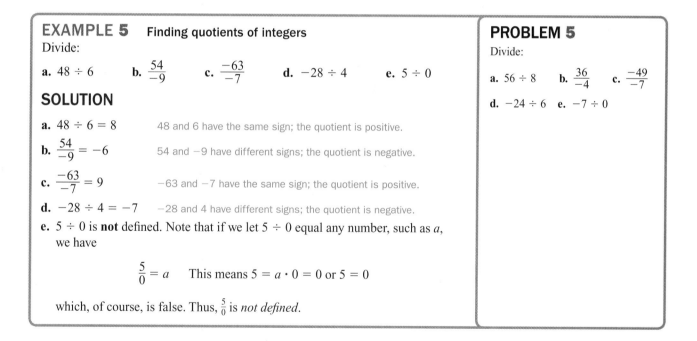

EXAMPLE 5   Finding quotients of integers

Divide:

a. $48 \div 6$ **b.** $\dfrac{54}{-9}$ **c.** $\dfrac{-63}{-7}$ **d.** $-28 \div 4$ **e.** $5 \div 0$

SOLUTION

a. $48 \div 6 = 8$ 48 and 6 have the same sign; the quotient is positive.

b. $\dfrac{54}{-9} = -6$ 54 and -9 have different signs; the quotient is negative.

c. $\dfrac{-63}{-7} = 9$ -63 and -7 have the same sign; the quotient is positive.

d. $-28 \div 4 = -7$ -28 and 4 have different signs; the quotient is negative.

e. $5 \div 0$ is **not** defined. Note that if we let $5 \div 0$ equal any number, such as a, we have

$$\frac{5}{0} = a$$ This means $5 = a \cdot 0 = 0$ or $5 = 0$

which, of course, is false. Thus, $\frac{5}{0}$ is *not defined*.

PROBLEM 5

Divide:

a. $56 \div 8$ **b.** $\dfrac{36}{-4}$ **c.** $\dfrac{-49}{-7}$

d. $-24 \div 6$ **e.** $-7 \div 0$

If the division involves real numbers written as fractions, we use the following procedure.

PROCEDURE

Dividing Fractions

To divide $\frac{a}{b}$ by $\frac{c}{d}$, multiply $\frac{a}{b}$ by the reciprocal of $\frac{c}{d}$, that is,

$$\frac{a}{b} \div \frac{c}{d} = \frac{a}{b} \cdot \frac{d}{c} = \frac{ad}{bc} \qquad (b, c, \text{ and } d \neq 0)$$

Of course, the rules of signs still apply!

Answers to PROBLEMS

5. a. 7   **b.** -9 **c.** 7

  d. -4 **e.** Not defined

EXAMPLE 6   Finding quotients of fractions

Divide:

a. $\dfrac{2}{5} \div \left(-\dfrac{3}{4}\right)$ **b.** $-\dfrac{5}{6} \div \left(-\dfrac{7}{2}\right)$ **c.** $-\dfrac{3}{7} \div \dfrac{6}{7}$

SOLUTION

a. $\dfrac{2}{5} \div \left(-\dfrac{3}{4}\right) = \dfrac{2}{5} \cdot \left(-\dfrac{4}{3}\right) = -\dfrac{8}{15}$

  Different signs   Negative quotient

b. $-\dfrac{5}{6} \div \left(-\dfrac{7}{2}\right) = -\dfrac{5}{6} \cdot \left(-\dfrac{2}{7}\right) = \dfrac{10}{42} = \dfrac{5}{21}$

  Same sign   Positive quotient

c. $-\dfrac{3}{7} \div \dfrac{6}{7} = -\dfrac{3}{7} \cdot \dfrac{7}{6} = -\dfrac{21}{42} = -\dfrac{1}{2}$

  Different signs   Negative quotient

PROBLEM 6

Divide:

a. $\dfrac{4}{5} \div \left(-\dfrac{3}{4}\right)$ **b.** $-\dfrac{5}{6} \div \left(-\dfrac{7}{4}\right)$

c. $-\dfrac{4}{7} \div \dfrac{8}{7}$

D ⟩ Applications Involving Multiplying and Dividing Real Numbers

When you are driving and push down or let up on the gas or brake pedal, your car changes speed. This change in speed over a period of time is called *acceleration* and is given by

Final speed   Starting speed

$$a = \dfrac{f - s}{t}$$

Acceleration   Time period

We use this idea in Example 7.

Acceleration   Deceleration

EXAMPLE 7   Acceleration and deceleration

You are driving at 55 miles per hour (mi/hr), and over the next 10 seconds,

a. you *increase* your speed to 65 miles per hour. What is your acceleration?

b. you *decrease* your speed to 40 miles per hour. What is your acceleration?

SOLUTION   Your starting speed is $s = 55$ miles per hour and the time period is $t = 10$ seconds (sec).

PROBLEM 7

a. Find your acceleration if you increase your speed from 55 miles per hour to 70 miles an hour over the next 5 seconds.

b. Find your acceleration if you decrease your speed from 50 miles per hour to 40 miles per hour over the next 5 seconds.

Answers to PROBLEMS

6. a. $-\dfrac{16}{15}$ **b.** $\dfrac{10}{21}$ **c.** $-\dfrac{1}{2}$ **7. a.** $3 \dfrac{\text{mi/hr}}{\text{sec}}$ **b.** $-2 \dfrac{\text{mi/hr}}{\text{sec}}$

a. Your final speed is $f = 65$ miles per hour, so

$$a = \frac{(65 - 55) \text{ mi/hr}}{10 \text{ sec}} = \frac{10 \text{ mi/hr}}{10 \text{ sec}} = 1 \frac{\text{mi/hr}}{\text{sec}}$$

Thus, your acceleration is 1 mile per hour each second.

b. Here the final speed is 40, so

$$a = \frac{(40 - 55) \text{ mi/hr}}{10 \text{ sec}} = \frac{-15 \text{ mi/hr}}{10 \text{ sec}} = -1\frac{1}{2} \frac{\text{mi/hr}}{\text{sec}}$$

Thus, your acceleration is $-1\frac{1}{2}$ miles per hour every second. When acceleration is *negative,* it is called *deceleration.* Deceleration can be thought of as negative acceleration, so your deceleration is $1\frac{1}{2}$ miles per hour each second.

MathZone ⊕

Boost your grade at mathzone.com!

> Practice Problems   > Self-Tests
> NetTutor   > e-Professors
                         > Videos

> **Exercises 1.4**

⟨ **A** ⟩ **Multiplying Real Numbers**   In Problems 1–20, perform the indicated operation.

1. $4 \cdot 9$

2. $16 \cdot 2$

3. $-10 \cdot 4$

4. $-6 \cdot 8$

5. $-9 \cdot 9$

6. $-2 \cdot 5$

7. $-6 \cdot (-3)(-2)$

8. $-4 \cdot (-5)(-3)$

9. $-9 \cdot (-2)(-3)$

10. $-7 \cdot (-10)(-2)$

11. $-2.2(3.3)$

12. $-1.4(3.1)$

13. $-1.3(-2.2)$

14. $-1.5(-1.1)$

15. $\frac{5}{6}\left(-\frac{5}{7}\right)$

16. $\frac{3}{8}\left(-\frac{5}{7}\right)$

17. $-\frac{3}{5}\left(-\frac{5}{12}\right)$

18. $-\frac{4}{7}\left(-\frac{21}{8}\right)$

19. $-\frac{6}{7}\left(\frac{35}{8}\right)$

20. $-\frac{7}{5}\left(\frac{15}{28}\right)$

⟨ **B** ⟩ **Evaluating Expressions Involving Exponents**   In Problems 21–30, perform the indicated operation.

21. -9^2

22. $(-9)^2$

23. $(-5)^2$

24. -5^2

25. -5^3

26. $(-5)^3$

27. $(-6)^4$

28. -6^4

29. $-\left(\frac{1}{2}\right)^5$

30. $\left(-\frac{1}{2}\right)^5$

⟨ **C** ⟩ **Dividing Real Numbers**   In Problems 31–60, perform the indicated operation.

31. $\frac{14}{2}$

32. $10 \div 2$

33. $-50 \div 10$

34. $\frac{-20}{5}$

35. $\frac{-30}{10}$

36. $-40 \div 8$

37. $\frac{-0}{3}$

38. $-0 \div 8$

39. $-5 \div 0$

40. $\frac{-8}{0}$

41. $\frac{0}{7}$

42. $0 \div (-7)$

43. $-15 \div (-3)$

44. $-20 \div (-4)$

45. $\frac{-25}{-5}$

46. $\frac{-16}{-2}$

47. $\frac{18}{-9}$

48. $\frac{35}{-7}$

49. $30 \div (-5)$

50. $80 \div (-10)$

51. $\frac{3}{5} \div \left(-\frac{4}{7}\right)$

52. $\frac{4}{9} \div \left(-\frac{1}{7}\right)$

53. $-\frac{2}{3} \div \left(-\frac{7}{6}\right)$

54. $-\frac{5}{6} \div \left(-\frac{25}{18}\right)$

55. $-\frac{5}{8} \div \frac{7}{8}$

56. $-\frac{4}{5} \div \frac{8}{15}$

57. $\frac{-3.1}{6.2}$

58. $\frac{1.2}{-4.8}$

59. $\frac{-1.6}{-9.6}$

60. $\frac{-9.8}{-1.4}$

⟨ **D** ⟩   **Applications Involving Multiplying and Dividing Real Numbers**   Use the following information in Problems 61–65:

1 all-beef frank	+45 calories
1 slice of bread	+65 calories
Running (1 min)	−15 calories
Swimming (1 min)	−7 calories

61. *Caloric gain or loss*   If a person eats 2 beef franks and runs for 5 minutes, what is the caloric gain or loss?

62. *Caloric gain or loss*   If a person eats 2 beef franks and runs for 30 minutes, what is the caloric gain or loss?

63. *Caloric gain or loss*   If a person eats 2 beef franks with 2 slices of bread and then runs for 15 minutes, what is the caloric gain or loss?

64. *Caloric gain or loss*   If a person eats 2 beef franks with 2 slices of bread and then runs for 15 minutes and swims for 30 minutes, what is the caloric gain or loss?

65. *"Burning off" calories*   If a person eats 2 beef franks, how many minutes does the person have to run to "burn off" the calories? (*Hint:* You must *spend* the calories contained in the 2 beef franks.)

66. *Automobile acceleration*   The highest road-tested acceleration for a standard production car is from 0 to 60 miles per hour in 3.275 seconds for a Ford RS 200 Evolution. What was the acceleration of this car? Give your answer to one decimal place.

67. *Automobile acceleration*   The highest road-tested acceleration for a street-legal car is from 0 to 60 miles per hour in 3.89 seconds for a Jankel Tempest. What was the acceleration of this car? Give your answer to one decimal place.

68. *Cost of saffron*   Which food do you think is the most expensive? It is saffron, which comes from Spain. It costs $472.50 to buy 3.5 ounces at Harrods, a store in Great Britain. What is the cost of 1 ounce of saffron?

69. *Long-distance telephone charges*   The price of a long-distance call from Tampa to New York is $3.05 for the first 3 minutes and $0.70 for each additional minute or fraction thereof. What is the cost of a 5-minute long-distance call from Tampa to New York?

70. *Long-distance telephone charges*   The price of a long-distance call from Tampa to New York using a different phone company is $3 for the first 3 minutes and $0.75 for each additional minute or fraction thereof. What is the cost of a 5-minute long-distance call from Tampa to New York using this phone company?

Toothpaste   The number of brushings in each Aquafresh tube is shown. How many uses are in each Aquafresh® tube or pump? Using 1 to 1.2 grams of toothpaste on the brush (numbers are approximate).

(a) Find the grams of toothpaste used per day written as a mixed number.
(b) Find the number of days the tube will last if you brush the indicated number of times.

Tube Content	Brushings
4.3 oz (120 grams)	100 brushings
4.6 oz (130 grams)	120 brushings
6.0 oz (170 grams)	150 brushings
6.4 oz (180 grams)	170 brushings
7.6 oz (210 grams)	200 brushings
8.2 oz (230 grams)	220 brushings

71. 100 brushings using the 120-gram tube and brushing four times a day. (a) _____ (b) _____

72. 150 brushings using the 170-gram tube and brushing three times a day. (a) _____ (b) _____

73. 200 brushings using the 210-gram tube and brushing four times a day. (a) _____ (b) _____

74. 220 brushings using the 230-gram tube and brushing four times a day. (a) _____ (b) _____

Source: www.DentalProfessional.com.

⟩ ⟩ ⟩   ***Using Your Knowledge***

Have a Decidedly Lovable Day   Have you met anybody *nice* today or did you have an *unpleasant* experience? Perhaps the person you met was *very nice* or your experience *very unpleasant*. Psychologists and linguists have a numerical way to indicate the difference between *nice* and *very nice* or between *unpleasant* and *very unpleasant*. Suppose you assign a positive number (+2, for example) to the adjective *nice,* a negative number (say, −2) to *unpleasant,* and a positive number greater than 1 (say +1.75) to *very*. Then, *very nice* means

Very   nice
  ↓   ↓
$(1.75) \cdot (2) = 3.50$

and *very unpleasant* means

Very   unpleasant

$$(1.75) \cdot (-2) = -3.50$$

Here are some adverbs and adjectives and their average numerical values, as rated by a panel of college students. (Values differ from one panel to another.)

Adverbs		Adjectives	
Slightly	0.54	Wicked	−2.5
Rather	0.84	Disgusting	−2.1
Decidedly	0.16	Average	−0.8
Very	1.25	Good	3.1
Extremely	1.45	Lovable	2.4

Find the value of each.

75. Slightly wicked

76. Decidedly average

77. Extremely disgusting

78. Rather lovable

79. Very good

By the way, if you got all the answers correct, you are 4.495!

〉 〉 〉 *Write On*

80. Why is $-a^2$ always negative for any nonzero value of a?

81. Why is $(-a)^2$ always positive for any nonzero value of a?

82. Is $(-1)^{100}$ positive or negative? What about $(-1)^{99}$?

83. Explain why $\frac{a}{0}$ is not defined.

〉 〉 〉 *Concept Checker*

Fill in the blank(s) with the correct word(s), phrase, or mathematical statement.

84. When two numbers with **the *same (like)*** sign are **multiplied,** the **product** is _____.

positive

85. When two numbers with ***different (unlike)*** signs are **multiplied,** the **product** is _____.

negative

86. When dividing two numbers with the ***same (like)*** sign, give the **quotient** a _____ sign.

87. When dividing two numbers with ***different (unlike)*** signs, give the **quotient** a _____ sign.

〉 〉 〉 *Mastery Test*

Perform the indicated operations

88. $\frac{-3.6}{1.2}$

89. $\frac{-3.1}{-12.4}$

90. $\frac{6.5}{-1.3}$

91. $-3 \cdot 11$

92. $9 \cdot (-10)$

93. $-5 \cdot (-11)$

94. -9^2

95. $(-8)^2$

96. $\left(-\frac{1}{5}\right)^2$

97. $-2.2(3.2)$

98. $-1.3(-4.1)$

99. $-\frac{3}{7}\left(-\frac{4}{5}\right)$

100. $\frac{6}{7}\left(-\frac{2}{3}\right)$

101. $\frac{1}{5} \div \left(-\frac{4}{7}\right)$

102. $-\frac{6}{7} \div \left(-\frac{3}{5}\right)$

103. $-\frac{4}{5} \div \frac{8}{5}$

104. The driver of a car traveling at 50 miles per hour slams on the brakes and slows down to 10 miles per hour in 5 seconds. What is the acceleration of the car?

〉 〉 〉 *Skill Checker*

Perform the indicated operations.

105. $\frac{22 + 4}{8 - 10}$

106. $\frac{4 - 8}{6 - 4}$

107. $\frac{20 \cdot 7}{10}$

108. $\frac{30 \cdot 4}{10}$

1.5

Order of Operations

▶ Objectives

A ⟩ Evaluate expressions using the correct order of operations.

B ⟩ Evaluate expressions with more than one grouping symbol.

C ⟩ Solve an application involving the order of operations.

▶ To Succeed, Review How To . . .

1. Add, subtract, multiply, and divide real numbers (pp. 52–54, 60–62).

2. Evaluate expressions containing exponents (pp. 61–62).

▶ Getting Started
Collecting the Rent

Now that we know how to do the fundamental operations with real numbers, we need to know *in what order* we should do them. Let's suppose all rooms in a motel are taken. (For simplicity, we won't include extra persons in the rooms.) How can we figure out how much money we should collect? To do this, we first multiply the price of each room by the number of rooms available at that price. Next, we add all these figures to get the final answer. The calculations look like this:

$$44 \cdot \$44 = \$1936$$
$$150 \cdot \$48 = \$7200$$
$$45 \cdot \$58 = \$2610$$
$$\$1936 + \$7200 + \$2610 = \$11{,}746 \quad \text{Total}$$

Note that we *multiplied* before we *added.* This is the correct order of operations. As you will see, changing the order of operations can change the answer!

MOTEL GUIDEBOOK 21

Chicago Area (cont.)

# Rooms	Single
44	$44
150	$48
45	$58

- OD pool
- Bus line
- Restaurant
- Non-smoking rooms
- Airport - 15 min.

Peoria Area

# Rms	SGL

A ⟩ Using the Order of Operations

If we want to find the answer to $3 \cdot 4 + 5$, do we

(1) add 4 and 5 first and then multiply by 3? That is, $3 \cdot 9 = 27$?

or

(2) multiply 3 by 4 first and then add 5? That is, $12 + 5 = 17$?

In (1), the answer is 27. In (2), the answer is 17. What if we write $3 \cdot (4 + 5)$ or $(3 \cdot 4) + 5$? What do the parentheses mean? To obtain an answer we can agree upon, we need the following rules.

> **RULES**
>
> **Order of Operations**
>
> Operations are always performed in the following order:
>
> **1.** Do all calculations inside **grouping symbols** such as **p**arentheses ( ) or brackets [ ].
>
> **2.** **E**valuate all exponents.
>
> **3.** Do **m**ultiplications and **d**ivisions as they occur from *left to right.*
>
> **4.** Do **a**dditions and **s**ubtractions as they occur from *left to right.*

You can remember the order by remembering the underlined letters PEMDAS.

With these conventions

$$3 \cdot 4 + 5$$

$$= \quad 12 \quad + 5 \qquad \text{First multiply.}$$

$$= \quad 17 \qquad \text{Then add.}$$

Similarly,

$$3 \cdot (4 + 5)$$

$$= 3 \cdot \quad 9 \qquad \text{First add inside parentheses.}$$

$$= \quad 27 \qquad \text{Then multiply.}$$

But

$$(3 \cdot 4) + 5$$

$$= \quad 12 \quad + 5 \qquad \text{First multiply inside parentheses.}$$

$$= \quad 17 \qquad \text{Then add.}$$

Note that multiplications and divisions are done **in order,** from left to right. This means that sometimes you do multiplications first, sometimes you do divisions first, depending on the order in which they occur. Thus,

$$12 \div 4 \cdot 2$$

$$= \quad 3 \quad \cdot 2 \qquad \text{First divide.}$$

$$= \quad 6 \qquad \text{Then multiply.}$$

But

$$12 \cdot 4 \div 2$$

$$= \quad 48 \quad \div 2 \qquad \text{First multiply.}$$

$$= \quad 24 \qquad \text{Then divide.}$$

EXAMPLE 1   Evaluating expressions

Find the value of each expression.

a. $8 \cdot 9 - 3$ **b.** $27 + 3 \cdot 5$

SOLUTION

a.

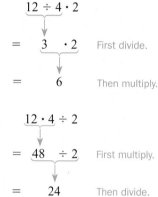

$$8 \cdot 9 - 3$$

$$= 72 - 3 \qquad \text{Do multiplications and divisions in order from left to right } (8 \cdot 9 = 72).$$

$$= \quad 69 \qquad \text{Then do additions and subtractions in order from left to right } (72 - 3 = 69).$$

b.

$$27 + 3 \cdot 5$$

$$= 27 + 15 \qquad \text{Do multiplications and divisions in order from left to right } (3 \cdot 5 = 15).$$

$$= \quad 42 \qquad \text{Then do additions and subtractions in order from left to right } (27 + 15 = 42).$$

PROBLEM 1

Find the value of each expression.

a. $3 \cdot 8 - 9$ **b.** $22 + 4 \cdot 9$

Answers to PROBLEMS

1. a. 15   **b.** 58

EXAMPLE 2   Expressions with grouping symbols and exponents

Find the value of each expression.

a. $63 \div 7 - (2 + 3)$

b. $8 \div 2^3 + 3 - 1$

SOLUTION

a.
$$63 \div 7 - (2 + 3)$$

$= 63 \div 7 - \quad 5$ First do the operation inside the parentheses.

$= \quad 9 \quad - \quad 5$ Next do the division.

$= \quad\quad 4$ Then do the subtraction.

b.
$$8 \div 2^3 + 3 - 1$$

$= 8 \div 8 + 3 - 1$ First do the exponentation.

$= \quad 1 \quad + 3 - 1$ Next do the division.

$= \quad\quad 4 \quad - 1$ Then do the addition.

$= \quad\quad 3$ Do the final subtraction.

EXAMPLE 3   Expression with grouping symbols

Find the value of: $8 \div 4 \cdot 2 + 3(5 - 2) - 3 \cdot 2$

SOLUTION

$$8 \div 4 \cdot 2 + 3(5 - 2) - 3 \cdot 2$$

$= 8 \div 4 \cdot 2 + 3 \quad (3) \quad - 3 \cdot 2$ First do the operations inside the parentheses. Now do the multiplications and divisions in order from left to right:

$= \quad 2 \cdot 2 + 3(3) \quad - 3 \cdot 2$ This means do $8 \div 4 = 2$ first.

$= \quad 4 \quad + 3(3) \quad - 3 \cdot 2$ Then do $2 \cdot 2 = 4$.

$= \quad 4 \quad + \quad 9 \quad - 3 \cdot 2$ Next do $3(3) = 9$.

$= \quad 4 \quad + \quad 9 \quad - \quad 6$ And finally, do $3 \cdot 2 = 6$.

$= \quad\quad 13 \quad\quad - \quad 6$ We are through with multiplications and divisions. Now do the addition.

$= \quad\quad\quad 7$ The final operation is a subtraction.

PROBLEM 2

Find the value of each expression.

a. $56 \div 8 - (3 + 1)$

b. $54 \div 3^3 + 5 - 2$

PROBLEM 3

Find the value of:

$10 \div 5 \cdot 2 + 4(5 - 3) - 4 \cdot 2$

B ⟩ Evaluating Expressions with More than One Grouping Symbol

Suppose a local department store is having a great sale. The advertised items are such good deals that you decide to buy two bedspreads and two mattresses. The price of one bedspread and one mattress is $14 + $88. Thus, the price of two of each is $2 \cdot (14 + 88)$. If you then decide to buy a lamp, the total price is

$$[2 \cdot (14 + 88)] + 12$$

We have used two types of grouping symbols in [2 · (14 + 88)] + 12, parentheses ( ) and brackets [ ]. There is one more grouping symbol, braces { }. Typically, the order of these grouping symbols is {[( )]}. Here is the rule we use to handle grouping symbols.

RULE

Grouping Symbols

When grouping symbols occur within other grouping symbols, perform the computations in the *innermost* grouping symbols first.

Thus, to find the value of [2 · (14 + 88)] + 12, we first add 14 and 88 (the operation inside parentheses, the innermost grouping symbols), then multiply by 2, and finally add 12. Here is the procedure:

$$[2 \cdot (14 + 88)] + 12 \qquad \text{Given}$$

$$= \; [2 \cdot (102)] \; + 12 \qquad \text{Add 14 and 88 inside the parentheses.}$$

$$= \quad 204 \qquad + 12 \qquad \text{Multiply 2 by 102 inside the brackets.}$$

$$= \qquad 216 \qquad \text{Do the final addition.}$$

EXAMPLE 4   Expressions with three grouping symbols

Find the value of: $20 \div 4 + \{2 \cdot 9 - [3 + (6 - 2)]\}$

SOLUTION   The innermost grouping symbols are the parentheses, so we do the operations inside the parentheses, then inside the brackets, and, finally, inside the braces. Here are the details.

$$20 \div 4 + \{2 \cdot 9 - [3 + (6 - 2)]\} \qquad \text{Given}$$

$$= 20 \div 4 + \{2 \cdot 9 - [3 + 4]\} \qquad \text{Subtract inside the parentheses } (6 - 2 = 4).$$

$$= 20 \div 4 + \{2 \cdot 9 - 7\} \qquad \text{Add inside the brackets } (3 + 4 = 7).$$

$$= 20 \div 4 + \{18 - 7\} \qquad \text{Multiply inside the braces } (2 \cdot 9 = 18).$$

$$= 20 \div 4 + \quad 11 \qquad \text{Subtract inside the braces } (18 - 7 = 11).$$

$$= \quad 5 \; + \quad 11 \qquad \text{Divide } (20 \div 4 = 5).$$

$$= \qquad 16 \qquad \text{Do the final addition.}$$

PROBLEM 4

Find the value of:

$$50 \div 5 + \{2 \cdot 7 - [4 + (5 - 2)]\}$$

Answers to PROBLEMS

4. 17

Fraction bars are sometimes used as grouping symbols to indicate an expression representing a single number. To find the value of such expressions, simplify above and below the fraction bars following the order of operations. Thus,

$$\frac{2(3 + 8) + 4}{2(4) - 10}$$

$$= \frac{2(11) + 4}{2(4) - 10} \qquad \text{Add inside the parentheses in the numerator } (3 + 8 = 11).$$

$$= \frac{22 + 4}{8 - 10} \qquad \text{Multiply in the numerator } [2(11) = 22] \text{ and in the denominator } [2(4) = 8].$$

$$= \frac{26}{-2} \qquad \text{Add in the numerator } (22 + 4 = 26), \text{ subtract in the denominator } (8 - 10 = -2).$$

$$= -13 \qquad \text{Do the final division. (Remember to use the rules of signs.)}$$

EXAMPLE 5   Using a fraction bar as a grouping symbol

Find the value of:

$$-5^2 + \frac{3(4 - 8)}{2} + 10 \div 5$$

SOLUTION $-5^2 + \dfrac{3(4 - 8)}{2} + 10 \div 5 \qquad$ Given

$$= -5^2 + \frac{3(-4)}{2} + 10 \div 5 \qquad \text{Subtract inside the parentheses } (4 - 8 = -4).$$

$$= -25 + \frac{3(-4)}{2} + 10 \div 5 \qquad \text{Do the exponentiation } (5^2 = 25, \text{ so } -5^2 = -25).$$

$$= -25 + \frac{-12}{2} + 10 \div 5 \qquad \text{Multiply above the division bar } [3(-4) = -12].$$

$$= -25 + (-6) + 10 \div 5 \qquad \text{Divide } \left(\frac{-12}{2} = -6\right).$$

$$= -25 + (-6) + \quad 2 \qquad \text{Divide } (10 \div 5 = 2).$$

$$= \quad -31 \quad + \quad 2 \qquad \text{Add } [-25 + (-6) = -31].$$

$$= \quad -29 \qquad \text{Do the final addition.}$$

PROBLEM 5

Find the value of:

$$-3^2 + \frac{2(3 - 6)}{3} + 20 \div 5$$

C ⟩ Applications Involving the Order of Operations

EXAMPLE 6   Ideal heart rate

Your ideal heart rate while exercising is found by subtracting your age A from 205, multiplying the result by 7, and dividing the answer by 10. In symbols, this is

$$\text{Ideal rate} = [(205 - A) \cdot 7] \div 10$$

Suppose you are 25 years old ($A = 25$). What is your ideal heart rate?

SOLUTION

$$\text{Ideal rate} = [(205 - A) \cdot 7] \div 10$$

$$= [(205 - 25) \cdot 7] \div 10 \qquad \text{Substitute 25 for } A.$$

$$= [180 \cdot 7] \div 10 \qquad \text{Subtract } (205 - 25 = 180).$$

$$= 1260 \div 10 \qquad \text{Multiply inside brackets } (180 \cdot 7 = 1260).$$

$$= 126 \qquad \text{Do the final division.}$$

PROBLEM 6

Use the formula to find your ideal heart rate assuming you are 20 years old. Round the answer to the nearest whole number.

Answers to PROBLEMS

5. -7 **6.** 130 (rounded from 129.5)

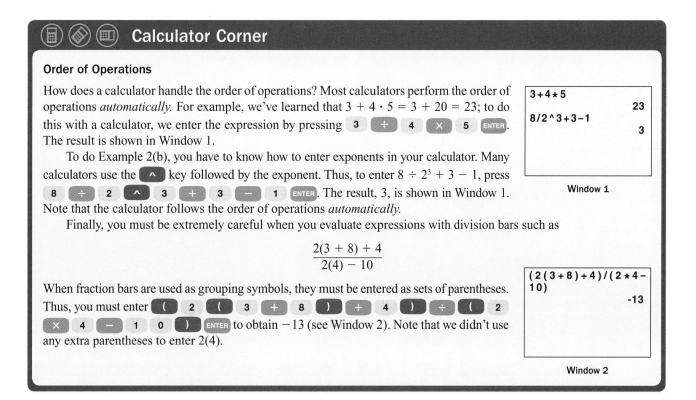

Calculator Corner

Order of Operations

How does a calculator handle the order of operations? Most calculators perform the order of operations *automatically*. For example, we've learned that $3 + 4 \cdot 5 = 3 + 20 = 23$; to do this with a calculator, we enter the expression by pressing [3] [+] [4] [×] [5] [ENTER]. The result is shown in Window 1.

To do Example 2(b), you have to know how to enter exponents in your calculator. Many calculators use the [^] key followed by the exponent. Thus, to enter $8 \div 2^3 + 3 - 1$, press [8] [÷] [2] [^] [3] [+] [3] [−] [1] [ENTER]. The result, 3, is shown in Window 1. Note that the calculator follows the order of operations *automatically*.

Finally, you must be extremely careful when you evaluate expressions with division bars such as

$$\frac{2(3+8)+4}{2(4)-10}$$

When fraction bars are used as grouping symbols, they must be entered as sets of parentheses. Thus, you must enter [(] [2] [(] [3] [+] [8] [)] [+] [4] [)] [÷] [(] [2] [×] [4] [−] [1] [0] [)] [ENTER] to obtain -13 (see Window 2). Note that we didn't use any extra parentheses to enter $2(4)$.

```
3+4*5
                    23
8/2^3+3-1
                     3
```
Window 1

```
(2(3+8)+4)/(2*4-
10)
                   -13
```
Window 2

Exercises 1.5

> Boost your grade at mathzone.com!
> Practice Problems   Self-Tests
> NetTutor   e-Professors
>   Videos

‹ **A** › **Using the Order of Operations**   In Problems 1–20, find the value of the given expression.

1. $4 \cdot 5 + 6$

2. $3 \cdot 4 + 6$

3. $7 + 3 \cdot 2$

4. $6 + 9 \cdot 2$

5. $7 \cdot 8 - 3$

6. $6 \cdot 4 - 9$

7. $20 - 3 \cdot 5$

8. $30 - 6 \cdot 5$

9. $48 \div 6 - (3 + 2)$

10. $81 \div 9 - (4 + 5)$

11. $3 \cdot 4 \div 2 + (6 - 2)$

12. $3 \cdot 6 \div 2 + (5 - 2)$

13. $36 \div 3^2 + 4 - 1$

14. $16 \div 2^3 + 3 - 2$

15. $8 \div 2^3 - 3 + 5$

16. $9 \div 3^2 - 8 + 5$

17. $10 \div 5 \cdot 2 + 8 \cdot (6 - 4) - 3 \cdot 4$

18. $15 \div 3 \cdot 3 + 2 \cdot (5 - 2) + 8 \div 4$

19. $4 \cdot 8 \div 2 - 3(4 - 1) + 9 \div 3$

20. $6 \cdot 3 \div 3 - 2(3 - 2) - 8 \div 2$

‹ **B** › **Evaluating Expressions with More than One Grouping Symbol**   In Problems 21–40, find the value of the given expression.

21. $20 \div 5 + \{3 \cdot 4 - [4 + (5 - 3)]\}$

22. $30 \div 6 + \{4 \div 2 \cdot 3 - [3 + (5 - 4)]\}$

23. $(20 - 15) \cdot [20 \div 2 - (2 \cdot 2 + 2)]$

24. $(30 - 10) \cdot [52 \div 4 - (3 \cdot 3 + 3)]$

25. $\{4 \div 2 \cdot 6 - (3 + 2 \cdot 3) + [5(3 + 2) - 1]\}$

26. $-6^2 + \dfrac{4(6 - 8)}{2} + 15 \div 3$

27. $-7^2 + \dfrac{3(8 - 4)}{4} + 10 \div 2 \cdot 3$

28. $-5^2 + \dfrac{6(3 - 7)}{4} + 9 \div 3 \cdot 2$

29. $(-6)^2 \cdot 4 \div 4 - \dfrac{3(7 - 9)}{2} - 4 \cdot 3 \div 2^2$

30. $(-4)^2 \cdot 3 \div 8 - \dfrac{4(6 - 10)}{2} - 3 \cdot 8 \div 2^3$

Web IT go to **mhhe.com/bello** for more lessons

31. $\left[\dfrac{7 - (-3)}{8 - 6}\right]\left[\dfrac{3 + (-8)}{7 - 2}\right]$

32. $\left[\dfrac{4 - (-3)}{8 - 1}\right]\left[\dfrac{-4 - (-9)}{8 - 3}\right]$

33. $\dfrac{(-3)(-2)(-4)}{3(-2)} - (-3)^2$

34. $\dfrac{(-2)(-4)(-5)}{10(-2)} - (-4)^2$

35. $\dfrac{(-10)(-6)}{(-2)(-3)(-5)} - (-3)^3$

36. $\dfrac{(-8)(-10)}{(-3)(-4)(-2)} - (-2)^3$

37. $\dfrac{(-2)^3(-3)(-9)}{(-2)^2(-3)^2} - 3^2$

38. $\dfrac{(-2)(-3)^2(-10)}{(-2)^2(-3)} - 3^2$

39. $-5^2 - \dfrac{(-2)(-3)(-4)}{(-12)(-2)}$

40. $-6^2 - \dfrac{(-3)(-5)(-8)}{(-3)(-4)}$

⟨ C ⟩ Applications Involving the Order of Operations

41. *Gasoline octane rating* Have you noticed the octane rating of gasoline at the gas pump? This octane rating is given by the equation

$$\frac{R + M}{2}$$

where R is a number measuring the performance of gasoline using the Research Method and M is a number measuring the performance of gasoline using the Motor Method. If a certain gasoline has $R = 92$ and $M = 82$, what is its octane rating?

42. *Gasoline octane rating* If a gasoline has $R = 97$ and $M = 89$, what is its octane rating?

43. *Exercise pulse rate* If A is your age, the minimum pulse rate you should maintain during aerobic activities is $0.72(220 - A)$. What is the minimum pulse rate you should maintain if you are the specified age?

  a. 20 years old   **b.** 45 years old

44. *Exercise pulse rate* If A is your age, the maximum pulse rate you should maintain during aerobic activities is $0.88(220 - A)$. What is the maximum pulse rate you should maintain if you are the specified age?

  a. 20 years old   **b.** 45 years old

45. *Weight* Your weight depends on many factors like gender and height. For example, if you are a woman more than 5 ft (60 inches) tall, your weight W (in pounds) should be

  $W = 105 + 5(h - 60)$, where h is your height in inches

  a. What should the weight of a woman measuring 62 inches be?

  b. What should the weight of a woman measuring 65 inches be?

46. *Weight for a man* A man of medium frame measuring h inches should weigh

  $W = 140 + 3(h - 62)$ pounds

  a. What should the weight of a man measuring 72 inches be?

  b. What should the weight of a man measuring 68 inches be?

47. *Cell phone rates* At the present time, the Verizon America Choice 900 plan costs $59.99 per month and gives you 900 anytime minutes with unlimited night and weekend minutes. After that, you pay $0.40 per minute. The monthly cost C is

  $C = \$59.99 + 0.40(m - 900)$, where m is the number of minutes used

  a. Find the monthly cost for a talker that used 1000 minutes.

  b. Find the monthly cost for a talker that used 945 minutes.

48. *Cell phone rates* The $39.99 Cingular 450 plan provides 450 anytime minutes, 450 roll over minutes, and 5000 night and weekend minutes. After that, you pay $0.45 per minute. The monthly cost C is

  $C = \$39.99 + 0.45(m - 450)$, where m is the number of minutes

  a. Find the monthly cost for a talker that used 500 minutes.

  b. Find the monthly cost for a talker that used 645 minutes.

49. *Parking rates* The short-term garage at Tampa International Airport gives you the first hour of parking free! After that, you pay $1.25 per half hour with an $18 daily maximum. The formula for the cost C is

  $C = 2.50(h - 1.5) + 1.25$

  a. What is the cost for parking 3 hours in the short-term garage?

  b. Horacio had to wait 10 hours for his mother's plane; how much was the parking fee?

50. *Parking rates* The long-term garage at Tampa International Airport gives you the first hour of parking free! After that you pay $1.25 per hour up to a $14 daily maximum. The formula for the cost C is

  $C = 1.25(h - 11) + 12.50$

  a. What is the cost for parking 11 hours in the long-term garage?

  b. What is the cost for parking 19 hours in the long-term garage?

❯❯❯ *Using Your Knowledge*

Children's Dosages   What is the corresponding dose (amount) of medication for children when the adult dosage is known? There are several formulas that tell us.

51. *Friend's rule* (for children under 2 years):

(Age in months · Adult dose) ÷ 150 = Child's dose

Suppose a child is 10 months old and the adult dose of aspirin is a 75-milligram tablet. What is the child's dose? [*Hint:* Simplify (10 · 75) ÷ 150.]

53. *Young's rule* (for children between 3 and 12):

(Age · Adult dose) ÷ (Age + 12) = Child's dose

Suppose a child is 6 years old and the adult dose of an antibiotic is 4 tablets every 12 hours. What is the child's dose? [*Hint:* Simplify (6 · 4) ÷ (6 + 12).]

52. *Clarke's rule* (for children over 2 years):

(Weight of child · Adult dose) ÷ 150 = Child's dose

If a 7-year-old child weighs 75 pounds and the adult dose is 4 tablets a day, what is the child's dose? [*Hint:* Simplify (75 · 4) ÷ 150.]

❯❯❯ *Write On*

54. State the rules for the order of operations, and explain why they are needed.

55. Explain whether the parentheses are needed when finding

    a. $2 + (3 \cdot 4)$ **b.** $2 \cdot (3 + 4)$

56. **a.** When evaluating an expression, do you *always* have to do multiplications before divisions? Give examples to support your answer.

    b. When evaluating an expression, do you *always* have to do additions before subtractions? Give examples to support your answer.

❯❯❯ *Concept Checker*

Fill in the blank(s) with the correct word(s), phrase, or mathematical statement.

57. The **acronym** (abbreviation) to remember the **order of operations** is _____.

58. In the acronym of Problem 57, the **P** means _____.

59. In the acronym of Problem 57, the **E** means _____.

60. In the acronym of Problem 57, the **M** means _____.

61. In the acronym of Problem 57, the **D** means _____.

62. In the acronym of Problem 57, the **A** means _____.

63. In the acronym of Problem 57, the **S** means _____.

multiplication	innermost
division	outside
addition	exponents
subtraction	parentheses
PEMDAS	

64. When **grouping symbols** occur within **other grouping symbols,** perform the computations in the _____ grouping symbols **first.**

❯❯❯ *Mastery Test*

Find the value of the expressions.

65. $63 \div 9 - (2 + 5)$

66. $-64 \div 8 - (6 - 2)$

67. $16 + 2^3 + 3 - 9$

68. $3 \cdot 4 - 18$

69. $18 + 4 \cdot 5$

70. $12 \div 4 \cdot 2 + 2(5 - 3) - 3 \cdot 4$

71. $15 \div 3 + \{2 \cdot 4 - [6 + (2 - 5)]\}$

72. $-6^2 + \dfrac{4(6 - 12)}{3} + 8 \div 2$

73. The ideal heart rate while exercising for a person A years old is $[(205 - A) \cdot 7] \div 10$. What is the ideal heart rate for a 35-year-old person?

Perform the indicated operations:

74. $\frac{1}{9} \cdot 9$

75. $11 \cdot \frac{1}{11}$

76. $17 \cdot \frac{1}{17}$

77. $-2.3 + 2.3$

78. $-1.7 + 1.7$

79. $-2.5 \cdot \frac{1}{-2.5}$

80. $-3.7 \cdot \frac{1}{-3.7}$

1.6   Properties of the Real Numbers

▶ Objectives

A〉 Identify which of the properties (associative or commutative) are used in a statement.

B〉 Use the commutative and associative properties to simplify expressions.

C〉 Identify which of the properties (identity or inverse) are used in a statement.

D〉 Use the properties to simplify expressions.

E〉 Use the distributive property to remove parentheses in an expression.

▶ To Succeed, Review How To . . .

Add, subtract, multiply, and divide real numbers (pp. 52, 54, 60, 62).

▶ Getting Started
Clarifying Statements

Look at the sign we found hanging outside a lawn mower repair shop. What does it mean? Do the owners want

     a small (engine mechanic)?
         or
     a (small engine) mechanic?

The second meaning is the intended one. Do you see why? The manner in which you *associate* the words makes a difference. Now use parentheses to show how you think the following words should be associated:

         Guaranteed used cars

           Huge tire sale

If you wrote

       Guaranteed (used cars)   and   Huge (tire sale)

you are well on your way to understanding the associative property. In algebra, if we are multiplying or adding, the way in which we associate (combine) the numbers makes *no* difference in the answer. This is the associative property, and we shall study it and several other properties of real numbers in this section.

A〉 Identifying the Associative and Commutative Properties

How would you add $17 + 98 + 2$? You would probably add 98 and 2, get 100, and then add 17 to obtain 117. Even though the order of operations tells us to add from left to right, we can group the *addends* (the numbers we are adding) any way we want without changing the sum. Thus, $(17 + 98) + 2 = 17 + (98 + 2)$. This fact can be stated as follows.

ASSOCIATIVE PROPERTY OF ADDITION	For any real numbers *a*, *b*, and *c*, changing the *grouping* of two addends does not change the sum. In symbols, $$a + (b + c) = (a + b) + c$$

The associative property of addition tells us that the *grouping* does not matter in addition.

What about multiplication? Does the grouping matter? For example, is $2 \cdot (3 \cdot 4)$ the same as $(2 \cdot 3) \cdot 4$? Since both calculations give 24, the manner in which we group these numbers doesn't matter either. This fact can be stated as follows.

ASSOCIATIVE PROPERTY OF MULTIPLICATION	For any real numbers *a*, *b*, and *c*, changing the *grouping* of two factors does not change the product. In symbols, $$a \cdot (b \cdot c) = (a \cdot b) \cdot c$$

The associative property of multiplication tells us that the *grouping* does not matter in multiplication.

We have now seen that grouping numbers differently in addition and multiplication yields the same answer. What about order? As it turns out, the order in which we do additions or multiplications doesn't matter either. For example, $2 + 3 = 3 + 2$ and $5 \cdot 4 = 4 \cdot 5$. In general, we have the following properties.

COMMUTATIVE PROPERTY OF ADDITION	For any numbers *a* and *b*, changing the *order* of two addends does not change the sum. In symbols, $$a + b = b + a$$

COMMUTATIVE PROPERTY OF MULTIPLICATION	For any numbers *a* and *b*, changing the *order* of two factors does not change the product. In symbols, $$a \cdot b = b \cdot a$$

The commutative properties of addition and multiplication tell us that *order* doesn't matter in addition or multiplication.

EXAMPLE 1   Identifying properties

Name the property illustrated in each of the following statements:

a. $a \cdot (b \cdot c) = (b \cdot c) \cdot a$ **b.** $(5 + 9) + 2 = 5 + (9 + 2)$
c. $(5 + 9) + 2 = (9 + 5) + 2$ **d.** $(6 \cdot 2) \cdot 3 = 6 \cdot (2 \cdot 3)$

SOLUTION

a. We changed the *order* of multiplication. The commutative property of multiplication was used.

b. We changed the *grouping* of the numbers. The associative property of addition was used.

c. We changed the *order* of the 5 and the 9 within the parentheses. The commutative property of addition was used.

d. Here we changed the *grouping* of the numbers. We used the associative property of multiplication.

PROBLEM 1

Name the property illustrated in each of the following statements:

a. $5 \cdot (4 \cdot 6) = (5 \cdot 4) \cdot 6$

b. $(x \cdot y) \cdot z = z \cdot (x \cdot y)$

c. $(4 + 8) + 7 = 4 + (8 + 7)$

d. $(4 + 3) + 5 = (3 + 4) + 5$

Answers to PROBLEMS

1. a. Associative prop. of mult.   **b.** Commutative prop. of mult.
  c. Associative prop. of addition   **d.** Commutative prop. of addition

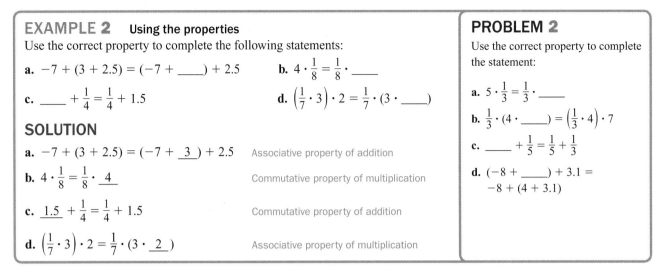

EXAMPLE 2   Using the properties

Use the correct property to complete the following statements:

a. $-7 + (3 + 2.5) = (-7 + \underline{\quad}) + 2.5$

b. $4 \cdot \frac{1}{8} = \frac{1}{8} \cdot \underline{\quad}$

c. $\underline{\quad} + \frac{1}{4} = \frac{1}{4} + 1.5$

d. $\left(\frac{1}{7} \cdot 3\right) \cdot 2 = \frac{1}{7} \cdot (3 \cdot \underline{\quad})$

SOLUTION

a. $-7 + (3 + 2.5) = (-7 + \underline{3}) + 2.5$ Associative property of addition

b. $4 \cdot \frac{1}{8} = \frac{1}{8} \cdot \underline{4}$ Commutative property of multiplication

c. $\underline{1.5} + \frac{1}{4} = \frac{1}{4} + 1.5$ Commutative property of addition

d. $\left(\frac{1}{7} \cdot 3\right) \cdot 2 = \frac{1}{7} \cdot (3 \cdot \underline{2})$ Associative property of multiplication

PROBLEM 2

Use the correct property to complete the statement:

a. $5 \cdot \frac{1}{3} = \frac{1}{3} \cdot \underline{\quad}$

b. $\frac{1}{3} \cdot (4 \cdot \underline{\quad}) = \left(\frac{1}{3} \cdot 4\right) \cdot 7$

c. $\underline{\quad} + \frac{1}{5} = \frac{1}{5} + \frac{1}{3}$

d. $(-8 + \underline{\quad}) + 3.1 = -8 + (4 + 3.1)$

B › Using the Associative and Commutative Properties to Simplify Expressions

The associative and commutative properties are also used to simplify expressions. For example, suppose x is a number: $(7 + x) + 8$ can be simplified by adding the numbers together as follows.

$$(7 + x) + 8 = 8 + (7 + x)$$ Commutative property of addition
$$= (8 + 7) + x$$ Associative property of addition
$$= 15 + x$$ You can also write the answer as $x + 15$ using the commutative property of addition.

Of course, you normally just add the 7 and 8 without going through all these steps, but this example shows *why* and *how* your shortcut can be done.

EXAMPLE 3   Using the properties to simplify expressions

Simplify: $6 + 4x + 8$

SOLUTION

$$6 + 4x + 8 = (6 + 4x) + 8$$ Order of operations
$$= 8 + (6 + 4x)$$ Commutative property of addition
$$= (8 + 6) + 4x$$ Associative property of addition
$$= 14 + 4x$$ Add (parentheses aren't needed).

You can also write the answer as $4x + 14$ using the commutative property of addition.

PROBLEM 3

Simplify: $3 + 5x + 8$

C › Identifying Identities and Inverses

The properties we've just mentioned are applicable to all real numbers. We now want to discuss two special numbers that have unique properties, the numbers 0 and 1. If we add 0 to a number, the number is unchanged; that is, the number 0 preserves the **identity** of all numbers under addition. Thus, 0 is called the *identity element for addition*. This definition can be stated as follows.

| IDENTITY ELEMENT FOR ADDITION | Zero is the **identity element for addition;** that is, for any number *a*, $$a + 0 = 0 + a = a$$ |

Answers to PROBLEMS

2. a. 5   **b.** 7   **c.** $\frac{1}{3}$ **d.** 4

3. $11 + 5x$ or $5x + 11$

The number 1 preserves the **identity** of all numbers under multiplication; that is, if we multiply a number by 1, the number remains unchanged. Thus, 1 is called the *identity element for multiplication,* as stated here.

IDENTITY ELEMENT FOR MULTIPLICATION	The number **1** is the **identity element for multiplication;** that is, for any number a, $$a \cdot 1 = 1 \cdot a = a$$

We complete our list of properties by stating two ideas that we will discuss fully later.

ADDITIVE INVERSE (OPPOSITE)	For every number a, there exists another number $-a$ called its **additive inverse** (or **opposite**) such that $a + (-a) = 0$.

When dealing with multiplication, the *multiplicative inverse* is called the *reciprocal.*

MULTIPLICATIVE INVERSE (RECIPROCAL)	For every number a (except 0), there exists another number $\frac{1}{a}$ called the **multiplicative inverse** (or **reciprocal**) such that $a \cdot \frac{1}{a} = 1$

EXAMPLE 4   Identifying properties

Name the property illustrated in each of the following statements:

a. $5 + (-5) = 0$ **b.** $\frac{8}{3} \cdot \frac{3}{8} = 1$ **c.** $0 + 9 = 9$ **d.** $7 \cdot 1 = 7$

SOLUTION

a. $5 + (-5) = 0$ Additive inverse

b. $\frac{8}{3} \cdot \frac{3}{8} = 1$ Multiplicative inverse

c. $0 + 9 = 9$ Identity element for addition

d. $7 \cdot 1 = 7$ Identity element for multiplication

PROBLEM 4

Name the property illustrated:

a. $8 \cdot 1 = 8$ **b.** $\frac{7}{5} \cdot \frac{5}{7} = 1$

c. $3 + 0 = 3$ **d.** $(-4) + 4 = 0$

EXAMPLE 5   Using the properties

Fill in the blank so that the result is a true statement:

a. ___ $\cdot 0.5 = 0.5$ **b.** $\frac{3}{4} +$ ___ $= 0$ **c.** ___ $+ 2.5 = 0$

d. ___ $\cdot \frac{3}{5} = 1$ **e.** $1.8 \cdot$ ___ $= 1.8$ **f.** $\frac{3}{4} +$ ___ $= \frac{3}{4}$

SOLUTION

a. __1__ $\cdot 0.5 = 0.5$ Identity element for multiplication

b. $\frac{3}{4} + \left(-\frac{3}{4}\right) = 0$ Additive inverse property

c. $\underline{-2.5} + 2.5 = 0$ Additive inverse property

d. $\frac{5}{3} \cdot \frac{3}{5} = 1$ Multiplicative inverse property

e. $1.8 \cdot \underline{1} = 1.8$ Identity element for multiplication

f. $\frac{3}{4} + \underline{0} = \frac{3}{4}$ Identity element for addition

PROBLEM 5

Fill in the blank so that the result is a true statement:

a. $\frac{7}{3} +$ ___ $= \frac{7}{3}$

b. $\frac{4}{9} \cdot$ ___ $= 1$

c. $\frac{5}{8} +$ ___ $= 0$

d. $2.4 \cdot$ ___ $= 2.4$

e. $9.1 +$ ___ $= 0$

f. $0.2 \cdot$ ___ $= 0.2$

Answers to PROBLEMS

4. a. Identity element for mult.   **b.** Multiplicative inverse   **c.** Identity element for add.   **d.** Additive inverse

5. a. 0   **b.** $\frac{9}{4}$ **c.** $-\frac{5}{8}$ **d.** 1   **e.** -9.1 **f.** 1

D ⟩ Using the Identity and Inverse Properties to Simplify Expressions

EXAMPLE 6   Using the properties of addition

Use the properties of addition to simplify: $-3x + 5 + 3x$

SOLUTION

$$
\begin{aligned}
-3x + 5 + 3x &= (-3x + 5) + 3x && \text{Order of operations} \\
&= 3x + (-3x + 5) && \text{Commutative property of addition} \\
&= [3x + (-3x)] + 5 && \text{Associative property of addition} \\
&= 0 + 5 && \text{Additive inverse property} \\
&= 5 && \text{Identity element for addition}
\end{aligned}
$$

PROBLEM 6

Use the addition properties to simplify: $-5x + 7 + 5x$

Keep in mind that after you get enough practice, you won't need to go through all the steps we show here. We have given you all the steps to make sure you understand *why* and *how* we simplify expressions.

E ⟩ Using the Distributive Property to Remove Parentheses

The properties we've discussed all contain a single operation. Now suppose you wish to multiply a number, say 7, by the sum of 4 and 5. As it turns out, $7 \cdot (4 + 5)$ can be obtained in two ways:

$$7 \cdot (4 + 5)$$
$$7 \cdot \quad 9$$
$$63$$

Add within the parentheses first.

or

$$(7 \cdot 4) + (7 \cdot 5)$$
$$28 \quad + \quad 35$$
$$63$$

Multiply and then add.

Thus,

$$7 \cdot (4 + 5) = (7 \cdot 4) + (7 \cdot 5)$$ *First* multiply 4 by 7 and *then* 5 by 7.

The parentheses in $(7 \cdot 4) + (7 \cdot 5)$ can be omitted since, by the order of operations, multiplications must be done first. Thus, multiplication distributes over addition as follows:

DISTRIBUTIVE PROPERTY

For any numbers *a*, *b*, and *c*,

$$a(b + c) = ab + ac$$

Note that $a(b + c)$ means $a \cdot (b + c)$, ab means $a \cdot b$, and ac means $a \cdot c$. The distributive property can also be extended to more than two numbers inside the parentheses. Thus, $a(b + c + d) = ab + ac + ad$. Moreover, $(b + c)a = ba + ca$.

EXAMPLE 7   Using the distributive property

Use the distributive property to multiply the following (x, y, and z are real numbers):

a. $8(2 + 4)$ **b.** $3(x + 5)$ **c.** $3(x + y + z)$ **d.** $(x + y)z$

SOLUTION

a. $8(2 + 4) = 8 \cdot 2 + 8 \cdot 4$

$\qquad = 16 + 32$ Multiply first.

$\qquad = 48$ Add next.

b. $3(x + 5) = 3x + 3 \cdot 5$

$\qquad = 3x + 15$ This expression cannot be simplified further since we don't know the value of x.

c. $3(x + y + z) = 3x + 3y + 3z$

d. $(x + y)z = xz + yz$

PROBLEM 7

Use the distributive property to multiply:

a. $(a + b)c$ **b.** $5(a + b + c)$

c. $4(a + 7)$ **d.** $3(2 + 9)$

EXAMPLE 8   Using the distributive property

Use the distributive property to multiply:

a. $-2(a + 7)$ **b.** $-3(x - 2)$ **c.** $-(a - 2)$

SOLUTION

a. $-2(a + 7) = -2a + (-2)7$

$\qquad = -2a - 14$

b. $-3(x - 2) = -3x + (-3)(-2)$

$\qquad = -3x + 6$

c. What does $-(a - 2)$ mean? Since

$$-a = -1 \cdot a$$

we have

$$-(a - 2) = -1 \cdot (a - 2)$$
$$= -1 \cdot a + (-1) \cdot (-2)$$
$$= -a + 2$$

PROBLEM 8

Use the distributive property to multiply:

a. $-3(x + 5)$

b. $-5(a - 2)$

c. $-(x - 4)$

Finally, for easy reference, here is a list of all the properties we've studied.

PROPERTIES OF THE REAL NUMBERS

If a, b, and c are real numbers, then the following properties hold.

Addition	Multiplication	Property
$a + b = b + a$	$a \cdot b = b \cdot a$	Commutative property
$a + (b + c) = (a + b) + c$	$a \cdot (b \cdot c) = (a \cdot b) \cdot c$	Associative property
$a + 0 = 0 + a = a$	$a \cdot 1 = 1 \cdot a = a$	Identity property
$a + (-a) = 0$	$a \cdot \frac{1}{a} = 1 \ (a \neq 0)$	Inverse property
	$a(b + c) = ab + ac$	Distributive property
	$(a + b)c = ac + bc$	

Answers to PROBLEMS

7. a. $ac + bc$ **b.** $5a + 5b + 5c$ **c.** $4a + 28$ **d.** 33

8. a. $-3x - 15$ **b.** $-5a + 10$ **c.** $-x + 4$

> **CAUTION**
>
> The commutative and associative properties apply to addition and multiplication but **not** to subtraction and division. For example,
>
> $$3 - 5 \neq 5 - 3 \quad \text{and} \quad 6 \div 2 \neq 2 \div 6$$
>
> Also,
>
> $$5 - (4 - 2) \neq (5 - 4) - 2 \quad \text{and} \quad 6 \div (3 \div 3) \neq (6 \div 3) \div 3$$

> **MathZone** +x
>
> **Boost your grade at** mathzone.com!
> > Practice   > Self-Tests
> > Problems   > e-Professors
> > NetTutor   > Videos

〉 Exercises 1.6

〈 A 〉 Identifying the Associative, Commutative, and Distributive Properties   In Problems 1–10, name the property illustrated in each statement.

1. $9 + 8 = 8 + 9$

2. $b \cdot a = a \cdot b$

3. $4 \cdot 3 = 3 \cdot 4$

4. $(a + 4) + b = a + (4 + b)$

5. $3(x + 6) = 3x + 3 \cdot 6$

6. $8(2 + x) = 8 \cdot 2 + 8x$

7. $a \cdot (b \cdot c) = a \cdot (c \cdot b)$

8. $a \cdot (b \cdot c) = (a \cdot b) \cdot c$

9. $a + (b + 3) = (a + b) + 3$

10. $(a + 3) + b = (3 + a) + b$

〈 B 〉 Using the Associative and Commutative Properties to Simplify Expressions   In Problems 11–18, name the property illustrated in each statement, then find the missing number that makes the statement correct.

11. $-3 + (5 + 1.5) = (-3 + \underline{\quad}) + 1.5$

12. $(-4 + \underline{\quad}) + \frac{1}{5} = -4 + \left(3 + \frac{1}{5}\right)$

13. $7 \cdot \frac{1}{8} = \frac{1}{8} \cdot \underline{\quad}$

14. $\frac{9}{4} \cdot \underline{\quad} = 3 \cdot \frac{9}{4}$

15. $\underline{\quad} + \frac{3}{4} = \frac{3}{4} + 6.5$

16. $7.5 + \underline{\quad} = \frac{5}{8} + 7.5$

17. $\left(\frac{3}{7} \cdot 5\right) \cdot \underline{\quad} = \frac{3}{7} \cdot (5 \cdot 2)$

18. $\left(\frac{2}{9} \cdot \underline{\quad}\right) \cdot 4 = \frac{2}{9} \cdot (2 \cdot 4)$

〈 B 〉 Using the Associative and Commutative Properties to Simplify Expressions
〈 D 〉 Using the Identity and Inverse Properties to Simplify Expressions

In Problems 19–24, simplify.

19. $5 + 2x + 8$

20. $-10 + 2y + 12$

21. $\frac{1}{5}a - 3 + 3 - \frac{1}{5}a$

22. $\frac{2}{3}b + 4 - 4 - \frac{2}{3}b$

23. $5c + (-5c) + \frac{1}{5} \cdot 5$

24. $-\frac{1}{7} \cdot 7 - 4x + 4x$

〈 C 〉 Identifying Identities and Inverses   In Problems 25–28, name the property illustrated in each statement.

25. $9 \cdot \frac{1}{9} = 1$

26. $10 \cdot 1 = 10$

27. $8 + 0 = 8$

28. $-6 + 6 = 0$

〈 E 〉 Using the Distributive Property to Remove Parentheses   In Problems 29–38, use the distributive property to fill in the blank.

29. $8(9 + \underline{\quad}) = 8 \cdot 9 + 8 \cdot 3$

30. $7(\underline{\quad} + 10) = 7 \cdot 4 + 7 \cdot 10$

31. $\underline{\quad}(3 + b) = 15 + 5b$

32. $3(\_\_\_ + b) = 3 \cdot 8 + 3b$ **33.** $\_\_\_(b + c) = ab + ac$ **34.** $3(x + 4) = 3x + 3 \cdot \_\_\_$

35. $-4(x + \_\_\_) = -4x + 8$ **36.** $-3(x + 5) = -3x + \_\_\_$ **37.** $-2(5 + c) = -2 \cdot 5 + \_\_\_ \cdot c$

38. $-3(\_\_\_ + a) = 6 + (-3)a$

In Problems 39–78, use the distributive property to multiply.

39. $6(4 + x)$ **40.** $3(2 + x)$ **41.** $8(x + y + z)$ **42.** $5(x + y + z)$

43. $6(x + 7)$ **44.** $7(x + 2)$ **45.** $(a + 5)b$ **46.** $(a + 2)c$

47. $6(5 + b)$ **48.** $3(7 + b)$ **49.** $-4(x + y)$ **50.** $-3(a + b)$

51. $-9(a + b)$ **52.** $-6(x + y)$ **53.** $-3(4x + 2)$ **54.** $-2(3a + 9)$

55. $-\left(\dfrac{3a}{2} - \dfrac{6}{7}\right)$ **56.** $-\left(\dfrac{2x}{3} - \dfrac{1}{5}\right)$ **57.** $-(2x - 6y)$ **58.** $-(3a - 6b)$

59. $-(2.1 + 3y)$ **60.** $-(5.4 + 4b)$ **61.** $-4(a + 5)$ **62.** $-6(x + 8)$

63. $-x(6 + y)$ **64.** $-y(2x + 3)$ **65.** $-8(x - y)$ **66.** $-9(a - b)$

67. $-3(2a - 7b)$ **68.** $-4(3x - 9y)$ **69.** $0.5(x + y - 2)$ **70.** $0.8(a + b - 6)$

71. $\dfrac{6}{5}(a - b + 5)$ **72.** $\dfrac{2}{3}(x - y + 4)$ **73.** $-2(x - y + 4)$ **74.** $-4(a - b + 8)$

75. $-0.3(x + y - 6)$ **76.** $-0.2(a + b - 3)$ **77.** $-\dfrac{5}{2}(a - 2b + c - 1)$ **78.** $-\dfrac{4}{7}(2a - b + 3c - 5)$

⟩⟩⟩ Applications

79. *Distance traveled*   Tyrone drove from San Luis Obispo to Los Angeles via Santa Barbara $(101 + 96)$ miles. He then drove 62 more miles to San Bernardino. Write an expression that models the situation and then find the distance from San Luis Obispo to San Bernardino.

80. *Distance traveled*   Maria started from San Luis Obispo but first stopped at Santa Barbara, 101 miles. She then drove to San Bernardino via Los Angeles, $(96 + 62)$ miles. Write an expression that models the situation and then find the driving distance for Maria.

81. *Distance traveled*   Referring to Problems 79 and 80, which property tells us that the driving distances for Tyrone and Maria are the same?

82. *Price of a car*   The price P of a car is its base price (B) plus destination charges D, that is, $P = B + D$. Tran bought a Nissan in Smyrna, Tennessee, and there was no destination charge.

   a. What is D?

   b. Fill in the blank in the equation $P = B + _____$

   c. What property tells you that the equation in part **b** is correct?

83. *Area*   The area of a rectangle is found by multiplying its length L times its width W.

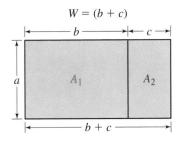

$$W = (b + c)$$

a. The length of the entire rectangle is a and its width is $(b + c)$. What is the area A of the rectangle?

b. The length of rectangle A_1 is a and its width b. What is the area of A_1?

c. The length of rectangle A_2 is a and its width c. What is the area of A_2?

d. The total area A of the rectangle is made up of the areas of rectangles A_1 and A_2, that is, $A = A_1 + A_2$ or $a(b + c) = ab + ac$, since $A = a(b + c)$, $A_1 = ab$ and $A_2 = ac$. Which property does this illustrate?

84. *Area*   The length of the entire rectangle is a and its width is b.

a. What is the area A of the entire rectangle?

b. What is the area of the smaller rectangle A_1?

c. The area of A_1 is the area A of the entire rectangle minus the area of A_2. Write an expression that models this situation.

d. Substitute the results obtained in **a** and **b** to rewrite the equation in part **c**.

85. *Weight*   **a.** If you are a woman more than 5 feet (60 inches) tall, your weight W (in pounds) should be

$$W = 105 + 5(h - 60),\text{ where } h \text{ is your height in inches}$$

Simplify this expression and find the weight of a woman 68 inch tall

b. The formula for the weight W of a man of medium frame measuring h inches is

$$W = 140 + 3(h - 62) \text{ pounds}$$

Simplify this expression and find the weight of a man 72 inches tall.

86. *Parking costs*   **a.** The formula for the cost C for parking in the short-term garage at Tampa International airport is

$$C = 2.50(h - 1.5) + 1.25$$

Simplify this formula and find the cost C of parking for 4 hours.

b. The formula for the cost C in the long-term parking garage is

$$C = 1.25(h - 11) + 12.50$$

Simplify this formula and find the cost C of parking for 12 hours.

❯❯❯ Using Your Knowledge

Multiplication Made Easy   The distributive property can be used to simplify certain multiplications. For example, to multiply 8 by 43, we can write

$$
\begin{aligned}
8(43) &= 8(40 + 3) \\
&= 320 + 24 \\
&= 344
\end{aligned}
$$

Use this idea to multiply the numbers.

87. $7(38)$ **88.** $8(23)$ **89.** $6(46)$ **90.** $9(52)$

❯❯❯ Write On

91. Explain why zero has no reciprocal.

92. We have seen that addition is commutative since $a + b = b + a$. Is subtraction commutative? Explain.

93. The associative property of addition states that $a + (b + c) = (a + b) + c$.
Is $a - (b - c) = (a - b) - c$? Explain.

94. Is $a + (b \cdot c) = (a + b)(a + c)$? Give examples to support your conclusion.

95. Multiplication is distributive over addition since $a \cdot (b + c) = a \cdot b + a \cdot c$. Is multiplication distributive over subtraction? Explain.

❯❯❯ Concept Checker

Fill in the blank(s) with the correct word(s), phrase, or mathematical statement.

96. The **associative property of addition** states that for any numbers a, b, and c,

_____.

97. The **associative property of multiplication** states that for any numbers a, b, and c,

_____.

98. The **commutative property of addition** states that for any numbers a, b, and c,

_____.

99. The **commutative property of multiplication** states that for any numbers a, b, and c,

_____.

100. The **identity element for addition** states that for any number a, _____.

101. The **identity element for multiplication** states that for any number a, _____.

102. The **additive inverse property** states that for any number a, there is an additive inverse _____.

$a(b + c) = ab + c$

$a(b + c) = ab + ac$

reciprocal

opposite

$\dfrac{1}{a}$

$-a$

$\dfrac{1}{a} \cdot a = a$

$a \cdot 1 = 1 \cdot a = a$

$a \cdot 0 = 0 \cdot a = a$

103. The **multiplicative inverse property** states that for any nonzero number a, there is multiplicative inverse ____.

104. The **multiplicative inverse of a** is also called the _____ of a.

105. The **distributive property** states that for any numbers a, b, and c, _____.

$$a + 0 = 0 + a = a$$
$$a \cdot b = b \cdot a$$
$$a + b = b + a$$
$$a \cdot (b \cdot c) = (a \cdot b) \cdot c$$
$$a + (b + c) = (a + b) + c$$

〉〉〉 *Mastery Test*

Use the distributive property to multiply the expression.

106. $-3(a + 4)$ **107.** $-4(a - 5)$ **108.** $-(a - 2)$

109. $4(a + 6)$ **110.** $2(x + y + z)$

Simplify:

111. $-2x + 7 + 2x$ **112.** $-4 - 2x + 2x + 4$ **113.** $7 + 4x + 9$ **114.** $-3 - 4x + 2$

Fill in the blank so that the result is a true statement.

115. $\frac{2}{5} + \underline{\ \ } = \frac{2}{5}$ **116.** $\frac{1}{4} + \underline{\ \ } = 0$ **117.** $-3.2 \cdot \underline{\ \ } = -3.2$

118. $\underline{\ \ } + 4.2 = 4.2$ **119.** $3(x + \underline{\ \ }) = 3x + 6$ **120.** $-2(x + 5) = -2x + \underline{\ \ }$

Name the property illustrated in each of the following statements.

121. $3 \cdot 1 = 3$ **122.** $\frac{5}{6} \cdot \frac{6}{5} = 1$ **123.** $2 + 0 = 2$

124. $-3 + 3 = 0$ **125.** $(6 \cdot x) \cdot 2 = 6 \cdot (x \cdot 2)$ **126.** $3 \cdot (4 \cdot 5) = 3 \cdot (5 \cdot 4)$

127. $(x + 2) + 5 = x + (2 + 5)$ **128.** $-3 \cdot 1 = -3$

Name the property illustrated in each statement, then find the missing number that makes the statement correct.

129. $\left(\frac{1}{5} \cdot 4\right) \cdot 2 = \frac{1}{5} \cdot (4 \cdot \underline{\ \ })$ **130.** $\frac{1}{3} + 2.4 = \underline{\ \ } + \frac{1}{3}$ **131.** $\frac{2}{5} \cdot \underline{\ \ } = 1 \cdot \frac{2}{5}$

132. $\frac{2}{11} + \underline{\ \ } = 0$ **133.** $-2 + (3 + 1.4) = (-2 + \underline{\ \ }) + 1.4$

〉〉〉 *Skill Checker*

Find:

134. $-5 + 3$ **135.** $-2 + (-5)$ **136.** $7 + (-9)$ **137.** $2(x + 1)$

1.7   Simplifying Expressions

▶ Objectives

A › Add and subtract like terms.

B › Use the distributive property to remove parentheses and then combine like terms.

C › Translate words into algebraic expressions and solve applications.

▶ To Succeed, Review How To . . .

1. Add and subtract real numbers (pp. 52–55).

2. Use the distributive property to simplify expressions (pp. 80–82).

▶ Getting Started
Combining Like Terms

If 3 tacos cost \$1.39 and 6 tacos cost \$2.39, then 3 tacos + 6 tacos will cost \$1.39 + \$2.39; that is, 9 tacos will cost \$3.78. Note that

$$3 \text{ tacos} + 6 \text{ tacos} = 9 \text{ tacos} \quad \text{and}$$
$$\$1.39 + \$2.39 = \$3.78$$

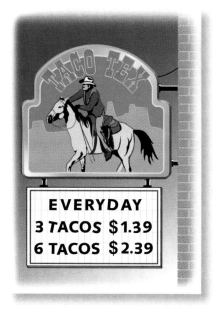

EVERYDAY
3 TACOS \$1.39
6 TACOS \$2.39

A › Adding and Subtracting Like Terms

In algebra, an **expression** is a collection of numbers and letters representing numbers (**variables**) connected by operation signs. The parts to be added or subtracted in these expressions are called **terms.** Thus, xy^2 is an expression with one term; $x + y$ is an expression with two terms, x and y; and $3x^2 - 2y + z$ has three terms, $3x^2$, $-2y$, and z.

The term "3 tacos" uses the number 3 to tell "how many." The number 3 is called the **numerical coefficient** (or simply the **coefficient**) of the term. Similarly, the terms $5x$, y, and $-8xy$ have numerical coefficients of 5, 1, and -8, respectively.

When two or more terms are *exactly alike* (except possibly for their coefficients or the order in which the factors are multiplied), they are called **like terms.** Thus, like terms contain the *same variables* with the *same exponents,* but their coefficients may be different. So 3 tacos and 6 tacos are like terms, $3x$ and $-5x$ are like terms, and $-xy^2$ and $7xy^2$ are like terms. On the other hand, $2x$ and $2x^2$ or $2xy^2$ and $2x^2y$ are *not* like terms. In this section we learn how to simplify expressions using like terms.

In an algebraic expression, like terms can be combined into a single term just as 3 tacos and 6 tacos can be combined into the single term 9 tacos. To combine like terms, make sure that the variable parts of the terms to be combined are identical (only the coefficients may be different), and then add (or subtract) the coefficients and keep the variables. This can be done using the distributive property:

$$3x \quad + \quad 5x \quad = \quad (3 + 5)x \quad = 8x$$
$$\underbrace{(x + x + x)} \quad + \quad \underbrace{(x + x + x + x + x)} \quad = \quad \overbrace{x + x + x + x + x + x + x + x}$$

$$2x^2 \quad + \quad 4x^2 \quad = \quad (2 + 4)x^2 \quad = 6x^2$$
$$\underbrace{(x^2 + x^2)} \quad + \quad \underbrace{(x^2 + x^2 + x^2 + x^2)} \quad = \quad \overbrace{x^2 + x^2 + x^2 + x^2 + x^2 + x^2}$$

$$3ab \quad + \quad 2ab \quad = \quad (3 + 2)ab \quad = 5ab$$
$$\underbrace{(ab + ab + ab)} \quad + \quad \underbrace{(ab + ab)} \quad = \quad \overbrace{ab + ab + ab + ab + ab}$$

But what about another situation, such as combining $-3x$ and $-2x$? We first write

$$-3x \quad + \quad (-2x)$$
$$\underbrace{(-x) + (-x) + (-x)} \quad + \quad \overbrace{(-x) + (-x)}$$

Thus,

$$-3x + (-2x) = [-3 + (-2)]x = -5x$$

Note that we write the addition of $-3x$ and $-2x$ as $-3x + (-2x)$, using parentheses around the $-2x$. We do this to avoid the confusion of writing

$$-3x + -2x$$

Never use two signs of operation together without parentheses.

As you can see, if both quantities to be added are preceded by minus signs, the result is preceded by a minus sign. If they are both preceded by plus signs, the result is preceded by a plus sign. But what about this expression, $-5x + (3x)$? You can visualize this expression as

$$-5x \quad + \quad 3x$$
$$\underbrace{(-x) + (-x) + (-x) + (-x) + (-x)} \quad + \quad \overbrace{x + x + x}$$

Since we have two more negative x's than positive x's, the result is $-2x$. Thus,

$$-5x + (3x) = (-5 + 3)x = -2x \quad \text{Use the distributive property.}$$

On the other hand, $5x + (-3x)$ can be visualized as

$$5x \quad + \quad (-3x)$$
$$\underbrace{x + x + x + x + x} \quad + \quad \overbrace{(-x) + (-x) + (-x)}$$

and the answer is $2x$; that is,

$$5x + (-3x) = [5 + (-3)]x = 2x \quad \text{Use the distributive property.}$$

Now here is an easy one: What is $x + x$? Since $1 \cdot x = x$, the coefficient of x is assumed to be 1. Thus,

$$x + x = 1 \cdot x + 1 \cdot x$$
$$= (1 + 1)x$$
$$= 2x$$

In all the examples that follow, make sure you use the fact that

$$ac + bc = (a + b)c$$

to combine the numerical coefficients.

EXAMPLE 1   Combining like terms using addition

Combine like terms:

a. $-7x + 2x$ **b.** $-4x + 6x$

c. $-2x + (-5x)$ **d.** $x + (-5x)$

SOLUTION

a. $-7x + 2x = (-7 + 2)x = -5x$

b. $-4x + 6x = (-4 + 6)x = 2x$

c. $-2x + (-5x) = [-2 + (-5)]x = -7x$

d. First, recall that $x = 1 \cdot x$. Thus,

$$x + (-5x) = 1x + (-5x)$$
$$= [1 + (-5)]x = -4x$$

PROBLEM 1

Combine like terms:

a. $-9x + 3x$ **b.** $-6x + 8x$

c. $-3x + (-7x)$ **d.** $x + (-6x)$

Subtraction of like terms is defined in terms of addition, as stated here.

SUBTRACTION

To subtract a number b from another number a, add the *additive inverse* (*opposite*) of b to a; that is,

$$a - b = a + (-b)$$

As before, to subtract like terms, we use the fact that $ac + bc = (a + b)c$. Thus,

Additive inverse

$$3x - 5x = 3x + (-5x) = [3 + (-5)]x = -2x$$
$$-3x - 5x = -3x + (-5x) = [(-3) + (-5)]x = -8x$$
$$3x - (-5x) = 3x + (5x) = (3 + 5)x = 8x$$
$$-3x - (-5x) = -3x + (5x) = (-3 + 5)x = 2x$$

Note that

$$-3x - (-5x) = -3x + 5x$$

In general, we have the following:

PROCEDURE

Subtracting $-b$

$$a - (-b) = a + b \qquad -(-b) \text{ is replaced by } +b, \text{ since } -(-b) = +b.$$

We can now combine like terms involving subtraction.

EXAMPLE 2   Combining like terms using subtraction

Combine like terms:

a. $-7ab - 9ab$ **b.** $8x^2 - 3x^2$

c. $-5ab^2 - (-8ab^2)$ **d.** $-6a^2b - (-2a^2b)$

PROBLEM 2

Combine like terms:

a. $3xy - 8xy$

b. $7x^2 - 5x^2$

c. $-3xy^2 - (-7xy^2)$

d. $-5x^2y - (-4x^2y)$

Answers to PROBLEMS

1. a. $-6x$ **b.** $2x$ **c.** $-10x$ **d.** $-5x$ **2. a.** $-5xy$ **b.** $2x^2$ **c.** $4xy^2$ **d.** $-x^2y$

SOLUTION

a. $7ab - 9ab = 7ab + (-9ab) = [7 + (-9)]ab = -2ab$

b. $8x^2 - 3x^2 = 8x^2 + (-3x^2) = [8 + (-3)]x^2 = 5x^2$

c. $-5ab^2 - (-8ab^2) = -5ab^2 + 8ab^2 = (-5 + 8)ab^2 = 3ab^2$

d. $-6a^2b - (-2a^2b) = -6a^2b + 2a^2b = (-6 + 2)a^2b = -4a^2b$

B ⟩ Removing Parentheses

Sometimes it's necessary to remove parentheses before combining like terms. For example, to combine like terms in

$$(3x + 5) + (2x - 2)$$

we have to remove the parentheses *first*. If there is a plus sign (or no sign) in front of the parentheses, we can simply remove the parentheses; that is,

$$+(a + b) = a + b \quad \text{and} \quad +(a - b) = a - b$$

With this in mind, we have

$$
\begin{aligned}
(3x + 5) + (2x - 2) &= 3x + 5 + 2x - 2 \\
&= 3x + 2x + 5 - 2 && \text{Use the commutative property.} \\
&= (3x + 2x) + (5 - 2) && \text{Use the associative property.} \\
&= 5x + 3 && \text{Simplify.}
\end{aligned}
$$

Note that we used the properties we studied to group like terms together.

Once you understand the use of these properties, you can then see that the simplification of $(3x + 5) + (2x - 2)$ consists of just adding $3x$ to $2x$ and 5 to -2. The computation can then be shown like this:

Like terms

$$(3x + 5) + (2x - 2) = 3x + 5 + 2x - 2$$

Like terms

$$= 5x + 3$$

We use this idea in Example 3.

EXAMPLE 3   Removing parentheses preceded by a plus sign

Remove parentheses and combine like terms:

a. $(4x - 5) + (7x - 3)$

b. $(3a + 5b) + (4a - 9b)$

SOLUTION   We first remove parentheses; then we add like terms.

a. $(4x - 5) + (7x - 3) = 4x - 5 + 7x - 3$

$$= 11x - 8$$

b. $(3a + 5b) + (4a - 9b) = 3a + 5b + 4a - 9b$

$$= 7a - 4b$$

PROBLEM 3

Remove parentheses and combine like terms:

a. $(7a - 5) + (8a - 2)$

b. $(4x + 6y) + (3x - 8y)$

Answers to PROBLEMS

3. a. $15a - 7$ **b.** $7x - 2y$

To simplify

$$4x + 3(x - 2) - (x + 5)$$

recall that $-(x + 5) = -1 \cdot (x + 5) = -x - 5$. We proceed as follows:

$4x + \underbrace{3(x - 2)} - (x + 5)$	Given
$= 4x + 3x - 6 - \underbrace{(x + 5)}$	Use the distributive property.
$= 4x + 3x - 6 - x - 5$	Remove parentheses.
$= (4x + 3x - x) + (-6 - 5)$	Combine like terms.
$= 6x - 11$	Simplify.

EXAMPLE 4   Removing parentheses preceded by a minus sign

Remove parentheses and combine like terms:

$$8x - 2(x - 1) - (x + 3)$$

SOLUTION

$8x \underbrace{- 2(x - 1)} - (x + 3)$	
$= 8x - 2x + 2 \underbrace{- (x + 3)}$	Use the distributive property.
$= 8x - 2x + 2 - x - 3$	Remove parentheses.
$= (8x - 2x - x) + (2 - 3)$	Combine like terms.
$= 5x + (-1)$ or $5x - 1$	Simplify.

PROBLEM 4

Remove parentheses and combine like terms:

$$9a - 3(a - 2) - (a + 4)$$

Sometimes parentheses occur within other parentheses. To avoid confusion, we use different grouping symbols. Thus, we usually don't write $((x + 5) + 3)$. Instead, we write $[(x + 5) + 3]$. To combine like terms in such expressions, remove the *innermost grouping symbols first*, as we illustrate in Example 5.

EXAMPLE 5   Removing grouping symbols

Remove grouping symbols and simplify:

$$[(x^2 - 1) + (2x + 5)] + [(x - 2) - (3x^2 + 3)]$$

SOLUTION   We first remove the innermost parentheses and then combine like terms. Thus,

$[(x^2 - 1) + (2x + 5)] + [(x - 2) - (3x^2 + 3)]$

$= [x^2 - 1 + 2x + 5] + [x - 2 - 3x^2 - 3]$	Remove parentheses. Note that $-(3x^2 + 3) = -3x^2 - 3$.
$= [x^2 + 2x + 4] + [-3x^2 + x - 5]$	Combine like terms inside the brackets.
$= x^2 + 2x + 4 - 3x^2 + x - 5$	Remove brackets.
$= -2x^2 + 3x - 1$	Combine like terms.

PROBLEM 5

Remove grouping symbols and simplify:

$$[(a^2 - 2) + (3a + 4)] + [(a - 5) - (4a^2 + 2)]$$

C › Applications: Translating Words into Algebraic Expressions

As we mentioned at the beginning of this section, we express ideas in algebra using expressions. In most applications, problems are first stated in words and have to be translated into algebraic expressions using mathematical symbols. Here's a short mathematics dictionary that will help you translate word problems.

Mathematics Dictionary

Addition (+)	Subtraction (−)
Write: $a + b$ (read "a plus b")	Write: $a - b$ (read "a minus b")
Words: Plus, sum, increase, more than, more, added to	Words: Minus, difference, decrease, less than, less, subtracted from
Examples: a plus b	Examples: a minus b
The sum of a and b	The difference of a and b
a increased by b	a decreased by b
b more than a	b less than a
b added to a	b subtracted from a
Multiplication (× or ·)	**Division (÷ or the bar −)**
Write: $a \cdot b$, ab, $(a)b$, $a(b)$, or $(a)(b)$ (read "a times b" or simply ab)	Write: $a \div b$ or $\frac{a}{b}$ (read "a divided by b")
Words: Times, of, product	Words: Divided by, quotient
Examples: a times b	Examples: a divided by b
The product of a and b	The quotient of a and b

The words and phrases contained in our mathematics dictionary involve the four fundamental operations of arithmetic. We use these words and phrases to translate sentences into *equations*. An **equation** is a sentence stating that two expressions are equal. Here are some words that in mathematics mean "equals":

Equals (=) means is, the same as, yields, gives, is obtained by

We use these ideas in Example 6.

EXAMPLE 6   **Translating and finding the number of terms**

Write in symbols and indicate the number of terms to the right of the equals sign.

a. The area A of a circle is obtained by multiplying π by the square of the radius r.

b. The perimeter P of a rectangle is obtained by adding twice the length L to twice the width W.

c. The current I in a circuit is given by the quotient of the voltage V and the resistance R.

SOLUTION

a. The area A of a circle   is obtained by   multiplying π by the square of the radius r.

$$A \qquad = \qquad \pi r^2$$

There is one term, πr^2, to the right of the equals sign.

b. The perimeter P of a rectangle   is obtained by   adding twice the length L to twice the width W.

$$P \qquad = \qquad 2L + 2W$$

There are two terms, $2L$ and $2W$, to the right of the equals sign.

c. The current I in a circuit   is given by   the quotient of the voltage V and the resistance R.

$$I \qquad = \qquad \frac{V}{R}$$

There is one term to the right of the equals sign.

PROBLEM 6

Write in symbols and indicate the number of terms to the right of the equals sign.

a. The surface area S of a sphere of radius r is obtained by multiplying 4π by the square of the radius r.

b. The perimeter P of a triangle is obtained by adding the lengths a, b, and c of each of the sides.

c. If P dollars are invested for one year at a rate r, the principal P is the quotient of the interest I and the rate r.

Answers to PROBLEMS

6. a. $S = 4\pi r^2$; one term

  b. $P = a + b + c$; three terms

  c. $P = \frac{I}{r}$; one term

EXAMPLE 7   Number game

Ready for some fun? Let's play a game. Pick a number. Add 2 to it. Triple the result. Subtract 6. Divide by 3. What do you get? The original number! How do we know? Translate and simplify!

SOLUTION   The steps are on the left, and the translation is on the right.

Pick a number. n

Add 2 to it. $n + 2$

Triple the result. $3(n + 2)$

Subtract 6. $3(n + 2) - 6$

Divide by 3. $\dfrac{3(n + 2) - 6}{3}$

What is the answer? Simplify!

Distributive law. $\dfrac{3n + 6 - 6}{3}$

Simplify numerator. $\dfrac{3n}{3}$

Reduce. n

Thus, you always get the original number n. Try it using 4, -2, and $\frac{1}{3}$.

PROBLEM 7

You try this one and find out what the answer will be: Pick a number. Add 3 to it. Double the result. Subtract twice the original number. Divide the result by 2. What is the answer?

🖩 ◈ 🖩 Calculator Corner

You can follow the steps in Example 7 with a calculator. Let's do it for the number 4. Here are the keystrokes:

4 + 2 ENTER × 3 − 6 ENTER ÷ 3 ENTER

The answer is the original number, 4. Your window should look like this:

```
4+2
                    6
Ans*3-6
                   12
Ans/3
                    4
```

If you have a calculator, try it for -2. The keystrokes are identical and your final answer should be -2.

Now that you know how to translate sentences into equations, let's translate symbols into words. For example, in the equation

$$W_{\text{T}} = W_{\text{b}} + W_{\text{s}} + W_{\text{e}}$$ Read "W sub T equals W sub b plus W sub s plus W sub e."

the letters T, b, s, and e are called **subscripts.** They help us represent the total weight W_{T} of a McDonald's® biscuit, which is composed of the weight of the biscuit W_{b}, the weight of the sausage W_{s}, and the weight of the eggs W_{e}. With this information, how would you translate the equation $W_{\text{T}} = W_{\text{b}} + W_{\text{s}} + W_{\text{e}}$? Here is one way: The total weight W_{T} of a McDonald's biscuit with sausage and eggs equals the weight W_{b} of the biscuit plus the weight W_{s} of the sausage plus the weight W_{e} of the eggs. (If you look at *The Fast-Food Guide,* you will see that $W_{\text{b}} = 75$ grams, $W_{\text{s}} = 43$ grams, and $W_{\text{e}} = 57$ grams.) Now try Example 8.

Answers to PROBLEMS

7. $\dfrac{2(n + 3) - 2n}{2} = 3$, so you always get 3.

EXAMPLE 8   Translating symbols to words

Translate into words: $P_T = P_A + P_B + P_C$, where P_T is the total pressure exerted by a gas, P_A is the partial pressure exerted by gas A, P_B is the partial pressure exerted by gas B, and P_C is the partial pressure exerted by gas C.

SOLUTION   The total pressure P_T exerted by a gas is the sum of the partial pressures P_A, P_B, and P_C exerted by gases A, B, and C, respectively.

PROBLEM 8

Translate into words:

$T_K = T_C + 273.15$, where T_K is the temperature in degrees Kelvin and T_C is the temperature in degrees Celsius.

You will find some problems like this in the *Using Your Knowledge.*

A NOTE ABOUT SUBSCRIPTS

Subscripts are used to distinguish individual variables that together represent a whole set of quantities. For example, W_1, W_2, and W_3 could be used to represent different weights; t_1, t_2, t_3, and t_4 to represent different times; and d_1 and d_2 to represent different distances.

We gave you a mathematics dictionary. Let's use it on a contemporary topic: extra pounds. Most formulas for your ideal weight W based on your height h contain the phrase: *for each additional inch over 5 feet (60 inches)*. You translate this as $(h - 60)$. Check it out. If you are 5 foot 3 in or 63 inches, your additional inches over 60 are $(h - 60) = (63 - 60)$ or 3. Every time you see the phrase *for each additional inch over 5 feet (60 inches)* translate it as $(h - 60)$

Source: http://www.halls.md.

TRANSLATE THIS

1. The history of the formulas for calculating ideal body weight W began in **1871** when Dr. P. P. Broca (a French surgeon) created this formula known as Broca's index. The ideal weight W (in pounds) for a woman h inches tall is 100 pounds for the first 5 feet and 5 pounds for each additional inch over 60.

2. The ideal weight W (in pounds) for men h inches tall is 110 pounds for the first 5 feet and 5 pounds for each additional inch over 60.

3. In 1974, Dr. B. J. Devine suggested a formula for the weight W in kilograms (kg) of men h inches tall: 50 plus 2.3 kilograms per inch over 5 feet (60 inches).

4. For women h inches tall, the formula for W is 45.5 plus 2.3 kilograms per inch over 5 feet. By the way, a kilogram (kg) is about 2.2 pounds.

5. In 1983, Dr. J. D. Robinson published a modification of the formula. For men h inches tall, the weight W should be 52 kilograms and 1.9 kilograms for each inch over 60.

In Problems 1–10 **TRANSLATE** the sentence and match the correct translation with one of the equations **A–O.**

A. $W = 50 + 2.3h - 60$

B. $W = 49 + 1.7(h - 60)$

C. $LBW = B + M + O$

D. $W = 100h + 5(h - 60)$

E. $W = 110 + 5(h - 60)$

F. $LBW = 0.32810C + 0.33929W - 29.5336$

G. $LBW = 0.32810W + 0.33929C - 29.5336$

H. $W = 100 + 5(h - 60)$

I. $LBW = 0.29569W + 0.41813C - 43.2933$

J. $W = 50h + 2.3(h - 60)$

K. $W = 110h + 5(h - 60)$

L. $W = 56.2 + 1.41(h - 60)$

M. $W = 50 + 2.3(h - 60)$

N. $W = 52 + 1.9(h - 60)$

O. $W = 45.5 + 2.3(h - 60)$

6. The Robinson formula W for women h inches tall is 49 kilograms and 1.7 kilograms for each inch over 5 feet.

7. A minor modification of Robinson formula is Miller's formula which defines the weight W for a man h inches tall as 56.2 kilograms added to 1.41 kilograms for each inch over 5 feet

8. There are formulas that suggest your lean body weight **(LBW)** is the sum of the weight of your bones **(B)**, muscles **(M)**, and organs **(O)**. Basically the sum of everything other than fat in your body.

9. For men over the age of 16, C centimeters tall and with weight W kilograms, the lean body weight **(LBW)** is the product of W and 0.32810, plus the product of C and 0.33929, minus 29.5336.

10. For women over the age of 30, C centimeters tall and weighting W kilograms the lean body weight **(LBW)** is the product of 0.29569 and W, plus the product of 0.41813 and C, minus 43.2933

Try some of the formulas before you go on and see how close to your ideal weight you are!

Answers to PROBLEMS

8. The temperature T_K in degrees Kelvin is the sum of the temperature T_C in degrees Celsius and 273.15.

> Exercises **1.7**

> **Boost your grade at** mathzone.com!
> > Practice   > Self-Tests
> > Problems   > e-Professors
> > NetTutor   > Videos

MathZone

‹ **A** › **Adding and Subtracting Like Terms**   In Problems 1–30, combine like terms (simplify).

1. $19a + (-8a)$

2. $-2b + 5b$

3. $-8c + 3c$

4. $-5d + (-7d)$

5. $4n^2 + 8n^2$

6. $3x^2 + (-9x^2)$

7. $-3ab^2 + (-4ab^2)$

8. $9ab^2 + (-3ab^2)$

9. $-4abc + 7abc$

10. $6xyz + (-9xyz)$

11. $0.7ab + (-0.3ab) + 0.9ab$

12. $-0.5x^2y + 0.8x^2y + 0.3x^2y$

13. $-0.3xy^2 + 0.2x^2y + (-0.6xy^2)$

14. $0.2x^2y + (-0.3xy^2) + 0.4xy^2$

15. $8abc^2 + 3ab^2c + (-8abc^2)$

16. $3xy + 5ab + 2xy + (-3ab)$

17. $-8ab + 9xy + 2ab + (-2xy)$

18. $7 + \frac{1}{2}x + 3 + \frac{1}{2}x$

19. $\frac{1}{5}a + \frac{3}{7}a^2b + \frac{2}{5}a + \frac{1}{7}a^2b$

20. $\frac{4}{9}ab + \frac{4}{5}ab^2 + \left(-\frac{1}{9}ab\right) + \left(-\frac{1}{5}a^2b\right)$

21. $13x - 2x$

22. $8x - (-2x)$

23. $6ab - (-2ab)$

24. $4.2xy - (-3.7xy)$

25. $-4a^2b - (3a^2b)$

26. $-8ab^2 - 4a^2b$

27. $3.1t^2 - 3.1t^2$

28. $-4.2ab - 3.8ab$

29. $0.3x^2 - 0.3x^2$

30. $0 - (-0.8xy^2)$

‹ **B** › **Removing Parentheses**   In Problems 31–55, remove parentheses and combine like terms.

31. $(3xy + 5) + (7xy - 9)$

32. $(8ab - 9) + (7 - 2ab)$

33. $(7R - 2) + (8 - 9R)$

34. $(5xy - 3ab) + (9ab - 8xy)$

35. $(5L - 3W) + (W - 6L)$

36. $(2ab - 2ac) + (ab - 4ac)$

37. $5x - (8x + 1)$

38. $3x - (7x + 2)$

39. $\frac{2x}{9} - \left(\frac{x}{9} - 2\right)$

40. $\frac{5x}{7} - \left(\frac{2x}{7} - 3\right)$

41. $4a - (a + b) + 3(b + a)$

42. $8x - 3(x + y) - (x - y)$

43. $7x - 3(x + y) - (x + y)$

44. $4(b - a) + 3(b + a) - 2(a + b)$

45. $-(x + y - 2) + 3(x - y + 6) - (x + y - 16)$

46. $[(a^2 - 4) + (2a^3 - 5)] + [(4a^3 + a) + (a^2 + 9)]$

47. $(x^2 + 7 - x) + [-2x^3 + (8x^2 - 2x) + 5]$

48. $[(0.4x - 7) + 0.2x^2] + [(0.3x^2 - 2) - 0.8x]$

49. $\left[\left(\frac{1}{4}x^2 + \frac{1}{5}x\right) - \frac{1}{8}\right] + \left[\left(\frac{3}{4}x^2 - \frac{3}{5}x\right) + \frac{5}{8}\right]$

50. $3[3(x + 2) - 10] + [5 + 2(5 + x)]$

51. $2[3(2a - 4) + 5] - [2(a - 1) + 6]$

52. $-2[6(a - b) + 2a] - [3b - 4(a - b)]$

53. $-3[4a - (3 + 2b)] - [6(a - 2b) + 5a]$

54. $-[-(x + y) + 3(x - y)] - [4(x + y) - (3x - 5y)]$

55. $-[-(0.2x + y) + 3(x - y)] - [2(x + 0.3y) - 5]$

‹ **C** › **Applications: Translating Words into Algebraic Expressions**   In Problems 56–70, translate the sentences into equations and indicate the number of terms to the right of the equals sign.

56. *Sleep*   The number of hours H a growing child should sleep is the difference between 17 and one-half the child's age A.

57. *Reciprocals*   The reciprocal of f is the sum of the reciprocal of u and the reciprocal of v.

58. *Rocket height*   The height h attained by a rocket is the sum of its height a at burnout and

$$\frac{r^2}{20}$$

where r is the speed at burnout.

59. *Temperature*   The Fahrenheit temperature F can be obtained by adding 40 to the quotient of n and 4, where n is the number of cricket chirps in 1 min.

60. *Circles*   The radius r of a circle is the quotient of the circumference C and 2π.

61. *Interest*   The interest received I is the product of the principal P, the rate r, and the time t.

62. *Profit*   The total profit P_T equals the total revenue R_T minus the total cost C_T.

63. *Perimeter*   The perimeter P is the sum of the lengths of the sides a, b, and c.

64. *Volume*   The volume V of a certain gas is the pressure P divided by the temperature T.

65. *Area*   The area A of a triangle is the product of the length of the base b and the height h, divided by 2.

66. *Area*   The area A of a circle is the product of π and the square of the radius r.

67. *Kinetic energy*   The kinetic energy K of a moving object is the product of a constant C, the mass m of the object, and the square of its velocity v.

68. *Averages*   The arithmetic mean m of two numbers a and b is found by dividing their sum by 2.

69. *Distance*   The distance in feet d that an object falls is the product of 16 and the square of the time in seconds t that the object has been falling.

70. *Horse power*   The horse power (hp) that a shaft can safely transmit is the product of a constant C, the speed s of the shaft, and the cube of the shaft's diameter d.

Just before the beginning of these exercises we gave some guidelines to find your ideal weight W based on your height h. How can you lose some pounds? Do Exercises 71–74 to find out! The secret is the calories you consume and spend each day.

Source: http://www.annecollins.com.

71. *Calories*   For women, the daily calories C needed to *lose* about 1 pound per week can be found by multiplying your weight W (in pounds) by 12 and deducting 500 calories. Translate this sentence and indicate the number of terms to the right of the equal sign.

72. *Calories*   For men, the daily calories C needed to *lose* about 1 pound per week can be found by multiplying your weight W (in pounds) by 14 and deducting 500 calories. Translate this sentence and indicate the number of terms to the right of the equal sign.

73. *Calories*   How can you "deduct" 500 calories each day? Exercising! Here are some exercises and the number of calories they use per hour.

Type of Exercise	Calories/Hour
Sleeping	55
Eating	85
Housework, moderate	160+
Dancing, ballroom	260
Walking, 3 mph	280
Aerobics	450+

a. How many calories C would you use in h hours by sleeping?

b. How many calories C would you use in h hours by eating?

c. How many calories C would you use in h hours by doing moderate housework?

74. *Calories*   Use the table in Problem 73 to find:

a. How many calories C would you use in h hours by ballroom dancing?

b. How many calories C would you use in h hours by walking at 3 miles per hour?

c. How many calories C would you use in h hours by doing aerobics?

〉〉〉   *Using Your Knowledge*

Geometric Formulas   The ideas in this section can be used to simplify formulas. For example, the **perimeter** (distance around) of the given rectangle is found by following the blue arrows. The perimeter is

$$P = W + L + W + L$$
$$= 2W + 2L$$

Use this idea to find the perimeter P of the given figure; then write a formula for P in symbols and in words.

75. The square of side S

S

76. The parallelogram of base b and side s

s

b

To obtain the actual measurement of certain perimeters, we have to add like terms if the measurements are given in feet and inches. For example, the perimeter of the rectangle shown here is

2 ft, 7 in.

4 ft, 1 in.

$$P = (2 \text{ ft} + 7 \text{ in.}) + (4 \text{ ft} + 1 \text{ in.}) + (2 \text{ ft} + 7 \text{ in.}) + (4 \text{ ft} + 1 \text{ in.})$$
$$= 12 \text{ ft} + 16 \text{ in.}$$

Since 16 inches = 1 foot + 4 inches,

$$P = 12 \text{ ft} + (1 \text{ ft} + 4 \text{ in.})$$
$$= 13 \text{ ft} + 4 \text{ in.}$$

Use these ideas to obtain the perimeter of the given rectangles.

77.

3 ft, 1 in.

6 ft, 2 in.

78.

4 ft, 5 in.

8 ft, 2 in.

79. The U.S. Postal Service has a regulation stating that "the sum of the length and girth of a package may be no more than 108 inches." What is the sum of the length and girth of the rectangular package below? (*Hint:* The girth of the package is obtained by measuring the length of the red line.)

L

h

w

80. Write in simplified form the height of the step block shown.

$3x$

$2x$

x

81. Write in simplified form the length of the metal plate shown.

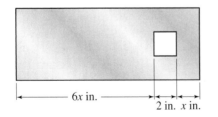

$6x$ in.

2 in. x in.

〉〉〉 *Write On*

82. Explain the difference between a factor and a term.

83. Write the procedure you use to combine like terms.

84. Explain how to remove parentheses when no sign or a plus sign precedes an expression within parentheses.

85. Explain how to remove parentheses when a minus sign precedes an expression within parentheses.

⟩⟩⟩ *Concept Checker*

Fill in the blank(s) with the correct word(s), phrase, or mathematical statement.

86. To **subtract** a number **b** from another number **a**, add the _____ of **b** to **a**.

87. $a - (-b) =$ _____ .

88. An _____ is a **sentence** stating that the expressions are **equal.**

89. We use the _____ sign to indicate that two expressions are equal.

reciprocal	equation
$a - b$	=
$a + b$	opposite

⟩⟩⟩ *Mastery Test*

Write in symbols and indicate the number of terms to the right of the equals sign:

90. When P dollars are invested at r percent, the amount of money A in an account at the end of 1 year is the sum of P and the product of P and r.

91. The area A of a rectangle is obtained by multiplying the length L by the width W.

Remove parentheses and combine like terms:

92. $9x - 3(x - 2) - (2x + 7)$

93. $(8y - 7) + 2(3y - 2)$

94. $(4t + u) + 4(5t - 7u)$

95. $-6ab^3 - (-3ab^3)$

Remove grouping symbols and simplify:

96. $[(2x^2 - 3) + (3x + 1)] + 2[(x - 1) - (x^2 + 2)]$

97. $3[(5 - 2x^2) + (2x - 1)] - 3[(5 - 2x) - (3 + x^2)]$

⟩⟩⟩ *Skill Checker*

Simplify the expression.

98. $\frac{1}{2}(12) - 6$

99. $2(9) - 7$

100. $3(z - 2)$

101. $-2(x + 5)$

⟩ Collaborative Learning

Form three groups. The steps to a number trick are shown. Group 1 starts with number 8, Group 2 with -3, and Group 3 with -1.2.

Steps	Group 1	Group 2	Group 3
Pick a number:	8	-3	-1.2
1. _____	10	-1	0.8
2. _____	30	-3	2.4
3. _____	24	-9	-3.6
4. _____	8	-3	-1.2
5. _____	0	0	0

Each group should fill in the blanks in the first column and find out how the trick works. The first group that discovers what the steps are wins! Of course, the other groups can challenge the results. Compare the five steps. Do all groups get the same steps? What are the steps?

‹CONTINUED›

The steps given here will give you a different final answer depending on the number you start with.

Number + 2
Ans · 3
Ans − 9
Ans/6

Are you ready for the race? All groups should start with the number 5. What answers do they get? The group that finishes first wins! Then do it again. Start with the number 7 this time. What answer do the groups get now?

Here is the challenge: Write an algebraic expression representing the steps in the table. Again, the group that finishes first wins.

Ultimate challenge: The groups write down the steps that describe the expression $3\left(\frac{x+2}{3} + 4\right) - x + 5$. ◪

Research Questions

1. In the *Human Side of Algebra* at the beginning of this chapter, we mentioned the Hindu numeration system. The Egyptians and Babylonians also developed numeration systems. Write a report about each of these numeration systems, detailing the symbols used for the digits 1–9, the base used, and the manner in which fractions were written.

2. Write a report on the life and works of Mohammed al-Khowarizmi, with special emphasis on the books he wrote.

3. We have now studied the four fundamental operations. But do you know where the symbols used to indicate these operations originated?

   a. Write a report about Johann Widmann's *Mercantile Arithmetic* (1489), indicating which symbols of operation were found in the book for the first time and the manner in which they were used.

   b. Introduced in 1557, the original equals sign used longer lines to indicate equality. Why were the two lines used to denote equality, what was the name of the person who introduced the symbol, and in what book did the notation first appear?

4. In this chapter we discussed mathematical expressions. Two of the signs of operation were used for the first time in the earliest-known treatise on algebra to write an algebraic expression. What is the name of this first treatise on algebra, and what is the name of the Dutch mathematician who used these two symbols in 1514?

5. The *Codex Vigilanus* written in Spain in 976 contains the Hindu-Arabic numerals 1–9. Write a report about the notation used for these numbers and the evolution of this notation.

⟩ Summary Chapter 1

Section	Item	Meaning	Example
1.1 A	Arithmetic expressions	Expressions containing numbers and operation signs	$3 + 4 - 8, 9 \cdot 4 \div 6$ are arithmetic expressions.
1.1 A	Algebraic expressions	Expressions containing numbers, operation signs, and variables	$3x - 4y + 3z, 2x \div y + 9z$, and $7x - 9y \div 3z$ are algebraic expressions.

Section	Item	Meaning	Example
1.1 A	Factor	The items to be multiplied	In the expression $4xy$, the factors are 4, x, and y.
1.1 B	Evaluate	To substitute a value for one or more of the variables in an expression	Evaluating $3x - 4y + 3z$ when $x = 1$, $y = 2$, and $z = 3$ yields $3 \cdot 1 - 4 \cdot 2 + 3 \cdot 3$ or 4.
1.2 A	Additive inverse (opposite)	The additive inverse of any integer a is $-a$.	The additive inverse of 5 is -5, and the additive inverse of -8 is 8.
1.2 B	Absolute value of n, denoted by $|n|$	The absolute value of a number n is the distance from n to 0.	$|-7| = 7$, $|13| = 13$, $|0.2| = 0.2$, and $\left|-\frac{1}{4}\right| = \frac{1}{4}$.
1.2 C	Set of natural numbers	$\{1, 2, 3, 4, 5, \ldots\}$	4, 13, and 497 are natural numbers.
	Set of whole numbers	$\{0, 1, 2, 3, 4, \ldots\}$	0, 92, and 384 are whole numbers.
	Set of integers	$\{\ldots, -2, -1, 0, 1, 2, \ldots\}$	-98, 0, and 459 are integers.
	Rational numbers	Numbers that can be written in the form $\frac{a}{b}$, a and b integers and $b \neq 0$	$\frac{1}{5}$, $\frac{-3}{4}$, and $\frac{0}{5}$ are rational numbers.
	Irrational numbers	Numbers that cannot be written as the ratio of two integers	$\sqrt{2}$, $-\sqrt[3]{2}$, and π are irrational numbers.
	Real numbers	The rationals and the irrationals	3, 0, -9, $\sqrt{2}$, and $\sqrt[3]{5}$ are real numbers.
1.3 A	Addition of real numbers	If both numbers have the *same* sign, add their absolute values and give the sum the common sign. If the numbers have *different* signs, subtract their absolute values and give the difference the sign of the number with the larger absolute value.	$-3 + (-5) = -8$ $-3 + 5 = +2$ $-3 + 1 = -2$
1.3 B	Subtraction of real numbers	$a - b = a + (-b)$	$3 - 5 = 3 + (-5) = -2$ and $4 - (-2) = 4 + 2 = 6$
1.4 A, C	Multiplication and division of real numbers	When multiplying or dividing two numbers with the *same* sign, the answer is *positive;* with *different* signs, the answer is *negative.*	$(-3)(-5) = 15$ $\frac{-15}{-3} = 5$ $(-3)(5) = -15$ $\frac{-15}{3} = -5$
1.4 B	Exponent	In the expression 3^2, 2 is the exponent.	3^2 means $3 \cdot 3$
	Base	In the expression 3^2, 3 is the base.	
1.5	Order of operations	1. Operations inside grouping symbols 2. Exponents 3. Multiplications and divisions as they occur from *left* to *right* 4. Additions and subtractions as they occur from *left* to *right*	$\begin{aligned} & 4^2 \div 2 \cdot 3 - 4 + [2(6 + 1) - 4] \\ =\ & 4^2 \div 2 \cdot 3 - 4 + [2(7) - 4] \\ =\ & 4^2 \div 2 \cdot 3 - 4 + [14 - 4] \\ =\ & 4^2 \div 2 \cdot 3 - 4 + [10] \\ =\ & 16 \div 2 \cdot 3 - 4 + [10] \\ =\ & 8 \cdot 3 \quad\ - 4 + 10 \\ =\ & \quad 24 \quad\ - 4 + 10 \\ =\ & \qquad\ 20 + 10 \\ =\ & \qquad\ 30 \end{aligned}$

(continued)

Section	Item	Meaning	Example
1.6 A	Associative property of addition	For any numbers a, b, c, $a + (b + c) = (a + b) + c$.	$3 + (4 + 9) = (3 + 4) + 9$
	Associative property of multiplication	For any numbers a, b, c, $a \cdot (b \cdot c) = (a \cdot b) \cdot c$.	$5 \cdot (2 \cdot 8) = (5 \cdot 2) \cdot 8$
	Commutative property of addition	For any numbers a and b, $a + b = b + a$.	$2 + 9 = 9 + 2$
	Commutative property of multiplication	For any numbers a and b, $a \cdot b = b \cdot a$.	$18 \cdot 5 = 5 \cdot 18$
1.6C	Identity element for addition	0 is the identity element for addition.	$3 + 0 = 0 + 3 = 3$
	Identity element for multiplication	1 is the identity element for multiplication.	$1 \cdot 7 = 7 \cdot 1 = 7$
	Additive inverse (opposite)	For any number a, its additive inverse is $-a$.	The additive inverse of 5 is -5 and that of -8 is 8.
	Multiplicative inverse (reciprocal)	The multiplicative inverse of a is $\frac{1}{a}$ if a is not 0 (0 has no reciprocal).	The multiplicative inverse of $\frac{3}{4}$ is $\frac{4}{3}$.
1.6 E	Distributive property	For any numbers a, b, c, $a(b + c) = ab + ac$.	$3(4 + x) = 3 \cdot 4 + 3 \cdot x$
1.7	Expression	A collection of numbers and letters connected by operation signs	xy^3, $x + y$, $x + y - z$, and $xy^3 - y$ are expressions.
	Terms	The parts that are to be added or subtracted in an expression	In the expression $3x^2 - 4x + 5$, the terms are $3x^2$, $-4x$, and 5.
	Numerical coefficient or coefficient	The part of the term indicating "how many"	In the term $3x^2$, 3 is the coefficient. In the term $-4x$, -4 is the coefficient. The coefficient of x is 1.
	Like terms	Terms with the same variables and exponents	$3x$ and $-4x$ are like terms. $-5x^2$ and $8x^2$ are like terms. $-xy^2$ and $3xy^2$ are like terms. *Note:* $-x^2y$ and $-xy^2$ **are not** like terms.
1.7C	Sums and differences	The *sum* of a and b is $a + b$.	The sum of 3 and x is $3 + x$.
	Product	The *difference* of a and b is $a - b$.	The difference of 6 and x is $6 - x$.
	Quotient	The *product* of a and b is $a \cdot b$, $(a)(b)$, $a(b)$, $(a)b$, or ab.	The product of 8 and x is $8x$.
		The *quotient* of a and b is $\frac{a}{b}$.	The quotient of 7 and x is $\frac{7}{x}$.

❯Review Exercises **Chapter 1**

(If you need help with these exercises, look in the section indicated in brackets.)

1. ❬ **1.1A** ❭ *Write in symbols:*

   a. The sum of a and b

   b. a minus b

   c. $7a$ plus $2b$ minus 8

2. ❬ **1.1A** ❭ *Write using juxtaposition:*

   a. 3 times m **b.** $-m$ times n times r

   c. $\frac{1}{7}$ of m **d.** The product of 8 and m

3. ⟨ **1.1A** ⟩ *Write in symbols:*
   a. The quotient of m and 9

   b. The quotient of 9 and n

4. ⟨ **1.1A** ⟩ *Write in symbols:*
   a. The quotient of $(m + n)$ and r

   b. The sum of m and n, divided by the difference of m and n

5. ⟨ **1.1B** ⟩ *For m = 9 and n = 3, evaluate:*
   a. $m + n$ **b.** $m - n$ **c.** $4m$

6. ⟨ **1.1B** ⟩ *For m = 9 and n = 3, evaluate:*
   a. $\dfrac{m}{n}$ **b.** $2m - 3n$ **c.** $\dfrac{2m + n}{n}$

7. ⟨ **1.2A** ⟩ *Find the additive inverse (opposite) of:*
   a. -5 **b.** $\dfrac{2}{3}$ **c.** 0.37

8. ⟨ **1.2B** ⟩ *Find the absolute value of:*
   a. $\left|-8\right|$ **b.** $\left|3\dfrac{1}{2}\right|$ **c.** $-\left|0.76\right|$

9. ⟨ **1.2C** ⟩ *Classify each of the numbers as natural, whole, integer, rational, irrational, or real. (Note: More than one category may apply.)*
   a. -7 **b.** 0

   c. $0.666\ldots = 0.\overline{6}$ **d.** $0.606006000\ldots$

   e. $\sqrt{41}$ **f.** $-1\dfrac{1}{8}$

10. ⟨ **1.3A** ⟩ *Add:*
    a. $7 + (-5)$ **b.** $(-0.3) + (-0.5)$

    c. $-\dfrac{3}{4} + \dfrac{1}{5}$ **d.** $3.6 + (-5.8)$

    e. $\dfrac{3}{4} + \left(-\dfrac{1}{2}\right)$

11. ⟨ **1.3B** ⟩ *Subtract:*
    a. $-16 - 4$ **b.** $-7.6 - (-5.2)$

    c. $\dfrac{5}{6} - \dfrac{9}{4}$

12. ⟨ **1.3C** ⟩ *Find:*
    a. $20 - (-12) + 15 - 12 - 5$

    b. $-17 + (-7) + 10 - (-7) - 8$

13. ⟨ **1.4A** ⟩ *Multiply:*
    a. $-5 \cdot 7$ **b.** $8(-2.3)$

    c. $-6(3.2)$ **d.** $-\dfrac{3}{7}\left(-\dfrac{4}{5}\right)$

14. ⟨ **1.4B** ⟩ *Find:*
    a. $(-4)^2$ **b.** -3^2

    c. $\left(-\dfrac{1}{3}\right)^3$ **d.** $-\left(-\dfrac{1}{3}\right)^3$

15. ⟨ **1.4C** ⟩ *Divide:*
    a. $\dfrac{-40}{10}$ **b.** $-8 \div (-4)$

    c. $\dfrac{3}{8} \div \left(-\dfrac{9}{16}\right)$ **d.** $-\dfrac{3}{4} \div \left(-\dfrac{1}{2}\right)$

16. ⟨ **1.5A** ⟩ *Find:*
    a. $64 \div 8 - (3 - 5)$

    b. $27 \div 3^2 + 5 - 8$

17. ⟨ **1.5B** ⟩ *Find:*
    a. $20 \div 5 + \{2 \cdot 3 - [4 + (7 - 9)]\}$

    b. $-6^2 + \dfrac{2(3 - 6)}{2} + 8 \div (-4)$

18. ⟨ **1.5C** ⟩ *The maximum pulse rate you should maintain during aerobic activities is 0.80(220 − A), where A is your age. What is the maximum pulse rate you should maintain if you are 30 years old?*

19. ⟨ **1.6A** ⟩ *Name the property illustrated in each statement.*
    a. $x + (y + z) = x + (z + y)$

    b. $6 \cdot (8 \cdot 7) = (6 \cdot 8) \cdot 7$

    c. $x + (y + z) = (x + y) + z$

20. ⟨ **1.6B** ⟩ *Simplify:*
    a. $6 + 4x - 10$

    b. $-8 + 7x + 10 - 15x$

21. ⟨ **1.6C, E** ⟩ *Fill in the blank so that the result is a true statement.*

   a. ___ $\cdot \frac{1}{5} = \frac{1}{5}$ **b.** $\frac{3}{4} +$ ___ $= 0$

   c. ___ $+ 3.7 = 0$ **d.** ___ $\cdot \frac{3}{2} = 1$

   e. ___ $+ \frac{2}{7} = \frac{2}{7}$ **f.** $3(x +$ ___ $) = 3x + (-15)$

22. ⟨ **1.6E** ⟩ *Use the distributive property to multiply.*

   a. $-3(a + 8)$

   b. $-4(x - 5)$

   c. $-(x - 4)$

23. ⟨ **1.7A** ⟩ *Combine like terms.*

   a. $-7x + 2x$

   b. $-2x + (-8x)$

   c. $x + (-9x)$

   d. $-6a^2b - (-9a^2b)$

24. ⟨ **1.7B** ⟩ *Remove parentheses and combine like terms.*

   a. $(4a - 7) + (8a + 2)$

   b. $9x - 3(x + 2) - (x + 3)$

25. ⟨ **1.7C** ⟩ *Write in symbols and indicate the number of terms to the right of the equals sign.*

   a. The number of minutes m you will wait in line at the bank is equal to the number of people p ahead of you divided by the number n of tellers, times 2.75.

$$m = \frac{p}{n}(2.75)$$

   b. The normal weight W of an adult (in pounds) can be estimated by subtracting 220 from the product of $\frac{11}{2}$ and h, where h is the person's height (in inches).

$$W = \frac{11h}{2} - 220$$

⟩ Practice Test **Chapter 1**

(Answers on page 104)

Visit www.mhhe.com/bello to view helpful videos that provide step-by-step solutions to several of the problems below.

1. Write in symbols:

   a. The sum of g and h **b.** g minus h

   c. $6g$ plus $3h$ minus 8

2. Write using juxtaposition:

   a. 3 times g **b.** $-g$ times h times r

   c. $\frac{1}{5}$ of g **d.** The product of 9 and g

3. Write in symbols:

   a. The quotient of g and 8

   b. The quotient of 8 and h

4. Write in symbols:

   a. The quotient of $(g + h)$ and r

   b. The sum of g and h, divided by the difference of g and h

5. For $g = 4$ and $h = 3$, evaluate:

   a. $g + h$ **b.** $g - h$ **c.** $5g$

6. For $g = 8$ and $h = 4$, evaluate:

   a. $\frac{g}{h}$ **b.** $2g - 3h$ **c.** $\frac{2g + h}{h}$

7. Find the additive inverse (opposite) of:

   a. -9 **b.** $\frac{3}{5}$ **c.** $0.222\ldots$

8. Find the absolute value.

   a. $\left| -\frac{1}{4} \right|$ **b.** $|13|$ **c.** $-|0.92|$

9. Consider the set $\{-8, \frac{5}{3}, \sqrt{2}, 0, 3.4, 0.333\ldots, -3\frac{1}{2}, 7, 0.123\ldots\}$. List the numbers in the set that are:

   a. Natural numbers   **b.** Whole numbers

   c. Integers   **d.** Rational numbers

   e. Irrational numbers   **f.** Real numbers

10. Add:

   a. $9 + (-7)$ **b.** $(-0.9) + (-0.8)$

   c. $-\frac{3}{4} + \frac{2}{5}$ **d.** $1.7 + (-3.8)$

   e. $\frac{4}{5} + \left(-\frac{1}{2} \right)$

11. Subtract:

    a. $-18 - 6$ **b.** $-9.2 - (-3.2)$

    c. $\dfrac{1}{6} - \dfrac{5}{4}$

12. Find:

    a. $10 - (-15) + 12 - 15 - 6$

    b. $-15 + (-8) + 12 - (-8) - 9$

13. Multiply:

    a. $-6 \cdot 8$ **b.** $9(-2.3)$

    c. $-7(8.2)$ **d.** $-\dfrac{2}{7}\left(-\dfrac{3}{5}\right)$

14. Find:

    a. $(-7)^2$ **b.** -9^2

    c. $\left(-\dfrac{2}{3}\right)^3$ **d.** $-\left(-\dfrac{2}{3}\right)^3$

15. Divide:

    a. $\dfrac{-50}{10}$ **b.** $-14 \div (-7)$

    c. $\dfrac{5}{8} \div \left(-\dfrac{11}{16}\right)$ **d.** $-\dfrac{3}{4} \div \left(-\dfrac{7}{9}\right)$

16. Find:

    a. $56 \div 7 - (4 - 9)$

    b. $36 \div 2^2 + 7$

17. Find the value of:

    a. $30 \div 6 + \{3 \cdot 4 - [2 + (8 - 10)]\}$

    b. $-7^2 + \dfrac{4(3 - 9)}{2} + 12 \div (-4)$

18. The handicap H of a bowler with average A is $H = 0.80(200 - A)$. What is the handicap of a bowler whose average is 180?

19. Name the property illustrated in each statement.

    a. $x + (y + z) = (x + y) + z$

    b. $6 \cdot (8 \cdot 7) = 6 \cdot (7 \cdot 8)$

    c. $x + (y + z) = (y + z) + x$

20. Simplify:

    a. $7 - 4x - 10$

    b. $-9 + 6x + 12 - 13x$

21. Fill in the blank so that the result is a true statement.

    a. $9.2 + \underline{\quad} = 0$ **b.** $\underline{\quad} \cdot \dfrac{5}{2} = 1$

    c. $\underline{\quad} \cdot 0.3 = 0.3$ **d.** $\dfrac{1}{4} + \underline{\quad} = 0$

    e. $\underline{\quad} + \dfrac{3}{7} = \dfrac{3}{7}$

    f. $-3(x + \underline{\quad}) = -3x + (-15)$

22. Use the distributive property to multiply.

    a. $-3(x + 9)$ **b.** $-5(a - 4)$

    c. $-(x - 8)$

23. Combine like terms.

    a. $-3x + (-9x)$

    b. $-9ab^2 - (-6ab^2)$

24. Remove parentheses and combine like terms.

    a. $(5a - 6) - (7a + 2)$

    b. $8x - 4(x - 2) - (x + 2)$

25. Write in symbols and indicate the number of terms to the right of the equals sign:

The temperature F (in degrees Fahrenheit) can be found by finding the sum of 37 and one-quarter the number of chirps C a cricket makes in 1 min.

›Answers to Practice Test Chapter 1

Answer	If You Missed	Review		
	Question	Section	Examples	Page
1. a. $g + h$ **b.** $g - h$ **c.** $6g + 3h - 8$	1	1.1	1	37
2. a. $3g$ **b.** $-ghr$ **c.** $\frac{1}{5}g$ **d.** $9g$	2	1.1	2	37
3. a. $\frac{g}{8}$ **b.** $\frac{8}{h}$	3	1.1	3a, b	38
4. a. $\frac{g+h}{r}$ **b.** $\frac{g+h}{g-h}$	4	1.1	3c, d	38
5. a. 7 **b.** 1 **c.** 20	5	1.1	4a, b, c	38
6. a. 2 **b.** 4 **c.** 5	6	1.1	4d, e	38
7. a. 9 **b.** $-\frac{3}{5}$ **c.** $-0.222\ldots$	7	1.2	1, 2	43–44
8. a. $\frac{1}{4}$ **b.** 13 **c.** -0.92	8	1.2	3, 4	45
9. a. 7 **b.** 0, 7 **c.** $-8, 0, 7$	9	1.2	5	46–47
d. $-8, \frac{5}{3}, 0, 3.4, 0.333\ldots, -3\frac{1}{2}, 7$				
e. $\sqrt{2}, 0.123\ldots$ **f.** All				
10. a. 2 **b.** -1.7 **c.** $-\frac{7}{20}$	10	1.3	1–5	52–54
d. -2.1 **e.** $\frac{3}{10}$				
11. a. -24 **b.** -6 **c.** $-\frac{13}{12}$	11	1.3	6, 7	55
12. a. 16 **b.** -12	12	1.3	8	56
13. a. -48 **b.** -20.7 **c.** -57.4 **d.** $\frac{6}{35}$	13	1.4	1, 2	60–61
14. a. 49 **b.** -81 **c.** $-\frac{8}{27}$ **d.** $\frac{8}{27}$	14	1.4	3, 4	62
15. a. -5 **b.** 2 **c.** $-\frac{10}{11}$ **d.** $\frac{27}{28}$	15	1.4	5, 6	63–64
16. a. 13 **b.** 16	16	1.5	1, 2	69–70
17. a. 17 **b.** -64	17	1.5	3–5	70–72
18. 16	18	1.5	6	72
19. a. Associative property of addition	19	1.6	1, 2	77–78
b. Commutative property of multiplication				
c. Commutative property of addition				
20. a. $-4x - 3$ **b.** $3 - 7x$	20	1.6	3	78
21. a. -9.2 **b.** $\frac{2}{5}$ **c.** 1	21	1.6	4, 5, 7, 8	79, 81
d. $-\frac{1}{4}$ **e.** 0 **f.** 5				
22. a. $-3x - 27$ **b.** $-5a + 20$ **c.** $-x + 8$	22	1.6	7, 8	81
23. a. $-12x$ **b.** $-3ab^2$	23	1.7	1, 2	88–89
24. a. $-2a - 8$ **b.** $3x + 6$	24	1.7	3–5	89–90
25. $F = 37 + \frac{1}{4}C$; two terms	25	1.7	6, 8	91, 93

Chapter

2

two

▶ Equations, Problem Solving, and Inequalities

The Human Side of Algebra

Most of the mathematics used in ancient Egypt is preserved in the Rhind papyrus, a document bought in 1858 in Luxor, Egypt, by Henry Rhind. Problem 24 in this document reads:

> A quantity and its $\frac{1}{7}$ added become 19. What is the quantity?

If we let q represent the quantity, we can translate the problem as

$$q + \frac{1}{7}q = 19$$

Unfortunately, the Egyptians were unable to simplify $q + \frac{1}{7}q$ because the sum is $\frac{8}{7}q$, and they didn't have a notation for the fraction $\frac{8}{7}$. How did they attempt to find the answer? They used the method of "false position." That is, they *assumed* the answer was 7, which yields $7 + \frac{1}{7} \cdot 7$ or 8. But the sum as stated in the problem is not 8, it's 19. How can you make the 8 into a 19? By multiplying by $\frac{19}{8}$, that is, by finding $8 \cdot \frac{19}{8} = 19$. But if you multiply the 8 by $\frac{19}{8}$, you should also multiply the assumed answer, 7, by $\frac{19}{8}$ to obtain the true answer, $7 \cdot \frac{19}{8}$. If you solve the equation $\frac{8}{7}q = 19$, you will see that $7 \cdot \frac{19}{8}$ is indeed the correct answer!

105

2.1 The Addition and Subtraction Properties of Equality

Objectives

A Determine whether a number satisfies an equation.

B Use the addition and subtraction properties of equality to solve equations.

C Use both properties together to solve an equation.

To Succeed, Review How To . . .

1. Add and subtract real numbers (pp. 52, 54).
2. Follow the correct order of operations (pp. 68–73).
3. Simplify expressions (pp. 78, 88–89).

Getting Started

A Lot of Garbage!

In this section, we study some ideas that will enable us to solve equations. Do you know what an equation is? Here is an example. How much waste do you generate every day? According to Franklin Associates, LTD, the average American generates about 2.7 pounds of waste daily if we *exclude* paper products! If w and p represent the total amount of waste and paper products (respectively) generated daily by the average American, $w - p = 2.7$. Further research indicates that $p = 1.7$ pounds; thus, $w - 1.7 = 2.7$. The statement

$$w - 1.7 = 2.7$$

is an *equation,* a statement indicating that two expressions are equal. Some equations are *true* ($1 + 1 = 2$), some are *false* ($2 - 5 = 3$), and some ($w - 1.7 = 2.7$) are neither true nor false. The equation $w - 1.7 = 2.7$ is a *conditional* equation that is true for certain values of the *variable* or unknown w. To find the total amount of waste generated daily (w), we have to *solve* $w - 1.7 = 2.7$; that is, we must find the value of the variable that makes the equation a true statement. We learn how to do this next.

A ⟩ Finding Solutions to an Equation

In the equation $w - 1.7 = 2.7$ in the *Getting Started,* the variable w can be replaced by many numbers, but only *one* number will make the resulting statement *true*. This number is called the *solution* of the equation. Can you find the solution of $w - 1.7 = 2.7$? Since w is the total amount of waste, $w = 1.7 + 2.7 = 4.4$, and 4.4 is the solution of the equation.

SOLUTIONS

The **solutions** of an equation are the replacements of the variable that make the equation a *true* statement. When we find the solution of an equation, we say that we have **solved** the equation.

How do we know whether a given number solves (or *satisfies*) an equation? We write the number in place of the variable in the given equation and see whether the result is true. For example, to decide whether 4 is a solution of the equation

$$x + 1 = 5$$

we replace x by 4. This gives the true statement:

$$4 + 1 = 5$$

so 4 is a solution of the equation $x + 1 = 5$.

EXAMPLE 1   Verifying a solution

Determine whether the given number is a solution of the equation:

a. 9; $x - 4 = 5$ **b.** 8; $5 = 3 - y$ **c.** 10; $\frac{1}{2}z - 5 = 0$

SOLUTION   Remember the rule: if you substitute the number for the variable and the result is *true,* the number is a solution.

a. If x is 9, $x - 4 = 5$ becomes $9 - 4 = 5$, which is a *true* statement. Thus, 9 is a solution of the equation.

b. If y is 8, $5 = 3 - y$ becomes $5 = 3 - 8$, which is a *false* statement. Hence, 8 is *not* a solution of the equation.

c. If z is 10, $\frac{1}{2}z - 5 = 0$ becomes $\frac{1}{2}(10) - 5 = 0$, which is a *true* statement. Thus, 10 is a solution of the equation.

PROBLEM 1

Determine whether the given number is a solution of the equation:

a. 7; $x - 5 = 3$

b. 4; $1 = 5 - y$

c. 6; $\frac{1}{3}z - 2 = 0$

We have learned how to determine whether a number satisfies an equation. Now to find such a number, we must find an *equivalent* equation whose solution is obvious.

EQUIVALENT EQUATIONS

Two equations are **equivalent** if their solutions are the same.

How do we find these equivalent equations? We use the properties of equality.

B ⟩ Using the Addition and Subtraction Properties of Equality

Look at the ad in the illustration. It says that $5 has been cut from the price of a gallon of paint and that the sale price is $6.69. What was the old price p of the paint? Since the old price p was cut by $5, the new price is $p - 5$. Since the new price is $6.69,

$$p - 5 = 6.69$$

To find the old price p, we add back the $5 that was cut. That is,

$$p - 5 + 5 = 6.69 + 5 \qquad \text{We add 5 to both sides of the equation to obtain an equivalent equation.}$$

$$p = 11.69$$

Thus, the old price was $11.69; this can be verified, since $11.69 - 5 = 6.69$. Note that by adding 5 to both sides of the equation $p - 5 = 6.69$, we produced an equivalent equation,

$p = 11.69$, whose solution is obvious. This example illustrates the fact that we can *add* the same number on both sides of an equation and produce an *equivalent* equation—that is, an equation whose solution is identical to the solution of the original one. Here is the property.

THE ADDITION PROPERTY OF EQUALITY	For any number **c**, the equation **a = b** is equivalent to $$a + c = b + c$$

We use this property in Example 2.

EXAMPLE 2   Using the addition property

Solve:

a. $x - 3 = 9$

b. $x - \frac{1}{7} = \frac{5}{7}$

SOLUTION

a. This problem is similar to our example of paint prices. To solve the equation, we need x by itself on one side of the equation. We can achieve this by adding 3 (the additive inverse of -3) on both sides of the equation.

$$x - 3 = 9$$
$$x - 3 + 3 = 9 + 3 \quad \text{Add 3 to both sides.}$$
$$x = 12$$

Thus, 12 is the solution of $x - 3 = 9$.

CHECK   Substituting 12 for x in the original equation, we have $12 - 3 = 9$, a true statement.

b.

$$x - \frac{1}{7} = \frac{5}{7}$$
$$x - \frac{1}{7} + \frac{1}{7} = \frac{5}{7} + \frac{1}{7} \quad \text{Add } \tfrac{1}{7} \text{ to both sides.}$$
$$x = \frac{6}{7} \quad \text{Simplify.}$$

Thus, $\frac{6}{7}$ is the solution of $x - \frac{1}{7} = \frac{5}{7}$.

CHECK $\frac{6}{7} - \frac{1}{7} = \frac{5}{7}$ is a true statement.

PROBLEM 2

Solve:

a. $x - 5 = 7$ **b.** $x - \frac{1}{5} = \frac{3}{5}$

Sometimes it's necessary to simplify an equation before we isolate x on one side. For example, to solve the equation

$$3x + 5 - 2x - 9 = 6x + 5 - 6x$$

we first simplify both sides of the equation by collecting like terms:

$$3x + 5 - 2x - 9 = 6x + 5 - 6x$$

$$(3x - 2x) + (5 - 9) = (6x - 6x) + 5 \quad \text{Group like terms.}$$
$$x + (-4) = 0 + 5 \quad \text{Combine like terms.}$$
$$x - 4 = 5 \quad \text{Rewrite } x + (-4) \text{ as } x - 4.$$
$$x - 4 + 4 = 5 + 4 \quad \text{Now add 4 to both sides.}$$
$$x = 9$$

Thus, 9 is the solution of the equation.

Answers to PROBLEMS

2. **a.** 12   **b.** $\frac{4}{5}$

CHECK   We substitute 9 for x in the original equation. To save time, we use the following diagram where $\overset{?}{=}$ means "are they equal?"

$3x + 5 - 2x - 9 \overset{?}{=} 6x + 5 - 6x$	
$3(9) + 5 - 2(9) - 9$	$6(9) + 5 - 6(9)$
$27 + 5 - 18 - 9$	$54 + 5 - 54$
$32 - 18 - 9$	5
$32 - 27$	
5	

Since both sides yield 5, our result is correct.

Now suppose that the price of an article is increased by \$3 and the article currently sells for \$8. What was its old price, p? The equation here is

$$\underbrace{\text{Old price}}_{p} \quad \underbrace{\text{went up}}_{+} \quad \underbrace{\$3}_{3} \quad \underbrace{\text{and is now}}_{=} \quad \underbrace{\$8.}_{8}$$

To solve this equation, we have to bring the price down; that is, we need to subtract 3 on both sides of the equation:

$$p + 3 - 3 = 8 - 3$$
$$p = 5$$

Thus, the old price was \$5. We have *subtracted* 3 on both sides of the equation. Here is the property that allows us to do this.

THE SUBTRACTION PROPERTY OF EQUALITY	For any number c, the equation $a = b$ is equivalent to $$a - c = b - c$$

This property tells us that we can *subtract* the same number on both sides of an equation to produce an *equivalent* equation. Note that since $a - c = a + (-c)$, you can think of *subtracting* c as *adding* $(-c)$.

EXAMPLE 3   **Using the subtraction property**

Solve:

a. $2x + 4 - x + 2 = 10$ **b.** $-3x + \dfrac{5}{7} + 4x - \dfrac{3}{7} = \dfrac{6}{7}$

SOLUTION

a. To solve the equation, we need to get x by itself on the left; that is, we want $x = \square$, where $\square$ is a number. We proceed as follows:

$$2x + 4 - x + 2 = 10$$
$$x + 6 = 10 \qquad \text{Simplify.}$$
$$x + 6 - 6 = 10 - 6 \qquad \text{Subtract 6 from both sides.}$$
$$x = 4$$

Thus, 4 is the solution of the equation.

CHECK

$2x + 4 - x + 2 \overset{?}{=} 10$	
$2(4) + 4 - 4 + 2$	10
$8 + 2$	
10	

PROBLEM 3

Solve:

a. $5y + 2 - 4y + 3 = 17$

b. $-5z + \dfrac{3}{8} + 6z - \dfrac{1}{8} = \dfrac{5}{8}$

(continued)

Answers to PROBLEMS

3. a. 12   **b.** $\dfrac{3}{8}$

b.
$$-3x + \frac{5}{7} + 4x - \frac{3}{7} = \frac{6}{7}$$

$$x + \frac{2}{7} = \frac{6}{7} \qquad \text{Simplify.}$$

$$x + \frac{2}{7} - \frac{2}{7} = \frac{6}{7} - \frac{2}{7} \qquad \text{Subtract } \tfrac{2}{7} \text{ from both sides.}$$

$$x = \frac{4}{7}$$

Thus, $\frac{4}{7}$ is the solution of the equation.

CHECK

$$\begin{array}{c|c} -3x + \frac{5}{7} + 4x - \frac{3}{7} \overset{?}{=} \frac{6}{7} \\ \hline -3\left(\frac{4}{7}\right) + \frac{5}{7} + 4\left(\frac{4}{7}\right) - \frac{3}{7} & \frac{6}{7} \\ -\frac{12}{7} + \frac{5}{7} + \frac{16}{7} - \frac{3}{7} & \\ -\frac{7}{7} + \frac{13}{7} & \\ \frac{6}{7} & \end{array}$$

C ⟩ Using the Addition and Subtraction Properties Together

Can we solve the equation $2x - 7 = x + 2$? Let's try. If we add 7 to both sides, we obtain

$$2x - 7 + 7 = x + 2 + 7$$
$$2x = x + 9$$

But this is not yet a solution. To solve this equation, we must get x by itself on the left—that is, $x = \square$, where $\square$ is a number. How do we do this? We want variables on one side of the equation (and we have them: $2x$) but only specific numbers on the other (here we are in trouble because we have an x on the right). To "get rid of" this x, we subtract x from both sides:

$$2x - x = x - x + 9 \qquad \text{Remember, } x = 1x.$$
$$x = 9$$

Thus, 9 is the solution of the equation.

CHECK

$$\begin{array}{c|c} 2x - 7 \overset{?}{=} x + 2 \\ \hline 2(9) - 7 & 9 + 2 \\ 11 & 11 \end{array}$$

By the way, you do not have to have the variable by itself on the *left* side of the equation. You may have the variables on the *right* side of the equation and your solution may be of the form $\square = x$, where $\square$ is a number.

A recommended procedure is to isolate the variable on the side of the equation that contains the highest coefficient of variables after simplification. Thus, when solving $7x + 3 = 4x + 6$, isolate the variables on the left side of the equation. On the other hand, when solving $4x + 6 = 7x + 3$, isolate the variables on the right side of the equation. Obviously, your solution would be the same in either case. [See Example 4(c) where the variables are on the right.]

Now let's review our procedure for solving equations by adding or subtracting.

> **PROCEDURE**
>
> **Solving Equations by Adding or Subtracting**
>
> **1.** Simplify both sides if necessary.
> **2.** Add or subtract the same numbers on both sides of the equation so that one side contains only variables.
> **3.** Add or subtract the same expressions on both sides of the equation so that the other side contains only numbers.

We use these three steps to solve Example 4.

EXAMPLE 4   Solving equations by adding or subtracting

Solve:

a. $3 = 8 + x$

c. $0 = 3(z - 2) + 4 - 2z$

b. $4y - 3 = 3y + 8$

d. $2(x + 1) = 3x + 5$

SOLUTION

a.

$$3 = 8 + x \quad \text{Given}$$

1. Both sides of the equation are already simplified.

$$3 = 8 + x$$

2. Subtract 8 on both sides.

$$3 - 8 = 8 - 8 + x$$
$$-5 = x$$

Step 3 is not necessary here, and the solution is -5.

$$x = -5$$

CHECK

$$\begin{array}{c|c} 3 \stackrel{?}{=} 8 + x \\ \hline 3 & 8 + (-5) \\ & 3 \end{array}$$

b.

$$4y - 3 = 3y + 8 \quad \text{Given}$$

1. Both sides of the equation are already simplified.

$$4y - 3 = 3y + 8$$

2. Add 3 on both sides.

$$4y - 3 + 3 = 3y + 8 + 3$$
$$4y = 3y + 11$$

3. Subtract $3y$ on both sides.

$$4y - 3y = 3y - 3y + 11$$
$$y = 11$$

The solution is 11.

CHECK

$$\begin{array}{c|c} 4y - 3 \stackrel{?}{=} 3y + 8 \\ \hline 4(11) - 3 & 3(11) + 8 \\ 44 - 3 & 33 + 8 \\ 41 & 41 \end{array}$$

c.

$$0 = 3(z - 2) + 4 - 2z \quad \text{Given}$$

1. Simplify by using the distributive property and combining like terms.

$$0 = 3z - 6 + 4 - 2z$$
$$0 = z - 2$$

PROBLEM 4

Solve:

a. $5 = 7 + x$

b. $5x - 2 = 4x + 3$

c. $0 = 3(y - 3) + 7 - 2y$

d. $3(z + 1) = 4z + 8$

(continued)

> **Answers to PROBLEMS**
>
> **4. a.** -2 **b.** 5   **c.** 2   **d.** -5

2. Add 2 on both sides. $0 + 2 = z - 2 + 2$

$$2 = z$$

Step 3 is not necessary, and $z = 2$
the solution is 2.

CHECK $0 \stackrel{?}{=} 3(z - 2) + 4 - 2z$

$$
\begin{array}{c|l}
0 & 3(2 - 2) + 4 - 2(2) \\
 & 3(0) + 4 - 4 \\
 & 0 + 4 - 4 \\
 & 0
\end{array}
$$

d. $2(x + 1) = 3x + 5$ Given

1. Simplify. $2x + 2 = 3x + 5$

2. Subtract 2 on both sides. $2x + 2 - 2 = 3x + 5 - 2$

$$2x = 3x + 3$$

3. Subtract $3x$ on both sides so all $2x - 3x = 3x - 3x + 3$
the variables are on the left. $-x = 3$

If $-x = 3$, then $x = -3$ because the opposite of a number is the number with
its sign changed; that is, if the opposite of x is 3, then x itself must be -3.
Thus, the solution is -3.

CHECK $2(x + 1) \stackrel{?}{=} 3x + 5$

$$
\begin{array}{c|c}
2(-3 + 1) & 3(-3) + 5 \\
2(-2) & -9 + 5 \\
-4 & -4
\end{array}
$$

Keep in mind the following rule; we will use it in Example 5.

> **PROCEDURE**
>
> Solving $-x = a$
> If a is a real number and $-x = a$, then $x = -a$.

EXAMPLE 5   Solving equations by adding or subtracting
Solve: $8x + 7 = 9x + 3$

SOLUTION

1. The equation is already $8x + 7 = 9x + 3$ Given
simplified.

2. Subtract 7 on both sides. $8x + 7 - 7 = 9x + 3 - 7$

$$8x = 9x - 4$$

3. Subtract $9x$ on both sides. $8x - 9x = 9x - 9x - 4$

$$-x = -4$$

Since $-x = -4$, then $x = -(-4) = 4$, and the solution is 4.

CHECK $8x + 7 \stackrel{?}{=} 9x + 3$

$$
\begin{array}{c|c}
8(4) + 7 & 9(4) + 3 \\
32 + 7 & 36 + 3 \\
39 & 39
\end{array}
$$

PROBLEM 5

Solve: $6y + 5 = 7y + 2$

Answers to PROBLEMS

5. 3

The equations in Examples 2–5 each had exactly one solution. For an equation that can be written as $ax + b = c$, there are three possibilities for the solution:

1. The equation has **one** solution. This is a **conditional** equation.
2. The equation has **no** solution. This is a **contradictory** equation.
3. The equation has **infinitely many** solutions. This is an **identity.**

EXAMPLE 6   Solving a contradictory equation

Solve: $3 + 8(x + 1) = 5 + 8x$

SOLUTION

$$3 + 8(x + 1) = 5 + 8x \quad \text{Given}$$

1. Simplify by using the distributive property and combining like terms.

$$3 + 8x + 8 = 5 + 8x$$
$$11 + 8x = 5 + 8x$$

2. Subtract 5 on both sides.

$$11 - 5 + 8x = 5 - 5 + 8x$$
$$6 + 8x = 8x$$

3. Subtract $8x$ on both sides.

$$6 + 8x - 8x = 8x - 8x$$
$$6 = 0$$

The statement "$6 = 0$" is a *false* statement. When this happens, it indicates that the equation has *no* solution—that is, it's a contradictory equation and we write "no solution."

PROBLEM 6

Solve: $5 + 3(z - 1) = 4 + 3z$

EXAMPLE 7   Solving an equation with infinitely many solutions

Solve: $7 + 2(x + 1) = 9 + 2x$

SOLUTION

$$7 + 2(x + 1) = 9 + 2x \quad \text{Given}$$

1. Simplify by using the distributive property and combining like terms.

$$7 + 2x + 2 = 9 + 2x$$
$$9 + 2x = 9 + 2x$$

You could stop here. Since both sides are *identical,* this equation is an identity. Every real number is a solution. But what happens if you go on? Let's see.

2. Subtract 9 on both sides.

$$9 - 9 + 2x = 9 - 9 + 2x$$
$$2x = 2x$$

3. Subtract $2x$ on both sides.

$$2x - 2x = 2x - 2x$$
$$0 = 0$$

The statement "$0 = 0$" is a true statement. When this happens, it indicates that *any* real number is a solution. (Try $x = -1$ or $x = 0$ in the original equation.) The equation has infinitely many solutions, and we write "all real numbers" for the solution.

PROBLEM 7

Solve: $3 + 4(y + 2) = 11 + 4y$

Answers to PROBLEMS

6. No solution

7. All real numbers

The graph shows the cost of tuition and fees from 1995 to 2005 (red bars). The cost has been very steady at about $2000 a year. Can you tell what the cost will be next year? What about in 2010? We will give you a formula, based on this graph, to be able to predict.

Source: Trends in College Pricing, the College Board. http://www.ed.gov.

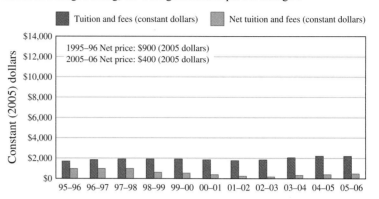

| ■ Tuition and fees (constant dollars) | ■ Net tuition and fees (constant dollars) |

1995–96 Net price: $900 (2005 dollars)
2005–06 Net price: $400 (2005 dollars)

Constant (2005) dollars

$14,000
$12,000
$10,000
$8,000
$6,000
$4,000
$2,000
$0

95–96 96–97 97–98 98–99 99–00 00–01 01–02 02–03 03–04 04–05 05–06

EXAMPLE 8   Application: College tuition and fees

How much is your tuition this year? It is estimated that annual tuition and required fees for in-state students at 2-year colleges is given by the expression **126x + 1512,** where x is the number of years after 2000.

a. Determine the estimated tuition and fees for 2000 ($x = 0$).

b. What will you predict the tuition and fees will be in the year 2010?

SOLUTION

a. When $x = 0$,

$$126x + 1512 = 126(0) + 1512 = 1512$$

Thus, estimated tuition and fees for 2000 ($x = 0$) amounted to $1512.

b. The year 2010 is 10 years after 2000, so $x = 10$

When $x = 10$,

$$126x + 1512 = 126(10) + 1512 = 1260 + 1512 = 2772$$

Thus, estimated tuition and fees for 2010 ($x = 10$) will amount to $2772.

PROBLEM 8

Use the expression $126x + 1512$ to answer the following questions.

a. What will the estimated annual tuition and fees be in the year 2004?

b. What will you predict the tuition and fees will be in the year 2020?

MathZone+

Boost your grade at mathzone.com!

> Practice Problems   > Self-Tests
> NetTutor   > e-Professors
                     > Videos

› Exercises **2.1**

‹ A › Verifying Solutions to an Equation In Problems 1–10, determine whether the given number is a solution of the
equation. (Do not solve.)

1. $x = 3$; $x - 1 = 2$

2. $x = 4$; $6 = x - 10$

3. $y = -2$; $3y + 6 = 0$

4. $z = -3$; $-3z + 9 = 0$

5. $n = 2$; $12 - 3n = 6$

6. $m = 3\frac{1}{2}$; $3\frac{1}{2} + m = 7$

7. $d = 10$; $\frac{2}{5}d + 1 = 3$

8. $c = 2.3$; $3.4 = 2c - 1.4$

9. $a = 2.1$; $4.6 = 11.9 - 3a$

10. $x = \frac{1}{10}$; $0.2 = \frac{7}{10} - 5x$

Answers to PROBLEMS

8. a. $2016   **b.** $4032

〈 **B** 〉 **Using the Addition and Subtraction Properties of Equality**   In Problems 11–30, solve and check the given equations.

11. $x - 5 = 9$

12. $y - 3 = 6$

13. $11 = m - 8$

14. $6 = n - 2$

15. $y - \frac{2}{3} = \frac{8}{3}$

16. $R - \frac{4}{3} = \frac{35}{3}$

17. $2k - 6 - k - 10 = 5$

18. $3n + 4 - 2n - 6 = 7$

19. $\frac{1}{4} = 2z - \frac{2}{3} - z$

20. $\frac{7}{2} = 3v - \frac{1}{5} - 2v$

21. $0 = 2x - \frac{3}{2} - x - 2$

22. $0 = 3y - \frac{1}{4} - 2y - \frac{1}{2}$

23. $\frac{1}{5} = 4c + \frac{1}{5} - 3c$

24. $0 = 6b + \frac{19}{2} - 5b - \frac{1}{2}$

25. $-3x + 3 + 4x = 0$

26. $-5y + 4 + 6y = 0$

27. $\frac{3}{4}y + \frac{1}{4} + \frac{1}{4}y = \frac{1}{4}$

28. $\frac{1}{2}y + \frac{2}{3} + \frac{1}{2}y = \frac{1}{3}$

29. $3.4 = -3c + 0.8 + 2c + 0.1$

30. $1.7 = -3c + 0.3 + 4c + 0.4$

〈 **C** 〉 **Using the Addition and Subtraction Properties Together**   In Problems 31–55, solve and check the given equations.

31. $6p + 9 = 5p$

32. $7q + 4 = 6q$

33. $3x + 3 + 2x = 4x$

34. $2y + 4 + 6y = 7y$

35. $4(m - 2) + 2 - 3m = 0$

36. $3(n + 4) + 2 = 2n$

37. $5(y - 2) = 4y + 8$

38. $3(z - 1) = 4z + 1$

39. $3a - 1 = 2(a - 4)$

40. $4(b + 1) = 5b - 3$

41. $5(c - 2) = 6c - 2$

42. $-4R + 6 = 5 - 3R + 8$

43. $3x + 5 - 2x + 1 = 6x + 4 - 6x$

44. $6f - 2 - 4f = -2f + 5 + 3f$

45. $-2g + 4 - 5g = 6g + 1 - 14g$

46. $-2x + 3 + 9x = 6x - 1$

47. $6(x + 4) + 4 - 2x = 4x$

48. $6(y - 1) - 2 + 2y = 8y + 4$

49. $10(z - 2) + 10 - 2z = 8(z + 1) - 18$

50. $7(a + 1) - 1 - a = 6(a + 1)$

51. $3b + 6 - 2b = 2(b - 2) + 4$

52. $3b + 2 - b = 3(b - 2) + 5$

53. $2p + \frac{2}{3} - 5p = -4p + 7\frac{1}{3}$

54. $4q + \frac{2}{7} - 6q = -3q + 2\frac{2}{7}$

55. $5r + \frac{3}{8} - 9r = -5r + 1\frac{1}{2}$

〉〉〉 **Applications**

56. *Price increases*   The price of an item is increased by $7; it now sells for $23. What was the old price of the item?

57. *Average hourly earnings*   In a certain year, the average hourly earnings were $9.81, an increase of 40¢ over the previous year. What were the average hourly earnings the previous year?

58. *Consumer Price Index*   The Consumer Price Index for housing in a recent year was 169.6, a 5.7-point increase over the previous year. What was the Consumer Price Index for housing the previous year?

59. *Medical costs*   The cost of medical care increased 142.2 points in a 6-year period. If the cost of medical care reached the 326.9 mark, what was it 6 years ago?

60. *SAT scores*   In the last 10 years, mathematics scores in the Scholastic Aptitude Test (SAT) have declined 16 points, to 476. What was the mathematics score 10 years ago?

61. *Waste generation*   From 1960 to 2003, the amount of waste generated each year increased by a whopping 148.1 million tons, ultimately reaching 236.2 million tons! How much waste was generated in 1960? The figure reached 234 million tons in 2000. What was the increase (in million tons) from 1960?

62. *Materials recovery*   From 1960 to 2003, the amount of materials recovered each year increased by 66.7 million tons to 72.3 million tons. How much waste was recovered in 1960? By 2003, 3.4 million more tons were recovered than in 2000. How many million tons were recovered in 2000?

63. *Popularity of exercise-walking*   What is the most popular sports activity in the United States? The answer is exercise-walking! According to the Census Bureau, 38% of the females 7 years and older engage in this activity. If this percent is 17% more than the percent of males who exercise-walk, what percent of males engage in exercise-walking?

64. *Popularity of swimming* The most popular sports activity among males 7 years and older is swimming, which is enjoyed by 26% of males. If this is 1% more than the percent of females swimming, what percent of females participate in swimming?

65. *Popularity of cultural activities* According to the Census Bureau, among people 18 years old and older, 8% more attended historic parks than art museums. If 35% of the population attended historic parks, what percent attended art museums?

66. *Red meat consumption* According to the U.S. Department of Agriculture, from 1970 to 2000, the per capita consumption of red meat declined by 16 pounds to 113.5 pounds. What was the per capita consumption of red meat in 1970? In 2000, consumption went up 1.1 pounds from the 1999 level. What was the per capita consumption in 1999?

What's your ideal weight? In general, it depends on your sex and height but there are many opinions and we can approximate all of them!

Source: http://www.halls.md/

67. *Ideal weight* According to Broca's index, the ideal weight W (in pounds) for a woman h inches tall is $W = 100 + 5(h - 60)$.

  a. A woman 62 inches tall weighs 120 pounds. Does her weight satisfy the equation?

  b. What should her weight be?

68. *Ideal weight* The ideal weight W (in pounds) for a man h inches tall is $W = 110 + 5(h - 60)$.

  a. A man 70 inches tall weighs 160 pounds. Does his weight satisfy the equation?

  b. Verify that 160 pounds is the correct weight for this man.

69. *Ideal weight* The B. J. Devine formula for the weight W (in kilograms) of a man h inches tall is $W = 50 + 2.3(h - 60)$.

  a. A man 70 inches tall weighs 75 kilograms. Does his weight satisfy the equation?

  b. How many kilograms overweight is this man?

70. *Ideal weight* For a woman h inches tall, the Devine formula (in kilograms) is $W = 45.5 + 2.3(h - 60)$.

  a. A woman 70 inches tall weighs 68.5 kilograms. Does this weight satisfy the equation?

  b. Verify that 68.5 kilograms is her correct weight.

71. *Ideal weight* The J. D. Robinson formula suggests a weight $W = 52 + 1.9(h - 60)$ (in kilograms) for a man h inches tall.

  a. A 70-inch-tall man weighs 70 kilograms. Does this weight satisfy the equation?

  b. How many kilograms underweight is this man?

72. *Ideal weight* The Robinson formula for the weight W of women h inches tall is $W = 49 + 1.7(h - 60)$.

  a. A 70-inch-tall woman weighs 65 kilograms. Does that weight satisfy the equation?

  b. Is she over or under weight, and by how much?

73. *Ideal weight* Miller's formula defines the weight W of a man h inches tall as $W = 56.2 + 1.41(h - 60)$ kilograms.

  a. A 70-inch-tall man weighs 70.3 kilograms. Does his weight satisfy the equation?

  b. Verify that 70.3 kilograms is the correct weight for this man.

How many calories can you eat and still lose weight?

Source: http://www.annecollins.com.

74. *Caloric intake* The daily caloric intake C needed for a woman to lose 1 pound per week is given by $C = 12W - 500$, where W is her weight in pounds.

  a. A woman weighs 120 pounds and consumes 940 calories each day. Does the caloric intake satisfy the equation?

  b. Verify that the 940 calories per day satisfy the equation.

75. *Caloric intake* The daily caloric intake C needed for a man to lose 1 pound per week is given by $C = 14W - 500$, where W is his weight in pounds.

  a. A man weighs 150 pounds and eats 1700 calories each day. Does the caloric intake satisfy the equation?

  b. How many calories over or under is he?

76. *Caloric intake* The daily caloric intake C for a very active person to lose one pound per week is given by $C = 17W - 500$, where W is the weight of the person.

  a. A person weighs 200 pounds and consumes 2900 calories daily. Does the caloric intake satisfy the equation?

  b. Verify that the 2900 calories per day satisfy the equation.

〉〉〉 *Using Your Knowledge*

Some Detective Work In this section we learned how to determine whether a given number *satisfies* an equation. Let's use this idea to do some detective work!

77. Suppose the police department finds a femur bone from a human female. The relationship between the length f of the femur and the height H of a female (in centimeters) is given by

$$H = 1.95f + 72.85$$

If the length of the femur is 40 centimeters and a missing female is known to be 120 centimeters tall, can the bone belong to the missing female?

78. If the length of the femur in Problem 77 is 40 centimeters and a missing female is known to be 150.85 centimeters tall, can the bone belong to the missing female?

80. Would you believe the driver in Problem 79 if he says he was going 90 miles per hour?

79. Have you seen police officers measuring the length of a skid mark after an accident? There's a formula for this. It relates the velocity V_a at the time of an accident and the length L_a of the skid mark at the time of the accident to the velocity and length of a test skid mark. The test skid mark is obtained by driving a car at a predetermined speed V_t, skidding to a stop, and then measuring the length of the skid L_t. The formula is

$$V_a^2 = \frac{L_a V_t^2}{L_t}$$

If $L_t = 36$, $L_a = 144$, $V_t = 30$, and the driver claims that at the time of the accident his velocity V_a was 50 miles per hour, can you believe him?

⟩⟩⟩ Write On

81. Explain what is meant by the solution of an equation.

82. Explain what is meant by equivalent equations.

83. Make up an equation that has no solution and one that has infinitely many solutions.

84. If the next-to-last step in solving an equation is $-x = -5$, what is the solution of the equation? Explain.

⟩⟩⟩ Concept Checker

Fill in the blank(s) with the correct word(s), phrase, or mathematical statement.

85. According to the **Addition Property of Equality** for any numbers a, b, and c the equation $a = b$ is **equivalent** to the equation _____.

86. According to the **Subtraction Property of Equality** for any numbers a, b, and c the equation $a = b$ is **equivalent** to the equation _____.

$a - c = b - c$	$a + c = b + c$
$c - a = b - a$	$b + a = c + a$

⟩⟩⟩ Mastery Test

Solve.

87. $5 + 4(x + 1) = 3 + 4x$

88. $2 + 4(x + 1) = 5x + 3$

89. $x - 5 = 4$

90. $x - \frac{1}{5} = \frac{3}{5}$

91. $x - 2.3 = 3.4$

92. $x - \frac{1}{7} = \frac{1}{4}$

93. $2x + 6 - x + 2 = 12$

94. $3x + 5 - 2x + 3 = 7$

95. $-5x + \frac{2}{9} + 6x - \frac{4}{9} = \frac{5}{9}$

96. $5y - 2 = 4y + 1$

97. $0 = 4(z - 3) + 5 - 3z$

98. $2 - (4x + 1) = 1 - 4x$

99. $3(x + 2) + 3 = 2 - (1 - 3x)$

100. $3(x + 1) - 3 = 2x - 5$

Determine whether the given number is a solution of the equation.

101. $-7; \frac{1}{7}z - 1 = 0$

102. $-\frac{3}{5}; \frac{5}{3}x + 1 = 0$

⟩⟩⟩ Skill Checker

Perform the indicated operation.

103. $4(-5)$

104. $6(-3)$

105. $-\frac{2}{3}\left(\frac{3}{4}\right)$

106. $-\frac{5}{7}\left(\frac{7}{10}\right)$

Find the reciprocal of each number.

107. $\frac{3}{2}$

108. $\frac{2}{5}$

Find the LCM of each pair of numbers.

109. 6 and 16

110. 9 and 12

111. 10 and 8

112. 30 and 18

2.2

The Multiplication and Division Properties of Equality

▶ Objectives

A ▷ Use the multiplication and division properties of equality to solve equations.

B ▷ Multiply by reciprocals to solve equations.

C ▷ Multiply by LCMs to solve equations.

D ▷ Solve applications involving percents.

▶ To Succeed, Review How To . . .

1. Multiply and divide signed numbers (pp. 60–61, 62–64).

2. Find the reciprocal of a number (pp. 63, 79).

3. Find the LCM of two or more numbers (pp. 13, 15).

4. Write a fraction as a percent, and vice versa (pp. 27–28).

▶ Getting Started

How Good a Deal Is This?

The tire in the ad is on sale at half price. It now costs $28. What was its old price, p? Since you are paying half price for the tire, the new price is $\frac{1}{2}$ of p—that is, $\frac{1}{2}p$ or $\frac{p}{2}$. Since this price is $28, we have

$$\frac{p}{2} = 28$$

S&K TIRES *1/2 Price Sale*
(50% off single tire price)
YAMOTO STEEL BELTED
$28⁰⁰*
*plus tax

But what was the old price? Twice as much, of course. Thus, to obtain the old price p, we multiply both sides of the equation by 2, the reciprocal of $\frac{1}{2}$, to obtain

$$2 \cdot \frac{p}{2} = 2 \cdot 28 \qquad \text{Note that } 2 \cdot \frac{p}{2} = 1p \text{ or } p.$$

$$p = 56$$

Hence, the old price was $56, as can be easily checked, since $\frac{56}{2} = 28$.

This example shows how you can *multiply* both sides of an equation by a nonzero number and obtain an *equivalent* equation—that is, an equation whose solution is the same as the original one. This is the multiplication property, one of the properties we will study in this section.

A ▷ Using the Multiplication and Division Properties of Equality

The ideas discussed in the *Getting Started* can be generalized as the following property.

THE MULTIPLICATION PROPERTY OF EQUALITY	For any nonzero number c, the equation $a = b$ is equivalent to $$ac = bc$$

This means that we can multiply both sides of an equation by the same nonzero number and obtain an equivalent equation. We use this property in Example 1.

EXAMPLE 1   Using the multiplication property

Solve:

a. $\dfrac{x}{3} = 2$ **b.** $\dfrac{y}{5} = -3$

SOLUTION

a. We multiply both sides of $\dfrac{x}{3} = 2$ by 3, the reciprocal of $\dfrac{1}{3}$.

$$3 \cdot \dfrac{x}{3} = 3 \cdot 2$$

$$\overset{1}{3} \cdot \dfrac{x}{\underset{1}{3}} = 6 \qquad \text{Note that } 3 \cdot \tfrac{1}{3} = 1, \text{ since 3 and } \tfrac{1}{3} \text{ are reciprocals.}$$

$$x = 6$$

Thus, the solution is 6.

CHECK $\dfrac{x}{3} \overset{?}{=} 2$

$$\begin{array}{c|c} \dfrac{6}{3} & 2 \\[2mm] \hline 2 & \end{array}$$

b. We multiply both sides of $\dfrac{y}{5} = -3$ by 5, the reciprocal of $\dfrac{1}{5}$.

$$5 \cdot \dfrac{y}{5} = 5(-3)$$

$$\overset{1}{5} \cdot \dfrac{y}{\underset{1}{5}} = -15 \qquad \text{Recall that } 5 \cdot (-3) = -15.$$

$$y = -15$$

Thus, the solution is -15.

CHECK $\dfrac{y}{5} \overset{?}{=} -3$

$$\begin{array}{c|c} \dfrac{-15}{5} & -3 \\[2mm] \hline -3 & \end{array}$$

PROBLEM 1

Solve:

a. $\dfrac{x}{5} = 3$ **b.** $\dfrac{y}{4} = -5$

Suppose the price of an article is doubled, and it now sells for $50. What was its original price, p? Half as much, right? Here is the equation:

$$2p = 50$$

We solve it by dividing both sides by 2 (to find half as much):

$$\dfrac{2p}{2} = \dfrac{50}{2}$$

$$\dfrac{\overset{1}{2}p}{\underset{1}{2}} = 25$$

Thus, the original price p is $25, as you can check:

$$2 \cdot 25 = 50$$

Note that dividing both sides by 2 (the *coefficient* of p) is the same as *multiplying* by $\frac{1}{2}$. Thus, you can also solve $2p = 50$ by multiplying by $\frac{1}{2}$ (the reciprocal of 2) to obtain

$$\tfrac{1}{2} \cdot 2p = \tfrac{1}{2} \cdot 50$$

$$p = 25$$

This example suggests that, just as the addition property of equality lets us *add* a number on each side of an equation, the division property lets us *divide* each side of an equation by a (nonzero) number to obtain an *equivalent* equation. We now state this property and use it in the next example.

THE DIVISION PROPERTY OF EQUALITY	For any nonzero number c, the equation $a = b$ is equivalent to $$\frac{a}{c} = \frac{b}{c}$$

This means that we can divide both sides of an equation by the same nonzero number and obtain an equivalent equation. Note that we can also *multiply* both sides of $a = b$ by the *reciprocal* of c—that is, by $\frac{1}{c}$—to obtain

$$\frac{1}{c} \cdot a = \frac{1}{c} \cdot b \quad \text{or} \quad \frac{a}{c} = \frac{b}{c} \quad \text{Same result!}$$

We shall solve equations by multiplying by reciprocals later in this section.

EXAMPLE 2   Using the division property

Solve:

a. $8x = 24$ **b.** $5x = -20$ **c.** $-3x = 7$

SOLUTION

a. We need to get x by itself on the left. That is, we need $x = \square$, where $\square$ is a number. Thus, we divide both sides of the equation by 8 (the coefficient of x) to obtain.

$$\frac{8x}{8} = \frac{24}{8}$$

$$\frac{\overset{1}{8}x}{\cancel{8}} = 3$$

$$x = 3$$

The solution is 3.

CHECK $8x \overset{?}{=} 24$

$$\begin{array}{c|c} 8 \cdot 3 & 24 \\ \hline 24 & \end{array}$$

You can also solve this problem by *multiplying* both sides of $8x = 24$ by $\frac{1}{8}$, the reciprocal of 8.

$$\frac{1}{8} \cdot 8x = \frac{1}{8} \cdot 24$$

$$1x = \frac{1}{\cancel{8}} \cdot \frac{\overset{3}{24}}{1}$$

$$x = 3$$

b. Here, we divide both sides of $5x = -20$ by 5 (the coefficient of x) so that we have x by itself on the left. Thus,

$$\frac{5x}{5} = \frac{-20}{5}$$

$$x = -4$$

The solution is -4.

CHECK $5x \overset{?}{=} -20$

$$\begin{array}{c|c} 5 \cdot (-4) & -20 \\ \hline -20 & \end{array}$$

Of course, you can also solve this problem by *multiplying* both sides by $\frac{1}{5}$, the reciprocal of 5. When solving these types of equations, you always have the option

PROBLEM 2

Solve:

a. $3x = 12$

b. $7x = -21$

c. $-5x = 20$

of dividing both sides of the equation by a specific number or multiplying both sides of the equation by the reciprocal of the number.

$$\frac{1}{5} \cdot 5x = \frac{1}{5} \cdot (-20)$$

$$1x = \frac{1}{\cancel{5}} \cdot \frac{\cancel{-20}^{-4}}{1}$$

$$x = -4$$

c. In this case we divide both sides of $-3x = 7$ by -3 (the coefficient of x). We then have

$$\frac{\cancel{-3}x}{\cancel{-3}} = \frac{7}{-3}$$

$$x = -\frac{7}{3}$$ Recall that the quotient of two numbers with different signs is negative.

The solution is $-\frac{7}{3}$.

CHECK $-3x \stackrel{?}{=} 7$

$$-3\left(-\frac{7}{3}\right) \,\Big|\, 7$$

$$7 \,\Big|\,$$

As you know, this problem can also be solved by multiplying both sides by $-\frac{1}{3}$, the reciprocal of -3.

$$-\frac{1}{3} \cdot (-3x) = -\frac{1}{3} \cdot (7)$$

$$1x = -\frac{1}{3} \cdot \frac{7}{1}$$

$$x = -\frac{7}{3}$$

B › Multiplying by Reciprocals

In Example 2, the coefficients of the variables were integers. In such cases, it's easy to divide each side of the equation by this coefficient. When the coefficient of the variable is a fraction, it's easier to multiply each side of the equation by the reciprocal of the coefficient. Thus, to solve $-3x = 7$, divide each side by -3, but to solve $\frac{3}{4}x = 18$, multiply each side by the reciprocal of the coefficient of x, that is, by $\frac{4}{3}$, as shown next.

EXAMPLE 3   Solving equations by multiplying by reciprocals

Solve:

a. $\frac{3}{4}x = 18$ **b.** $-\frac{2}{5}x = 8$ **c.** $-\frac{3}{8}x = -15$

SOLUTION

a. Multiply both sides of $\frac{3}{4}x = 18$ by the reciprocal of $\frac{3}{4}$, that is, by $\frac{4}{3}$. We then get

$$\frac{4}{3}\left(\frac{3}{4}x\right) = \frac{4}{3}(18)$$

$$1 \cdot x = \frac{4}{3} \cdot \frac{\cancel{18}^{6}}{1} = \frac{24}{1}$$

$$x = 24$$

Hence the solution is 24.

PROBLEM 3

Solve:

a. $\frac{3}{5}x = 12$

b. $-\frac{2}{5}x = 6$

c. $-\frac{4}{5}x = -8$

(continued)

Answers to PROBLEMS

3. a. 20   **b.** -15 **c.** 10

CHECK

$$\frac{3}{4}x \stackrel{?}{=} 18$$

$$\begin{array}{c|c} \frac{3}{4}(24) & 18 \\ \hline \frac{3}{4} \cdot \frac{\overset{6}{24}}{1} & \\ {}_{1} & \\ 18 & \end{array}$$

b. Multiplying both sides of $-\frac{2}{5}x = 8$ by the reciprocal of $-\frac{2}{5}$, that is, by $-\frac{5}{2}$, we have

$$-\frac{5}{2}\left(-\frac{2}{5}x\right) = -\frac{5}{2}(8)$$

$$1 \cdot x = \frac{-5}{\overset{}{2}} \cdot \frac{\overset{4}{8}}{1} = -\frac{20}{1}$$

$$x = -20$$

The solution is -20.

CHECK

$$-\frac{2}{5}x \stackrel{?}{=} 8$$

$$\begin{array}{c|c} -\frac{2}{5}(-20) & 8 \\ \hline -\frac{2}{5}\left(\frac{\overset{-4}{-20}}{1}\right) & \\ {}_{1} & \\ 8 & \end{array}$$

c. Here we multiply both sides of $-\frac{3}{8}x = -15$ by $-\frac{8}{3}$, the reciprocal of $-\frac{3}{8}$, to obtain

$$-\frac{8}{3}\left(-\frac{3}{8}x\right) = -\frac{8}{3}(-15)$$

$$1 \cdot x = \frac{-8(-15)}{3} = \frac{-8(\overset{-5}{-15})}{\overset{}{3}}$$

$$= 40$$

The solution is 40.

CHECK

$$-\frac{3}{8}x \stackrel{?}{=} -15$$

$$\begin{array}{c|c} -\frac{3}{8}(40) & -15 \\ \hline -\frac{3}{8}(\overset{5}{40}) & \\ {}_{1} & \\ -15 & \end{array}$$

C › Multiplying by the LCM

Finally, if the equation we are solving contains sums or differences of fractions, we first eliminate these fractions by multiplying each term in the equation by the smallest number that is a multiple of each of the denominators. This number is called the *least common multiple* (or LCM for short) of the denominators. If you forgot about LCMs and fractions, review Sections R.1 and R.2 at the beginning of the book.

Which equation would you rather solve?

$$3x + 2x = 1 \quad \text{or} \quad \frac{x}{2} + \frac{x}{3} = \frac{1}{6}$$

Probably the first! But if you multiply each term in the second equation by the LCM of 2, 3, and 6 (which is 6), you obtain the first equation!

Do you remember how to find the LCM of two numbers? If you don't, here's a quick way to do it. Suppose you wish to solve the equation

$$\frac{x}{6} + \frac{x}{16} = 22$$

To find the LCM of 6 and 16, write the denominators in a horizontal row (see step 1 to the right) and divide each of them by the largest number that will divide *both* of them. In this case, the number is 2. The quotients are 3 and 8, as shown in step 3. Since there are no numbers other than 1 that will divide both 3 and 8, the LCM is the product of 2 and the final quotients 3 and 8, as indicated in step 4.

1. $\begin{array}{|cc}6 & 16\end{array}$

2. $2\begin{array}{|cc}6 & 16\end{array}$

3. $2\begin{array}{|cc}6 & 16\\3 & 8\end{array}$

4. $2\begin{array}{|cc}6 & 16\\3 & 8\end{array} \rightarrow 2 \cdot 3 \cdot 8 = 48$
is the LCM.

For an even quicker method, write the multiples of 16 (the larger of the two denominators) until you find a multiple of 16 that is *divisible* by 6. Like this: 16 32 ④⑧

↑
divisible
by 6

The LCM is 48, as before.

Now we can multiply each side of the equation by this LCM (48):

$$48\left(\frac{x}{6} + \frac{x}{16}\right) = 48 \cdot 22 \qquad \text{To avoid confusion, place parentheses around } \frac{x}{6} + \frac{x}{16}.$$

$$\overset{8}{\cancel{48}} \cdot \frac{x}{\cancel{6}} + \overset{3}{\cancel{48}} \cdot \frac{x}{\cancel{16}} = 48 \cdot 22 \qquad \begin{array}{l}\text{Use the distributive property. Don't multiply}\\48 \cdot 22 \text{ yet, we'll simplify this later.}\end{array}$$

$$8x + 3x = 48 \cdot 22 \qquad \text{Simplify the left side.}$$

$$11x = 48 \cdot 22 \qquad \text{Combine like terms.}$$

$$\frac{\cancel{11}x}{\cancel{11}} = \frac{48 \cdot \overset{2}{\cancel{22}}}{\cancel{11}} \qquad \begin{array}{l}\text{Divide both sides by 11. Do you see why we waited}\\\text{to multiply } 48 \cdot 22?\end{array}$$

$$x = 96$$

The solution is 96.

CHECK

$$\frac{x}{6} + \frac{x}{16} \overset{?}{=} 22$$

$$\begin{array}{c|c}\frac{96}{6} + \frac{96}{16} & 22\\16 + 6 & \\22 & \end{array}$$

Note that if we wish to clear fractions in

$$\frac{a}{b} + \frac{c}{d} = \frac{e}{f}$$

we can use the following procedure.

PROCEDURE

Clearing Fractions

To clear fractions in an equation, multiply both sides of the equation by the LCM of the denominators, or, equivalently, multiply each term by the LCM.

Thus, if we multiply both sides of $\frac{a}{b} + \frac{c}{d} = \frac{e}{f}$ by the LCM of b, d, and f (which we shall call L), we get

$$L\left(\frac{a}{b} + \frac{c}{d}\right) = L\frac{e}{f} \qquad \text{(Where } b, d, \text{ and } f \neq 0)$$

Note the added parentheses.

or

$$\frac{La}{b} + \frac{Lc}{d} = \frac{Le}{f} \qquad \text{Where } b, d, \text{ and } f \neq 0$$

Thus, to clear the fractions here, we multiply each term by L (using the distributive property).

EXAMPLE 4   Solving equations by multiplying by the LCM

Solve:

a. $\frac{x}{10} + \frac{x}{8} = 9$ **b.** $\frac{x}{3} - \frac{x}{8} = 10$

SOLUTION

a. The LCM of 10 and 8 is 40 (since the first four multiples of 10 are 10, 20, 30, and 40, and 8 divides 40). You can also find the LCM by writing

$$2 \quad \underline{|10 \quad 8} \\ \quad \rule{0pt}{0pt}\quad 5 - 4 \rightarrow 2 \cdot 5 \cdot 4 = 40$$

Multiplying each term by 40, we have

$$40 \cdot \frac{x}{10} + 40 \cdot \frac{x}{8} = 40 \cdot 9$$
$$4x + 5x = 40 \cdot 9 \qquad \text{Simplify.}$$
$$9x = 40 \cdot 9 \qquad \text{Combine like terms.}$$
$$x = 40 \qquad \text{Divide by 9.}$$

The solution is 40.

CHECK $\frac{x}{10} + \frac{x}{8} \overset{?}{=} 9$

$$\begin{array}{c|c} \frac{40}{10} + \frac{40}{8} & 9 \\ 4 + 5 & \\ 9 & \end{array}$$

b. The LCM of 3 and 8 is $3 \cdot 8 = 24$, since the largest number that divides 3 and 8 is 1.

$$1 \quad \underline{|3 \quad 8} \\ \quad \rule{0pt}{0pt}\quad 3 - 8 \rightarrow 1 \cdot 3 \cdot 8 = 24$$

Multiplying each term by 24 yields

$$24 \cdot \frac{x}{3} - 24 \cdot \frac{x}{8} = 24 \cdot 10$$
$$8x - 3x = 24 \cdot 10 \qquad \text{Simplify.}$$
$$5x = 24 \cdot 10 \qquad \text{Combine like terms.}$$
$$\frac{\cancel{5}x}{\cancel{5}} = \frac{24 \cdot \overset{2}{\cancel{10}}}{\underset{1}{\cancel{5}}} \qquad \text{Divide by 5.}$$
$$x = 48$$

The solution is 48.

CHECK $\frac{x}{3} - \frac{x}{8} \overset{?}{=} 10$

$$\begin{array}{c|c} \frac{48}{3} - \frac{48}{8} & 10 \\ 16 - 6 & \\ 10 & \end{array}$$

PROBLEM 4

Solve:

a. $\frac{x}{10} + \frac{x}{6} = 8$

b. $\frac{x}{4} - \frac{x}{5} = 1$

Answers to PROBLEMS

4. a. 30   **b.** 20

In some cases, the numerators of the fractions involved contain more than one term. However, the procedure for solving the equation is still the same, as we illustrate in Example 5.

EXAMPLE 5   Solving equations by multiplying by the LCM

Solve:

a. $\dfrac{x+1}{3} + \dfrac{x-1}{10} = 5$ **b.** $\dfrac{x+1}{3} - \dfrac{x-1}{8} = 4$

SOLUTION

a. The LCM of 3 and 10 is $3 \cdot 10 = 30$, since 3 and 10 don't have any common factors. Multiplying each term by 30, we have

$$\overset{10}{\cancel{30}}\left(\frac{x+1}{\cancel{3}}\right) + \overset{3}{\cancel{30}}\left(\frac{x-1}{\cancel{10}}\right) = 30 \cdot 5 \qquad \text{Note the parentheses!}$$

$$10(x+1) + 3(x-1) = 150$$

$$10x + 10 + 3x - 3 = 150 \qquad \text{Use the distributive property.}$$

$$13x + 7 = 150 \qquad \text{Combine like terms.}$$

$$13x = 143 \qquad \text{Subtract 7.}$$

$$x = 11 \qquad \text{Divide by 13.}$$

We leave the check for you to verify.

b. Here the LCM is $3 \cdot 8 = 24$. Multiplying each term by 24, we obtain

$$\overset{8}{\cancel{24}}\left(\frac{x+1}{\cancel{3}}\right) - \overset{3}{\cancel{24}}\left(\frac{x-1}{\cancel{8}}\right) = 24 \cdot 4 \qquad \text{Note the parentheses!}$$

$$8(x+1) - 3(x-1) = 96$$

$$8x + 8 - 3x + 3 = 96 \qquad \text{Use the distributive property.}$$

$$5x + 11 = 96 \qquad \text{Combine like terms.}$$

$$5x = 85 \qquad \text{Subtract 11.}$$

$$x = 17 \qquad \text{Divide by 5.}$$

Be sure you check this answer in the original equation.

PROBLEM 5

Solve:

a. $\dfrac{x+2}{3} + \dfrac{x-2}{4} = 6$

b. $\dfrac{x+2}{5} - \dfrac{x-2}{3} = 0$

D › Applications Involving Percent Problems

Percent problems are among the most common types of problems, not only in mathematics, but also in many other fields. Basically, there are three types of percent problems.

Type 1   asks you to find a number that is a given percent of a specific number.

Example: 20% (read "20 percent") of 80 is what number?

Type 2   asks you what percent of a number is another given number.

Example: What percent of 20 is 5?

Type 3   asks you to find a number when it's known that a given number equals a percent of the unknown number.

Example: 10 is 40% of what number?

To do these problems, you need only recall how to translate words into equations and how to write percents as fractions (pp. 91–93 and 27–28 respectively).

Now do you remember what 20% means? The symbol % is read as "percent," which means "per hundred." Recall that

$$20\% = \frac{20}{100} = \frac{\overset{1}{\cancel{20}}}{\underset{5}{\cancel{100}}} = \frac{1}{5}$$

Similarly,

$$60\% = \frac{60}{100} = \frac{\overset{3}{\cancel{60}}}{\underset{5}{\cancel{100}}} = \frac{3}{5}$$

$$17\% = \frac{17}{100}$$

We are now ready to solve some percent problems.

EXAMPLE 6 Finding a percent of a number (type 1)

Twenty percent of 80 is what number?

SOLUTION

Let's translate this.

20%	of	80	is	what number?
↓	↓	↓	↓	↓
$\frac{20}{100}$	$\cdot$	80	$=$	n

$$\frac{1}{5} \cdot 80 = n \qquad \text{Since } \frac{20}{100} = \frac{1}{5}$$

$$\frac{80}{5} = n \qquad \text{Multiply } \frac{1}{5} \text{ by } \frac{80}{1}.$$

$$n = 16 \qquad \text{Reduce } \frac{80}{5}.$$

Thus, 20% of 80 is 16.

PROBLEM 6

Forty percent of 30 is what number?

EXAMPLE 7 Finding a percent (type 2)

What percent of 20 is 5?

SOLUTION

Let's translate this.

What percent	of	20	is	5?
↓	↓	↓	↓	↓
x	$\cdot$	20	$=$	5

$$\frac{x \cdot 20}{20} = \frac{5}{20} \qquad \text{Divide by 20.}$$

$$x = \frac{1}{4} \qquad \text{Reduce } \frac{5}{20}.$$

But x represents a percent, so we must change $\frac{1}{4}$ to a percent.

$$x = \frac{1}{4} = \frac{25}{100} \quad \text{or} \quad 25\%$$

Thus, 5 is 25% of 20.

PROBLEM 7

What percent of 30 is 6?

EXAMPLE 8 Finding a number (type 3)

Ten is 40% of what number?

SOLUTION

First we translate:

10	is	40%	of	what number?
↓	↓	↓	↓	↓
10	$=$	$\frac{40}{100}$	$\cdot$	n

PROBLEM 8

Twenty is 40% of what number?

Answers to PROBLEMS

6. 12   **7.** 20%   **8.** 50

$$\frac{40}{100} \cdot n = 10 \qquad \text{Rearrange.}$$

$$\frac{2}{5} \cdot n = 10 \qquad \text{Reduce } \tfrac{40}{100}.$$

$$\overset{5}{\cancel{\frac{5}{2}}} \cdot \underset{1}{\cancel{\frac{2}{5}}} \cdot n = \frac{5}{2} \cdot \cancel{10} \qquad \text{Multiply by } \tfrac{5}{2}.$$

$$n = 25$$

Thus, ten is 40% of 25.

EXAMPLE 9   Application: Angioplasty versus TPA

The study cited in the illustration claims that you can save more lives with angioplasty (a procedure in which a balloon-tipped instrument is inserted in your arteries, the balloon is inflated, and the artery is unclogged!) than with a blood-clot-breaking drug called TPA. Of the 451 patients studied, 226 were randomly assigned for TPA and 225 for angioplasty. After 6 months the results were as follows:

a. 6.2% of the angioplasty patients died. To the nearest whole number, how many patients is that?

b. 7.1% of the drug therapy patients died. To the nearest whole number, how many patients is that?

c. 2.2% of the angioplasty patients had strokes. To the nearest whole number, how many patients is that?

d. 4% of the drug therapy group patients had strokes. To the nearest whole number, how many patients is that?

SOLUTION

a. We need to take 6.2% of 225 (the number of angioplasty patients).

$$6.2\% \text{ of } 225 \quad \text{Means} \quad 0.062 \cdot 225 = 13.95$$

or 14 when rounded to the nearest whole number.

b. Here, we need 7.1% of 226 = 0.071 · 226 = 16.046, or 16 when rounded to the nearest whole number.

c. 2.2% of 225 = 0.022 · 225 = 4.95, or 5 when rounded to the nearest whole number.

d. 4% of 226 = 0.04 · 226 = 9.04, or 9 when rounded to the nearest whole number.

Now, look at the answers 14, 16, 5, and 9 for parts **a–d,** respectively. Are those answers faithfully depicted in the graphs? What do you think happened? (You will revisit this study in Exercises 71–80.)

PROBLEM 9

The study also said that after six months, 5.3% of the angioplasty patients had a heart attack.

a. To the nearest whole number, how many patients is that?

b. In addition, 10.6% of the drug therapy group had a heart attack. To the nearest whole number, how many patients is that?

Are the answers for **a** and **b** faithfully depicted in the graph?

Saving Lives

Community hospitals without on-site cardiac units can save more lives with angioplasty than with drug treatment, a new study of 451 heart attack victims shows.

■ Clot-breaking drug
■ Angioplasty

Number of patients with the following outcome, six weeks after treatment:

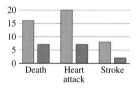

Number of patients with the following outcome, six months after treatment:

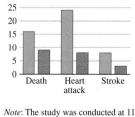

Note: The study was conducted at 11 community hospitals without on-site cardiac surgery units.

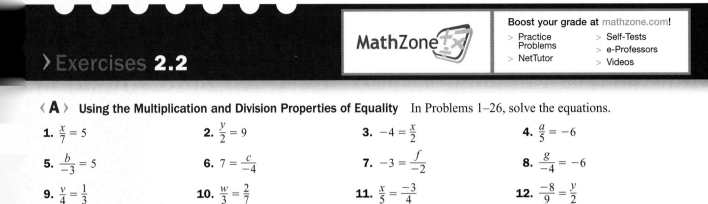

> Exercises **2.2**

⟨ A ⟩ Using the Multiplication and Division Properties of Equality   In Problems 1–26, solve the equations.

1. $\frac{x}{7} = 5$

2. $\frac{y}{2} = 9$

3. $-4 = \frac{x}{2}$

4. $\frac{a}{5} = -6$

5. $\frac{b}{-3} = 5$

6. $7 = \frac{c}{-4}$

7. $-3 = \frac{f}{-2}$

8. $\frac{g}{-4} = -6$

9. $\frac{v}{4} = \frac{1}{3}$

10. $\frac{w}{3} = \frac{2}{7}$

11. $\frac{x}{5} = \frac{-3}{4}$

12. $\frac{-8}{9} = \frac{y}{2}$

13. $3z = 33$

14. $4y = 32$

15. $-42 = 6x$

16. $7b = -49$

17. $-8c = 56$

18. $-5d = 45$

19. $-5x = -35$

20. $-12 = -3x$

21. $-3y = 11$

22. $-5z = 17$

23. $-2a = 1.2$

24. $-3b = 1.5$

25. $3t = 4\frac{1}{2}$

26. $4r = 6\frac{2}{3}$

⟨ B ⟩ Multiplying by Reciprocals   In Problems 27–40, solve the equations.

27. $\frac{1}{3}x = -0.75$

28. $\frac{1}{4}y = 0.25$

29. $-6 = \frac{3}{4}C$

30. $-2 = \frac{2}{9}F$

31. $\frac{5}{6}a = 10$

32. $24 = \frac{2}{7}z$

33. $-\frac{4}{5}y = 0.4$

34. $0.5x = \frac{-1}{4}$

35. $\frac{-2}{11}p = 0$

36. $\frac{-4}{9}q = 0$

37. $-18 = \frac{3}{5}t$

38. $-8 = \frac{2}{7}R$

39. $\frac{7x}{0.02} = -7$

40. $-6 = \frac{6y}{0.03}$

⟨ C ⟩ Multiplying by the LCM   In Problems 41–60, solve the equations.

41. $\frac{y}{2} + \frac{y}{3} = 10$

42. $\frac{a}{4} + \frac{a}{3} = 14$

43. $\frac{x}{7} + \frac{x}{3} = 10$

44. $\frac{z}{6} + \frac{z}{4} = 20$

45. $\frac{x}{5} + \frac{x}{10} = 6$

46. $\frac{r}{2} + \frac{r}{6} = 8$

47. $\frac{t}{6} + \frac{t}{8} = 7$

48. $\frac{f}{9} + \frac{f}{12} = 14$

49. $\frac{x}{2} + \frac{x}{5} = \frac{7}{10}$

50. $\frac{a}{3} + \frac{a}{7} = \frac{20}{21}$

51. $\frac{c}{3} - \frac{c}{5} = 2$

52. $\frac{F}{4} - \frac{F}{7} = 3$

53. $\frac{W}{6} - \frac{W}{8} = \frac{5}{12}$

54. $\frac{m}{6} - \frac{m}{10} = \frac{4}{3}$

55. $\frac{x}{5} - \frac{3}{10} = \frac{1}{2}$

56. $\frac{3y}{7} - \frac{1}{14} = \frac{1}{14}$

57. $\frac{x+4}{4} - \frac{x+2}{3} = -\frac{1}{2}$

58. $\frac{w-1}{2} + \frac{w}{8} = \frac{7w+1}{16}$

59. $\frac{x}{6} + \frac{3}{4} = x - \frac{7}{4}$

60. $\frac{x}{6} + \frac{4}{3} = x - \frac{1}{3}$

⟨ D ⟩ Applications Involving Percent Problems   Translate into an equation and solve.

61. 30% of 40 is what number?

62. 17% of 80 is what number?

63. 40% of 70 is what number?

64. What percent of 40 is 8?

65. What percent of 30 is 15?

66. What percent of 40 is 4?

67. 30 is 20% of what number?

68. 10 is 40% of what number?

69. 12 is 60% of what number?

70. 24 is 50% of what number?

Medical Study   Refer to the study described in Example 9. It was determined that not all 451 patients actually received treatment. As a matter of fact, only 211 patients received the clot-breaking drug (TPA) and 171 received angioplasty. Exercises 71–76 refer to the chart shown here. Give your answers to one decimal place.

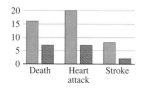

Clot-breaking drug (211)
Angioplasty   (171)

Number of patients with the following outcome, six weeks after treatment:

Source: Data from *Journal of the American Medical Association.*

71. As you can see, 16 of the patients receiving the clot-breaking drug died. What percent of the 211 patients is that?

72. What percent of the 171 patients receiving angioplasty died?

73. What percent of the 211 patients receiving the clot-breaking drug had heart attacks?

74. What percent of the 171 patients receiving angioplasty had heart attacks?

75. What percent of the 211 patients receiving the clot-breaking drug had a stroke?

76. What percent of the 171 patients receiving angioplasty had a stroke?

Medical Study   Exercises 77–80 refer to the chart shown here. Give your answers to one decimal place.

Clot-breaking drug (211)
Angioplasty   (171)

Number of patients with the following outcome, six months after treatment:

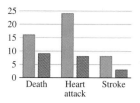

Source: Data from *Journal of the American Medical Association.*

77. As you can see, 9 of the patients receiving angioplasty died. What percent of the 171 patients is that?

78. What percent of the 171 patients receiving angioplasty had a heart attack?

79. What percent of the 211 patients receiving the clot-breaking drug had a stroke?

80. What percent of the 171 patients receiving angioplasty had a stroke?

Type of Exercise	Calories/Hour
Sleeping	55
Eating	85
Housework, moderate	160+
Dancing, ballroom	260
Walking, 3 mph	280
Aerobics	450+

Source: http://www.annecollins.com/.

Burning Calories   The table shows the number of calories used in 1 hour of different activities. To lose 1 pound, you need to burn 3500 calories. If you want to do it by *sleeping* h hours at 55 calories per hour (line 1), you have the equation $55h = 3500$, and $h = 63\frac{7}{11}$ hours, so you have to sleep more than 63 hours to lose 1 pound!

In Problems 81–85:

a. Use the chart to write an equation indicating the number h of hours you need to lose 1 pound (3500 calories) by doing the indicated exercise.

b. Solve the equation to find the number of hours needed to lose one pound.

81. Eating

82. Moderate housework

83. Ballroom dancing

84. Walking at 3 mph

85. Aerobics

By the way, if you eat just one McDonald's hamburger (280 calories), you need to walk for exactly 1 hour to burn the calories (see the chart)!

> > > *Using Your Knowledge*

The Cost of a Strike Have you ever been on strike? If so, after negotiating a salary increase, did you come out ahead? How can you find out? Let's take an example. If an employee earns $10 an hour and works 250 8-hour days a year, her salary would be $10 \cdot 8 \cdot 250 = \$20,000$. What percent increase in salary is needed to make up for the lost time during the strike? First, we must know how many days' wages were lost (the average strike in the United States lasts about 100 days). Let's assume that 100 days were lost. The wages for 100 days would be $10 \cdot 8 \cdot 100 = \$8000$. The employee needs $20,000 plus $8000 or $28,000 to make up the lost wages. Let x be the multiple of the old wages needed to make up the loss within 1 year. We then have

$$20,000 \cdot x = 28,000$$

$$\frac{20,000 \cdot x}{20,000} = \frac{28,000}{20,000} \qquad \text{Divide by 20,000.}$$

$$x = 1.40 \quad \text{or} \quad 140\%$$

Thus, the employee needs a 40% increase to recuperate her lost wages within 1 year!

86. Assume that an employee works 8 hours a day for 250 days. If the employee makes $20 an hour, what is her yearly salary?

87. If the employee goes on strike for 100 days, find the amount of her lost wages.

88. What percent increase will be necessary to make up for the lost wages within 1 year?

> > > **Write On**

89. a. What is the difference between an expression and an equation?

   b. What is the difference between simplifying an expression and solving an equation?

91. When solving the equation $-\frac{3}{4}x = 15$, would it be easier to divide by $-\frac{3}{4}$ or to multiply by the reciprocal of $-\frac{3}{4}$? Explain your answer and solve the equation.

90. When solving the equation $-3x = 18$, would it be easier to divide by -3 or to multiply by the reciprocal of -3? Explain your answer and solve the equation.

> > > *Concept Checker*

Fill in the blank(s) with the correct word(s), phrase, or mathematical statement.

92. According to the **Multiplication Property of Equality,** for any nonzero number c, the equation $a = b$ is equivalent to _____.

93. According to the **Division Property of Equality,** for any nonzero number c, the equation $a = b$ is equivalent to _____.

94. The **smallest number** that is a **multiple** of each of the **denominators** in an equation is called the _____ of the denominators.

95. To **clear fractions** in an equation, **multiply** both sides of the equation by the _____ of the **denominators.**

LCM	$\frac{a}{c} = \frac{b}{c}$
multiple	$\frac{a}{b} = \frac{c}{a}$
$\boldsymbol{ab = ac}$	**LCD**
$\boldsymbol{ac = bc}$	

> > > *Mastery Test*

96. 15 is 30% of what number?

97. What percent of 45 is 9?

98. 40% of 60 is what number?

Solve.

99. $\dfrac{x+2}{4} + \dfrac{x-1}{5} = 3$

100. $\dfrac{x+3}{2} - \dfrac{x-2}{3} = 5$

101. $\dfrac{y}{6} + \dfrac{y}{10} = 8$

102. $\dfrac{y}{5} - \dfrac{y}{8} = 3$

103. $\dfrac{4}{5}y = 8$

104. $-\dfrac{3}{4}y = 6$

105. $-\frac{2}{7}y = -4$ **106.** $-7y = 16$ **107.** $\frac{x}{2} = -7$

108. $-\frac{y}{4} = -3$

〉〉〉 *Skill Checker*

Use the distributive property to multiply.

109. $4(x - 6)$ **110.** $3(6 - x)$ **111.** $5(8 - y)$ **112.** $6(8 - 2y)$

113. $9(6 - 3y)$ **114.** $-3(4x - 2)$ **115.** $-5(3x - 4)$

Find:

116. $20 \cdot \frac{3}{4}$ **117.** $24 \cdot \frac{1}{6}$ **118.** $-5 \cdot \left(-\frac{4}{5}\right)$ **119.** $-7 \cdot \left(-\frac{3}{7}\right)$

2.3   Linear Equations

▶ Objectives

A 〉 Solve linear equations in one variable.

B 〉 Solve a literal equation for one of the unknowns.

▶ To Succeed, Review How To . . .

1. Find the LCM of three numbers (p. 15).
2. Add, subtract, multiply, and divide real numbers (pp. 52, 54, 60, 62).
3. Use the distributive property to remove parentheses (pp. 81, 89–90).

▶ Getting Started
Getting the Best Deal

Suppose you want to rent a midsize car. Which is the better deal, paying $39.99 per day *plus* $0.20 per mile or paying $49.99 per day with unlimited mileage? Well, it depends on how many miles you travel! First, let's see how many miles you have to travel before the costs are the same. The total daily cost, C, based on traveling m miles consists of two parts: $39.99 plus the additional charge at $0.20 per mile. Thus,

The total daily cost C	consists of	fixed cost + mileage.
C	$=$	$39.99 + 0.20m$

If the costs are identical, C must be $49.99, that is,

The total daily cost C	is the same as	the flat rate.
$39.99 + 0.20m$	$=$	$49.99

Solving for m will tell us how many miles we must go for the mileage rate and the flat rate to be the same.

$$39.99 + 0.20m = 49.99 \qquad \text{Given.}$$

$$39.99 - 39.99 + 0.20m = 49.99 - 39.99 \qquad \text{Subtract 39.99.}$$

$$0.20m = 10$$

$$\frac{0.20m}{0.20} = \frac{10}{0.20} \qquad \text{Divide by 0.20.}$$

$$m = 50$$

Thus, if you plan to travel fewer than 50 miles, the mileage rate is better. If you travel 50 miles, the cost is the same for both, and if you travel more than 50 miles the flat rate is better. In this section we shall learn how to solve equations such as $39.99 + 0.20m = 49.99$, a type of equation called a *linear equation in one variable.*

A ⟩ Solving Linear Equations in One Variable

Let's concentrate on the idea we used to solve $39.99 + 0.20m = 49.99$. Our main objective is for our solution to be in the form of $m = \Box$, where $\Box$ is a number. Because of this, we first subtracted 39.99 and then divided by 0.20. This technique works for *linear equations.* Here is the definition.

LINEAR EQUATION	A **linear equation** is an equation that can be written in the form $$ax + b = c$$ where $a \neq 0$ and a, b, and c are real numbers.

How do we solve linear equations? The same way we solved for m in the *Getting Started.* We use the properties of equality that we studied in Sections 2.1 and 2.2. Remember the steps?

1. The equation is already simplified. $ax + b = c$

2. Subtract b. $ax + b - b = c - b$

$ax = c - b$

3. Divide by a. $\dfrac{ax}{a} = \dfrac{c - b}{a}$

$x = \dfrac{c - b}{a}$

Linear equations may have

a. *One* solution

b. *No* solution—no value of the variable will make the equation true. ($x + 1 = x + 2$ has no solution.)

c. *Infinitely many* solutions—any value of the variable will make the equation true. [$2x + 2 = 2(x + 1)$ has infinitely many solutions.]

EXAMPLE 1   Solving linear equations

Solve:

a. $3x + 7 = 13$ **b.** $-5x - 3 = 1$

SOLUTION

a. **1.** The equation is already simplified. $3x + 7 = 13$

  2. Subtract 7. $3x + 7 - 7 = 13 - 7$

                                          $3x = 6$

  3. Divide by 3 (or multiply by $\dfrac{3x}{3} = \dfrac{6}{3}$
     the reciprocal of 3).

                                          $x = 2$

The solution is 2.

CHECK $\dfrac{3x + 7 \overset{?}{=} 13}{}$
$$3(2) + 7 \;\big|\; 13$$
$$6 + 7$$
$$13$$

b. **1.** The equation is already simplified. $-5x - 3 = 1$

  2. Add 3. $-5x - 3 + 3 = 1 + 3$

                                          $-5x = 4$

  3. Divide by -5 (or multiply by the $\dfrac{-5x}{-5} = \dfrac{4}{-5}$
     reciprocal of -5).

                                          $x = -\dfrac{4}{5}$

The solution is $-\dfrac{4}{5}$.

CHECK $\dfrac{-5x - 3 \overset{?}{=} 1}{}$
$$-5\left(-\dfrac{4}{5}\right) - 3 \;\big|\; 1$$
$$4 - 3$$
$$1$$

PROBLEM 1

Solve:

a. $4x + 5 = 17$

b. $-3x - 5 = 2$

Note that the main idea is to place the variables on one side of the equation so you can write the solution in the form $x = \square$ (or $\square = x$), where $\square$ is a number (a constant). Can we solve the equation $5(x + 2) = 3(x + 1) + 9$? This equation is *not* of the form $ax + b = c$, but we can write it in this form if we use the distributive property to remove parentheses, as shown in Example 2.

EXAMPLE 2   Using the distributive property to solve a linear equation

Solve: $5(x + 2) = 3(x + 1) + 9$

SOLUTION

1. There are no fractions to clear. $5(x + 2) = 3(x + 1) + 9$

2. Use the distributive property. $5x + 10 = 3x + 3 + 9$

                                     $5x + 10 = 3x + 12$ Add 3 and 9.

3. Subtract 10 on both sides. $5x + 10 - 10 = 3x + 12 - 10$

                                     $5x = 3x + 2$

4. Subtract $3x$ on both sides. $5x - 3x = 3x - 3x + 2$

                                     $2x = 2$

PROBLEM 2

Solve: $7(x + 1) = 4(x + 2) + 5$

(continued)

Answers to PROBLEMS

1. a. 3   **b.** $-\dfrac{7}{3}$ **2.** 2

5. Divide both sides by 2 (or multiply by the reciprocal of 2).

$$\frac{2x}{2} = \frac{2}{2}$$

$$x = 1$$

The solution is 1.

6. CHECK

$$\frac{5(x + 2) \stackrel{?}{=} 3(x + 1) + 9}{\begin{array}{c|c} 5(1 + 2) & 3(1 + 1) + 9 \\ 5(3) & 3(2) + 9 \\ 15 & 6 + 9 \\ & 15 \end{array}}$$

What if we have fractions in the equation? We can clear the fractions by multiplying by the *least common multiple* (LCM), as we did in Section 2.2. For example, to solve

$$\frac{3}{4} + \frac{x}{10} = 1$$

we first multiply each term by 20, the LCM of 4 and 10. We can obtain the LCM by noting that 20 is the first multiple of 10 that is divisible by 4, or by writing

$$\begin{array}{c} 2 \underline{|4 \qquad 10} \\ \underline{}2 \longrightarrow 5 \rightarrow 2 \cdot 2 \cdot 5 = 20 \quad \text{is the LCM} \end{array}$$

Now, solve $\frac{3}{4} + \frac{x}{10} = 1$ as follows:

1. Clear the fractions by multiplying by 20, the LCM of 4 and 10.

$$\overset{5}{\cancel{20}} \cdot \frac{3}{\cancel{4}} + \overset{2}{\cancel{20}} \cdot \frac{x}{\cancel{10}} = 20 \cdot 1$$

2. Simplify.

$$15 + 2x = 20$$

3. Subtract 15.

$$15 - 15 + 2x = 20 - 15$$

4. The right side has numbers only.

$$2x = 5$$

5. Divide by 2 (or multiply by the reciprocal of 2).

$$\frac{2x}{2} = \frac{5}{2}$$

The solution is $\frac{5}{2}$.

$$x = \frac{5}{2}$$

6. CHECK

$$\frac{\dfrac{3}{4} + \dfrac{x}{10} \stackrel{?}{=} 1}{\begin{array}{c|c} \dfrac{3}{4} + \dfrac{\frac{5}{2}}{10} & 1 \\[2ex] \dfrac{3}{4} + \dfrac{5}{2} \cdot \dfrac{1}{10} & \\[2ex] \dfrac{3}{4} + \dfrac{5}{20} & \\[2ex] \dfrac{3}{4} + \dfrac{1}{4} & \\[2ex] 1 & \end{array}}$$

Here is a shortcut for finding the LCM of two or more numbers; say 4 and 10.

1. Check if one is a multiple of another. (If it is, the larger number is the LCM. For example, the LCM of 7 and 14 is 14 and the LCM of 12 and 24 is 24.)

2. Select the largest number (10).

3. Double it (20), triple it (30), and so on until the other number (4) exactly divides into the doubled or tripled quantity. Since 4 exactly divides into 20, the LCM is 20.

Note: If you were looking for the LCM of 15 and 25, select the 25. Double it (50) but 15 does not exactly divide into 50. Triple the 25 (75). Now 15 exactly divides into 75, so the LCM of 15 and 25 is 75.

The procedure we have used to solve the preceding examples can be used to solve any linear equation. As before, what we need to do is isolate the variables on one side of the equation and the numbers on the other so that we can write the solution in the form $x = \square$ or $\square = x$. Here's how we accomplish this, with a step-by-step example shown on the right. (If you have forgotten how to find the LCM for three numbers, see Section R.2.)

PROCEDURE

Solving Linear Equations

$$\frac{x}{4} - \frac{1}{6} = \frac{7}{12}(x - 2) \quad \text{Given}$$

This is one term.

1. Clear any fractions by multiplying each term on both sides of the equation by the LCM of the denominators, 12.

1. $12 \cdot \dfrac{x}{4} - 12 \cdot \dfrac{1}{6} = 12 \cdot \overbrace{\left[\dfrac{7}{12}(x - 2)\right]}$

2. Remove parentheses and collect like terms (simplify) if necessary.

2. $\quad 3x - 2 = 7(x - 2)$
$\quad 3x - 2 = 7x - 14$

3. Add or subtract the same number on both sides of the equation so that the numbers are isolated on one side.

3. $\quad 3x - 2 + 2 = 7x - 14 + 2$
$\quad 3x = 7x - 12$

4. Add or subtract the same term or expression on both sides of the equation so that the variables are isolated on the other side.

4. $\quad 3x - 7x = 7x - 7x - 12$
$\quad -4x = -12$

5. If the coefficient of the variable is not 1, divide both sides of the equation by the coefficient (or, equivalently, multiply by the reciprocal of the coefficient of the variable).

5. $\quad \dfrac{-4x}{-4} = \dfrac{-12}{-4}$
$\quad x = 3 \qquad \text{The solution}$

6. Be sure to check your answer in the original equation.

6. CHECK

EXAMPLE 3   Solving linear equations

Solve:

a. $\dfrac{7}{24} = \dfrac{x}{8} + \dfrac{1}{6}$

b. $\dfrac{1}{5} - \dfrac{x}{4} = \dfrac{7(x + 3)}{10}$

SOLUTION   We use the six-step procedure.

a.

$$\frac{7}{24} = \frac{x}{8} + \frac{1}{6} \quad \text{Given}$$

1. Clear the fractions; the LCM is 24.

$$\cancel{24} \cdot \frac{7}{\cancel{24}} = \overset{3}{\cancel{24}} \cdot \frac{x}{\cancel{8}} + \overset{4}{\cancel{24}} \cdot \frac{1}{\cancel{6}}$$

2. Simplify.

$$7 = 3x + 4$$

3. Subtract 4.

$$7 - 4 = 3x + 4 - 4$$

4. The left side has numbers only.

$$3 = 3x$$

PROBLEM 3

Solve:

a. $\dfrac{20}{21} = \dfrac{x}{7} + \dfrac{x}{3}$

b. $\dfrac{1}{4} - \dfrac{x}{5} = \dfrac{17(x + 4)}{20}$

(continued)

5. Divide by 3 (or multiply by the reciprocal of 3).

$$\frac{3}{3} = \frac{3x}{3}$$

$$1 = x$$

$$x = 1 \quad \text{The solution}$$

6. CHECK

$$\frac{7}{24} \overset{?}{=} \frac{x}{8} + \frac{1}{6}$$

$$\begin{array}{c|c} \dfrac{7}{24} & \dfrac{1}{8} + \dfrac{1}{6} \\[2ex] & \dfrac{3}{24} + \dfrac{4}{24} \\[2ex] & \dfrac{7}{24} \end{array}$$

b.

$$\frac{1}{5} - \frac{x}{4} = \frac{7(x+3)}{10} \quad \text{Given}$$

1. Clear the fractions; the LCM is 20. Remember, you can get the LCM by selecting 10 (the largest of 5, 4, and 10) and doubling it, which gives 20. Since 5 and 4 exactly divide into 20, 20 is the LCM of 5, 4, and 10.

$$\overset{4}{\cancel{20}} \cdot \frac{1}{5} - \overset{5}{\cancel{20}} \cdot \frac{x}{4} = \overset{2}{\cancel{20}} \cdot \frac{7(x+3)}{\cancel{10}}$$

2. Simplify and use the distributive law.

$$4 - 5x = 14(x + 3)$$
$$4 - 5x = 14x + 42$$

3. Subtract 4.

$$4 - 4 - 5x = 14x + 42 - 4$$
$$-5x = 14x + 38$$

4. Subtract 14x.

$$-5x - 14x = 14x - 14x + 38$$
$$-19x = 38$$

5. Divide by −19 (or multiply by the reciprocal of −19).

$$\frac{-19x}{-19} = \frac{38}{-19}$$

$$x = -2 \quad \text{The solution}$$

6. CHECK

$$\frac{1}{5} - \frac{x}{4} \overset{?}{=} \frac{7(x+3)}{10}$$

$$\begin{array}{c|c} \dfrac{1}{5} - \dfrac{-2}{4} & \dfrac{7(-2+3)}{10} \\[2ex] \dfrac{1}{5} + \dfrac{1}{2} & \dfrac{7(1)}{10} \\[2ex] \dfrac{2}{10} + \dfrac{5}{10} & \dfrac{7}{10} \\[2ex] \dfrac{7}{10} & \end{array}$$

B 〉 Solving Literal Equations

The procedures for solving linear equations that we've just described can also be used to solve some *literal equations*. A **literal equation** is an equation that contains known quantities expressed by means of letters.

In business, science, and engineering, literal equations are usually given as formulas such as the area A of a circle of radius r ($A = \pi r^2$), the simple interest earned on a principal P at a given rate r for a given time t ($I = Prt$), and so on. Unfortunately, these formulas are not always in the form we need to solve the problem at hand. However, as it turns out, we can use the same methods we've just learned to solve for a particular variable in such a formula. For example, let's solve for P in the formula $I = Prt$. To keep track of the variable P, we first circle it in color:

$$I = \textcircled{P}rt$$

$$\frac{I}{rt} = \frac{\textcircled{P}rt}{rt} \qquad \text{Divide by } rt.$$

$$\frac{I}{rt} = \textcircled{P} \qquad \text{Simplify.}$$

$$\textcircled{P} = \frac{I}{rt} \qquad \text{Rewrite.}$$

Now let's look at another example.

EXAMPLE 4   Solving a literal equation

A trapezoid is a four-sided figure in which only two of the sides are parallel. The area of the trapezoid shown here is

$$A = \frac{h}{2}(b_1 + b_2)$$

where h is the altitude and b_1 and b_2 are the bases. Solve for b_2.

SOLUTION   We will circle b_2 to remember that we are solving for it. Now we use our six-step procedure.

$$A = \frac{h}{2}(b_1 + \textcircled{b_2}) \qquad \text{Given}$$

1. Clear the fraction; the LCM is 2.

$$2 \cdot A = \cancel{2} \cdot \frac{h}{2}(b_1 + \textcircled{b_2})$$
$$2A = h(b_1 + \textcircled{b_2})$$

2. Remove parentheses.

$$2A = hb_1 + h\textcircled{b_2}$$

3. There are no numbers to isolate.

4. Subtract the same term, hb_1, on both sides.

$$2A - hb_1 = hb_1 - hb_1 + h\textcircled{b_2}$$
$$2A - hb_1 = h\textcircled{b_2}$$

5. Divide both sides by the coefficient of b_2, h.

$$\frac{2A - hb_1}{h} = \frac{h\textcircled{b_2}}{h}$$

$$\frac{2A - hb_1}{h} = \textcircled{b_2}$$

$$b_2 = \frac{2A - hb_1}{h} \qquad \text{The solution}$$

6. No check is necessary.

PROBLEM 4

The speed S of an ant in centimeters per second is $S = \frac{1}{6}(C - 4)$, where C is the temperature in degrees Celsius. Solve for C.

EXAMPLE 5   Using a pattern to solve a literal equation

The cost of renting a car is $40 per day, plus 20¢ per mile m *after* the first 150 miles. (The first 150 miles are free.)

a. What is the formula for the total daily cost C based on the miles m traveled?

b. Solve for m in the equation.

PROBLEM 5

Some people claim that the relationship between shoe size S and foot length L is $S = 3L - 24$, where L is the length of the foot in inches.

a. If a person wears size 12 shoes, what is the length L of their foot?

b. Solve for L.

(continued)

Answers to PROBLEMS

4. $C = 6S + 4$

5. a. 12 in.   **b.** $L = \dfrac{S + 24}{3}$

SOLUTION

a. Let's try to find a pattern to the general formula.

Miles Traveled	Daily Cost	Per-Mile Cost	Total Cost
50	$40	0	$40
100	$40	0	$40
150	$40	0	$40
200	$40	$0.20(200 - 150) = 10$	$40 + $10 = $50
300	$40	$0.20(300 - 150) = 30$	$40 + $30 = $70

Yes, there is a pattern; can you see that the daily cost when traveling more than 150 miles is $C = 40 + 0.20(m - 150)$?

b. Again, we circle the variable we want to isolate.

$$C = 40 + 0.20(\textcircled{m} - 150) \qquad \text{Given}$$

$$C = 40 + 0.20\,\textcircled{m} - 30 \qquad \text{Use the distributive property.}$$

$$C = 10 + 0.20\,\textcircled{m} \qquad \text{Simplify.}$$

$$C - 10 = 10 - 10 + 0.20\,\textcircled{m} \qquad \text{Subtract 10.}$$

$$C - 10 = 0.20\,\textcircled{m}$$

$$\frac{C - 10}{0.20} = \frac{0.20\,\textcircled{m}}{0.20} \qquad \text{Divide by 0.20.}$$

$$\frac{C - 10}{0.20} = \textcircled{m}$$

As you know, you can use your calculator to graph certain equations. Let's look at Example 5 again. Suppose you want to find the number m of miles you can travel for different costs (say, when $C = \$50, \$100, \$150$, and so on). You are given $C = 40 + 0.20(m - 150)$ and you solve for m, as before, to obtain

$$\frac{C - 10}{0.20} = m$$

Using x instead of C for the cost and y instead of m for the miles (it's easier to work with x's and y's on the calculator), you graph

$$y = \frac{x - 10}{0.20}$$

The graph is shown in Figure 2.1. Note that the cost x starting at $50 (shown on the horizontal axis) occurs in $50 increments, while the number of miles traveled (vertical scale) is in 100-mile increments. You can also use the trace feature offered by most calculators to automatically find the number of miles y you can travel when the cost x is $50. Do you see the answer in Figure 2.2? It says you can travel 200 miles when you pay $50.

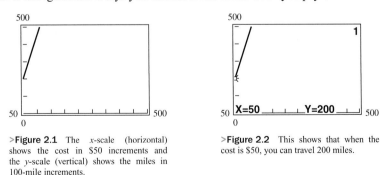

>**Figure 2.1** The x-scale (horizontal) shows the cost in $50 increments and the y-scale (vertical) shows the miles in 100-mile increments.

>**Figure 2.2** This shows that when the cost is $50, you can travel 200 miles.

But what does this have to do with solving literal equations? The point is that if you want to see the graph of $2x + 3y = 6$ using a calculator, you have to **know how to solve for y.** Let's do that next.

EXAMPLE 6   Solving for a specified variable

Solve for y: $2x + 3y = 6$

SOLUTION   Remember, we want to isolate the y.

$$2x + 3y = 6 \qquad \text{Given}$$

$$2x - 2x + 3y = 6 - 2x \qquad \text{Subtract 2x.}$$

$$\frac{3y}{3} = \frac{6 - 2x}{3} \qquad \text{Divide by 3.}$$

$$y = \frac{6 - 2x}{3} \qquad \text{The solution}$$

PROBLEM 6

Solve for x: $3x + 4y = 7$

Here is a graph that may be of more interest to you: the annual cost of tuition and fees at a private four year college or university.

As you can see, the cost C was more than \$20,000 in 2005. What would the tuition be in 2006–2007? What about in 2010? In how many years will annual tuition and fees reach \$25,000? We will answer all these questions next.

EXAMPLE 7   Costs at private 4-year colleges and universities

The tuition costs C at private colleges and universities is as shown in the graph and can be approximated by $C = 992x + 16{,}145$, where x is the number of years after 2000.

a. What was the estimated cost C in the year 2000? ($x = 0$)

b. What will be the estimated cost C in 2010?

c. When will the cost C be \$25,000? (Answer to the nearest year.)

SOLUTION

a. In the year 2000, $x = 0$ (0 years after 2000) and

$$C = 992(0) + 16{,}145 = \$16{,}145$$

This means that according to the formula, the tuition and costs was \$16,145 in the year 2000 (compare with the graph).

b. In the year 2010, $x = 10$ (2010 − 2000, or 10 years after 2000)

$$C = 992(10) + 16{,}145 = 9920 + 16{,}145$$

$$= \$26{,}065$$

c. We want to find when will the cost $C = 992x + 16{,}145$ will be \$25,000.

This will happen when $992x + 16{,}145 = 25{,}000$

We now solve for x $992x + 16{,}145 = 25{,}000$

Subtract $16{,}145$ $992x + 16{,}145 - 16{,}145 = 25{,}000 - 16{,}145$

Simplify $992x = 8855$

PROBLEM 7

Use the formula
$C = 992x + 16{,}145$ to find

a. The estimated cost in 2001.

b. The estimated cost in 2005.

c. When cost will be \$20,000.

(continued)

Divide by 992 $\dfrac{992x}{992}$ $= \dfrac{8855}{992}$

Simplify x $\approx \quad 9$

This means that in about 9 years after 2000, or in 2009, tuition and costs at 4-year private colleges and universities will be about \$25,000. You can check this by letting $x = 9$ in the formula $C = 992x + 16{,}145$ obtaining $C = 992(9) + 16{,}145$ or \$25,073, which is a little more than \$25,000.

› Exercises **2.3**

〈 **A** 〉 **Solving Linear Equations in One Variable**   In Problems 1–50, solve the equation.

1. $3x - 12 = 0$

2. $5a + 10 = 0$

3. $2y + 6 = 8$

4. $4b - 5 = 3$

5. $-3z - 4 = -10$

6. $-4r - 2 = 6$

7. $-5y + 1 = -13$

8. $-3x + 1 = -9$

9. $3x + 4 = x + 10$

10. $4x + 4 = x + 7$

11. $5x - 12 = 6x - 8$

12. $5x + 7 = 7x + 19$

13. $4v - 7 = 6v + 9$

14. $8t + 4 = 15t - 10$

15. $6m - 3m + 12 = 0$

16. $10k + 15 - 5k = 25$

17. $10 - 3z = 8 - 6z$

18. $8 - 4y = 10 + 6y$

19. $5(x + 2) = 3(x + 3) + 1$

20. $y - (4 - 2y) = 7(y - 1)$

odd

21. $5(4 - 3a) = 7(3 - 4a)$

22. $\dfrac{3}{4}y - 4.5 = \dfrac{1}{4}y + 1.3$

23. $-\dfrac{7}{8}c + 5.6 = -\dfrac{5}{8}c - 3.3$

24. $x + \dfrac{2}{3}x = 10$

25. $-2x + \dfrac{1}{4} = 2x + \dfrac{4}{5}$

26. $6x + \dfrac{1}{7} = 2x - \dfrac{2}{7}$

27. $\dfrac{x - 1}{2} + \dfrac{x - 2}{2} = 3$

28. $\dfrac{3x + 5}{3} + \dfrac{x + 3}{3} = 12$

29. $\dfrac{x}{5} - \dfrac{x}{4} = 1$

30. $\dfrac{x}{3} - \dfrac{x}{2} = 1$

31. $\dfrac{x + 1}{4} - \dfrac{2x - 2}{3} = 3$

32. $\dfrac{z + 4}{3} = \dfrac{z + 6}{4}$

33. $\dfrac{2h - 1}{3} = \dfrac{h - 4}{12}$

34. $\dfrac{5 - 6y}{7} - \dfrac{-7 - 4y}{3} = 2$

35. $\dfrac{2w + 3}{2} - \dfrac{3w + 1}{4} = 1$

36. $\dfrac{7r + 2}{6} + \dfrac{1}{2} = \dfrac{r}{4}$

37. $\dfrac{8x - 23}{6} + \dfrac{1}{3} = \dfrac{5}{2}x$

38. $\dfrac{x + 1}{2} + \dfrac{x + 2}{3} + \dfrac{x + 4}{4} = -8$

39. $\dfrac{x - 5}{2} - \dfrac{x - 4}{3} = \dfrac{x - 3}{2} - (x - 2)$

40. $\dfrac{x + 1}{2} + \dfrac{x + 2}{3} + \dfrac{x + 3}{4} = 16$

41. $-4x + \dfrac{1}{2} = 4\left(\dfrac{1}{8} - x\right)$

42. $-6x + \dfrac{2}{3} = 4\left(\dfrac{1}{5} - x\right)$

43. $\dfrac{1}{2}(8x + 4) - 5 = \dfrac{1}{4}(4x + 8) + 1$

44. $\dfrac{1}{3}(3x + 9) + 2 = \dfrac{1}{9}(9x + 18) + 3$

45. $x + \dfrac{x}{2} - \dfrac{3x}{5} = 9$

46. $\dfrac{5x}{3} - \dfrac{3x}{4} + \dfrac{11}{6} = 0$

47. $\dfrac{4x}{9} - \dfrac{3}{2} = \dfrac{5x}{6} - \dfrac{3x}{2}$

48. $\dfrac{7x}{2} - \dfrac{4x}{3} + \dfrac{2x}{5} = -\dfrac{11}{6}$

49. $\dfrac{3x + 4}{2} - \dfrac{1}{8}(19x - 3) = 1 - \dfrac{7x + 18}{12}$

50. $\dfrac{11x - 2}{3} - \dfrac{1}{2}(3x - 1) = \dfrac{17x + 7}{6} - \dfrac{2}{9}(7x - 2)$

〈 **B** 〉 **Solving Literal Equations**   In Problems 51–60, solve each equation for the indicated variable.

51. $C = 2\pi r$; solve for r.

52. $A = bh$; solve for h.

53. $3x + 2y = 6$; solve for y.

54. $5x - 6y = 30$; solve for y.

55. $A = \pi(r^2 + rs)$; solve for s.

56. $T = 2\pi(r^2 + rh)$; solve for h.

57. $\dfrac{V_2}{V_1} = \dfrac{P_1}{P_2}$; solve for V_2.

58. $\dfrac{a}{b} = \dfrac{c}{d}$; solve for a.

59. $S = \dfrac{f}{H - h}$; solve for H.

60. $I = \dfrac{E}{R + nr}$; solve for R.

〉〉〉 *Applications*

61. *Shoe size and length of foot* The relationship between a person's shoe size S and the length of the person's foot L (in inches) is given by (a) $S = 3L - 22$ for a man and (b) $S = 3L - 21$ for a woman. Solve for L for parts (a) and (b).

62. *Man's weight and height* The relationship between a man's weight W (in pounds) and his height H (in inches) is given by $W = 5H - 190$. Solve for H.

63. *Woman's weight and height* The relationship between a woman's weight W (in pounds) and her height H (in inches) is given by $W = 5H - 200$. Solve for H.

64. *Sleep hours and child's age* The number H of hours a growing child A years old should sleep is $H = 17 - \dfrac{A}{2}$. Solve for A.

Recall from Exercise 61 that the relationship between shoe size S and the length of a person's foot L (in inches) is given by the equations

$$S = 3L - 22 \quad \text{for men}$$

$$S = 3L - 21 \quad \text{for women}$$

65. *Shoe size* If Tyrone wears size 11 shoes, what is the length L of his foot?

66. *Shoe size* If Maria wears size 7 shoes, what is the length L of her foot?

67. *Shoe size* Sam's size 7 tennis shoes fit Sue perfectly! What size women's tennis shoe does Sue wear?

68. *Package delivery charges* The cost of first-class mail is 41 cents for the first ounce and 17 cents for each additional ounce. A delivery company will charge \$5.51 for delivering a package weighing up to 2 lb (32 oz). When would the U.S. Postal Service price, $P = 0.41 + 0.17(x - 1)$, where x is the weight of the package in ounces, be the same as the delivery company's price?

69. *Cost of parking* The parking cost at a garage is modeled by the equation $C = 1 + 0.75(h - 1)$, where h is the number of hours you park and C is the cost in dollars. When is the cost C equal to \$11.50?

70. *Baseball run production* Do you follow major-league baseball? Did you know that the average number of runs per game was highest in 1996? The average number of runs scored per game for the National League can be approximated by the equation $N = 0.165x + 4.68$, where x is the number of years after 1996. For the American League, the approximation is $A = -0.185x + 5.38$. When will $N = A$? That is, when will the National League run production be the same as that of the American League?

Who tends to waste the most time at work, older or younger people? Look at the table and judge for yourself.

Year of Birth	Time Wasted per Day
1950–1959	0.68 hr
1960–1969	1.19 hr
1970–1979	1.61 hr
1980–1985	1.95 hr

Source: www.salary.com http://www.salary.com/.

The number of hours H wasted each day by a person born in the 50s (5), 60s (6), 70s (7), or 80s (8) can be approximated by

$$H = 0.423x - 1.392, \text{ where } x \text{ is the decade in which the person was born.}$$

71. *Wasted work time* According to the formula, **a.** how many hours were wasted per day by a person born in the 1950–1959 decade? **b.** What about according to the table?

72. *Wasted work time* According to the formula, **a.** how many hours were wasted per day by a person born in the 1960–1969 decade? **b.** What about according to the table?

73. *Wasted work time*   According to the formula, **a.** how many hours were wasted per day by a person born in the 1970–1979 decade? **b.** What about according to the table?

74. *Wasted work time*   Which of the "decades" in the table does not really cover 10 years?

75. *Wasted work time*   According to the formula, **a.** how many hours were wasted per day by a person born in the 1980–1985 decade? **b.** What about according to the table? **c.** Based on the results of Problems 71–75, who tends to waste the most time at work?

〉〉〉 *Using Your Knowledge*

Car Deals!   In the *Getting Started* at the beginning of this section, we discussed a method that would tell us which was the better deal: using the mileage rate or the flat rate to rent a car. In Problems 76–78, we ask similar questions based on the rates given.

76. Suppose you wish to rent a subcompact that costs $30.00 per day, plus $0.15 per mile. Write a formula for the cost *C* based on traveling *m* miles.

77. How many miles do you have to travel so that the mileage rate, cost *C* in Problem 76, is the same as the flat rate, which is $40 per day?

78. If you were planning to travel 300 miles during the week, would you use the mileage rate or the flat rate given in Problems 76 and 77?

79. In a free-market economy, merchants sell goods to make a profit. These profits are obtained by selling goods at a *markup* in price. This markup *M* can be based on the cost *C* or on the selling price *S*. The formula relating the selling price *S*, the cost *C*, and the markup *M* is

$$S = C + M$$

A merchant plans to have a 20% markup on cost.

a. What would the selling price *S* be?

b. Use the formula obtained in part **a** to find the selling price of an article that cost the merchant $8.

80. **a.** If the markup on an article is 25% of the selling price, use the formula $S = C + M$ to find the cost *C* of the article.

b. Use the formula obtained in part **a** to find the cost of an article that sells for $80.

〉〉〉 *Write On*

81. The definition of a linear equation in one variable states that *a* cannot be zero. What happens if $a = 0$?

82. In this section we asked you to solve a formula for a specified variable. Write a paragraph explaining what that means.

83. The simplification of a linear equation led to the following step:

$$3x = 2x$$

If you divide both sides by *x*, you get $3 = 2$, which indicates that the equation has no solution. What's wrong with this reasoning?

〉〉〉 *Concept Checker*

Fill in the blank(s) with the correct word(s), phrase, or mathematical statement.

84. A **linear equation** is an equation which can be written in the form _____.

85. An **equation** that contains **known quantities expressed by means of letters** is called a _____ equation.

unknown   ***x = c***

literal   ***ax + b = c***

〉〉〉 *Mastery Test*

86. Solve for b_1 in the equation $A = \frac{h}{2}(b_1 + b_2)$.

87. Solve for *m* in the equation $50 = 40 + 0.20(m - 100)$.

Solve.

88. $\frac{7}{12} = \frac{x}{4} + \frac{x}{3}$

89. $\frac{1}{3} - \frac{x}{5} = \frac{8(x + 2)}{15}$

90. $10(x + 2) = 6(x + 1) + 18$

91. $-5(x + 2) = -3(x + 1) - 9$

92. $-4x - 5 = 2$

93. $3x + 8 = 11$

⟩ ⟩ ⟩ *Skill Checker*

Translate each statement into a mathematical expression.

94. The sum of n and $2n$

95. The quotient of $(a + b)$ and c

96. The quotient of a and $(b - c)$

97. The product of a and the sum of b and c

98. The difference of a and b, times c

99. The difference of a and the product of b and c

2.4   Problem Solving: Integer, General, and Geometry Problems

▶ Objectives

Use the RSTUV method to solve:

A ⟩ Integer problems

B ⟩ General word problems

C ⟩ Geometry word problems

▶ To Succeed, Review How To . . .

1. Translate sentences into equations (pp. 36–38).
2. Solve linear equations (pp. 131–136).

▶ Getting Started
Twin Problem Solving

Now that you've learned how to solve equations, you are ready to apply this knowledge to solve real-world problems. These problems are usually stated in words and consequently are called **word** or **story problems.** This is an area in which many students encounter difficulties, but don't panic; we are about to give you a surefire method of tackling word problems.

To start, let's look at a problem that may be familiar to you. Look at the photo of the twins, Mary and Margaret. At birth, Margaret was 3 ounces heavier than Mary, and together they weighed 35 ounces. Can you find their weights? There you have it, a word problem! In this section we shall study an effective way to solve these problems.

Here's how we solve word problems. It's as easy as 1-2-3-4-5.

> **PROCEDURE**
>
> **RSTUV Method for Solving Word Problems**
>
> **1.** **R**ead the problem carefully and decide what is asked for (the unknown).
> **2.** **S**elect a variable to represent this unknown.
> **3.** **T**hink of a plan to help you write an equation.
> **4.** **U**se algebra to solve the resulting equation.
> **5.** **V**erify the answer.

If you really want to learn how to do word problems, this is the method you have to master. Study it carefully, and then use it. *It works!* How do we remember all of these steps? Easy. Look at the first letter in each sentence. We call this the **RSTUV method.**

> **NOTE**
>
> We will present problem solving in a five-step (RSTUV) format. Follow the instructions at each step until you understand them. Now, cover the Example you are studying (a 3 by 5 index card will do) and work the corresponding margin problem. Check the answer and see if you are correct; if not, study the steps in the Example again.

The mathematics dictionary in Section 1.7 also plays an important role in the solution of word problems. The words contained in the dictionary are often the **key** to translating the problem. But there are other details that may help you. Here's a restatement of the RSTUV method that offers hints and tips.

HINTS AND TIPS   Our problem-solving procedure (RSTUV) contains five steps. The numbered steps are listed, with hints and tips following each.

1. Read the problem. Mathematics is a language. As such, you have to learn how to read it. You may not understand or even get through reading the problem the first time. That's OK. Read it again and as you do, pay attention to key words or instructions such as *compute, draw, write, construct, make, show, identify, state, simplify, solve,* and *graph.* (Can you think of others?)

2. Select the unknown. How can you answer a question if you don't know what the question is? One good way to find the unknown (variable) is to look for the question mark "?" and read the material to its left. Try to determine what is given and what is missing.

3. Think of a plan. (Translate!) Problem solving requires many skills and strategies. Some of them are *look for a pattern; examine a related problem; use a formula; make tables, pictures, or diagrams; write an equation; work backward; and make a guess.* In algebra, the plan should lead to writing an equation or an inequality.

4. Use algebra to solve the resulting equation. If you are studying a mathematical technique, it's almost certain that you will have to use it in solving the given problem. Look for ways the technique you're studying could be used to solve the problem.

5. Verify the answer. Look back and check the results of the original problem. Is the answer reasonable? Can you find it some other way?

Are you ready to solve the problem in the *Getting Started* now? Let's restate it for you:

At birth, Mary and Margaret (the Stimson twins) together weighed 35 ounces. Margaret was 3 ounces heavier than Mary. Can you find their weights?

Basically, this problem is about sums. Let's put our RSTUV method to use.

SUMMING THE WEIGHTS OF MARY AND MARGARET

1. Read the problem. Read the problem slowly—not once, but two or three times. (Reading algebra is *not* like reading a magazine; you may have to read algebra problems several times before you understand them.)

2. Select the unknown. (*Hint:* Let the unknown be the quantity you know nothing about.) Let w (in ounces) represent the weight for Mary (which makes Margaret $w + 3$ ounces, since she was 3 ounces heavier).

3. Think of a plan. Translate the sentence into an equation: Together

$$
\underbrace{\text{Margaret}}_{(w+3)} \quad \underbrace{\text{and}}_{+} \quad \underbrace{\text{Mary}}_{w} \quad \underbrace{\text{weighed}}_{=} \quad \underbrace{\text{35 oz.}}_{35}
$$

4. Use algebra to solve the resulting equation.

$$(w + 3) + w = 35 \qquad \text{Given}$$
$$w + 3 + w = 35 \qquad \text{Remove parentheses.}$$
$$2w + 3 = 35 \qquad \text{Combine like terms.}$$
$$2w + 3 - 3 = 35 - 3 \qquad \text{Subtract 3.}$$
$$2w = 32 \qquad \text{Simplify.}$$
$$\frac{2w}{2} = \frac{32}{2} \qquad \text{Divide by 2.}$$
$$w = 16 \qquad \text{Mary's weight}$$

Thus, Mary weighed 16 ounces, and Margaret weighed $16 + 3 = 19$ ounces.

5. Verify the solution. Do they weigh 35 ounces together? Yes, $16 + 19$ is 35.

A ⟩ Solving Integer Problems

Sometimes a problem uses words that may be new to you. For example, a popular type of word problem in algebra is the **integer** problem. You remember the integers—they are the positive integers 1, 2, 3, and so on; the negative integers $-1, -2, -3$, and so on; and 0. Now, if you are given any integer, can you find the integer that comes right after it? Of course, you simply add 1 to the given integer and you have the answer. For example, the integer that comes after 4 is $4 + 1 = 5$ and the one that comes after -4 is $-4 + 1 = -3$. In general, if n is any integer, the integer that follows n is $n + 1$. We usually say that n and $n + 1$ are **consecutive integers.** We have illustrated this idea here.

Now suppose you are given an *even* integer (an integer divisible by 2) such as 8 and you are asked to find the next even integer (which is 10). This time, you must add 2 to 8 to obtain the answer. What is the next even integer after 24? $24 + 2 = 26$. If n is even, the next even integer after n is $n + 2$. What about the next *odd* integer after 5? First, recall that the odd integers are $\ldots, -3, -1, 1, 3, \ldots$. The next odd integer after 5 is $5 + 2 = 7$. Similarly, the next odd integer after 21 is $21 + 2 = 23$. In general, if n is odd, the next odd integer after n is $n + 2$. Thus, if you need two consecutive even or odd integers and you call the first one n, the next one will be $n + 2$. We shall use all these ideas as well as the RSTUV method in Example 1.

EXAMPLE 1   A consecutive integer problem

The sum of three consecutive even integers is 126. Find the integers.

SOLUTION   We use the RSTUV method.

1. Read the problem. We are asked to find three consecutive *even* integers.

2. Select the unknown. Let n be the first of the integers. Since we want three consecutive even integers, we need to find the next two consecutive even integers. The next even integer after n is $n + 2$, and the one after $n + 2$ is $(n + 2) + 2 = n + 4$. Thus, the three consecutive even integers are n, $n + 2$, and $n + 4$.

3. Think of a plan. Translate the sentence into an equation.

The sum of 3 consecutive even integers   is   126.

$$n + (n + 2) + (n + 4) \qquad = \qquad 126$$

PROBLEM 1

The sum of three consecutive odd integers is 129. Find the integers.

(continued)

Answers to PROBLEMS

1. 41, 43, 45

4. Use algebra to solve the resulting equation.

$$n + (n + 2) + (n + 4) = 126 \qquad \text{Given}$$
$$n + n + 2 + n + 4 = 126 \qquad \text{Remove parentheses.}$$
$$3n + 6 = 126 \qquad \text{Combine like terms.}$$
$$3n + 6 - 6 = 126 - 6 \qquad \text{Subtract 6.}$$
$$3n = 120 \qquad \text{Simplify.}$$
$$\frac{3n}{3} = \frac{120}{3} \qquad \text{Divide by 3.}$$
$$n = 40$$

Thus, the three consecutive even integers are 40, $40 + 2 = 42$, and $40 + 4 = 44$.

5. Verify the solution. Since $40 + 42 + 44 = 126$, our result is correct.

The RSTUV method works for numbers other than consecutive integers. We solve a different type of integer problem in the next example.

EXAMPLE 2   An integer problem

Twelve less than 5 times a number is the same as the number increased by 4. Find the number.

SOLUTION   We use the RSTUV method.

1. Read the problem. We are asked to find a certain number.

2. Select the unknown. Let n represent the number.

3. Think of a plan. Translate the sentence into an equation.

Twelve less than	5 times a number	is the same as	the number increased by 4,
	$5n - 12$	$=$	$n + 4$

4. Use algebra to solve the resulting equation.

$$5n - 12 = n + 4 \qquad \text{Given}$$
$$5n - 12 + 12 = n + 4 + 12 \qquad \text{Add 12.}$$
$$5n = n + 16 \qquad \text{Simplify.}$$
$$5n - n = n - n + 16 \qquad \text{Subtract } n.$$
$$4n = 16 \qquad \text{Simplify.}$$
$$\frac{4n}{4} = \frac{16}{4} \qquad \text{Divide by 4.}$$
$$n = 4$$

Thus, the number is 4.

5. Verify the solution. Is 12 less than 5 times 4 the same as 4 increased by 4? That is, is $5(4) - 12 = 4 + 4$?

$$5(4) - 12 = 4 + 4$$
$$20 - 12 = 8$$

Yes, this sentence is true.

PROBLEM 2

Fourteen less than 3 times a number is the same as the number increased by 2. Find the number.

B › General Word Problems

Many interesting problems can be solved using the RSTUV method. The next example is typical.

Answers to PROBLEMS

2. 8

EXAMPLE 3   Solving a general word problem

Have you eaten at a fast-food restaurant lately? If you eat a cheeseburger and fries, you would consume 1070 calories. As a matter of fact, the fries contain 30 more calories than the cheeseburger. How many calories are there in each food?

SOLUTION   Again, we use the RSTUV method.

1. Read the problem. We are asked to find the number of calories in the cheeseburger and in the fries.

2. Select the unknown. Let c represent the number of calories in the cheeseburger. This makes the number of calories in the fries $c + 30$, that is, 30 more calories. (We could instead let f be the number of calories in the fries; then $f - 30$ would be the number of calories in the cheeseburger.)

3. Think of a plan. Translate the problem.

The calories in fries	and	the calories in cheeseburger	total	1070.
$(c + 30)$	$+$	c	$=$	1070

4. Use algebra to solve the equation.

$$(c + 30) + c = 1070 \qquad \text{Given}$$
$$c + 30 + c = 1070 \qquad \text{Remove parentheses.}$$
$$2c + 30 = 1070 \qquad \text{Combine like terms.}$$
$$2c + 30 - 30 = 1070 - 30 \qquad \text{Subtract 30.}$$
$$2c = 1040 \qquad \text{Simplify.}$$
$$\frac{2c}{2} = \frac{1040}{2} \qquad \text{Divide by 2.}$$
$$c = 520$$

Thus, the cheeseburger has 520 calories. Since the fries have 30 calories more than the cheeseburger, the fries have $520 + 30 = 550$ calories.

5. Verify the solution. Does the total number of calories—that is, $520 + 550$—equal 1070? Since $520 + 550 = 1070$, our results are correct.

PROBLEM 3

A McDonald's® cheeseburger and small fries together contain 540 calories. The cheeseburger has 120 more calories than the fries. How many calories are in each food?

30 more calories than the cheeseburger

1070 calories

C › Geometry Word Problems

The last type of problem we discuss here deals with an important concept in geometry: angle measure. Angles are measured using a unit called a **degree** (symbolized by °).

Types of Angles

Angle of Measure 1°

One complete revolution around a circle is 360° and $\frac{1}{360}$ of a complete revolution is 1°.

Complementary Angles

Two angles whose sum is 90° are called **complementary angles.** Example: a 30° angle and a 60° angle are complementary since $30° + 60° = 90°$.

$90° - x$

x

(continued)

Types of Angles *(continued)*

Supplementary Angles

Two angles whose sum is 180° are called **supplementary angles.** Example: a 50° angle and a 130° angle are supplementary since 50° + 130° = 180°.

As you can see from the table, if x represents the measure of an angle,

$$90 - x \quad \text{is the measure of its } \textbf{complement}$$

and

$$180 - x \quad \text{is the measure of its } \textbf{supplement}$$

Let's use this information to solve a problem.

EXAMPLE 4   Complementary and supplementary angles

Find the measure of an angle whose supplement is 30° less than 3 times its complement.

SOLUTION   As usual, we use the RSTUV method.

1. Read the problem. We are asked to find the measure of an angle. Make sure you understand the ideas of complement and supplement. If you don't understand them, review those concepts.

2. Select the unknown. Let m be the measure of the angle. By definition, we know that

$$90 - m \quad \text{is its complement}$$
$$180 - m \quad \text{is its supplement}$$

3. Think of a plan. We translate the problem statement into an equation:

| The supplement | is | 30° less than | 3 times its complement. |

$$180 - m \quad = \quad 3(90 - m) - 30$$

4. Use algebra to solve the equation.

$180 - m = 3(90 - m) - 30$	Given
$180 - m = 270 - 3m - 30$	Remove parentheses.
$180 - m = 240 - 3m$	Combine like terms.
$180 - 180 - m = 240 - 180 - 3m$	Subtract 180.
$-m = 60 - 3m$	Simplify.
$+3m - m = 60 - 3m + 3m$	Add 3m.
$2m = 60$	Simplify.
$m = 30$	Divide by 2.

Thus, the measure of the angle is 30°, which makes its complement $90° - 30° = 60°$ and its supplement $180° - 30° = 150°$.

5. Verify the solution. Is the supplement (150°) 30° less than 3 times the complement; that is, is $150 = 3 \cdot 60 - 30$? Yes! Our answer is correct.

PROBLEM 4

Find the measure of an angle whose supplement is 45° less than 4 times its complement.

Answers to PROBLEMS

4. 45°

TRANSLATE THIS

1. Since 1980 **the amount *A* of garbage generated has increased by 50% of *A* to 236 million tons** per year in 2003.

2. The number *n* of landfills has **declined by 6157 to 1767.**

3. Since 1990 **the amount of garbage *g* sent to landfills has decreased by 9 million tons to 131 million tons** in 2003.

4. In 2003 the net per capita discard rate (that's how much garbage you discard) was **3.09 pounds (per person per day) down by 0.05 pounds from the *p* pounds** discarded in 2002.

5. The total materials recycled *R* (in millions of tons) increased by 66.7 million tons to 72.3 million tons from 1960 to 2003.

The third step in the RSTUV procedure is to **TRANSLATE** the information into an equation. In Problems 1–10 **TRANSLATE** the sentence and match the correct translation with one of the equations A–O.

A. $9 - g = 131$
B. $R + 66.7 = 72.3$
C. $33 = 0.14g$
D. $A + 0.50A = 236$
E. $3.09 = p - 0.05$
F. $M + 49.8 = 55.4$
G. $131 = 0.55g$
H. $n - 6157 = 1767$
I. $g - 9 = 131$
J. $35.2\% = t + 23.1\%$
K. $0.93b = 2$
L. $3.09 - p = 0.05$
M. $0.93 = 2b$
N. $35.2\%y = 23.1\%$
O. $6157 - n = 1767$

6. The materials recovered for recycling *M* (in millions of tons) increased by **49.8 tons to 55.4 tons** from 1960 to 2003.

7. In 2003 about **33 million tons (14%) of the total garbage *g*** was burned.

8. In 2003 about **131 million tons (55%) of the total garbage *g*** generated went to landfills.

9. In 2003, the largest category in all garbage generated, 35.2% was paper. This **35.2% exceeded the amount of yard trimmings *t*** (the second largest category) **by 23.1%.**

10. In 2003, about **93% of the *b* billion pounds of lead in recycled batteries yielded 2 billion pounds** of lead.

Source: Municipal Solid Waste Generation, Recycling and Disposal in the United States. http://www.epa.gov/.

You will have an opportunity to solve and finish these problems in Exercises 41–50.

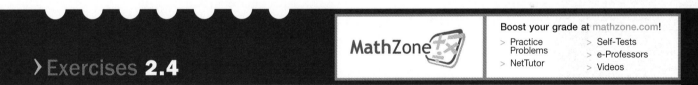

> Exercises **2.4**

MathZone ⁺ˣ

Boost your grade at mathzone.com!
> Practice Problems
> NetTutor
> Self-Tests
> e-Professors
> Videos

〈 **A** 〉 **Solving Integer Problems**   In Problems 1–20, solve the given problem.

1. The sum of three consecutive even integers is 138. Find the integers.

2. The sum of three consecutive odd integers is 135. Find the integers.

3. The sum of three consecutive even integers is −24. Find the integers.

4. The sum of three consecutive odd integers is −27. Find the integers.

5. The sum of two consecutive integers is −25. Find the integers.

6. The sum of two consecutive integers is −9. Find the integers.

7. Find three consecutive integers (n, $n + 1$, and $n + 2$) such that the last added to twice the first is 23.

8. Maria, Latasha, and Kim live in apartments numbered consecutively (n, $n + 1$, $n + 2$). Kim's apartment number added to twice Maria's apartment number gives 47. What are their apartment numbers?

9. Three lockers are numbered consecutively (n, $n + 1$, $n + 2$) in such a way that the sum of the first and last lockers is the same as twice the middle locker number. What are the locker numbers?

10. Pedro spent 27 more dollars for his math book than for his English book. If his total bill was $141, what was the price of each of the books?

11. Tyrone bought two used books. One book was $24 more than the other. If his total purchase was $64, what was the price of each of the books?

12. The total number of points in a basketball game was 179. The winning team scored 5 more points than the losing team. What was the score?

13. Another basketball game turned out to be a rout in which the winning team scored 55 more points than the losing team. If the total number of points in the game was 133, what was the final score?

14. Sandra and Mida went shopping. Their total bill was $210, but Sandra spent only two-fifths as much as Mida. How much did each one spend?

15. The total of credit charges on one card is $4 more than three-eighths the charges on the other. If the total charges are $147, how much was charged on each card?

16. Marcus, Liang, and Mourad went to dinner. Liang's bill was $2 more than Marcus's bill, and Mourad's bill was $2 more than Liang's bill. If the total bill was $261 before tip, find the amount for each individual bill.

17. The sum of three numbers is 254. The second is 3 times the first, and the third is 5 less than the second. Find the numbers.

18. The sum of three consecutive integers is 17 less than 4 times the smallest of the three integers. Find the integers.

19. The larger of two numbers is 6 times the smaller. Their sum is 147. Find the numbers.

20. Five times a certain fraction yields the same as 3 times 1 more than the fraction. Find the fraction.

‹ **B** › **General Word Problems**   In Problems 21–34, solve the following problems.

21. Do you have Internet access? Polls indicate that 66% of all adults do. About 25% more adults access the Internet from home than from work. If 15% go online from other locations (not work or home), what is the percent of adults accessing the Internet from home? What is the percent of adults accessing the Internet from work?

Source: Harris Interactive poll.

22. The Internet market audience (in thousands) in New York, the leading market, exceeds that in Los Angeles by 727. If their combined market audience is 7955, what are the Internet market audiences (in thousands) in New York and in Los Angeles?

23. How much money do you make per week? The salary difference between men over 25 years old and men 16–24 years old is $330. If men over 25 make $722 per week, how much do men between 16 and 24 make per week?

24. For women, the difference in weekly salary is $188 but women over 25 make only $542 per week. (Compare this with the figure for men in Problem 23.) How much do women between 16 and 24 make per week?

Source: U.S. Bureau of Labor Statistics Household Data.

25. The height of the Empire State Building including its antenna is 1472 feet. The building is 1250 feet tall. How tall is the antenna?

26. The cost C for renting a car is given by $C = 0.10m + 10$, where m is the number of miles traveled. If the total cost amounted to $17.50, how many miles were traveled?

27. A toy rocket goes vertically upward with an initial velocity of 96 feet per second. After t seconds, the velocity of the rocket is given by the formula $v = 96 - 32t$, neglecting air resistance. In how many seconds will the rocket reach its highest point? (*Hint:* At the highest point, $v = 0$.)

28. Refer to Problem 27 and find the number of seconds t that must elapse before the velocity decreases to 16 feet per second.

29. In a recent school election, 980 votes were cast. The winner received 372 votes more than the loser. How many votes did each of the two candidates receive?

30. The combined annual cost of the U.S. and Russian space programs has been estimated at $71 billion. The U.S. program is cheaper; it costs $19 billion less than the Russian program. What is the cost of each program?

31. A direct-dialed telephone call from Tampa to New York is $3.05 for the first 3 minutes and $0.70 for each additional minute or fraction thereof. If Juan's call cost him $7.95, how many minutes was the call?

32. A parking garage charges $1.25 for the first hour and $0.75 for each additional hour or fraction thereof.

    a. If Sally paid $7.25 for parking, for how many hours was she charged?

    b. If Sally has only $10 with her, what is the maximum number of hours she can park in the garage? (*Hint:* Check your answer!)

33. The cost of a taxicab is $0.95 plus $1.25 per mile.

    a. If the fare was $28.45, how long was the ride?

    b. If a limo service charges $15 for a 12-mile ride to the airport, which one is the better deal, the cab or the limo?

34. According to the Department of Health and Human Services, the number of heavy alcohol users in the 35-and-older category exceeds the number of heavy alcohol users in the 18–25 category by 844,000. If the total number of persons in these two categories is 7,072,000, how many persons are there in each of these categories?

‹ **C** ›   **Geometry Word Problems**   In Problems 35–40, find the measure of the given angles.

35. An angle whose supplement is 20° more than twice its complement

36. An angle whose supplement is 10° more than three times its complement

37. An angle whose supplement is 3 times the measure of its complement

38. An angle whose supplement is 4 times the measure of its complement

39. An angle whose supplement is 54° less than 4 times its complement

40. An angle whose supplement is 41° less than 3 times its complement

〉〉〉   Applications

Problems 41–50 are based on the translations you already did in the *Translate This* at the beginning of the section. Now, you only have to solve the resulting equations! Use the RSTUV procedure to solve Problems 41–50. Remember, look at the *Translate This* for the equation!

41. *Garbage*   Since 1980 **the amount A of garbage generated has increased by 50% of A to 236 million tons** per year in 2003. Find A to the nearest million ton.

42. *Landfills*   The number n of landfills has **declined** by 6157 to 1767. Find n.

43. *Garbage*   Since 1990 **the amount of garbage g sent to landfills has decreased by 9 million tons to 131 million tons** in 2003. Find g.

44. *Garbage*   In 2003 the net per capita discard rate (that's how much garbage you discard) was **3.09 pounds (per person per day) down by 0.05 pounds from the p pounds** discarded in 2002. Find p.

45. *Recycling*   The total materials recycled R (in millions of tons) **increased by 66.7 million tons to 72.3 million tons** from 1960 to 2003. Find R.

46. *Recycling*   The materials recovered for recycling M **(in millions of tons) increased by 49.8 tons to 55.4 tons** from 1960 to 2003. Find M.

47. *Garbage*   In 2003 about **33 million tons (14%) of the total garbage g** was burned. Find g to the nearest million ton.

48. *Landfills*   In 2003 about **131 million tons (55%) of the total garbage g** generated went to landfills. Find g to the nearest million ton.

49. *Garbage*   In 2003, the largest category in all garbage generated, 35.2%, was paper. This **35.2% exceeded the amount of yard trimmings t** (the second largest category) **by 23.1%.** Find t.

50. *Recycling*   In 2003, about **93% of the b billion pounds of lead in recycled batteries yielded 2 billion pounds** of lead. Find b to two decimal places.

〉〉〉   Using Your Knowledge

Diophantus's Equation   If you are planning to become an algebraist (an expert in algebra), you may not enjoy much fame. As a matter of fact, very little is known about one of the best algebraists of all time, the Greek Diophantus. According to a legend, the following problem is in the inscription on his tomb:

One-sixth of his life God granted him youth. After a twelfth more, he grew a beard. After an additional seventh, he married, and 5 years later, he had a son. Alas, the unfortunate son's life span was only one-half that of his father, who consoled his grief in the remaining 4 years of his life.

51. Use your knowledge to find how many years Diophantus lived. (*Hint:* Let x be the number of years Diophantus lived.)

〉〉〉 *Write On*

52. When reading a word problem, what is the first thing you try to determine?

53. How do you verify your answer in a word problem?

54. The "T" in the RSTUV method means that you must "Think of a plan." Some strategies you can use in this plan include *look for a pattern* and *make a picture*. Can you think of three other strategies?

〉〉〉 *Concept Checker*

Fill in the blank(s) with the correct word(s), phrase, or mathematical statement.

55. The procedure used to solve word problems is called the _____ procedure.

56. n and $n + 1$ are called _____ integers.

57. Two angles whose sum is 90° are called _____ angles.

58. Two angles whose sum is 180° are called _____ angles.

right	**supplementary**
acute	**RSTUV**
complementary	**consecutive**
obtuse	**even**

〉〉〉 *Mastery Test*

59. The sum of three consecutive odd integers is 249. Find the integers.

60. Three less than 4 times a number is the same as the number increased by 9. Find the number.

61. If you eat a single slice of a 16-inch mushroom pizza and a 10-ounce chocolate shake, you have consumed 530 calories. If the shake has 70 more calories than the pizza, how many calories are there in each?

62. Find the measure of an angle whose supplement is 47° less than 3 times its complement.

〉〉〉 *Skill Checker*

Solve.

63. $55T = 100$

64. $88R = 3240$

65. $15T = 120$

66. $81T = 3240$

67. $-75x = -600$

68. $-45x = -900$

69. $-0.02P = -70$

70. $-0.05R = -100$

71 $-0.04x = -40$

72. $-0.03x = 30$

2.5

Problem Solving: Motion, Mixture, and Investment Problems

▶ Objectives

Use the RSTUV method to solve:

A ❭   Motion problems

B ❭   Mixture problems

C ❭   Investment problems

▶ To Succeed, Review How To . . .

1. Translate sentences into equations (pp. 36–38, 91–93).
2. Solve linear equations (pp. 131–136).

▶ Getting Started

Birds in Motion

As we have mentioned previously, some of the strategies we can use to help us think of a plan include *use a formula, make a table,* and *draw a diagram.* We will use these strategies as we solve problems in this section. For example, in the cartoon, the bird is trying an impossible task, unless he turns around! If he does, how does Curls know that it will take him less than 2 hours to fly 100 miles? Because there's a formula to figure this out!

By Permission of John L. Hart FLP, and Creators Syndicate, Inc.

If an object moves at a constant rate R for a time T, the distance D traveled by the object is given by

$$D = RT$$

The object could be your car moving at a constant rate R of 55 miles per hour for $T = 2$ hours. In this case, you would have traveled a distance $D = 55 \times 2 = 110$ miles. Here the rate is in miles per *hour* and the time is in *hours.* (Units have to be consistent!) Similarly, if you jog at a constant rate of 5 miles per hour for 2 hours, you would travel $5 \times 2 = 10$ miles. Here the units are in miles per hour, so the time must be in hours. In working with the formula for distance, and especially in more complicated problems, it is often helpful to write the formula in a chart. For example, to figure out how far you jogged, you would write:

	R	×	**T**	=	**D**
Jogger	5 mi/hr		2 hr		10 mi

As you can see, we used the formula $D = RT$ to solve this problem and organized our information in a chart. We shall continue to use formulas and charts when solving the rest of the problems in this section.

A ⟩ Motion Problems

Let's go back to the bird problem, which is a **motion problem.** If the bird turns around and is flying at 5 miles per hour with a tail wind of 50 miles per hour, its rate R would be $50 + 5 = 55$ miles per hour. The wind is helping the bird, so the wind speed must be added to his rate. Clumsy Carp wants to know how long it would take the bird to fly a distance of 100 miles; that is, he wants to find the time T. We can write this information in a table, substituting 55 for R and 100 for D.

	R	×	T	=	D
Bird	55		T		100

Since

$$R \times T = D$$

we have

$$55T = 100$$

$$T = \frac{100}{55} = \frac{20}{11} = 1\frac{9}{11} \text{ hr} \qquad \text{Divide by 55.}$$

Curls was right; it does take less than 2 hours!
Now let's try some other motion problems.

EXAMPLE 1   Finding the speed

The longest regularly scheduled bus route is Greyhound's "Supercruiser" Miami-to-San Francisco route, a distance of 3240 miles. If this distance is covered in about 82 hours, what is the average speed R of the bus rounded to the nearest tenth?

SOLUTION   We use our RSTUV method.

1. Read the problem. We are asked to find the rate of the bus.

2. Select the unknown. Let R represent this rate.

3. Think of a plan. What type of information do you need? How can you enter it so that $R \times T = D$? Translate the problem and enter the information in a chart.

	R	×	T	=	D
Supercruiser	R		82		3240

The equation is

$$R \times 82 = 3240 \quad \text{or} \quad 82R = 3240$$

4. Use algebra to solve the equation.

$$82R = 3240 \qquad \text{Given}$$

$$\frac{82R}{82} = \frac{3240}{82} \qquad \text{Divide by 82.}$$

$$R \approx 39.5 \qquad 82\overline{)3240.00} \qquad \text{By long division (to the nearest tenth)}$$

```
      39.51
82)3240.00
   246
   ---
   780
   738
   ---
   42 0
   41 0
   ----
   1 00
     82
   ----
     18
```

The bus's average speed is 39.5 miles per hour.

PROBLEM 1

Example 1 is about bus routes in the United States. There is a 6000-mile bus trip from Caracas to Buenos Aires that takes 214 hours, including a 12-hour stop in Santiago and a 24-hour stop in Lima. What is the average speed of the bus rounded to the nearest tenth? *Hint:* Do not count rest time.

Answers to PROBLEMS

1. 33.7 mi/hr

5. Verify the solution. Check the arithmetic by substituting 39.5 into the formula $R \times T = D$.

$$39.5 \times 82 = 3239$$

Although the actual distance is 3240, this difference is acceptable because we rounded our answer to the nearest tenth.

Some motion problems depend on the relationship between the distances traveled by the objects involved. When you think of a plan for these types of problems, a diagram can be very useful.

EXAMPLE 2   Finding the time it takes to overtake an object

The Supercruiser bus leaves Miami traveling at an average rate of 40 miles per hour. Three hours later, a car leaves Miami for San Francisco traveling on the same route at 55 miles per hour. How long does it take for the car to overtake the bus?

SOLUTION   Again, we use the RSTUV method.

1. Read the problem. We are asked to find how many hours it takes the car to overtake the bus.

2. Select the unknown. Let T represent this number of hours.

3. Think of a plan. Translate the given information and enter it in a chart. Note that if the car goes for T hours, the bus goes for $(T + 3)$ hours (since it left 3 hours earlier).

	R	×	T	=	D
Car	55		T		$55T$
Bus	40		$T + 3$		$40(T + 3)$

Since the vehicles are traveling in the same direction, the diagram of this problem looks like this:

When the car overtakes the bus, they will have traveled the *same* distance. According to the chart, the car has traveled $55T$ miles and the bus $40(T + 3)$ miles. Thus,

$$\underbrace{55T}_{\text{Distance traveled by car}} \underbrace{=}_{\text{overtakes}} \underbrace{40(T + 3)}_{\text{distance traveled by Supercruiser}}$$

4. Use algebra to solve the equation.

$$55T = 40(T + 3) \qquad \text{Given}$$
$$55T = 40T + 120 \qquad \text{Simplify.}$$
$$55T - 40T = 40T - 40T + 120 \qquad \text{Subtract } 40T.$$

PROBLEM 2

How long does it take if the car in Example 2 is traveling at 60 miles per hour?

(continued)

$$15T = 120 \qquad \text{Simplify.}$$

$$\frac{15T}{15} = \frac{120}{15} \qquad \text{Divide by 15.}$$

$$T = 8$$

It takes the car 8 hours to overtake the bus.

5. Verify the solution. The car travels for 8 hours at 55 miles per hour; thus it travels $55 \times 8 = 440$ miles, whereas the bus travels at 40 miles per hour for 11 hours, a total of $40 \times 11 = 440$ miles. Since the car traveled the same distance, it overtook the bus in 8 hours.

In Example 2, the two vehicles moved in the *same* direction. A variation of this type of problem involves motion toward each other, as shown in Example 3.

EXAMPLE 3   Two objects moving toward each other

The Supercruiser leaves Miami for San Francisco 3240 miles away, traveling at an average speed of 40 miles per hour. At the same time, a slightly faster bus leaves San Francisco for Miami traveling at 41 miles per hour. How many hours will it take for the buses to meet?

SOLUTION   We use the RSTUV method.

1. Read the problem. We are asked to find how many hours it takes for the buses to meet.

2. Select the unknown. Let T represent the hours each bus travels before they meet.

3. Think of a plan. Translate the information and enter it in a chart.

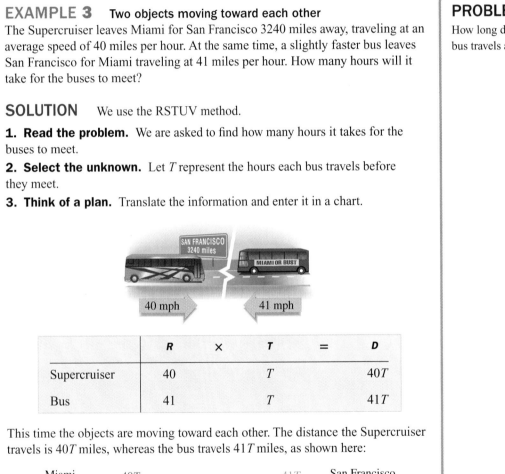

	R	×	T	=	D
Supercruiser	40		T		$40T$
Bus	41		T		$41T$

This time the objects are moving toward each other. The distance the Supercruiser travels is $40T$ miles, whereas the bus travels $41T$ miles, as shown here:

When they meet, the combined distance traveled by *both* buses is 3240 miles. This distance is also $40T + 41T$. Thus, we have

Distance traveled by Supercruiser	and	distance traveled by other bus	is	total distance.
$40T$	$+$	$41T$	$=$	3240

PROBLEM 3

How long does it take if the faster bus travels at 50 miles per hour?

---

4. Use algebra to solve the equation.

$$40T + 41T = 3240 \qquad \text{Given}$$
$$81T = 3240 \qquad \text{Combine like terms.}$$
$$\frac{81T}{81} = \frac{3240}{81} \qquad \text{Divide by 81.}$$
$$T = 40$$

Thus, each bus traveled 40 hours before they met.

5. Verify the solution. The Supercruiser travels $40 \times 40 = 1600$ miles in 40 hours, whereas the other bus travels $40 \times 41 = 1640$ miles. Clearly, the total distance traveled is $1600 + 1640 = 3240$ miles.

B ⟩ Mixture Problems

Another type of problem that can be solved using a chart is the **mixture problem.** In a mixture problem, two or more things are combined to form a mixture. For example, dental-supply houses mix pure gold and platinum to make white gold for dental fillings. Suppose one of these houses wishes to make 10 troy ounces of white gold to sell for $415 per ounce. If pure gold sells for $400 per ounce and platinum sells for $475 per ounce, how much of each should the supplier mix?

We are looking for the amount of each material needed to make 10 ounces of a mixture selling for $415 per ounce. If we let x be the total number of ounces of gold, then $10 - x$ (the balance) must be the number of ounces of platinum. Note that if a quantity T is split into two parts, one part may be x and the other $T - x$. This can be checked by adding $T - x$ and x to obtain $T - x + x = T$. We then enter all the information in a chart. The top line of the chart tells us that if we multiply the price of the item by the ounces used, we will get the total price:

	Price/Ounce	×	Ounces	=	Total Price
Gold	400		x		$400x$
Platinum	475		$10 - x$		$475(10 - x)$
Mixture	415		10		4150

The first line (following the word *gold*) tells us that the price of gold ($400) times the amount being used (x ounces) gives us the total price ($400x$). In the second line, the price of platinum ($475) times the amount being used ($10 - x$ ounces) gives us the total price of $475(10 - x)$. Finally, the third line tells us that the price of the mixture is $415, that we need 10 ounces of it, and that its total price will be $4150.

Since the sum of the total prices of gold and platinum in the last column must be equal to the total price of the mixture, it follows that

$$400x + 475(10 - x) = 4150$$
$$400x + 4750 - 475x = 4150 \qquad \text{Simplify.}$$
$$4750 - 75x = 4150$$
$$4750 - 4750 - 75x = 4150 - 4750 \qquad \text{Subtract 4750.}$$
$$-75x = -600 \qquad \text{Simplify.}$$
$$\frac{-75x}{-75} = \frac{-600}{-75} \qquad \text{Divide by } -75.$$
$$x = 8$$

Thus, the supplier must use 8 ounces of gold and $10 - 8 = 2$ ounces of platinum. You can verify that this is correct! (8 ounces of gold at $400 per ounce and 2 ounces of platinum at $475 per ounce make 10 ounces of the mixture that costs $4150.)

EXAMPLE **4**   **A mixture problem**

How many ounces of a 50% acetic acid solution should a photographer add to 32 ounces of a 5% acetic acid solution to obtain a 10% acetic acid solution?

SOLUTION   We use the RSTUV method.

1. Read the problem. We are asked to find the number of ounces of the 50% solution (a solution consisting of 50% acetic acid) that should be added to make a 10% solution.

2. Select the unknown. Let x stand for the number of ounces of 50% solution to be added.

3. Think of a plan. Remember to write the percents as decimals. To translate the problem, we first use a chart. In this case, the headings should include the percent of acetic acid and the amount to be mixed. The product of these two numbers will then give us the amount of pure acetic acid. Note that the percents have been converted to decimals.

	%	×	Ounces	=	Amount of Pure Acid in Final Mixture
50% solution	0.50		x		$0.50x$
5% solution	0.05		32		1.60
10% solution	0.10		$x + 32$		$0.10(x + 32)$

Since we have x ounces of one solution and 32 ounces of the other, we have $(x + 32)$ ounces of the final mixture.

Since the sum of the amounts of pure acetic acid should be the same as the total amount of pure acetic acid in the final mixture, we have

$$0.50x + 1.60 = 0.10(x + 32) \quad \text{Given}$$

4. Use algebra to solve the equation.

$$10 \cdot 0.50x + 10 \cdot 1.60 = 10 \cdot [0.10(x + 32)] \quad \text{Multiply by 10 to clear the decimals.}$$

$$5x + 16 = 1(x + 32)$$

$$5x + 16 = x + 32$$

$$5x + 16 - 16 = x + 32 - 16 \quad \text{Subtract 16.}$$

$$5x = x + 16 \quad \text{Simplify.}$$

$$5x - x = x - x + 16 \quad \text{Subtract } x.$$

$$4x = 16 \quad \text{Simplify.}$$

$$\frac{4x}{4} = \frac{16}{4} \quad \text{Divide by 4.}$$

$$x = 4$$

Thus, the photographer must add 4 ounces of the 50% solution.

5. Verify the solution. We leave the verification to you.

PROBLEM 4

What if we want to obtain a 30% acetic acid solution?

C ⟩ Investment Problems

Finally, there's another problem that's similar to the mixture problem—the **investment problem.** Investment problems depend on the fact that the simple interest I you can earn (or pay) on principal P invested at rate r for 1 year is given by the formula

$$I = Pr$$

Now suppose you have a total of $10,000 invested. Part of the money is invested at 6% and the rest at 8%. If the bank tells you that you have earned $730 interest for 1 year, how much do you have invested at each rate?

In this case, we need to know how much is invested at each rate. Thus, if we say that we have invested P dollars at 6%, the rest of the money, that is, $10,000 - P$, would be invested at 8%. Note that for this to work, the sum of the amount invested at 6%, P dollars, plus the rest, $10,000 - P$, must equal $10,000. Since $P + 10,000 - P = 10,000$, we have done it correctly. This information is entered in a chart, as shown here:

	P	×	r	=	I
6% investment	P		0.06		$0.06P$
8% investment	$10,000 - P$		0.08		$0.08(10,000 - P)$

This column must add to 10,000.

The total interest is the sum of the two expressions.

From the chart we can see that the total interest is the sum of the expressions in the last column, $0.06P + 0.08(10,000 - P)$. Recall that the bank told us that our total interest is $730. Thus,

$$0.06P + 0.08(10,000 - P) = 730$$

You can solve this equation by first multiplying each term by 100 to clear the decimals:

$$100 \cdot 0.06P + 100 \cdot 0.08(10,000 - P) = 100 \cdot 730$$
$$6P + 8(10,000 - P) = 73,000$$
$$6P + 80,000 - 8P = 73,000$$
$$-2P = -7000$$
$$P = 3500$$

You could also solve

$$0.06P + 0.08(10,000 - P) = 730$$

as follows:

$$0.06P + 800 - 0.08P = 730 \qquad \text{Simplify.}$$
$$800 - 0.02P = 730$$
$$800 - 800 - 0.02P = 730 - 800 \qquad \text{Subtract 800.}$$
$$-0.02P = -70 \qquad \text{Simplify.}$$
$$\frac{-0.02P}{-0.02} = \frac{-70}{-0.02} \qquad \text{Divide by } -0.02.$$
$$P = 3500$$

$$\begin{array}{r} 35\ 00 \\ 0.02\overline{)70.00} \\ \hline 6 \\ 10 \\ \underline{10} \\ 0 \end{array}$$

Thus, $3500 is invested at 6% and the rest, $10,000 - 3500$, or $6500, is invested at 8%. You can verify that 6% of $3500 added to 8% of $6500 yields $730.

EXAMPLE 5   An investment portfolio

A woman has some stocks that yield 5% annually and some bonds that yield 10%. If her investment totals $6000 and her annual income from the investments is $500, how much does she have invested in stocks and how much in bonds?

SOLUTION   As usual, we the RSTUV method.

1. Read the problem. We are asked to find how much is invested in stocks and how much in bonds.

2. Select the unknown. Let s be the amount invested in stocks. This makes the amount invested in bonds $(6000 - s)$.

3. Think of a plan. A chart is a good way to visualize this problem. Remember that the headings will be the formula $P \times r = I$ and that the percents must be written as decimals. Now we enter the information:

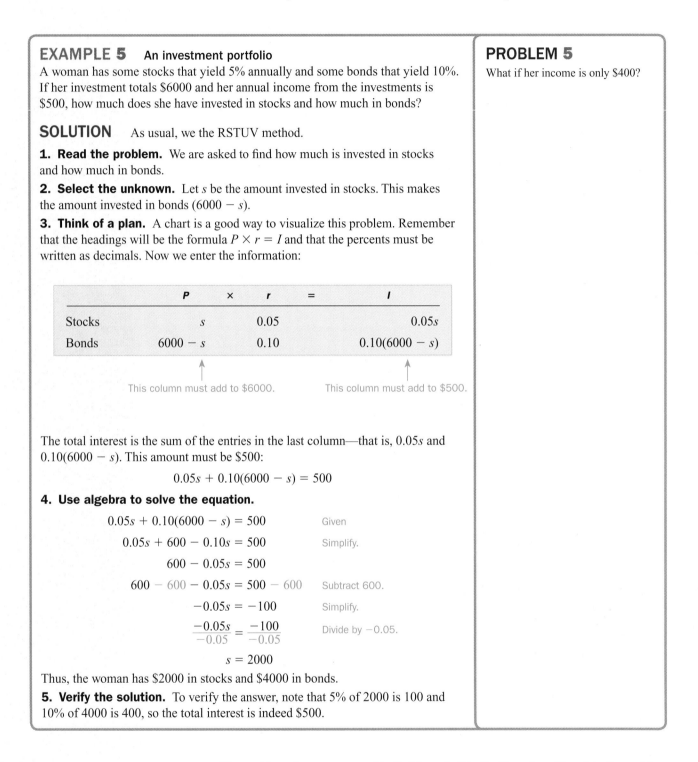

	P	$\times$	r	$=$	I
Stocks	s		0.05		$0.05s$
Bonds	$6000 - s$		0.10		$0.10(6000 - s)$

This column must add to $6000.   This column must add to $500.

The total interest is the sum of the entries in the last column—that is, $0.05s$ and $0.10(6000 - s)$. This amount must be $500:

$$0.05s + 0.10(6000 - s) = 500$$

4. Use algebra to solve the equation.

$0.05s + 0.10(6000 - s) = 500$	Given
$0.05s + 600 - 0.10s = 500$	Simplify.
$600 - 0.05s = 500$	
$600 - 600 - 0.05s = 500 - 600$	Subtract 600.
$-0.05s = -100$	Simplify.
$\dfrac{-0.05s}{-0.05} = \dfrac{-100}{-0.05}$	Divide by -0.05.
$s = 2000$	

Thus, the woman has $2000 in stocks and $4000 in bonds.

5. Verify the solution. To verify the answer, note that 5% of 2000 is 100 and 10% of 4000 is 400, so the total interest is indeed $500.

PROBLEM 5
What if her income is only $400?

The problems here correspond to Problems 1–9 in the Exercises. Translate the problem and place the information in a box with the heading: $\boxed{R \times T = D}$ Do not solve!

TRANSLATE THIS

1. The distance from Los Angeles to Sacramento is about 400 miles. A bus covers this distance in about 8 hours.

2. The distance from Boston to New Haven is 120 miles. A car leaves Boston at 7 A.M. and gets to New Haven just in time for lunch, at exactly 12 noon.

3. A 120-VHS Memorex videotape contains 246 meters of tape. When played at standard speed (SP), it will play for 120 minutes.

4. A Laser Writer Select prints one 12-inch page in 6 seconds. What is the rate of output for this printer?

5. The air distance from Miami to Tampa is about 200 miles. A jet flies at an average speed of 400 miles per hour.

The third step in the RSTUV procedure is to **TRANSLATE** the information into an equation. In Problems 1–10 **TRANSLATE** the sentence and match the correct translation with one of the boxes A–O.

A.

R	×	T	=	D
?		120		246

B.

R	×	T	=	D
30		T		$30T$

C.

R	×	T	=	D
260		5		?

D.

R	×	T	=	D
60		$T-1$		$60(T-1)$

E.

R	×	T	=	D
15		T		$15T$

F.

R	×	T	=	D
?		5		120

G.

R	×	T	=	D
?		8		400

H.

R	×	T	=	D
60		$T+1$		$60(T+1)$

I.

R	×	T	=	D
?		6		12

J.

R	×	T	=	D
60		T		$60T$

K.

R	×	T	=	D
60		$T-\frac{1}{2}$		$60(T-\frac{1}{2})$

L.

R	×	T	=	D
120		?		246

M.

R	×	T	=	D
400		?		200

N.

R	×	T	=	D
60		$T+\frac{1}{2}$		$60(T+\frac{1}{2})$

O.

R	×	T	=	D
50		T		$50T$

6. A freight train leaves the station traveling at 30 miles per hour for T hours.

7. One hour later (see Problem 6), a passenger train leaves the same station traveling at 60 miles per hour in the same direction.

8. An accountant catches a train that travels at 50 miles per hour for T hours.

9. The basketball coach at a local high school left for work on his bicycle traveling at 15 miles per hour for T hours

10. Half an hour later (see Problem 9), his wife noticed that he had forgotten his lunch. She got in her car and took his lunch to him. Luckily, she got to school at exactly the same time as her husband. She made her trip traveling at 60 miles per hour.

MathZone ✚✗

Boost your grade at mathzone.com!
> Practice Problems   > Self-Tests
> NetTutor   > e-Professors
> Videos

› Exercises **2.5**

‹ A › Motion Problems   In Problems 1–16, use the RSTUV method to solve the motion problems.

1. The distance from Los Angeles to Sacramento is about 400 miles. A bus covers this distance in about 8 hours. What is the average speed of the bus?

2. The distance from Boston to New Haven is 120 miles. A car leaves Boston at 10 A.M. and gets to New Haven just in time for lunch, at exactly 12 noon. What is the speed of the car?

3. A 120-VHS Memorex videotape contains 246 meters of tape. When played at standard speed (SP), it will play for 120 minutes. What is the rate of play of the tape? Answer to the nearest whole number.

4. A Laser Writer Select prints one 12-inch page in 6 seconds.

  a. What is the rate of output for this printer?

  b. How long would it take to print 60 pages at this rate?

5. The air distance from Miami to Tampa is about 200 miles. If a jet flies at an average speed of 400 miles per hour, how long does it take to go from Tampa to Miami?

6. A freight train leaves the station traveling at 30 miles per hour. One hour later, a passenger train leaves the same station traveling at 60 miles per hour in the same direction. How long does it take for the passenger train to overtake the freight train?

7. A bus leaves the station traveling at 60 kilometers per hour. Two hours later, a student shows up at the station with a briefcase belonging to her absent-minded professor who is riding the bus. If she immediately starts after the bus at 90 kilometers per hour, how long will it be before she reunites the briefcase with her professor?

8. An accountant catches a train that travels at 50 miles per hour, whereas his boss leaves 1 hour later in a car traveling at 60 miles per hour. They had decided to meet at the train station in the next town and, strangely enough, they get there at exactly the same time! If the train and the car traveled in a straight line on parallel paths, how far is it from one town to the other?

9. The basketball coach at a local high school left for work on his bicycle traveling at 15 miles per hour. Half an hour later, his wife noticed that he had forgotten his lunch. She got in her car and took his lunch to him. Luckily, she got to school at exactly the same time as her husband. If she made her trip traveling at 60 miles per hour, how far is it from their house to the school?

10. A jet traveling 480 miles per hour leaves San Antonio for San Francisco, a distance of 1632 miles. An hour later another plane, going at the same speed, leaves San Francisco for San Antonio. How long will it be before the planes pass each other?

11. A car leaves town A going toward B at 50 miles per hour. At the same time, another car leaves B going toward A at 55 miles per hour. How long will it be before the two cars meet if the distance from A to B is 630 miles?

12. A contractor has two jobs that are 275 kilometers apart. Her headquarters, by sheer luck, happen to be on a straight road between the two construction sites. Her first crew left headquarters for one job traveling at 70 kilometers per hour. Two hours later, she left headquarters for the other job, traveling at 65 kilometers per hour. If the contractor and her first crew arrived at their job sites simultaneously, how far did the first crew have to drive?

13. A plane has 7 hours to reach a target and come back to base. It flies out to the target at 480 miles per hour and returns on the same route at 640 miles per hour. How many miles from the base is the target?

14. A man left home driving at 40 miles per hour. When his car broke down, he walked home at a rate of 5 miles per hour; the entire trip (driving and walking) took him $2\frac{1}{4}$ hours. How far from his house did his car break down?

15. The space shuttle *Discovery* made a historical approach to the Russian space station *Mir*, coming to a point 366 feet (4392 inches) away from the station. For the first 180 seconds of the final 366-foot approach, the *Discovery* traveled 20 times as fast as during the last 60 seconds of the approach. How fast was the shuttle approaching during the first 180 seconds and during the last 60 seconds?

16. If, in Problem 15, the commander decided to approach the *Mir* traveling only 13 times faster for the first 180 seconds as for the last 60 seconds, what would be the shuttle's final rate of approach?

⟨ **B** ⟩ **Mixture Problems** In Problems 17–27, use the RSTUV method to solve these mixture problems.

17. How many liters of a 40% glycerin solution must be mixed with 10 liters of an 80% glycerin solution to obtain a 65% solution?

18. How many parts of glacial acetic acid (99.5% acetic acid) must be added to 100 parts of a 10% solution of acetic acid to give a 28% solution? (Round your answer to the nearest whole part.)

19. If the price of copper is 65¢ per pound and the price of zinc is 30¢ per pound, how many pounds of copper and zinc should be mixed to make 70 pounds of brass selling for 45¢ per pound?

20. Oolong tea sells for $19 per pound. How many pounds of Oolong should be mixed with another tea selling at $4 per pound to produce 50 pounds of tea selling for $7 per pound?

21. How many pounds of Blue Jamaican coffee selling at $5 per pound should be mixed with 80 pounds of regular coffee selling at $2 per pound to make a mixture selling for $2.60 per pound? (The merchant cleverly advertises this mixture as "Containing the incomparable Blue Jamaican coffee"!)

22. How many ounces of vermouth containing 10% alcohol should be added to 20 ounces of gin containing 60% alcohol to make a pitcher of martinis that contains 30% alcohol?

23. Do you know how to make manhattans? They are mixed by combining bourbon and sweet vermouth. How many ounces of manhattans containing 40% vermouth should a bartender mix with manhattans containing 20% vermouth so that she can obtain a half gallon (64 ounces) of manhattans containing 30% vermouth?

24. A car radiator contains 30 quarts of 50% antifreeze solution. How many quarts of this solution should be drained and replaced with pure antifreeze so that the new solution is 70% antifreeze?

25. A car radiator contains 30 quarts of 50% antifreeze solution. How many quarts of this solution should be drained and replaced with water so that the new solution is 30% antifreeze?

26. A 12-ounce can of frozen orange juice concentrate is mixed with 3 cans of cold water to obtain a mixture that is 10% juice. What is the percent of pure juice in the concentrate?

27. The instructions on a 12-ounce can of Welch's Orchard fruit juice state: "Mix with 3 cans cold water" and it will "contain 30% juice when properly reconstituted." What is the percent of pure juice in the concentrate? What does this mean to you?

⟨ **C** ⟩   **Investment Problems**   In Problems 28–36, use the RSTUV method to solve these investment problems.

28. Two sums of money totaling $15,000 earn, respectively, 5% and 7% annual interest. If the total interest from both investments amounts to $870, how much is invested at each rate?

29. An investor invested $20,000, part at 6% and the rest at 8%. Find the amount invested at each rate if the annual income from the two investments is $1500.

30. A woman invested $25,000, part at 7.5% and the rest at 6%. If her annual interest from these two investments amounted to $1620, how much money did she invest at each rate?

31. A man has a savings account that pays 5% annual interest and some certificates of deposit paying 7% annually. His total interest from the two investments is $1100, and the total amount of money in the two investments is $18,000. How much money does he have in the savings account?

32. A woman invested $20,000 at 8%. What additional amount must she invest at 6% so that her annual income is $2200?

33. A sum of $10,000 is split, and the two parts are invested at 5% and 6%, respectively. If the interest from the 5% investment exceeds the interest from the 6% investment by $60, how much is invested at each rate?

34. An investor receives $600 annually from two investments. He has $500 more invested at 8% than at 6%. Find the amount invested at each rate.

35. How can you invest $40,000, part at 6% and the remainder at 10%, so that the interest earned in each of the accounts is the same?

36. An inheritance of $10,000 was split into two parts and invested. One part lost 5% and the second gained 11%. If the 5% loss was the same as the 11% gain, how much was each investment?

⟩ ⟩ ⟩   *Using Your Knowledge*

Gesselmann's Guessing   Some of the mixture problems given in this section can be solved using the *guess-and-correct* procedure developed by Dr. Harrison A. Gesselmann of Cornell University. The procedure depends on taking a guess at the answer and then using a calculator to correct this guess. For example, to solve the very first mixture problem presented in this section (p. 157), we have to mix gold and platinum to obtain 10 ounces of a mixture selling for $415 per ounce. Our first guess is to use *equal amounts* (5 ounces each) of gold and platinum. This gives a mixture with a price per pound equal to the average price of gold ($400) and platinum ($475), that is,

$$\frac{400 + 475}{2} = \$437.50 \text{ per pound}$$

As you can see from the following figure, more gold must be used in order to bring the $437.50 average down to the desired $415:

5 oz	Desired	Average	5 oz
$400	$415	$437.50	$475

22.50

37.50

Thus, the correction for the additional amount of gold that must be used is

$$\frac{22.50}{37.50} \times 5 \text{ oz}$$

This expression can be obtained by the keystroke sequence

22.50 [÷] 37.50 [×] 5 [ENTER]

which gives the correction 3. The correct amount is

$$
\begin{aligned}
\text{First guess} \quad & \text{5 oz of gold} \\
+\text{Correction} \quad & \underline{\text{3 oz of gold}} \\
\text{Total} = \; & \text{8 oz of gold}
\end{aligned}
$$

and the remaining 2 ounces is platinum.

If your instructor permits, use this method to work Problems 19 and 23.

〉〉〉 *Write On*

37. Ask your pharmacist or your chemistry instructor if they mix products of different concentrations to make new mixtures. Write a paragraph on your findings.

38. Most of the problems involved have precisely the information you need to solve them. In real life, however, irrelevant information (called *red herrings*) may be present. Find some problems with red herrings and point them out.

39. The *guess-and-correct* method explained in the *Using Your Knowledge* also works for investment problems. Write the procedure you would use to solve investment problems using this method.

〉〉〉 *Concept Checker*

Fill in the blank(s) with the correct word(s), phrase, or mathematical statement.

40. The formula for the distance D traveled at a rate R in time T is given by _____.

41. The formula for the annual interest I earned (or paid) on a principal P at a rate r is given by _____.

$R = DT$	$I = Pr$
$T = DR$	$P = Ir$
$D = RT$	

〉〉〉 *Mastery Test*

42. Billy has two investments totaling $8000. One investment yields 5% and the other 10%. If the total annual interest is $650, how much money is invested at each rate?

43. How many gallons of a 10% salt solution should be added to 15 gallons of a 20% salt solution to obtain a 16% solution?

44. Two trains are 300 miles apart, traveling toward each other on adjacent tracks. One is traveling at 40 miles per hour and the other at 35 miles per hour. After how many hours do they meet?

45. A bus leaves Los Angeles traveling at 50 miles per hour. An hour later a car leaves at 60 miles per hour to try to catch the bus. How long does it take the car to overtake the bus?

46. The distance from South Miami to Tampa is 250 miles. This distance can be covered in 5 hours by car. What is the average speed on the trip?

〉〉〉 *Skill Checker*

Find:

47. $2.9(30) + 71$

48. $\frac{5}{9}(95 - 32)$

49. $\frac{1}{2}(20)(10)$

50. $2(38) - 10$

51. $\frac{22 + 1}{3}$

52. $\frac{21 + 1}{3}$

2.6   Formulas and Geometry Applications

▶ Objectives

A〉 Solve a formula for one variable and use the result to solve a problem.

B〉 Solve problems involving geometric formulas.

C〉 Solve geometric problems involving angle measurement.

D〉 Solve an application using formulas.

▶ To Succeed, Review How To . . .

1. Reduce fractions (pp. 4–6).
2. Perform the fundamental operations using decimals (pp. 22, 23).
3. Evaluate expressions using the correct order of operations (pp. 61–62, 69–72).

▶ Getting Started
Do You Want Fries with That?

One way to solve word problems is to use a formula. Here's an example in which an incorrect formula has been used. The ad claims that the new Monster Burger has 50% more beef than the Lite Burger because its diameter, 6, is 50% more than 4. But is that the right way to measure them?

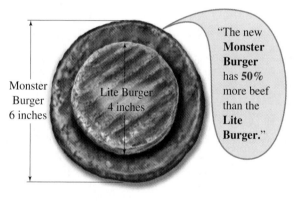

Monster Burger 6 inches

Lite Burger 4 inches

"The new **Monster Burger** has **50%** more beef than the **Lite Burger.**"

We should compare the two hamburgers by comparing the *volume* of the beef in each burger. The volume V of a burger is $V = Ah$, where A is the area and h is the height. The area of the Lite Burger is $L = \pi r^2$, where r is the radius (half the distance across the middle) of the burger. For the Lite Burger, $r = 2$ inches, so its area is

$$L = \pi(2 \text{ in.})^2 = 4\pi \text{ in.}^2$$

For the Monster Burger, $r = 3$ inches, so its area M is given by

$$M = \pi(3 \text{ in.})^2 = 9\pi \text{ in.}^2$$

The volume of the Lite Burger is $V = Ah = 4\pi h$, and for the Monster Burger the volume is $9\pi h$. If the height is the same for both burgers, we can simplify the volume calculation. The difference in volumes is $9\pi h - 4\pi h = 5\pi h$, and the *percent* increase for the Monster Burger is given by

$$\text{Percent increase} = \frac{\text{increase}}{\text{base}} = \frac{5\pi h}{4\pi h} = 1.25 \text{ or } 125\%$$

Thus, based on the discussion, the Monster Burgers actually have 125% more beef than the Lite Burgers. What other assumptions must you make for this to be so? (You'll have an opportunity in the *Write On* to give your opinion.) In this section we shall use formulas (like that for the volume or area of a burger) to solve problems.

Problems in many fields of endeavor can be solved if the proper formula is used. For example, why aren't you electrocuted when you hold the two terminals of your car battery?

You may know the answer. It's because the voltage V in the battery (12 volts) is too small. We know this from a formula in physics that tells us that the voltage V is the product of the current I and the resistance R; that is, $V = IR$. To find the current I going through your body when you touch both terminals, solve for I by dividing both sides by R to obtain

$$I = \frac{V}{R}$$

A car battery carries 12 volts and $R = 20{,}000$ ohms, so the current is

$$I = \frac{12}{20{,}000} = \frac{6}{10{,}000} = 0.0006 \text{ amp} \qquad \text{Volts divided by ohms yields amperes (amp).}$$

Since it takes about 0.001 ampere to give you a slight shock, your car battery should pose no threat.

A ⟩ Using Formulas

EXAMPLE 1   Solving problems in anthropology

Anthropologists know how to estimate the height of a man (in centimeters, cm) by using a bone as a clue. To do this, they use the formula

$$H = 2.89h + 70.64$$

where H is the height of the man and h is the length of his humerus.

a. Estimate the height of a man whose humerus bone is 30 centimeters long.

b. Solve for h.

c. If a man is 163.12 centimeters tall, how long is his humerus?

SOLUTION

a. We write 30 in place of h (in parentheses to indicate multiplication):

$$H = 2.89(30) + 70.64$$
$$= 86.70 + 70.64$$
$$= 157.34 \text{ cm}$$

Thus, a man with a 30-centimeter humerus should be about 157 centimeters tall.

b. We circle h to track the variable we are solving for.

$$H = 2.89 \,\widehat{h} + 70.64 \qquad \text{Given}$$
$$H - 70.64 = 2.89 \,\widehat{h} \qquad \text{Subtract 70.64.}$$
$$\frac{H - 70.64}{2.89} = \frac{2.89 \,\widehat{h}}{2.89} \qquad \text{Divide by 2.89.}$$

Thus,

$$\widehat{h} = \frac{H - 70.64}{2.89}$$

c. This time, we substitute 163.12 for H in the preceding formula to obtain

$$h = \frac{163.12 - 70.64}{2.89} = 32$$

Thus, the length of the humerus of a 163.12-centimeter-tall man is 32 centimeters.

EXAMPLE 2   Comparing temperatures

The formula for converting degrees Fahrenheit (°F) to degrees Celsius (°C) is

$$C = \frac{5}{9}(F - 32)$$

a. On a hot summer day, the temperature is 95°F. How many degrees Celsius is that?

PROBLEM 1

The estimated height of a woman (in inches) is given by $H = 2.8h + 28.1$, where h is the length of her humerus (in inches).

a. Estimate the height of a woman whose humerus bone is 15 inches long.

b. Solve for h.

c. If a woman is 61.7 inches tall, how long is her humerus?

PROBLEM 2

The formula for the speed of an ant is $S = \frac{1}{6}(C - 4)$ centimeters per second, where C is the temperature in degrees Celsius.

a. If the temperature is 22°C, how fast is the ant moving?

Answers to PROBLEMS

1. a. 70.1 in.   **b.** $h = \frac{H - 28.1}{2.8}$ **c.** 12 in.   **2. a.** 3 cm/sec   **b.** $C = 6S + 4$ **c.** 16°C

b. Solve for F.

c. What is F when $C = 10$?

b. Solve for C.

c. What is C when $S = 2$ cm/sec?

SOLUTION

a. In this case, $F = 95$. So

$$C = \frac{5}{9}(F - 32) = \frac{5}{9}(95 - 32)$$

$$= \frac{5}{9}(63) \qquad \text{Subtract inside the parentheses first.}$$

$$= \frac{5 \cdot 63}{9}$$

$$= 35$$

Thus, the temperature is 35°C.

b. We circle F to track the variable we are solving for.

$$C = \frac{5}{9}(\text{\textcircled{F}} - 32) \qquad \text{Given}$$

$$9 \cdot C = 9 \cdot \frac{5}{9}(\text{\textcircled{F}} - 32) \qquad \text{Multiply by 9.}$$

$$9C = 5(\text{\textcircled{F}} - 32) \qquad \text{Simplify.}$$

$$9C = 5\text{\textcircled{F}} - 160 \qquad \text{Use the distributive property.}$$

$$9C + 160 = 5\text{\textcircled{F}} - 160 + 160 \qquad \text{Add 160.}$$

$$\frac{9C + 160}{5} = \frac{5\text{\textcircled{F}}}{5} \qquad \text{Divide by 5.}$$

Thus,

$$\text{\textcircled{F}} = \frac{9C + 160}{5}$$

c. Substitute 10 for C:

$$F = \frac{9 \cdot 10 + 160}{5}$$

$$= \frac{250}{5}$$

$$= 50$$

Thus, $F = 50$.

EXAMPLE 3   Solving retail problems

The retail selling price R of an item is obtained by adding the original cost C and the markup M on the item.

a. Write a formula for the retail selling price.

b. Find the markup M of an item that originally cost $50.

SOLUTION

a. Write the problem in words and then translate it.

The retail selling price   is obtained   by adding the original cost C and the markup M.

$$R \qquad = \qquad C + M$$

b. Here $C = \$50$ and we must solve for M. Substituting 50 for C in $R = C + M$, we have

$$R = 50 + M$$

$$R - 50 = M \qquad \text{Subtract 50.}$$

Thus, $M = R - 50$.

PROBLEM 3

The final cost F of an item is obtained by adding the original cost C and the tax T on the item.

a. Write a formula for the final cost F of the item.

b. Find the tax T on an item that originally cost $10.

Answers to PROBLEMS

3. a. $F = C + T$ **b.** $T = F - 10$

B ❭ Using Geometric Formulas

Many of the formulas we encounter in algebra come from geometry. For example, to find the *area* of a figure, we must find the number of square units contained in the figure. Unit squares look like these:

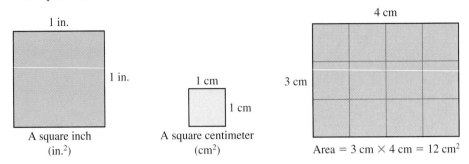

A square inch
(in.²)

A square centimeter
(cm²)

Area = 3 cm × 4 cm = 12 cm²

Now, to find the **area** of a figure, say a rectangle, we must find the number of square units it contains. For example, the area of a rectangle 3 centimeters by 4 centimeters is 3 cm × 4 cm = 12 cm² (read "12 square centimeters"), as shown in the diagram.

In general, we can find the area A of a rectangle by multiplying its length L by its width W, as given here.

AREA OF A RECTANGLE	The area **A** of a rectangle of length **L** and width **W** is $$A = LW$$

What about a rectangle's **perimeter** (distance around)? We can find the perimeter by adding the lengths of the four sides. Since we have two sides of length L and two sides of length W, the perimeter of the rectangle is

$$W + L + W + L = 2L + 2W$$

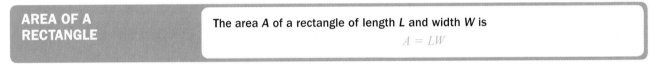

In general, we have the following formula.

PERIMETER OF A RECTANGLE	The perimeter **P** of a rectangle of length **L** and width **W** is $$P = 2L + 2W$$

EXAMPLE 4   Finding areas and perimeters

Find:

a. The area of the rectangle shown in the figure.

b. The perimeter of the rectangle shown in the figure.

c. If the perimeter of a rectangle 30 inches long is 110 inches, what is the width of the rectangle?

2.3 in.

1.4 in.

SOLUTION

a. The area:
$$A = LW$$
$$= (2.3 \text{ in.}) \cdot (1.4 \text{ in.})$$
$$= 3.22 \text{ in.}^2$$

b. The perimeter:
$$P = 2L + 2W$$
$$= 2(2.3 \text{ in.}) + 2(1.4 \text{ in.})$$
$$= 4.6 \text{ in.} + 2.8 \text{ in.}$$
$$= 7.4 \text{ in.}$$

PROBLEM 4

a. What is the area of a rectangle 2.4 by 1.2 inches?

b. What is the perimeter of the rectangle in part **a?**

c. If the perimeter of a rectangle 20 inches long is 100 inches, what is the width of the rectangle?

Answers to PROBLEMS

4. a. 2.88 in.² **b.** 7.2 in.

  c. 30 in.

c. The perimeter:

$$P = 2L + 2W$$
$$110 \text{ in.} = 2 \cdot (30 \text{ in.}) + 2W \qquad \text{Substitute 30 for } L \text{ and 110 for } P.$$
$$110 \text{ in.} = 60 \text{ in.} + 2W \qquad \text{Simplify.}$$
$$110 \text{ in.} - 60 \text{ in.} = 2W \qquad \text{Subtract 60.}$$
$$50 \text{ in.} = 2W$$
$$25 \text{ in.} = W \qquad \text{Divide by 2.}$$

Thus, the width of the rectangle is 25 inches.

Note that the area is given in square units, whereas the perimeter is a length and is given in linear units.

If we know the area of a rectangle, we can always calculate the area of the shaded triangle:

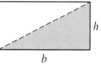

The area of the triangle is $\frac{1}{2}$ the area of the rectangle, which is bh. Thus, we have the following formula.

AREA OF A TRIANGLE	If a triangle has base *b* and perpendicular height *h*, its area *A* is $$A = \frac{1}{2}bh$$

Note that this time we used b and h instead of L and W. This formula holds true for any type of triangle.

EXAMPLE 5   Finding areas of a triangle

a. Find the area of a triangular piece of cloth 20 centimeters long and 10 centimeters high.

b. The area of the triangular sail on a toy boat is 250 square centimeters. If the base of the sail is 20 centimeters long, how high is the sail?

SOLUTION

a. $A = \frac{1}{2}bh$

$\quad = \frac{1}{2}(20 \text{ cm}) \cdot (10 \text{ cm})$

$\quad = 100 \text{ cm}^2$

b. Substituting 250 for A and 20 for b, we obtain

$$A = \frac{1}{2}bh \qquad \text{Becomes}$$
$$250 = \frac{1}{2} \cdot 20 \cdot h$$
$$250 = 10h \qquad \text{Simplify.}$$
$$25 = h \qquad \text{Divide by 10.}$$

Thus, the height of the sail is 25 centimeters.

PROBLEM 5

a. Find the area of a triangle 30 inches long and 15 inches high.

b. The area of the sail on a boat is 300 square feet. If the base of the sail is 20 ft long, how high is the sail?

Answers to PROBLEMS

5. a. 225 in.^{2} **b.** 30 ft

The area of a circle can easily be found if we know the **radius** r of the circle, which is the distance from the center of the circle to its edge. Here is the formula.

AREA OF A CIRCLE	The area A of a circle of radius r is $$A = \pi \cdot r \cdot r$$ $$= \pi r^2$$

As you can see, the formula for finding the area of a circle involves the number π (read "pie"). The number π is irrational; it cannot be written as a terminating or repeating decimal or a fraction, but it can be *approximated*. In most of our work we shall say that π is about 3.14 or $\frac{22}{7}$; that is, $\pi \approx 3.14$ or $\pi \approx \frac{22}{7}$. The number π is also used in finding the perimeter (distance around) a circle. This perimeter of the circle is called the **circumference** C and is found by using the following formula.

Radius r

The area A of a circle of radius r is
$$A = \pi \cdot r \cdot r = \pi r^2$$

CIRCUMFERENCE OF A CIRCLE	The circumference C of a circle of radius r is $$C = 2\pi r$$

Since the radius of a circle is half the diameter, another formula for the circumference of a circle is $C = \pi d$, where d is the diameter of the circle.

EXAMPLE 6 A CD's recorded area and circumference

The circumference C of a CD is 12π centimeters.

a. What is the radius r of the CD?

b. Every CD has a circular region in the center with no grooves, as can be seen in the figure. If the radius of this circular region is 2 centimeters and the rest of the CD has grooves, what is the area of the grooved (recorded) region?

SOLUTION

a. The circumference of a circle is

$$C = 2\pi r$$
$$12\pi = 2\pi r \qquad \text{Substitute } 12\pi \text{ for } C.$$
$$6 = r \qquad \text{Divide by } 2\pi.$$

Thus, the radius of the CD is 6 centimeters.

No sound

Recorded

b. To find the grooved (shaded) area, we find the area of the entire CD and subtract the area of the ungrooved (white) region. The area of the entire CD is

$$A = \pi r^2$$
$$A = \pi(6)^2 = 36\pi \qquad \text{Substitute } r = 6.$$

The area of the white region is

$$A = \pi(2)^2 = 4\pi$$

Thus, the area of the shaded region is $36\pi - 4\pi = 32\pi$ square centimeters.

PROBLEM 6

The circumference C of a CD is 9π inches.

a. What is the radius r of the CD?

b. The radius of the nonrecorded part of this CD is 0.75 inches. What is the area of the recorded region?

C ⟩ Solving for Angle Measurements

We've already mentioned complementary angles (two angles whose sum is 90°) and supplementary angles (two angles whose sum is 180°). Now we introduce the idea of **vertical angles.** The figure shows two intersecting lines with angles numbered ①, ❷, ③, and ❹.

Angles ① and ③ are placed "vertically"; they are called *vertical angles.* Another pair of vertical angles is ❷ and ❹.

VERTICAL ANGLES	**Vertical angles** have equal measures.

Note that if you add the measures of angles ① and ❷, you get 180°, a **straight angle.** Similarly, the sums of the measures of angles ❷ and ③, ③ and ❹, and ❹ and ① yield straight angles.

EXAMPLE 7   Finding the measures of angles

Find the measures of the marked angles in each figure:

a.

$(3x)°$

$(2x - 10)°$

b.

$(8x - 15)°$ $(3x - 5)°$

c.

$(3x - 3)°$

$(2x + 18)°$

PROBLEM 7

a. Find the measures if the angle in the diagram for part **a** is $(8x)°$ instead of $(3x)°$.

b. Find the measures if the angle in the diagram for part **b** is $(4x - 3)°$ instead of $(3x - 5)°$.

c. Find the measures if the angle in the diagram for part **c** is $(x + 17)°$ instead of $(2x + 18)°$.

SOLUTION

a. The sum of the measures of the two marked angles must be 180°, since angles are supplementary (they form a straight angle). Thus,

$$(2x - 10) + 3x = 180$$
$$5x - 10 = 180 \qquad \text{Simplify.}$$
$$5x = 190 \qquad \text{Add 10.}$$
$$x = 38 \qquad \text{Divide by 5.}$$

To find the measure of each of the angles, replace x with 38 in $2x - 10$ and in $3x$ to obtain

$$2(38) - 10 = 66 \quad \text{and} \quad 3(38) = 114$$

Thus, the measures are 66° and 114°, respectively. Note that $114 + 66 = 180$, so our result is correct.

(continued)

Answers to PROBLEMS

7. a. 28° and 152°   **b.** Both are 9°

c. 54° and 36°

b. The two marked angles are vertical angles, so their measures must be equal. Thus,

$$8x - 15 = 3x - 5$$

$$8x = 3x + 10 \quad \text{Add 15.}$$

$$5x = 10 \quad \text{Subtract } 3x.$$

$$x = 2 \quad \text{Divide by 5.}$$

Now, replace x with 2 in $8x - 15$ to obtain $8 \cdot 2 - 15 = 1$. Since the angles are vertical, their measures are both $1°$.

c. The sum of the measures of the two marked angles must be $90°$, since the angles are complementary. Thus,

$$(3x - 3) + (2x + 18) = 90$$

$$5x + 15 = 90 \quad \text{Simplify.}$$

$$5x = 75 \quad \text{Subtract 15.}$$

$$x = 15 \quad \text{Divide by 5.}$$

Replacing x with 15 in $(3x - 3)$ and $(2x + 18)$, we obtain $3 \cdot 15 - 3 = 42$ and $2 \cdot 15 + 18 = 48$. Thus, the measures of the angles are $42°$ and $48°$, respectively. Note that the sum of the measures of the two angles is $90°$, as expected.

D › Solving Applications by Using Formulas

Many formulas find uses in our daily lives. For example, do you know the *exact* relationship between your shoe size S and the length L of your foot in inches? Here are the formulas used in the United States:

$$L = \frac{22 + S}{3} \quad \text{For men}$$

$$L = \frac{21 + S}{3} \quad \text{For women}$$

EXAMPLE 8   Sizing shoes

Use the shoe sizing formulas to find:

a. The length of the foot corresponding to a size 1 shoe for men.

b. The length of the foot corresponding to a size 1 shoe for women.

c. Solve for S and determine the size shoe needed by Matthew McGrory (the man with the largest feet), whose left foot is 18 inches long.

SOLUTION

a. To find the length of the foot corresponding to a size 1 shoe for men, substitute 1 for S in

$$L = \frac{22 + S}{3}$$

$$= \frac{22 + 1}{3}$$

$$= \frac{23}{3} = 7\frac{2}{3} \text{ in.}$$

PROBLEM 8

a. Find the length of the foot corresponding to a size 2 shoe for men.

b. Find the length of the foot corresponding to a size 3 shoe for women.

c. McGrory's "smaller" right foot is 17 inches long. What shoe size fits his right foot?

Answers to PROBLEMS

8. a. 8 in.   **b.** 8 in.   **c.** 29

b. This time we substitute 1 for S in

$$L = \frac{21 + S}{3}$$

$$L = \frac{21 + 1}{3} = \frac{22}{3} = 7\frac{1}{3} \text{ in.}$$

c. $L = \dfrac{22 + S}{3}$ Given

$3L = 3 \cdot \dfrac{22 + S}{3}$ Multiply by the LCM 3.

$3L = 22 + S$ Simplify.

$3L - 22 = S$ Subtract 22.

Since the length of McGrory's foot is 18 in., substitute 18 for L to obtain

$$S = 3 \cdot 18 - 22 = 32$$

Thus, McGrory needs a size 32 shoe!

TRANSLATE THIS

1. The height H of a man (in centimeters) is the sum of 70.64 and the product of 2.89 and h, where h is the length of the man's humerus bone.

2. The formula C for converting degrees Fahrenheit F to degrees Celsius C is $\frac{5}{9}$ times the difference of F and **32.**

3. The area A of a square whose side is S units is S squared.

4. The area A of a rectangle of length L and width W is the product of L and W.

5. The perimeter P of a rectangle of length L and width W is the sum of twice the length L and twice the length W.

The third step in the RSTUV procedure is to **TRANSLATE** the information into an equation. In Problems 1–10 **TRANSLATE** the sentence and match the correct translation with one of the equations A–O.

A. $\frac{5}{9}F - 32$

B. $C = 2\pi r$

C. $H = 70.64 + 2.89h$

D. $L = \dfrac{S + 22}{3}$

E. $C = \frac{5}{9}(F - 32)$

F. $P = 2L + 2W$

G. $L = S + \dfrac{21}{3}$

H. $C = \frac{5}{9}(32 - F)$

I. $A = S^2$

J. $L = S + \dfrac{22}{3}$

K. $A = \pi r^2$

L. $L = \dfrac{S + 21}{3}$

M. $A = \frac{1}{2}bh$

N. $H = (70.64 + 2.89)f$

O. $A = LW$

6. The area A of a triangle with base b and height h is one-half the product of b and h.

7. The area A of a circle of radius r is the product of π and the square of r.

8. The circumference C of a circle of radius r is twice the product of π and r.

9. The length L of a man's foot is obtained by finding the sum of S and **22,** and dividing the result by 3.

10. The length L of a woman's foot is the quotient of the sum S plus 21, and 3.

> Exercises **2.6**

< A > **Using Formulas**   In Problems 1–10, use the formulas to find the solution.

1. The number of miles D traveled in T hours by an object moving at a rate R (in miles per hour) is given by $D = RT$.

  a. Find D when $R = 30$ and $T = 4$.

  b. Find the distance traveled by a car going 55 miles per hour for 5 hours.

  c. Solve for R in $D = RT$.

  d. If you travel 180 miles in 3 hours, what is R?

2. The rate of travel R of an object moving a distance D in time T is given by

$$R = \frac{D}{T}$$

  a. Find R when $D = 240$ miles and $T = 4$ hours.

  b. Find the rate of travel of a train that traveled 140 miles in 4 hours.

  c. Solve for T in $R = \frac{D}{T}$.

  d. How long would it take the train of part **b** to travel 105 miles?

3. The height H of a man (in inches) is related to his weight W (in pounds) by the formula $W = 5H - 190$.

  a. If a man is 60 inches tall, what should his weight be?

  b. Solve for H.

  c. If a man weighs 200 pounds, how tall should he be?

4. The number of hours H a growing child should sleep is

$$H = 17 - \frac{A}{2}$$

where A is the age of the child in years.

  a. How many hours should a 6-year-old sleep?

  b. Solve for A in $H = 17 - \frac{A}{2}$.

  c. At what age would you expect a child to sleep 11 hours?

5. The formula for converting degrees Celsius to degrees Fahrenheit is

$$F = \frac{9}{5}C + 32$$

  a. If the temperature is 15°C, what is the corresponding Fahrenheit temperature?

  b. Solve for C.

  c. What is the corresponding Celsius temperature when $F = 50$?

6. The profit P a business makes is obtained by subtracting the expenses E from the income I.

  a. Write a formula for the profit P.

  b. Find the profit of a business with $2700 in income and $347 in expenses.

  c. Solve for E in the formula you wrote in part **a.**

  d. If the profit is $750 and the income is $1300, what are the expenses?

7. The energy efficiency ratio (EER) for an air conditioner is obtained by dividing the British thermal units (Btu) per hour by the watts w.

  a. Write a formula that will give the EER of an air conditioner.

  b. Find the EER of an air conditioner with a capacity of 9000 British thermal units per hour and a rating of 1000 watts.

  c. Solve for the British thermal units in your formula for EER.

  d. How many British thermal units does a 2000-watt air conditioner produce if its EER is 10?

8. The capital C of a business is the difference between the assets A and the liabilities L.

  a. Write a formula that will give the capital of a business.

  b. If a business has $4800 in assets and $2300 in liabilities, what is the capital of the business?

  c. Solve for L in your formula for C.

  d. If a business has $18,200 in capital and $30,000 in assets, what are its liabilities?

9. The selling price S of an item is the sum of the cost C and the margin (markup) M.

  a. Write a formula for the selling price of a given item.

  b. A merchant wishes to have a $15 margin (markup) on an item costing $52. What should be the selling price of this item?

  c. Solve for M in your formula for S.

  d. If the selling price of an item is $18.75 and its cost is $10.50, what is the markup?

10. The tip speed S_T of a propeller is equal to π times the diameter d of the propeller times the number N of revolutions per second.

  a. Write a formula for S_T.

  b. If a propeller has a 2-meter diameter and it is turning at 100 revolutions per second, find S_T. (Use $\pi \approx 3.14$.)

  c. Solve for N in your formula for S_T.

  d. What is N when $S_T = 275\pi$ and $d = 2$?

⟨ **B** ⟩   **Using Geometric Formulas**   In Problems 11–15 use the geometric formulas to find the solution.

11. The perimeter P of a rectangle of length L and width W is $P = 2L + 2W$.

   a. Find the perimeter of a rectangle 10 centimeters by 20 centimeters.

   b. What is the length of a rectangle with a perimeter of 220 centimeters and a width of 20 centimeters?

13. The circumference C of a circle of radius r is $C = 2\pi r$.

   a. Find the circumference of a circle with a radius of 10 inches. (Use $\pi \approx 3.14$.)

   b. Solve for r in $C = 2\pi r$.

   c. What is the radius of a circle whose circumference is 20π inches?

12. The perimeter P of a rectangle of length L and width W is $P = 2L + 2W$.

   a. Find the perimeter of a rectangle 15 centimeters by 30 centimeters.

   b. What is the width of a rectangle with a perimeter of 180 centimeters and a length of 60 centimeters?

14. If the circumference C of a circle of radius r is $C = 2\pi r$, what is the radius of a tire whose circumference is 26π inches?

15. The area A of a rectangle of length L and width W is $A = LW$.

   a. Find the area of a rectangle 4.2 meters by 3.1 meters.

   b. Solve for W in the formula $A = LW$.

   c. If the area of a rectangle is 60 square meters and its length is 10 meters, what is the width of the rectangle?

⟨ **C** ⟩   **Solving for Angle Measurements**   In Problems 16–29, find the measure of each marked angle.

16.

17.

18.

19.

20.

21.

22.

23.

24.

25.

26.

27.

28.

29.

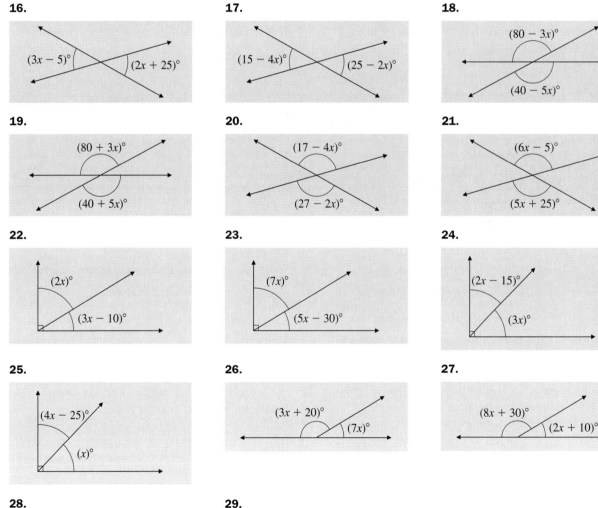

⟨ D ⟩ Solving Applications by Using Formulas

30. *Supersized omelet* One of the largest rectangular omelets ever cooked was 30 feet long and had an 80-foot perimeter. How wide was it?

31. *Largest pool* If you were to walk around the largest rectangular pool in the world, in Casablanca, Morocco, you would walk more than 1 kilometer. To be exact, you would walk 1110 meters. If the pool is 480 meters long, how wide is it?

32. *Football field dimensions* The playing surface of a football field is 120 yards long. A player jogging around the perimeter of this surface jogs 346 yards. How wide is the playing surface of a football field?

33. *CD diameter* A point on the rim of a CD record travels 14.13 inches each revolution. What is the diameter of this CD? (Use $\pi \approx 3.14$.)

34. *Gigantic pizza* One of the largest pizzas ever made had a 251.2-foot circumference! What was its diameter? (Use $\pi \approx 3.14$.)

35. *Continental suit sizes* Did you know that clothes are sized differently in different countries? If you want to buy a suit in Europe and you wear a size A in America, your continental (European) size C will be $C = A + 10$.

   a. Solve for A.

   b. If you wear a continental size 50 suit, what would be your American size?

36. *Continental dress sizes* Your continental dress size C is given by $C = A + 30$, where A is your American dress size.

   a. Solve for A.

   b. If you wear a continental size 42 dress, what is your corresponding American size?

37. *Recreational boats data* According to the National Marine Manufacturers Association, the number N of recreational boats (in millions) has been steadily increasing since 1975 and is given by $N = 9.74 + 0.40t$, where t is the number of years after 1975.

   a. What was the number of recreational boats in 1985?

   b. Solve for t in $N = 9.74 + 0.40t$.

   c. In what year would you expect the number of recreational boats to reach 17.74 million?

38. *NCAA men's basketball teams* The number N of NCAA men's college basketball teams has been increasing since 1980 according to the formula $N = 720 + 5t$, where t is the number of years after 1980.

   a. How many teams would you expect in the year 2000?

   b. Solve for t.

   c. In what year would you expect the number of teams to reach 795?

39. *Vehicle insurance.* The amount A spent on vehicle insurance by *persons under 25* can be approximated by the equation $A = 28x + 420$, where x is the number of years after 2000.

   a. According to the formula, how much would a person under 25 spend on vehicle insurance in 2010?

   b. Solve for x.

   c. In how many years would vehicle insurance be $1000? Answer to the nearest year.

   Source: Bureau of Labor Consumer Expenditure Survey

40. *Vehicle insurance.* The amount A spent on vehicle insurance by *persons aged 25–54* can be approximated by the equation $A = 40x + 786$, where x is the number of years after 2000.

   a. According to the formula, how much would a person aged 25–54 spend on vehicle insurance in 2010?

   b. Solve for x.

   c. In how many years would vehicle insurance be $1000? Answer to the nearest year.

   Source: Bureau of Labor Consumer Expenditure Survey

41. *Height of a man* The height H of a man (in inches) can be estimated by the equation $H = 3.3r + 34$, where r is the length of the radius bone (the bone from the wrist to the elbow).

   a. Estimate the height of a man whose radius bone is 10 inches.

   b. Solve for r.

   c. If a man is 70.3 inches tall, how long is his radius?

   Source: Science Safari: The First People

42. *Height of a woman* The height H of a woman (in inches) can be estimated by $H = 3.3r + 32$, where r is the length of the radius bone (the bone from the wrist to the elbow).

   a. Estimate the height of a woman whose radius bone is 10 inches.

   b. Solve for r.

   c. If a woman is 61.7 inches tall, how long is her radius?

   Source: Science Safari: The First People

43. *Height of a woman* The height H of a woman (in centimeters) can be estimated by $H = 2.9t + 62$, where t is the length of the tibia bone (the bone between the knee and ankle).

   a. Estimate the height of a woman whose tibia bone is 30 centimeters.

   b. Solve for t.

   c. If a woman is 151.9 inches tall, how long is her radius?

   Source: Science Safari: The First People

〉〉〉 *Using Your Knowledge*

Living with Algebra   Many practical problems around the house require some knowledge of the formulas we've studied. For example, let's say that you wish to carpet your living room. You need to use the formula for the area A of a rectangle of length L and width W, which is $A = LW$.

44. Carpet sells for $8 per square yard. If the cost of carpeting a room was $320 and the room is 20 feet long, how wide is it? (Note that one square yard is nine square feet.)

45. If you wish to plant new grass in your yard, you can buy sod squares of grass that can simply be laid on the ground. Each sod square is approximately 1 square foot. If you have 5400 sod squares and you wish to sod an area that is 60 feet wide, what is the length?

46. If you wish to fence your yard, you need to know its perimeter (the distance around the yard). If the yard is W feet by L feet, the perimeter P is given by $P = 2W + 2L$. If your rectangular yard needs 240 feet of fencing and your yard is 70 feet long, what is the width?

〉〉〉 *Write On*

47. Write an explanation of what is meant by the *perimeter* of a geometric figure.

48. Write an explanation of what is meant by the *area* of a geometric figure.

49. Remember the hamburgers in the *Getting Started?* Write two explanations of how the Monster burger can have 50% more beef than the Lite burger and still look like the one in the picture.

50. To make a fair comparison of the amount of beef in two hamburgers, should you compare the circumferences, areas, or volumes? Explain.

51. The Lite burger has a 4-inch diameter whereas the Monster burger has a 6-inch diameter. How much bigger (in percent) is the circumference of the Monster burger? Can you now explain the claim in the ad? Is the claim correct? Explain.

〉〉〉 *Concept Checker*

Fill in the blank(s) with the correct word(s), phrase, or mathematical statement.

52. The **area** A of a **rectangle** of length L and width W is _____.

53. The **perimeter** P of a **rectangle** of length L and width W is _____.

54. The **area** A of a **triangle** with base b and height h is _____.

55. The **area** A of a **circle** of radius r is _____.

56. The **circumference** C of a **circle** of radius r is _____.

57. Vertical angles are angles of _____ measure.

LW	**πr^2**
2L + 2W	**$2\pi r$**
bh	**$2\pi r^2$**
$\frac{1}{2}bh$	**unequal**
πr	**equal**

〉〉〉 *Mastery Test*

Find the measures of the marked angles.

58.

59.

60.

61.

$(7x + 30)°$

$(3x + 10)°$

62.

$(4x + 11)°$

$(4x + 9)°$

63.

$(6x + 14)°$

$(2x + 6)°$

64. A circle has a radius of 10 inches.

   a. Find its area and its circumference.

   b. If the circumference of a circle is 40π inches, what is its radius?

65. The formula for estimating the height H (in centimeters) of a woman using the length h of her humerus as a clue is given by the equation $H = 2.75h + 71.48$.

   a. Estimate the height of a woman whose humerus bone is 20 centimeters long.

   b. Solve for h.

   c. If a woman is 140.23 centimeters tall, how long is her humerus?

66. The total cost T of an item is obtained by adding its cost C and the tax t on the item.

   a. Write a formula for the total cost T.

   b. Find the tax t on an item that cost \$8 if the total after adding the tax is \$8.48.

67. The area A of a triangle is $A = \frac{1}{2}bh$, where b is the base of the triangle and h is its height.

   a. What is the area of a triangle 15 inches long and with a 10-inch base?

   b. Solve for b in $A = \frac{1}{2}bh$.

   c. The area of a triangle is 18 square inches, and its height is 9 inches. How long is the base of the triangle?

68. According to the Motion Picture Association of America, the number N of motion picture theaters (in thousands) has been growing according to the formula $N = 15 + 0.60t$, where t is the number of years after 1975.

   a. How many theaters were there in 1985? (*Hint:* $t = 10$.)

   b. Solve for t in $N = 15 + 0.60t$.

   c. In what year did the number of theaters total 27,000?

69. The formula for converting degrees Fahrenheit F to degrees Celsius C is

$$C = \frac{5}{9}F - \frac{160}{9}$$

   a. Find the Celsius temperature on a day in which the thermometer reads 41°F.

   b. Solve for F.

   c. What is F when $C = 20$?

〉〉〉 *Skill Checker*

Solve:

70. $3x - 2 = 2(x - 2)$

71. $2x - 1 = x + 3$

72. $3x - 2 = 2(x - 1)$

73. $4(x + 1) = 3x + 7$

74. $\frac{-x}{4} + \frac{x}{6} = \frac{x - 3}{6}$

75. $\frac{x}{3} - \frac{x}{2} = 1$

2.7   Properties of Inequalities

▷ Objectives

A ▷ Determine which of two numbers is greater.

B ▷ Solve and graph linear inequalities.

C ▷ Write, solve, and graph compound inequalities.

D ▷ Solve an application involving inequalities.

▷ To Succeed, Review How To . . .

1. Add, subtract, multiply, and divide real numbers (pp. 52, 54, 60, 62).
2. Solve linear equations (pp. 131–136).

▷ Getting Started
Savings on Sandals

After learning to solve linear equations, we need to learn to solve *linear inequalities*. Fortunately, the rules are very similar, but the notation is a little different. For example, the ad says that the price you will pay for these sandals will be cut \$1 to \$3. That is, you will save \$1 to \$3. If x is the amount of money you can save, what can x

be? Well, it is at least \$1 and can be as much as \$3; that is, $x = 1$ or $x = 3$ or x is between 1 and 3. In the language of algebra, we write this fact as

$$1 \le x \le 3$$ Read "1 is less than or equal to x and x is less than or equal to 3" or "x is between 1 and 3, inclusive."

The statement $1 \le x \le 3$ is made up of two parts:

$1 \le x$ which means that 1 is *less than or equal to x* (or that x is *greater than or equal to* 1), and

$x \le 3$ which means that x is *less than or equal to* 3 (or that 3 is *greater than or equal to x*).

These statements are examples of *inequalities,* which we will learn how to solve in this section.

In algebra, an **inequality** is a statement with $>$, $<$, $\ge$, or $\le$ as its verb. Inequalities can be represented on a number line. Here's how we do it. As you recall, a number line is constructed by drawing a line, selecting a point on this line, and calling it zero (the origin):

We then locate equally spaced points to the right of the origin on the line and label them with the *positive* integers 1, 2, 3, and so on. The corresponding points to the left of zero are labeled -1, -2, -3, and so on (the *negative* integers). This construction allows the *association of numbers with points on the line*. The number associated with a point is called the **coordinate** of that point. For example, there are points associated with the numbers -2.5, $-1\frac{1}{2}$, $\frac{3}{4}$, and 2.5:

All these numbers are real numbers. As we have mentioned, the real numbers include natural (counting) numbers, whole numbers, integers, fractions, and decimals as well as the irrational numbers (which we discuss in more detail later). Thus, the real numbers can all be represented on the number line.

A ❭ Order of Numbers

As you can see, numbers are placed in order on the number line. *Greater* numbers are always to the *right* of *smaller* ones. (The farther to the *right,* the *greater* the number.) Thus, any number to the *right* of a second number is said to be **greater than** ($>$) the second number. We also say that the second number is **less than** ($<$) the first number. For example, since 3 is to the right of 1, we write

$$3 \quad > \quad 1 \quad \text{or} \quad 1 \quad < \quad 3$$

Similarly,

$$-1 > -3 \quad \text{or} \quad -3 < -1$$
$$0 > -2 \quad \text{or} \quad -2 < 0$$
$$3 > -1 \quad \text{or} \quad -1 < 3$$

Note that the inequality signs $>$ and $<$ always point to the smaller number.

EXAMPLE 1   Writing inequalities

Fill in the blank with $>$ or $<$ so that the resulting statement is true.

a. 3 _____ 4   **b.** -4 _____ -3 **c.** -2 _____ -3

SOLUTION   We first construct a number line containing these numbers. (Of course, we could just think about the number line without actually drawing one.)

$$\begin{array}{c}\longleftarrow\!\!+\!\!+\!\!+\!\!+\!\!+\!\!+\!\!+\!\!+\!\!+\!\!+\!\!+\!\!\longrightarrow \\ {\scriptstyle -5 \;\; -4 \;\; -3 \;\; -2 \;\; -1 \;\; 0 \;\; 1 \;\; 2 \;\; 3 \;\; 4 \;\; 5}\end{array}$$

a. Since 3 is to the left of 4, $3 < 4$.

b. Since -4 is to the left of -3, $-4 < -3$.

c. Since -2 is to the right of -3, $-2 > -3$.

PROBLEM 1

Fill in the blank with $>$ or $<$ so that the resulting statement is true.

a. 5 _____ 3   **b.** -1 _____ -4

c. -5 _____ -4

B ❭ Solving and Graphing Inequalities

Just as we solved equations, we can also solve inequalities. We do this by extending the addition and multiplication properties of equality (Sections 2.1 and 2.2) to include inequalities. We say that we have *solved* a given inequality when we obtain an inequality equivalent to the one given and in the form $x < \square$ or $x > \square$. (Note that the variable is on the **left** side of the inequality.) For example, consider the inequality

$$x < 3$$

$x < 3$
$2 < 3$
$1\frac{1}{2} < 3$
$1 < 3$
$0 < 3$
$-\frac{1}{2} < 3$

There are many real numbers that will make this inequality a true statement. A few of them are shown in the accompanying table. Thus, 2 is a solution of $x < 3$ because $2 < 3$. Similarly, $-\frac{1}{2}$ is a solution of $x < 3$ because $-\frac{1}{2} < 3$.

As you can see from this table, 2, $1\frac{1}{2}$, 1, 0, and $-\frac{1}{2}$ are *solutions* of the inequality $x < 3$. Of course, we can't list all the real numbers that satisfy the inequality $x < 3$ because there are infinitely many of them, but we can certainly show all the solutions of $x < 3$ *graphically* by using a number line:

$$x < 3$$
$$\begin{array}{c}\longleftarrow\!\!+\!\!+\!\!+\!\!+\!\!+\!\!+\!\!\circ\!\!+\!\!+\!\!+\!\!\longrightarrow \\ {\scriptstyle -5 \;\; -4 \;\; -3 \;\; -2 \;\; -1 \;\; 0 \;\; 1 \;\; 2 \;\; 3 \;\; 4 \;\; 5}\end{array}$$

Answers to PROBLEMS

1. a. $>$ **b.** $>$ **c.** $<$

> **NOTE**
>
> $x < 3$ tells you to draw your heavy line to the *left* of 3 because, in this case, the symbol $<$ points left.

This representation is called the **graph** of the solutions of $x < 3$, which are indicated by the heavy line. Note that there is an open circle at $x = 3$ to indicate that 3 is *not* part of the graph of $x < 3$ (since 3 is not less than 3). Also, the colored arrowhead points to the left (just as the $<$ in $x < 3$ points to the left) to indicate that the heavy line continues to the left without end. On the other hand, the graph of $x \geq 2$ should continue to the right without end:

$x \geq 2$

> **NOTE**
>
> $x \geq 2$ tells you to draw your heavy line to the *right* of 2 because, in this case, the symbol $\geq$ points right.

Moreover, since $x = 2$ is included in the graph, a solid dot appears at the point $x = 2$.

EXAMPLE 2   Graphing inequalities

Graph the inequality on a number line.

a. $x \geq -1$ **b.** $x < -2$

SOLUTION

a. The numbers that satisfy the inequality $x \geq -1$ are the numbers that are *greater than or equal to* -1, that is, the number -1 and all the numbers to the right of -1 (remember, $\geq$ points to the right and the dot must be solid). The graph is shown here:

$x \geq -1$

$$\begin{array}{ccccccccccc} -5 & -4 & -3 & -2 & -1 & 0 & 1 & 2 & 3 & 4 & 5 \end{array}$$

b. The numbers that satisfy the inequality $x < -2$ are the numbers that are *less than* -2, that is, the numbers to the *left of but not including* -2 (note that $<$ points to the left and that the dot is open). The graph of these points is shown here:

$x < -2$

$$\begin{array}{ccccccccccc} -5 & -4 & -3 & -2 & -1 & 0 & 1 & 2 & 3 & 4 & 5 \end{array}$$

PROBLEM 2

Graph the inequality on a number line.

a. $x \leq -2$ **b.** $x > -3$

We solve more complicated inequalities just as we solve equations, by finding an equivalent inequality whose solution is obvious. Remember, we have solved a given inequality when we obtain an inequality in the form $x < \square$ or $x > \square$ which is equivalent to the one given. Thus, to solve the inequality $2x - 1 < x + 3$, we try to find an equivalent inequality of the form $x < \square$ or $x > \square$. As before, we need some properties. The first of these are the addition and subtraction properties.

If $3 < 4$, then

$$3 + 5 < 4 + 5 \qquad \text{Add 5.}$$
$$8 < 9 \qquad \text{True.}$$

Similarly, if $3 > -2$, then

$$3 + 7 > -2 + 7 \qquad \text{Add 7.}$$
$$10 > 5 \qquad \text{True.}$$

Answers to PROBLEMS

2. a.

b.

Also, if $3 < 4$, then

$$3 - 1 < 4 - 1 \qquad \text{Subtract 1.}$$
$$2 < 3 \qquad \text{True.}$$

Similarly, if $3 > -2$, then

$$3 - 5 > -2 - 5 \qquad \text{Subtract 5.}$$
$$-2 > -7 \qquad \text{True because } -2 \text{ is to the right of } -7.$$

In general, we have the following properties.

ADDITION AND SUBTRACTION PROPERTIES OF INEQUALITIES

You can *add* or *subtract* the same number a on both sides of an inequality and obtain an equivalent inequality. In symbols,

If	$x < y$	If	$x > y$
then	$x + a < y + a$	then	$x + a > y + a$
or	$x - b < y - b$	or	$x - b > y - b$

Note: These properties also hold when the symbols $\leq$ and $\geq$ are used.

NOTE

Since $x - b = x + (-b)$, subtracting b from both sides is the same as adding the inverse of b, $(-b)$, so you can think of subtracting b as adding $(-b)$.

Now let's return to the inequality $2x - 1 < x + 3$. To solve this inequality, we need the variables by themselves (isolated) on one side, so we proceed as follows:

$$2x - 1 < x + 3 \qquad \text{Given}$$
$$2x - 1 + 1 < x + 3 + 1 \qquad \text{Add 1.}$$
$$2x < x + 4 \qquad \text{Simplify.}$$
$$2x - x < x - x + 4 \qquad \text{Subtract } x.$$
$$x < 4 \qquad \text{Simplify.}$$

Any number less than 4 is a solution. The graph of this inequality is as follows:

$$2x - 1 < x + 3 \text{ or, equivalently, } x < 4$$

You can check that this solution is correct by selecting any number from the graph (say 0) and replacing x with that number in the original inequality. For $x = 0$, we have $2(0) - 1 < 0 + 3$, or $-1 < 3$, a true statement. Of course, this is only a "partial" check, since we didn't try *all* the numbers in the graph. You can check a little further by selecting a number *not* on the graph to make sure the result is false. For example, when $x = 5$,

$$2x - 1 < x + 3 \quad \text{becomes} \quad 2(5) - 1 < 5 + 3$$
$$10 - 1 < 8$$
$$9 < 8 \qquad \text{False.}$$

EXAMPLE 3 Using the addition and subtraction properties to solve and graph inequalities

Solve and graph the inequality on a number line:

a. $3x - 2 < 2(x - 2)$

b. $4(x + 1) \geq 3x + 7$

PROBLEM 3

Solve and graph:

a. $4x - 3 < 3(x - 2)$

b. $3(x + 2) \geq 2x + 5$

Answers to PROBLEMS

3. a. $x < -3$

b. $x \geq -1$

SOLUTION

a.
$$3x - 2 < 2(x - 2) \qquad \text{Given}$$
$$3x - 2 < 2x - 4 \qquad \text{Simplify.}$$
$$3x - 2 + 2 < 2x - 4 + 2 \qquad \text{Add 2.}$$
$$3x < 2x - 2 \qquad \text{Simplify.}$$
$$3x - 2x < 2x - 2x - 2 \qquad \text{Subtract 2x (or add } -2x\text{).}$$
$$x < -2 \qquad \text{Simplify.}$$

Any number less than -2 is a solution. The graph of this inequality is as follows:

$$3x - 2 < 2(x - 2) \text{ or, equivalently, } x < -2$$

b.
$$4(x + 1) \geq 3x + 7 \qquad \text{Given}$$
$$4x + 4 \geq 3x + 7 \qquad \text{Simplify.}$$
$$4x + 4 - 4 \geq 3x + 7 - 4 \qquad \text{Subtract 4.}$$
$$4x \geq 3x + 3 \qquad \text{Simplify.}$$
$$4x - 3x \geq 3x - 3x + 3 \qquad \text{Subtract 3x (or add } -3x\text{).}$$
$$x \geq 3 \qquad \text{Simplify.}$$

Any number greater than or equal to 3 is a solution. The graph of this inequality is as follows:

$$4(x + 1) \geq 3x + 7 \text{ or, equivalently, } x \geq 3$$

How do we solve an inequality such as $\frac{x}{2} < 3$? If half a number is less than 3, the number must be less than 6. This suggests that you can *multiply (or divide)* both sides of an inequality by a *positive* number and obtain an equivalent inequality.

$$\frac{x}{2} < 3 \qquad \text{Given}$$
$$2 \cdot \frac{x}{2} < 2 \cdot 3 \qquad \text{Multiply by 2.}$$
$$x < 6 \qquad \text{Simplify.}$$

And to solve

$$2x < 8$$
$$\frac{2x}{2} < \frac{8}{2} \qquad \text{Divide by 2 (or multiply by the reciprocal of 2).}$$
$$x < 4 \qquad \text{Simplify.}$$

Any number less than 4 is a solution. Let's try some more examples.
If $3 < 4$, then

$$5 \cdot 3 < 5 \cdot 4$$
$$15 < 20 \qquad \text{True.}$$

Note that we are multiplying or dividing *both* sides of the inequality by a **positive** number. In such cases, we do not change the inequality symbol.
If $-2 > -10$, then

$$5 \cdot (-2) > 5 \cdot (-10)$$
$$-10 > -50 \qquad \text{True.}$$

Also, if $6 < 8$, then

$$\frac{6}{2} < \frac{8}{2}$$
$$3 < 4 \qquad \text{True.}$$

Similarly, if $-6 > -10$, then

$$-\frac{6}{2} > -\frac{10}{2}$$
$$-3 > -5 \quad \text{True.}$$

Here are the properties we've just used.

MULTIPLICATION AND DIVISION PROPERTIES OF INEQUALITIES FOR POSITIVE NUMBERS	You can *multiply* or *divide* both sides of an inequality by any *positive* number a and obtain an equivalent inequality. In symbols,

If	$x < y$ (and a is positive)	If	$x > y$ (and a is positive)
then	$ax < ay$	then	$ax > ay$
and	$\dfrac{x}{a} < \dfrac{y}{a}$	and	$\dfrac{x}{a} > \dfrac{y}{a}$

Note: These properties also hold when the symbols $\leq$ and $\geq$ are used.

> **NOTE**
>
> Since dividing x by a is the same as multiplying x by the reciprocal of a, you can think of dividing by a as multiplying by the reciprocal of a.

EXAMPLE 4   **Using the multiplication and division properties with positive numbers**

Solve and graph the inequality on a number line:

a. $5x + 3 \leq 2x + 9$ **b.** $4(x - 1) > 2x + 6$

SOLUTION

a.

$5x + 3 \leq 2x + 9$	Given
$5x + 3 - 3 \leq 2x + 9 - 3$	Subtract 3 (or add -3).
$5x \leq 2x + 6$	Simplify.
$5x - 2x \leq 2x - 2x + 6$	Subtract $2x$ (or add $-2x$).
$3x \leq 6$	Simplify.
$\dfrac{3x}{3} \leq \dfrac{6}{3}$	Divide by 3 (or multiply by the reciprocal of 3).
$x \leq 2$	Simplify.

Any number less than or equal to 2 is a solution. The graph is as follows:

$5x + 3 \leq 2x + 9$ or, equivalently, $x \leq 2$

b.

$4(x - 1) > 2x + 6$	Given
$4x - 4 > 2x + 6$	Simplify.
$4x - 4 + 4 > 2x + 6 + 4$	Add 4.
$4x > 2x + 10$	Simplify.
$4x - 2x > 2x - 2x + 10$	Subtract $2x$ (or add $-2x$).
$2x > 10$	Simplify.
$\dfrac{2x}{2} > \dfrac{10}{2}$	Divide by 2 (or multiply by the reciprocal of 2).
$x > 5$	Simplify.

Any number greater than 5 is a solution. The graph is as follows:

$4(x - 1) > 2x + 6$ or, equivalently, $x > 5$

PROBLEM 4

Solve and graph:

a. $4x + 3 \leq 2x + 5$

b. $5(x - 1) > 3x + 1$

Answers to PROBLEMS

4. a.

$x \leq 1$

b.

$x > 3$

You may have noticed that the multiplication (or division) property allows us to multiply or divide only by a *positive* number. However, to solve the inequality $-2x < 4$, we need to divide by -2 or multiply by $-\frac{1}{2}$, a number that is *not* positive.

Let's first see what happens when we divide both sides of an inequality by a *negative* number. Consider the inequality

$$2 < 4$$

If we divide both sides of this inequality by -2, we get

$$\frac{2}{-2} < \frac{4}{-2} \quad \text{or} \quad -1 < -2$$

which is *not* true. To obtain a true statement, we must *reverse the inequality sign* and write:

$$-1 > -2$$

Similarly, consider

$$-6 > -8$$
$$\frac{-6}{-2} > \frac{-8}{-2}$$
$$3 > 4$$

which again is not true. However, the statement becomes true when we reverse the inequality sign and write $3 < 4$. So if we *divide* both sides of an inequality by a *negative* number, we must reverse the inequality sign to obtain an equivalent inequality. Similarly, if we *multiply* both sides of an inequality by a *negative* number, we must reverse the inequality sign to obtain an equivalent inequality:

$$2 < 4 \qquad \text{Given}$$
$$-3 \cdot 2 > -3 \cdot 4 \qquad \text{If we multiply by } -3, \text{ we reverse the inequality sign.}$$
$$-6 > -12$$

Note that now we are multiplying or dividing *both* sides of the inequality by a **negative** number. In such cases, we reverse the inequality symbol.

Here are some more examples.

$$3 < 12 \qquad\qquad\qquad\qquad 8 > 4$$
$$-2 \cdot 3 > -2 \cdot 12 \qquad\qquad \frac{8}{-2} < \frac{4}{-2}$$
$$\text{Reverse the sign.} \qquad\qquad \text{Reverse the sign.}$$
$$-6 > -24 \qquad\qquad\qquad\qquad -4 < -2$$

These properties are stated here.

MULTIPLICATION AND DIVISION PROPERTIES OF INEQUALITIES FOR NEGATIVE NUMBERS	You can *multiply* or *divide* both sides of an inequality by a *negative* number a and obtain an equivalent inequality provided you *reverse* the inequality sign. In symbols,

If	$x < y$ (and a is negative)	**If**	$x > y$ (and a is negative)
then	$ax > ay$ Reverse the sign.	**then**	$ax < ay$ Reverse the sign.
and	$\dfrac{x}{a} > \dfrac{y}{a}$ Reverse the sign.	**and**	$\dfrac{x}{a} < \dfrac{y}{a}$ Reverse the sign.

Note: These properties also hold when the symbols $\leq$ and $\geq$ are used.

We use these properties in Example 5.

EXAMPLE 5   Using the multiplication and division properties with negative numbers

Solve:

a. $-3x < 15$ **b.** $\frac{-x}{4} > 2$ **c.** $3(x - 2) \leq 5x + 2$

PROBLEM 5

Solve:

a. $-4x < 20$

b. $\frac{-x}{3} > 2$

c. $2(x - 1) \leq 4x + 1$

(continued)

Answers to PROBLEMS

5. a. $x > -5$ **b.** $x < -6$ **c.** $x \geq -\frac{3}{2}$

SOLUTION

a. To solve this inequality, we need the x by itself on the left; that is, we have to divide both sides by -3. Of course, when we do this, we must reverse the inequality sign.

$$-3x < 15 \qquad \text{Given}$$

$$\frac{-3x}{-3} > \frac{15}{-3} \qquad \text{Divide by } -3 \text{ and reverse the sign.}$$

$$x > -5 \qquad \text{Simplify.}$$

Any number greater than -5 is a solution.

b. Here we multiply both sides by -4 and reverse the inequality sign.

$$\frac{-x}{4} > 2 \qquad \text{Given}$$

$$-4\left(\frac{-x}{4}\right) < -4 \cdot 2 \qquad \text{Multiply by } -4 \text{ and reverse the inequality sign.}$$

$$x < -8 \qquad \text{Simplify.}$$

Any number less than -8 is a solution.

c.

$$3(x - 2) \leq 5x + 2 \qquad \text{Given}$$

$$3x - 6 \leq 5x + 2 \qquad \text{Simplify.}$$

$$3x - 6 + 6 \leq 5x + 2 + 6 \qquad \text{Add 6.}$$

$$3x \leq 5x + 8 \qquad \text{Simplify.}$$

$$3x - 5x \leq 5x - 5x + 8 \qquad \text{Subtract 5x (or add } -5x\text{).}$$

$$-2x \leq 8 \qquad \text{Simplify.}$$

$$\frac{-2x}{-2} \geq \frac{8}{-2} \qquad \text{Divide by } -2 \text{ (or multiply by the reciprocal of } -2\text{) and reverse the inequality sign.}$$

$$x \geq -4 \qquad \text{Simplify.}$$

Thus, any number greater than or equal to -4 is a solution.

Of course, to solve more complicated inequalities (such as those involving fractions), we simply follow the six-step procedure we use for solving linear equations (p.135).

EXAMPLE 6   **Using the six-step procedure to solve an inequality**

Solve:

$$\frac{-x}{4} + \frac{x}{6} < \frac{x - 3}{6}$$

SOLUTION   We follow the six-step procedure for linear equations.

$$\frac{-x}{4} + \frac{x}{6} < \frac{x - 3}{6} \qquad \text{Given}$$

1. Clear the fractions; the LCM is 12.

$$12 \cdot \left(\frac{-x}{4}\right) + 12 \cdot \frac{x}{6} < 12\left(\frac{x - 3}{6}\right)$$

2. Remove parentheses (use the distributive property).

$$-3x + 2x < 2(x - 3)$$

Collect like terms.

$$-x < 2x - 6$$

3. There are no numbers on the left, only the variable $-x$.

4. Subtract $2x$.

$$-x - 2x < 2x - 2x - 6$$

$$-3x < -6$$

PROBLEM 6

Solve:

$$\frac{-x}{3} + \frac{x}{4} < \frac{x - 8}{4}$$

Answers to **PROBLEMS**

6. $x > 6$

5. Divide by the coefficient of x, -3, and reverse the inequality sign.

$$\frac{-3x}{-3} > \frac{-6}{-3}$$

Remember that when you divide both sides of the equation by -3, which is negative, you have to reverse the inequality sign from $<$ to $>$.

$$x > 2$$

Thus, any number greater than 2 is a solution.

6. CHECK   Try $x = 12$. (It's a good idea to try the LCM. Do you see why?)

$$\frac{-x}{4} + \frac{x}{6} \overset{?}{<} \frac{x-3}{6}$$

$$\frac{-12}{4} + \frac{12}{6} \,\Big|\, \frac{12-3}{6}$$

$$-3 + 2 \,\Big|\, \frac{9}{6}$$

$$-1 \,\Big|\, \frac{3}{2}$$

Since $-1 < \frac{3}{2}$, the inequality is true. Of course, this only partially *verifies* the answer, since we're unable to try *every* solution.

C ⟩ Solving and Graphing Compound Inequalities

What about inequalities like the one at the beginning of this section? Inequalities such as

$$1 \le x \le 3$$

are called **compound inequalities** because they are equivalent to two other inequalities; that is, $1 \le x \le 3$ means $1 \le x$ *and* $x \le 3$. Thus, if we are asked to solve the inequalities

$$2 \le x \quad \text{and} \quad x \le 4$$

we write

$$2 \le x \le 4 \qquad \text{$2 \le x \le 4$ is a compound inequality.}$$

The graph of this inequality consists of all the points between 2 and 4, inclusive, as shown here:

$$2 \le x \le 4$$

<img: number line from -5 to 5 with solid dots at 2 and 4 and segment between>

You can also write the solution as   [2, 4]

include 2   start   end   include 4

> **NOTE**
>
> To graph $2 \le x \le 4$, place a solid dot at 2, a solid dot at 4, and draw the line segment *between* 2 and 4. In interval notation [2, 4]

To graph $2 < x < 4$ use the same procedure but place an open dot at 2, and at 4

$$2 < x < 4$$

<img: number line from -3 to 5 with open dots at 2 and 4>

or (2, 4)

do not include 2   start   end   do not include 4

The key to solving compound inequalities is to try writing the inequality in the form $a \le x \le b$ (or $a < x < b$), where a and b are real numbers. This form is called the *required* solution. Of course, if we try to write $x < 3$ and $x > 7$ in this form, we get $7 < x < 3$, which is not true.

In general, the notation used to write the solution of compound inequalities and their resulting graphs are as follows:

Symbols	Graph	Interval Notation
$a < x < b$		(a, b)
$a \leq x < b$		$[a, b)$
$a < x \leq b$		$(a, b]$
$a \leq x \leq b$		$[a, b]$

Note that when an open circle ○ or a parentheses like (or) are used, the endpoints are **not** included. When a closed circle ● or a bracket like [or] are used, the endpoints **are** included.

EXAMPLE 7   Solving compound inequalities
Solve and graph on a number line:

a. $1 < x$ and $x < 3$

b. $5 \geq -x$ and $x \leq -3$

c. $x + 1 \leq 5$ and $-2x < 6$

SOLUTION

a. The inequalities $1 < x$ *and* $x < 3$ are written as $1 < x < 3$. Thus, the solution consists of the numbers *between* 1 and 3, as shown here.

Draw an open circle at 1, an open circle at 3, and draw the line segment between 1 and 3.

$$1 < x < 3$$

or $(1, 3)$

b. Since we wish to write $5 \geq -x$ and $x \leq -3$ in the form $a \leq x \leq b$, we multiply both sides of $5 \geq -x$ by -1 to obtain

$$-1 \cdot 5 \leq -1 \cdot (-x)$$

$$-5 \leq x$$

We now have

$$-5 \leq x \quad \text{and} \quad x \leq -3$$

that is,

$$-5 \leq x \leq -3$$

Thus, the solution consists of all the numbers between -5 and -3 *inclusive*.

Note that -5 and -3 have solid dots.

$$-5 \leq x \leq -3$$

or $[-5, -3]$

c. We solve $x + 1 \leq 5$ by subtracting 1 from both sides to obtain

$$x + 1 - 1 \leq 5 - 1$$

$$x \leq 4$$

We now have

$$x \leq 4 \quad \text{and} \quad -2x < 6$$

We then divide both sides of $-2x < 6$ by -2 to obtain $x > -3$. We now have

$$x \leq 4 \quad \text{and} \quad x > -3$$

PROBLEM 7
Solve and graph:

a. $2 < x$ and $x < 4$

b. $3 \geq -x$ and $x \leq -1$

c. $x + 2 \leq 6$ and $-3x < 6$

Answers to PROBLEMS

7. a. (number line graph)   **b.** (number line graph)   **c.** (number line graph)

Rearranging these inequalities, we write

$$-3 < x \quad \text{and} \quad x \le 4$$

that is,

$$-3 < x \le 4$$

Here the solution consists of all numbers between -3 and 4 and the number 4 itself:

−3 not included   4 included

$$-3 < x \le 4$$

−5 −4 −3 −2 −1   0   1   2   3   4   5

$$(-3, 4]$$

CAUTION

When graphing **linear inequalities** (inequalities that can be written in the form $ax + b \le c$, where a, b, and c are real numbers and a is not 0), the graph is usually a *ray* pointing in the same direction as the inequality and with the variable on the left-hand side as shown here:

$$x \ge a$$
$$a$$

$$x \le a$$
$$a$$

On the other hand, when graphing a **compound inequality,** the graph is usually a *line segment:*

$$a \le x \le b$$
$$a \qquad\qquad b$$

D ❭ Solving an Application Involving Inequalities

EXAMPLE 8   Is smoking decreasing?

Do you think the number of smokers is increasing or decreasing? According to the National Health Institute Survey, the percent of smokers t years after 1965 is given by the equation $P = 43 - 0.64t$. This means that in 1965 ($t = 0$), the percent P of smokers was $P = 43 - 0.64(0) = 43\%$. If the trend continues, in what year would you expect the percent of smokers to be less than 23%? (Assume t, the number of years after 1965, is an integer.)

SOLUTION   We want to find the values of t for which $P = 43 - 0.64t < 23$.

$43 - 0.64t < 23$	Given
$43 - 43 - 0.64t < 23 - 43$	Subtract 43.
$-0.64t < -20$	Simplify.
$\dfrac{-0.64t}{-0.64} > \dfrac{-20}{-0.64}$	Divide by -0.64. (Don't forget to reverse the inequality!)
$t > 31.25$	Simplify.

This means that 31.25 or 32 years after 1965—that is, by $1965 + 32 = 1997$—the number of smokers should be less than 23%.

PROBLEM 8

Tobacco use is decreasing in the 12-to-17-year age group. The percent of smokers in that group is $P = 20 - 2x$, where x is the number of years after 1997 ($x = 0$). Using this information, when would the percent of smokers in this group be less than 10%?

Answers to PROBLEMS

8. $x > 5$, or more than 5 years after 1997, that is, in 2003

❯ Exercises **2.7**

⟨ **A** ⟩ **Order of Numbers**   In Problems 1–10, fill in the blank with $>$ or $<$ so that the resulting statement is true.

1. 8 _____ 9

2. -8 _____ -9

3. -4 _____ -9

4. 7 _____ 3

5. $\frac{1}{4}$ _____ $\frac{1}{3}$

6. $\frac{1}{5}$ _____ $\frac{1}{2}$

7. $-\frac{2}{3}$ _____ -1

8. $-\frac{1}{5}$ _____ -1

9. $-3\frac{1}{4}$ _____ -3

10. $-4\frac{1}{5}$ _____ -4

⟨ **B** ⟩ **Solving and Graphing Inequalities**   In Problems 11–30, solve and graph the inequalities on a number line.

11. $2x + 6 \leq 8$

12. $4y - 5 \leq 3$

13. $-3y - 4 \geq -10$

14. $-4z - 2 \geq 6$

15. $-5x + 1 < -14$

16. $-3x + 1 < -8$

17. $3a + 4 \leq a + 10$

18. $4b + 4 \leq b + 7$

19. $5z - 12 \geq 6z - 8$

20. $5z + 7 \geq 7z + 19$

21. $10 - 3x \leq 7 - 6x$

22. $8 - 4y \leq -12 + 6y$

23. $5(x + 2) < 3(x + 3) + 1$

24. $5(4 - 3x) < 7(3 - 4x) + 12$

25. $-2x + \frac{1}{4} \geq 2x + \frac{4}{5}$

26. $6x + \frac{1}{7} \geq 2x - \frac{2}{7}$

27. $\frac{x}{5} - \frac{x}{4} \leq 1$

28. $\frac{x}{3} - \frac{x}{2} \leq 1$

29. $\frac{7x + 2}{6} + \frac{1}{2} \geq \frac{3}{4}x$

30. $\frac{8x - 23}{6} + \frac{1}{3} \geq \frac{5}{2}x$

⟨ **C** ⟩ **Solving and Graphing Compound Inequalities**   In Problems 31–40, solve and graph the inequalities on a number line.

31. $x < 3$ and $-x < -2$

32. $-x < 5$ and $x < 2$

33. $x + 1 < 4$ and $-x < -1$

34. $x - 2 < 1$ and $-x < 2$

35. $x - 2 < 3$ and $2 > -x$

36. $x - 3 < 1$ and $1 > -x$

37. $x + 2 < 3$ and $-4 < x + 1$

38. $x + 4 < 5$ and $-1 \leq x + 2$

39. $x - 1 \geq 2$ and $x + 7 < 12$

40. $x - 2 > 1$ and $-x \geq -5$

In Problems 41–50, write the given information as an inequality.

41. The temperature t in your refrigerator is between 20°F and 40°F.

42. The height h (in feet) of any mountain is always less than or equal to that of Mount Everest, 29,029 feet.

43. Joe's salary s for this year will be between $12,000 and $13,000.

44. Your gas mileage m (in miles per gallon) is between 18 and 22, depending on your driving.

45. The number of possible eclipses e in a year varies from 2 to 7, inclusive.

46. The assets a of the Du Pont family are in excess of $150 billion.

47. The cost c of ordinary hardware (tools, mowers, and so on) is between $3.50 and $4.00 per pound.

48. The range r (in miles) of a rocket is always less than 19,000 miles.

49. The altitude a (in feet) attained by the first liquid-fueled rocket was less than 41 feet.

50. The number of days d a person remained in the weightlessness of space before 1988 did not exceed 370.

⟨ **D** ⟩ **Solving Applications Involving Inequalities**

51. *NCAA men's basketball* The number N of NCAA men's college basketball teams t years after 1980 is given by $N = 720 + 5t$. When would you expect that the NCAA would have more than 800 teams?

52. *Heart disease* According to the U.S. National Center for Health Statistics, the number D of deaths per 100,000 population attributed to heart disease has been decreasing for t years after 1999 and is given by $D = -10t + 260$. After what year would you expect the number of deaths per 100,000 to be less than 200?

53. *Cost of a hospital room* From 1980 on, the daily room charge C of a hospital room has been increasing and is given by $C = 127 + 17t$, where t is the number of years after 1980. After what year would you expect the average hospital room rate to surpass the $300 per day mark?

54. *Poultry consumption* The annual per capita consumption of poultry products (in pounds) t years after 1985 is given by $C = 45 + 2t$. When would you expect the per capita consumption of poultry products to exceed 60 pounds per year? (See the Collaborative Learning at the end of the chapter.)

❯❯❯ *Using Your Knowledge*

A Question of Inequality Can you solve the problem in the cartoon? Let

$$J = \text{Joe's height}$$
$$B = \text{Bill's height}$$
$$F = \text{Frank's height}$$
$$S = \text{Sam's height}$$

Translate each statement into an equation or an inequality.

55. Joe is 5 feet (60 inches) tall.

56. Bill is taller than Frank.

57. Frank is 3 inches shorter than Sam.

58. Frank is taller than Joe.

59. Sam is 6 feet 5 inches (77 inches) tall.

60. According to the statement in Problem 56, Bill is taller than Frank, and according to the statement in Problem 58, Frank is taller than Joe. Write these two statements as an inequality of the form $a > b > c$.

61. Based on the answer to Problem 60 and the fact that you can obtain Frank's height by using the results of Problems 57 and 59, what can you really say about Bill's height?

〉〉〉 *Write On*

62. Write the similarities and differences in the procedures used to solve equations and inequalities.

63. As you solve an inequality, when do you have to change the direction of the inequality?

64. A student wrote "$2 < x < -5$" to indicate that x was between 2 and -5. Why is this wrong?

65. Write the steps you would use to solve the inequality $-3x < 15$.

〉〉〉 *Concept Checker*

Fill in the blank(s) with the correct word(s), phrase, or mathematical statement.

66. According to the **Addition Principle of Inequality** if $x < y$ and a is a **real** number, _____.

67. According to the **Subtraction Principle of Inequality** if $x < y$ and a is a **real** number, _____.

68. According to the **Multiplication Principle of Inequality** if $x < y$ and a is a **positive** number, _____.

69. According to the **Multiplication Principle of Inequality** if $x < y$ and a is a **negative** number, _____.

70. According to the **Division Principle of Inequality** if $x < y$ and a is a **positive** number, _____.

71. According to the **Division Principle of Inequality** if $x < y$ and a is a **negative** number, _____.

$x + a < y + a$ $\dfrac{-x}{a} < \dfrac{-y}{a}$

$a - x < a - y$ $\dfrac{x}{a} < \dfrac{y}{a}$

$x - a < y - a$ $ax < ay$

$ax > ay$ $\dfrac{-x}{a} > \dfrac{-y}{a}$

$\dfrac{x}{a} > \dfrac{y}{a}$

〉〉〉 *Mastery Test*

Solve and graph on a number line:

72. $x + 2 \leq 6$ and $-3x \leq 9$

73. $3 \geq -x$ and $x \leq -1$

74. $2 < x$ and $x < 4$

75. $\dfrac{-x}{3} + \dfrac{x}{4} < \dfrac{x-4}{4}$

76. $4x + 5 < x + 11$

77. $3(x - 1) > x + 3$

78. $4x - 7 < 3(x - 2)$

79. $3(x + 1) \geq 2x + 5$

80. $x \geq -2$

81. $x < 1$

Fill in the blank with $>$ or $<$ so that the resulting statement is true:

82. 5 _____ 7

83. -2 _____ -1

84. -4 _____ -5

85. $\dfrac{1}{3}$ _____ -3

86. According to the U.S. Department of Agriculture, the total daily grams of fat F consumed per person is modeled by the equation $F = 181.5 + 0.8t$, where t is the number of years after 2000. After what year would you expect the daily consumption of grams of fat to exceed 189.5?

〉〉〉 *Skill Checker*

In Problems 87–95 graph the number on a number line.

87. 1

88. 4

89. -2

90. -3	**91.** 0	**92.** 3
93. 5	**94.** -1	**95.** -4

› Collaborative Learning

Average Annual Per Capita Consumption
(in pounds)

Source: American Meat Institute.

Form four groups: Beefies, porkies, chickens, and turkeys.

1. In Problem 54 of Section 2.7, we gave a formula for the consumption C of poultry in the United States. Use the formula to estimate the U.S. consumption of poultry in 2001. How close is the estimate to the one given in the table? (2001 is the last year shown.)

2. From 1990 on, the consumption of beef (B), pork (P), and turkey (T) has been rather steady. Write a consumption equation that approximates the annual per capita consumption of *your product*.

3. The consumption of chicken (C) has been increasing about two pounds per year since 1980 so that $C = 45 + 2t$. In what year is C highest? In what year is it lowest?

4. Write an inequality comparing the annual consumption of *your product* and C.

5. In what year(s) was the annual consumption of *your product* less than C, equal to C, and greater than C?

› Research Questions

1. What does the word *papyrus* mean? Explain how the Rhind papyrus got its name.

2. Write a report on the contents and origins of the Rhind papyrus.

3. The Rhind papyrus is one of two documents detailing Egyptian mathematics. What is the name of the other document, and what type of material does it contain?

4. Write a report about the rule of false position and the rule of double false position.

5. Find out who invented the symbols for greater than ($>$) and less than ($<$).

6. What is the meaning of the word *geometry*? Give an account of the origin of the subject.

7. Problem 50 in the Rhind papyrus gives the method for finding the area of a circle. Write a description of the problem and the method used.

❭ Summary **Chapter 2**

Section	Item	Meaning	Example
2.1	Equation	A statement indicating that two expressions are **equal**	$x - 8 = 9$, $\frac{1}{2} - 2x = \frac{3}{4}$, and $0.2x + 8.9 = \frac{1}{2} + 6x$ are equations.
2.1A	Solutions	The solutions of an equation are the replacements of the variable that make the equation a **true** statement.	4 is a solution of $x + 1 = 5$.
	Equivalent equations	Two equations are **equivalent** if their solutions are the **same.**	$x + 1 = 4$ and $x = 3$ are equivalent.
2.1B	The addition property of equality	$a = b$ is equivalent to $a + c = b + c$.	$x - 1 = 2$ is equivalent to $x - 1 + 1 = 2 + 1$.
	The subtraction property of equality	$a = b$ is equivalent to $a - c = b - c$.	$x + 1 = 2$ is equivalent to $x + 1 - 1 = 2 - 1$.
2.1C	Conditional equation	An equation with **one** solution	$x + 7 = 9$ is a conditional equation whose solution is 2.
	Contradictory equation	An equation with **no** solution	$x + 1 = x + 2$ is a contradictory equation.
	Identity	An equation with **infinitely** many solutions	$2(x + 1) - 5 = 2x - 3$ is an identity. Any real number is a solution.
2.2A	The multiplication property of equality	$a = b$ is equivalent to $ac = bc$ if c is not 0.	$\frac{x}{2} = 3$ is equivalent to $2 \cdot \frac{x}{2} = 2 \cdot 3$.
	The division property of equality	$a = b$ is equivalent to $\frac{a}{c} = \frac{b}{c}$ if c is not 0.	$2x = 6$ is equivalent to $\frac{2x}{2} = \frac{6}{2}$.
2.2B	Reciprocal	The reciprocal of $\frac{a}{b}$ is $\frac{b}{a}$.	The reciprocal of $\frac{5}{2}$ is $\frac{2}{5}$.
2.2C	LCM (least common multiple)	The smallest number that is a multiple of each of the given numbers	The LCM of 3, 8, and 9 is 72.
2.3A	Linear equation	An equation that can be written in the form $ax + b = c$	$5x + 5 = 2x + 6$ is a linear equation (it can be written as $3x + 5 = 6$).
2.3B	Literal equation	An equation that contains letters other than the variable for which we wish to solve	$I = Prt$ and $C = 2\pi r$ are literal equations.
2.4	RSTUV method	To solve word problems, **R**ead, **S**elect a variable, **T**ranslate, **U**se algebra, and **V**erify your answer.	
2.4A	Consecutive integers	If n is an integer, the next consecutive integer is $n + 1$. If n is an even (odd) integer, the next consecutive even (odd) integer is $n + 2$.	4, 5, and 6 are three consecutive integers. 2, 4, and 6 are three consecutive even integers.

Section	Item	Meaning	Example
2.4C	Complementary angles	Two angles whose sum measures 90°	Two angles with measures 50° and 40° are complementary.
	Supplementary angles	Two angles whose sum measures 180°	Two angles with measures 35° and 145° are supplementary.
2.6B	Perimeter	The distance around a geometric figure	The perimeter P of a rectangle with length L and width W is $P = 2L + 2W$.
	Area of a rectangle	The area A of a rectangle of length L and width W is $A = LW$.	The area of a rectangle 8 inches long and 4 inches wide is $A = 8 \text{ in.} \cdot 4 \text{ in.} = 32 \text{ in.}^2$
	Area of a triangle	The area A of a triangle with base b and height h is $A = \frac{1}{2}bh$.	The area of a triangle with base 5 cm and height 10 cm is $A = \frac{1}{2} \cdot 5 \text{ cm} \cdot 10 \text{ cm} = 25 \text{ cm}^2$.
	Area of a circle	The area A of a circle of radius r is $A = \pi r^2$.	The area of a circle whose radius is 5 inches is $A = \pi(5)^2$, that is, $25\pi \text{ in.}^2$
	Circumference of a circle	The circumference of a circle with radius r is $C = 2\pi r$.	The circumference of a circle whose radius is 10 inches is $C = 2\pi(10)$, that is, $20\pi \text{ in.}$
2.6C	Vertical angles	Angles ① and ② are vertical angles.	
2.7	Inequality	A statement with $>$, $<$, $\geq$, or $\leq$ for its verb	$2x + 1 > 5$ and $3x - 5 \leq 7 - x$ are inequalities.
2.7B	The addition property of inequalities	$a < b$ is equivalent to $a + c < b + c$.	$x - 1 < 2$ is equivalent to $x - 1 + 1 < 2 + 1$.
	The subtraction property of inequalities	$a < b$ is equivalent to $a - c < b - c$.	$x + 1 < 2$ is equivalent to $x + 1 - 1 < 2 - 1$.
	The multiplication property of inequalities	$a < b$ is equivalent to $ac < bc$ if $c > 0$ or $ac > bc$ if $c < 0$	$\frac{x}{2} < 3$ is equivalent to $2 \cdot \frac{x}{2} < 2 \cdot 3$.
	The division property of inequalities	$a < b$ is equivalent to $\frac{a}{c} < \frac{b}{c}$, if $c > 0$ or $\frac{a}{c} > \frac{b}{c}$, if $c < 0$	$-2x < 6$ is equivalent to $\frac{-2x}{-2} > \frac{6}{-2}$.

>Review Exercises **Chapter 2**

(If you need help with these exercises, look in the section indicated in brackets.)

1. ⟨**2.1A**⟩ *Determine whether the given number satisfies the equation.*

  a. $5; 7 = 14 - x$ **b.** $4; 13 = 17 - x$

  c. $-2; 8 = 6 - x$

2. ⟨**2.1B**⟩ *Solve the given equation.*

  a. $x - \frac{1}{3} = \frac{1}{3}$ **b.** $x - \frac{5}{7} = \frac{2}{7}$

  c. $x - \frac{5}{9} = \frac{1}{9}$

3. ⟨**2.1B**⟩ *Solve the given equation.*

  a. $-3x + \frac{5}{9} + 4x - \frac{2}{9} = \frac{5}{9}$

  b. $-2x + \frac{4}{7} + 3x - \frac{2}{7} = \frac{6}{7}$

  c. $-4x + \frac{5}{6} + 5x - \frac{1}{6} = \frac{5}{6}$

4. ⟨**2.1C**⟩ *Solve the given equation.*

  a. $3 = 4(x - 1) + 2 - 3x$

  b. $4 = 5(x - 1) + 9 - 4x$

  c. $5 = 6(x - 1) + 8 - 5x$

5. ⟨**2.1C**⟩ *Solve the given equation.*

  a. $6 + 3(x + 1) = 2 + 3x$

  b. $-2 + 4(x - 1) = -7 - 4x$

  c. $-1 - 2(x + 1) = 3 - 2x$

6. ⟨**2.1C**⟩ *Solve the given equation.*

  a. $5 + 2(x + 1) = 2x + 7$

  b. $-2 + 3(x - 1) = -5 + 3x$

  c. $-3 - 4(x - 1) = 1 - 4x$

7. ⟨**2.2A**⟩ *Solve the given equation.*

  a. $\frac{1}{5}x = -3$ **b.** $\frac{1}{7}x = -2$

  c. $5x = -10$

8. ⟨**2.2B**⟩ *Solve the given equation.*

  a. $-\frac{3}{4}x = -9$ **b.** $-\frac{3}{5}x = -9$

  c. $-\frac{2}{3}x = -6$

9. ⟨**2.2C**⟩ *Solve the given equation.*

  a. $\frac{x}{3} + \frac{2x}{4} = 5$ **b.** $\frac{x}{4} + \frac{3x}{2} = 6$

  c. $\frac{x}{5} + \frac{3x}{10} = 10$

10. ⟨**2.2C**⟩ *Solve the given equation.*

  a. $\frac{x}{3} - \frac{x}{4} = 1$ **b.** $\frac{x}{2} - \frac{x}{7} = 10$

  c. $\frac{x}{4} - \frac{x}{5} = 2$

11. ⟨**2.2C**⟩ *Solve the given equation.*

  a. $\frac{x - 1}{4} - \frac{x + 1}{6} = 1$

  b. $\frac{x - 1}{6} - \frac{x + 1}{8} = 0$

  c. $\frac{x - 1}{8} - \frac{x + 1}{10} = 0$

12. ⟨**2.2D**⟩ *Solve.*

  a. What percent of 30 is 6?

  b. What percent of 40 is 4?

  c. What percent of 50 is 10?

13. ⟨**2.2D**⟩ *Solve.*

  a. 20 is 40% of what number?

  b. 30 is 90% of what number?

  c. 25 is 75% of what number?

14. ⟨**2.3A**⟩ *Solve.*

  a. $\frac{1}{5} - \frac{x}{4} = \frac{19(x + 4)}{20}$

  b. $\frac{1}{5} - \frac{x}{4} = \frac{6(x + 5)}{5}$

  c. $\frac{1}{5} - \frac{x}{4} = \frac{29(x + 6)}{20}$

15. ⟨ **2.3B** ⟩ *Solve.*

   a. $A = \frac{1}{2}bh$; solve for h.

   b. $C = 2\pi r$; solve for r.

   c. $V = \frac{bh}{3}$; solve for b.

16. ⟨ **2.4A** ⟩ *Find the numbers described.*

   a. The sum of two numbers is 84 and one of the numbers is 20 more than the other.

   b. The sum of two numbers is 47 and one of the numbers is 19 more than the other.

   c. The sum of two numbers is 81 and one of the numbers is 23 more than the other.

17. ⟨ **2.4B** ⟩ *If you eat a fried chicken breast and a 3-ounce piece of apple pie, you have consumed 578 calories.*

   a. If the pie has 22 more calories than the chicken breast, how many calories are in each?

   b. Repeat part **a** where the number of calories consumed is 620 and the pie has 38 more calories than the chicken breast.

   c. Repeat part **a** where the number of calories consumed is 650 and the pie has 42 more calories than the chicken breast.

18. ⟨ **2.4C** ⟩ *Find the measure of an angle whose supplement is:*

   a. 20 degrees less than 3 times its complement.

   b. 30 degrees less than 3 times its complement.

   c. 40 degrees less than 3 times its complement.

19. ⟨ **2.5A** ⟩ *Solve.*

   a. A car leaves a town traveling at 40 miles per hour. An hour later, another car leaves the same town traveling at 50 miles per hour in the same direction. How long does it take the second car to overtake the first one?

   b. Repeat part **a** where the first car travels at 30 miles per hour and the second one at 50 miles per hour.

   c. Repeat part **a** where the first car travels at 40 miles per hour and the second one at 60 miles per hour.

20. ⟨ **2.5B** ⟩ *Solve.*

   a. How many pounds of a product selling at $1.50 per pound should be mixed with 15 pounds of another product selling at $3 per pound to obtain a mixture selling at $2.40 per pound?

   b. Repeat part **a** where the products sell for $2, $3, and $2.50, respectively.

   c. Repeat part **a** where the products sell for $6, $2, and $4.50, respectively.

21. ⟨ **2.5C** ⟩ *Solve.*

   a. A woman invests $30,000, part at 5% and part at 6%. Her annual interest amounts to $1600. How much does she have invested at each rate?

   b. Repeat part **a** where the rates are 7% and 9%, respectively, and her annual return amounts to $2300.

   c. Repeat part **a** where the rates are 6% and 10%, respectively, and her annual return amounts to $2000.

22. ⟨ **2.6A** ⟩ *The cost C of a long-distance call is $C = 3.05m + 3$, where m is the number of minutes the call lasts.*

   a. Solve for m and then find the length of a call that cost $27.40.

   b. Repeat part **a** where $C = 3.15m + 3$ and the call cost $34.50.

   c. Repeat part **a** where $C = 3.25m + 2$ and the call cost $21.50.

23. ⟨ **2.6C** ⟩ *Find the measures of the marked angles.*

   a.

   b.

   c.

24. ⟨**2.7A**⟩ *Fill in the blank with the symbol $<$ or $>$ to make the resulting statement true.*

  a. -8 _____ -7

  b. $\frac{1}{2}$ _____ -3

  c. 4 _____ $4\frac{1}{3}$

25. ⟨**2.7B**⟩ *Solve and graph the given inequality.*

  a. $4x - 2 < 2(x + 2)$

  b. $5x - 4 < 2(x + 1)$

  c. $7x - 1 < 3(x + 1)$

26. ⟨**2.7B**⟩ *Solve and graph the given inequality.*

  a. $6(x - 1) \geq 4x + 2$

  b. $5(x - 1) \geq 2x + 1$

  c. $4(x - 2) \geq 2x + 2$

27. ⟨**2.7B**⟩ *Solve and graph the given inequality.*

  a. $-\dfrac{x}{3} + \dfrac{x}{6} \leq \dfrac{x - 1}{6}$

  b. $-\dfrac{x}{4} + \dfrac{x}{7} \leq \dfrac{x - 1}{7}$

  c. $-\dfrac{x}{5} + \dfrac{x}{3} \leq \dfrac{x - 1}{3}$

28. ⟨**2.7C**⟩ *Solve and graph the compound inequality.*

  a. $x + 2 \leq 4$ and $-2x < 6$

  b. $x + 3 \leq 5$ and $-3x < 9$

  c. $x + 1 \leq 2$ and $-4x < 8$

⟩ Practice Test **Chapter 2**

(Answers on page 200)

Visit www.mhhe.com/bello to view helpful videos that provide step-by-step solutions to several of the problems below.

1. Does the number 3 satisfy the equation $6 = 9 - x$?

2. Solve $x - \frac{2}{7} = \frac{3}{7}$.

3. Solve $-2x + \frac{7}{8} + 3x - \frac{5}{8} = \frac{5}{8}$.

4. Solve $2 = 3(x - 1) + 5 - 2x$.

5. Solve $2 + 5(x + 1) = 8 + 5x$.

6. Solve $-3 - 2(x - 1) = -1 - 2x$.

7. Solve $\frac{2}{3}x = -4$.

8. Solve $-\frac{2}{3}x = -6$.

9. Solve $\frac{x}{4} + \frac{2x}{3} = 11$.

10. Solve $\frac{x}{3} - \frac{x}{5} = 2$.

11. Solve $\frac{x - 2}{5} - \frac{x + 1}{8} = 0$.

12. What percent of 55 is 11?

13. Nine is 36% of what number?

14. Solve $\frac{1}{5} - \frac{x}{3} = \frac{23(x + 5)}{15}$.

15. Solve for h in $S = \frac{1}{3}\pi r^2 h$.

16. The sum of two numbers is 75. If one of the numbers is 15 more than the other, what are the numbers?

17. A man has invested a certain amount of money in stocks and bonds. His annual return from these investments is $840. If the stocks produce $230 more in returns than the bonds, how much money does he receive annually from each investment?

18. Find the measure of an angle whose supplement is 50° less than 3 times its complement.

19. A freight train leaves a station traveling at 30 miles per hour. Two hours later, a passenger train leaves the same station traveling in the same direction at 42 miles per hour. How long does it take for the passenger train to catch the freight train?

20. How many pounds of coffee selling for $1.10 per pound should be mixed with 30 pounds of coffee selling for $1.70 per pound to obtain a mixture that sells for $1.50 per pound?

21. An investor bought some municipal bonds yielding 5% annually and some certificates of deposit yielding 7% annually. If her total investment amounts to $20,000 and her annual return is $1160, how much money is invested in bonds and how much in certificates of deposit?

22. The cost C of riding a taxi is $C = 1.95 + 0.85m$, where m is the number of miles (or fraction) you travel.
   a. Solve for m.
   b. How many miles did you travel if the cost of the ride was $20.65?

23. Find x and the measures of the marked angles.

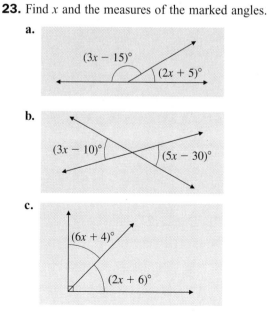

   a.

   b.

   c.

24. Fill in the blank with $<$ or $>$ to make the resulting statement true.
   a. -3 _____ -5
   b. $-\frac{1}{3}$ _____ 3

25. Solve and graph the inequality.
   a. $-\frac{x}{2} + \frac{x}{4} \leq \frac{x + 2}{4}$
   b. $x + 1 \leq 3$ and $-2x < 6$

❯Answers to Practice Test **Chapter 2**

Answer	If You Missed	Review		
	Question	Section	Examples	Page
1. Yes	1	2.1	1	107
2. $x = \dfrac{5}{7}$	2	2.1	2	108
3. $x = \dfrac{3}{8}$	3	2.1	3	109–110
4. $x = 0$	4	2.1	4, 5	111–112
5. No solution	5	2.1	6	113
6. All real numbers	6	2.1	7	113
7. $x = -6$	7	2.2	1, 2, 3	119–122
8. $x = 9$	8	2.2	1, 2, 3	119–122
9. $x = 12$	9	2.2	4	124
10. $x = 15$	10	2.2	4	124
11. $x = 7$	11	2.2	5	125
12. 20%	12	2.2	7	126
13. 25	13	2.2	8	126–127
14. $x = -4$	14	2.3	1, 2, 3	133–136
15. $h = \dfrac{3S}{\pi r^2}$	15	2.3	4, 5, 6	137–139
16. 30 and 45	16	2.4	1, 2	145–146
17. $305 from bonds and $535 from stocks	17	2.4	3	147
18. 20°	18	2.4	4	148
19. 5 hours	19	2.5	1, 2, 3	154–157
20. 15 pounds	20	2.5	4	158
21. $12,000 in bonds, $8000 in certificates	21	2.5	5	160
22. a. $m = \dfrac{C - 1.95}{0.85}$ **b.** 22 miles	22	2.6	1, 2	166–167
23. a. $x = 38$; 99° and 81° **b.** $x = 10$; both are 20° **c.** $x = 10$; 64° and 26°	23	2.6	3, 4, 5, 6, 7	167–172
24. a. $>$ **b.** $<$	24	2.7	1	180
25. a. $x \geq -1$ **b.** $-3 < x \leq 2$	25	2.7	2, 3, 4, 5, 6, 7	181–189

25. a. $x \geq -1$

b. $-3 < x \leq 2$

›Cumulative Review **Chapters 1–2**

1. Find the additive inverse (opposite) of -7.

2. Find: $\left|-9\frac{9}{10}\right|$

3. Find: $-\frac{2}{7} + \left(-\frac{2}{9}\right)$

4. Find: $-0.7 - (-8.9)$

5. Find: $(-2.4)(3.6)$

6. Find: $-(2^4)$

7. Find: $-\frac{7}{8} \div \left(-\frac{5}{24}\right)$

8. Evaluate $y \div 5 \cdot x - z$ for $x = 6, y = 60, z = 3$.

9. Which property is illustrated by the following statement?

$9 \cdot (8 \cdot 5) = 9 \cdot (5 \cdot 8)$

10. Multiply: $6(5x + 7)$

11. Combine like terms: $-5cd - (-6cd)$

12. Simplify: $2x - 2(x + 4) - 3(x + 1)$

13. Write in symbols: The quotient of $(a - 4b)$ and c

14. Does the number 4 satisfy the equation $11 = 15 - x$?

15. Solve for x: $5 = 4(x - 3) + 4 - 3x$

16. Solve for x: $-\frac{7}{3}x = -21$

17. Solve for x: $\frac{x}{3} - \frac{x}{5} = 2$

18. Solve for x: $4 - \frac{x}{4} = \frac{2(x + 1)}{9}$

19. Solve for b in the equation $S = 6a^2b$.

20. The sum of two numbers is 155. If one of the numbers is 35 more than the other, what are the numbers?

21. Maria has invested a certain amount of money in stocks and bonds. The annual return from these investments is $595. If the stocks produce $105 more in returns than the bonds, how much money does Maria receive annually from each type of investment?

22. Train A leaves a station traveling at 40 mph. Six hours later, train B leaves the same station traveling in the same direction at 50 mph. How long does it take for train B to catch up to train A?

23. Arlene purchased some municipal bonds yielding 12% annually and some certificates of deposit yielding 14% annually. If Arlene's total investment amounts to $5000 and the annual income is $660, how much money is invested in bonds and how much is invested in certificates of deposit?

24. Solve and graph: $-\frac{x}{6} + \frac{x}{5} \leq \frac{x - 5}{5}$

Chapter

3

three

▶ **Graphs of Linear Equations, Inequalities, and Applications**

The Human Side of Algebra

René Descartes, "the reputed founder of modern philosophy," was born March 31, 1596, near Tours, France. His frail health caused his formal education to be delayed until he was 8, when his father enrolled him at the Royal College at La Fleche. It was soon noticed that the boy needed more than normal rest, and he was advised to stay in bed as long as he liked in the morning. Descartes followed this advice and made a lifelong habit of staying in bed late whenever he could.

The idea of analytic geometry came to Descartes while he watched a fly crawl along the ceiling near a corner of his room. Descartes described the path of the fly in terms of its distance from the adjacent walls by developing the *Cartesian coordinate system,* which we study in Section 3.1. Impressed by his knowledge of philosophy and mathematics, Queen Christine of Sweden engaged Descartes as a private tutor. He arrived in Sweden to discover that she expected him to teach her philosophy at 5 o'clock in the morning in the ice-cold library of her palace. Deprived of his beloved morning rest, Descartes caught "inflammation of the lungs," from which he died on February 11, 1650, at age 53.

Later on, the first person to use the word *graph* was James Joseph Sylvester, a teacher born in London in 1814, who used the word in an article published in 1878. Sylvester passed the knowledge of graphs to his students, writing articles in *Applied Mechanics* and admonishing his students that they would "do well to graph on squared paper some curves like the following." Unfortunately for his students, Sylvester had a dangerous temper and attacked a student with a sword cane at the University of Virginia. The infraction? The student was reading a newspaper during his class. So now you can learn from this lesson: concentrate on your graphs and do not read newspapers in the classroom!

3.1 Line Graphs, Bar Graphs, and Applications

Objectives

A ▷ Graph (plot) ordered pairs of numbers.

B ▷ Determine the coordinates of a point in the plane.

C ▷ Read and interpret ordered pairs on a line graph.

D ▷ Read and interpret ordered pairs on a bar graph.

E ▷ Find the quadrant in which a point lies.

F ▷ Given a chart or ordered pairs, create the corresponding line graph.

▷ To Succeed, Review How To . . .

1. Evaluate an expression (pp. 61–62, 69–72).
2. Solve linear equations (pp. 132–136).

▷ Getting Started

Hurricanes and Graphs

The map shows the position of Hurricane Desi. The hurricane is near the intersection of the vertical line indicating 90° longitude and the horizontal line indicating 25° latitude. This point can be identified by assigning to it an **ordered pair** of numbers, called **coordinates,** showing the longitude first and the latitude second. Thus, the hurricane would have the coordinates

(91, 25)

This is the longitude (units right or left).   This is the latitude (units up or down).

This method is used to give the position of cities, islands, ships, airplanes, and so on. For example, the coordinates of New Orleans on the map are (90, 30), whereas those of Pensacola are approximately (88, 31). In mathematics we use a system very similar to this one to locate points in a plane. In this section we learn how to graph points in a *Cartesian plane* and then examine the relationship of these points to linear equations.

Here is the way we construct a **Cartesian coordinate system** (also called a rectangular coordinate system).

1. Draw a number line (Figure 3.1).
2. Draw another number line perpendicular to the first one and crossing it at 0 (the origin) (see Figure 3.2).

On the number line, each point on the graph is a **number.** On a coordinate plane, each point is the **graph** of an ordered pair. The individual numbers in an ordered pair are called **coordinates.** For example, the point *P* in Figure 3.3 is associated with the ordered pair (2, 3). The first coordinate of *P* is 2 and the second coordinate is 3. The point $Q(-1, 2)$ has a first coordinate of -1 and a second coordinate of 2. We call the horizontal number line the **x-axis** and label it with the letter *x*; the vertical number line is called the **y-axis** and is labeled with the letter *y*. We can now say that the point $P(2, 3)$ has *x*-coordinate (**abscissa**) 2, and *y*-coordinate (**ordinate**) 3.

>Figure 3.1

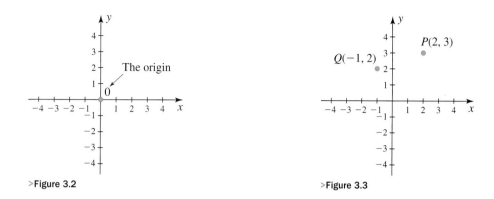

>Figure 3.2

>Figure 3.3

A > Graphing Ordered Pairs

In general, if a point P has coordinates (x, y), we can always locate or graph the point in the coordinate plane. We start at the origin and go x units to the *right* if x is *positive;* we go to the *left* if x is *negative*. We then go y units *up* if y is *positive*, *down* if y is *negative*.

x-coordinate ┐   ┌ y-coordinate

(x, y)

Tells us to go right or left ┘   └ Tells us to go up or down

For example, to graph the point $R(3, -2)$, we start at the origin and go 3 units right (since the x-coordinate 3 is positive) and 2 units down (since the y-coordinate -2 is negative). The point is graphed in Figure 3.4.

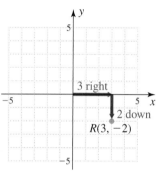

>Figure 3.4

> **NOTE**
>
> All points on the x-axis have y-coordinate 0 (zero units up or down); all points on the y-axis have x-coordinate 0 (zero units right or left).

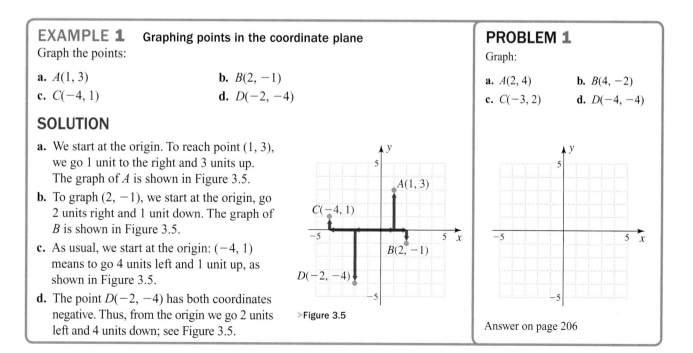

EXAMPLE 1   **Graphing points in the coordinate plane**

Graph the points:

a. $A(1, 3)$ **b.** $B(2, -1)$

c. $C(-4, 1)$ **d.** $D(-2, -4)$

SOLUTION

a. We start at the origin. To reach point $(1, 3)$, we go 1 unit to the right and 3 units up. The graph of A is shown in Figure 3.5.

b. To graph $(2, -1)$, we start at the origin, go 2 units right and 1 unit down. The graph of B is shown in Figure 3.5.

c. As usual, we start at the origin: $(-4, 1)$ means to go 4 units left and 1 unit up, as shown in Figure 3.5.

d. The point $D(-2, -4)$ has both coordinates negative. Thus, from the origin we go 2 units left and 4 units down; see Figure 3.5.

>Figure 3.5

PROBLEM 1

Graph:

a. $A(2, 4)$ **b.** $B(4, -2)$

c. $C(-3, 2)$ **d.** $D(-4, -4)$

Answer on page 206

B › Finding Coordinates

Since every point P in the plane is associated with an ordered pair (x, y), we should be able to find the coordinates of any point as shown in Example 2.

EXAMPLE 2   Finding coordinates

Determine the coordinates of each of the points in Figure 3.6.

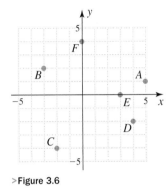

> Figure 3.6

PROBLEM 2

Determine the coordinates of each of the points.

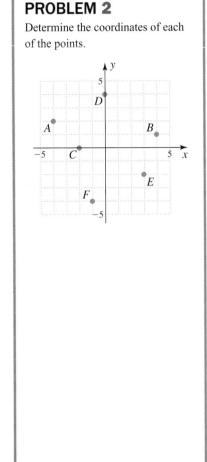

SOLUTION   Point A is 5 units to the right of the origin and 1 unit above the horizontal axis. The ordered pair corresponding to A is $(5, 1)$. The coordinates of the other four points can be found in a similar manner. Here is the summary.

Point	Start at the origin, move:	Coordinates
A	5 units *right*, 1 unit *up*	$(5, 1)$
B	3 units *left*, 2 units *up*	$(-3, 2)$
C	2 units *left*, 4 units *down*	$(-2, -4)$
D	4 units *right*, 2 units *down*	$(4, -2)$
E	3 units *right*, 0 units *up*	$(3, 0)$
F	0 units *right*, 4 units *up*	$(0, 4)$

C › Applications: Line Graphs

Now that you know how to find the coordinates of a point, we learn how to read and interpret line graphs.

Answers to PROBLEMS

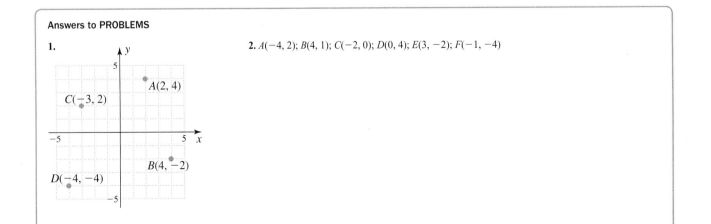

1.

2. $A(-4, 2)$; $B(4, 1)$; $C(-2, 0)$; $D(0, 4)$; $E(3, -2)$; $F(-1, -4)$

EXAMPLE 3   Reading and interpreting line graphs

Figure 3.7 gives the age coversion from human years to dog years. The ordered pair (1, 12) means that 1 human year is equivalent to about 12 dog years.

>Figure 3.7

Source: Data from Cindy's K9 Clips.

a. What does the ordered pair (3, 30) represent?

b. If a dog is 9 years old in human years, how old is it in dog years? Write the ordered pair corresponding to this situation.

c. If retirement age is 65, what is the retirement age for dogs in human years? That is, how many human years correspond to 65 dog years? Write the ordered pair corresponding to this situation.

SOLUTION

a. Three human years are equivalent to 30 dog years.

b. Start at (0, 0) and move right to 9 on the horizontal axis. (See Figure 3.8.) Now, go up until you reach the graph as shown. The point is 60 units high (the y-coordinate is 60). Thus, 9 years old in human years is equivalent to 60 years old in dog years. The ordered pair corresponding to this situation is (9, 60).

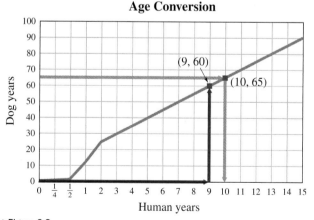

>Figure 3.8

c. We have to find how many human years are represented by 65 dog years. This time, we go to the point at which y is 65 units high, then move right until we reach the graph. At the point for which y is 65 on the graph, x is 10. Thus, the equivalent retirement age for dogs is 10 human years. At that age, they are entitled to Canine Security benefits! The ordered pair corresponding to this situation is (10, 65).

PROBLEM 3

a. What does the ordered pair (5, 40) in Figure 3.7 represent?

b. If a dog is 11 human years old, how old is it in dog years?

c. If the drinking age for humans is 21, what is the equivalent drinking age for dogs in human years? (Answer to the nearest whole number.)

Answers to PROBLEMS

3. a. 5 human years are equivalent to 40 dog years.   **b.** 70   **c.** 2

D ⟩ Applications: Bar Graphs

Another popular use of ordered pairs is **bar graphs,** in which certain *categories* are paired with certain numbers. For example, if you have a $1000 balance at 18% interest and are making the minimum $25 payment each month on your credit card, it will take you forever (actually 5 years) to pay it off. If you decide to pay it off in 12 months, how much would your payment be? The bar graph in Figure 3.9 tells you, provided you know how to read it! First, start at the 0 point and move right horizontally until you get to the **category** labeled 12 months (blue arrow), then go up vertically to the end of the bar (red arrow). According to the vertical scale labeled Monthly payment (the frequency), the arrow is 92 units long, meaning that the monthly payment will be $92 per month. The ordered pair corresponding to this situation is (12, 92).

>**Figure 3.9**

Source: Data from KJE Computer Solutions, LLC.

EXAMPLE **4** Reading and interpreting bar graphs	PROBLEM **4**

a. Referring to the graph, what would your payment be if you decide to pay off the $1000 in 24 months?

b. How much would you save if you do so?

SOLUTION

a. To find the payment corresponding to 24 months, move right on the horizontal axis to the category labeled 24 months and then vertically to the end of the bar in Figure 3.10. According to the vertical scale, the monthly payment will be $50.

>**Figure 3.10**

Source: Data from KJE Computer Solutions, LLC.

b. If you pay the minimum $25 payment for 60 months, you would pay 60 × $25 = $1500. If you pay $50 for 24 months, you would pay 24 × $50 = $1200 and would save $300 ($1500 − $1200).

PROBLEM 4

a. What would your payment be if you decide to pay off the $1000 in 48 months?

b. How much would you save if you do so?

E ⟩ Quadrants

Figure 3.11 shows that the *x*- and *y*-axes divide the plane into four regions called **quadrants.** These quadrants are numbered in counterclockwise order using Roman numerals and starting in the upper right-hand region. What can we say about the coordinates of the points in each quadrant?

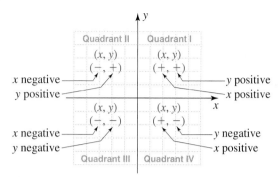

>Figure 3.11

In Quadrant I: Both coordinates positive

In Quadrant II: First coordinate negative, second positive

In Quadrant III: Both coordinates negative

In Quadrant IV: First coordinate positive, second negative

EXAMPLE 5   **Find the quadrant in which each ordered pair lies**

In which quadrant, if any, are the points located?

a. $A(-3, 2)$ **b.** $B(1, -2)$ **c.** $C(3, 4)$

d. $D(-4, -1)$ **e.** $E(0, 3)$ **f.** $F(2, 0)$

SOLUTION   The points are shown on the graph.

a. The point $A(-3, 2)$ is in the second quadrant (QII).

b. The point $B(1, -2)$ is in the fourth quadrant (QIV).

c. The point $C(3, 4)$ is in the first quadrant (QI).

d. The point $D(-4, -1)$ is in the third quadrant (QIII).

e. The point $E(0, 3)$ is on the y-axis (*no quadrant*).

f. The point $F(2, 0)$ is on the x-axis (*no quadrant*).

PROBLEM 5

In which quadrant, if any, are the points located?

a. $A(1, 3)$ **b.** $B(-2, -2)$

c. $C(-4, 2)$ **d.** $D(2, -1)$

e. $E(0, 1)$ **f.** $F(3, 0)$

You may be wondering: Why do we need to know about quadrants? There are two reasons:

1. If you are using a graphing calculator, you have to specify the (WINDOW) (basically, the quadrants) you want to display.

2. If you are graphing ordered pairs, you need to adjust your graph to show the quadrants in which your ordered pairs lie. We will illustrate how this works next.

F › Applications: Creating Line Graphs

Suppose you want to graph the ordered pairs in the table on page 210 (we give equal time to cats!). The ordered pair (c, h) would pair the actual age c of a cat (in "regular" time) to the equivalent human age h. Since both c and h are positive, the graph would have to be in

Quadrant I. Values of c would be from 0 to 21 and values of h from 10 to 100, suggesting a scale of 10 units for each equivalent human year. Let us do all this in Example 6.

Cat's Actual Age	Equivalent Human Age	Cat's Actual Age	Equivalent Human Age
6 months	10 years	10 years	56 years
8 months	13 years	12 years	64 years
1 year	15 years	14 years	72 years
2 years	24 years	16 years	80 years
4 years	32 years	18 years	88 years
6 years	40 years	20 years	96 years
8 years	48 years	21 years	100 years

It was once thought that 1 year in the life of a cat was equivalent to 7 years of a human life. Recently, a new scale has been accepted: after the first 2 years, the cat's life proceeds more slowly in relation to human life and each feline year is approximately 4 human years. The general consensus is that at about age 7 a cat can be considered "middle-aged," and age 10 and beyond "old."

Source: Data from X Mission Internet.

EXAMPLE 6   Creating and interpreting line graphs

a. Make a coordinate grid to graph the ordered pairs in the table above starting with 1 year.

b. Graph the ordered pairs starting with (1, 15).

c. What does (21, 100) mean?

d. Study the pattern (2, 24), (4, 32), (6, 40). What would be the number h in (8, h)?

e. What is the number c in (c, 56)?

f. The oldest cat, Spike, lived to an equivalent human age of 140 years. What was Spike's actual age?

SOLUTION   Parts **a** and **b** are shown in the graph.

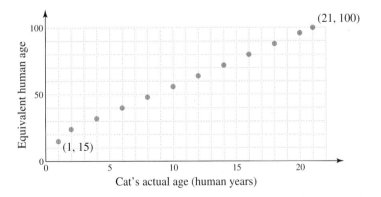

c. (21, 100) means that if a cat's actual age is 21, its equivalent human age is 100 years.

PROBLEM 6

a. What does the ordered pair (16, 80) in the preceding table mean?

b. What is h in (12, h)?

c. What is c in (c, 72)?

d. CNN reports that there is a cat living in Texas whose equivalent human age is 148 years. What is the cat's actual age?

Source: CNN.

---

d. The numbers for the second coordinate are 24, 32, and 40 (increasing by 8). The next number would be 48, so the next ordered pair would be (8, 48). You can verify this from the table or the graph.

e. From the table or the graph, the ordered pair whose second coordinate is 56 is (10, 56), so $c = 10$.

f. Neither the table nor the graph goes to 140 (the graph stops at 100). But we can look at the last two ordered pairs in the table and follow the pattern, or we can extend the graph and find the answer. Here are the last two ordered pairs in the table:

(20, 96)

(21, 100) ⌐

     •

     •   To go from 100 to 140 we need

     • $10 \left(\frac{40}{4} = 10\right)$ increments of 4 for the

               second coordinate and 10 increments

(?, 140) ◄─┘ of 1 for the first coordinate. Thus,

(31, 140) is the next ordered pair. This means Spike was 31 human years old.

As you can see, *line graphs* can be used to show a certain change over a period of time. If we were to connect the points in the graph of Example 6 starting with (2, 24), we would have a *straight line*. We shall study linear equations and their graphs in Section 3.2.

Calculator Corner

Graphing Points

We can do most of the work in this section with our calculators. To do Example 1, first adjust the **viewing screen** or **window** of the calculator. Since the values of x range from -5 to 5, denoted by $[-5, 5]$, and the values of y also range from -5 to 5, we have a $[-5, 5]$ by $[-5, 5]$ viewing rectangle, or window, as shown in Window 1.

Window 1   Window 2

To graph the points A, B, C, and D of Example 1, set the calculator on the statistical graph mode (![2nd] ![Y=] 1), turn the plot on (![ENTER]), select the type of plot, and the list of numbers you are going to use for the x-coordinates (L_1) and the y-coordinates (L_2) as well as the type of mark you want the calculator to make (Window 2). Now, press ![STAT] 1 and enter the x-coordinates of A, B, C, and D under L_1 and the y-coordinates of A, B, C, and D under L_2. Finally, press ![GRAPH] to obtain the points A, B, C, and D in Window 3. Use these ideas to do Problems 1–5 in the exercise set.

Window 3

> Exercises **3.1**

⟨ A ⟩ Graphing Ordered Pairs   In Problems 1–5, graph the points.

1. a. $A(1, 2)$ **b.** $B(-2, 3)$

 c. $C(-3, 1)$ **d.** $D(-4, -1)$

2. a. $A\left(-2\frac{1}{2}, 3\right)$ **b.** $B\left(-1, 3\frac{1}{2}\right)$

 c. $C\left(-\frac{1}{2}, -4\frac{1}{2}\right)$ **d.** $D\left(\frac{1}{3}, 4\right)$

3. a. $A(0, 2)$ **b.** $B(-3, 0)$

 c. $C\left(3\frac{1}{2}, 0\right)$ **d.** $D\left(0, -1\frac{1}{4}\right)$

4. a. $A(20, 20)$ **b.** $B(-10, 20)$

 c. $C(35, -15)$ **d.** $D(-25, -45)$

5. a. $A(0, 40)$ **b.** $B(-35, 0)$

 c. $C(-40, -15)$ **d.** $D(0, -25)$

⟨ B ⟩ Determining Coordinates
⟨ E ⟩ Quadrants

In Problems 6–10, give the coordinates of the points and the quadrant in which each point lies.

9.

10.

⟨ **C** ⟩ **Applications: Line Graphs**   Use the following graph to work Problems 11–14.

11. What is the lower limit pulse rate for a 20-year-old? Write the answer as an ordered pair.

12. What is the upper limit pulse rate for a 20-year-old? Write the answer as an ordered pair.

13. What is the upper limit pulse rate for a 45-year-old? Write the answer as an ordered pair.

14. What is the lower limit pulse rate for a 50-year-old? Write the answer as an ordered pair.

Use the following graph to work Problems 15–24.

The green graph shows the pounds of waste generated per capita (per person) each day and the blue graph shows the total waste generated in millions of tons.

Municipal Solid Waste Generation Rates 1960–2003

Source: www.epa.gov.

15. How many pounds of waste were generated per person each day in 1960?

16. How many pounds of waste were generated per person each day in 2003?

17. How many more pounds were generated per person each day in 2003 than in 1960?

18. How many more pounds were generated per person each day in 2003 than in 1980?

19. Based on the graph, how many pounds of waste would you expect to be generated per person each day in 2008?

20. What was the total waste (in millions of tons) generated in 1960?

21. What was the total waste (in millions of tons) generated in 2003?

22. How many more million tons of waste were generated in 2003 than in 1960?

23. How many more million tons of waste were generated in 2003 than in 2000?

24. Based on your answer to Problem 23, how many million tons of waste would you expect to be generated in 2006?

Use the following graph to work Problems 25–28.

The graph shows the amount owed on a $1000 debt at an 18% interest rate when the minimum $25 payment is made or when a new monthly payment of $92 is made. Thus, at the current monthly payment of $25, it will take 60 months to pay the $1000 balance (you got to $0!). On the other hand, with a new $92 monthly payment, you pay off the $1000 balance in 12 months.

Source: Data from KJE Computer Solutions, LLC.

25. What is your balance after 6 months if you are paying $25 a month?

26. What is your balance after 6 months if you are paying $92 per month?

27. What is your balance after 18 months if you are paying $25 a month?

28. What is your balance after 48 months if you are paying $25 a month?

Use the following graph to work Problems 29–33.

Reduction in Young Americans Returning Through El Paso Border Crossing After Juarez Bars Closing Time Change

Source: Data from Institute for Public Strategies.

The graph shows the number of young Americans crossing the border at El Paso after the bar closing hour in Juarez, Mexico, changed from 3 A.M. to 2 A.M.

29. About how many people crossed the border at 12 A.M. when the bar closing time was 2 A.M.?

30. About how many people crossed the border at 12 A.M. when the bar closing time was 3 A.M.?

31. At what time was the difference in the number of border crossers greatest? Can you suggest an explanation for this?

32. At what time was the number of border crossers the same?

33. What was the difference in the number of border crossers at 5 A.M.?

⟨ **D** ⟩ **Applications: Bar Graphs** Use the following graph to work Problems 34–38.

Change in Number of Crossers with BAC* Over 0.08 per Weekend Night from the Juarez Bar Closing Time Shift

Source: Data from Institute for Public Strategies.

The numbers of crossers before 3 A.M. and after 3 A.M. are shown in the graph.

34. What was the total number of crossers before the change?

35. How many of those were before 3 A.M. and how many after 3 A.M.?

36. How many crossers were there after the change?

37. How many of those were before and how many after 3 A.M.?

38. Why do you think there were so many crossers with high BAC before the change?

Use the following graph to work Problems 39–42.

The graph shows the percent of selected materials and the number of tons recycled in each category in a recent year. The total amount of recycled materials was 70 million tons.

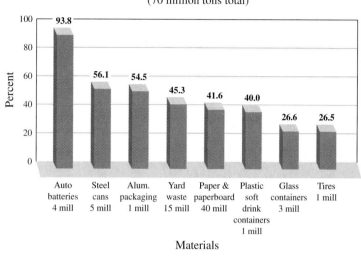

Recycling Rates of Selected Materials
(70 million tons total)

Source: Data from U.S. Environmental Protection Agency.

39. a. Which material was recycled the most.

   b. Which material was recycled the least.

41. What percent of the tires were recycled? What percent of the 70 million tons total were tires?

40. What percent of the auto batteries were recycled? What percent of the 70 million tons total were auto batteries?

42. How many more tons of glass containers than aluminum packaging were recycled?

Use the following graph to work Problems 43–46.

How does the down payment affect the monthly payment?

Purchase price	20,000
Sales tax +	1,050
Fees	40
Total price =	**21,090**
Cash down −	1,500
Net trade in −	1,000
Loan amount =	**18,590**

How the Down Payment Affects my Payment

Source: Data from KJE Computer Solutions, LLC.

43. What is the monthly payment if the down payment is $0?

45. What is the difference in the monthly payment if you increase the down payment from $1500 to $2500?

44. What is the monthly payment if the down payment is $2000?

46. What is the difference in the annual amount paid if you increase the down payment from $1500 to $2500?

⟨ **F** ⟩   **Applications: Creating Line Graphs**   The chart shows the dog's actual age and the equivalent human age for dogs of different weights and will be used in Problems 47–55.

Dog's Age	Weight (pounds)			
	15–30	30–49	50–74	75–100
1	12	12	14	16
2	19	21	23	23
3	25	25	26	29
4	32	32	35	35
5	36	36	37	39
6	40	42	43	45
7	44	45	46	48
8	48	50	52	52
9	52	53	54	56
10	56	56	58	61
11	60	60	62	65
12	62	63	66	70
13	66	67	70	75
14	70	70	75	80
15	72	76	80	85

Equivalent Human Age

Using different colors, graph the equivalent human age for a dog weighing:

47. Between 15 and 30 pounds

48. Between 30 and 49 pounds

49. Between 50 and 74 pounds

50. Between 75 and 100 pounds

51. If a dog's age is 15, which of the weight scales will make it appear oldest?

52. If a dog's age is 15, which of the weight scales will make it appear youngest?

53. Based on your answers to Problems 51 and 52, the _____ (less, more) a dog weighs, the _____ (younger, older) it appears on the human age scale.

54. If a dog is 6 years old, what is its equivalent human age? *Hint:* There are four answers!

55. From the table, at which dog's age(s) is there no agreement as to what the equivalent human age would be?

56. The red graph shows the monthly balance on a 30-year $100,000 mortgage at 6.25% paying $615.72 a month. If you make biweekly payments of $307.06 you save $27,027 in interest!

   a. What is the approximate balance of the 30-year (red) mortgage after 12 months?

   b. What is the approximate balance of the accelerated (blue) mortgage after 12 months?

   c. In how many months do you pay off the mortgage under the accelerated plan?

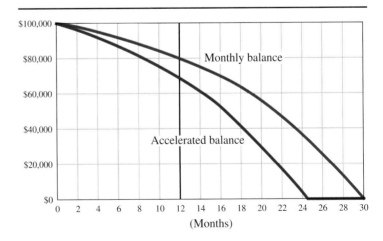

57. The red graph shows the monthly balance on a 30-year $200,000 mortgage at 7.5% paying $1398.43 a month. If you make biweekly payments of $699.22 you save $80,203 in interest!

    a. What is the approximate balance of the 30-year (red) mortgage after 20 months?

    b. What is the approximate balance of the accelerated (blue) mortgage after 20 months?

    c. In how many months do you pay off the mortgage under the accelerated plan?

58. The environmental lapse is the rate of decrease of temperature with altitude (elevation): the *higher* you are, the *lower* the temperature. The temperature drops about 4°F (minus 4 degrees Fahrenheit) for each 1000 feet of altitude.

    Source: www.answers.com.

Altitude in (1000 ft)	Temperature Change	Ordered Pair
1	−4°F	(1, −4)
2	−8°F	(2, −8)
3		
4		
5		

    a. Complete the table.

    b. Make a graph of the ordered pairs.

59. In the metric system, the environmental lapse is about **7°C** (minus 7 degrees Celsius) for each kilometer of altitude.

Altitude (kilometers)	Temperature Change	Ordered Pair
1	−7°C	(1, −7)
2	−14°C	(2, −14)
3		
4		
5		

    a. Complete the table.

    b. Make a graph of the ordered pairs.

60. The graph shows the annual percent of renters among people under age 25.

    a. What percent were renters in 2000?

    b. What percent were renters in 2004?

    c. In what years was the percent of renters decreasing?

    d. In what years was the percent of renters unchanged?

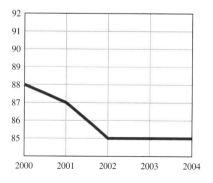

〉〉〉   ***Using Your Knowledge***

Graphing the Risks of "Hot" Exercise   The ideas we presented in this section are vital for understanding graphs. For example, do you exercise in the summer? To determine the risk of exercising in the heat, you must know how to read the graph on the right.

This can be done by first finding the temperature (the *y*-axis) and then reading across from it to the right, stopping at the vertical line representing the relative humidity. Thus, on a 90°F day, if the humidity is less than 30%, the weather is in the safe zone.

Use your knowledge to answer the following questions about exercising in the heat.

61. If the humidity is 50%, how high can the temperature go and still be in the safe zone for exercising? (Answer to the nearest degree.)

62. If the humidity is 70%, at what temperature will the danger zone start?

63. If the temperature is 100°F, what does the humidity have to be so that it is safe to exercise?

64. Between what temperatures should you use caution when exercising if the humidity is 80%?

65. Suppose the temperature is 86°F and the humidity is 60%. How many degrees can the temperature rise before you get to the danger zone?

〉〉〉   **Write On**

66. Suppose the point (a, b) lies in the first quadrant. What can you say about a and b?

67. Suppose the point (a, b) lies in the second quadrant. What can you say about a and b? Is it possible that $a = b$?

68. In what quadrant(s) can you have points (a, b) such that $a = b$? Explain.

69. Suppose the ordered pairs (a, b) and (c, d) have the same point as their graphs—that is, $(a, b) = (c, d)$. What is the relationship between a, b, c, and d?

70. Now suppose $(a + 5, b - 7)$ and $(3, 5)$ have the same point as their graphs. What are the values of a and b?

〉〉〉   **Concept Checker**

Fill in the blank(s) with the correct word(s), phrase, or mathematical statement.

71. In the **ordered pair** (x, y) the *x* is called the _____.

72. In the **ordered pair** (x, y) the *y* is called the _____.

coordinate   ordinance

abscissa   ordinate

〉〉〉   **Mastery Test**

Graph the points:

73. $A(4, 2)$

74. $B(3, -2)$

75. $C(-2, 1)$

76. $D(0, -3)$

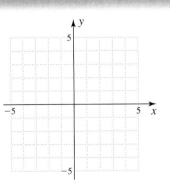

77. The projected cost (in cents per minute) for wireless phone use for the next four years is given in the table. Graph these points.

Year	cost (cents/min)
1	25
2	23
3	22
4	20

78. Referring to the graph, find an ordered pair giving the equivalent dog age (dog years) for a dog that is actually 7 years old (human years).

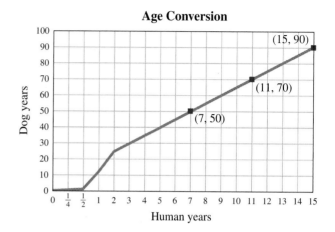

Age Conversion

79. Referring to the graph in Problem 78, find the age (human years) of a dog that is 70 years old (dog years).

80. What does the ordered pair (15, 90) in the graph in Problem 78 mean?

81. In which quadrant are the following points?

    a. $(-2, 3)$

    b. $(3, -2)$

    c. $(2, 2)$

    d. $(-1, -1)$

82. Give the coordinates of each of the points shown in the graph.

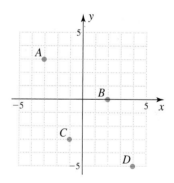

〉〉〉 *Skill Checker*

83. Solve: $y = 3x - 2$ when $y = 7$

84. Solve: $C = 0.10m + 10$ when $m = 30$

85. Solve: $3x + y = 9$ when $x = 0$

86. Solve: $y = 50x + 450$ when $x = 2$

87. Solve: $3x + y = 6$ when $x = 1$

88. Solve: $3x + y = 6$ when $y = 0$

3.2 Graphing Linear Equations in Two Variables

▶ Objectives

A ▷ Determine whether a given ordered pair is a solution of an equation.

B ▷ Find ordered pairs that are solutions of a given equation.

C ▷ Graph linear equations of the form $y = mx + b$ and $Ax + By = C$.

D ▷ Solve applications involving linear equations.

▶ To Succeed, Review How To . . .

1. Graph ordered pairs (p. 205).

2. Read and interpret ordered pairs on graphs (pp. 206–207).

▶ Getting Started

The graph shows the 72-hour forecast period from the National Hurricane Center. In the blue line, the x-coordinate shows the number of **hours** (0 to 72), whereas the y-coordinate gives the **wind speed.**

$$(x, y)$$

x is the hours ⤒ ⤒ y is the wind speed

Thus, at the beginning of the advisory ($x = 0$) the wind speed was 150 mph and 24 hours later) $x = 24$, the wind increased to 165 mph. What ordered pair corresponds to the wind speed $x = 48$ hours later? The answer is (48, 120).

NHC Maximum 1-Minute Wind Speed Forecast and Probabilities

Forecast period (hours)
Wilma advisory 23 10.00 pm CDT Oct 20 2005

A ▷ Solutions of Equations

How can we construct our own graph showing the wind speed of the hurricane for the first 24 hours? First, we label the x-axis with the hours from 0 to 24 and the y-axis with the wind speed from 0 to 200. We then graph the ordered pairs (x, y) representing the hours x and the wind speed y starting with (0, 150) and (24, 165), as shown in Figure 3.12. Does this line have any relation to algebra? Of course! The line representing the wind speed y can be approximated by the equation

$$y = \frac{5}{8}x + 150$$

Hurricane Intensity

>Figure 3.12

Note that at the beginning of the forecast (when $x = 0$) the wind speed was

$$y = \frac{5}{8} \cdot 0 + 150 = 150 \text{ (mph)}$$

If we write $x = 0$ and $y = 150$ as the ordered pair $(0, 150)$, we say that the ordered pair $(0, 150)$ **satisfies** or is a **solution** of the equation $y = \frac{5}{8}x + 150$. What about $x = 24$ hours after the beginning of the forecast? At that point the wind speed $y = \frac{5}{8} \cdot 24 + 150 = 165$ (mph), hence the ordered pair $(24, 165)$ also satisfies the equation $y = \frac{5}{8}x + 150$ since $165 = \frac{5}{8} \cdot 24 + 150$. Note that the equation $y = \frac{5}{8}x + 150$ is different from all other equations we have studied because it has two variables, x and y. Such equations are called **equations in two variables.**

DETERMINING IF AN ORDERED PAIR IS A SOLUTION

To determine whether an ordered pair is a solution of an equation in two variables, we substitute the x-coordinate for x and the y-coordinate for y in the given equation. If the resulting statement is *true,* then the ordered pair is a solution of, or *satisfies,* the equation.

Thus, $(2, 5)$ is a solution of $y = x + 3$, since in the ordered pair $(2, 5)$, $x = 2$, $y = 5$, and

$$y = x + 3$$

becomes

$$5 = 2 + 3$$

which is a true statement.

EXAMPLE 1   **Determining whether an ordered pair is a solution**
Determine whether the given ordered pairs are solutions of $2x + 3y = 10$.

a. $(2, 2)$ **b.** $(-3, 4)$ **c.** $(-4, 6)$

SOLUTION

a. In the ordered pair $(2, 2)$, $x = 2$ and $y = 2$. Substituting in

$$2x + 3y = 10$$

we get

$$2(2) + 3(2) = 10 \quad \text{or} \quad 4 + 6 = 10$$

which is true. Thus, $(2, 2)$ is a solution of

$$2x + 3y = 10$$

You can summarize your work as shown here:

$$\begin{array}{c|c} 2x \;+\; 3y \stackrel{?}{=} 10 & \\ \hline 2(2) + 3(2) & 10 \\ 4 \;+\; 6 & \\ 10 & \text{True} \end{array}$$

PROBLEM 1

Determine whether the ordered pairs are solutions of $3x + 2y = 10$.

a. $(2, 2)$
b. $(-3, 4)$
c. $(-4, 11)$

(continued)

Answers to PROBLEMS

1. a. Yes; $3(2) + 2(2) = 10$ **b.** No; $3(-3) + 2(4) \neq 10$ **c.** Yes; $3(-4) + 2(11) = 10$

b. In the ordered pair $(-3, 4)$, $x = -3$ and $y = 4$. Substituting in

$$2x + 3y = 10$$

we get

$$2(-3) + 3(4) = 10$$
$$-6 + 12 = 10$$

which is *not* true. Thus, $(-3, 4)$ is not a solution of the given equation. You can summarize your work as shown here:

$$\begin{array}{c|c} 2x \quad + \quad 3y \quad \overset{?}{=} \quad 10 & \\ \hline 2(-3) + 3(4) & \\ & 10 \\ -6 \ + 12 & \\ 6 & \text{False} \end{array}$$

c. In the ordered pair $(-4, 6)$, the x-coordinate is -4 and the y-coordinate is 6. Substituting these numbers for x and y, respectively, we have

$$2x + 3y = 10$$
$$2(-4) + 3(6) = 10$$

or

$$-8 + 18 = 10$$

which is a true statement. Thus, $(-4, 6)$ satisfies, or is a solution of, the given equation. You can summarize your work as shown here.

$$\begin{array}{c|c} 2x \quad + \quad 3y \quad \overset{?}{=} \quad 10 & \\ \hline 2(-4) + 3(6) & \\ & 10 \\ -8 \ + 18 & \\ 10 & \text{True} \end{array}$$

B ⟩ Finding Missing Coordinates

In some cases, rather than verifying that a certain ordered pair *satisfies* an equation, we actually have to *find* ordered pairs that are solutions of the given equation. For example, we might have the equation

$$y = 2x + 5$$

and would like to find several ordered pairs that satisfy this equation. To do this we substitute any number for x in the equation and then find the corresponding y-value. A good number to choose is $x = 0$. In this case,

$$y = 2 \cdot 0 + 5 = 5 \qquad \text{Zero is easy to work with because the value of } y \text{ is easily found when 0 is substituted for } x.$$

Thus, the ordered pair $(0, 5)$ satisfies the equation $y = 2x + 5$. For $x = 1$, $y = 2 \cdot 1 + 5 = 7$; hence, $(1, 7)$ also satisfies the equation.

We can let x be any number in an equation and then find the corresponding y-value. Conversely, we can let y be any number in the given equation and then find the x-value. For example, if we are given the equation

$$y = 3x - 2$$

and we are asked to find the value of x in the ordered pair $(x, 7)$, we simply let y be 7 and obtain

$$7 = 3x - 2$$

We then solve for x by rewriting the equation as

$$3x - 2 = 7$$
$$3x = 9 \quad \text{Add 2.}$$
$$x = 3 \quad \text{Divide by 3.}$$

Thus, $x = 3$ and the ordered pair satisfying $y = 3x - 2$ is $(3, 7)$, as can be verified since $7 = 3(3) - 2$.

EXAMPLE 2 Finding the missing coordinate

Complete the given ordered pairs so that they satisfy the equation $y = 4x + 3$.

a. $(x, 11)$　　　**b.** $(-2, y)$

SOLUTION

a. In the ordered pair $(x, 11)$, y is 11. Substituting 11 for y in the given equation, we have

$$11 = 4x + 3$$
$$4x + 3 = 11 \quad \text{Rewrite with } 4x + 3 \text{ on the left.}$$
$$4x = 8 \quad \text{Subtract 3.}$$
$$x = 2 \quad \text{Divide by 4.}$$

Thus, $x = 2$ and the ordered pair is $(2, 11)$.

b. Here $x = -2$. Substituting this value in $y = 4x + 3$ yields

$$y = 4(-2) + 3 = -8 + 3 = -5$$

Thus, $y = -5$ and the ordered pair is $(-2, -5)$.

PROBLEM 2

Complete the ordered pairs so that they satisfy the equation $y = 3x + 4$.

a. $(x, 7)$　　**b.** $(-2, y)$

EXAMPLE 3 Approximating cholesterol levels

Figure 3.13 shows the decrease in cholesterol with exercise over a 12-week period. If C is the cholesterol level and w is the number of weeks elapsed, the line shown can be approximated by $C = -3w + 215$.

>Figure 3.13

a. According to the graph, what is the cholesterol level at the end of 1 week and at the end of 12 weeks?

b. What is the cholesterol level at the end of 1 week and at the end of 12 weeks using the equation $C = -3w + 215$?

PROBLEM 3

a. According to the graph, what is the cholesterol level at the end of 4 weeks?

b. What is the cholesterol level at the end of 4 weeks using the equation in Example 3?

SOLUTION

a. The cholesterol level at the end of 1 week corresponds to the point $(1, 211)$ on the graph (the point of intersection of the vertical line above 1 and the graph). Thus, the cholesterol level at the end of 1 week is 211. Similarly, the cholesterol level at the end of 12 weeks corresponds to the point $(12, 175)$. Thus, the cholesterol level at the end of 12 weeks is 175.

(continued)

Answers to **PROBLEMS**

2. a. $(1, 7)$　**b.** $(-2, -2)$

3. a. About 202　**b.** 203

b. From the equation $C = -3w + 215$, the cholesterol level at the end of 1 week ($w = 1$) is given by $C = -3(1) + 215 = 212$ (close to the 211 on the graph!) and the cholesterol level at the end of 12 weeks ($w = 12$) is $C = -3(12) + 215 = -36 + 215$, or 179.

C › Graphing Linear Equations by Plotting Points

Suppose you want to rent a car that costs $30 per day plus $0.20 per mile traveled. If we have the equation for the daily cost C based on the number m of miles traveled, we can graph this equation. The equation is

$$C = \underbrace{0.20m}_{\text{20¢ per mile}} + \underbrace{30}_{\text{\$30 each day}}$$

Now remember that a solution of this equation must be an ordered pair of numbers of the form (m, C). For example, if you travel 10 miles, $m = 10$ and the cost is

$$C = 0.20(10) + 30 = 2 + 30 = \$32$$

Thus, $(10, 32)$ is an ordered pair satisfying the equation; that is, $(10, 32)$ is a *solution* of the equation. If we go 20 miles, $m = 20$ and

$$C = 0.20(20) + 30 = 4 + 30 = \$34$$

Hence, $(20, 34)$ is also a solution.

As you see, we can go on forever finding solutions. It's much better to organize our work, list these two solutions and some others, and then graph the points obtained in the Cartesian coordinate system. In this system the number of miles m will appear on the horizontal axis, and the cost C on the vertical axis. The corresponding points given in the table appear in the accompanying figure.

m	C
0	30
10	32
20	34
30	36

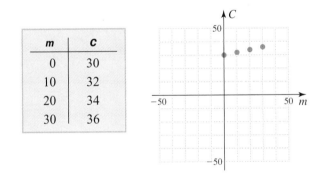

Note that m must be *positive* or *zero*. We have selected values for m that make the cost easy to compute—namely, 0, 10, 20, and 30.

It seems that if we join the points appearing in the graph, we obtain a straight line. Of course, if we knew this for sure, we could have saved time! Why? Because if we know that the graph of an equation is a straight line, we simply find *two* solutions of the equation, graph the two points, and then join them with a straight line. As it turns out, the graph of $C = 0.20m + 30$ *is* a straight line.

The procedure used to graph $C = 0.20m + 30$ can be generalized to graph other lines. Here are the steps.

PROCEDURE

Graphing Lines

1. Choose a value for one variable, calculate the value of the other variable, and graph the resulting ordered pair.

2. Repeat step 1 to obtain at least two ordered pairs.

3. Graph the ordered pairs and draw a line passing through the points. (You can use a third ordered pair as a check.)

This procedure is called **graphing,** and the following rule lets us know that the graph is indeed a *straight line*.

RULE

Straight-Line Graphs

The graph of a linear equation of the form

$$Ax + By = C \quad \text{or} \quad y = mx + b$$

where A, B, C, m, and b are constants (A and B are not both 0) is a **straight line,** and every straight line has an equation that can be written in one of these forms.

Thus, the graph of $C = 0.20m + 30$ is a straight line and the graph of $3x + y = 6$ is also a straight line, as we show next.

EXAMPLE 4 Graphing lines of the form $Ax + By = C$

Graph: $3x + y = 6$

SOLUTION The equation is of the form $Ax + By = C$, and thus the graph is a straight line. Since two points determine a line, we shall graph two points and join them with a straight line, the graph of the equation. Two easy points to use occur when we let $x = 0$ and find y, and let $y = 0$ and find x. For $x = 0$,

$$3x + y = 6$$

becomes

$$3 \cdot 0 + y = 6 \quad \text{or} \quad y = 6$$

Thus, $(0, 6)$ is on the graph.

When $y = 0$, $3x + y = 6$ becomes

$$3x + 0 = 6$$
$$3x = 6$$
$$x = 2$$

Hence, $(2, 0)$ is also on the graph.

We chose $x = 0$ and $y = 0$ because the calculations are easy. There are infinitely many points that satisfy the equation $3x + y = 6$, but the points $(0, 6)$ and $(2, 0)$—the y- and x-intercepts—are easy to find.

It's a good idea to pick a *third* point as a *check*. For example, if we let $x = 1$, $3x + y = 6$ becomes

$$3 \cdot 1 + y = 6$$
$$3 + y = 6$$
$$y = 3$$

PROBLEM 4

Graph: $2x + y = 4$

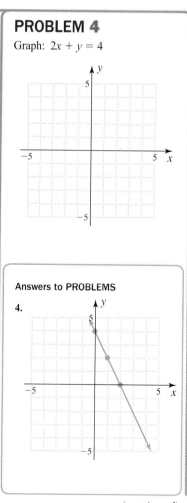

Answers to PROBLEMS

4.

(continued)

Now we have our third point, (1, 3), as shown in the following table. The points (0, 6), (2, 0), and (1, 3), as well as the completed graph of the line, are shown in Figure 3.14.

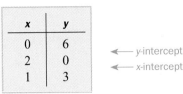

x	y
0	6	← y-intercept
2	0	← x-intercept
1	3

Note that the point (0, 6) where the line crosses the y-axis is called the **y-intercept,** and the point (2, 0) where the line crosses the x-axis is called the **x-intercept.**

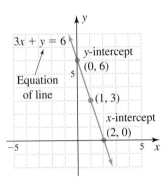

The graph in Figure 3.14 cannot show the entire line, which extends indefinitely in both directions; that is, the graph shown is a part of a line that continues without end in both directions, as indicated by the arrows.

>Figure 3.14

EXAMPLE 5   **Graphing lines of the form $y = mx + b$**

Graph: $y = -2x + 6$

SOLUTION   The equation is of the form $y = mx + b$, which is a straight line. Thus, we follow the procedure for graphing lines.

1. Let $x = 0$.

$$y = -2x + 6 \qquad \text{becomes}$$
$$y = -2(0) + 6 = 6$$

Thus, the point (0, 6) is on the graph. (See Figure 3.15.)

2. Let $y = 0$.

$$y = -2x + 6 \qquad \text{becomes}$$
$$0 = -2x + 6$$
$$2x = 6 \qquad \text{Add } 2x.$$
$$x = 3 \qquad \text{Divide by 2.}$$

Hence, (3, 0) is also on the line.

3. Graph (0, 6) and (3, 0), and draw a line passing through both points.
To make sure, we will use a third point.
Let $x = 1$.

$$y = -2x + 6 \qquad \text{becomes}$$
$$y = -2(1) + 6 = 4$$

Thus, the point (1, 4) is also on the line, as shown in Figure 3.15.

>Figure 3.15

PROBLEM 5

Graph: $y = -3x + 6$

Answers to PROBLEMS

5.

D › Applications Involving Linear Equations

Have you heard the term "wind chill factor"? It is the temperature you actually *feel* because of the wind. Thus, for example, if the temperature is 10 degrees Fahrenheit (10°F) and the wind is blowing at 15 miles per hour, you will *feel* as if the temperature is −7°F. If the wind is blowing at a *constant* 15 miles per hour, the wind chill factor W can be roughly approximated by the equation

$$W = 1.3t - 20, \quad \text{where } t \text{ is the temperature in } °F$$

EXAMPLE 6   Graphing a linear equation

Graph: $W = 1.3t - 20$

SOLUTION   First, note that we have to label the axes differently. Instead of y we have W, and instead of x we have t. Next, note that the wind chill factors we will get seem to be negative (try $t = -5$, $t = 0$, and $t = 5$). Thus, we will concentrate on points in the third and fourth quadrants, as shown in the grid. Now let's obtain some points.

For $t = -5$, $W = 1.3(-5) - 20 = -26.5$.   Graph $(-5, -26.5)$.
For $t = 0$, $W = 1.3(0) - 20 = -20$.   Graph $(0, -20)$.
For $t = 5$, $W = 1.3(5) - 20 = -13.5$.   Graph $(5, -13.5)$.

Note that the equation $W = 1.3t - 20$ applies only when the wind is blowing at 15 miles per hour! Moreover, you should recognize that the graph is a line (it is of the form $y = mx + b$ with W instead of y and t instead of x) and that the graph of the line belongs mostly in the third and fourth quadrants.

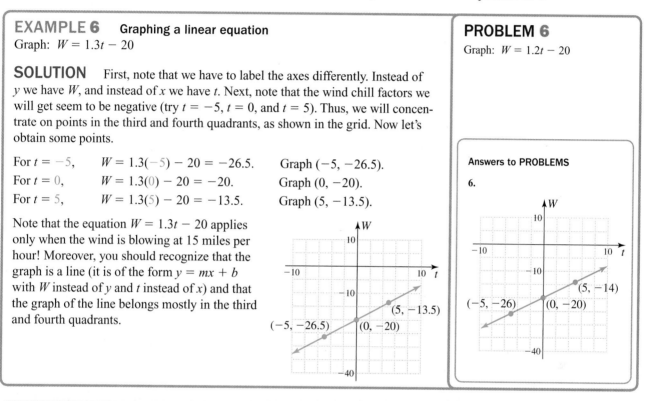

PROBLEM 6

Graph: $W = 1.2t - 20$

Answers to PROBLEMS

6.

▣ ◈ ▥ Calculator Corner

Graphing Lines

You can satisfy three of the objectives of this section using a calculator. We will show you how by graphing the equation $2x + 3y = 10$, determining whether the point $(2, 2)$ satisfies the equation, and finding the x-coordinate in the ordered pair $(x, -6)$ using a calculator.

The graph of $2x + 3y = 10$ consists of all points satisfying $2x + 3y = 10$. To graph this equation, first solve for y obtaining

$$y = \frac{10}{3} - \frac{2x}{3}$$

If you want to determine whether $(2, 2)$ is a solution of the equation, turn the statistical plot off (press 1 and select OFF), then set the window for integers (ZOOM 8 ENTER), and press Y= . Then enter

$$Y_1 = \frac{10}{3} - \frac{2x}{3}$$

Window 1

Press GRAPH .

Use TRACE to move the cursor around. In Window 1, $x = 2$ and $y = 2$ are shown; thus, $(2, 2)$ satisfies the equation. To find x in $(x, -6)$, use TRACE until you are at the point where $y = -6$; then read the value of x, which is 14. Try it! You can use these ideas to do Problems 1–16.

We do not want to leave you with the idea that all equations you will encounter are linear equations. Later in the book we shall study:

1. Absolute-value equations of the form $y = |x|$
2. Quadratic equations of the form $y = x^2$
3. Cubic equations of the form $y = x^3$

These types of equations can be graphed using the same procedure as that for graphing lines, with one notable exception: in step (3), you draw a smooth curve passing through the points. The results are shown here.

Boost your grade at mathzone.com!
> Practice Problems
> NetTutor
> Self-Tests
> e-Professors
> Videos

> Exercises **3.2**

⟨ **A** ⟩   **Solutions of Equations**   In Problems 1–6, determine whether the ordered pair is a solution of the equation.

1. $(3, 2); x + 2y = 7$

2. $(4, 2); x - 3y = 2$

3. $(5, 3); 2x - 5y = -5$

4. $(-2, 1); -3x = 5y + 1$

5. $(2, 3); -5x = 2y + 4$

6. $(-1, 1); 4y = -2x + 2$

⟨ **B** ⟩   **Finding Missing Coordinates**   In Problems 7–16, find the missing coordinate.

7. $(3, \underline{\quad})$ is a solution of $2x - y = 6$.

8. $(-2, \underline{\quad})$ is a solution of $-3x + y = 8$.

9. $(\underline{\quad}, 2)$ is a solution of $3x + 2y = -2$.

10. $(\underline{\quad}, -5)$ is a solution of $x - y = 0$.

11. $(0, \underline{\quad})$ is a solution of $3x - y = 3$.

12. $(0, \underline{\quad})$ is a solution of $x - 2y = 8$.

13. $(\underline{\quad}, 0)$ is a solution of $2x - y = 6$.

14. $(\underline{\quad}, 0)$ is a solution of $-2x - y = 10$.

15. $(-3, \underline{\quad})$ is a solution of $-2x + y = 8$.

16. $(-5, \underline{\quad})$ is a solution of $-3x - 2y = 9$.

⟨ **C** ⟩   **Graphing Linear Equations by Plotting Points**   In Problems 17–40, graph the equation.

17. $2x + y = 4$

18. $y + 3x = 3$

19. $-2x - 5y = -10$

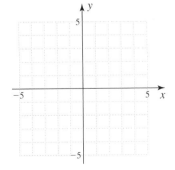

20. $-3x - 2y = -6$

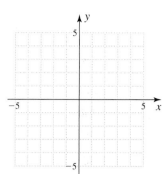

21. $y + 3 = 3x$

22. $y - 4 = -2x$

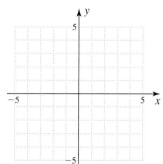

23. $6 = 3x - 6y$

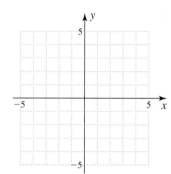

24. $6 = 2x - 3y$

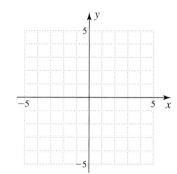

25. $-3y = 4x + 12$

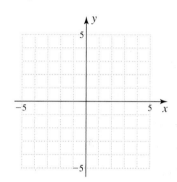

26. $-2x = 5y + 10$

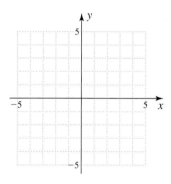

27. $-2y = -x + 4$

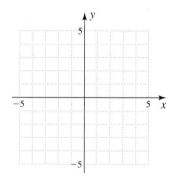

28. $-3y = -x + 6$

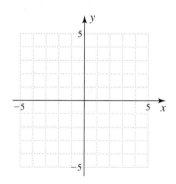

29. $-3y = -6x + 3$

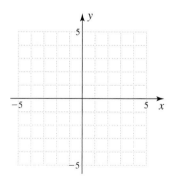

30. $-4y = -2x + 4$

31. $y = 2x + 4$

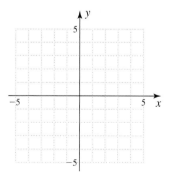

32. $y = 3x + 6$

33. $y = -2x + 4$

34. $y = -3x + 6$

35. $y = -3x - 6$

36. $y = -2x - 4$

37. $y = \frac{1}{2}x - 2$

38. $y = \frac{1}{3}x - 1$

39. $y = -\frac{1}{2}x - 2$

40. $y = -\frac{1}{3}x - 1$

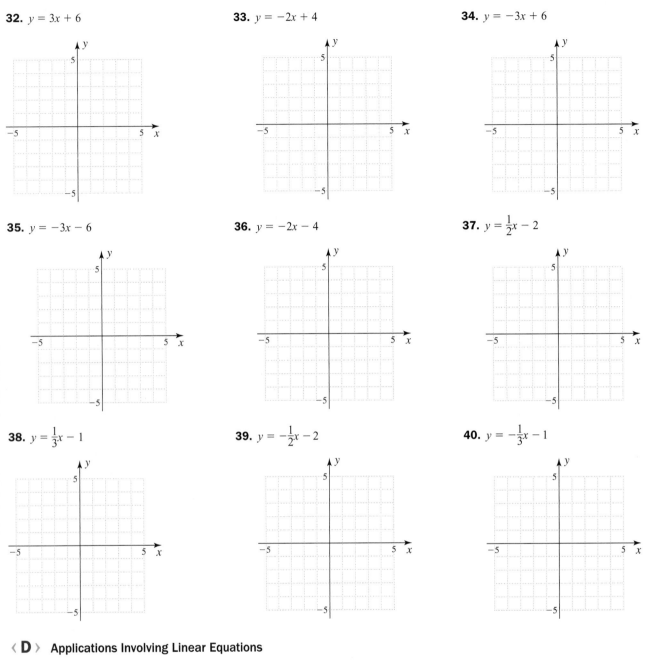

⟨ D ⟩ Applications Involving Linear Equations

41. *Wind chill factor* When the wind is blowing at a constant 5 miles per hour, the wind chill factor W can be approximated by $W = 1.1t - 9$, where t is the temperature in °F and W is the wind chill factor.

  a. Find W when $t = 0$.

  b. Find W when $t = 10$.

  c. Graph W.

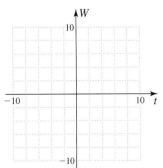

42. *Wind chill factor* When the wind is blowing at a constant 20 miles per hour, the wind chill factor W can be approximated by $W = 1.3t - 21$, where t is the temperature in °F and W is the wind chill factor.

  a. Find W when $t = 0$.

  b. Find W when $t = 10$.

  c. Graph W.

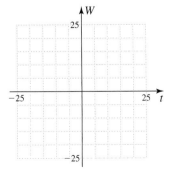

43. *College tuition and fees*　Do you want a rough approximation of the annual cost C (in dollars) of tuition and required fees at a two-year college? This cost can be approximated by $C = 200t + 7000$, where t the number of years after 1996.

   a. What was the cost of tuition in 1996?

   b. What was it in 2001?

   c. What would it be in 2006? In 2010?

   d. Make a graph of the line representing C.

44. *Cost of textbooks*　The annual average cost C (in dollars) of textbooks at a two-year college is approximately $C = 30t + 680$, where t is the number of years after 2000.

   a. What was the cost of textbooks in 2000?

   b. What was it in 2001?

   c. What would it be in 2005? In 2010?

   d. Make a graph of the line representing C.

45. *Environmental lapse*　Suppose the temperature is 60° Fahrenheit (60°F). As your altitude increases, the temperature y at x (thousand) feet above sea level is given by the equation

$$y = -4x + 60$$

   a. What is the temperature at **1** (thousand) feet?

   b. What is the temperature at **10** (thousand) feet?

   c. What is the temperature at sea level ($x = 0$)?

   d. Graph $y = -4x + 60$ using the points you obtained in parts **a, b,** and **c**.

   e. What happens to the temperature if your altitude is more than **15** (thousand) feet?

Temperature above Sea Level

46. *Environmental lapse*　Suppose the temperature is 16° Celsius (16°C). As your altitude increases, the temperature y at x kilometers above sea level is given by the equation

$$y = -7x + 14$$

   a. What is the temperature when the altitude is **1** kilometer?

   b. What is the temperature when the altitude is **2** kilometers?

   c. What is the temperature at sea level ($x = 0$)?

   d. Graph $y = -7x + 14$ using the points you obtained in parts **a, b,** and **c**.

   e. What happens to the temperature if your altitude is more than **2** kilometers?

Temperature above Sea Level

47. *Catering prices*   A caterer charges $40 per person plus a $200 setup fee. The total cost C for an event can be represented by the equation

$$C = 40x + 200$$

where x is the number of people attending. Assume that fewer than 100 people will be attending.

   a. What is the cost when 30 people are attending?

   b. What is the cost when 50 people are attending?

   c. What is the cost for just setting up? (x = 0)

   d. Graph C = 40x + 200 using the points you obtained in parts **a**, **b**, and **c**.

Catering Costs

48. *Service staff cost*   The service staff for a reception charges $30 per hour, with a 5-hour minimum ($150). The cost C can be represented by the equation

$$C = 30h + 150$$

where h is the number of hours beyond 5. Assume your reception is shorter than 8 hours.

   a. What is the cost from 0 to 5 hours?

   b. What is the cost for 6 hours?

   c. What is the cost for 8 hours?

   d. Graph C = 30h + 150 using the points obtained in parts **a**, **b**, and **c**.

Service Staff Costs

〉〉〉   *Using Your Knowledge*

Anthropology   You already know how to determine whether an ordered pair satisfies an equation. Anthropological detectives use this knowledge to estimate the living height of a person using one dried bone as a clue. Suppose a detective finds a 17.9-inch femur bone from a male. To find the height H (in inches) of its owner, use the formula H = 1.88 f + 32.010 and determine that the owner's height H must have been H = 1.88(17.9) + 32.010 ≈ 66 inches. Of course, the ordered pair (17.9, 66) satisfies the equation. Now, for the fun part. Suppose you find a 17.9-inch femur bone, but this time you are looking for a missing female 66 inches tall. Can this femur belong to her? No! How do we know? The ordered pair (17.9, 66) *does not* satisfy the equation for female femur bones, which is H = 1.945 f + 28. Try it! Now, use the formulas to work Problems 49–54.

Living Height (inches)	
Male	**Female**
$H = 1.880 f + 32.010$	$H = 1.945 f + 28.379$
$H = 2.894h + 27.811$	$H = 2.754h + 28.140$
$H = 3.271r + 33.829$	$H = 3.343r + 31.978$
$H = 2.376t + 30.970$	$H = 2.352t + 29.439$
where *f*, *h*, *r*, and *t* represent the length of the femur, humerus, radius, and tibia bones, respectively.	

49. An 18-inch humerus bone from a male subject has been found. Can the bone belong to a 6'8" basketball player missing for several weeks?

50. A 16-inch humerus bone from a female subject has been found. Can the bone belong to a missing 5'2" female student?

51. A radius bone measuring 14 inches is found. Can it belong to Sandy Allen, the world's tallest living woman? (She is 7 feet tall!) How long to the nearest inch should the radius bone be for it to be Sandy Allen's bone?
   Source: Guinness World Records.

52. The tallest woman in medical history was Zeng Jinlian. A tibia bone measuring 29 inches is claimed to be hers. If the claim proves to be true, how tall was she, and what ordered pair would satisfy the equation relating the length of the tibia and the height of a woman?

53. A 10-inch radius of a man is found. Can it belong to a man 66.5 inches tall?

54. The longest recorded bone—an amazing 29.9 inches long—is the femur of the German giant Constantine who died in 1902. How tall was he, and what ordered pair would satisfy the equation relating the length of his femur and his height?

〉〉〉 *Write On*

55. Write the procedure you use to show that an ordered pair (a, b) satisfies an equation.

56. Write the procedure you use to determine whether the graph of an equation is a straight line.

57. Write the procedure you use to graph a line.

58. In your own words, write what "wind chill factor" means.

〉〉〉 *Concept Checker*

Fill in the blank(s) with the correct word(s), phrase, or mathematical statement.

59. The **graph** of $Ax + By = C$, where A, B, and C are constants, is the graph of a _____ line.   **straight**

60. The **point** at which a line **crosses the y-axis** is called the y-_____ and the **point** at which a line **crosses the x-axis** is called the x-_____.   **intercept**

〉〉〉 *Mastery Test*

61. Determine whether the ordered pair is a solution of $3x + 2y = 10$.

  a. $(1, 2)$

  b. $(2, 1)$

62. Graph: $2x + y = 6$

63. Graph: $y = -3x + 6$

64. Complete the ordered pairs so they satisfy the equation $y = 3x + 2$.

  a. $(x, 5)$

  b. $(-3, y)$

65. When the wind is blowing at a constant 10 miles per hour, the wind chill factor W can be approximated by $W = 1.3t - 18$, where t is the temperature in $°F$ and W is the wind chill factor.

  a. Find W when $t = 0$.

  b. Find W when $t = 10$.

  c. Graph W.

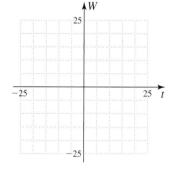

〉〉〉 *Skill Checker*

66. Solve: $2x - 6 = 0$

67. Solve: $2x + 4 = 0$

68. Solve: $3 + 3y = 0$

69. Solve: $0 = 0 - 0.1t$

70. Solve: $0 = 8 - 0.1t$

3.3   Graphing Lines Using Intercepts: Horizontal and Vertical Lines

▶ Objectives

A ⟩ Graph lines using intercepts.

B ⟩ Graphs lines passing through the origin.

C ⟩ Graph horizontal and vertical lines.

D ⟩ Solve applications involving graphs of lines.

▶ To Succeed, Review How To . . .

1. Solve linear equations (pp. 132–135).

2. Graph ordered pairs of numbers (p. 205).

▶ Getting Started

How Much Money Do You Have?

What is your household income? As you can see from the graphs, some of the median household incomes were slowly increasing, whereas others were almost flat over the time period shown. For example, the median household income h for Hispanics was slightly under $30,000 in 1972 and slightly over $30,000 in 1999. The graph of h can be approximated by the horizontal line $h = 30,000$. In this section, we learn how to graph horizontal and vertical lines, and lines whose graphs go through the origin. How do we know which lines are horizontal, which ones are vertical, and which ones pass through the origin? We will tell you in this section!

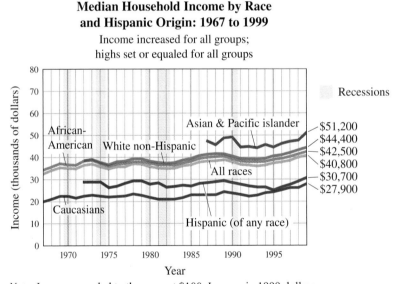

Median Household Income by Race and Hispanic Origin: 1967 to 1999

Income increased for all groups; highs set or equaled for all groups

Note: Income rounded to the nearest $100. Income in 1999 dollars.

Source: Data from U.S. Census Bureau, Current Population Survey, March 1968 to 2000.

A ⟩ Graphing Lines Using Intercepts

In Section 3.2 we mentioned that the graph of a linear equation of the form $Ax + By = C$ is a *straight line,* and we graphed such lines by finding ordered pairs that satisfied the equation of the line. But there is a quicker and simpler way to graph these equations by using the idea of *intercepts.* Here are the definitions we need.

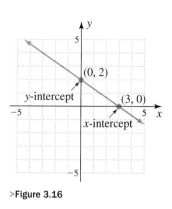

>**Figure 3.16**

PROCEDURE

1. The **x-intercept** $(a, 0)$ of a line is the point at which the line crosses the x-axis. To find the x-intercept, let $y = 0$ and solve for x.

2. The **y-intercept** $(0, b)$ of a line is the point at which the line crosses the y-axis. To find the y-intercept, let $x = 0$ and solve for y.

The graph in Figure 3.16 shows a line with y-intercept $(0, 2)$ and x-intercept $(3, 0)$. How do we use these ideas to graph lines? Here is the procedure.

PROCEDURE

To Graph a Line Using the Intercepts

1. Find the x-intercept $(a, 0)$.

2. Find the y-intercept $(0, b)$.

3. Graph the points $(a, 0)$ and $(0, b)$, and connect them with a line.

4. Find a third point to use as a check (make sure the point is on the line!).

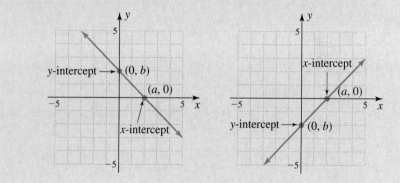

EXAMPLE 1   Graphing lines using intercepts

Graph: $5x + 2y = 10$

SOLUTION   First we find the x- and y-intercepts.

1. Let $x = 0$ in $5x + 2y = 10$. Then

$$5(0) + 2y = 10$$
$$2y = 10$$
$$y = 5$$

Hence, $(0, 5)$ is the y-intercept.

2. Let $y = 0$ in $5x + 2y = 10$. Then

$$5x + 2(0) = 10$$
$$5x = 10$$
$$x = 2$$

Hence, $(2, 0)$ is the x-intercept.

3. Now we graph the points $(0, 5)$ and $(2, 0)$ and connect them with a line as shown in Figure 3.17.

PROBLEM 1

Graph: $2x + 5y = 10$

Answers to PROBLEMS

1.

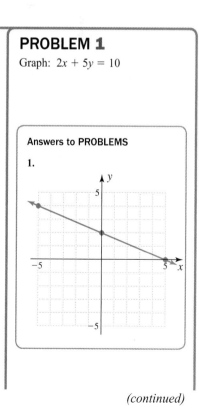

(continued)

4. We need to find a *third* point to use as a *check:* we let $x = 4$ and replace x with 4 in

$$5x + 2y = 10$$
$$5(4) + 2y = 10$$
$$20 + 2y = 10$$
$$2y = -10 \quad \text{Subtract 20.}$$
$$y = -5$$

The three points $(0, 5)$, $(2, 0)$, and $(4, -5)$ as well as the completed graph are shown in Figure 3.17.

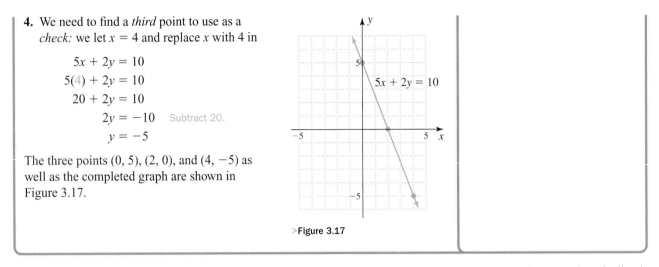

> Figure 3.17

The procedure used to graph a line using the intercepts works even when the line is *not* written in the form $Ax + By = C$, as we will show in Example 2.

EXAMPLE 2   Graphing lines using intercepts

Graph: $2x - 3y - 6 = 0$

SOLUTION

1. Let $y = 0$. Then

$$2x - 3(0) - 6 = 0$$
$$2x - 6 = 0$$
$$2x = 6 \quad \text{Add 6.}$$
$$x = 3 \quad \text{Divide by 2.}$$

$(3, 0)$ is the x-intercept.

2. Let $x = 0$. Then

$$2(0) - 3y - 6 = 0$$
$$-3y - 6 = 0$$
$$-3y = 6 \quad \text{Add 6.}$$
$$y = -2 \quad \text{Divide by } -3.$$

$(0, -2)$ is the y-intercept.

3. Connect $(0, -2)$ and $(3, 0)$ with a line, the graph of $2x - 3y - 6 = 0$, see Figure 3.18.

4. Use a third point as a check. Examine the line. It seems that for $x = -3$, y is -4. Let us check.

If $x = -3$, we have

$$2(-3) - 3y - 6 = 0$$
$$-6 - 3y - 6 = 0$$
$$-3y - 12 = 0 \quad \text{Simplify.}$$
$$-3y = 12 \quad \text{Add 12.}$$
$$y = -4 \quad \text{Divide by } -3.$$

As we suspected, the resulting point $(-3, -4)$ is on the line. Do you see why we picked $x = -3$? We did so because -3 seemed to be the only x-coordinate that yielded a y-value that was an integer.

> Figure 3.18

PROBLEM 2

Graph: $3x - 2y - 6 = 0$

Answers to PROBLEMS

2.

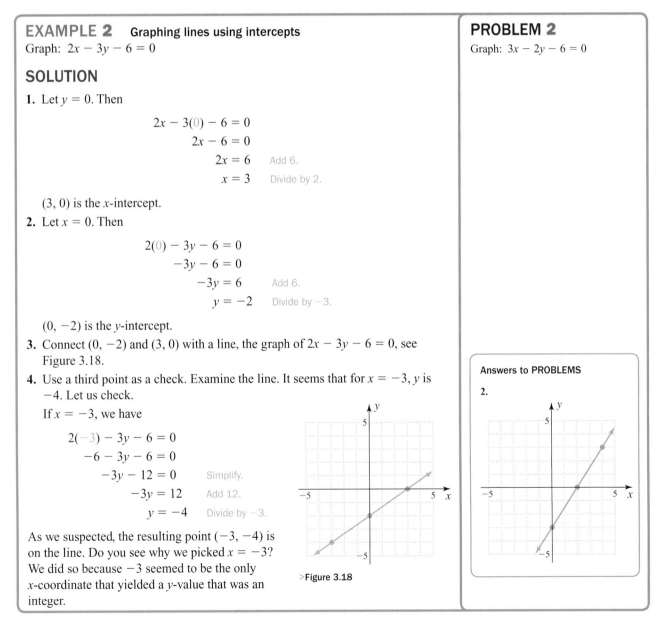

B › Graphing Lines Through the Origin

Sometimes it isn't possible to use the *x*- and *y*-intercepts to graph an equation. For example, to graph $x + 5y = 0$, we can start by letting $x = 0$ to obtain $0 + 5y = 0$ or $y = 0$. This means that $(0, 0)$ is part of the graph of the line $x + 5y = 0$. If we now let $y = 0$, we get $x = 0$, which is the same ordered pair. The line $x + 5y = 0$ goes through the origin $(0, 0)$, as seen in Figure 3.19, so we need to find another point. An easy one is $x = 5$. When 5 is substituted for *x* in $x + 5y = 0$, we obtain

$$5 + 5y = 0 \quad \text{or} \quad y = -1$$

Thus, a second point on the graph is $(5, -1)$. We join the points $(0, 0)$ and $(5, -1)$ to get the graph of the line shown in Figure 3.19. To use a third point as a check, let $x = -5$, which gives $y = 1$ and the point $(-5, 1)$, also shown on the line.

Here is the procedure for recognizing and graphing lines that go through the origin.

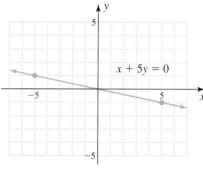

>Figure 3.19

STRAIGHT-LINE GRAPH THROUGH THE ORIGIN	The graph of an equation of the form $Ax + By = 0$, where *A* and *B* are constants and not equal to zero, is a straight line that goes through the **origin.**

PROCEDURE

Graphing Lines Through the Origin

To graph a line through the origin, use the point $(0, 0)$, find another point, and draw the line passing through $(0, 0)$ and this other point. Find a third point and verify that it is on the graph of the line.

EXAMPLE 3   Graphing lines passing through the origin

Graph: $2x + y = 0$

SOLUTION   The line $2x + y = 0$ is of the form $Ax + By = 0$ so it goes through the origin $(0, 0)$. To find another point on the line, we let $y = 4$ in $2x + y = 0$:

$$2x + 4 = 0$$
$$x = -2$$

Now we join the points $(0, 0)$ and $(-2, 4)$ with a line, the graph of $2x + y = 0$, as shown in Figure 3.20. Note that the check point $(2, -4)$ is on the graph.

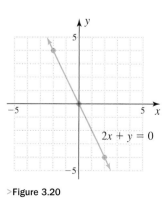

>Figure 3.20

PROBLEM 3

Graph: $3x + y = 0$

Answers to PROBLEMS

3.

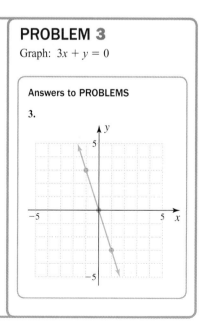

Here is a summary of the techniques used to graph lines that are *not* vertical or horizontal.

To graph $Ax + By = C$ (A, B, and C not 0)	To graph $Ax + By = 0$ (A and B not 0)
Find two points on the graph (preferably the x- and y-intercepts) and join them with a line. The result is the graph of $Ax + By = C$.	The graph goes through $(0, 0)$. Find another point and join $(0, 0)$ and the point with a line. The result is the graph of $Ax + By = 0$.
To graph $2x + y = -4$:	To graph $x + 4y = 0$, let $x = -4$ in
Let $x = 0$, find $y = -4$, and graph $(0, -4)$. Let $y = 0$, find $x = -2$, and graph $(-2, 0)$.	$$x + 4y = 0$$ $$-4 + 4y = 0$$ $$y = 1$$
Join $(0, -4)$ and $(-2, 0)$ to obtain the graph. Use an extra point to check. For example, if $x = -4$, $2(-4) + y = -4$, and $y = 4$. The point $(-4, 4)$ is on the line, so our graph is correct!	Join $(0, 0)$ and $(-4, 1)$ to obtain the graph. Use $x = 4$ as a check. When $x = 4$, $4 + 4y = 0$, and $y = -1$. The point $(4, -1)$ is on the line, so our graph is correct!

C › Graphing Horizontal and Vertical Lines

Not all linear equations are written in the form $Ax + By = C$. For example, consider the equation $y = 3$. It may seem that this equation is not in the form $Ax + By = C$. However, we can write the equation $y = 3$ as

$$0 \cdot x + y = 3$$

which is an equation written in the desired form. How do we graph the equation $y = 3$? Since it doesn't matter what value we give x, the result is always $y = 3$, as can be seen in the following table:

x	y
0	3
1	3
2	3

In the equation $0 \cdot x + y = 3$, x can be any number; you always get $y = 3$.

These ordered pairs, as well as the graph of $y = 3$, appear in Figure 3.21.

>Figure 3.21   >Figure 3.22   >Figure 3.23

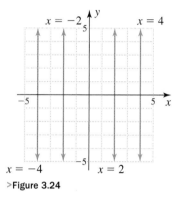

>Figure 3.24

Note that the equation $y = 3$ has for its graph a *horizontal* line crossing the y-axis at $y = 3$. The graphs of some other horizontal lines, all of which have equations of the form $y = k$ (k a constant), appear in Figure 3.22.

If the graph of any equation $y = k$ is a horizontal line, what would the graph of the equation $x = 3$ be? A vertical line, of course! We first note that the equation $x = 3$ can be written as

$$x + 0 \cdot y = 3$$

Thus, the equation is of the form $Ax + By = C$ so that its graph is a straight line. Now for any value of y, the value of x remains 3. Three values of x and the corresponding y-values appear in the following table. These three points, as well as the completed graph, are shown in Figure 3.23. The graphs of other vertical lines $x = -4$, $x = -2$, $x = 2$, and $x = 4$ are given in Figure 3.24.

x	y
3	0
3	1
3	2

Here are the formal definitions.

GRAPH OF $y = k$

The graph of any equation of the form

$$y = k \qquad \text{where } k \text{ is a constant}$$

is a **horizontal line** crossing the y-axis at k.

GRAPH OF $x = k$

The graph of any equation of the form

$$x = k \qquad \text{where } k \text{ is a constant}$$

is a **vertical line** crossing the x-axis at k.

EXAMPLE 4 Graphing vertical and horizontal lines

Graph: **a.** $2x - 4 = 0$ **b.** $3 + 3y = 0$

SOLUTION

a. We first solve for x.

$$2x - 4 = 0 \quad \text{Given}$$
$$2x = 4 \quad \text{Add 4.}$$
$$x = 2 \quad \text{Divide by 2.}$$

The graph of $x = 2$ is a vertical line crossing the x-axis at 2, as shown in Figure 3.25.

b. In this case, we solve for y.

$$3 + 3y = 0 \quad \text{Given}$$
$$3y = -3 \quad \text{Subtract 3.}$$
$$y = -1 \quad \text{Divide by 3.}$$

The graph of $y = -1$ is a horizontal line crossing the y-axis at -1, as shown in Figure 3.26.

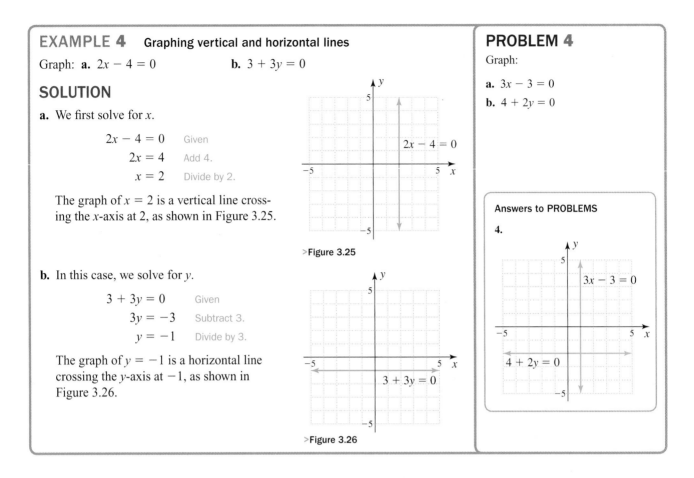

>Figure 3.25

>Figure 3.26

PROBLEM 4

Graph:

a. $3x - 3 = 0$

b. $4 + 2y = 0$

Answers to PROBLEMS

4.

D ⟩ Applications Involving Graphs of Lines

As we saw in the *Getting Started,* graphing a linear equation in two variables can be very useful in solving real-world problems. Here is another example.

EXAMPLE 5 Gender differences in hospital stays

Figure 3.27 shows the average stay in the hospital (in days) for males and females using data from the U.S. National Center for Health Statistics. If t represents the number of years after 1970 and d represents the number of days, the equations can be approximated as follows:

For males: $d = 9 - 0.1t$
For females: $d = 8 - 0.1t$

a. Graph these two equations.

b. Can t be negative?

c. Can d be negative?

d. Based on your answers to parts **b** and **c**, in what quadrant should the two graphs lie?

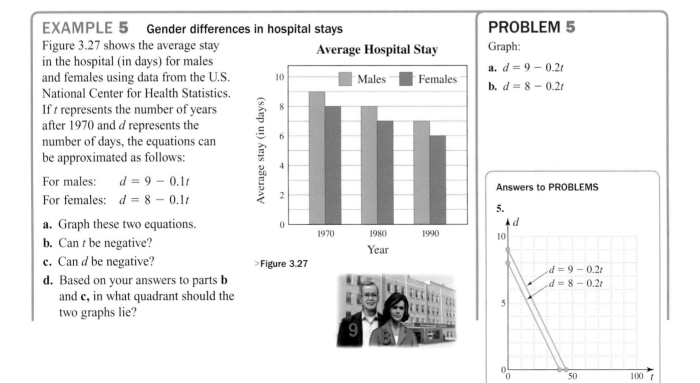

>Figure 3.27

PROBLEM 5

Graph:

a. $d = 9 - 0.2t$

b. $d = 8 - 0.2t$

Answers to PROBLEMS

5.

SOLUTION

a. We find the t- and d-intercepts for males and females, graph them, and connect them with a line.

For males, when $t = 0$,

$$d = 9 - 0.1(0) = 9$$

Thus, $(0, 9)$ is the d-intercept. To find the t-intercept, let $d = 0$:

$$0 = 9 - 0.1t$$
$$0.1t = 9$$
$$t = 90$$

Thus, $(90, 0)$ is the t-intercept. Graph the intercepts $(0, 9)$ and $(90, 0)$ for males and join them with a line:

For females, when $t = 0$,

$$d = 8 - 0.1(0) = 8$$

Thus, $(0, 8)$ is the d-intercept. To find the t-intercept, let $d = 0$:

$$0 = 8 - 0.1t$$
$$0.1t = 8$$
$$t = 80$$

Thus, $(80, 0)$ is the t-intercept. Graph the intercepts $(0, 8)$ and $(80, 0)$ for females and join them with a line:

Average stay for males

Average stay for females

Note that the scale on the t-axis is not the same as that on the d-axis.

b. Because t represents the number of years after 1970, it cannot be negative.

c. Because d represents the number of days, it cannot be negative.

d. Since t and d cannot be negative, the graphs should lie only in quadrant I.

🖩 ◈ ▦ Calculator Corner

Graphing Lines

Your calculator can easily do the examples in this section, but you have to *know* how to solve for y to enter the given equations. (Even with a calculator, you have to know algebra!) Then to graph $3x + y = 6$ (Example 4 in Section 3.2), we first solve for y to obtain $y = 6 - 3x$; thus, we enter $Y_1 = 6 - 3x$ and press GRAPH to obtain the result shown in Window 1. We used the *default* or *standard* window, which is a $[-10, 10]$ by $[-10, 10]$ rectangle. If you have a calculator, do Examples 2 and 3 of Section 3.3 now. Note that you can graph $3 + 3y = 0$ (Example 4b) by solving for y to obtain $y = -1$, but you *cannot* graph $2x - 4 = 0$ with most calculators. Can you see why?

   To graph the equations in Example 5, you have to select the appropriate window. Let's see why. Suppose you simply enter $d = 9 - 0.1t$ by using y instead of d and x instead of t. The result is shown in Window 2. To see more of the graph, you have to adjust the window. Algebraically, we know that the t-intercept is $(90, 0)$ and the d-intercept is $(0, 9)$; thus, an appropriate window might be $[0, 90]$ by $[0, 10]$. Now you only have to select the scales for x and y. Since the x's go from 0 to 90, make Xscl $= 10$. The y's go from 0 to 10, so we can let Yscl $= 1$. (See Window 3.) The completed graph appears with a scale that allows us to see most of the graph in Window 4. Press GRAPH to see it! Now do you see how the equation $d = 9 - 0.1t$ can be graphed and why it's graphed in quadrant I?

(continued)

Window 1   Window 2   Window 3   Window 4

WINDOW
Xmin=0
Xmax=90
Xscl=10
Ymin=0
Ymax=10
Yscl=1

You've gotten this far, so you should be rewarded! Suppose you want to find out how long the projected average hospital stay would be for males in the year 2000. Some calculators find the value of y when given x. Press [2nd] [TRACE] [ENTER] 30 [ENTER] to find the value of $y = 9 - 0.1x$ when $X = 30$ ($2000 - 1970 = 30$). (If your calculator doesn't have this feature, you can use [ZOOM] and [TRACE] to find the answer.) Window 5 shows the answer. Can you see what it is? Can you do it without your calculator? Now see if you can find the projected average hospital stay for females in the year 2000.

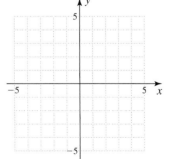

Window 5

> Exercises **3.3**

‹ A › Graphing Lines Using Intercepts   In Problems 1–10, graph the equations.

1. $x + 2y = 4$

2. $y + 2x = 2$

3. $-5x - 2y = -10$

4. $-2x - 3y = -6$

5. $y - 3x - 3 = 0$

6. $y + 2x - 4 = 0$

7. $6 = 6x - 3y$

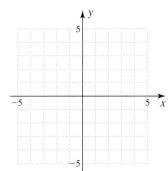

8. $6 = 3y - 2x$

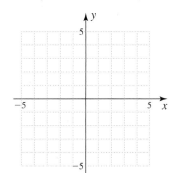

9. $3x + 4y + 12 = 0$

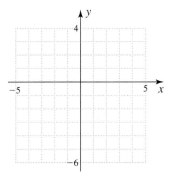

10. $5x + 2y + 10 = 0$

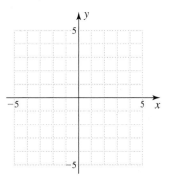

⟨ **B** ⟩ **Graphing Lines Through the Origin**   In Problems 11–20, graph the equations.

11. $3x + y = 0$

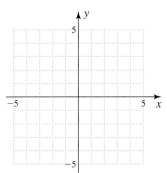

12. $4x + y = 0$

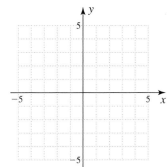

13. $2x + 3y = 0$

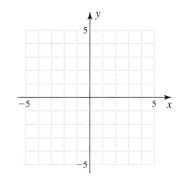

14. $3x + 2y = 0$

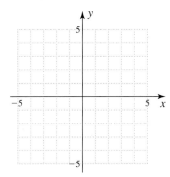

15. $-2x + y = 0$

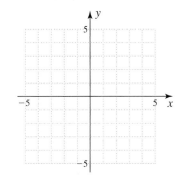

16. $-3x + y = 0$

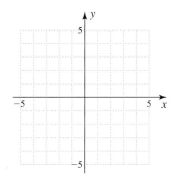

17. $2x - 3y = 0$

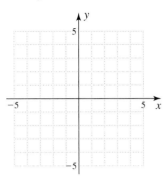

18. $3x - 2y = 0$

19. $-3x = -2y$

20. $-2x = -3y$

⟨**C**⟩ **Graphing Horizontal and Vertical Lines**   In Problems 21–30, graph the equations.

21. $y = -4$

22. $y = -\dfrac{3}{2}$

23. $2y + 6 = 0$

24. $-3y + 9 = 0$

25. $x = -\dfrac{5}{2}$

26. $x = \dfrac{7}{2}$

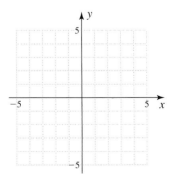

27. $2x + 4 = 0$

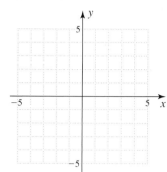

28. $3x - 12 = 0$

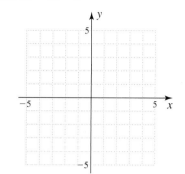

29. $2x - 9 = 0$

30. $-2x + 7 = 0$

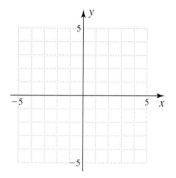

⟨ **D** ⟩   **Applications Involving Graphs of Lines**

31. *Average stay in hospital*   According to the U.S. National Center for Health Statistics, the total average stay in the hospital, d (in days), can be approximated by $d = 5.6 - 0.08t$, where t is the number of years after 2000.

   a. What is the d-intercept?

   b. What is the t-intercept?

   c. Graph the equation.

32. *Sales of jazz recordings*   According to the Recording Industry Association, the percent p of U.S. dollar sales for jazz recordings can be approximated by $p = 6 - 0.6t$, where t is the number of years after 2000.

   a. Find the t- and p-intercepts for this equation.

   b. Graph the equation.

33. *Daily fat intake*   According to the U.S. Department of Agriculture, the total daily fat intake g (in grams) per person can be approximated by $g = 140 + t$, where t is the number of years after 1950.

   a. What was the daily fat intake per person in 1950?

   b. What was the daily fat intake per person in 1990?

   c. What would you project the daily fat intake to be in the year 2000?

   d. Use the information from parts **a**–**c** to graph $g = 140 + t$.

34. *Death rates*   According to the U.S. Health and Human Services, the number D of deaths from heart disease per 100,000 population can be approximated by $D = -5t + 290$, where t is the number of years after 1998.

   a. How many deaths per 100,000 population were there in 1998?

   b. How many deaths per 100,000 population would you expect in 2008?

   c. Find the t- and D-intercepts for $D = -5t + 290$ and graph the equation.

35. *Studying and grades*   If you are enrolled for *c* credit hours, how many hours per week *W* should you spend outside class studying? For best results, the suggestion is

$$W = 3c$$

a. Suppose you are taking 6 credit hours. How many hours should you spend outside the class studying?

b. Suppose you are a full-time student taking 12 credit hours. How many hours should you spend outside class studying?

c. Graph $y = 3c$ using the points you obtained in parts **a** and **b**.

Source: University of Michigan, Flint

Study Hours vs. Credit Hours

36. *Studying and grades*   You may even get away with studying only 2 hours per week *W* outside class for each credit hour *c* you are taking.

a. Suppose you are taking 6 credit hours. How many hours should you spend outside the class studying?

b. Suppose you are a full time student taking 12 credit hours. How many hours should you spend outside class studying?

c. Graph $y = 2c$ using the points you obtained in parts **a** and **b**.

Study Hours vs. Credit Hours

37. *Water pressure*   Suppose you take scuba diving. The water pressure *P* (in pounds per cubic foot) as you descend to a depth of *f* feet is given by $P = 62.4\,f$.

a. What would the pressure *P* be at sea level ($f = 0$)?

b. What would the pressure *P* be at a depth of 10 feet?

c. According to PADI (Professional Association of Diving Instructors) the maximum depth for beginners is 40 feet. What would the pressure *P* be at 40 feet?

d. Graph $P = 62.4\,f$ using the points you obtained in parts **a–c**.

Pressure at Depth of *f* Feet

38. *Temperature*   When you fly in a balloon, the temperature *T* decreases 7°F for each 1000 feet *f* of altitude. Thus, at 3 (thousand) feet, the temperature would decrease by 21°F. If you started at 70°F, now the temperature is 70°F − 21°F, or 49°F.

a. What would the decrease in temperature be when you are 1 (thousand) feet high?

b. What about when you are 5 (thousand) feet high?

c. Graph $T = -7\,f$ using the points you obtained in parts **a** and **b**.

Decrease in Temperature

〉〉〉 *Using Your Knowledge*

Cholesterol Level   In Example 3 of Section 3.2 (pp. 223–224), we mentioned that the decrease in the cholesterol level C after w weeks elapsed could be approximated by $C = -3w + 215$.

39. According to this equation, what was the cholesterol level initially?

40. What was the cholesterol level at the end of 12 weeks?

41. Use the results of Problems 39 and 40 to graph the equation $C = -3w + 215$.

42. According to the graph shown in Example 3 (pp. 223–224), the initial cholesterol level was 215. If we assume that the initial cholesterol level is 230, we can approximate the reduction in cholesterol by $C = -3w + 230$.

    a. Find the intercepts and graph the equation on the same coordinate axes as $C = -3w + 215$.

    b. How many weeks does it take for a person with an initial cholesterol level of 230 to reduce it to 175?

〉〉〉 *Write On*

43. If in the equation $Ax + By = C$, $A = 0$ and B and C are not zero, what type of graph will result?

44. If in the equation $Ax + By = C$, $B = 0$ and A and C are not zero, what type of graph will result?

45. If in the equation $Ax + By = C$, $C = 0$ and A and B are not zero, what type of graph will result?

46. What are the intercepts of the line $Ax + By = C$, and how would you find them?

47. How many points are needed to graph a straight line?

〉〉〉 *Concept Checker*

Fill in the blank(s) with the correct word(s), phrase, or mathematical statement.

48. The *x*-intercept of a line is the point at which the line **crosses the *x*-axis** and has **coordinates** _____.

49. The *y*-intercept of a line is the point at which the line **crosses the *y*-axis** and has **coordinates** _____.

50. The **coordinates** of the **origin** are _____.

51. The **graph of** $Ax + By = 0$, where A and B are constants and not equal to 0, is a **straight line that passes through the** _____.

52. The **graph of** $y = k$, where k a constant, is a _____ **line.**

53. The **graph of** $x = k$, where k a constant, is a _____ **line.**

(0, 0)	**(*x*, 0)**
origin	**(0, *x*)**
horizontal	**(*y*, 0)**
vertical	**(0, *y*)**
(*x*, *y*)	

⟩⟩⟩ *Mastery Test*

Graph:

54. $x + 2y = 4$

55. $-2x + y = 4$

56. $3x - 6 = 0$

57. $4 + 2y = 0$

58. $-4x + y = 0$

59. $-x + 4y = 0$

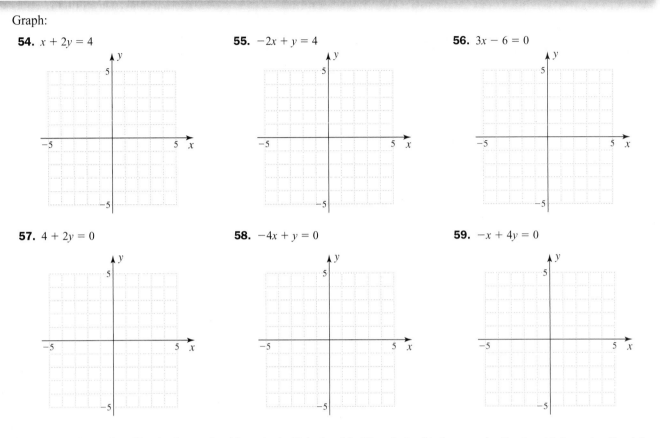

60. The number N (in millions) of recreational boats in the United States can be approximated by $N = 10 + 0.4t$, where t is the number of years after 1975.

   a. How many recreational boats were there in 1975?

   b. How many would you expect in 1995?

   c. Find the t- and N-intercepts of $N = 10 + 0.4t$ and graph the equation.

61. The relationship between the Continental dress size C and the American dress size A is $C = A + 30$.

   a. Find the intercepts for $C = A + 30$ and graph the equation.

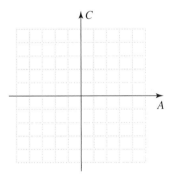

   b. In what quadrant should the graph lie?

⟩⟩⟩ *Skill Checker*

Find:

62. $13.5 - 3.5$

63. $3 - (-6)$

64. $5 - (-7)$

65. $-6 - 3$

66. $-5 - 7$

3.4 The Slope of a Line: Parallel and Perpendicular Lines

▶ Objectives

A ⟩ Find the slope of a line given two points.

B ⟩ Find the slope of a line given the equation of the line.

C ⟩ Determine whether two lines are parallel, perpendicular, or neither.

D ⟩ Solve applications involving slope.

▶ To Succeed, Review How To . . .

1. Add, subtract, multiply, and divide signed numbers (pp. 52–56, 60–64).
2. Solve an equation for a specified variable (pp. 136–140).

▶ Getting Started

Online Services Are Sloping Upward

Can you tell from the graph when the number of subscribers to commercial online services soared? After year 1. The annual increase in subscribers the first year was

$$\frac{\text{Difference in subscribers}}{\text{Difference in years}} = \frac{3.5 - 3.0}{1 - 0} = \frac{0.5}{1} = 0.5 \text{ (million per year)}$$

The annual increase in subscribers in years 1–5 was

$$\frac{13.5 - 3.5}{5 - 1} = \frac{10}{4} = 2.5 \text{ (million per year)}$$

Source: ClickZ Network.

As you can see from the graph, the line gets "steeper" in years 1 to 5 than during the first year, 0 to 1. The "steepness" of the lines (0 to 1 and 1 to 5) was calculated by comparing the *vertical* change of the line (the **rise**) to the *horizontal* change (the **run**). This measure of steepness is called the **slope** of the line. In this section, we shall learn how to find the slope of a line when two points are given or when the equation of the line is given and also how to determine when lines are parallel, perpendicular, or neither.

A ⟩ Finding Slopes from Two Points

The steepness of a line can be measured by using the ratio of the **vertical rise** (or **fall**) to the corresponding **horizontal run.** This ratio is called the *slope.* For example, a staircase that rises 3 feet in a horizontal distance of 4 feet is said to have a slope of $\frac{3}{4}$. The definition of slope is as follows.

SLOPE	The **slope** m of the line going through the points (x_1, y_1) and (x_2, y_2), where $x_1 \neq x_2$, is given by $$m = \frac{y_2 - y_1}{x_2 - x_1} = \frac{\text{rise} \uparrow}{\text{run} \rightarrow}$$ This means that the slope of a line is the change in y (Δy) over the change in x (Δx), that is, $$m = \frac{\Delta y}{\Delta x} = \frac{y_2 - y_1}{x_2 - x_1}$$

The slope for a vertical line such as $x = 3$ is not defined because all points on this line have the same x-value, 3, and hence,

$$m = \frac{y_2 - y_1}{3 - 3} = \frac{y_2 - y_1}{0}$$

which is undefined because division by zero is undefined (see Example 3). The slope of a horizontal line is zero because all points on such a line have the same y-values. Thus, for the line $y = 7$,

$$m = \frac{7 - 7}{x_2 - x_1} = \frac{0}{x_2 - x_1} = 0 \qquad \text{See Example 3.}$$

EXAMPLE 1   Finding the slope given two points: Positive slope

Find the slope of the line passing through the points $(0, -6)$ and $(3, 3)$ in Figure 3.28.

SOLUTION   Suppose we choose $(x_1, y_1) = (0, -6)$ and $(x_2, y_2) = (3, 3)$. Then we use the equation for slope to obtain

$$m = \frac{3 - (-6)}{3 - 0} = \frac{9}{3} = 3$$

If we choose $(x_1, y_1) = (3, 3)$ and $(x_2, y_2) = (0, -6)$, then

$$m = \frac{-6 - 3}{0 - 3} = \frac{-9}{-3} = 3$$

As you can see, it makes no difference which point is labeled (x_1, y_1) and which is labeled (x_2, y_2). Since an interchange of the two points simply changes the sign of both the numerator and the denominator in the slope formula, the result is the same in both cases.

>**Figure 3.28**   Line with positive slope: $\frac{\text{Rise}}{\text{Run}} = \frac{9}{3} = 3$

PROBLEM 1

Find the slope of the line going through the points $(0, -6)$ and $(2, 2)$.

EXAMPLE 2   Finding the slope given two points: Negative slope

Find the slope of the line that passes through the points $(3, -4)$ and $(-2, 3)$ in Figure 3.29.

SOLUTION   We take $(x_1, y_1) = (-2, 3)$ so that $(x_2, y_2) = (3, -4)$. Then

$$m = \frac{-4 - 3}{3 - (-2)} = -\frac{7}{5}$$

>**Figure 3.29**   Line with negative slope: $\frac{\text{Rise}}{\text{Run}} = -\frac{7}{5}$

PROBLEM 2

Find the slope of the line going through the points $(2, -4)$ and $(-3, 2)$.

Answers to PROBLEMS

1. 4   **2.** $-\frac{6}{5}$

Examples 1 and 2 are illustrations of the fact that a line that rises from left to right has a *positive slope* and one that falls from left to right has a *negative slope*. What about vertical and horizontal lines? We shall discuss them in Example 3.

EXAMPLE 3   Finding the slopes of vertical and horizontal lines

Find the slope of the line passing through the given points.

a. $(-4, -2)$ and $(-4, 3)$ **b.** $(1, 4)$ and $(4, 4)$

SOLUTION

a. Substituting $(-4, -2)$ for (x_1, y_1) and $(-4, 3)$ for (x_2, y_2) in the equation for slope, we obtain

$$m = \frac{3 - (-2)}{-4 - (-4)} = \frac{3 + 2}{-4 + 4} = \frac{5}{0}$$

which is undefined. Thus, the slope of a vertical line is *undefined* (see Figure 3.30).

b. This time $(x_1, y_1) = (1, 4)$ and $(x_2, y_2) = (4, 4)$, so

$$m = \frac{4 - 4}{4 - 1} = \frac{0}{3} = 0$$

Thus, the slope of a horizontal line is zero (see Figure 3.30).

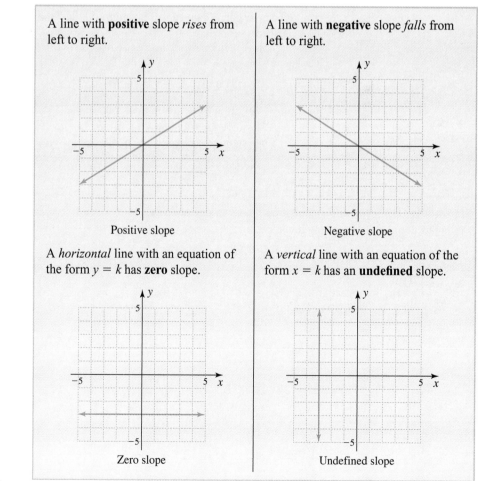

> **Figure 3.30**

PROBLEM 3

Find the slope of the line passing through the given points.

a. $(4, 1)$ and $(-3, 1)$

b. $(-2, 4)$ and $(-2, 1)$

Let's summarize our work with slopes so far.

A line with **positive** slope *rises* from left to right.

Positive slope

A line with **negative** slope *falls* from left to right.

Negative slope

A *horizontal* line with an equation of the form $y = k$ has **zero** slope.

Zero slope

A *vertical* line with an equation of the form $x = k$ has an **undefined** slope.

Undefined slope

B ⟩ Finding Slopes from Equations

In the *Getting Started,* we found the slope of the line by using a ratio. The slope of a line can also be found from its equation. Thus, if we approximate the number of subscribers to commercial online services between years 1 and 5 by the equation

$$y = 2.5x + 1 \quad \text{(millions)}$$

where x is the number of years after year 0 and y is the number of subscribers, we can find the slope of $y = 2.5x + 1$ by using two values for x and then finding the corresponding y-values.

For $x = 1$, $y = 2.5(1) + 1 = 3.5$, so $(1, 3.5)$ is on the line.
For $x = 2$, $y = 2.5(2) + 1 = 6$, so $(2, 6)$ is on the line.

The slope of $y = 2.5x + 1$ passing through $(1, 3.5)$ and $(2, 6)$ is

$$m = \frac{6 - 3.5}{2 - 1} = 2.5$$

which is simply the **coefficient** of x in the equation $y = 2.5x + 1$. This idea can be generalized as follows.

SLOPE OF $y = mx + b$

The slope of the line defined by the equation $y = mx + b$ is n.

Thus, if you are given the equation of a line and you want to find its slope, you would use this procedure.

PROCEDURE

Finding a slope
1. Solve the equation for y.
2. The slope is m, the coefficient of x.

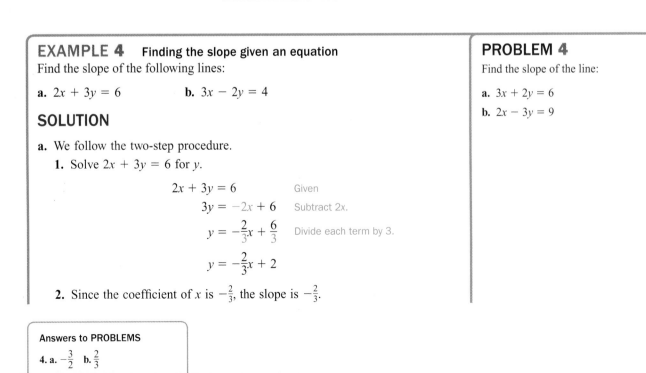

EXAMPLE 4   Finding the slope given an equation
Find the slope of the following lines:

a. $2x + 3y = 6$ **b.** $3x - 2y = 4$

SOLUTION

a. We follow the two-step procedure.
   1. Solve $2x + 3y = 6$ for y.

$$2x + 3y = 6 \qquad \text{Given}$$
$$3y = -2x + 6 \qquad \text{Subtract } 2x.$$
$$y = -\frac{2}{3}x + \frac{6}{3} \qquad \text{Divide each term by 3.}$$
$$y = -\frac{2}{3}x + 2$$

   2. Since the coefficient of x is $-\frac{2}{3}$, the slope is $-\frac{2}{3}$.

PROBLEM 4
Find the slope of the line:

a. $3x + 2y = 6$

b. $2x - 3y = 9$

Answers to PROBLEMS
4. a. $-\frac{3}{2}$ b. $\frac{2}{3}$

b. We follow the steps.

  1. Solve for y.

$$3x - 2y = 4 \qquad \text{Given}$$
$$-2y = -3x + 4 \qquad \text{Subtract 3x.}$$
$$y = \frac{-3}{-2}x + \frac{4}{-2} \qquad \text{Divide each term by } -2.$$
$$y = \frac{3}{2}x - 2$$

  2. The slope is the coefficient of x, so the slope is $\frac{3}{2}$.

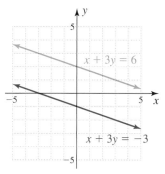

>Figure 3.31

C ⟩ Finding Parallel and Perpendicular Lines

Parallel lines are lines in the plane that never intersect. In Figure 3.31, the lines $x + 3y = 6$ and $x + 3y = -3$ appear to be parallel lines. How can we be sure? By solving each equation for y and determining whether both lines have the same slope.

For $x + 3y = 6$: $y = -\frac{1}{3}x + 2$ The slope is $-\frac{1}{3}$.

For $x + 3y = -3$: $y = -\frac{1}{3}x - 1$ The slope is $\frac{1}{3}$.

Since the two lines have the same slope but different y-intercepts, they are parallel lines.

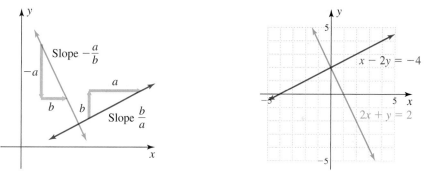

>Figure 3.32   >Figure 3.33

The two lines in Figure 3.32 appear to be **perpendicular**; that is, they meet at a 90° angle. Note that their slopes are *negative reciprocals* and that the product of their slopes is

$$-\frac{a}{b} \cdot \frac{b}{a} = -1$$

The lines $2x + y = 2$ and $x - 2y = -4$ have graphs that also appear to be perpendicular (see Figure 3.33). To show that this is the case, we check their slopes.

For $2x + y = 2$: $y = -2x + 2$ The slope is -2.

For $x - 2y = -4$: $y = \frac{1}{2}x + 2$ The slope is $\frac{1}{2}$.

Since the product of the slopes is $-2 \cdot \frac{1}{2} = -1$, the lines are perpendicular.

SLOPES OF PARALLEL AND PERPENDICULAR LINES

Two lines with the **same** slope but different y-intercepts are **parallel**.

Two lines whose slopes have a product of **−1** are **perpendicular**.

EXAMPLE 5   Finding whether two given lines are parallel, perpendicular, or neither

Decide whether the pair of lines are parallel, perpendicular, or neither:

a. $x - 3y = 6$
$2x - 6y = -12$

b. $2x + y = 6$
$x + y = 4$

c. $2x + y = 5$
$x - 2y = 4$

SOLUTION

a. We find the slope of each line by solving the equations for y.

$x - 3y = 6$	Given	$2x - 6y = -12$	Given
$-3y = -x + 6$	Subtract x.	$-6y = -2x - 12$	Subtract 2x.
$y = \frac{1}{3}x - 2$	Divide by −3.	$y = \frac{1}{3}x + 2$	Divide by −6.
The slope is $\frac{1}{3}$.		The slope is $\frac{1}{3}$.	

Since both slopes are $\frac{1}{3}$, the slopes are equal and the y-intercepts are different, the lines are parallel, as shown in Figure 3.34.

b. We find the slope of each line by solving the equations for y.

$2x + y = 6$	Given	$x + y = 4$	Given
$y = -2x + 6$	Subtract 2x.	$y = -x + 4$	Subtract x.
The slope is -2.		The slope is -1.	

Since the slopes are -2 and -1, their product is not -1, so the lines are neither parallel nor perpendicular, as shown in Figure 3.35.

c. We again solve both equations for y to find their slopes.

$2x + y = 5$	Given	$x - 2y = 4$	Given
$y = -2x + 5$	Subtract 2x.	$-2y = -x + 4$	Subtract x.
The slope is -2.		$y = \frac{1}{2}x - 2$	Divide by −2.
		The slope is $\frac{1}{2}$.	

Since the product of the slopes is $-2 \cdot \frac{1}{2} = -1$, the lines are perpendicular, as shown in Figure 3.36.

PROBLEM 5

Decide whether the pair of lines are parallel, perpendicular, or neither.

a. $x - 2y = 3$
$2x - 4y = 8$

b. $3x + y = 6$
$x + y = 2$

c. $3x + y = 6$
$x - 3y = 5$

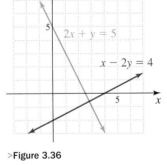

>Figure 3.34

>Figure 3.35

>Figure 3.36

Answers to PROBLEMS

5. a. Parallel   **b.** Neither
   c. Perpendicular

D › Applications Involving Slope

In the *Getting Started,* we saw that the number of commercial online subscribers *increased* at a rate of 2.5 million per year. The slope of a line can be used to describe a rate of change. For example, the number N of deaths (per 100,000 population) due to heart disease can be approximated by $N = -5t + 300$, where t is the number of years after 1960. Since the slope of the line $N = -5t + 300$ is -5, this means that the number of deaths per 100,000 population due to heart disease is *decreasing* by 5 every year.

EXAMPLE 6   Escalating health care costs

As you have probably heard, health care costs have been soaring for several years. According to the Health Insurance Association of America, the average daily hospital room charge C can be approximated by $C = 17t + 127$ (dollars), where t is the number of years after 1980.

a. What is the slope of $C = 17t + 127$?

b. What does the slope represent?

SOLUTION

a. The slope of the line $C = 17t + 127$ is 17.

b. The slope 17 represents the annual *increase* (in dollars) for the average daily room charge.

PROBLEM 6

The decrease in cholesterol level C over w weeks is given by $C = -3w + 215$.

a. What is the slope of $C = -3w + 215$?

b. What does the slope represent?

▣ ◈ ▦   Calculator Corner

Using STAT **to find** $y = ax + b$

In this section we learned how to find the slope of a line given two points. Your calculator can do better! Let's try Example 1, where we are given the points $(0, -6)$ and $(3, 3)$ and asked to find the slope. Press STAT 1 and enter the two x-coordinates (0 and 3) under L_1 and the two y-coordinates (-6 and 3) under L_2. Your calculator is internally programmed to use a process called *regression* to find the equation of the line passing through these two points. To do so, press STAT , move the cursor right to reach CALC, enter 4 for LinReg (Linear Regression), and then press ENTER . The resulting line shown in Window 1 is written in the slope-intercept form $y = ax + b$. Clearly, the slope is 3 and the y-intercept is -6. See what happens when you use your calculator to do Example 2.

What happens when you do Example 3a? You get the warning shown in Window 2, indicating that the slope is *undefined*. Here you have to *know* some algebra to conclude that the resulting line is a *vertical* line. (Your calculator can't do that for you!)

```
LinReg
y=ax+b
a=3
b=-6
r =1
```
Window 1

```
ERR:DOMAIN
1:Quit
2:Goto
```
Window 2

Answers to PROBLEMS

6. a. -3 **b.** The cholesterol is dropping 3 points each week.

⟩ Exercises **3.4**

⟨ **A** ⟩ **Finding Slopes from Two Points**   In Problems 1–14, find the slope of the line that passes through the two given points.

1. $(1, 2)$ and $(3, 4)$

2. $(1, -2)$ and $(-3, -4)$

3. $(0, 5)$ and $(5, 0)$

4. $(3, -6)$ and $(5, -6)$

5. $(-1, -3)$ and $(7, -4)$

6. $(-2, -5)$ and $(-1, -6)$

7. $(0, 0)$ and $(12, 3)$

8. $(-1, -1)$ and $(-10, -10)$

9. $(3, 5)$ and $(-2, 5)$

10. $(4, -3)$ and $(2, -3)$

11. $(4, 7)$ and $(-5, 7)$

12. $(-3, -5)$ and $(-2, -5)$

13. $\left(-\frac{1}{2}, -\frac{1}{3}\right)$ and $\left(-\frac{1}{2}, \frac{1}{3}\right)$

14. $\left(-\frac{1}{5}, 2\right)$ and $\left(-\frac{1}{5}, 1\right)$

⟨ **B** ⟩ **Finding Slopes from Equations**   In Problems 15–26, find the slope of the given line.

15. $y = 3x + 7$

16. $y = -4x + 6$

17. $-3y = 2x - 4$

18. $4y = 6x + 3$

19. $x + 3y = 6$

20. $-x + 2y = 3$

21. $-2x + 5y = 5$

22. $3x - y = 6$

23. $y = 6$

24. $x = 7$

25. $2x - 4 = 0$

26. $2y - 3 = 0$

⟨ **C** ⟩ **Finding Parallel and Perpendicular Lines**   In Problems 27–37, determine whether the given lines are parallel, perpendicular, or neither.

27. $y = 2x + 5$ and $4x - 2y = 7$

28. $y = 4 - 5x$ and $15x + 3y = 3$

29. $2x + 5y = 8$ and $5x - 2y = -9$

30. $3x + 4y = 4$ and $2x - 6y = 7$

31. $x + 7y = 7$ and $2x + 14y = 21$

32. $y - 5x = 12$ and $y - 3x = 8$

33. $2x + y = 7$ and $-2x - y = 9$

34. $2x - 4 = 0$ and $x - 1 = 0$

35. $2y - 4 = 0$ and $3y - 6 = 0$

36. $3y = 6$ and $2x = 6$

37. $3x = 7$ and $2y = 7$

⟨ **D** ⟩ **Applications Involving Slope**

38. *Daily fat consumption*   According to the U.S. Department of Agriculture, the daily fat consumption F per person can be approximated by the equation

$$F = 190 + 0.8t \text{ (grams)}$$

where t is the number of years after 2000.

   a. What is the slope of this line?

   b. What does the slope represent?

39. *Average hospital stay*   According to the National Center for Health Statistics, the average hospital stay S can be approximated by the equation

$$S = 5 - 0.1t \text{ (days)}$$

where t is the number of years after 2000.

   a. What is the slope of this line?

   b. Is the average hospital stay increasing or decreasing?

   c. What does the slope represent?

40. *Daily seafood consumption* According to the U.S. Department of Agriculture, the daily consumption T of tuna can be approximated by the equation

$$T = 3.87 - 0.13t \text{ (pounds)}$$

and the daily consumption F of fish and shellfish can be approximated by the equation

$$F = -0.29t + 15.46$$

a. Is the consumption of tuna increasing or decreasing?

b. Is the consumption of fish and shellfish increasing or decreasing?

c. Which consumption (tuna or fish and shellfish) is decreasing faster?

42. *Life expectancy of men* The average life span y of an American man is given by the equation $y = 0.15t + 74$, where t is the number of years after 2000.

a. What is the slope of this line?

b. Is the life span of American men increasing or decreasing?

c. What does the slope represent?

Source: U.S. National Center for Health Statistics, *Statistical Abstract of the United States.*

44. *Velocity of a thrown ball* The speed v of a ball thrown up with an initial velocity of 15 meters per second is given by the equation $v = 15 - 5t$, where v is the velocity (in meters per second) and t is the number of seconds after the ball is thrown.

a. What is the slope of this line?

b. Is the velocity of the ball increasing or decreasing?

c. What does the slope represent?

46. *Milk products consumption* The number of gallons of milk products g consumed annually by the average American can be approximated by the equation $g = 24 - 0.2t$, where t is the number of years after 2000.

a. What is the slope of this line?

b. Is the consumption of milk products increasing or decreasing?

41. *Life expectancy of women* The average life span (life expectancy) y of an American woman is given by the equation $y = 0.15t + 80$, where t is the number of years after 2000.

a. What is the slope of this line?

b. Is the life span of American women increasing or decreasing?

c. What does the slope represent?

Source: U.S. National Center for Health Statistics, *Statistical Abstract of the United States.*

43. *Velocity of a thrown ball* The speed v of a ball thrown up with an initial velocity of 128 feet per second is given by the equation $v = 128 - 32t$, where v is the velocity (in feet per second) and t is the number of seconds after the ball is thrown.

a. What is the slope of this line?

b. Is the velocity of the ball increasing or decreasing?

c. What does the slope represent?

45. *Daily fat consumption* The number of fat grams f consumed daily by the average American can be approximated by the equation $f = 165 + 0.4t$, where t is the number of years after 2000.

a. What is the slope of this line?

b. Is the consumption of fat increasing or decreasing?

c. What does the slope represent?

Source: U.S. Dept. of Agriculture, *Statistical Abstract of the United States.*

c. What does the slope represent?

Source: U.S. Dept. of Agriculture, *Statistical Abstract of the United States.*

Footwear For Problems 47–49, use the information below. The graph shows the annual expenditure on footwear in five consecutive years by persons under 25 years of age.

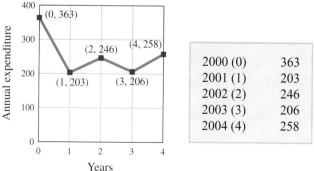

2000 (0)	363
2001 (1)	203
2002 (2)	246
2003 (3)	206
2004 (4)	258

Years

Source: U.S. Department of Labor

47. a. In what year did the expenditures on footwear decrease the most?

b. What was the decrease from year 0 to year 1?

c. Find the slope m of the line in part **b.**

d. What does the slope m represent?

48. a. In what years did the expenditures on footwear decrease?

   b. What was the decrease during year 2?

   c. Find the slope *m* of the line for year 2

   d. What does the slope *m* represent?

49. a. In what year did the expenditures on footwear increase the most?

   b. Find the slope *m* of the line from year 3 to year 4.

   c. Find the slope *m* of the line from year 1 to year 2.

   d. Which slope is larger, the slope for year 3 to 4 or the slope for year 1 to 2?

〉〉〉 *Using Your Knowledge*

Up, Down, or Away!   The slope of a line can be positive, negative, zero, or undefined. Use your knowledge to sketch the following lines.

50. A line that has a negative slope

51. A line that has a positive slope

52. A line with zero slope

53. A line with an undefined slope

〉〉〉 *Write On*

54. Write in your own words what is meant by the *slope* of a line.

55. Explain why the slope of a horizontal line is zero.

56. Explain why the slope of a vertical line is undefined.

〉〉〉 *Concept Checker*

Fill in the blank(s) with the correct word(s), phrase, or mathematical statement.

57. The **slope** of the line through (x_1, y_1) and (x_2, y_2) is _____ .

58. The **slope** of the line $y = mx + b$ is _____ .

59. Two **lines** with the **same slope** and **different** *y*-intercepts are _____ lines.

60. Two **lines** with **slopes whose product is** -1 are _____ lines.

$\dfrac{x_2 - x_1}{y_2 - y_1}$ **perpendicular**

b m

parallel $\dfrac{y_2 - y_1}{x_2 - x_1}$

〉〉〉 *Mastery Test*

Find the slope of the line passing through the given points:

61. $(2, -3)$ and $(4, -5)$

62. $(-1, 2)$ and $(4, -2)$

63. $(-4, 2)$ and $(-4, 5)$

64. $(-3, 4)$ and $(-5, 4)$

65. Find the slope of the line $3x + 2y = 6$.

66. Find the slope of the line $-3x + 4y = 12$.

Determine whether the lines are parallel, perpendicular, or neither:

67. $-x + 3y = -6$ and $2x - 6y = -7$

68. $2x + 3y = 5$ and $3x - 2y = 5$

69. $3x - 2y = 6$ and $-2x - 3y = 6$

70. The U.S. population P (in millions) can be approximated by $P = 2.2t + 180$, where t is the number of years after 1960.

   a. What is the slope of this line?

   b. How fast is the U.S. population growing each year? (State your answer in millions.)

Simplify:

71. $2[x - (-4)]$

72. $3[x - (-6)]$

73. $-2[x - (-1)]$

3.5 Graphing Lines Using Points and Slopes

▶ Objectives

Find and graph an equation of a line given:

A ▶ Its slope and a point on the line.

B ▶ Its slope and y-intercept.

C ▶ Two points on the line.

▶ To Succeed, Review How To . . .

1. Add, subtract, multiply, and divide signed numbers (pp. 52–56, 60–64).

2. Solve a linear equation for a specified variable (pp. 136–140).

▶ Getting Started

The Formula for Cholesterol Reduction

As we discussed in Example 3, Section 3.2 (pp. 223–224), the cholesterol level C can be approximated by $C = -3w + 215$, where w is the number of weeks elapsed. How did we get this equation? You can see by looking at the graph that the cholesterol level decreases *about* 3 points each week, so the slope of the line is -3. You can make this approximation more exact by using the points $(0, 215)$ and $(12, 175)$ to find the slope

$$m = \frac{215 - 175}{0 - 12} = -\frac{40}{12} \approx -3.3$$

Cholesterol Level Reduction

Since the y-intercept is at 215, you can reason that the cholesterol level starts at 215 and decreases about 3 points each week. Thus, the cholesterol level C based on the number of weeks elapsed is

$$C = 215 - 3w$$

or, equivalently,

$$C = -3w + 215$$

If you want a more exact approximation, you can also write

$$C = -3.3w + 215$$

Thus, if you are given the slope m and y-intercept b of a line, the equation of the line will be $y = mx + b$.

   In this section, we shall learn how to find and graph a linear equation when we are given the slope and a point on the line, the slope and the y-intercept, or two points on the line.

A ▶ Using the Point-Slope Form of a Line

We can use the slope of a line to obtain the equation of the line, provided we are given one point on the line. Thus, suppose a line has slope m and passes through the point (x_1, y_1).

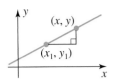

>Figure 3.37

If we let (x, y) be a second point on the line, the slope of the line shown in Figure 3.37 is given by

$$\frac{y - y_1}{x - x_1} = m$$

Multiplying both sides by $(x - x_1)$, we get the *point-slope form* of the line.

POINT-SLOPE FORM

The **point-slope form** of the equation of the line going through (x_1, y_1) and having slope *m* is

$$y - y_1 = m(x - x_1)$$

EXAMPLE 1   Finding an equation for a line given a point and the slope

a. Find an equation of the line that passes through the point $(2, -3)$ and has slope $m = -4$.

b. Graph the line.

SOLUTION

a. Using the point-slope form, we get

$$y - (-3) = -4(x - 2)$$
$$y + 3 = -4x + 8$$
$$y = -4x + 5$$

b. To graph this line, we start at the point $(2, -3)$. Since the slope of the line is -4 and, by definition, the slope is

$$\frac{\text{Rise}}{\text{Run}} = -\frac{4}{1}$$

we go 4 units *down* (the rise) and 1 unit *right* (the run), ending at the point $(3, -7)$. We then join the points $(2, -3)$ and $(3, -7)$ with a line, which is the graph of $y = -4x + 5$ (see Figure 3.38). As a final check, does the point $(3, -7)$ satisfy the equation $y = -4x + 5$? When $x = 3$, $y = -4x + 5$ becomes

$$y = -4(3) + 5 = -7 \quad \text{True!}$$

Thus, the point $(3, -7)$ is on the line $y = -4x + 5$. Note that the equation $y + 3 = -4x + 8$ can be written in the standard form $Ax + By = C$.

$$y + 3 = -4x + 8 \quad \text{Given}$$
$$4x + y + 3 = 8 \quad \text{Add } 4x.$$
$$4x + y = 5 \quad \text{Subtract 3.}$$

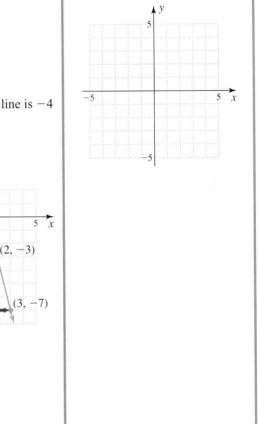

>Figure 3.38

PROBLEM 1

a. Find an equation of the line that goes through the point $(2, -4)$ and has slope $m = -3$.

b. Graph the line.

Answers to PROBLEMS

1. a. $y = -3x + 2$

b.

B › The Slope-Intercept Form of a Line

An important special case of the point-slope form is that in which the given point is the point where the line intersects the *y*-axis. Let this point be denoted by $(0, b)$. Then b is called the *y-intercept* of the line. Using the point-slope form, we obtain

$$y - b = m(x - 0)$$

or

$$y - b = mx$$

By adding b to both sides, we get the *slope-intercept form* of the equation of the line.

SLOPE-INTERCEPT FORM

The **slope-intercept form** of the equation of the line having slope m and y-intercept b is

$$y = mx + b$$

EXAMPLE 2   Finding an equation of a line given the slope and the y-intercept

a. Find an equation of the line having slope 5 and y-intercept -4.
b. Graph the line.

SOLUTION

a. In this case $m = 5$ and $b = -4$. Substituting in the slope-intercept form, we obtain

$$y = 5x + (-4) \quad \text{or} \quad y = 5x - 4$$

b. To graph this line, we start at the y-intercept $(0, -4)$. Since the slope of the line is 5 and by definition the slope is

$$\frac{\text{Rise}}{\text{Run}} = \frac{5}{1}$$

go 5 units *up* (the rise) and 1 unit *right* (the run), ending at $(1, 1)$. We join the points $(0, -4)$ and $(1, 1)$ with a line, which is the graph of $y = 5x - 4$ (see Figure 3.39). Now check that the point $(1, 1)$ is on the line $y = 5x - 4$.

When $x = 1$, $y = 5x - 4$ becomes

$$y = 5(1) - 4 = 1 \quad \text{True!}$$

Thus, the point $(1, 1)$ is on the line $y = 5x - 4$.

>Figure 3.39

PROBLEM 2
a. Find an equation of the line with slope 3 and y-intercept -6.
b. Graph the line.

C › The Two-Point Form of a Line

If a line passes through two given points, you can find and graph the equation of the line by using the point-slope form as shown in Example 3.

EXAMPLE 3   Finding an equation of a line given two points

a. Find an equation of the line passing through the points $(-2, 3)$ and $(1, -3)$.
b. Graph the line.

PROBLEM 3
a. Find an equation of the line going through $(5, -6)$ and $(-3, 4)$.
b. Graph the line.

Answers to PROBLEMS

2. a. $y = 3x - 6$
b.

Answer on page 262

(continued)

SOLUTION

a. We first find the slope of the line.

$$m = \frac{3 - (-3)}{-2 - 1} = \frac{6}{-3} = -2$$

Now we can use either the point $(-2, 3)$ or $(1, -3)$ and the point-slope form to find the equation of the line. Using the point $(-2, 3) = (x_1, y_1)$ and $m = -2$,

$$y - y_1 = m(x - x_1)$$

becomes

$$y - 3 = -2[x - (-2)]$$
$$y - 3 = -2(x + 2)$$
$$y - 3 = -2x - 4$$
$$y = -2x - 1$$

or, in standard form, $2x + y = -1$.

b. To graph this line, we simply plot the given points $(-2, 3)$ and $(1, -3)$ and join them with a line, as shown in Figure 3.40. To check our results, we make sure that both points satisfy the equation $2x + y = -1$. For $(-2, 3)$, let $x = -2$ and $y = 3$ in $2x + y = -1$ to obtain

$$2(-2) + 3 = -4 + 3 = -1$$

Thus, $(-2, 3)$ satisfies $2x + y = -1$, and our result is correct.

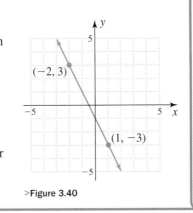

>Figure 3.40

At this point, many students ask, "How do we know which form to use?" The answer depends on what information is given in the problem. The following table helps you make this decision. It's a good idea to examine this table closely before you attempt the problems in Exercises 3.5.

Finding the Equation of a Line

Given	Use
A point (x_1, y_1) and the slope m	Point-slope form: $y - y_1 = m(x - x_1)$
The slope m and the y-intercept b	Slope-intercept form: $y = mx + b$
Two points (x_1, y_1) and (x_2, y_2), $x_1 \neq x_2$	Two-point form: $y - y_1 = m(x - x_1)$, where $m = \dfrac{y_2 - y_1}{x_2 - x_1}$

Note that the resulting equation can always be written in the standard form: $Ax + By = C$.

Answers to PROBLEMS

3. a. $y = -\frac{5}{4}x + \frac{1}{4}$

b.

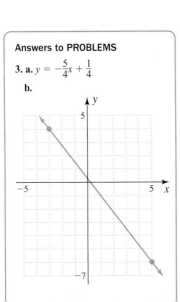

Boost your grade at mathzone.com!

MathZone

> Practice Problems   > Self-Tests
> NetTutor   > e-Professors
   > Videos

› Exercises **3.5**

‹ **A** ›   **Using the Point-Slope Form of a Line**   In Problems 1–6, find the equation of the line that has the given properties (*m* is the slope), then graph the line.

1. Goes through $(1, 2)$; $m = \frac{1}{2}$ **2.** Goes through $(-1, -2)$; $m = -2$ **3.** Goes through $(2, 4)$; $m = -1$

4. Goes through $(-3, 1)$; $m = \frac{3}{2}$ **5.** Goes through $(4, 5)$; $m = 0$ **6.** Goes through $(3, 2)$; slope is not defined (does not exist)

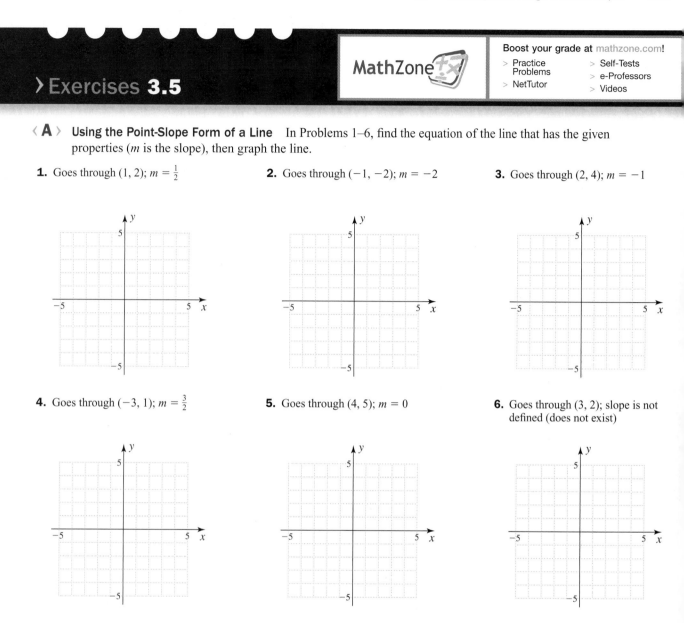

‹ **B** ›   **The Slope–Intercept Form of a Line**   In Problems 7–16, find an equation of the line with the given slope and *y*-intercept.

7. Slope, 2; *y*-intercept, -3 **8.** Slope, 3; *y*-intercept, -5

9. Slope, -4; *y*-intercept, 6   **10.** Slope, -6; *y*-intercept, -7

11. Slope, $\frac{3}{4}$; *y*-intercept, $\frac{7}{8}$ **12.** Slope $\frac{7}{8}$; *y*-intercept, $\frac{3}{8}$

13. Slope, 2.5; *y*-intercept, -4.7 **14.** Slope, 2.8; *y*-intercept, -3.2

15. Slope, -3.5; *y*-intercept, 5.9   **16.** Slope, -2.5; *y*-intercept, 6.4

In Problems 17–20, find an equation of the line having slope m and y-intercept b, and then graph the line.

17. $m = \frac{1}{4}, b = 3$

18. $m = -\frac{2}{5}, b = 1$

19. $m = -\frac{3}{4}, b = -2$

20. $m = -\frac{1}{3}, b = -1$

⟨ **C** ⟩ **The Two-Point Form of a Line**   In Problems 21–30, find an equation of the line passing through the given points, write the equation in standard form, and then graph the equation.

21. $(2, 3)$ and $(7, 8)$

22. $(-2, -3)$ and $(1, -6)$

23. $(2, 2)$ and $(1, -1)$

24. $(-3, 4)$ and $(-2, 0)$

25. $(3, 0)$ and $(0, 4)$

26. $(0, -3)$ and $(4, 0)$

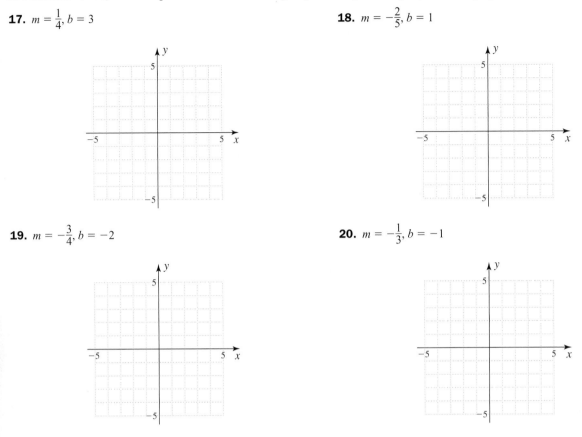

27. $(3, 0)$ and $(3, 2)$

28. $(-4, 2)$ and $(-4, 0)$

29. $(-2, -3)$ and $(1, -3)$

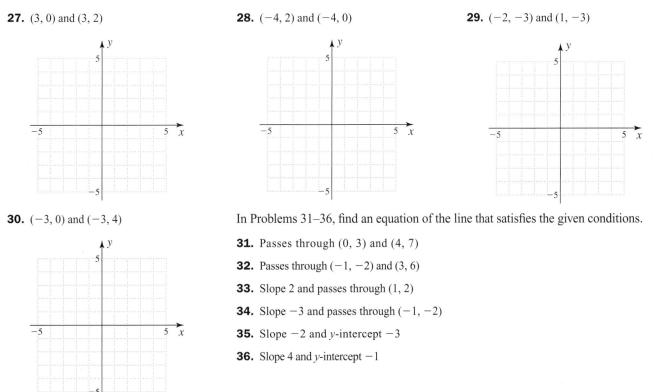

30. $(-3, 0)$ and $(-3, 4)$

In Problems 31–36, find an equation of the line that satisfies the given conditions.

31. Passes through $(0, 3)$ and $(4, 7)$

32. Passes through $(-1, -2)$ and $(3, 6)$

33. Slope 2 and passes through $(1, 2)$

34. Slope -3 and passes through $(-1, -2)$

35. Slope -2 and y-intercept -3

36. Slope 4 and y-intercept -1

〉〉〉 *Using Your Knowledge*

The Business of Slopes and Intercepts   The slope m and the y-intercept of an equation play important roles in economics and business. Let's see how.

Suppose you wish to go into the business of manufacturing fancy candles. First, you have to buy some ingredients such as wax, paint, and so on. Assume that all these ingredients cost you $100. This is the *fixed cost*. Now suppose it costs $2 to manufacture each candle. This is the *marginal cost*. What would be the total cost y if the marginal cost is $2, x units are produced, and the fixed cost is $100? The answer is

$$\underbrace{y}_{\substack{\text{Total cost}\\ \text{(in dollars)}}} = \underbrace{2x}_{\substack{\text{Cost for}\\ x\ \text{units}}} + \underbrace{100}_{\substack{\text{Fixed}\\ \text{cost}}}$$

In general, an equation of the form

$$y = mx + b$$

gives the total cost y of producing x units, when m is the cost of producing 1 unit and b is the fixed cost.

37. Find the total cost y of producing x units of a product costing $2 per unit if the fixed cost is $50.

38. Find the total cost y of producing x units of a product whose production cost is $7 per unit if the fixed cost is $300.

39. The total cost y of producing x units of a certain product is given by

$$y = 2x + 75$$

   a. What is the production cost for each unit?

   b. What is the fixed cost?

〉〉〉 *Write On*

40. If you are given the equation of a nonvertical line, describe the procedure you would use to find the slope of the line.

41. If you are given a vertical line whose x-intercept is the origin, what is the name of the line?

42. If you are given a horizontal line whose y-intercept is the origin, what is the name of the line?

43. How would you write the equation of a vertical line in the standard form $Ax + By = C$?

44. How would you write the equation of a horizontal line in the standard form $Ax + By = C$?

45. Write an explanation of why a vertical line cannot be written in the form $y = mx + b$.

> > > **Concept Checker**

Fill in the blank(s) with the correct word(s), phrase, or mathematical statement.

46. The **point-slope form** of the equation of the line passing through (x_1, y_1) and having slope m is _____.

47. The **slope-intercept** form of the equation of the line having slope m and y-intercept b is _____.

48. The **two-point** form of the equation of the line passing through the points (x_1, y_1) and (x_2, y_2) is _____.

49. The **standard form** of a straight line equation in two variables is _____.

$$x - x_1 = m(y - y_1) \qquad x - x_1 = \frac{y_2 - y_1}{x_2 - x_1}(y - y_1)$$

$$y = mx + b \qquad\qquad y = ax + m$$

$$x + y = C \qquad\qquad y - y_1 = m(x - x_1)$$

$$Ax + By = C$$

$$y - y_1 = \frac{y_2 - y_1}{x_2 - x_1}(x - x_1)$$

> > > **Mastery Test**

50. Find an equation of the line with slope 3 and y-intercept -6, and graph the line.

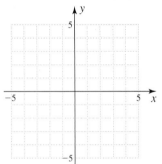

51. Find an equation of the line with slope $-\frac{3}{4}$ and y-intercept 2, and graph the line.

52. Find an equation of the line passing through the point $(-2, 3)$ and with slope -3, and graph the line.

53. Find an equation of the line passing through the point $(3, -1)$ and with slope $\frac{2}{3}$, and graph the line.

54. Find an equation of the line passing through the points $(-3, 4)$ and $(-2, -6)$, write it in standard form, and graph the line.

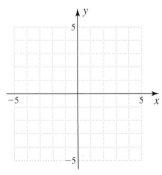

55. Find an equation of the line passing through the points $(-1, -6)$ and $(-3, 4)$, write it in standard form, and graph the line.

> > > **Skill Checker**

56. Find the point-slope form of the equation of a line with slope $m = 2.5$ and passing through the point $(0, 1.75)$.

57. Find the slope intercept form of the equation of a line with y-intercept 50 and slope 0.35.

58. Find the slope m of the line passing through $(6, 12)$ and $(8, 20)$.

59. Write in standard form the equation of the line passing through $(6, 12)$ and $(8, 20)$.

3.6

Applications of Equations of Lines

▶ Objectives

A ▷ Solve applications involving the point-slope formula.

B ▷ Solve applications involving the slope-intercept formula.

C ▷ Solve applications involving the two-point formula.

▶ To Succeed, Review How To . . .

1. Find the equation of a line given a point and the slope (p. 260).
2. Find the equation of a line given a point and the y-intercept (pp. 260–261).
3. Find the equation of a line given two points (pp. 261–262).

▶ Getting Started

Taxi! Taxi!

How much is a taxi ride in your city? In Boston, it is $1.50 for the first $\frac{1}{4}$ mile and $2 for each additional mile. As a matter of fact, it costs $21 for a 10-mile ride. Can we find a simple formula based on the number of miles m you ride? If C represents the total cost of the ride and we know that a 10-mile ride costs $21, we are given one point—namely, (10, 21). We also know that the slope is 2, since the costs are $2 per additional mile. Using the point-slope form, we have

$$C - 21 = 2(m - 10)$$

or $$C = 2m - 20 + 21$$

That is, $$C = 2m + 1$$

Note that this is consistent with the fact that a 10-mile ride costs $21, since

$$C = 2(10) + 1 = 21$$

In the preceding section, we learned how to find and graph the equation of a line when given a point and the slope, a point and the y-intercept, or two points. In this section we are going to use the appropriate formulas to solve application problems.

A ▷ Applications Involving the Point-Slope Formula

EXAMPLE 1   Taxicab fare calculations

Taxi fares in Key West are $2.25 for the first $\frac{1}{5}$ mile and then $2.50 for each additional mile. If a 10-mile ride costs $26.75, find an equation for the total cost C of an m-mile ride. What will be the cost for a 30-mile ride?

SOLUTION   Since we know that a 10-mile ride costs $26.75, and each additional mile costs $2.50, we are given the point (10, 26.75) and the slope 2.5.

PROBLEM 1

A taxicab company charges $2.50 for the first $\frac{1}{4}$ mile and $2.00 for each additional mile. If a 10-mile ride costs $22, find an equation for the total cost C of an m-mile ride. What is the cost of a 25-mile ride?

(continued)

Using the point-slope form, we have

$$C - 26.75 = 2.5(m - 10)$$

or

$$C = 2.5m - 25 + 26.75$$

That is,

$$C = 2.5m + 1.75$$

For a 30-mile ride, $m = 30$ and $C = 2.5(30) + 1.75 = \$76.75$.

B › Applications Involving the Slope-Intercept Formula

Do you have a cell phone? Which plan do you have? Most plans have a set number of free minutes for a set fee, after which you pay for additional minutes used. For example, at the present time, Verizon Wireless® has a plan that allows 900 free minutes for $50 with unlimited weekend minutes. After that, you pay $0.35 for each additional minute. We will consider such a plan in Example 2.

EXAMPLE 2   Cell phone plan charges

Maria subscribed to a cell phone plan with 900 free minutes, a $50 monthly fee, and $0.35 for each additional minute. Find an equation for the total cost C of her plan when she uses m minutes after the first 900. What is her cost when she uses 1200 minutes?

SOLUTION   Let m be the number of minutes Maria uses after the first 900. If she does not go over the limit, she will pay $50. The y-intercept will then be 50. Since she pays $0.35 for each additional minute, the slope is 0.35. Thus, the total cost C when m additional minutes are used is

$$C = 0.35m + 50$$

If she uses 1200 minutes, she has to pay for 300 of them ($1200 - 900$ free $= 300$ paid), and the cost will then be

$$C = 0.35(300) + 50$$
$$= \quad 105 \quad + 50$$
$$= \$155$$

Thus, Maria will pay $155 for her monthly payment.

PROBLEM 2

Find the cost of the plan if the monthly fee is $40 and each minute after the first 900 costs $0.50. What is the cost when she uses 1000 minutes?

C › Applications Involving the Two-Point Formula

To be on the Internet you need an Internet service provider (ISP), a company that provides Internet access to individuals and companies. There are many ISP plans and many have unlimited use; that is, subscribers can stay online as much time as they wish. Economical plans involve a set fee for a set number of hours and an additional charge for extra hours. For example, at the present time, America Online® has a "light usage" plan and a "limited" plan. We discuss the cost for the limited plan in Example 3.

Answers to PROBLEMS

2. $C = 0.50m + 40$; $90

EXAMPLE 3   On-line subscriber costs

An online limited plan costs $9.95 for up to 5 hours of use, $12.90 for 6 hours, and $18.80 for 8 hours. Find an equation for the total cost C of this plan when h hours are used ($h > 5$). What is the cost for 10 hours?

SOLUTION   We will examine ordered pairs of the form (h, C), where C is the cost and h is the number of hours used. The ordered pair corresponding to 6 hours of use costing $12.90 is (6, 12.90), and the ordered pair corresponding to 8 hours of use costing $18.80 is (8, 18.80). We now find the slope. Using the slope formula, we have

$$m = \frac{y_2 - y_1}{x_2 - x_1} = \frac{18.80 - 12.90}{8 - 6} = \frac{5.9}{2} = \$2.95$$

To find the equation for C we use the point-slope form with $m = 2.95$ and (6, 12.90) as our point, obtaining

$$C - 12.90 = 2.95(h - 6)$$

or

$$C = 2.95h - 17.70 + 12.90$$

or

$$C = 2.95h - 4.80$$

For 10 hours of use, $h = 10$ and

$$C = 2.95(10) - 4.80$$
$$= 29.50 - 4.80$$
$$= \$24.70$$

PROBLEM 3

There is another plan that costs $14.95 for up to 5 hours, $17.45 for 6 hours, and $24.95 for 9 hours. Find an equation for the total cost C of this plan when h hours are used ($h > 5$). What is the cost when 10 hours are used?

MathZone +x

> Exercises 3.6

⟨ **A** ⟩   **Applications Involving the Point-Slope Formula**   In Problems 1–10, solve the application.

1. *San Francisco taxi fares*   Taxi fares in San Francisco are $2 for the first mile and $1.70 for each additional mile. If a 10-mile ride costs $17.30, find an equation for the total cost C of an m-mile ride. What would the price be for a 30-mile ride?

2. *San Francisco taxi fares*   A different taxicab company in San Francisco charges $3 for the first mile and $1.50 for each additional mile. If a 20-mile ride costs $31.50, find an equation for the total cost C of an m-mile ride. What would be the price of a 10-mile ride? Which company is cheaper, this one or the company in Problem 1?

3. *San Francisco taxi fares*   Pedro took a cab in San Francisco and paid the fare quoted in Problem 1. Tyrone paid the fare quoted in Problem 2. Amazingly, they paid the same amount! How far did they ride?

4. *New York taxi fares*   In New York, a 20-mile cab ride is $32 and consists of an initial set charge and $1.50 per mile. Find an equation for the total cost C of an m-mile ride. How much would you have to pay for a 30-mile ride?

5. *San Francisco taxi fares*   The cost C for San Francisco fares (Problem 1) is $2 for the first mile and $1.70 for each mile thereafter. We can find C by following these steps:

   a. What is the cost of the first mile?

   b. If the whole trip is m miles, how many miles do you travel after the first mile?

   c. How much do you pay per mile after the first mile?

   d. What is the cost of all the miles after the first?

   e. The total cost C is the sum of the cost of the first mile and the cost of all the miles after the first. What is that cost? Is your answer the same as that in Problem 1?

6. *New York taxi fares*   New York fares are easier to compute. They are simply $2 for the initial set charge and $1.50 for each mile. Use the procedure in Problem 5 to find the total cost C for an m-mile trip. Do you get the same answer as you did in Problem 4?

Answers to PROBLEMS

3. $C = 2.5h + 2.45$; $27.45

7. *Cell phone rental overseas*   Did you know that you can rent cell phones for your overseas travel? If you are in Paris, the cost for a 1-week rental, including 60 minutes of long-distance calls to New York, is $175.

   a. Find a formula for the total cost C of a rental phone that includes m minutes of long-distance calls to New York.

   b. What is the weekly charge for the phone?

   c. What is the per-minute usage charge?

9. *Wind chill temperatures*   The table shows the relationship between the actual temperature (x) in degrees Fahrenheit and the wind chill temperature (y) when the wind speed is 5 miles per hour.

Wind Chill (Wind Speed 5 mi/hr)				
Temperature	(x)	20	25	30
Wind Chill	(y)	13	19	25

   a. Find the slope of the line (the rate of change of the wind chill temperature) using the points (20, 13) and (25, 19).

   b. Find the slope of the line using the points (25, 19) and (30, 25).

   c. Are the two slopes the same?

   d. Use the point-slope form to find y using the points (20, 13) and (25, 19).

   e. Use the point-slope form to find y using the points (25, 19) and (30, 25). Do you get an equation equivalent to the one in part **d**?

   f. Use your formula to find the wind chill when the temperature is 5°F and the wind speed is 5 miles per hour.

8. *International long-distance rates*   Long-distance calls from the Hilton Hotel in Paris to New York cost $7.80 per minute.

   a. Find a formula for the cost C of m minutes of long-distance calls from Paris to New York.

   b. How many minutes can you use so that the charges are identical to those you would pay when renting the phone of Problem 7? Answer to the nearest minute.

10. *Heat index values*   The table shows the relationship between the actual temperature (x) and the heat index or apparent temperature (y) when the relative humidity is 50%. *Note:* Temperatures above 105°F can cause severe heat disorders with continued exposure!

Relative Humidity (50%)				
Temperature	(x)	100	102	104
Heat Index	(y)	118	124	130

   a. Find the slope of the line using the points (100, 118) and (102, 124).

   b. Find the slope of the line using the points (102, 124) and (104, 130).

   c. Are the two slopes the same?

   d. Use the point-slope form of the line to find y using the points (100, 118) and (102, 124).

   e. Use the point-slope form of the line to find y using the points (102, 124) and (104, 130). Do you get an equation equivalent to the one in part **d**?

   f. Use your formula to find the heat index when the temperature is 106°F and the relative humidity is 50%.

⟨ B ⟩   Applications Involving the Slope-Intercept Formula

In Problems 11–22, solve the application.

11. *Electrician's charges*   According to Microsoft Home Advisor®, if you have an electrical failure caused by a blown fuse or circuit breaker you may have to call an electrician. "Plan on spending at least $100 for a service call, plus the electrician's hourly rate." Assume that the service call is $100 and the electrician charges $37.50 an hour.

   a. Find an equation for the total cost C of a call lasting h hours.

   b. If your bill amounts to $212.50, for how many hours were you charged?

12. *Appliance technician's charges*   Microsoft Home Advisor suggests that the cost C for fixing a faulty appliance is about $75 for the service call plus the technician's hourly rate. Assume that this rate is $40 per hour.

   a. Find an equation for the total cost C of a call lasting h hours.

   b. If your bill amounts to $195, for how many hours were you charged?

13. *International phone rental charges*   The total cost C for renting a phone for 1 week in London, England, is $40 per week plus $2.30 per minute for long-distance charges when calling the United States.

   a. Find a formula for the cost C of renting a phone and using m minutes of long-distance charges.

   b. If the total cost C amounted to $201, how many minutes were used?

   c. The Dorchester Hotel in London charges $7.20 per minute for long-distance charges to the United States. Find a formula for the cost C of m minutes of long-distance calls from the hotel to the United States.

   d. How many minutes can you use so that the charges are identical to those you would pay when renting the phone of part **a** (round to nearest minute)?

14. *Phone rental charges*   The rate for incoming calls for a rental phone is $1.20 per minute. If the rental charge is $50 per week, find a formula for the total cost C of renting a phone for m incoming minutes. If you paid $146 for the rental, for how many incoming minutes were you billed?

15. *Cell phone costs*   Verizon Wireless has a plan that costs $50 per month for 900 anytime minutes and unlimited weekend minutes. The charge for each additional minute is $0.35.

   a. Find a formula for the cost C when m additional minutes are used.

   b. If your bill was for $88.50, how many additional minutes did you use?

17. *Cell phone costs*   How many additional minutes do you have to use so that the costs for the plans in Problems 15 and 16 are identical? After how many additional minutes is the plan in Problem 15 cheaper?

19. *Estimating a female's height*   The height y for a female (in inches) can be estimated by multiplying the length x of her femur bone by 1.95 and adding 29 to the result.

   a. Write an equation for a woman's height y in slope-intercept form.

   b. What is the slope?

   c. What is the y-intercept?

21. *Average hospital stay*   Since 1970, the number y of days the average person stays in the hospital has steadily decreased from 8 days at the rate of 0.1 day per year. If x represents the number of years after 1970, write a slope-intercept equation representing the average number y of days a person stays in the hospital. What would the average stay in the year 2000 be? In 2010?

16. *Cell phone costs*   A Verizon competitor has a similar plan, but it charges $45 per month and $0.40 for each additional minute.

   a. Find a formula for the cost C when m additional minutes are used.

   b. If your bill was for $61, how many additional minutes did you use?

18. *Estimating a man's height*   You can estimate a man's height y (in inches) by multiplying the length x of his femur bone by 1.88 and adding 32 to the result.

   a. Write an equation for a man's estimated height y in slope-intercept form.

   b. What is the slope?

   c. What is the y-intercept?

20. *Estimating height from femur length*   Refer to Problems 18 and 19.

   a. What would be the length x of a femur that would yield the same height y for a male and a female?

   b. What would the estimated height of a person having such a femur bone be?

22. *Average hospital stay for females*   Since 1970, the number y of days the average female stays in the hospital has steadily decreased from 7.5 days at the rate of 0.1 day per year. If x represents the number of years after 1970, write a slope-intercept equation representing the average number y of days a female stays in the hospital. What would the average stay in the year 2000 be? In 2010?

⟨ **C** ⟩   **Applications Involving the Two-Point Formula**   In Problems 23–26, solve the application.

23. *Dog years vs. human years*   It is a common belief that 1 human year is equal to 7 dog years. That is not very accurate, since dogs reach adulthood within the first couple of years. A more accurate formula indicates that when a dog is 3 years old in human years, it would be 30 years old in dog years. Moreover, a 9-year-old dog is 60 years old in dog years.

   a. Form the ordered pairs (h, d), where h is the age of the dog in human years and d is the age of the dog in dog years. What does (3, 30) mean? What does (9, 60) mean?

   b. Find the slope of the line using the points (3, 30) and (9, 60).

   c. Find an equation for the age d of a dog based on the dog's age h in human years $(h > 1)$.

   d. If a dog is 4 human years old, how old is it in dog years?

   e. If a human could retire at age 65, what is the equivalent retirement age for a dog (in human years)?

   f. The drinking age for humans is usually 21 years. What is the equivalent drinking age for dogs?

24. *Cat years vs. human years*   When a cat is $h = 2$ years old in human years, it is $c = 24$ years old in cat years, and when a cat is $h = 6$ years old in human years, it is $c = 40$ years old in cat years.

   a. Form the ordered pairs (h, c), where h is the age of the cat in human years and c is the age of the cat in cat years. What does (2, 24) mean? What does (6, 40) mean?

   b. Find the slope of the line using the points (2, 24) and (6, 40).

   c. Find an equation for the age c of a cat based on the cat's age h in human years $(h > 1)$.

   d. If a cat is 4 human years old, how old is it in cat years?

   e. If a human could retire at age 60, what is the equivalent retirement age for a cat in human years?

   f. The drinking age for humans is usually 21 years. What is the equivalent drinking age for cats?

25. *Blood alcohol concentration* Your blood alcohol level is dependent on the amount of liquor you consume, your weight, and your gender. Suppose you are a 150-pound male and you consume 3 beers (5% alcohol) over a period of 1 hour. Your blood alcohol concentration (BAC) would be 0.052. If you have 5 beers, then your BAC would be 0.103. (You are legally drunk then! Most states regard a BAC of 0.08 as legally drunk.) Consider the ordered pairs (3, 0.052) and (5, 0.103).

a. Find the slope of the line passing through the two points.

b. If *b* represents the number of beers you had in 1 hour and *c* represents your BAC, find an equation for *c*.

c. Find your BAC when you have had 4 beers in 1 hour. Find your BAC when you have had 6 beers in 1 hour.

d. How many beers do you have to drink in 1 hour to be legally drunk (BAC = 0.08)? Answer to the nearest whole number.

27. *Medical applications* A rough estimate of a child's weight *W* (in kilograms) can be made from the child's age *A* (1 to 6) using the formula:

Weight in kilograms = (age in years) · 2 + 8

a. Write the equation in symbols using *W* for weight and *A* for age.

b. What is the slope *m* of $W = 2A + 8$?

c. What is the intercept *b* of $W = 2A + 8$?

d. Find the estimated weight *W* when $A = 1$. Graph the point.

e. Find the estimated weight *W* when $A = 6$. Graph the point.

f. Use the two points from parts **d** and **e** to graph $W = 2A + 8$.

Weight (kg)

Source: http://www.medal.org.

26. *Blood alcohol concentration* Suppose you are a 125-pound female and you consume 3 beers (5% alcohol) over a period of 1 hour. Your blood alcohol concentration (BAC) would be 0.069. If you have 5 beers, then your BAC would be 0.136. (You are really legally drunk then!) Consider the ordered pairs (3, 0.069) and (5, 0.136).

a. Find the slope of the line passing through the two points.

b. If *b* represents the number of beers you had in 1 hour and *c* represents your BAC, find an equation for *c*.

c. Find your BAC when you have had 4 beers in 1 hour. Find your BAC when you have had 6 beers in 1 hour.

d. How many beers do you have to drink in 1 hour to be legally drunk (BAC = 0.08)? Answer to the nearest whole number. You can check this information by going to link 7-2-5 on the Bello Website at mhhe.com/bello.

28. *Medical applications* A rough estimate of a child's weight *W* (in kilograms), from age 7 to 12, can be made from the formula $(7 \leq A \leq 12)$:

$$\text{Weight in kilograms} = \frac{(\text{age in years}) \cdot 7 - 5}{2}$$

a. Write the equation in symbols using *W* for weight and *A* for age.

b. What is the slope *m* of $W = \frac{7A - 5}{2}$?

c. What is the intercept *b* of $W = \frac{7A - 5}{2}$?

d. Find the estimated weight *W* when $A = 7$. Graph the point.

e. Find the estimated weight *W* when $A = 12$. Graph the point.

f. Use the two points from parts **d** and **e** to graph $W = \frac{7A - 5}{2}$.

For convenience, use a *y*-scale with 10 unit increments.

Weight (kg)

Source: http://www.medal.org.

29. *Medical applications*   A child's height *H* (in centimeters) can be estimated using the age *A* (2–12 years) by using the formula:

Height in centimeters = (age in years) · 6 + 77

  a. Write the equation in symbols using *H* for height and *A* for age.
  b. What is the slope *m* of $H = 6A + 77$?
  c. What is the intercept *b* of $H = 6A + 77$?
  d. Find the estimated height *H* when $A = 2$. Graph the point.

  e. Find the estimated height *H* when $A = 12$. Graph the point.

  f. Use the two points from parts **d** and **e** to graph $H = 6A + 77$.

For convenience, use a *y*-scale with 50 unit increments.

Height (cm)

Source: http://www.medal.org.

30. *Medical applications*   According to Munoz et al., the most reliable estimate for the height *H* (in centimeters) of a female is obtained by using the tibia bone and is given by

Height in centimeters = 76.53 + 2.41(length of tibia in cm)

  a. Write the equation in symbols using *H* for height and *t* for the length of the tibia in centimeters.
  b. What is the slope *m* of $H = 76.53 + 2.41t$?
  c. What is the intercept *b* of $H = 76.53 + 2.41t$?
  Source: http://www.medal.org.

32. *Medical applications*   If the gender of the person is unknown, the height *H* (in centimeters) of the person can be estimated by using the length of the humerus bone and is given by

Height in centimeters = 49.84 + 3.83 (length of humerus in cm)

31. *Medical applications*   According to Munoz et al., the femur bone gives the most reliable height estimate *H* (in centimeters) for the height of a male. The equation is

Height in centimeters = 62.92 + 2.39 (length of femur in cm)

  a. Write the equation in symbols using *H* for height and *f* for the length of the femur in centimeters.
  b. What is the slope *m* of $H = 62.92 + 2.39f$?
  c. What is the intercept *b* of $H = 62.92 + 2.39f$?
  Source: http://www.medal.org.

___

  a. Write the equation in symbols using *H* for height and *h* for the length of the humerus in centimeters.
  b. What is the slope *m* of $H = 49.84 + 3.83h$?
  c. What is the intercept *b* of $H = 49.84 + 3.83h$?
  Source: http://www.medal.org.

〉〉〉 Write On

33. In general, write the steps you use to find the formula for a linear equation in two variables similar to the ones in Problems 11–16.

35. What do the slopes you obtained in Problems 25 and 26 mean?

34. What is the minimum number of points you need to find the formula for a linear equation in two variables like the ones in Problems 23–26?

36. Would the slopes in Problems 25 and 26 be different if the weights of the male and female are different? Explain.

〉〉〉 Using Your Knowledge

Internet   We have so far neglected the graphs of the applications we have studied, but graphs are especially useful when comparing different situations. For example, the America Online "Light Usage" plan costs $4.95 per month for up to 3 hours and $2.95 for each additional hour. If *C* is the cost and *h* is the number of hours:

37. Graph *C* when *h* is between 0 and 3 hours, inclusive.

38. Graph *C* when *h* is more than 3 hours.

39. A different AOL plan costs $14.95 for unlimited hours. Graph the cost for this plan.

40. When is the plan in Problems 37 and 38 cheaper? When is the plan in Problem 39 cheaper?

⟩⟩⟩ *Mastery Test*

41. Irena rented a car and paid $40 a day and $0.15 per mile. Find an equation for the total daily cost C when she travels m miles. What is her cost if she travels 450 miles in a day?

42. In 1990 about 186 million tons of trash were produced in the United States. The amount increases by 3.4 million tons each year after 1990. Use the point-slope formula to find an equation for the total amount of trash y (millions of tons) produced x years after 1990:

    a. What would be the slope?

    b. What point would you use to find the equation?

    c. What is the equation?

43. Somjit subscribes to an Internet service with a flat monthly rate for up to 20 hours of use. For each hour over this limit, there is an additional per-hour fee. The table shows Somjit's first two bills.

Month	Hours of Use	Monthly Fee
March	30	$24
April	33	$26.70

    a. What does the point (30, 24) mean?

    b. What does the point (33, 26.70) mean?

    c. Find an equation for the cost C of x hours ($x > 20$) of use.

⟩⟩⟩ *Skill Checker*

Fill in the blank with $<$ or $>$ so that the result is a true statement:

44. 0 ____ 600

45. 0 ____ -8

46. 0 ____ 6

47. 10 ____ 0

48. -5 ____ 0

49. -1 ____ 0

50. -3 ____ -1

3.7   Graphing Inequalities in Two Variables

▶ Objective

A ⟩ Graph linear inequalities in two variables.

▶ To Succeed, Review How To . . .

1. Use the symbols $>$ and $<$ to compare numbers (p. 180).
2. Graph lines (pp. 209–211, 235–240).
3. Evaluate an expression (pp. 61–62, 69–72).

▶ Getting Started
Renting Cars and Inequalities

Suppose you want to rent a car costing $30 a day and $0.20 per mile. The total cost T depends on the number x of days the car is rented and the number y of miles traveled and is given by

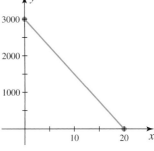

If you want the cost to be *exactly* $600 ($T = 600$), we graph the equation $600 = 30x + 0.20y$ by finding the intercepts. For $x = 0$,

$$600 = 30(0) + 0.20y$$

$$\frac{600}{0.20} = y \qquad \text{Divide both sides by 0.20.}$$

$$y = 3000$$

Thus, $(0, 3000)$ is the y-intercept. Now for $y = 0$,

$$600 = 30x + 0.20(0)$$

$$20 = x. \qquad \text{Divide both sides by 30.}$$

The shaded region represents the points for which $30x + 0.02y < 600$.

The line is not part of the graph, so it's shown dashed.

Thus, $(20, 0)$ is the x-intercept. Join the two intercepts with a line to obtain the graph shown on the previous page.

If we want to spend *less* than $600, the total cost T must satisfy (be a solution of) the inequality

$$30x + 0.20y < 600$$

All points satisfying the equation $30x + 0.20y = 600$ are *on* the line. Where are the points so that $30x + 0.20y < 600$? The line $30x + 0.20y = 600$ divides the plane into three regions:

   1. The points *below* the line
   2. The points *on* the line
   3. The points *above* the line

The test point $(0, 0)$ is *below* the line $30x + 0.20y = 600$ and satisfies $30x + 0.20y < 600$. Thus, all the other points *below* the line also satisfy the inequality and are shown shaded in the second graph. The line is *not* part of the answer, so it's shown **dashed.** In this section we shall learn how to solve linear inequalities in two variables, and we shall also examine why their solutions are regions of the plane.

A › Graphing Linear Inequalities in Two Variables

The procedure we used in the *Getting Started* can be generalized to graph any linear inequality that can be written in the form $Ax + By < C$. Here are the steps.

PROCEDURE

Graphing a Linear Inequality

   1. Determine the line that is the boundary of the region. If the inequality involves $\leq$ or $\geq$, draw the line **solid;** if it involves $<$ or $>$, draw a **dashed** line. The points on the *solid* line are part of the solution set.
   2. Use any point (a, b) not on the line as a test point. Substitute the values of a and b for x and y in the inequality. If a true statement results, shade the side of the line containing the test point. If a false statement results, shade the other side.

EXAMPLE 1 Graphing linear inequalities where the line is not part of the solution set

Graph: $2x - 4y < -8$

SOLUTION   We use the two-step procedure for graphing inequalities.

1. We first graph the boundary line $2x - 4y = -8$.

   When $x = 0$, $-4y = -8$,   and $y = 2$

   When $y = 0$, $2x = -8$,   and $x = -4$

x	y
0	2
-4	0

   Since the inequality involved is $<$, join the points $(0, 2)$ and $(-4, 0)$ with a dashed line as shown in Figure 3.41.

2. Select an easy test point and see whether it satisfies the inequality. If it does, the solution lies on the same side of the line as the test point; otherwise, the solution is on the other side of the line.

   An easy point is $(0, 0)$, which is *below* the line. If we substitute $x = 0$ and $y = 0$ in the inequality $2x - 4y < -8$, we obtain

   $$2 \cdot 0 - 4 \cdot 0 < -8 \quad \text{or} \quad 0 < -8$$

   which is *false*. Thus, the point $(0, 0)$ is not part of the solution. Because of this, the solution consists of the points *above* (on the other side of) the line $2x - 4y = -8$, as shown shaded in Figure 3.42. Note that the line itself is shown **dashed** to indicate that it isn't part of the solution.

>Figure 3.41

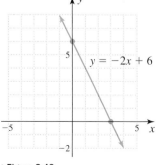

>Figure 3.42

PROBLEM 1

Graph: $3x - 2y < -6$

Answers to PROBLEMS

1.

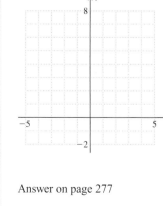

In Example 1, the line was not part of the solution for the given inequality. Next, we give an example in which the line is part of the solution for the inequality.

EXAMPLE 2 Graphing linear inequalities where the line is part of the solution set

Graph: $y \leq -2x + 6$

SOLUTION   As usual, we use the two-step procedure for graphing inequalities.

1. We first graph the line $y = -2x + 6$.

   When $x = 0$, $y = 6$

   When $y = 0$, $0 = -2x + 6$ or $x = 3$

x	y
0	6
3	0

   Since the inequality involved is $\leq$, the graph of the line is shown solid in Figure 3.43.

>Figure 3.43

PROBLEM 2

Graph: $y \leq -3x + 6$

Answer on page 277

2. Select the point (0, 0) or any other point not on the line as a test point. When $x = 0$ and $y = 0$, the inequality

$$y \leq -2x + 6$$

becomes

$$0 \leq -2 \cdot 0 + 6 \quad \text{or} \quad 0 \leq 6$$

which is *true.* Thus, all the points on the same side of the line as (0, 0)—that is, the points *below* the line—are solutions of $y \leq -2x + 6$. These solutions are shown shaded in Figure 3.44.

The line $y = -2x + 6$ is shown solid because it is part of the solution, since $y \leq -2x + 6$ allows $y = -2x + 6$. [For example, the point (3, 0) satisfies the inequality $y \leq -2x + 6$ because $0 \leq -2 \cdot 3 + 6$ yields $0 \leq 0$, which is true.]

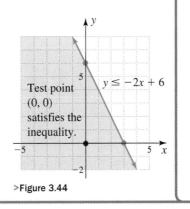

>Figure 3.44

As you recall from Section 3.3, a line with an equation of the form $Ax + By = 0$ passes through the origin, so we cannot use the point (0, 0) as a test point. This is not a problem! Just use any other convenient point, as shown in Example 3.

EXAMPLE 3   **Graphing linear inequalities where the boundary line passes through the origin**

Graph: $y + 2x > 0$

SOLUTION   We follow the two-step procedure.

1. We first graph the boundary line $y + 2x = 0$:

When $x = 0$, $y = 0$
When $y = -4$, $-4 + 2x = 0$ or $x = 2$

Join the points (0, 0) and (2, −4) with a dashed line as shown in Figure 3.45 since the inequality involved is >.

x	y
0	0
2	−4

>Figure 3.45

PROBLEM 3

Graph: $y + 3x > 0$

(continued)

Answers to PROBLEMS

2.

3.

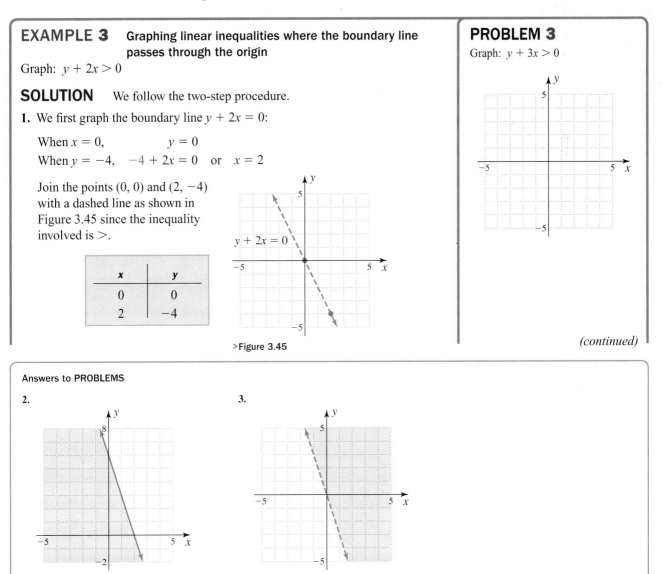

2. Because the line goes through the origin, we cannot use (0, 0) as a test point. A convenient point to use is (1, 1), which is *above* the line. Substituting $x = 1$ and $y = 1$ in $y + 2x > 0$, we obtain

$$1 + 2(1) > 0 \quad \text{or} \quad 3 > 0$$

which is true. Thus, we shade the points *above* the line $y + 2x = 0$, as shown in Figure 3.46.

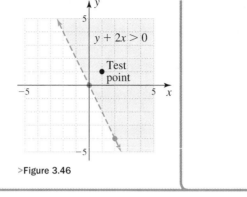

>Figure 3.46

Finally, if the inequalities involve horizontal or vertical lines, we also proceed in the same manner.

EXAMPLE 4   Graphing linear inequalities involving horizontal or vertical lines

Graph:

a. $x \geq -2$ **b.** $y - 2 < 0$

SOLUTION

a. The two-step procedure applies here also.

   1. Graph the vertical line $x = -2$ as solid, since the inequality involved is $\geq$.

   2. Use (0, 0) as a test point. When $x = 0$, $x \geq -2$ becomes $0 \geq -2$, a true statement. So we shade all the points to the *right* of the line $x = -2$, as shown in Figure 3.47.

b. We again follow the steps.

   1. First, we write the inequality as $y < 2$, and then we graph the horizontal line $y = 2$ as dashed, since the inequality involved is $<$.

   2. Use (0, 0) as a test point. When $y = 0$, $y < 2$ becomes $0 < 2$, a true statement. So we shade all the points *below* the line $y = 2$, as shown in Figure 3.48.

>Figure 3.47

>Figure 3.48

PROBLEM 4

Graph:

a. $x \geq 2$ **b.** $y + 2 > 0$

Answers to PROBLEMS

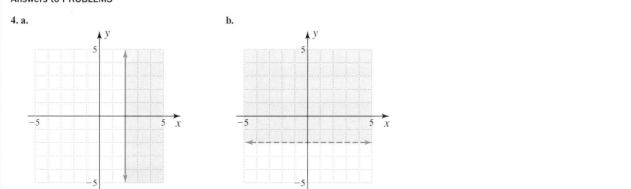

4. a.   b.

▦ ◈ ▥ Calculator Corner

Solving Linear Inequalities

Your calculator can help solve linear inequalities, but you still have to know the algebra. Thus, to do Example 1, you must first solve $2x - 4y = -8$ for y to obtain $y = \frac{1}{2}x + 2$. Now graph this equation using the Zdecimal window, a window in which the x- and y-coordinates are decimals between -4.7 and 4.7. You can do this by pressing ⟨ZOOM⟩ ⟨4⟩. Now graph $y = \frac{1}{2}x + 2$.

To decide whether you need to shade above or below the line, move the cursor *above* the line. In the decimal window, you can see the values for x and y. For $x = 2$ and $y = 5$, $2x - 4y < -8$ becomes

$$2(2) - 4(5) < -8 \quad \text{or} \quad 4 - 20 < -8$$

a true statement. Thus, you need to shade *above* the line. You can do this with the DRAW feature. Press ⟨2nd⟩ ⟨PRGM⟩ ⟨7⟩ and enter the line above which you want to shade—that is, $\frac{1}{2}x + 2$. Now press ⟨,⟩ and enter the y-range that you want to shade. Let the calculator shade below 5 and above -5 by entering ⟨5⟩ ⟨,⟩ ⟨(-)⟩ ⟨5⟩ ⟨,⟩. Finally, enter the value at which you want the shading to end, (say 5), close the parentheses, and press ⟨ENTER⟩.

The instructions we asked you to enter are shown in Window 1, and the resulting graph is shown in Window 2. What is missing? *You* should know the line itself is not part of the graph, since the inequality $<$ is involved.

```
Shade(.5X+2,5,
-5,5)
```

Window 1

Window 2

MathZone ⁺ˣ

⟩ Exercises **3.7**

⟨ **A** ⟩ **Graphing Linear Inequalities in Two Variables**   In Problems 1–32, graph the inequalities.

1. $2x + y > 4$

2. $y + 3x > 3$

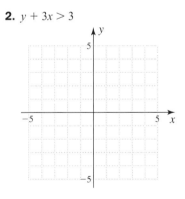

3. $-2x - 5y \leq 10$

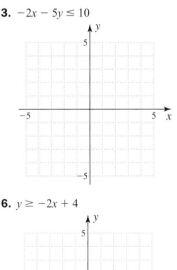

4. $-3x - 2y \leq -6$

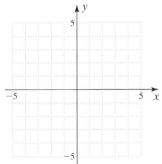

5. $y \geq 3x - 3$

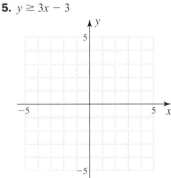

6. $y \geq -2x + 4$

7. $6 < 3x - 6y$

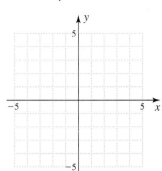

8. $6 < 2x - 3y$

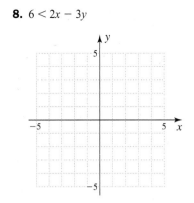

9. $3x + 4y \geq 12$

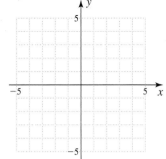

10. $-3y \geq 6x + 6$

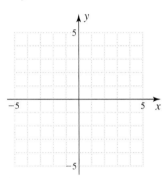

11. $10 < -2x + 5y$

12. $4 < x - y$

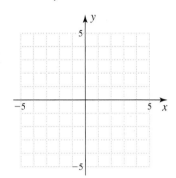

13. $x \geq 2y - 4$

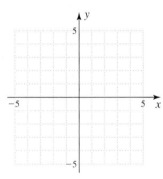

14. $2x \geq 4y + 2$

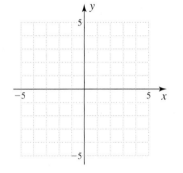

15. $y < -x + 5$

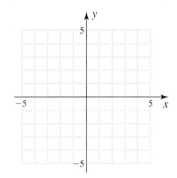

16. $2y < 4x - 8$

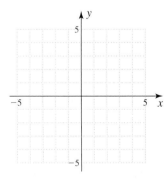

17. $2y < 4x + 5$

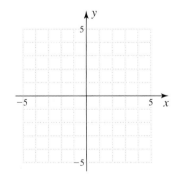

18. $2y \geq 3x + 5$

19. $x > 1$

20. $y \leq \dfrac{3}{2}$

21. $x \leq \dfrac{5}{2}$

22. $x \geq -\dfrac{2}{3}$

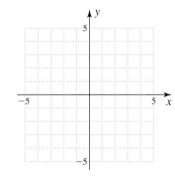

23. $y \leq -\dfrac{3}{2}$

24. $x - \dfrac{1}{3} \geq 0$

25. $x - \dfrac{2}{3} > 0$

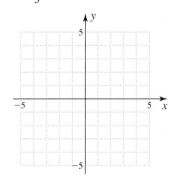

26. $y - \dfrac{5}{2} > \dfrac{1}{2}$

27. $y + \dfrac{1}{3} \geq \dfrac{2}{3}$

28. $x + \dfrac{1}{5} < \dfrac{6}{5}$

29. $2x + y < 0$

30. $2y + x \geq 0$

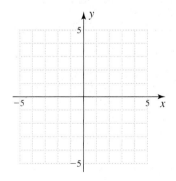

31. $y - 3x > 0$

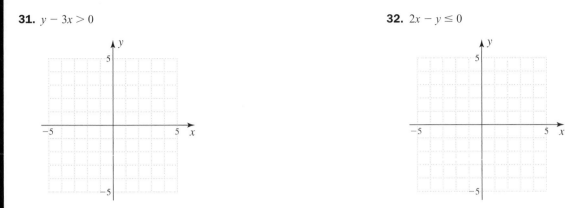

32. $2x - y \le 0$

In Problems 33 and 34, the graph of the given system of inequalities defines a geometric figure. Graph the inequalities on the same set of axes and identify the figure.

33. $-x \le -3$, $-y \ge -4$, $x \le 4$, $y \ge 2$

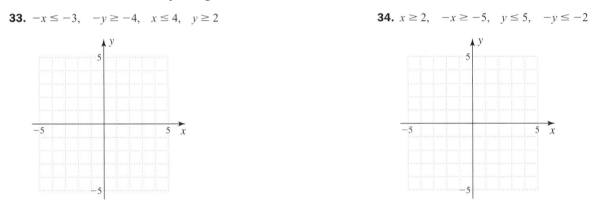

34. $x \ge 2$, $-x \ge -5$, $y \le 5$, $-y \le -2$

〉〉〉 **Applications**

Poverty Levels   For Problems 35–38, use the given graph. The graph shows the percent of people classified as poor by the Federal government in three age categories: 65 and older, under 18, and 18–64.

Note: The data points are placed at the midpoints of the respective years. Data for people 18 to 64 and 65 and older are not available from 1960 to 1965.

Source: U.S. Census Bureau, Current Population Survey. 1960 to 2006 Annual Social and Economic Supplements.

35. Which of the categories (65+, <18, 18–64) has the **highest** percent of persons in poverty in 2005? What is this percent?

36. Which of the three categories (65+, <18, 18–64) has the **lowest** percent of persons in poverty in 2005? What is this percent?

37. What is the first year in which the data is available for the poverty rates of persons in the 65+ and 18–64 categories?

38. a. What was the approximate percent of persons in the 65+, <18, and 18–64 categories in 1966?

   b. Determine if there was an increase, decrease, or almost no change from 1966 to 2006 in each of the categories 65+, <18, and 18–64?

〉〉〉 *Using Your Knowledge*

Savings on Rentals   The ideas we discussed in this section can save you money when you rent a car. Here's how.

Suppose you have the choice of renting a car from company A or from company B. The rates for these companies are as follows:

   Company A: $20 a day plus $0.20 per mile
   Company B: $15 a day plus $0.25 per mile

If x is the number of miles traveled in a day and y represents the cost for that day, the equations representing the cost for each company are

   Company A: $y = 0.20x + 20$
   Company B: $y = 0.25x + 15$

40. On the same coordinate axes you used in Problem 39, graph the equation representing the cost for company B.

42. When is the cost less for company A?

39. Graph the equation representing the cost for company A.

41. When is the cost the same for both companies?

43. When is the cost less for company B?

〉〉〉 *Write On*

44. Describe in your own words the graph of a linear inequality in two variables. How does it differ from the graph of a linear inequality in one variable?

46. Explain how you decide whether the boundary line is solid or dashed when graphing a linear inequality in two variables.

45. Write the procedure you use to solve a linear inequality in two variables. How does the procedure differ from the one you use to solve a linear inequality in one variable?

47. Explain why a point on the boundary line cannot be used as a test point when graphing a linear inequality in two variables.

〉〉〉 *Concept Checker*

Fill in the blank(s) with the correct word(s), phrase, or mathematical statement.

48. A **linear inequality** in two variables x and y involving < can be written in the form _____.

49. If a **linear inequality** involves ≤ or ≥ the boundary of the region is drawn as a _____ line.

50. If a **linear inequality** has a **boundary** that is a **solid line,** the line is in the _____ **set** of the inequality.

51. If a **linear inequality** involves < or > the **boundary** of the region is drawn as a _____ line.

solid

dashed

color

$x + y < C$

$Ax + By < C$

solution

⟩ ⟩ ⟩ *Mastery Test*

Graph:

52. $x - 4 \le 0$

53. $y \ge -4$

54. $x + 2 < 0$

55. $y - 3 > 0$

56. $3x - 2y < -6$

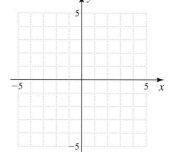

57. $y \le -4x + 8$

58. $y > 2x$

59. $y < 3x$

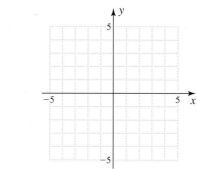

60. $x - 2y > 0$

61. $3x + y \le 0$

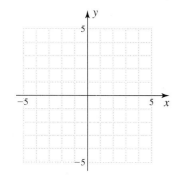

62. $y - 4x > 0$

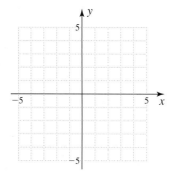

In Problems 63–65, find the product.

63. $-3 \cdot 8$ **64.** $(-2)(-7)$ **65.** $(7)(-3)$

In Problems 66–67, find the quotient.

66. $-\dfrac{24}{6}$ **67.** $\dfrac{24}{-6}$

〉Collaborative Learning

Hourly Earnings in Selected Industries

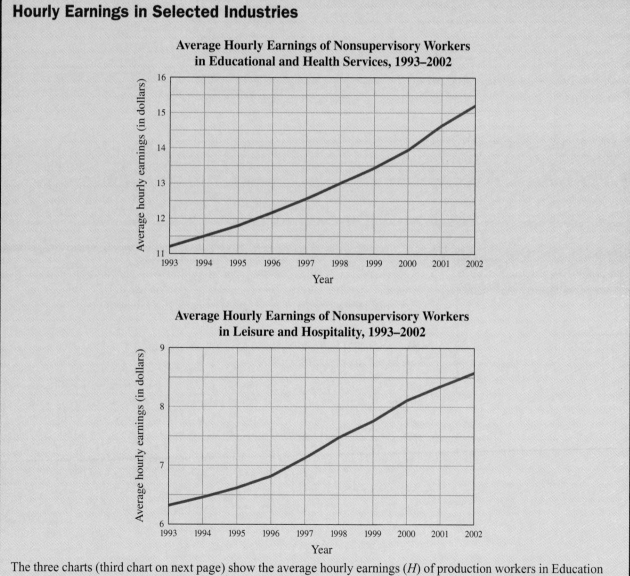

The three charts (third chart on next page) show the average hourly earnings (H) of production workers in Education and Health Services (E), Leisure and Hospitality (L), and Construction (C). Form three groups, E, L, and C.

1. What was your group salary (to the nearest dollar) in 1993?

2. What was your group salary (to the nearest dollar) in 2002?

〈CONTINUED〉

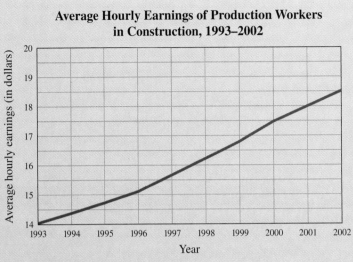

Average Hourly Earnings of Production Workers in Construction, 1993–2002

Source: U.S. Department of Labor, Bureau of Labor Statistics. Select the desired industry.

3. Use 1993 = 0. What is $(0, H)$, where H is the hourly earnings to the nearest dollar for each of the groups?

4. What is $(9, H)$, where H is the hourly earnings to the nearest dollar for each of the groups?

5. What is the slope of the line for each of the groups?

6. What is the equation of the line for the hourly earnings H for each of the groups?

7. What would be the predicted hourly earnings for each of the groups in 2010?

8. If you were making a career decision based on the equations obtained in Question 6, which career would you choose?

> **Research Questions**

1. Some historians claim that the official birthday of analytic geometry is November 10, 1619. Investigate and write a report on why this is so and on the event that led Descartes to the discovery of analytic geometry.

2. Find out what led Descartes to make his famous pronouncement "*Je pense, donc je suis*" (I think, therefore, I am) and write a report about the contents of one of his works, *La Geometrie.*

3. From 1629 to 1633, Descartes "was occupied with building up a cosmological theory of vortices to explain all natural phenomena." Find out the name of the treatise in which these theories were explained and why it wasn't published until 1654, after his death.

4. Inequalities are used in solving "linear programming" problems in mathematics, economics, and many other fields. Find out what linear programming is and write a short paper about it. Include the techniques involved and the mathematicians and scientists who cooperated in the development of this field.

5. Linear programming problems are sometimes solved using the "simplex" method. Write a few paragraphs describing the simplex method, the people who developed it, and its uses.

6. For many years, scientists have tried to improve on the simplex method. As far back as 1979, the Soviet Academy of Sciences published a paper that did just that. Find the name of the author of the paper as well as the name of the other mathematicians who have supplied and then improved on the proof contained in the paper.

Section	Item	Meaning	Example
3.1	x-axis	A horizontal number line on a coordinate plane	
	y-axis	A vertical number line on a coordinate plane	
	Abscissa	The first coordinate in an ordered pair	The abscissa in the ordered pair (3, 4) is 3.
	Ordinate	The second coordinate in an ordered pair	The ordinate in the ordered pair (3, 4) is 4.
	Quadrant	One of the four regions into which the axes divide the plane	
3.2A	Solution of an equation	The ordered pair (a, b) is a solution of an equation if when the values for a and b are substituted for the variables in the equation, the result is a true statement.	(4, 5) is a solution of the equation $2x + 3y = 23$ because if 4 and 5 are substituted for x and y in $2x + 3y = 23$, the result $2(4) + 3(5) = 23$ is true.
3.2C	Graph of a linear equation	The graph of a linear equation of the form $y = mx + b$ is a straight line.	The graph of the equation $y = 3x + 6$ is a straight line.
	Graph of a linear equation	The graph of a linear equation of the form $Ax + By = C$ is a straight line, and every straight line has an equation of this form.	The linear equation $3x + 6y = 12$ has a straight line for its graph.
3.3A	x-intercept	The point at which a line crosses the x-axis	The x-intercept of the line $3x + 6y = 12$ is (4, 0).
	y-intercept	The point at which a line crosses the y-axis	The y-intercept of the line $3x + 6y = 12$ is (0, 2).
3.3C	Horizontal line	A line whose equation can be written in the form $y = k$	The line $y = 3$ is a horizontal line.
	Vertical line	A line whose equation can be written in the form $x = k$	The line $x = -5$ is a vertical line.
3.4A	Slope of a line	The ratio of the vertical change to the horizontal change of a line $$m = \frac{y_2 - y_1}{x_2 - x_1}$$	
	Slope of a line through (x_1, y_1) and (x_2, y_2), $x_1 \neq x_2$		The slope of the line through (3, 5) and (6, 8) is $$m = \frac{8 - 5}{6 - 3} = 1.$$
3.4B	Slope of the line $y = mx + b$	The slope of the line $y = mx + b$ is m.	The slope of the line $y = \frac{3}{5}x + 7$ is $\frac{3}{5}$.

(continued)

Section	Item	Meaning	Example
3.4C	Parallel lines	Two lines are parallel if their slopes are equal and they have different y-intercepts.	The lines $y = 2x + 5$ and $3y - 6x = 8$ are parallel (both have a slope of 2).
	Perpendicular lines	Two lines are perpendicular if the product of their slopes is -1.	The lines defined by $y = 2x + 1$ and $y = -\frac{1}{2}x + 2$ are perpendicular since $2 \cdot (-\frac{1}{2}) = -1$.
3.5A	Point-slope form	The point-slope form of an equation for the line passing through (x_1, x_2) and with slope m is $y - y_1 = m(x - x_1)$.	The point-slope form of an equation for the line passing through $(2, -5)$ and with slope -3 is $y - (-5) = -3(x - 2)$.
3.5B	Slope-intercept form	The slope-intercept form of an equation for the line with slope m and y-intercept b is $y = mx + b$.	The slope-intercept form of an equation for the line with slope -5 and y-intercept 2 is $y = -5x + 2$.
3.5C	Two-point form	The two-point form of an equation for a line going through (x_1, y_1) and (x_2, y_2) is $y - y_1 = m(x - x_1)$, where $m = \dfrac{y_2 - y_1}{x_2 - x_1}$.	The two-point form of a line going through $(2, 3)$ and $(7, 13)$ is $y - 3 = 2(x - 2)$.
3.7A	Linear inequality	An inequality that can be written in the form $Ax + By < C$ or $Ax + By > C$ (substituting $\leq$ for $<$ or $\geq$ for $>$ also yields a linear inequality).	$3x < 2y - 6$ is a linear inequality because it can be written as $3x - 2y < -6$.
3.7A	Graph of a linear inequality	1. Find the boundary. Draw the line solid for $\leq$ or $\geq$, dashed for $<$ or $>$. 2. Use (a, b) as a test point and shade the side of the line containing the test point if the resulting inequality is a true statement; otherwise, shade the region that does not contain the test point.	Graph of $3x - 2y < 6$. Using $(0, 0)$ as the test point for graphing $3x - 2y < 6$, we get $$3(0) - 2(0) < 6,\ \text{true}$$ Shade the region containing the test point, which is the region above the dashed line.

> Review Exercises **Chapter 3**

(If you need help with these exercises, look in the section indicated in brackets.)

1. ⟨**3.1A**⟩ *Graph the points.*

  a. A: $(-1, 2)$

  b. B: $(-2, -1)$

  c. C: $(3, -3)$

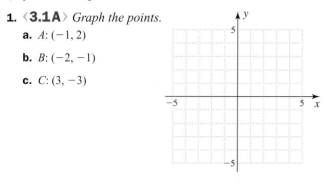

2. ⟨**3.1B**⟩ *Find the coordinates of the points shown on the graph.*

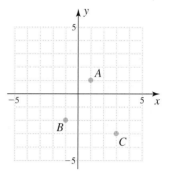

3. ⟨**3.1C**⟩ *The graph shows the new wind chill temperatures (top) and the old wind chill temperatures (bottom) for different wind speeds.*

  a. On the top graph, what does the ordered pair $(10, -10)$ represent?

  b. If the wind speed is 40 miles per hour, what is the approximate new wind chill temperature?

  c. If the wind speed is 40 miles per hour, what is the approximate old wind chill temperature?

Wind Chill Temperature Comparison

Source: Data from National Oceanic and Atmospheric Administration.

4. ⟨**3.1D**⟩ *The bar graph indicates the number of months and monthly payment needed to pay off a $500 loan at 18% annual interest.*

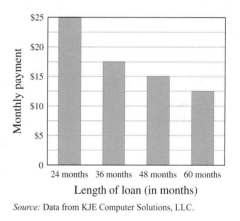

Source: Data from KJE Computer Solutions, LLC.

To the nearest dollar, what is the monthly payment if you want to pay off the loan in:

  a. 60 months?

  b. 48 months?

  c. 24 months?

5. ⟨**3.1E**⟩ *Determine the quadrant in which each of the points is located:*

  a. *A*

  b. *B*

  c. *C*

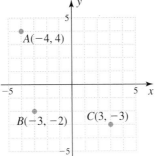

6. ⟨**3.1F**⟩ *The ordered pairs represent the wind chill temperature in degrees Fahrenheit when the wind speed in miles per hour is as indicated.*

Wind Speed	Wind Chill
10	−10
20	−15
30	−18

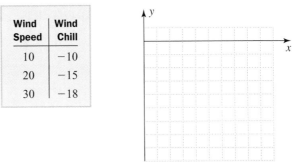

  a. Graph the ordered pairs.

  b. What does $(10, -10)$ mean?

  c. What is the number s in $(s, -15)$?

7. ⟨**3.2A**⟩ *Determine whether the given point is a solution of* $x - 2y = -3$.

  a. $(1, -2)$

  b. $(2, -1)$

  c. $(-1, 1)$

8. ⟨**3.2B**⟩ *Find x in the given ordered pair so that the pair satisfies the equation* $2x - y = 4$.

  a. $(x, 2)$

  b. $(x, 4)$

  c. $(x, 0)$

9. ⟨**3.2C**⟩ *Graph:*

  a. $x + y = 4$

  b. $x + y = 2$

  c. $x + 2y = 2$

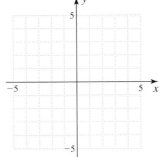

10. ⟨**3.2C**⟩ *Graph:*

  a. $y = \frac{3}{2}x + 3$

  b. $y = -\frac{3}{2}x + 3$

  c. $y = \frac{3}{4}x + 4$

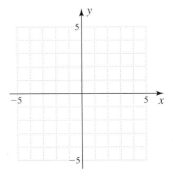

11. ⟨**3.2D**⟩ *The average annual consumption g of milk products (in gallons) per person can be approximated by* $g = 30 - 0.2t$, *where t is the number of years after 1980. Graph:*

  a. $g = 30 - 0.2t$

  b. $g = 20 - 0.2t$

  c. $g = 10 - 0.2t$

12. ⟨**3.3A**⟩ *Graph:*

  a. $2x - 3y - 12 = 0$

  b. $3x - 2y - 12 = 0$

  c. $2x + 3y + 12 = 0$

13. ⟨**3.3B**⟩ *Graph:*

   a. $3x + y = 0$

   b. $-2x + 3y = 0$

   c. $-3x + 2y = 0$

14. ⟨**3.3C**⟩ *Graph:*

   a. $2x - 6 = 0$

   b. $2x - 2 = 0$

   c. $2x - 4 = 0$

15. ⟨**3.3C**⟩ *Graph:*

   a. $y = -1$

   b. $y = -3$

   c. $y = -4$

16. ⟨**3.4A**⟩ *Find the slope of the line passing through the given points.*

   a. $(-5, 2)$ and $(-5, 4)$

   b. $(-3, 5)$ and $(3, 5)$

   c. $(-2, -1)$ and $(-2, -5)$

17. ⟨**3.4A**⟩ *Find the slope of the line passing through the given points.*

   a. $(1, -4)$ and $(2, -3)$

   b. $(5, -2)$ and $(8, 5)$

   c. $(3, -4)$ and $(4, -8)$

18. ⟨**3.4B**⟩ *Find the slope of the line.*

   a. $3x + 2y = 6$

   b. $x + 4y = 4$

   c. $-2x + 3y = 6$

19. ⟨**3.4C**⟩ *Decide whether the lines are parallel, perpendicular, or neither.*

   a. $2x + 3y = 6$
       $6x = 6 - 4y$

   b. $3x + 2y = 4$
      $-2x + 3y = 4$

   c. $2x + 3y = 6$
      $-2x - 3y = 6$

20. ⟨**3.4D**⟩ *The number N of theaters t years after 1975 can be approximated by*

$$N = 0.6t + 15 \text{ (thousand)}$$

   a. What is the slope of this line?

   b. What does the slope represent?

   c. How many theaters were added each year?

21. ⟨**3.5A**⟩ *Find an equation of the line going through the point $(3, -5)$ and with the given slope.*

   a. $m = -2$

   b. $m = -3$

   c. $m = -4$

22. ⟨**3.5B**⟩ *Find an equation of the line with the given slope and intercept.*

   a. Slope 5, y-intercept -2

   b. Slope 4, y-intercept 7

   c. Slope 6, y-intercept -4

23. ⟨**3.5C**⟩ *Find an equation for the line passing through the given points.*

   a. $(-1, 2)$ and $(4, 7)$

   b. $(-3, 1)$ and $(7, 6)$

   c. $(1, 2)$ and $(7, -2)$

24. ⟨**3.6A**⟩

   a. The cost C of a long-distance call that lasts for m minutes is \$3 plus \$0.20 for each minute. If a 10-minute call costs \$5, write an equation for the cost C and find the cost of a 15-minute call.

   b. Long-distance rates for m minutes are \$5 plus \$0.20 for each minute. If a 10-minute call costs \$7, write an equation for the cost C and find the cost of a 15-minute call.

   c. Long-distance rates for m minutes are \$5 plus \$0.30 for each minute. If a 10-minute call costs \$8, write an equation for the cost C and find the cost of a 15-minute call.

25. ⟨**3.6B**⟩

   a. A cell phone plan costs \$30 per month with 500 free minutes and \$0.40 for each additional minute. Find an equation for the total cost C of the plan when m minutes are used after the first 500. What is the cost when 800 minutes are used?

   b. Find an equation when the cost is \$40 per month and \$0.30 for each additional minute. What is the cost when 600 minutes are used?

   c. Find an equation when the cost is \$50 per month and \$0.20 for each additional minute. What is the cost when 900 minutes are used?

26. ⟨**3.6C**⟩

   a. The bills for two long-distance calls are \$3 for 5 minutes and \$5 for 10 minutes. Find an equation for the total cost C of the calls when m minutes are used.

   b. Repeat part **a** if the charges are \$4 for 5 minutes and \$6 for 10 minutes.

   c. Repeat part **a** if the charges are \$5 for 5 minutes and \$7 for 10 minutes.

27. ⟨**3.7A**⟩ *Graph.*

   a. $2x - 4y < -8$

   b. $3x - 6y < -12$

   c. $4x - 2y < -8$

28. ⟨**3.7A**⟩ *Graph.*

   a. $-y \le -2x + 2$

   b. $-y \le -2x + 4$

   c. $-y \le -x + 3$

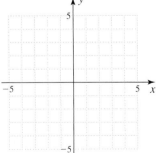

29. ⟨**3.7A**⟩ *Graph.*

a. $2x + y > 0$

b. $3x + y > 0$

c. $3x - y < 0$

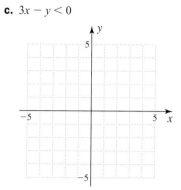

30. ⟨**3.7A**⟩ *Graph.*

a. $x \geq -4$

b. $y - 4 < 0$

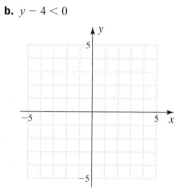

c. $2y - 4 \geq 0$

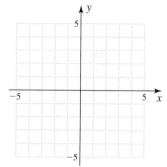

〉 Practice Test **Chapter 3**

(Answers on pages 298–302)

Visit www.mhhe.com/bello to view helpful videos that provide step-by-step solutions to several of the problems below.

1. Graph the point $(-2, 3)$.

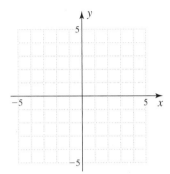

2. Find the coordinates of point A shown on the graph.

3. The graph shows the new wind chill temperatures (top) and the old wind chill temperatures (bottom) for different wind speeds.

Wind Chill Temperature Comparison

Source: Data from National Oceanic and Atmospheric Administration.

  a. On the top graph, what does the ordered pair $(20, -15)$ represent?

  b. If the wind speed is 90 miles per hour, what is the approximate new wind chill temperature?

  c. If the wind speed is 90 miles per hour, what is the approximate old wind chill temperature?

4. The bar graph indicates the number of months and monthly payment needed to pay off a $1000 loan at 8% annual interest.

Source: Data from KJE Computer Solutions, LLC.

What is the monthly payment if you want to pay off the loan in the specified number of months?

  a. 60 months

  b. 48 months

  c. 24 months

5. Determine the quadrant in which each of the points is located:

   a. *A*

   b. *B*

   c. *C*

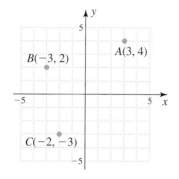

6. The ordered pairs represent the old wind chill temperature in degrees Fahrenheit when the wind speed in miles per hour is as indicated.

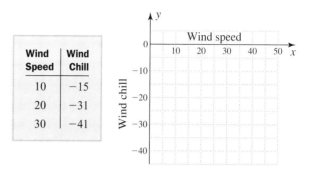

Wind Speed	Wind Chill
10	−15
20	−31
30	−41

   a. Graph the ordered pairs.

   b. What does $(10, -15)$ mean?

   c. What is the number *s* in $(s, -15)$?

7. Determine whether the ordered pair $(1, -2)$ is a solution of $2x - y = -2$.

8. Find *x* in the ordered pair $(x, 2)$ so that the ordered pair satisfies the equation $3x - y = 10$.

9. Graph $x + 2y = 4$.

10. Graph $y = -\frac{3}{2}x - 3$.

11. The average daily consumption *g* of protein (in grams) per person can be approximated by $g = 100 + 0.7t$, where *t* is the number of years after 2000. Graph $g = 100 + 0.7t$.

12. Graph $2x - 5y - 10 = 0$.

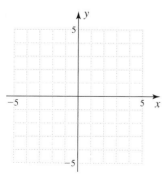

13. Graph $2x - 3y = 0$.

14. Graph $3x - 6 = 0$.

15. Graph $y = -4$.

16. Find the slope of the line passing through:

a. $(2, -8)$ and $(-4, -2)$

b. $(6, -8)$ and $(-4, 6)$

17. Find the slope of the line passing through:

a. $(-3, -2)$ and $(-3, 4)$

b. $(2, 4)$ and $(4, 4)$

18. Find the slope of the line

$y - 2x = 6$.

19. Decide whether the lines are parallel, perpendicular, or neither.

a. $3y = x + 5$

$2x - 6y = 6$

b. $3y = -x + 5$

$9x - 3y = 6$

20. The number N of stores in a city t years after 2000 can be approximated by

$$N = 0.8t + 15 \text{ (hundred)}$$

a. What is the slope of this line?

b. What does the slope represent?

c. How many stores were added each year?

21. Find an equation of the line passing through the point $(2, -6)$ with slope -5. Write the answer in point-slope form and then graph the line.

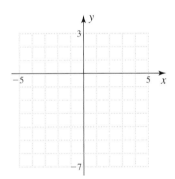

22. Find an equation of the line with slope 5 and y-intercept -4. Write the answer in slope-intercept form and then graph the line.

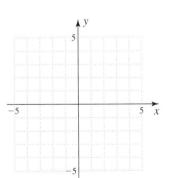

23. Find an equation of the line passing through the points $(2, -8)$ and $(-4, -2)$. Write the answer in standard form.

24. Long-distance rates for m minutes are \$10 plus \$0.20 for each minute. If a 10-minute call costs \$12, write an equation for the total cost C and find the cost of a 15-minute call.

25. A cell phone plan costs \$40 per month with 500 free minutes and \$0.50 for each additional minute. Find an equation for the total cost C of the plan when m minutes are used after the first 500. What is the cost when 800 total minutes are used?

26. The bills for two long-distance calls are \$7 for 5 minutes and \$10 for 10 minutes. Find an equation for the total cost C of calls when m minutes are used.

27. Graph $3x - 2y < -6$.

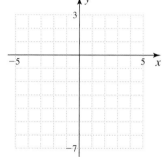

28. Graph $-y \le -3x + 3$.

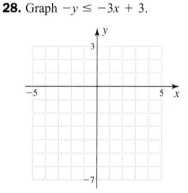

29. Graph $4x - y > 0$.

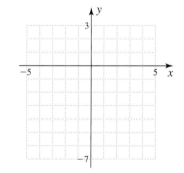

30. Graph $2y - 8 \ge 0$.

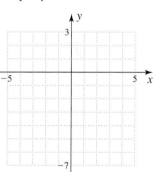

›Answers to Practice Test Chapter 3

Answer	If You Missed	Review		
	Question	Section	Examples	Page
1.	1	3.1	1	205
2. $(-2, -1)$	2	3.1	2	206
3. a. When the wind speed is 20 miles per hour, the wind chill temperature is -15 degrees Fahrenheit. **b.** -30 degrees Fahrenheit **c.** -40 degrees Fahrenheit	3	3.1	3	207
4. a. $20 **b.** $25 **c.** $45	4	3.1	4	208
5. a. Quadrant I **b.** Quadrant II **c.** Quadrant III	5	3.1	5	209
6. a. **b.** When the wind speed is 10 mi/hr, the wind chill temperature is $-15°$F. **c.** 10	6	3.1	6	210–211
7. No	7	3.2	1	221–222

Answer	If You Missed		Review	
	Question	**Section**	**Examples**	**Page**
8. $x = 4$	8	3.2	2	223
9.	9	3.2	4	225

$x + 2y = 4$

10.	10	3.2	5	226

$y = -\frac{3}{2}x - 3$

11.	11	3.2	6	227

$g = 100 + 0.7t$

12.	12	3.3	1, 2	235–236

$2x - 5y - 10 = 0$

Answer	If You Missed	Review		
	Question	Section	Examples	Page
13.	13	3.3	3	237
14.	14	3.3	4a	240
15.	15	3.3	4b	240
16. a. -1 **b.** $-\dfrac{7}{5}$	16	3.4	1, 2	250
17. a. Undefined **b.** 0	17	3.4	3	251
18. 2	18	3.4	4	252–253
19. a. Parallel **b.** Perpendicular	19	3.4	5	254
20. a. 0.8 **b.** Annual increase in the number of stores **c.** 80	20	3.4	6	255

Answer	If You Missed	Review		
	Question	Section	Examples	Page
21. $y + 6 = -5(x - 2)$	21	3.5	1	260

Answer	If You Missed	Review		
22. $y = 5x - 4$	22	3.5	2	261

Answer	If You Missed	Review		
23. $x + y = -6$	23	3.5	3	261–262
24. $C = 0.20m + 10$; $13	24	3.6	1	267–268
25. $C = 40 + 0.50m$; 190	25	3.6	2	268
26. $C = 0.60m + 4$	26	3.6	3	269
27.	27	3.7	1	276

Answer	If You Missed	Review		
	Question	Section	Examples	Page
28.	28	3.7	2	276–277
29.	29	3.7	3	277–278
30.	30	3.7	4	278

⟩Cumulative Review **Chapters 1–3**

1. Find the additive inverse (opposite) of -1.

2. Find the absolute value: $\left|-3\frac{1}{7}\right|$

3. Add: $-\frac{1}{6} + \left(-\frac{3}{8}\right)$

4. Subtract: $9.7 - (-3.3)$

5. Multiply: $(-2.6)(7.6)$

6. Multiply: $(-6)^4$

7. Divide: $-\frac{6}{7} \div \left(-\frac{1}{14}\right)$

8. Evaluate $y \div 2 \cdot x - z + 3$ for $x = 3$, $y = 8$, $z = 3$.

9. Which property is illustrated by the following statement?
$$(4 + 3) + 9 = (3 + 4) + 9$$

10. Multiply: $3(6x - 7)$

11. Combine like terms: $-4cd^2 - (-5cd^2)$

12. Simplify: $3x - 3(x + 4) - (x + 2)$

13. Translate into symbols: The quotient of $(m + n)$ and p.

14. Does the number -14 satisfy the equation $1 = 15 - x$?

15. Solve for x: $1 = 5(x - 2) + 5 - 4x$

16. Solve for x: $-\frac{8}{3}x = -24$

17. Solve for x: $\frac{x}{6} - \frac{x}{9} = 3$

18. Solve for x: $4 - \frac{x}{5} = \frac{2(x + 1)}{11}$

19. Solve for d in the equation $S = 7c^2d$.

20. The sum of two numbers is 170. If one of the numbers is 40 more than the other, what are the numbers?

21. Dave has invested a certain amount of money in stocks and bonds. The total annual return from these investments is $625. If the stocks produce $245 more in returns than the bonds, how much money does Dave receive annually from each type of investment?

22. Train A leaves a station traveling at 20 miles per hour. Two hours later, train B leaves the same station traveling in the same direction at 40 miles per hour. How long does it take for train B to catch up to train A?

23. Martin purchased some municipal bonds yielding 10% annually and some certificates of deposit yielding 11% annually. If Martin's total investment amounts to $25,000 and the annual income is $2630, how much money is invested in bonds and how much is invested in certificates of deposit?

24. Graph: $-\frac{x}{7} + \frac{x}{6} \leq \frac{x - 6}{6}$

25. Graph the point $C(-3, 4)$.

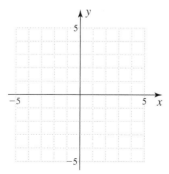

26. What are the coordinates of point A?

27. Determine whether the ordered pair $(-5, 1)$ is a solution of $4x + y = -21$.

28. Find x in the ordered pair $(x, 2)$ so that the ordered pair satisfies the equation $3x - y = -8$.

29. Graph: $2x + y = 8$

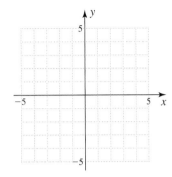

30. Graph: $4x - 8 = 0$

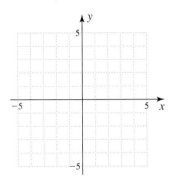

31. Find the slope of the line going through the points $(-5, 3)$ and $(-6, -6)$.

32. What is the slope of the line $4x - 2y = 18$?

33. Find the pair of parallel lines.

(**1**) $-5y = 4x + 7$

(**2**) $-15y + 12x = 7$

(**3**) $12x + 15y = 7$

34. Find the slope of the line passing through the points:

a. $(9, 2)$ and $(9, -5)$

b. $(-3, -7)$ and $(4, -7)$

35. Add: $-\frac{2}{9} + \left(-\frac{1}{8}\right)$

36. Subtract: $4.3 - (-3.9)$

37. Find: $(-2)^4$

38. Divide: $-\frac{1}{5} \div \left(-\frac{1}{10}\right)$

39. Evaluate $y \div 2 \cdot x - z$ for $x = 2$, $y = 8$, $z = 3$.

40. Simplify: $x + 4(x - 2) + (x - 3)$.

41. Write in symbols: The quotient of $(d + 4e)$ and f.

42. Solve for x: $5 = 3(x - 1) + 5 - 2x$

43. Solve for x: $\frac{x}{7} - \frac{x}{9} = 2$

44. The sum of two numbers is 110. If one of the numbers is 40 more than the other, what are the numbers?

45. Susan purchased some municipal bonds yielding 11% annually and some certificates of deposit yielding 14% annually. If Susan's total investment amounts to $9000 and the annual income is $1140, how much money is invested in bonds and how much is invested in certificates of deposit?

46. Graph: $-\frac{x}{7} + \frac{x}{4} \geq \frac{x - 4}{4}$

47. Graph the point $C(-2, -3)$.

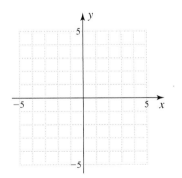

48. Determine whether the ordered pair $(-3, -4)$ is a solution of $5x - y = -19$.

49. Find x in the ordered pair $(x, -1)$ so that the ordered pair satisfies the equation $4x + 2y = -10$.

50. Graph: $3x + y = 3$

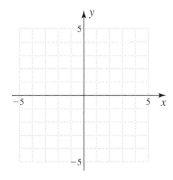

51. Graph: $3x + 9 = 0$

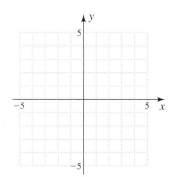

52. Find the slope of the line going through the points $(-4, -6)$ and $(-7, 1)$.

53. What is the slope of the line $12x - 4y = -14$?

54. Find the pair of parallel lines.

   (1) $20x - 5y = 2$

   (2) $5y + 20x = 2$

   (3) $-y = -4x + 2$

55. Find the slope of the line passing through:

   a. $(0, 8)$ and $(1, 8)$

   b. $(-1, 2)$ and $(-1, 7)$

56. Find an equation of the line that goes through the point $(2, 0)$ and has slope $m = -3$.

57. Find an equation of the line having slope 3 and y-intercept -1.

58. Graph: $6x - y < -6$

59. Graph: $-y \geq -6x - 6$

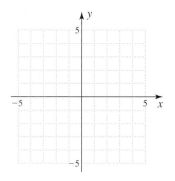

Section

Chapter

4
four

▶ # Exponents and Polynomials

The Human Side of Algebra

In the "Golden Age" of Greek mathematics, 300–200 B.C., three mathematicians "stood head and shoulders above all the others of the time." One of them was Apollonius of Perga in Southern Asia Minor. Around 262–190 B.C., Apollonius developed a method of "tetrads" for expressing large numbers, using an equivalent of exponents of the single myriad (10,000). It was not until about the year 250 that the *Arithmetica* of Diophantus advanced the idea of exponents by denoting the square of the unknown as Δ^Y, the first two letters of the word *dunamis,* meaning "power." Similarly, K^Y represented the cube of the unknown quantity. It was not until 1360 that Nicole Oresme of France gave rules equivalent to the product and power rules of exponents that we study in this chapter. Finally, around 1484, a manuscript written by the French mathematician Nicholas Chuquet contained the *denominacion* (or power) of the unknown quantity, so that our algebraic expressions $3x$, $7x^2$, and $10x^3$ were written as .3. and .7.^{2} and .10.^{3}. What about zero and negative exponents? $8x^0$ became .8.^{0} and $8x^{-2}$ was written as .8.$^{2.m}$, meaning ".8. *seconds moins,*" or 8 to the negative two power. Some things do change!

4.1 The Product, Quotient, and Power Rules for Exponents

▶ Objectives

A〉 Multiply expressions using the product rule for exponents.

B〉 Divide expressions using the quotient rule for exponents.

C〉 Use the power rules to simplify expressions.

▶ To Succeed, Review How To . . .

1. Multiply and divide integers (pp. 60–61, 62–64).
2. Use the commutative and associative properties (p. 78).

▶ Getting Started

Squares, Cubes, and Exponents

Exponential notation is used to indicate how many times a quantity is to be used as a *factor*. For example, the area A of the *square* in the figure can be written as

$$A = x \cdot x = x^2 \quad \text{Read "x squared."}$$

The exponent 2 indicates that the x is used as a factor twice. Similarly, the volume V of the *cube* is

$$V = x \cdot x \cdot x = x^3 \quad \text{Read "x cubed."}$$

This time, the exponent 3 indicates that the x is used as a factor three times. Can you think of a way to represent x^4 or x^5 using a geometric figure?

If a variable x (called the **base**) is to be used n times as a factor, we use the following definition:

DEFINITION OF x^n	$\underbrace{x \cdot x \cdot x \cdot \cdots \cdot x}_{n \text{ factors}} = \underset{\text{Base}}{x^n} \longleftarrow \text{Exponent}$

When n is a natural number, some of the powers of x are

$$x \cdot x \cdot x \cdot x \cdot x = x^5 \quad \text{x to the fifth power}$$
$$x \cdot x \cdot x \cdot x = x^4 \quad \text{x to the fourth power}$$
$$x \cdot x \cdot x = x^3 \quad \text{x to the third power (also read "x cubed")}$$
$$x \cdot x = x^2 \quad \text{x to the second power (also read "x squared")}$$
$$x = x^1 \quad \text{x to the first power (also read as x)}$$

Note that if the base carries no exponent, the exponent is assumed to be 1. Moreover, for $a \neq 0$, a^0 is defined to be 1.

DEFINITION OF a^1 AND a^0	**1.** $a = a^1, b = b^1,$ and $c = c^1$ **2.** $a^0 = 1 \quad (a \neq 0)$

In this section we shall learn how to multiply and divide expressions containing exponents by using the product and quotient rules.

A ⟩ Multiplying Expressions

We are now ready to multiply expressions involving exponents. For example, to multiply x^2 by x^3, we first write

$$\overbrace{\underbrace{x^2}_{x \cdot x} \cdot \underbrace{x^3}_{x \cdot x \cdot x}}^{x^5}$$

or

Here we've just added the exponents of x^2 and x^3, 2 and 3, to find the exponent of the result, 5. Similarly, $2^2 \cdot 2^3 = 2^{2+3} = 2^5$ and $3^2 \cdot 3^3 = 3^{2+3} = 3^5$. Thus,

$$a^3 \cdot a^4 = a^{3+4} = a^7$$

and

$$b^2 \cdot b^4 = b^{2+4} = b^6$$

From these and similar examples, we have the following product rule for exponents.

PRODUCT RULE FOR EXPONENTS

If m and n are positive integers,

$$x^m \cdot x^n = x^{m+n}$$

This rule means that to multiply expressions with the *same base,* we keep the base and *add* the exponents.

NOTE

Note that $x^m \cdot y^n \neq (x \cdot y)^{m+n}$ because the bases x and y are not equal.

NOTE

Before you apply the product rule, make sure that the *bases* are the same.

Of course, some expressions may have numerical coefficients other than 1. For example, the expression $3x^2$ has the numerical coefficient 3. Similarly, the numerical coefficient of $5x^3$ is 5. If we decide to multiply $3x^2$ by $5x^3$, we just multiply numbers by numbers (coefficients) and letters by letters. This procedure is possible because of the *commutative and associative properties of multiplication* we've studied. Using these two properties, we then write

$$(3x^2)(5x^3) = (3 \cdot 5)(x^2 \cdot x^3) \qquad \text{We use parentheses to indicate multiplication.}$$
$$= 15x^{2+3}$$
$$= 15x^5$$

and

$$(8x^2y)(4xy^2)(2x^5y^3)$$
$$= (8 \cdot 4 \cdot 2) \cdot (x^2 \cdot x^1 \cdot x^5)(y^1 \cdot y^2 \cdot y^3) \qquad \text{Note that } x = x^1 \text{ and } y = y^1.$$
$$= 64x^8y^6$$

NOTE

Be sure you understand the difference between adding and multiplying expressions. Thus,

$$5x^2 + 7x^2 = 12x^2$$

but

$$(5x^2)(7x^2) = 35x^{2+2} = 35x^4$$

EXAMPLE 1   Multiplying expressions with positive coefficients

Multiply:

a. $(5x^4)(3x^7)$

b. $(3ab^2c^3)(4a^2b)(2bc)$

SOLUTION

a. We use the commutative and associative properties to write the coefficients and the letters together:

$$(5x^4)(3x^7) = (5 \cdot 3)(x^4 \cdot x^7)$$
$$= 15x^{4+7}$$
$$= 15x^{11}$$

b. Using the commutative and associative properties, and the fact that $a = a^1$, $b = b^1$, and $c = c^1$, we write

$$(3ab^2c^3)(4a^2b)(2bc) = (3 \cdot 4 \cdot 2)(a^1 \cdot a^2)(b^2 \cdot b^1 \cdot b^1)(c^3 \cdot c^1)$$
$$= 24a^{1+2}b^{2+1+1}c^{3+1}$$
$$= 24a^3b^4c^4$$

PROBLEM 1

Multiply:

a. $(3x^5)(4x^7)$

b. $(4ab^3c^2)(3ab^3)(2bc)$

To multiply expressions involving signed coefficients, we recall the rule of signs for multiplication:

> **RULES**
>
> **Signs for Multiplication**
> **1.** When multiplying two numbers with the *same* (like) sign, the product is *positive* $(+)$.
> **2.** When multiplying two numbers with *different* (unlike) signs, the product is *negative* $(-)$.

Thus, to multiply $(-3x^5)$ by $(8x^2)$, we first note that the expressions have *different* signs. The product should have a *negative* coefficient; that is,

$$(-3x^5)(8x^2) = (-3 \cdot 8)(x^5 \cdot x^2)$$
$$= -24x^{5+2}$$
$$= -24x^7$$

Of course, if the expressions have the *same* sign, the product should have a *positive* coefficient. Thus,

$$(-2x^3)(-7x^5) = (-2)(-7)(x^3 \cdot x^5)$$ Note that the product of -2 and -7 was written as
$$= +14x^{3+5}$$ $(-2)(-7)$ and *not* as $(-2 \cdot -7)$ to avoid confusion.
$$= 14x^8$$ Recall that $+14 = 14$.

EXAMPLE 2   Multiplying expressions with negative coefficients

Multiply:

a. $(7a^2bc)(-3ac^4)$

b. $(-2xyz^3)(-4x^3yz^2)$

SOLUTION

a. Since the coefficients have *different* signs, the result must have a *negative* coefficient. Hence,

$$(7a^2bc)(-3ac^4) = (7)(-3)(a^2 \cdot a^1)(b^1)(c^1 \cdot c^4)$$
$$= -21a^{2+1}b^1c^{1+4}$$
$$= -21a^3bc^5$$

PROBLEM 2

Multiply:

a. $(2a^3bc)(-5ac^5)$

b. $(-3x^3yz^4)(-2xyz^4)$

b. The expressions have the *same* sign, so the result must have a *positive* coefficient. That is,

$$(-2xyz^3)(-4x^3yz^2) = [(-2)(-4)](x^1 \cdot x^3)(y^1 \cdot y^1)(z^3 \cdot z^2)$$
$$= +8x^{1+3}y^{1+1}z^{3+2}$$
$$= 8x^4y^2z^5$$

B ⟩ Dividing Expressions

We are now ready to discuss the division of one expression by another. As you recall, the same rule of signs that applies to the multiplication of integers applies to the division of integers. We write this rule for easy reference.

> **RULES**
>
> **Signs for Division**
>
> **1.** When dividing two numbers with the *same* (like) sign, the quotient is *positive* $(+)$.
>
> **2.** When dividing two numbers with *different* (unlike) signs, the quotient is *negative* $(-)$.

Now we know what to do with the numerical coefficients when we divide one expression by another. But what about the exponents? To divide expressions involving exponents, we need another rule. So to divide x^5 by x^3, we first use the definition of exponent and write

$$\frac{x^5}{x^3} = \frac{x \cdot x \cdot x \cdot x \cdot x}{x \cdot x \cdot x} \quad (x \neq 0) \qquad \text{Remember, division by zero is not allowed!}$$

Since $(x \cdot x \cdot x)$ is common to the numerator and denominator, we have

$$\frac{x^5}{x^3} = \frac{(x \cdot x \cdot x) \cdot x \cdot x}{(x \cdot x \cdot x)} = x \cdot x = x^2$$

Here the colored x's mean that we divided the numerator and denominator by the common factor $(x \cdot x \cdot x)$. Of course, you can immediately see that the exponent 2 in the answer is simply the difference of the original two exponents, 5 and 3; that is,

$$\frac{x^5}{x^3} = x^{5-3} = x^2 \quad (x \neq 0)$$

Similarly,

$$\frac{x^7}{x^4} = x^{7-4} = x^3 \quad (x \neq 0)$$

and

$$\frac{y^4}{y^1} = y^{4-1} = y^3 \quad (y \neq 0)$$

We can now state the rule for dividing expressions involving exponents.

> **QUOTIENT RULE FOR EXPONENTS**
>
> If m and n are positive integers and m is greater than n then,
>
> $$\frac{x^m}{x^n} = x^{m-n} \quad (x \neq 0)$$
>
> This means that to divide expressions with the same base, we keep the same base and subtract the exponents.

EXAMPLE 3   Dividing expressions with negative coefficients

Find the quotient:

$$\frac{24x^2y^6}{-6xy^4}$$

SOLUTION   We could write

$$\frac{24x^2y^6}{-6xy^4} = \frac{2 \cdot 2 \cdot (2 \cdot 3 \cdot x) \cdot x \cdot (y \cdot y \cdot y \cdot y) \cdot y \cdot y}{-(2 \cdot 3 \cdot x) \cdot (y \cdot y \cdot y \cdot y)}$$

$$= -2 \cdot 2 \cdot x \cdot y \cdot y$$

$$= -4xy^2$$

but to save time, it's easier to divide 24 by -6, x^2 by x, and y^6 by y^4, like this:

$$\frac{24x^2y^6}{-6xy^4} = \frac{24}{-6} \cdot \frac{x^2}{x} \cdot \frac{y^6}{y^4}$$

$$= -4 \cdot x^{2-1} \cdot y^{6-4}$$

$$= -4xy^2$$

PROBLEM 3

Find the quotient:

$$\frac{25a^3b^7}{-5ab^3}$$

C ⟩ Simplifying Expressions Using the Power Rules

Suppose we wish to find $(5^3)^2$. By definition, squaring a quantity means that we multiply the quantity by itself. Thus,

$$(5^3)^2 = 5^3 \cdot 5^3 = 5^{3+3} \quad \text{or} \quad 5^6$$

We could get this answer by multiplying exponents in $(5^3)^2$ to obtain $5^{3 \cdot 2} = 5^6$. Similarly,

$$(x^2)^3 = x^2 \cdot x^2 \cdot x^2 = x^{2 \cdot 3} = x^6$$

Again, we multiplied exponents in $(x^2)^3$ to get x^6. We use these ideas to state the following power rule for exponents.

POWER RULE FOR EXPONENTS

If m and n are positive integers.

$$(x^m)^n = x^{mn}$$

This means that when raising a *power* to a *power*, we keep the base and *multiply* the exponents.

EXAMPLE 4   Raising a power to a power

Simplify:

a. $(2^3)^4$ **b.** $(x^2)^3$ **c.** $(y^4)^5$

SOLUTION

a. $(2^3)^4 = 2^{3 \cdot 4} = 2^{12}$ **b.** $(x^2)^3 = x^{2 \cdot 3} = x^6$ **c.** $(y^4)^5 = y^{4 \cdot 5} = y^{20}$

PROBLEM 4

Simplify:

a. $(3^2)^4$ **b.** $(a^3)^5$ **c.** $(b^5)^3$

Answers to PROBLEMS

3. $-5a^2b^4$

4. a. 3^8 **b.** a^{15} **c.** b^{15}

Sometimes we need to raise several factors inside parentheses to a power, such as in $(x^2y^3)^3$. We use the definition of cubing (see the *Getting Started*) and write:

$$(x^2y^3)^3 = x^2y^3 \cdot x^2y^3 \cdot x^2y^3$$
$$= (x^2 \cdot x^2 \cdot x^2)(y^3 \cdot y^3 \cdot y^3)$$
$$= (x^2)^3(y^3)^3$$
$$= x^6y^9$$

Since we are cubing x^2y^3, we could get the same answer by multiplying each of the exponents in x^2y^3 by 3 to obtain $x^{2 \cdot 3}y^{3 \cdot 3} = x^6y^9$. Thus, to raise several factors inside parentheses to a power, we raise each factor to the given power, as stated in the following rule.

POWER RULE FOR PRODUCTS

If m, n, and k are positive integers,

$$(x^m y^n)^k = (x^m)^k(y^n)^k = x^{mk}y^{nk}$$

This means that to raise a product to a power, we raise each factor in the product to that power.

NOTE

The power rule applies to products only:

$$(x + y)^n \neq x^n + y^n$$

EXAMPLE 5   Using the power rule for products

Simplify:

a. $(3x^2y^2)^3$ **b.** $(-2x^2y^3)^3$

SOLUTION

a. $(3x^2y^2)^3 = 3^3(x^2)^3(y^2)^3$ Note that since 3 is a *factor* in $3x^2y^2$,
         $= 27x^6y^6$ 3 is also raised to the third power.

b. $(-2x^2y^3)^3 = (-2)^3(x^2)^3(y^3)^3$
            $= -8x^6y^9$

PROBLEM 5

Simplify:

a. $(2a^3b^2)^4$ **b.** $(-3a^3b^2)^3$

Just as we have a product and a quotient rule for exponents, we also have a power rule for products and for quotients. Since the quotient

$$\frac{a}{b} = a \cdot \frac{1}{b},$$

we can use the power rule for products and some of the real-number properties to obtain the following.

THE POWER RULE FOR QUOTIENTS

If m is a positive integer,

$$\left(\frac{x}{y}\right)^m = \frac{x^m}{y^m} \quad (y \neq 0)$$

This means that to raise a quotient to a power, we raise the numerator and denominator in the quotient to that power.

EXAMPLE 6 Using the power rule for quotients
Simplify:

a. $\left(\dfrac{4}{5}\right)^3$

b. $\left(\dfrac{2x^2}{y^3}\right)^4$

SOLUTION

a. $\left(\dfrac{4}{5}\right)^3 = \dfrac{4^3}{5^3} = \dfrac{64}{125}$

b. $\dfrac{(2x^2)^4}{(y^3)^4} = \dfrac{2^4 \cdot (x^2)^4}{(y^3)^4}$

$= \dfrac{16x^{2 \cdot 4}}{y^{3 \cdot 4}}$

$= \dfrac{16x^8}{y^{12}}$

PROBLEM 6
Simplify:

a. $\left(\dfrac{2}{3}\right)^3$ **b.** $\left(\dfrac{3a^2}{b^4}\right)^3$

Here is a summary of the rules we've just studied.

RULES FOR EXPONENTS

If m, n, and k are positive integers, the following rules apply:

Rule	Example
1. Product rule for exponents: $x^m x^n = x^{m+n}$	$x^5 \cdot x^6 = x^{5+6} = x^{11}$
2. Quotient rule for exponents: $\dfrac{x^m}{x^n} = x^{m-n} \ (m > n, x \neq 0)$	$\dfrac{p^8}{p^3} = p^{8-3} = p^5$
3. Power rule for products: $(x^m y^n)^k = x^{mk} y^{nk}$	$(x^4 y^3)^4 = x^{4 \cdot 4} y^{3 \cdot 4} = x^{16} y^{12}$
4. Power rule for quotients: $\left(\dfrac{x}{y}\right)^m = \dfrac{x^n}{y^m} \ (y \neq 0)$	$\left(\dfrac{a^3}{b^4}\right)^6 = \dfrac{a^{3 \cdot 6}}{b^{4 \cdot 6}} = \dfrac{a^{18}}{b^{24}}$

Now let's look at an example where several of these rules are used.

EXAMPLE 7 Using the power rule for products and quotients
Simplify:

a. $(3x^4)^3(-2y^3)^2$ **b.** $\left(\dfrac{3}{5}\right)^3 \cdot 5^4$

SOLUTION

a. $(3x^4)^3(-2y^3)^2 = (3)^3(x^4)^3(-2)^2(y^3)^2$ Use Rule 3.

$= 27x^{4 \cdot 3}(4)y^{3 \cdot 2}$ Use the Power Rule for Exponents.

$= (27 \cdot 4)x^{12}y^6$

$= 108x^{12}y^6$

b. $\left(\dfrac{3}{5}\right)^3 \cdot 5^4 = \dfrac{3^3}{5^3} \cdot 5^4$ Use Rule 4.

$= \dfrac{3^3}{5^3} \cdot \dfrac{5^4}{1}$ Since $5^4 = \dfrac{5^4}{1}$

$= \dfrac{3^3 \cdot 5^4}{5^3}$ Multiply.

$= 3^3 \cdot 5$ Since $\dfrac{5^4}{5^3} = 5^{4-3} = 5$

$= 27 \cdot 5$ Since $3^3 = 27$

$= 135$ Multiply.

PROBLEM 7
Simplify:

a. $(2a^3)^4(-3b^2)^3$ **b.** $\left(\dfrac{2}{3}\right)^3 \cdot 3^5$

Answers to PROBLEMS

6. **a.** $\dfrac{8}{27}$ **b.** $\dfrac{27a^6}{b^{12}}$

7. **a.** $-432a^{12}b^6$ **b.** 72

So far, we have discussed only the theory and rules for exponents. Does anybody need or use exponents? We will show one application in Example 8.

EXAMPLE 8   Application: Volume of an ice cream cone

The volume V of a cone is $V = \frac{1}{3}\pi r^2 h$, where r is the radius of the base (opening) and h is the height of the cone. The volume S of a hemisphere is $S = \frac{2}{3}\pi r^3$. The ice cream cone in the photo is 10 centimeters tall and has a radius of 2 centimeters; the mound of ice cream is 2 centimeters high. To make computation easier, assume that π is about 3; that is, use 3 for π.

a. How much ice cream does it take to fill the cone?

b. What is the volume of the mound of ice cream?

c. What is the total volume of ice cream (cone plus mound) in the photo?

SOLUTION

a. The volume V of the cone is $V = \frac{1}{3}\pi r^2 h$, where $r = 2$ and $h = 10$. Thus, $V = \frac{1}{3}\pi(2)^2(10) = \frac{40\pi}{3}$ cubic centimeters (cm³) or approximately 40 cubic centimeters.

b. The volume of the mound of ice cream is the volume of the hemisphere:

$$S = \frac{2}{3}\pi r^3 = \frac{2}{3}\pi(2)^3 = \frac{16\pi}{3} \text{ cm}^3 \approx 16 \text{ cm}^3$$

c. The total amount of ice cream is $\frac{40\pi}{3} + \frac{16\pi}{3}$ cubic centimeters $= \frac{56\pi}{3}$ cubic centimeters or about 56 cubic centimeters.

PROBLEM 8

Answer the same questions as in Example 8 if the cone is 4 inches tall and has a 1-inch radius. Can you see that the hemisphere must have a 1-inch radius?

Calculator Corner

Checking Exponents

Can we check the rules of exponents using a calculator? We can do it if we have only one variable and we agree that **two expressions are equivalent if their graphs are identical.** Thus, to check Example 1(a) we have to check that $(5x^4)(3x^7) = 15x^{11}$. We will consider $(5x^4)(3x^7)$ and $15x^{11}$ separately. As a matter of fact, let $Y_1 = (5x^4)(3x^7)$ and $Y_2 = 15x^{11}$; enter Y_1 in your calculator by pressing Y= and entering $(5x^4)(3x^7)$. (Remember that exponents are entered by pressing ^ .) Now, press GRAPH and the graph appears, as shown at the left. Now enter $Y_2 = 15x^{11}$. You get the same graph, indicating that the graphs actually coincide and thus are equivalent. Now, for a numerical quick surprise, press 2nd GRAPH. The screen shows the information on the right. What does it mean? It means that for the values of x in the table (0, 1, 2, 3, etc.), Y_1 and Y_2 have the same values.

X	Y₁	Y₂
0	0	0
1	15	15
2	30720	30720
3	2.66E6	2.66E6
4	6.29E7	6.29E7
5	7.32E8	7.32E8
6	5.44E9	5.44E9
X=0		

Answers to PROBLEMS

8. a. $\frac{4\pi}{3}$ in.³ ≈ 4 in.³

　b. $\frac{2\pi}{3}$ in.³ ≈ 2 in.³

　c. 2π in.³ ≈ 6 in.³

› Exercises **4.1**

‹ **A** › **Multiplying Expressions**   In Problems 1–16, find the product.

1. $(4x)(6x^2)$

2. $(2a^2)(3a^3)$

3. $(5ab^2)(6a^3b)$

4. $(-2xy)(x^2y)$

5. $(-xy^2)(-3x^2y)$

6. $(x^2y)(-5xy)$

7. $b^3\left(\dfrac{-b^2c}{5}\right)$

8. $-a^3b\left(\dfrac{ab^4c}{3}\right)$

9. $\left(\dfrac{-5xy^2z}{2}\right)\left(\dfrac{-3x^2yz^5}{5}\right)$

10. $(-2x^2yz^3)(4xyz)$

11. $(-2xyz)(3x^2yz^3)(5x^2yz^4)$

12. $(-a^2b)(-0.4b^2c^3)(1.5abc)$

13. $(a^2c^3)(-3b^2c)(-5a^2b)$

14. $(ab^2c)(-ac^2)(-0.3bc^3)(-2.5a^3c^2)$

15. $(-2abc)(-3a^2b^2c^2)(-4c)(-b^2c)$

16. $(xy)(yz)(xz)(yz)$

‹ **B** › **Dividing Expressions**   In Problems 17–34, find the quotient.

17. $\dfrac{x^7}{x^3}$

18. $\dfrac{8a^3}{4a^2}$

19. $\dfrac{-8a^4}{16a^2}$

20. $\dfrac{9y^5}{6y^2}$

21. $\dfrac{12x^5y^3}{6x^2y}$

22. $\dfrac{18x^6y^2}{9xy}$

23. $\dfrac{-6x^6y^3}{12x^3y}$

24. $\dfrac{8x^8y^4}{4x^5y^2}$

25. $\dfrac{-14a^8y^6}{-21a^5y^2}$

26. $\dfrac{-2a^5y^8}{-6a^2y^3}$

27. $\dfrac{-27a^2b^8c^3}{-36ab^5c^2}$

28. $\dfrac{-5x^6y^8z^5}{10x^2y^5z^2}$

29. $\dfrac{3a^3 \cdot a^5}{2a^4}$

30. $\dfrac{y^2 \cdot y^8}{y \cdot y^3}$

31. $\dfrac{(2x^2y^3)(-3x^5y)}{6xy^3}$

32. $\dfrac{(-3x^3y^2z)(4xy^3z)}{6xy^2z}$

33. $\dfrac{(-x^2y)(x^3y^2)}{x^3y}$

34. $\dfrac{(-8x^2y)(-7x^5y^3)}{-2x^2y^3}$

‹ **C** › **Simplifying Expressions Using the Power Rules**   In Problems 35–70, simplify.

35. $(2^2)^3$

36. $(3^1)^2$

37. $(3^2)^1$

38. $(2^3)^2$

39. $(x^3)^3$

40. $(y^2)^4$

41. $(y^3)^2$

42. $(x^4)^3$

43. $(-a^2)^3$

44. $(-b^3)^5$

45. $(2x^3y^2)^3$

46. $(3x^2y^3)^2$

47. $(2x^2y^3)^2$

48. $(3x^4y^4)^3$

49. $(-3x^3y^2)^3$

50. $(-2x^5y^4)^4$

51. $(-3x^6y^3)^2$

52. $(y^4z^3)^5$

53. $(-2x^4y^4)^3$

54. $(-3y^5z^3)^4$

55. $\left(\dfrac{2}{3}\right)^4$

56. $\left(\dfrac{-3}{4}\right)^3$

57. $\left(\dfrac{3x^2}{2y^3}\right)^3$

58. $\left(\dfrac{3a^2}{2b^3}\right)^4$

59. $\left(\dfrac{-2x^2}{3y^3}\right)^4$

60. $\left(\dfrac{-3x^2}{2y^3}\right)^4$

61. $(2x^3)^2(3y^3)$

62. $(3a)^3(2b)^2$

63. $(-3a)^2(-4b)^3$

64. $(-2x^2)^3(-3y^3)^2$

65. $-(4a^2)^2(-3b^3)^2$

66. $-(3x^3)^3(-2y^3)^3$

67. $\left(\dfrac{2}{3}\right)^5 \cdot 3^6$

68. $\left(\dfrac{4}{5}\right)^3 \cdot 5^4$

69. $\left(\dfrac{x}{y}\right)^5 \cdot y^7$

70. $\left(\dfrac{2a^2}{3b}\right)^5 \cdot 3^4b^7$

› › ›   **Applications**

Compound Interest   If you invest P dollars at rate r compounded annually, the compound amount A in your account at the end of n years is given by:

$$A = P(1 + r)^n$$

Use this formula in Problems 71–74.

71. If you invest $1000 at 8% compounded annually, how much will be in your account at the end of 3 years?

72. If you invest $500 at 10% compounded annually, how much will be in your account at the end of 3 years?

73. Suppose you invest $1000 at 10% compounded annually. How much will be in your account at the end of 4 years?

74. Which investment will produce more money: $500 invested at 8% compounded annually for 3 years or $600 invested at 7% compounded annually for 3 years? How much more money?

Cuban Sandwiches   Problems 75–80 refer to the photo (to the right, top).

75. A standard pickle container measures x by x by $\frac{2}{3}x$ inches. What is the volume V of the container?

76. Using the formula obtained in Problem 75, what is the volume of the container if $x = 6$ inches?

77. A Cuban sandwich takes about one cubic inch of pickles. How many sandwiches can be made with the contents of the container?

78. A container of pickles holds about 24 ounces. If pickles cost $0.045 per ounce, what is the cost of the pickles in the container?

79. What is the cost of the pickles in a Cuban sandwich (see Problems 77 and 78)?

80. How many ounces of pickles does a Cuban sandwich take (see Problems 77 and 78)?

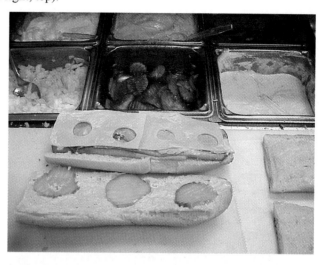

Volume   Problems 81–85 refer to the photo (to the right, middle).

81. The dimensions of the green bean container are x by $2x$ by $\frac{1}{2}x$. What is the volume V of the container?

82. Using the formula given in Problem 81, what is the volume of the container if $x = 6$?

83. Green beans are served in a small bowl that has the approximate shape of a hemisphere with a radius of 2 inches. If the volume of a hemisphere is $S = \frac{2}{3}\pi r^3$, what is the volume of one serving of green beans? Use 3 for π.

84. Based on your answers to Problems 82 and 83, how many servings of green beans does the whole container hold?

85. If each serving of green beans costs $0.24 and sells for $1.99, how much does the restaurant make per container?

Volume   Problems 86–90 refer to the photo (to the right, bottom).

86. The volume V of a cylinder is $V = \pi r^2 h$, where r is the radius and h is the height. If the pot is x inches tall and its radius is $\frac{1}{2}x$, what is the volume of the pot? Use 3 for π.

87. When the pot is used for cooking beans, it is only $\frac{3}{4}$ full. What is the volume of the beans in the pot? (Refer to the formula for the volume of a cylinder given in Problem 86.)

88. If the pot is $\frac{3}{4}$ full and the height of the pot is 10 inches, what is the volume of the beans? (Refer to the formula for the volume of a cylinder given in Problem 86.)

89. A serving of beans is about 1 cup, or 15 in.³, of beans. How many servings of beans are in a pot of beans?

90. How many pots of beans are needed to serve 150 people?

Volume of Solids   We can write the formula for the volume of many solids using exponents. In Problems 91–94 the volume of a solid is given in words. Write the formula in symbols.

91. The volume V of a cylinder of radius r and height h is the product of π, the square of r, and h.

Cylinder

92. The volume V of a sphere of radius r is the product of $\frac{4}{3}\pi$ and the cube of r.

Sphere

93. The volume V of a cone of height h and radius r is the product of $\frac{1}{3}\pi$, the square of r, and h.

Cone

94. The volume V of a cube is the length L of one of the sides cubed.

Cube

❯❯❯ Using Your Knowledge

Follow That Pattern   Many interesting patterns involve exponents. Use your knowledge to find the answers to the following problems.

95.
$$1^2 = 1$$
$$(11)^2 = 121$$
$$(111)^2 = 12,321$$
$$(1111)^2 = 1,234,321$$

a. Find $(11,111)^2$.

b. Find $(111,111)^2$.

96.
$$1^2 = 1$$
$$2^2 = 1 + 2 + 1$$
$$3^2 = 1 + 2 + 3 + 2 + 1$$
$$4^2 = 1 + 2 + 3 + 4 + 3 + 2 + 1$$

a. Use this pattern to write 5^2.

b. Use this pattern to write 6^2.

97.
$$1 + 3 = 2^2$$
$$1 + 3 + 5 = 3^2$$
$$1 + 3 + 5 + 7 = 4^2$$

a. Find $1 + 3 + 5 + 7 + 9$.

b. Find $1 + 3 + 5 + 7 + 9 + 11 + 13$.

98. Can you discover your own pattern? What is the largest number you can construct by using the number 9 three times? (It is not 999!)

❯❯❯ Write On

99. Explain why the product rule does not apply to the expression $x^2 \cdot y^3$.

100. Explain why the product rule does not apply to the expression $x^2 + y^3$.

101. What is the difference between the product rule and the power rule?

102. Explain why you cannot use the quotient rule stated in the text to conclude that
$$\frac{x^m}{x^m} = 1 = x^0, \quad x \neq 0$$

❯❯❯ Concept Checker

Fill in the blank(s) with the correct word(s), phrase, or mathematical statement.

103. If m and n are **positive integers**, $x^m \cdot x^n =$ _____.

104. When **multiplying** numbers with the **same (like) signs**, the **product** is _____.

105. When **multiplying** numbers with **different (unlike) signs**, the **product** is _____.

an integer x^{m+n}

0   **negative**

106. When **dividing** numbers with the **same (like) signs**, the quotient is _____.

107. When **dividing** numbers with **different (unlike) signs**, the quotient is _____.

108. If m and n are **positive integers** with $m > n$, $\left(\frac{x^m}{x^n}\right) =$ _____.

109. If m and n are **positive integers** $(x^m)^n =$ _____.

110. If m, n, and k are **positive integers** $(x^m y^n)^k =$ _____.

111. If m is a **positive integer**, $\left(\frac{x}{y}\right)^m =$ _____.

112. In the quotient $\left(\frac{x}{y}\right)^m$, the *denominator* y can not be _____.

positive	y^m
x^{m-n}	$x^{mk}y^{nk}$
x^{mn}	$\dfrac{x^m}{y^m}$
$\dfrac{x^m}{y^n}$	$y^{mk}y^{nk}$
x^m	

〉〉〉 *Mastery Test*

Simplify:

113. $(5a^3)(6a^8)$

114. $(3x^2yz)(2xy^2)(8xz^3)$

115. $(-3x^3yz)(4xz^4)$

116. $(-3ab^2c^4)(-5a^4bc^3)$

117. $\dfrac{15x^4y^7}{-3x^2y}$

118. $\dfrac{-5x^2y^6z^4}{15xy^4z^2}$

119. $\dfrac{-3a^2b^9c^2}{-12ab^2c}$

120. $(3^2)^2$

121. $(y^5)^4$

122. $(3x^2y^3)^2$

123. $\left(\dfrac{3y^2}{x^5}\right)^3$

124. $(-2x^2y^3)^2(3x^4y^5)^3$

125. $\left(\dfrac{1}{2^5}\right) \cdot 2^7$

126. $\left(\dfrac{x^4}{y^7}\right) \cdot y^9$

〉〉〉 *Skill Checker*

Find:

127. 4^3

128. 5^2

129. $\dfrac{1}{2^3}$

130. $\dfrac{1}{3^2}$

131. $\dfrac{3^4}{2^4}$

4.2   Integer Exponents

▶ Objectives

A〉 Write an expression with negative exponents as an equivalent one with positive exponents.

B〉 Write a fraction involving exponents as a number with a negative power.

C〉 Multiply and divide expressions involving negative exponents.

▶ To Succeed, Review How To . . .

1. Raise numbers to a power (p. 312).
2. Use the rules of exponents (p. 314).

▶ Getting Started
Negative Exponents in Science

In science and technology, negative numbers are used as exponents. For example, the diameter of the DNA molecule pictured on the next page is 10^{-8} meter, and the time it takes for an electron to go from source to screen in a TV tube is 10^{-6} second. So what do the numbers 10^{-8} and 10^{-6} mean? Look at the pattern obtained by *dividing by 10* in each step:

Exponents decrease by 1.	$10^3 = 1000$	Divide by 10 at each step.
	$10^2 = 100$	
	$10^1 = 10$	
↓	$10^0 = 1$	↓

Hence the following also holds true:

Exponents decrease by 1.

$$10^{-1} = \frac{1}{10} = \frac{1}{10}$$

$$10^{-2} = \frac{1}{100} = \frac{1}{10^2}$$

$$10^{-3} = \frac{1}{1000} = \frac{1}{10^3}$$

Divide by 10 at each step.

Thus, the diameter of a DNA molecule is

$$10^{-8} = \frac{1}{10^8} = 0.00000001 \text{ meter}$$

and the time elapsed between source and screen in a TV tube is

$$10^{-6} = \frac{1}{10^6} = 0.000001 \text{ second}$$

These are very small numbers and very cumbersome to write, so, for convenience, we use negative exponents to represent them. In this section we shall study expressions involving negative and zero exponents.

A ⟩ Negative Exponents

In the pattern used in the *Getting Started,* $10^0 = 1$. In general, we make the following definition.

ZERO EXPONENT

> For any nonzero number *x*,
> $$x^0 = 1$$
> This means that a nonzero number *x* raised to the zero power is **1**.

Thus, $5^0 = 1$, $8^0 = 1$, $6^0 = 1$, $(3y)^0 = 1$, and $(-2x^2)^0 = 1$.

Let us look again at the pattern in the *Getting Started;* as you can see,

$$10^{-1} = \frac{1}{10^1}, \quad 10^{-2} = \frac{1}{10^2}, \quad 10^{-3} = \frac{1}{10^3}, \quad \text{and} \quad 10^{-n} = \frac{1}{10^n}$$

We also have the following definition.

NEGATIVE EXPONENT

> If *n* is a positive integer, then
> $$x^{-n} = \frac{1}{x^n} \qquad (x \neq 0)$$

This definition says that x^{-n} and x^n are reciprocals, since

$$x^{-n} \cdot x^n = \frac{1}{x^n} \cdot x^n = 1$$

By definition, then,

$$5^{-2} = \frac{1}{5^2} = \frac{1}{5 \cdot 5} = \frac{1}{25} \quad \text{and} \quad 2^{-3} = \frac{1}{2^3} = \frac{1}{2 \cdot 2 \cdot 2} = \frac{1}{8}$$

When n is positive, we obtain this result:

$$\left(\frac{1}{x}\right)^{-n} = \frac{1}{\left(\frac{1}{x}\right)^n} = \frac{1}{\frac{1}{x^n}} = 1 \cdot \frac{x^n}{1} = x^n$$

which can be stated as follows.

nTH POWER OF A QUOTIENT	If n is a positive integer, then $$\left(\frac{1}{x}\right)^{-n} = x^n, \ (x \neq 0)$$

Thus,

$$\left(\frac{1}{4}\right)^{-3} = 4^3 = 64 \quad \text{and} \quad \left(\frac{1}{5}\right)^{-2} = 5^2 = 25$$

EXAMPLE 1   Rewriting with positive exponents

Use positive exponents to rewrite and simplify:

a. 6^{-2} b. 4^{-3} c. $\left(\frac{1}{7}\right)^{-3}$ d. $\left(\frac{1}{x}\right)^{-5}$

SOLUTION

a. $6^{-2} = \frac{1}{6^2} = \frac{1}{6 \cdot 6} = \frac{1}{36}$ b. $4^{-3} = \frac{1}{4^3} = \frac{1}{4 \cdot 4 \cdot 4} = \frac{1}{64}$

c. $\left(\frac{1}{7}\right)^{-3} = 7^3 = 343$ d. $\left(\frac{1}{x}\right)^{-5} = x^5$

PROBLEM 1

Rewrite and simplify:

a. 2^{-3} b. 3^{-3}

c. $\left(\frac{1}{2}\right)^{-3}$ d. $\left(\frac{1}{a}\right)^{-4}$

If we want to rewrite

$$\frac{x^{-2}}{y^{-3}}$$

without negative exponents, we use the definition of negative exponents to rewrite x^{-2} and y^{-3}:

$$\frac{x^{-2}}{y^{-3}} = \frac{\frac{1}{x^2}}{\frac{1}{y^3}} = \frac{1}{x^2} \cdot \frac{y^3}{1} = \frac{y^3}{x^2}$$

so that the expression is in the simpler form,

$$\frac{x^{-2}}{y^{-3}} = \frac{y^3}{x^2}$$

Here is the rule for changing fractions with negative exponents to equivalent ones with positive exponents.

SIMPLIFYING FRACTIONS WITH NEGATIVE EXPONENTS

For any nonzero numbers x and y and any positive integers m and n,

$$\frac{x^{-m}}{y^{-n}} = \frac{y^n}{x^m}$$

This means that to write an equivalent fraction without negative exponents, we interchange numerators and denominators and make the exponents positive.

EXAMPLE 2   Writing equivalent fractions with positive exponents

Write as an equivalent fraction without negative exponents and simplify:

a. $\dfrac{3^{-2}}{4^{-3}}$ **b.** $\dfrac{a^{-7}}{b^{-3}}$ **c.** $\dfrac{x^2}{y^{-4}}$

SOLUTION

a. $\dfrac{3^{-2}}{4^{-3}} = \dfrac{4^3}{3^2} = \dfrac{64}{9}$ **b.** $\dfrac{a^{-7}}{b^{-3}} = \dfrac{b^3}{a^7}$ **c.** $\dfrac{x^2}{y^{-4}} = x^2 y^4$

PROBLEM 2

Write as an equivalent expression without negative exponents and simplify:

a. $\dfrac{2^{-4}}{3^{-3}}$ **b.** $\dfrac{x^{-9}}{y^{-4}}$ **c.** $\dfrac{a^5}{b^{-8}}$

B › Writing Fractions Using Negative Exponents

As we saw in the *Getting Started,* we can write fractions involving powers in the denominator using negative exponents.

EXAMPLE 3   Writing equivalent fractions with negative exponents

Write using negative exponents:

a. $\dfrac{1}{5^4}$ **b.** $\dfrac{1}{7^5}$ **c.** $\dfrac{1}{x^5}$ **d.** $\dfrac{3}{x^4}$

SOLUTION   We use the definition of negative exponents.

a. $\dfrac{1}{5^4} = 5^{-4}$ **b.** $\dfrac{1}{7^5} = 7^{-5}$

c. $\dfrac{1}{x^5} = x^{-5}$ **d.** $\dfrac{3}{x^4} = 3 \cdot \dfrac{1}{x^4} = 3x^{-4}$

PROBLEM 3

Write using negative exponents:

a. $\dfrac{1}{7^6}$ **b.** $\dfrac{1}{8^5}$

c. $\dfrac{1}{a^9}$ **d.** $\dfrac{7}{a^6}$

C › Multiplying and Dividing Expressions with Negative Exponents

In Section 4.1, we multiplied expressions that contained positive exponents. For example,

$$x^5 \cdot x^8 = x^{5+8} = x^{13} \quad \text{and} \quad y^2 \cdot y^3 = y^{2+3} = y^5$$

Can we multiply expressions involving negative exponents using the same idea? Let's see.

$$x^5 \cdot x^{-2} = x^5 \cdot \dfrac{1}{x^2} = \dfrac{x^5}{x^2} = x^3$$

Adding exponents,

$$x^5 \cdot x^{-2} = x^{5+(-2)} = x^3 \qquad \text{Same answer!}$$

Similarly,

$$x^{-3} \cdot x^{-2} = \dfrac{1}{x^3} \cdot \dfrac{1}{x^2} = \dfrac{1}{x^{3+2}} = \dfrac{1}{x^5} = x^{-5}$$

Adding exponents,

$$x^{-3} \cdot x^{-2} = x^{-3+(-2)} = x^{-5} \qquad \text{Same answer again!}$$

So, we have the following rule.

> **PRODUCT RULE FOR EXPONENTS**
>
> If *m* and *n* are integers,
>
> $$x^m \cdot x^n = x^{m+n}$$
>
> This rule says that when we *multiply* expressions with the same base, we keep the base and *add* the exponents.

Note that the rule does *not* apply to $x^7 \cdot y^6$ because the bases are different.

Answers to PROBLEMS

2. a. $\dfrac{3^3}{2^4} = \dfrac{27}{16}$ **b.** $\dfrac{y^4}{x^9}$ **c.** $a^5 b^8$

3. a. 7^{-6} **b.** 8^{-5} **c.** a^{-9} **d.** $7a^{-6}$

EXAMPLE 4   Using the product rule
Multiply and simplify (that is, write the answer without negative exponents):

a. $2^6 \cdot 2^{-4}$ **b.** $4^3 \cdot 4^{-5}$ **c.** $y^{-2} \cdot y^{-3}$ **d.** $a^{-5} \cdot a^5$

SOLUTION

a. $2^6 \cdot 2^{-4} = 2^{6+(-4)} = 2^2 = 4$ **b.** $4^3 \cdot 4^{-5} = 4^{3+(-5)} = 4^{-2} = \dfrac{1}{4^2} = \dfrac{1}{16}$

c. $y^{-2} \cdot y^{-3} = y^{-2+(-3)} = y^{-5} = \dfrac{1}{y^5}$ **d.** $a^{-5} \cdot a^5 = a^{-5+5} = a^0 = 1$

PROBLEM 4
Multiply and simplify:

a. $3^{-4} \cdot 3^6$ **b.** $2^4 \cdot 2^{-6}$

c. $b^{-3} \cdot b^{-5}$ **d.** $x^{-7} \cdot x^7$

NOTE

We wrote the answers in Example 4 *without* using negative exponents. In algebra, it's customary to write answers *without* negative exponents.

In Section 4.1, we divided expressions with the same base. Thus,

$$\frac{7^5}{7^2} = 7^{5-2} = 7^3 \quad \text{and} \quad \frac{8^3}{8} = 8^{3-1} = 8^2$$

The rule used there can be extended to any exponents that are integers.

QUOTIENT RULE FOR EXPONENTS

If m and n are integers,

$$\frac{x^m}{x^n} = x^{m-n}, \quad x \neq 0$$

This rule says that when we *divide* expressions with the *same* base, we keep the base and *subtract* the exponents.

Note that

$$\frac{x^m}{x^n} = x^m \cdot \frac{1}{x^n} = x^m \cdot x^{-n} = x^{m-n}$$

EXAMPLE 5   Using the quotient rule
Simplify:

a. $\dfrac{6^5}{6^{-2}}$ **b.** $\dfrac{x}{x^5}$ **c.** $\dfrac{y^{-2}}{y^{-2}}$ **d.** $\dfrac{z^{-3}}{z^{-4}}$

SOLUTION

a. $\dfrac{6^5}{6^{-2}} = 6^{5-(-2)} = 6^{5+2} = 6^7$

b. $\dfrac{x}{x^5} = x^{1-5} = x^{-4} = \dfrac{1}{x^4}$

c. $\dfrac{y^{-2}}{y^{-2}} = y^{-2-(-2)} = y^{-2+2} = y^0 = 1$ Recall the definition for the zero exponent.

d. $\dfrac{z^{-3}}{z^{-4}} = z^{-3-(-4)} = z^{-3+4} = z^1 = z$

PROBLEM 5
Simplify:

a. $\dfrac{3^5}{3^{-3}}$ **b.** $\dfrac{a}{a^{-6}}$

c. $\dfrac{b^{-5}}{b^{-5}}$ **d.** $\dfrac{a^{-4}}{a^6}$

Answers to PROBLEMS

4. a. 9   **b.** $\dfrac{1}{4}$ **c.** $\dfrac{1}{b^8}$ **d.** 1   **5. a.** 3^8 **b.** a^7 **c.** 1   **d.** $\dfrac{1}{a^{10}}$

Here is a summary of the definitions and rules of exponents we've just discussed. Please, make sure you read and understand these rules and procedures before you go on.

RULES FOR EXPONENTS

If m, n, and k are integers, the following rules apply:

Rule		Example
Product rule for exponents:	$x^m x^n = x^{m+n}$	$x^{-2} \cdot x^6 = x^{-2+6} = x^4$
Zero exponent:	$x^0 = 1 \quad (x \neq 0)$	$9^0 = 1$, $y^0 = 1$, and $(3a)^0 = 1$
Negative exponent:	$x^{-n} = \dfrac{1}{x^n} \quad (x \neq 0)$	$3^{-4} = \dfrac{1}{3^4} = \dfrac{1}{81}$, $y^{-7} = \dfrac{1}{y^7}$
Quotient rule for exponents:	$\dfrac{x^m}{x^n} = x^{m-n} \quad (x \neq 0)$	$\dfrac{p^8}{p^3} = p^{8-3} = p^5$
Power rule for exponents:	$(x^m)^n = x^{mn}$	$(a^3)^9 = a^{3 \cdot 9} = a^{27}$
Power rule for products:	$(x^m y^n)^k = x^{mk} y^{nk}$	$(x^4 y^3)^4 = x^{4 \cdot 4} y^{3 \cdot 4} = x^{16} y^{12}$
Power rule for quotients:	$\left(\dfrac{x}{y}\right)^m = \dfrac{x^m}{y^m} \quad (y \neq 0)$	$\left(\dfrac{a^3}{b^4}\right)^8 = \dfrac{a^{3 \cdot 8}}{b^{4 \cdot 8}} = \dfrac{a^{24}}{b^{32}}$
Negative to positive exponent:	$\left(\dfrac{1}{x}\right)^{-n} = x^n \quad (x \neq 0)$	$\left(\dfrac{1}{a^2}\right)^{-n} = (a^2)^n = a^{2n}$
Negative to positive exponent:	$\dfrac{x^{-m}}{y^{-n}} = \dfrac{y^n}{x^m} \quad (x \neq 0)$	$\dfrac{x^{-7}}{y^{-9}} = \dfrac{y^9}{x^7}$

Now let's do an example that requires us to use several of these rules.

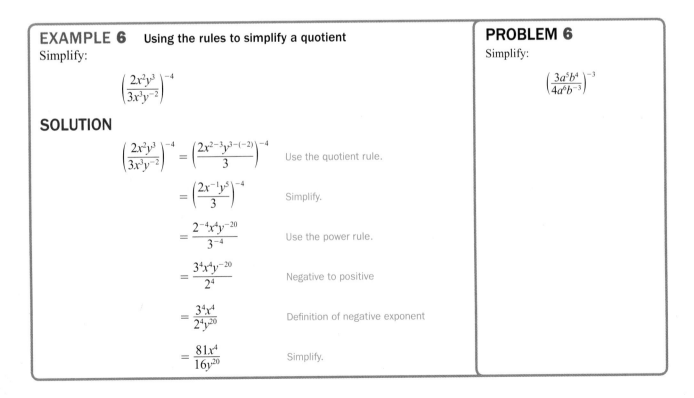

EXAMPLE 6   Using the rules to simplify a quotient

Simplify:

$$\left(\frac{2x^2 y^3}{3x^3 y^{-2}}\right)^{-4}$$

SOLUTION

$$\left(\frac{2x^2 y^3}{3x^3 y^{-2}}\right)^{-4} = \left(\frac{2x^{2-3} y^{3-(-2)}}{3}\right)^{-4}$$ Use the quotient rule.

$$= \left(\frac{2x^{-1} y^5}{3}\right)^{-4}$$ Simplify.

$$= \frac{2^{-4} x^4 y^{-20}}{3^{-4}}$$ Use the power rule.

$$= \frac{3^4 x^4 y^{-20}}{2^4}$$ Negative to positive

$$= \frac{3^4 x^4}{2^4 y^{20}}$$ Definition of negative exponent

$$= \frac{81x^4}{16y^{20}}$$ Simplify.

PROBLEM 6

Simplify:

$$\left(\frac{3a^5 b^4}{4a^6 b^{-3}}\right)^{-3}$$

Let's use exponents in a practical way. What are the costs involved when you are driving a car? Gas, repairs, the car loan payment. What else? Depreciation! The depreciation on a car is the difference between what you paid for the car and what the car is worth now. When you pay P dollars for your car it is then worth 100% of P, but here is what happens to the value as years go by and the car depreciates 10% each year:

Year	Value
0	100% of P
1	90% of $P = (1 - 0.10)P$
2	90% of $(1 - 0.10)P = (1 - 0.10)(1 - 0.10)P$
	$= (1 - 0.10)^2 P$
3	90% of $(1 - 0.10)^2 P = (1 - 0.10)(1 - 0.10)^2 P$
	$= (1 - 0.10)^3 P$

After n years, the value will be

$$(1 - 0.10)^n P$$

But what about x years *ago*? x years ago means $-x$, so the value C of the car x years ago would have been

$$C = (1 - 0.10)^{-x} P$$

There are more examples using exponents in the Using Your Knowledge Exercises.

EXAMPLE 7   **Car depreciation**

Lashonda bought a 3-year-old car for $10,000. If the car depreciates 10% each year:

a. What will be the value of the car in 2 years?

b. What was the price of the car when it was new?

SOLUTION

a. The value in 2 years will be

$$10{,}000(1 - 0.10)^2 = 10{,}000(0.90)^2 = \$8100$$

b. The car was new 3 years ago and it was then worth

$$10{,}000(1 - 0.10)^{-3} = 10{,}000(0.90)^{-3} = \$13{,}717.42$$

PROBLEM 7

a. Find the value of the car in 3 years.

b. Find the value of the car 2 years ago.

> Exercises **4.2**

MathZone+

Boost your grade at mathzone.com!
> Practice Problems
> NetTutor
> Self-Tests
> e-Professors
> Videos

< A >   **Negative Exponents**   In Problems 1–16, write using positive exponents and then simplify.

1. 4^{-2}

2. 2^{-3}

3. 5^{-3}

4. 7^{-2}

5. $\left(\frac{1}{8}\right)^{-2}$

6. $\left(\frac{1}{6}\right)^{-3}$

7. $\left(\frac{1}{x}\right)^{-7}$

8. $\left(\frac{1}{y}\right)^{-6}$

9. $\frac{4^{-2}}{3^{-3}}$

10. $\frac{2^{-4}}{3^{-2}}$

11. $\frac{5^{-2}}{3^{-4}}$

12. $\frac{6^{-2}}{5^{-3}}$

13. $\frac{a^{-5}}{b^{-6}}$

14. $\frac{p^{6}}{q^{5}}$

15. $\frac{x^{-9}}{y^{-9}}$

16. $\frac{t^{-3}}{s^{-3}}$

Answers to PROBLEMS

7. a. $7290   **b.** $12,345.68

⟨ **B** ⟩   **Writing Fractions Using Negative Exponents**   In Problems 17–22, write using negative exponents.

17. $\dfrac{1}{2^3}$ **18.** $\dfrac{1}{3^4}$ **19.** $\dfrac{1}{y^5}$ **20.** $\dfrac{1}{b^6}$

21. $\dfrac{1}{q^5}$ **22.** $\dfrac{1}{t^4}$

⟨ **C** ⟩   **Multiplying and Dividing Expressions with Negative Exponents**   In Problems 23–46, multiply and simplify.
(Remember to write your answers without negative exponents.)

23. $3^5 \cdot 3^{-4}$ **24.** $4^{-6} \cdot 4^8$ **25.** $2^{-5} \cdot 2^7$ **26.** $3^8 \cdot 3^{-5}$

27. $4^{-6} \cdot 4^4$ **28.** $5^{-4} \cdot 5^2$ **29.** $6^{-1} \cdot 6^{-2}$ **30.** $3^{-2} \cdot 3^{-1}$

31. $2^{-4} \cdot 2^{-2}$ **32.** $4^{-1} \cdot 4^{-2}$ **33.** $x^6 \cdot x^{-4}$ **34.** $y^7 \cdot y^{-2}$

35. $y^{-3} \cdot y^5$ **36.** $x^{-7} \cdot x^8$ **37.** $a^3 \cdot a^{-8}$ **38.** $b^4 \cdot b^{-7}$

39. $x^{-5} \cdot x^3$ **40.** $y^{-6} \cdot y^2$ **41.** $x \cdot x^{-3}$ **42.** $y \cdot y^{-5}$

43. $a^{-2} \cdot a^{-3}$ **44.** $b^{-5} \cdot b^{-2}$ **45.** $b^{-3} \cdot b^3$ **46.** $a^6 \cdot a^{-6}$

In Problems 47–60, divide and simplify.

47. $\dfrac{3^4}{3^{-1}}$ **48.** $\dfrac{2^2}{2^{-2}}$ **49.** $\dfrac{4^{-1}}{4^2}$ **50.** $\dfrac{3^{-2}}{3^3}$

51. $\dfrac{y}{y^3}$ **52.** $\dfrac{x}{x^4}$ **53.** $\dfrac{x}{x^{-2}}$ **54.** $\dfrac{y}{y^{-3}}$

55. $\dfrac{x^{-3}}{x^{-1}}$ **56.** $\dfrac{x^{-4}}{x^{-2}}$ **57.** $\dfrac{x^{-3}}{x^4}$ **58.** $\dfrac{y^{-4}}{y^5}$

59. $\dfrac{x^{-2}}{x^{-5}}$ **60.** $\dfrac{y^{-3}}{y^{-6}}$

In Problems 61–70, simplify.

61. $\left(\dfrac{a}{b^3}\right)^2$ **62.** $\left(\dfrac{a^2}{b}\right)^3$ **63.** $\left(\dfrac{-3a}{2b^2}\right)^{-3}$ **64.** $\left(\dfrac{-2a^2}{3b^0}\right)^{-2}$ **65.** $\left(\dfrac{a^{-4}}{b^2}\right)^{-2}$

66. $\left(\dfrac{a^{-2}}{b^3}\right)^{-3}$ **67.** $\left(\dfrac{x^5}{y^{-2}}\right)^{-3}$ **68.** $\left(\dfrac{x^6}{y^{-3}}\right)^{-2}$ **69.** $\left(\dfrac{x^{-4}y^3}{x^5y^5}\right)^{-3}$ **70.** $\left(\dfrac{x^{-2}y^0}{x^7y^2}\right)^{-2}$

⟩ ⟩ ⟩   *Applications*

71. *Watch out for fleas* Do you have a dog or a cat? Sometimes they can get fleas! A flea is 2^{-4} inches long. Write 2^{-4} using positive exponents and as a simplified fraction.

72. *Ants in your pantry!* The smallest ant is about $\frac{1}{25}$ of an inch. Write $\frac{1}{25}$ using positive exponents and using negative exponents.

73. *Micron* A **micron** is a unit of length equivalent to 0.001 millimeters.

   a. Write 0.001 as a fraction.

   b. Write the fraction obtained in part **a** with a denominator that is a power of 10.

   c. Write the fraction obtained in part **b** using negative expo-nents.

   d. Write 0.001 using negative exponents.

74. *Nanometer* A **nanometer** is a *billionth* of a meter.

   a. Write *one billionth* as a fraction

   b. Write the fraction of part **a** with a denominator that is a power of 10.

   c. Write the fraction of part **b** using negative exponents.

   d. Write *one billionth* using negative exponents.

⟩⟩⟩ *Using Your Knowledge*

Exponential Growth The idea of exponents can be used to measure population growth. Thus, if we assume that the world population is increasing about 2% each year (experts say the rate is between 1% and 3%), we can predict the population of the world next year by *multiplying* the present world population by 1.02 (100% + 2% = 102% = 1.02). If we let the world population be P, we have

$$\text{Population in 1 year} = 1.02P$$

$$\text{Population in 2 years} = 1.02(1.02P) = (1.02)^2 P$$

$$\text{Population in 3 years} = 1.02(1.02)^2 P = (1.02)^3 P$$

75. If the population P in 2000 was 6 billion people, what would it be in 2 years—that is, in 2002? (Round your answer to three decimal places.)

76. What will the population be 5 years after 2000? (Round your answer to three decimal places.)

To find the population 1 year from now, we multiply by 1.02. What should we do to find the population 1 year *ago?* We divide by 1.02. Thus, if the population today is P,

$$\text{Population 1 year ago} \;= \frac{P}{1.02} = P \cdot 1.02^{-1}$$

$$\text{Population 2 years ago} = \frac{P \cdot 1.02^{-1}}{1.02} = P \cdot 1.02^{-2}$$

$$\text{Population 3 years ago} = \frac{P \cdot 1.02^{-2}}{1.02} = P \cdot 1.02^{-3}$$

77. If the population P in 2000 was 6 billion people, what was it in 1990? (Round your answer to three decimal places.)

78. What was the population in 1985? (Round your answer to three decimal places.)

Do you know what inflation is? It is the tendency for prices and wages to rise, making your money worth less today than in the future. The formula for the cost C of an item n years from now if the present value (value now) is P dollars and the inflation rate is $r\%$ is

$$\boxed{C = P(1 + r)^n}$$

79. In 2002, the cheapest Super Bowl tickets were $400. If we assume a 3% (0.03) inflation rate, how much would tickets cost:

  a. in 2003

  b. in 2007

80. In 1967 a Super Bowl ticket was $12. Assuming a 3% inflation rate:

  a. How much should the tickets cost 35 years later? Is the cost more or less than the actual $400 price?

  b. If a ticket is $400 now, how much should it have been 35 years ago?

81. In 1967 the Consumer Price Index (CPI) was 35. Thirty-five years later, it was 180, so ticket prices should be scaled up by a factor of $\frac{180}{35} \approx 5.14$. Using this factor, how much should a $12 ticket price be 35 years later?

82. The average annual cost for tuition and fees at a 4-year public college is about $20,000 and rising at a 3.5% rate. How much would you have to pay in tuition and fees 4 years from now?

83. The average price of a house is $170,000. If the inflation rate is 2%, what would the average price be in 3 years?

84. If the average price of a house now is $170,000 and the inflation rate is 2%, what was the average price 3 years ago?

85. A credit card account is a loan for which you pay at least 9% annually. If you purchase $100 with your card, make no payments for 4 months, and your rate is 1.5% a month, how much would you owe at the end of the 4 months?

86. The number of bacteria in a sample after n hours growing at a 50% rate is modeled by the equation

$$100(1 + 0.50)^n$$

  a. What would the number of bacteria be after 6 hours?

  b. What was the number of bacteria 2 hours ago?

87. On a Monday morning, you count 40 ants in your room. On Tuesday, you find 60. By Wednesday, they number 90. If the population continues to grow at this rate:

  a. What is the percent rate of growth?

  b. Write an equation of the form $P(1 + r)^n$ that models the number A of ants after n days.

  c. How many ants would you expect on Friday?

  d. How many ants would you expect in 2 weeks (14 days)?

  e. Name at least one factor that would cause the ant population to slow down its growth.

88. Referring to Problem 87, how many ants (to the nearest whole number) were there on the previous Sunday? On the previous Saturday?

⟩⟩⟩ *Write On*

89. Give three different explanations for why $x^0 = 1$ $(x \neq 0)$.

90. By definition, if n is a positive integer, then

$$x^{-n} = \frac{1}{x^n} \quad (x \neq 0)$$

  a. Does this rule hold if n is *any* integer? Explain and give some examples.

  b. Why do we have to state $x \neq 0$ in this definition?

91. Does $x^{-2} + y^{-2} = (x + y)^{-2}$? Explain why or why not.

92. Does

$$x^{-1} + y^{-1} = \frac{1}{x + y}?$$

  Explain why or why not.

⟩⟩⟩ *Concept Checker*

Fill in the blank(s) with the correct word(s), phrase, or mathematical statement.

93. For any nonzero number x, $x^0 = $ _____.

94. If n is a **positive integer**, $x^{-n} = $ _____.

95. If n is a **positive integer**, $\left(\frac{1}{x}\right)^{-n} = $ _____.

96. For any **nonzero numbers** x and y and any **positive integers** m and n, $\frac{x^{-m}}{y^{-n}} = $ _____.

97. If m and n are **integers**, $x^m \cdot x^n = $ _____.

98. If m and n are **integers**, $\frac{x^m}{x^n} = $ _____.

x^{-n}	x^{m+n}
x^n	n
$\dfrac{y^n}{x^m}$	x^{n-m}
$\dfrac{x^n}{y^n}$	x^{m-n}
x^{mn}	$\dfrac{1}{x^n}$
1	

⟩⟩⟩ *Mastery Test*

Write using positive exponents and simplify:

99. $\dfrac{7^{-2}}{6^{-2}}$

100. $\dfrac{p^{-3}}{q^{-7}}$

101. $\left(\frac{1}{9}\right)^{-2}$

102. $\left(\frac{1}{r}\right)^{-4}$

Write using negative exponents:

103. $\dfrac{1}{7^6}$

104. $\dfrac{1}{w^5}$

Simplify and write the answer with positive exponents:

105. $5^6 \cdot 5^{-4}$

106. $x^{-3} \cdot x^{-7}$

107. $\dfrac{z^{-5}}{z^{-7}}$

108. $\dfrac{r}{r^8}$

109. $\left(\dfrac{2xy^3}{3x^3y^{-3}}\right)^{-4}$

110. $\left(\dfrac{3x^{-2}y^{-3}}{2x^3y^{-2}}\right)^{-5}$

111. The price of a used car is $5000. If the car depreciates (loses its value) by 10% each year:

   a. What will the value of the car be in 3 years?

   b. What was the value of the car 2 years ago.

〉〉〉 *Skill Checker*

Find:

112. 7.31×10^1

113. 8.39×10^2

114. 7.314×10^3

115. 8.16×10^{-2}

116. 3.15×10^{-3}

4.3   Application of Exponents: Scientific Notation

▶ Objectives

A 〉 Write numbers in scientific notation.

B 〉 Multiply and divide numbers in scientific notation.

C 〉 Solve applications involving scientific notation.

▶ To Succeed, Review How To . . .

1. Use the rules of exponents (p. 324).
2. Multiply and divide real numbers (pp. 60–61, 62–64).

▶ Getting Started
Sun Facts

How many facts do you know about the sun? Here is some information taken from an encyclopedia.

Mass: 2.19×10^{27} tons

Temperature: 1.8×10^6 degrees Fahrenheit

Energy generated per minute: 2.4×10^4 horsepower

All the numbers here are written as products of a number between 1 and 10 and an appropriate power of 10. This is called *scientific notation*. When written in standard notation, these numbers are

2,190,000,000,000,000,000,000,000,000

1,800,000

24,000

It's easy to see why so many technical fields use scientific notation. In this section we shall learn how to write numbers in scientific notation and how to perform multiplications and divisions using these numbers.

A 〉 Scientific Notation

We define scientific notation as follows:

SCIENTIFIC NOTATION	A number in **scientific notation** is written as $$M \times 10^n$$ where *M* is a number between **1** and **10** and *n* is an integer.

How do we write a number in scientific notation? First, recall that when we *multiply* a number by a power of 10 ($10^1 = 10$, $10^2 = 100$, and so on), we simply move the decimal point as many places to the *right* as indicated by the exponent of 10. Thus,

$$7.31 \times 10^1 = 7.31 = 73.1$$ Exponent 1; move decimal point 1 place right.

$$72.813 \times 10^2 = 72.813 = 7281.3$$ Exponent 2; move decimal point 2 places right.

$$160.7234 \times 10^3 = 160.7234 = 160{,}723.4$$ Exponent 3; move decimal point 3 places right.

On the other hand, if we *divide* a number by a power of 10, we move the decimal point as many places to the *left* as indicated by the exponent of 10. Thus,

$$\frac{7}{10} = 0.7 = 7 \times 10^{-1}$$

$$\frac{8}{100} = 0.08 = 8 \times 10^{-2}$$

and

$$\frac{4.7}{100{,}000} = 0.000047 = 4.7 \times 10^{-5}$$

Remembering the following procedure makes it easy to write a number in scientific notation.

PROCEDURE

Writing a Number in Scientific Notation ($M \times 10^n$)

1. Move the decimal point in the given number so that there is only one nonzero digit to its left. The resulting number is *M*.
2. Count the number of places you moved the decimal point in step 1. If the decimal point was moved to the *left*, *n* is *positive;* if it was moved to the *right*, *n* is *negative*.
3. Write $M \times 10^n$.

For example,

$$5.3 = 5.3 \times 10^0$$ The decimal point in 5.3 must be moved 0 places to get 5.3.

$$87 = 8.7 \times 10^1 = 8.7 \times 10$$ The decimal point in 87 must be moved 1 place *left* to get 8.7.

$$68{,}000 = 6.8 \times 10^4$$ The decimal point in 68,000 must be moved 4 places *left* to get 6.8.

$$0.49 = 4.9 \times 10^{-1}$$ The decimal point in 0.49 must be moved 1 place *right* to get 4.9.

$$0.072 = 7.2 \times 10^{-2}$$ The decimal point in 0.072 must be moved 2 places *right* to get 7.2.

> **NOTE**
>
> After completing step 1 in the procedure, decide whether you should make the number obtained *larger* (*n* positive) or *smaller* (*n* negative).

EXAMPLE 1   Writing a number in scientific notation

The approximate distance to the sun is 93,000,000 miles, and the wavelength of its ultraviolet light is 0.000035 centimeter. Write 93,000,000 and 0.000035 in scientific notation.

SOLUTION

$$93{,}000{,}000 = 9.3 \times 10^7$$

$$0.000035 = 3.5 \times 10^{-5}$$

PROBLEM 1

The approximate distance to the moon is 239,000 miles and its mass is 0.012 that of the Earth. Write 239,000 and 0.012 in scientific notation.

EXAMPLE 2   Changing scientific notation to standard notation

A jumbo jet weighs 7.75×10^5 pounds, whereas a house spider weighs 2.2×10^{-4} pound. Write these weights in standard notation.

7.75×10^5

2.2×10^{-4}

SOLUTION

$7.75 \times 10^5 = 775{,}000$ To multiply by 10^5, move the decimal point 5 places right.

$2.2 \times 10^{-4} = 0.00022$ To multiply by 10^{-4}, move the decimal point 4 places left.

PROBLEM 2

In an average year, Carnival Cruise™ line puts more than 10.1×10^6 mints, each weighing 5.25×10^{-2} ounce, on guest's pillows. Write 10.1×10^6 and 5.25×10^{-2} in standard notation.

B ⟩ Multiplying and Dividing Using Scientific Notation

Consider the product $300 \cdot 2000 = 600{,}000$. In scientific notation, we would write

$$(3 \times 10^2) \cdot (2 \times 10^3) = 6 \times 10^5$$

To find the answer, we can multiply 3 by 2 to obtain 6 and multiply 10^2 by 10^3, obtaining 10^5. To multiply numbers in scientific notation, we proceed in a similar manner; here's the procedure.

> **PROCEDURE**
>
> **Multiplying Using Scientific Notation**
>
> 1. Multiply the decimal parts first and write the result in scientific notation.
> 2. Multiply the powers of 10 using the product rule.
> 3. The answer is the product of the numbers obtained in steps 1 and 2 after simplification.

EXAMPLE 3   Multiplying numbers in scientific notation

Multiply:

a. $(5 \times 10^3) \times (8.1 \times 10^4)$ **b.** $(3.2 \times 10^2) \times (4 \times 10^{-5})$

SOLUTION

a. We multiply the decimal parts first, then write the result in scientific notation.

$$5 \times 8.1 = 40.5 = 4.05 \times 10$$

PROBLEM 3

Multiply:

a. $(6 \times 10^4) \times (5.2 \times 10^5)$

b. $(3.1 \times 10^3) \times (5 \times 10^{-6})$

(continued)

Answers to PROBLEMS

1. 2.39×10^5; 1.2×10^{-2} **2.** 10,100,000; 0.0525   **3. a.** 3.12×10^{10} **b.** 1.55×10^{-2}

Next we multiply the powers of 10.

$$10^3 \times 10^4 = 10^7 \quad \text{Add exponents 3 and 4 to obtain 7.}$$

The answer is $(4.05 \times 10) \times 10^7$, or 4.05×10^8.

b. Multiply the decimals and write the result in scientific notation.

$$3.2 \times 4 = 12.8 = 1.28 \times 10$$

Multiply the powers of 10.

$$10^2 \times 10^{-5} = 10^{2-5} = 10^{-3}$$

The answer is $(1.28 \times 10) \times 10^{-3}$, or $1.28 \times 10^{1+(-3)} = 1.28 \times 10^{-2}$.

Division is done in a similar manner. For example,

$$\frac{3.2 \times 10^5}{1.6 \times 10^2}$$

is found by dividing 3.2 by 1.6 (yielding 2) and 10^5 by 10^2, which is 10^3. The answer is 2×10^3.

EXAMPLE 4 **Dividing numbers in scientific notation**
Find the quotient: $(1.24 \times 10^{-2}) \div (3.1 \times 10^{-3})$

SOLUTION First divide 1.24 by 3.1 to obtain $0.4 = 4 \times 10^{-1}$. Now divide powers of 10:

$$10^{-2} \div 10^{-3} = 10^{-2-(-3)}$$
$$= 10^{-2+3}$$
$$= 10^1$$

The answer is $(4 \times 10^{-1}) \times 10^1 = 4 \times 10^0 = 4$.

PROBLEM 4

Find the quotient:

$(1.23 \times 10^{-3}) \div (4.1 \times 10^{-4})$

C > Applications Involving Scientific Notation

Applications involving scientific notation are common because large and small numbers are used in many different fields of study such as astronomy.

EXAMPLE 5 **Energy from the sun**
The total energy received from the sun each minute is 1.02×10^{19} calories. Since the area of the Earth is 5.1×10^{18} square centimeters, the amount of energy received per square centimeter of the Earth's surface every minute (the solar constant) is

$$\frac{1.02 \times 10^{19}}{5.1 \times 10^{18}}$$

Simplify this expression.

SOLUTION Dividing 1.02 by 5.1, we obtain $0.2 = 2 \times 10^{-1}$. Now, $10^{19} \div 10^{18} = 10^{19-18} = 10^1$. Thus, the final answer is

$$(2 \times 10^{-1}) \times 10^1 = 2 \times 10^0 = 2$$

This means that the Earth receives about 2 calories of energy per square centimeter each minute.

PROBLEM 5

The population density for a country is the number of people per square mile. Monaco's population density is

$$\frac{3.3 \times 10^4}{7.5 \times 10^{-1}}$$

Simplify this expression.

Now, let's talk about more "earthly" matters.

Answers to PROBLEMS

4. 3   **5.** 44,000/mi²

EXAMPLE 6   Printing money

In a recent year, the Treasury Department reported printing the following amounts of money in the specified denominations:

$3,500,000,000 in $1 bills $2,160,000,000 in $20 bills
$1,120,000,000 in $5 bills $250,000,000 in $50 bills
$640,000,000 in $10 bills $320,000,000 in $100 bills

a. Write these numbers in scientific notation.

b. Determine how much money was printed (in billions).

SOLUTION

a. $3,500,000,000 = 3.5 \times 10^9$
$1,120,000,000 = 1.12 \times 10^9$
$640,000,000 = 6.4 \times 10^8$
$2,160,000,000 = 2.16 \times 10^9$
$250,000,000 = 2.5 \times 10^8$
$320,000,000 = 3.2 \times 10^8$

b. Since we have to write all the quantities in billions (a billion is 10^9), we have to write all numbers using 9 as the exponent. First, let's consider 6.4×10^8. To write this number with an exponent of 9, we write

$$6.4 \times 10^8 = (0.64 \times 10) \times 10^8 = 0.64 \times 10^9$$

Similarly,

$$2.5 \times 10^8 = (0.25 \times 10) \times 10^8 = 0.25 \times 10^9$$

and

$$3.2 \times 10^8 = (0.32 \times 10) \times 10^8 = 0.32 \times 10^9$$

Writing the other numbers, we get   Add the entire column.

$3.5 \ \times 10^9$
1.12×10^9
2.16×10^9
$\overline{7.99 \times 10^9}$

Thus, 7.99 billion dollars were printed.

EXAMPLE 7   New planets and scientific notation

In 2006 the International Astronomical Union decided to reclassify Pluto and the newly discovered Eris as "dwarf" planets. Pluto and Eris are now "dwarfs." What makes these planets "dwarf"? Clearly, their size! Here are the sizes of the diameter of the four giant planets—Jupiter, Saturn, Uranus, Neptune—and the two "dwarf" planets.

Jupiter 8.8736×10^4 miles
Neptune 3.0775×10^4 miles
Saturn 7.4978×10^4 miles
Uranus 3.2193×10^4 miles
Pluto 1.422×10^3 miles
Eris 1.490×10^3 miles

a. Which of the two dwarf planets is bigger?

b. Which planet is bigger, Jupiter or Saturn?

PROBLEM 6

How much money was minted in coins? The amounts of money minted in the specified denominations are as follows:

Pennies: $1,025,740,000
Nickels: $66,183,600
Dimes: $233,530,000
Quarters: $466,850,000
Half-dollars: $15,355,000

a. Write these numbers in scientific notation.

b. Determine how much money was minted (in billions).

PROBLEM 7

a. Ceres is a dwarf planet with a diameter of 5.80×10^2 miles. Is Ceres bigger or smaller than Eris?

b. Which planet is bigger, Neptune or Uranus?

Photo of Eris with the sun in the background.

(continued)

SOLUTION

a. To find the bigger dwarf planet, we compare the diameters of Pluto and Eris: 1.422×10^3 miles and 1.490×10^3 miles.

To do this,

1. Compare the exponents.
2. Compare the decimal parts.

Both have the same exponents (3), but one of the decimal parts (1.490 the decimal part of Eris) is larger, so we can conclude that Eris is larger than Pluto.

b. Again the exponents for Jupiter and Saturn are the same (4) but the decimal part of Jupiter (8.8736) is larger, so Jupiter is larger.

Read more about the "demotion" of Pluto and the new planets at: http://www.gps.caltech.edu/~mbrown/eightplanets/.

> Exercises **4.3**

< A > **Scientific Notation** In Problems 1–10, write the given number in scientific notation.

1. 55,000,000 (working women in the United States)

2. 69,000,000 (working men in the United States)

3. 300,000,000 (U.S. population now)

4. 309,000,000 (estimated U.S. population in the year 2010)

5. 1,900,000,000 (dollars spent on water beds and accessories in 1 year)

6. 0.035 (ounces in a gram)

7. 0.00024 (probability of four-of-a-kind in poker)

8. 0.000005 (the gram-weight of an amoeba)

9. 0.000000002 (the gram-weight of one liver cell)

10. 0.00000009 (wavelength of an X ray in centimeters)

In Problems 11–20, write the given number in standard notation.

11. 1.53×10^2 (pounds of meat consumed per person per year in the United States)

12. 5.96×10^2 (pounds of dairy products consumed per person per year in the United States)

13. 171×10^6 (fresh bagels produced per year in the United States)

14. 2.01×10^6 (estimated number of jobs created in service industries between now and the year 2010)

15. 6.85×10^9 (estimated worth, in dollars, of the five wealthiest women)

16. 1.962×10^{10} (estimated worth, in dollars, of the five wealthiest men)

17. 2.3×10^{-1} (kilowatts per hour used by your TV)

18. 4×10^{-2} (inches in 1 millimeter)

19. 2.5×10^{-4} (thermal conductivity of glass)

20. 4×10^{-11} (energy, in joules, released by splitting one uranium atom)

< B > **Multiplying and Dividing Using Scientific Notation** In Problems 21–30, perform the indicated operations and give your answer in scientific notation.

21. $(3 \times 10^4) \times (5 \times 10^5)$

22. $(5 \times 10^2) \times (3.5 \times 10^3)$

23. $(6 \times 10^{-3}) \times (5.1 \times 10^6)$

24. $(3 \times 10^{-2}) \times (8.2 \times 10^5)$

25. $(4 \times 10^{-2}) \times (3.1 \times 10^{-3})$

26. $(3.1 \times 10^{-3}) \times (4.2 \times 10^{-2})$

27. $\dfrac{4.2 \times 10^5}{2.1 \times 10^2}$

28. $\dfrac{5 \times 10^6}{2 \times 10^3}$

29. $\dfrac{2.2 \times 10^4}{8.8 \times 10^6}$

30. $\dfrac{2.1 \times 10^3}{8.4 \times 10^5}$

〈 C 〉 Applications Involving Scientific Notation

31. *Vegetable consumption*   The average American eats about 170 pounds of vegetables each year. Since there are about 300 million Americans, the number of pounds of vegetables consumed each year should be $(1.7 \times 10^2) \times (3 \times 10^8)$.

    a. Write this number in scientific notation.

    b. Write this number in standard notation.

33. *Garbage production*   America produces 236.2 million tons of garbage each year. Since a ton is 2000 pounds, and there are about 360 days in a year and 300 million Americans, the number of pounds of garbage produced each day of the year for each man, woman, and child in America is

$$\frac{(2.362 \times 10^8) \times (2 \times 10^3)}{(3 \times 10^8) \times (3.6 \times 10^2)}$$

Write this number in standard notation.

35. *Nuclear fission*   Nuclear fission is used as an energy source. Do you know how much energy a gram of uranium-235 gives? The answer is

$$\frac{4.7 \times 10^9}{235} \text{ kilocalories}$$

Write this number in scientific notation.

37. *Internet e-mail projections*   An Internet travel site called Travelocity® sends about 730 million e-mails a year, and that number is projected to grow to 4.38 billion per year. If a year has 365 days, how many e-mails per day is Travelocity sending now, and how many are they projecting to send later?

    Source: Iconocast.com.

39. *E-mail proliferation*   In 2006, about 300 million people in the United States will be sending 300 million e-mails every day. How many e-mails will each person be sending per year?

40. *Planets*   Do you know how many "giant" planets we have? Here are the distances from the sun (in kilometers) of the four giant planets—Jupiter, Neptune, Saturn, and Uranus

       Jupiter 7.78×10^8 km

       Neptune 4.50×10^9 km

       Saturn 1.43×10^9 km

       Uranus 2.87×10^9 km

    a. Which planet is closest to the sun?

    b. Which planet is farthest from the sun?

    c. Order the four planets from least to greatest distance from the sun.

41. *Terrestrial planets*   Do you know the terrestrial planets? Why are they called terrestrial? Here are the distances from the sun (in kilometers) of the four *terrestrial* planets (planets composed primarily of rock and metal)—Earth, Mars, Mercury, and Venus.

       Earth 1.55×10^8 km

       Mars 2.28×10^8 km

       Mercury 5.80×10^7 km

       Venus 1.03×10^8 km

32. *Soft drink consumption*   The average American drinks 52.4 gallons of soft drinks each year. Since there are about 300 million Americans, the number of gallons of soft drinks consumed each year should be $(5.24 \times 10^1) \times (3 \times 10^8)$.

    a. Write this number in scientific notation.

    b. Write this number in standard notation.

34. *Velocity of light*   The velocity of light can be measured by dividing the distance from the sun to the Earth (1.47×10^{11} meters) by the time it takes for sunlight to reach the Earth (4.9×10^2 seconds). Thus, the velocity of light is

$$\frac{1.47 \times 10^{11}}{4.9 \times 10^2}$$

How many meters per second is that?

36. *U.S. national debt*   The national debt of the United States is about $\$8.5 \times 10^{12}$ (about \$8.5 trillion). If we assume that the U.S. population is 300 million, each citizen actually owes $\frac{8.5 \times 10^{12}}{3 \times 10^8}$ dollars. How much money is that? Approximate the answer to the nearest cent. You can see the answer for this very minute at http://www.brillig.com/debt_clock/.

38. *Product information e-mails*   Every year, 18.25 billion e-mails requesting product information or service inquiries are sent. If a year has 365 days, how many e-mails a day is that?

    Source: Warp 9, Inc.

    a. Which planet is closest to the sun?

    b. Which planet is farthest from the sun?

    c. Order the four planets from least to greatest distance from the sun.

42. *Pluto demoted!*   In August 2006, the International Astronomical Union revised the definition of a planet. Under this new classification Eris, Pluto, and Ceres were classified as "dwarf" planets. The distance from the sun (in kilometers) of each of these planets is shown:

Ceres	4.14×10^8 km
Eris	1.45×10^{10} km
Pluto	5.90×10^9 km

   a. Which planet is closest to the sun?

   b. Which planet is farthest from the sun?

   c. Order the three planets from least to greatest distance from the sun.

44. *Terrestrial planets*   The mass of the *terrestrial* planets (in kilograms) is given:

Earth	5.98×10^{24} kg
Mars	6.42×10^{23} km
Mercury	3.30×10^{23} km
Venus	4.87×10^{24} km

43. *Giant planets*   The masses of the "giant" planets (in kilograms) are given:

Jupiter	1.90×10^{27} kg
Neptune	1.02×10^{26} kg
Saturn	5.69×10^{26} km
Uranus	8.68×10^{25} km

   a. Which planet is heaviest?

   b. Which planet is lightest?

   c. Order the four planets from lightest to heaviest.

---

   a. Which planet is heaviest?

   b. Which planet is lightest?

   c. Order the four planets from lightest to heaviest.

〉〉〉 Using Your Knowledge

Astronomical Quantities   As we have seen, scientific notation is especially useful when very large quantities are involved. Here's another example. In astronomy, we find that the speed of light is 299,792,458 meters per second.

45. Write 299,792,458 in scientific notation.

Astronomical distances are so large that they are measured in astronomical units (AU). An astronomical unit is defined as the average separation (distance) of the Earth and the sun—that is, 150,000,000 kilometers.

46. Write 150,000,000 in scientific notation.

47. Distances in astronomy are also measured in *parsecs:* 1 parsec = 2.06×10^5 AU. Thus, 1 parsec = $(2.06 \times 10^5) \times (1.5 \times 10^8)$ kilometers. Written in scientific notation, how many kilometers is that?

48. Astronomers also measure distances in *light-years,* the distance light travels in 1 year: 1 light-year = 9.46×10^{12} kilometers. The closest star, Proxima Centauri, is 4.22 light-years away. In scientific notation using two decimal places, how many kilometers is that?

49. Since 1 parsec = 3.09×10^{13} kilometers (see Problem 47) and 1 light-year = 9.46×10^{12} kilometers, the number of light-years in a parsec is

$$\frac{3.09 \times 10^{13}}{9.46 \times 10^{12}}$$

Write this number in standard notation rounded to two decimal places.

〉〉〉 Write On

50. Explain why the procedure used to write numbers in scientific notation works.

51. What are the advantages and disadvantages of writing numbers in scientific notation?

〉〉〉 Concept Checker

Fill in the blank(s) with the correct word(s), phrase, or mathematical statement.

52. A number is written in **scientific notation** if it is written in the form _____.

53. When a number is written in **scientific notation** in the form $M \times 10^n$, the M is a number between 1 and _____ and n is a(n) _____.

54. The **first** step in **multiplying** numbers in **scientific notation** is to **multiply** the _____ **parts.**

55. The **second** step in multiplying numbers in **scientific notation** is to **multiply** the _____ **of 10** using the product rule.

0	**M × 10ⁿ**
10	**10ᴹ**
whole number	**decimal**
integer	**powers**

⟩⟩⟩ *Mastery Test*

56. The width of the asteroid belt is 2.8×10^8 kilometers. The speed of *Pioneer 10,* a U.S. space vehicle, in passing through this belt was 1.4×10^5 kilometers per hour. Thus, *Pioneer 10* took

$$\frac{2.8 \times 10^8}{1.4 \times 10^5}$$

hours to go through the belt. How many hours is that in scientific notation?

57. The distance to the moon is about 289,000 miles, and its mass is 0.12456 that of the Earth. Write these numbers in scientific notation.

58. The Concorde (a supersonic passenger plane) weighed 4.08×10^5 pounds, and a cricket weighs 3.125×10^{-4} pound. Write these weights in standard notation.

Perform the operations and write the answer in scientific and standard notation.

59. $(2.52 \times 10^{-2}) \div (4.2 \times 10^{-3})$

60. $(4.1 \times 10^2) \times (3 \times 10^{-5})$

61. $(6 \times 10^4) \times (2.2 \times 10^3)$

62. $(3.2 \times 10^{-2}) \div (1.6 \times 10^{-5})$

⟩⟩⟩ *Skill Checker*

Find:

63. $-16(1)^2 + 118$

64. $-16(2)^2 + 118$

65. $-8(3)^2 + 80$

66. $3(2^2) - 5(3) + 8$

67. $-4 \cdot 8 \div 2 + 20$

68. $-5 \cdot 6 \div 2 + 25$

4.4   Polynomials: An Introduction

▶ Objectives

A ⟩ Classify polynomials.

B ⟩ Find the degree of a polynomial.

C ⟩ Write a polynomial in descending order.

D ⟩ Evaluate polynomials.

▶ To Succeed, Review How To . . .

1. Evaluate expressions (pp. 61–62, 69–72).
2. Add, subtract, and multiply expressions (pp. 78, 88–89, 309–311).

▶ Getting Started

A Diving Polynomial

The diver in the photo jumped from a height of 118 feet. Do you know how many feet above the water the diver will be after t seconds? Scientists have determined a formula for finding the answer:

$$-16t^2 + 118 \quad \text{(feet above the water)}$$

The expression $-16t^2 + 118$ is an example of a *polynomial.* Here are some other polynomials:

$$5x, \quad 9x - 2, \quad -5t^2 + 18t - 4,$$
$$\text{and} \quad y^5 - 2y^2 + \frac{4}{5}y - 6$$

We construct these polynomials by adding or subtracting products of numbers and variables raised to whole-number exponents. Of course, if we use any other operations, the result may not be a polynomial. For example,

$$x^2 - \frac{3}{x} \quad \text{and} \quad x^{-7} + 4x$$

are *not* polynomials (we divided by the variable x in the first one and used negative exponents in the second one). In this section we shall learn how to classify polynomials in one variable, find their degrees, write them in descending order, and evaluate them.

A ⟩ Classifying Polynomials

Polynomials can be used to track and predict the amount of waste (in millions of tons) generated annually in the United States. The polynomial approximating this amount is:

$$-0.001t^3 + 0.06t^2 + 2.6t + 88.6$$

where t is the number of years after 1960. But how do we *predict* how much waste will be generated in the year 2010 using this polynomial? We will show you in Example 6! First, let's look at the definition of polynomials.

POLYNOMIAL	A **polynomial** is an algebraic expression formed by using the operations of addition and subtraction on products of numbers and variables raised to whole-number exponents.

The parts of a polynomial separated by plus signs are called the **terms** of the polynomial. If there are subtraction signs, we can rewrite the polynomial using addition signs, since we know that $a - b = a + (-b)$. Thus,

$5x$ has one term: $\qquad\qquad\qquad$ $5x$

$9x - 2$ has two terms: $\qquad\qquad$ $9x$ and -2 $\qquad$ Recall that $9x - 2 = 9x + (-2)$.

$-5t^2 + 18t - 4$ has three terms: $\quad -5t^2, 18t,$ and -4 $\quad$ $-5t^2 + 18t - 4 = -5t^2 + 18t + (-4)$

Polynomials are classified according to the number of terms they have. Thus,

$5x$ has *one* term; it is called a *monomial*. $\qquad\qquad$ *mono* $\quad$ means one.

$9x - 2$ has *two* terms; it is called a *binomial*. $\qquad\quad$ *bi* $\qquad$ means two.

$-5t^2 + 18t - 4$ has *three* terms; it is called a *trinomial*. $\quad$ *tri* $\qquad$ means three.

> **NOTE**
> **1.** A polynomial of one term is a **monomial;**
> **2.** A polynomial of two terms, a **binomial;**
> **3.** A polynomial of three terms is a **trinomial;**
> **4.** A polynomial of more than three terms is just a **polynomial.**

EXAMPLE 1 Classifying polynomials

Classify each of the following polynomials as a monomial, binomial, or trinomial.

a. $6x - 1$ $\qquad$ **b.** -8 $\qquad$ **c.** $-4 + 3y - y^2$ $\qquad$ **d.** $5(x + 2) - 3$

SOLUTION

a. $6x - 1$ has two terms; it is a binomial.

b. -8 has only one term; it is a monomial.

c. $-4 + 3y - y^2$ has three terms; it is a trinomial.

d. $5(x + 2) - 3$ is a binomial. Note that $5(x + 2)$ is *one* term.

PROBLEM 1

Classify as monomial, binomial, or trinomial.

a. -5

b. $-3 + 4y + 6y^2$

c. $8x - 3$

d. $8(x + 9) - 3(x - 1)$

Answers to **PROBLEMS**

1. a. Monomial $\quad$ **b.** Trinomial

$\quad$ **c.** Binomial $\quad$ **d.** Binomial

B › Finding the Degree of a Polynomial

All the polynomials we have seen contain only one variable and are called *polynomials in one variable*. Polynomials in one variable, such as $x^2 + 3x - 7$, can also be classified according to the *highest* exponent of the variable. The highest exponent of the variable is called the **degree** of the polynomial. To find the degree of a polynomial, you simply examine each term and find the highest exponent of the variable. Thus, the degree of $3x^2 + 5x^4 - 2$ is found by looking at the exponent of the variable in each of the terms.

The exponent in $3x^2$ is 2.

The exponent in $5x^4$ is 4.

The exponent in -2 is 0 because $-2 = -2x^0$.   (Recall that $x^0 = 1$.)

Thus, the degree of $3x^2 + 5x^4 - 2$ is 4, the highest exponent of the variable in the polynomial. Similarly, the degree of $4y^3 - 3y^5 + 9y^2$ is 5, since 5 is the highest exponent of the variable present in the polynomial. By convention, a number such as -4 or 7 is called a **polynomial of degree 0,** because if $a \neq 0$, $a = ax^0$. Thus, $-4 = -4x^0$ and $7 = 7x^0$ are polynomials of degree 0. The number 0 itself is called the **zero polynomial** and is *not* assigned a degree. (Note that $0 \cdot x^1 = 0$; $0 \cdot x^2 = 0$, $0 \cdot x^3 = 0$, and so on, so the zero polynomial cannot have a degree.)

EXAMPLE 2   Finding the degree of a polynomial

Find the degree:

a. $-2t^2 + 7t - 2 + 9t^3$ **b.** 8   **c.** $-3x + 7$ **d.** 0

SOLUTION

a. The highest exponent of the variable t in the polynomial $-2t^2 + 7t - 2 + 9t^3$ is 3; thus, the degree of the polynomial is 3.

b. The degree of 8 is, by convention, 0.

c. Since $x = x^1$, $-3x + 7$ can be written as $-3x^1 + 7$, making the degree of the polynomial 1.

d. 0 is the zero polynomial; it does not have a degree.

PROBLEM 2

Find the degree:

a. 9   **b.** $-5z^2 + 2z - 8$

c. 0   **d.** $-8y + 1$

C › Writing a Polynomial in Descending Order

The degree of a polynomial is easier to find if we agree to write the polynomial in **descending order;** that is, the term with the *highest* exponent is written *first*, the *second* highest is *next,* and so on. Fortunately, the associative and commutative properties of addition permit us to do this rearranging! Thus, instead of writing $3x^2 - 5x^3 + 4x - 2$, we rearrange the terms and write $-5x^3 + 3x^2 + 4x - 2$ with exponents in the terms arranged in *descending* order. Similarly, to write $-3x^3 + 7 + 5x^4 - 2x$ in descending order, we use the associative and commutative properties and write $5x^4 - 3x^3 - 2x + 7$. Of course, it would not be incorrect to write this polynomial in ascending order (or with no order at all); it is just that we *agree* to write polynomials in descending order for uniformity and convenience.

EXAMPLE 3   Writing polynomials in descending order

Write in descending order:

a. $-9x + x^2 - 17$ **b.** $-5x^3 + 3x - 4x^2 + 8$

PROBLEM 3

Write in descending order:

a. $-4x^2 + 3x^3 - 8 + 2x$

b. $-3y + y^2 - 1$

(continued)

Answers to PROBLEMS

2. a. 0   **b.** 2   **c.** No degree   **d.** 1

3. a. $3x^3 - 4x^2 + 2x - 8$ **b.** $y^2 - 3y - 1$

SOLUTION

a. $-9x + x^2 - 17 = x^2 - 9x - 17$

b. $-5x^3 + 3x - 4x^2 + 8 = -5x^3 - 4x^2 + 3x + 8$

D ⟩ Evaluating Polynomials

Now, let's return to the diver in the *Getting Started.* You may be wondering why his height above the water after t seconds was $-16t^2 + 118$ feet. This expression doesn't even look like a number! But polynomials represent numbers when they are *evaluated.* So, if our diver is $-16t^2 + 118$ feet above the water after t seconds, then after 1 second (that is, when $t = 1$), our diver will be

$$-16(1)^2 + 118 = -16 + 118 = 102 \text{ ft}$$

above the water.

After 2 seconds (that is, when $t = 2$), his height will be

$$-16(2)^2 + 118 = -16 \cdot 4 + 118 = 54 \text{ ft}$$

above the water.

Note that

$$\text{At } t = 1, \quad -16t^2 + 118 = 102$$
$$\text{At } t = 2, \quad -16t^2 + 118 = 54$$

and so on.

In algebra, polynomials in one variable can be represented by using symbols such as $P(t)$ (read "P of t"), $Q(x)$, and $D(y)$, where the symbol in parentheses indicates the variable being used. Thus, $P(t) = -16t^2 + 118$ is the polynomial representing the height of the diver above the water and $G(t)$ is the polynomial representing the amount of waste generated annually in the United States. With this notation, $P(1)$ represents the value of the polynomial $P(t)$ when 1 is substituted for t in the polynomial; that is,

$$P(1) = -16(1)^2 + 118 = 102$$

and

$$P(2) = -16(2)^2 + 118 = 54$$

and so on.

EXAMPLE 4   Evaluating polynomials

When $t = 3$, what is the value of $P(t) = -16t^2 + 118$?

SOLUTION   When $t = 3$,

$$P(t) = -16t^2 + 118$$

becomes

$$P(3) = -16(3)^2 + 118$$
$$= -16(9) + 118$$
$$= -144 + 118$$
$$= -26$$

PROBLEM 4

Find the value of $P(x) = -16t^2 + 90$ when $t = 2$.

Note that in this case, the answer is *negative,* which means that the diver should be *below* the water's surface. However, since he can't continue to free-fall after hitting the water, we conclude that it took him between 2 and 3 seconds to hit the water.

Answers to PROBLEMS

4. 26

EXAMPLE 5   Evaluating polynomials

Evaluate $Q(x) = 3x^2 - 5x + 8$ when $x = 2$.

SOLUTION   When $x = 2$,

$$Q(x) = 3x^2 - 5x + 8$$

becomes

$$Q(2) = 3(2^2) - 5(2) + 8$$
$$= 3(4) - 5(2) + 8 \quad \text{Multiply } 2 \cdot 2 = 2^2.$$
$$= 12 - 10 + 8 \quad \text{Multiply } 3 \cdot 4 \text{ and } 5 \cdot 2.$$
$$= 2 + 8 \quad \text{Subtract } 12 - 10.$$
$$= 10 \quad \text{Add } 2 + 8.$$

Note that to evaluate this polynomial, we followed the order of operations studied in Section 1.5.

PROBLEM 5

Evaluate $R(x) = 5x^2 - 3x + 9$ when $x = 3$.

EXAMPLE 6   Generated waste

a. If $G(t) = -0.001t^3 + 0.06t^2 + 2.6t + 88.6$ is the amount of waste (in millions of tons) generated annually in the United States and t is the number of years *after* 1960, how much waste was generated in 1960 ($t = 0$)?

b. How much waste is predicted to be generated in the year 2010?

SOLUTION

a. At $t = 0$, $G(t) = -0.001t^3 + 0.06t^2 + 2.6t + 88.6$ becomes

$G(0) = -0.001(0)^{-3} + 0.06(0)^2 + 2.6(0) + 88.6 = 88.6$ (million tons)

Thus, 88.6 million tons were generated in 1960.

243.6 tons — 2010
$G(t)$ 88.6 tons — 1960

b. The year 2010 is $2010 - 1960 = 50$ years *after* 1960. This means that $t = 50$ and

$$G(50) = -0.001(50)^3 + 0.06(50)^2 + 2.6(50) + 88.6$$
$$= -0.001(125,000) + 0.06(2500) + 2.6(50) + 88.6$$
$$= 243.6 \text{ (million tons)}$$

The prediction is that 243.6 million tons of waste will be generated in the year 2010.

PROBLEM 6

a. How much waste was generated in 1961?

b. How much waste was generated in 2000?

EXAMPLE 7   Blood alcohol level

Do you know how many drinks it takes before you are considered legally drunk? In many states you are drunk if you have a blood alcohol level (BAL) of 0.10 or even lower (0.08). The chart on the next page shows your BAL after consuming 3 ounces of alcohol (6 beers with 4% alcohol or 30 ounces of 10% wine or 7.5 ounces of vodka or whiskey) in the time period shown. The polynomial equation $y = -0.0226x + 0.1509$ approximates the BAL for a 150-pound male and $y = -0.0257x + 0.1663$ the BAL of a 150-pound female.

PROBLEM 7

a. Use the graph to find the BAL for a male after 3 hours.

b. Use the graph to find the BAL for a female after 3 hours.

c. Evaluate $y = -0.0226x + 0.1509$ for $x = 3$.

d. Evaluate $y = -0.0257x + 0.1663$ for $x = 3$.

(continued)

Answers to PROBLEMS

5. 45   **6. a.** 91.259 million tons   **b.** 224.60 million tons   **7. a.** 0.082   **b.** 0.09   **c.** 0.0831   **d.** 0.0892

a. Use the graph to find the BAL for a male after 0.5 hour.

b. Use the graph to find the BAL for a female after 1 hour.

c. Evaluate $y = -0.0226x + 0.1509$ for $x = 0.5$. Does your answer coincide with the answer to part **a?**

d. Evaluate $y = -0.0257x + 0.1663$ for $x = 1$. Does your answer coincide with the answer to part **b?**

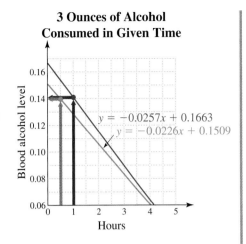

3 Ounces of Alcohol Consumed in Given Time

$y = -0.0257x + 0.1663$
$y = -0.0226x + 0.1509$

SOLUTION

a. First, locate 0.5 on the x-axis.
Move vertically until you reach the blue line and then horizontally (left) to the y-axis. The y-value at that point is approximately 0.14. This means that the BAL of a male 0.5 hour after consuming 3 ounces of alcohol is about 0.14 (legally drunk!).

b. This time, locate 1 on the x-axis, move vertically until you reach the red line, and then move horizontally to the y-axis. The y-value at that point is a little more than 0.14, so we estimate the answer to be 0.141 (legally drunk!).

c. When $x = 0.5$,

$$y = -0.0226x + 0.1509$$
$$= -0.0226(0.5) + 0.1509 = 0.1396$$

which is very close to the 0.14 from part **a.**

d. When $x = 1$,

$$y = -0.0257x + 0.1663$$
$$= -0.0257(1) + 0.1663 = 0.1406$$

which is also very close to the 0.141 from part **b.**

⬛ ◈ ⬛ Calculator Corner

Evaluating Polynomials

If you have a calculator, you can evaluate polynomials in several ways. One way is to make a picture (graph) of the polynomial and use the TRACE and ZOOM keys. Or, better yet, if your calculator has a "value" feature, it will automatically find the value of a polynomial for a given number. Thus, to find the value of $G(t) = -0.001t^3 + 0.06t^2 + 2.6t + 88.6$ when $t = 50$ in Example 6, first graph the polynomial. With a TI-83 Plus, press Y= and enter $-0.001X^3 + 0.06X^2 + 2.6X + 88.6$ for Y_1. (Note that we used X's instead of t's because X's are easier to enter.) If you then press GRAPH, nothing will show in your window! Why? Because a standard window gives values of X only between -10 and 10 and corresponding $Y_1 = G(x)$ values between -10 and 10. Adjust the X- and Y-values to those shown in Window 1 and press GRAPH again. To evaluate $G(X)$ at $X = 50$ with a TI-83 Plus, press 2nd TRACE 1. When the calculator prompts you by showing $X =$, enter 50 and press ENTER. The result is shown in Window 2 as $Y = 243.6$. This means that 50 years after 1960—that is, in the year 2010—243.6 million tons of waste will be generated.

You can also evaluate $G(X)$ by first storing the value you wish ($X = 50$) by pressing 50 STO▸ X,T,θ,n ENTER, then entering $-0.001X^3 + 0.06X^2 + 2.6X + 88.6$, and finally pressing ENTER again. The result is shown in Window 3.

WINDOW FORMAT
Xmin=0
Xmax=50
Xscl=10
Ymin=0
Ymax=200
Yscl=50

Window 1

X=50   Y= 243.6

Window 2

The beauty of the first method is that now you can evaluate $G(20)$, $G(50)$, or $G(a)$ for any number a in the chosen interval by simply entering the value of a and pressing ENTER. You don't have to reenter $G(X)$ or adjust the window again!

```
50→X
                50
-0.001X^3+0.06X^2
+2.6X+88.6
             243.6
```

Window 3

MathZone

Boost your grade at mathzone.com!
> Practice Problems   > Self-Tests
> NetTutor   > e-Professors
                          > Videos

⟩ Exercises **4.4**

⟨ **A** ⟩ Classifying Polynomials
⟨ **B** ⟩ Finding the Degree of a Polynomial

In Problems 1–10, classify each expression as a monomial, binomial, trinomial, or polynomial and give the degree.

1. $-5x + 7$

2. $8 + 9x^3$

3. $7x$

4. $-3x^4$

5. $-2x + 7x^2 + 9$

6. $-x + x^3 - 2x^2$

7. 18

8. 0

9. $9x^3 - 2x$

10. $-7x + 8x^6 + 3x^5 + 9$

⟨ **B** ⟩ Finding the Degree of a Polynomial
⟨ **C** ⟩ Writing a Polynomial in Descending Order

In Problems 11–20, write in descending order and give the degree of each polynomial.

11. $-3x + 8x^3$

12. $7 - 2x^3$

13. $4x - 7 + 8x^2$

14. $9 - 3x + x^3$

15. $5x + x^2$

16. $-3x - 7x^3$

17. $3 + x^3 - x^2$

18. $-3x^2 + 8 - 2x$

19. $4x^5 + 2x^2 - 3x^3$

20. $4 - 3x^3 + 2x^2 + x$

⟨ **D** ⟩ Evaluating Polynomials   In Problems 21–24, find the value of the polynomial when (a) $x = 2$ and (b) $x = -2$.

21. $3x - 2$

22. $x^2 - 3$

23. $2x^2 - 1$

24. $x^3 - 1$

25. If $P(x) = 3x^2 - x - 1$, find
  a. $P(2)$ **b.** $P(-2)$

26. If $Q(x) = 2x^2 + 2x + 1$, find
  a. $Q(2)$ **b.** $Q(-2)$

27. If $R(x) = 3x - 1 + x^2$, find
  a. $R(2)$ **b.** $R(-2)$

28. If $S(x) = 2x - 3 - x^2$, find
  a. $S(2)$ **b.** $S(-2)$

29. If $T(y) = -3 + y + y^2$, find
  a. $T(2)$ **b.** $T(-2)$

30. If $U(r) = -r - 4 - r^2$, find
  a. $U(2)$ **b.** $U(-2)$

⟩⟩⟩ Applications

31. *Height of dropped object*   If an object drops from an altitude of k feet, its height above ground after t seconds is given by $-16t^2 + k$ feet. If the object is dropped from an altitude of 150 feet, what would be the height of the object after the specified amount of time.

  a. t seconds

  b. 1 second   **c.** 2 seconds

32. *Velocity of dropped object* After t seconds have passed, the velocity of an object dropped from a height of 96 feet is $-32t$ feet per second. What would be the velocity of the object after the specified amount of time.

  a. 1 second

  b. 2 seconds

33. *Height of dropped object*   If an object drops from an altitude of k meters, its height above the ground after t seconds is given by $-4.9t^2 + k$ meters. If the object is dropped from an altitude of 200 meters, what would be the height of the object after the specified amount of time.

a. t seconds

b. 1 second   **c.** 2 seconds

34. *Velocity of dropped object*   After t seconds have passed, the velocity of an object dropped from a height of 300 meters is $-9.8t$ meters per second. What would be the velocity of the object after the specified amount of time.

a. 1 second

b. 2 seconds

35. *Annual number of robberies*   According to FBI data, the annual number of robberies (per 100,000 population) can be approximated by

$$R(t) = 1.76t^2 - 17.24t + 251$$

where t is the number of years after 1980.

a. What was the number of robberies (per 100,000) in 1980 ($t = 0$)?

b. How many robberies per 100,000 would you predict in the year 2000? In 2010?

36. *Annual number of assaults*   The number of aggravated assaults (per 100,000) can be approximated by

$$A(t) = -0.2t^3 + 4.7t^2 - 15t + 300$$

where t is the number of years after 2000.

a. What was the number of aggravated assaults (per 100,000) in 2000 ($t = 0$)?

b. How many aggravated assaults per 100,000 would you predict for the year 2020?

37. *Rock music sales*   What type of music do you buy? According to the Recording Industry Association of America, the percent of U.S. dollar music sales spent annually on rock music is given by

$$R(t) = 0.07t^3 - 1.15t^2 + 4t + 30$$

where t is the number of years after 1992. (Give answers to two decimals.)

a. What percent of music sales were spent on rock music during 1992?

b. What percent of music sales were spent on rock music during 2002?

c. Were rock music sales increasing or decreasing from 1992 to 2002?

d. During 1992, the value of all music sold amounted to $9.02 billion. How much was spent on rock music?

e. During 2002, the value of all music sold amounted to $13.0 billion. How much was spent on rock music?

38. *Country music sales*   The percent of U.S. dollar music sales spent annually on country music is given by

$$C(t) = -0.04t^2 - 0.6t + 18$$

where t is the number of years after 1992.

a. What percent of music sales were spent on country music during 1992?

b. What percent of music sales were spent on country music during 2002?

c. Were country music sales increasing or decreasing from 1992 to 2002?

d. During 1992, the value of all music sold amounted to $9.02 billion. How much was spent on country music? (Give answers to two decimals.)

e. During 2002, the value of all music sold amounted to $13.0 billion. How much was spent on country music? (Give answers to two decimals.)

Source: Encarta.msn.com

39. *Record low temperatures*   According to the *USA Today Weather Almanac,* the coldest city in the United States (based on average annual temperature) is International Falls, Minnesota. Record low temperatures (in °F) there can be approximated by $L(m) = -4m^2 + 57m - 175$, where m is the number of the month starting with March ($m = 3$) and ending with December ($m = 12$).

a. Find the record low during July.

b. If $m = 1$ were allowed, what would be the record low in January? Does the answer seem reasonable? Do you see why $m = 1$ is not one of the choices?

40. *Record high temperatures*   According to the *USA Today Weather Almanac,* the hottest city in the United States (based on average annual temperature) is Key West, Florida. Record high temperatures there can be approximated by $H(m) = -0.12m^2 + 2.9m + 77$, where m is the number of the month starting with January ($m = 1$) and ending with December ($m = 12$).

a. Find the record high during January. (Answer to the nearest whole number.)

b. In what two months would you expect the highest-ever temperature to have occurred? What is m for each of the two months?

c. Find $H(m)$ for each of the two months of part **b.** Which is higher?

d. The highest temperature ever recorded in Key West was 95°F and occurred in August 1957. How close was your approximation?

41. *Internet use in China*   The number of Internet users in China (in millions) is shown in the figure and can be approximated by $N(t) = 0.5t^2 + 4t + 2.1$, where t is the number of years after 1998.

   a. Use the graph to find the number of users in 2003.

   b. Evaluate $N(t)$ for $t = 5$. Is the result close to the approximation in part **a?**

Internet Use in China

Source: Data from *USA Today,* May 9, 2000.

42. *Stopping distances*   The stopping distance needed for a 3000-pound car to come to a complete stop when traveling at the indicated speeds is shown in the figure.

   a. Use the graph to estimate the number of feet it takes the car to stop when traveling at 85 miles per hour.

   b. Use the quadratic polynomial
   $D(s) = 0.05s^2 + 2.2s + 0.75$, where s is speed in miles per hour, to approximate the number of feet it takes for the car to stop when traveling at 85 miles per hour.

Auto Stopping Distance

Source: Data from *USA Today*/Foundation for Traffic Safety.

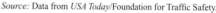

43. *Tuition and fees at 4-year private institutions*   The graph shows the tuition and fee charges for 4-year private institutions (blue graph). This cost can be approximated by the binomial $C(t) = 1033t + 16,072$, where t is the number of years after 2000.

   a. Use the graph to find the cost of tuition and fees in 2005.

   b. Evaluate $C(t)$ for $t = 5$. Is the result close to the value in part **a?**

   c. Estimate the cost of tuition and fees in 2010 ($t = 10$).

Source: http://www.ed.gov.

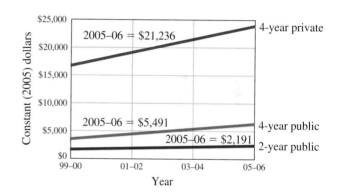

44. *Tuition and fees at 4-year public institutions*   The graph shows the tuition and fee charges for 4-year public institutions (red graph). This cost can be approximated by the binomial $C(t) = 397t + 3508$, where t is the number of years after 2000.

   a. Use the graph to find the cost of tuition and fees in 2005.

   b. Evaluate $C(t)$ for $t = 5$. Is the result close to the value in part **a?**

   c. Estimate the cost of tuition and fees in 2010 ($t = 10$).

45. *Tuition and fees at 2-year public institutions*   The graph shows the tuition and fee charges for 4-year public institutions (dark red graph). This cost can be approximated by the binomial $C(t) = 110t + 1642$, where t is the number of years after 2000.

   a. Use the graph to find the cost of tuition and fees in 2005.

   b. Evaluate $C(t)$ for $t = 5$. Is the result close to the value in part **a?**

   c. Estimate the cost of tuition and fees in 2010 ($t = 10$).

46. *Tuition, fees, and room and board for 4-year private institutions.* The graph shows the tuition, fees, and room and board charges at 4-year private institutions (blue graph). This cost can be approximated by the binomial $C(t) = 1357t + 22{,}240$, where t is the number of years after 2000.

   a. Use the graph to find the cost of tuition, fees, and room and board in 2005.

   b. Evaluate $C(t)$ for $t = 5$. Is the result close to the value in part **a?**

   c. Estimate the cost of tuition, fees, and room and board in 2010 ($t = 10$).

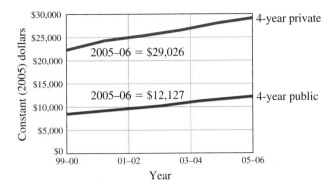

47. *Tuition, fees, and room and board for 4-year public institutions.* The graph shows the tuition, fees, and room and board charges at 4-year public institutions (red graph). This cost can be approximated by the binomial $C(t) = 738x + 8439$, where t is the number of years after 2000.

   a. Use the graph to find the cost of tuition, fees, and room and board in 2005.

   b. Evaluate $C(t)$ for $t = 5$. Is the result close to the value in part **a?**

   c. Estimate the cost of tuition, fees, and room and board in 2010 ($t = 10$).

48. *Tuition, fees, and room and board for 4-year public institutions* We can approximate the tuition, fees, and room and board charges at 4-year public institutions (red graph) by using the trinomial $C(t) = 29t^2 + 608t + 8411$, where t is the number of years after 2000.

   a. Evaluate $C(t)$ for $t = 5$. Is the result close to the $12,127 value given in the graph?

   b. Estimate the cost of tuition, fees, and room and board in 2010 ($t = 10$) and compare with the value obtained in part **c** of Problem 47.

   c. You can even approximate the tuition, fees, and room and board charges using $C(t) = -13t^3 + 126t^2 + 431t + 8450$, where t is the number of years after 2000. What will be $C(10)$ and how close is it to the values obtained in **b?**

   d. Which is the best approximation, the first-, the second-, or the third-degree polynomial?

〉〉〉 *Using Your Knowledge*

Faster and Faster Polynomials   We've already stated that if an object is simply *dropped* from a certain height, its velocity after t seconds is given by $-32t$ feet per second. What will happen if we actually *throw* the object down with an initial velocity, say v_0? Since the velocity $-32t$ is being helped by the velocity v_0, the new final velocity will be given by $-32t + v_0$ (v_0 is *negative* if the object is thrown *downward*).

49. Find the velocity after t seconds have elapsed of a ball thrown downward with an initial velocity of 10 feet per second.

50. What will the velocity of the ball in Problem 49 be after the specified amount of time?

   a. 1 second   **b.** 2 seconds

51. In the metric system, the velocity after t seconds of an object thrown downward with an initial velocity v_0 is given by the equation

$$-9.8t + v_0 \quad \text{(meters)}$$

What would be the velocity of a ball thrown downward with an initial velocity of 2 meters per second after the specified amount of time.

   a. 1 second

   b. 2 seconds

52. The height of an object after t seconds have elapsed depends on two factors: the initial velocity v_0 and the height s_0 from which the object is thrown. The polynomial giving this height is given by the equation

$$-16t^2 + v_0 t + s_0 \quad \text{(feet)}$$

where v_0 is the initial velocity and s_0 is the height from which the object is thrown. What would be the height of a ball thrown downward from a 300-foot tower with an initial velocity of 10 feet per second after the specified amount of time.

   a. 1 second   **b.** 2 seconds

〉〉〉 *Write On*

53. Write your own definition of a polynomial.

54. Is $x^2 + \frac{1}{x} + 2$ a polynomial? Why or why not?

55. Is $x^{-2} + x + 3$ a polynomial? Why or why not?

56. Explain how to find the degree of a polynomial in one variable.

57. The degree of x^4 is 4. What is the degree of 7^4? Why?

58. What does "evaluate a polynomial" mean?

〉〉〉 *Concept Checker*

Fill in the blank(s) with the correct word(s), phrase, or mathematical statement.

59. A _____ is an **algebraic expression** formed by using the operations of **addition and subtraction** on products of numbers and variables raised to whole number exponents.

60. A **polynomial** of **one** term is called a _____.

61. A **polynomial** of **two** terms is called a _____.

62. A **polynomial** of **three** terms is called a _____.

binomial

monomial

polynomial

trinomial

〉〉〉 *Mastery Test*

63. Evaluate $2x^2 - 3x + 10$ when $x = 2$.

64. If $P(x) = 3x^3 - 7x + 9$, find $P(3)$.

65. When $t = 2.5$, what is the value of $-16t^2 + 118$?

Find the degree of each polynomial.

66. $-5y - 3$

67. $4x^2 - 5x^3 + x^8$

68. -9

69. 0

Write each polynomial in descending order.

70. $-2x^4 + 5x - 3x^2 + 9$

71. $-8 + 5x^2 - 3x$

Classify as a monomial, binomial, trinomial, or polynomial.

72. $-4t + t^2 - 8$

73. $-5y$

74. $278 + 6x$

75. $2x^3 - x^2 + x - 1$

76. The amount of waste recovered (in millions of tons) in the United States can be approximated by $R(t) = 0.04t^2 - 0.59t + 7.42$, where t is the number of years after 1960.

  a. How many million tons were recovered in 1960?

  b. How many million tons would you predict will be recovered in the year 2010?

77. Refer to Example 7.

  a. Use the graph to find the BAL for a female after 2 hours.

  b. Use the graph to find the BAL for a male after 2 hours.

  c. Evaluate $y = -0.0257x + 0.1663$ for $x = 2$. Is the answer close to that of part **a?**

  d. Evaluate $y = -0.0226x + 0.1509$ for $x = 2$. Is the answer close to that of part **b?**

〉〉〉 *Skill Checker*

Find:

78. $-5ab + (2ab)$

79. $-3ab + (-4ab)$

80. $-8a^2b + (-5a^2b)$

81. $-3x^2y + 8x^2y - 2x^2y$

82. $-2xy^2 + 7xy^2 - 9xy^2$

83. $5xy^2 - (-3xy^2)$

84. $7x^2y - (-8x^2y)$

4.5

Addition and Subtraction of Polynomials

▶ Objectives

A ▷ Add polynomials.

B ▷ Subtract polynomials.

C ▷ Find areas by adding polynomials.

D ▷ Solve applications involving polynomials.

▶ To Succeed, Review How To . . .

1. Add and subtract like terms (pp. 78, 88–89).

2. Remove parentheses in expressions preceded by a minus sign (pp.89–90).

▶ Getting Started

Wasted Waste

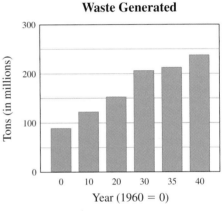

Waste Generated

The annual amount of waste (in millions of tons) generated in the United States is approximated by $G(t) = -0.001t^3 + 0.06t^2 + 2.6t + 88.6$, where t is the number of years after 1960. How much of this waste is recovered? That amount can be approximated by $R(t) = 0.06t^2 - 0.59t + 6.4$. From these two approximations, we can estimate that the amount of waste actually "wasted" (not recovered) is $G(t) - R(t)$. To find this difference, we simply subtract like terms. To make the procedure more familiar, we write it in columns:

$$G(t) = -0.001t^3 + 0.06t^2 + 2.6t \ + 88.6$$

$$\underline{(-)\,R(t) = (-) \qquad\quad 0.06t^2 - 0.59t + \ \ 6.4}$$

$$-0.001t^3 + \qquad\qquad + 3.19t + 82.2$$

└─ Note that $0.06t^2 - 0.06t^2 = 0$

Thus, the amount of waste generated and *not* recovered is

$$G(t) - R(t) = -0.001t^3 + 3.19t + 82.2.$$

Let's see what this means in millions of tons. Since t is the number of years after 1960, $t = 0$ in 1960, and the amount of waste generated, the amount of waste recycled, and the amount of waste not recovered are as follows:

$$G(0) = -0.001(0)^3 + 0.06(0)^2 + 2.6(0) + 88.6 = 88.6 \quad \text{(million tons)}$$

$$R(0) = 0.06(0)^2 - 0.59(0) + 6.4 \ = 6.4 \quad \text{(million tons)}$$

$$G(0) - R(0) = 88.6 - 6.4 = 82.2 \quad \text{(million tons)}$$

As you can see, there is much more material *not* recovered than material recovered. How can we find out whether the situation is changing? One way is to predict how much waste will be produced and how much recovered, say, in the year 2010. The amount can be approximated by $G(50) - R(50)$. Then we find out if, percentagewise, the situation is getting

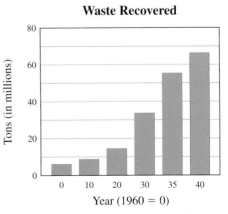

Waste Recovered

better. In 1960, the percent of materials recovered was 6.4/88.6, or about 7.2%. What percent would it be in the year 2010? In this section, we will learn how to add and subtract polynomials and use these ideas to solve applications.

A ⟩ Adding Polynomials

The addition of monomials and polynomials is a matter of combining like terms. For example, suppose we wish to add $3x^2 + 7x - 3$ and $5x^2 - 2x + 9$; that is, we wish to find

$$(3x^2 + 7x - 3) + (5x^2 - 2x + 9)$$

Using the commutative, associative, and distributive properties, we write

$$(3x^2 + 7x - 3) + (5x^2 - 2x + 9) = (3x^2 + 5x^2) + (7x - 2x) + (-3 + 9)$$
$$= (3 + 5)x^2 + (7 - 2)x + (-3 + 9)$$
$$= 8x^2 + 5x + 6$$

Similarly, the sum of $4x^3 + \frac{3}{7}x^2 - 2x + 3$ and $6x^3 - \frac{1}{7}x^2 + 9$ is written as

$$\left(4x^3 + \frac{3}{7}x^2 - 2x + 3\right) + \left(6x^3 - \frac{1}{7}x^2 + 9\right)$$
$$= (4x^3 + 6x^3) + \left(\frac{3}{7}x^2 - \frac{1}{7}x^2\right) + (-2x) + (3 + 9)$$
$$= 10x^3 + \frac{2}{7}x^2 - 2x + 12$$

In both examples, the polynomials have been written in descending order for convenience in combining like terms.

EXAMPLE 1 Adding polynomials

Add: $3x + 7x^2 - 7$ and $-4x^2 + 9 - 3x$

SOLUTION We first write both polynomials in descending order and then combine like terms to obtain

$$(7x^2 + 3x - 7) + (-4x^2 - 3x + 9) = (7x^2 - 4x^2) + (3x - 3x) + (-7 + 9)$$
$$= 3x^2 + 0 + 2$$
$$= 3x^2 + 2$$

PROBLEM 1

Add: $5x + 8x^2 - 3$ and $-3x^2 + 8 - 5x$

As in arithmetic, the addition of polynomials can be done by writing the polynomials in descending order and then placing like terms in columns. In arithmetic, you add 345 and 678 by writing the numbers in a column:

$$+345$$
$$+678$$

————— Units
————— Tens
————— Hundreds

Thus, to add $4x^3 + 3x - 7$ and $7x - 3x^3 + x^2 + 9$, we first write both polynomials in descending order with like terms in the same column, leaving space for any missing terms. We then add the terms in each of the columns:

The x^2 term is missing in $4x^3 + 3x - 7$.

$$\begin{array}{r} 4x^3 + 3x - 7 \\ -3x^3 + x^2 + 7x + 9 \\ \hline x^3 + x^2 + 10x + 2 \end{array}$$

EXAMPLE 2   Adding polynomials

Add: $-3x + 7x^2 - 2$ and $-4x^2 - 3 + 5x$

SOLUTION   We first write both polynomials in descending order, place like terms in a column, and then add as shown:

$$\begin{array}{r} 7x^2 - 3x - 2 \\ -4x^2 + 5x - 3 \\ \hline 3x^2 + 2x - 5 \end{array}$$

Horizontally, we write:

$$(7x^2 - 3x - 2) + (-4x^2 + 5x - 3)$$
$$= \underbrace{(7x^2 - 4x^2)}_{} + \underbrace{(-3x + 5x)}_{} + \underbrace{(-2 - 3)}_{}$$
$$= \quad 3x^2 \quad + \quad 2x \quad + \quad (-5)$$
$$= \quad 3x^2 \quad + \quad 2x \quad - \quad 5$$

PROBLEM 2

Add: $-5y + 8y^2 - 3$ and $-5y^2 - 4 + 6y$

B › Subtracting Polynomials

To subtract polynomials, we first recall that

$$\boxed{a - (b + c) = a - b - c}$$

To remove the parentheses from an expression preceded by a minus sign, we must change the sign of each term *inside* the parentheses. This is the same as multiplying each term inside the parentheses by -1. Thus,

$$(3x^2 - 2x + 1) - (4x^2 + 5x + 2) = 3x^2 - 2x + 1 - 4x^2 - 5x - 2$$
$$= (3x^2 - 4x^2) + (-2x - 5x) + (1 - 2)$$
$$= -x^2 + (-7x) + (-1)$$
$$= -x^2 - 7x - 1$$

Here's how we do it using columns:

$$\begin{array}{r} 3x^2 - 2x + 1 \\ (-)\,4x^2 + 5x + 2 \\ \hline \end{array} \quad \text{is written} \quad \longrightarrow \quad \begin{array}{r} 3x^2 - 2x + 1 \\ (+)-4x^2 - 5x - 2 \\ \hline -x^2 - 7x - 1 \end{array}$$

Note that we changed the sign of *every* term in $4x^2 + 5x + 2$ and wrote $-4x^2 - 5x - 2$.

NOTE

"Subtract b from a" means to find $a - b$.

EXAMPLE 3   Subtracting polynomials

Subtract $4x - 3 + 7x^2$ from $5x^2 - 3x$.

SOLUTION   We first write the problem in columns, then change the signs and add:

$$\begin{array}{r} 5x^2 - 3x \\ (-)\,7x^2 + 4x - 3 \\ \hline \end{array} \quad \text{is written} \quad \longrightarrow \quad \begin{array}{r} 5x^2 - 3x \\ (+)-7x^2 - 4x + 3 \\ \hline -2x^2 - 7x + 3 \end{array}$$

Thus, the answer is $-2x^2 - 7x + 3$.

PROBLEM 3

Subtract $5y - 4 + 8y^2$ from $6y^2 - 4y$.

Answers to PROBLEMS

2. $3y^2 + y - 7$

3. $-2y^2 - 9y + 4$

To do it horizontally, we write

$$(5x^2 - 3x) - (7x^2 + 4x - 3)$$

$$= 5x^2 - 3x - 7x^2 - 4x + 3 \qquad \text{Change the sign of every term in } 7x^2 + 4x - 3.$$

$$= (5x^2 - 7x^2) + (-3x - 4x) + 3 \qquad \text{Use the commutative and associative properties.}$$

$$= -2x^2 - 7x + 3$$

Just as in arithmetic, we can add or subtract more than two polynomials. For example, to add the polynomials $-7x + x^2 - 3$, $6x^2 - 8 + 2x$, and $3x - x^2 + 5$, we simply write each of the polynomials in descending order with like terms in the same column and add:

$$x^2 - 7x - 3$$
$$6x^2 + 2x - 8$$
$$\underline{-x^2 + 3x + 5}$$
$$6x^2 - 2x - 6$$

Or, horizontally, we write

$$(x^2 - 7x - 3) + (6x^2 + 2x - 8) + (-x^2 + 3x + 5)$$
$$= (x^2 + 6x^2 - x^2) + (-7x + 2x + 3x) + (-3 - 8 + 5)$$
$$= 6x^2 + (-2x) + (-6)$$
$$= 6x^2 - 2x - 6$$

EXAMPLE 4   Adding polynomials

Add: $x^3 + 2x - 3x^2 - 5$, $-8 + 2x - 5x^2$, and $7x^3 - 4x + 9$

SOLUTION   We first write all the polynomials in descending order with like terms in the same column and then add:

$$x^3 - 3x^2 + 2x - 5$$
$$-5x^2 + 2x - 8$$
$$\underline{7x^3 \qquad\quad - 4x + 9}$$
$$8x^3 - 8x^2 \qquad - 4$$

Horizontally, we have

$$(x^3 - 3x^2 + 2x - 5) + (-5x^2 + 2x - 8) + (7x^3 - 4x + 9)$$
$$= (x^3 + 7x^3) + (-3x^2 - 5x^2) + (2x + 2x - 4x) + (-5 - 8 + 9)$$
$$= 8x^3 + (-8x^2) + 0x + (-4)$$
$$= 8x^3 - 8x^2 - 4$$

PROBLEM 4

Add: $y^3 + 3y - 4y^2 - 6$,
$-9 + 3y - 6y^2$, and
$6y^3 - 5y + 8$

C ⟩ Finding Areas

Addition of polynomials can be used to find the sum of the areas of several rectangles. To find the total area of the shaded rectangles, add the individual areas.

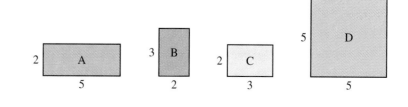

Answers to PROBLEMS

4. $7y^3 - 10y^2 + y - 7$

Since the area of a rectangle is the product of its length and its width, we have

Area of A + Area of B + Area of C + Area of D

$5 \cdot 2$ + $2 \cdot 3$ + $3 \cdot 2$ + $5 \cdot 5$

$= 10$ + 6 + 6 + 25

Thus, the total area is

$$10 + 6 + 6 + 25 = 47 \qquad \text{(square units)}$$

This same procedure can be used when some of the lengths are represented by variables, as shown in Example 5.

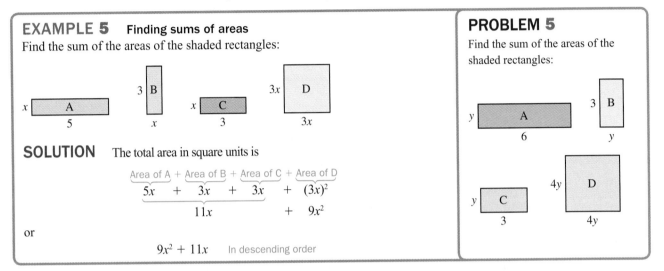

EXAMPLE 5   **Finding sums of areas**

Find the sum of the areas of the shaded rectangles:

SOLUTION   The total area in square units is

Area of A + Area of B + Area of C + Area of D

$5x$ + $3x$ + $3x$ + $(3x)^2$

$11x$ + $9x^2$

or

$$9x^2 + 11x \qquad \text{In descending order}$$

PROBLEM 5

Find the sum of the areas of the shaded rectangles:

D › Applications Involving Polynomials

The concepts we've just studied can be used to examine important issues in American life. Let's see how we can use these concepts to study health care.

EXAMPLE 6   **Health Care**

Based on Social Security Administration statistics, the annual amount of money spent on hospital care by the average American can be approximated by the equation

$$H(t) = 6.6t^2 + 30t + 682 \qquad \text{(dollars)}$$

while the annual amount spent for doctors' services is approximated by the equation

$$D(t) = 41t + 293$$

where t is the number of years after 1985.

a. Find the annual amount of money spent for doctors' services and hospital care.
b. How much was spent for doctors' services in 1985?
c. How much was spent on hospital care in 1985?
d. How much would you predict will be spent on hospital care and doctors' services in the year 2005?

PROBLEM 6

Using the equations given in Example 6, how much was spent on hospital care and doctors' services in 2000?

SOLUTION

a. To find the annual amount spent for doctors' services and hospital care, we find $D(t) + H(t)$:

$$
\begin{aligned}
D(t) = \quad & 41t + 293 \\
H(t) = 6.6t^2 + {} & 30t + 682 \\
\hline
6.6t^2 + {} & 71t + 975
\end{aligned}
$$

Answers to PROBLEMS

5. $16y^2 + 12y$ **6.** 3525

Thus, the annual amount spent for doctors' services and hospital care is $6.6t^2 + 71t + 975$.

b. To find how much was spent for doctors' services in 1985 ($t = 0$), find $D(0) = 41(0) + 293$. Thus, $293 was spent for doctors' services in 1985.

c. The amount spent on hospital care in 1985 ($t = 0$) was $H(0) = 6.6(0)^2 + 30(0) + 682 = 682$. Thus, $682 was spent for hospital care in 1985.

d. The year 2005 corresponds to $t = 2005 - 1985 = 20$. To predict the amount to be spent on hospital care and doctors' services in the year 2005, we find $6.6(20)^2 + 71(20) + 975 = \$5035$. Note that it's easier to evaluate $D(t) + H(t) = 6.6t^2 + 71t + 975$ for $t = 20$ than it is to evaluate $D(20)$ and $H(20)$ and then add.

EXAMPLE 7   Blood alcohol level

In Example 7 of Section 4.4, we introduced the polynomial equations $y = -0.0226x + 0.1509$ and $y = -0.0257x + 0.1663$ approximating the blood alcohol level (BAL) for a 150-pound male or female, respectively. In these equations x represents the time since consuming 3 ounces of alcohol.

It is known that the burn-off rate of alcohol is 0.015 per hour (that is, the BAL is reduced by 0.015 per hour if no additional alcohol is consumed). Find a polynomial that would approximate the BAL for a male x hours after consuming the 3 ounces of alcohol.

SOLUTION   The initial BAL is $-0.0226x + 0.1509$, but this level is *decreased* by 0.015 each hour. Thus, the actual BAL after x hours is $-0.0226x + 0.1509 - 0.015x$, or $-0.0376x + 0.1509$.

PROBLEM 7

Find a polynomial that would approximate the BAL for a female x hours after consuming 3 ounces of alcohol.

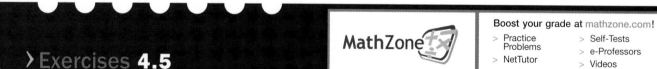

❭ Exercises 4.5

❬ A ❭ **Adding Polynomials**   In Problems 1–30, add as indicated.

1. $(5x^2 + 2x + 5) + (7x^2 + 3x + 1)$

2. $(3x^2 - 5x - 5) + (9x^2 + 2x + 1)$

3. $(-3x + 5x^2 - 1) + (-7 + 2x - 7x^2)$

4. $(3 - 2x^2 + 7x) + (-6 + 2x^2 - 5x)$

5. $(2x + 5x^2 - 2) + (-3 + 5x - 8x^2)$

6. $-3x - 2 + 3x^2$ and $-4 + 5x - 6x^2$

7. $-2 + 5x$ and $-3 - x^2 - 5x$

8. $-4x + 2 - 6x^2$ and $2 + 5x$

9. $x^3 - 2x + 3$ and $-2x^2 + x - 5$

10. $x^4 - 3 + 2x - 3x^3$ and $3x^4 - 2x^2 + 5 - x$

Answers to PROBLEMS

7. $-0.0407x + 0.1663$

11. $-6x^3 + 2x^4 - x$ and $2x^2 + 5 + 2x - 2x^3$

12. $\frac{1}{2}x^3 + x^2 - \frac{1}{5}x$ and $\frac{3}{5}x + \frac{1}{2}x^3 - 3x^2$

13. $\frac{1}{3} - \frac{2}{5}x^2 + \frac{3}{4}x$ and $\frac{1}{4}x - \frac{1}{5}x^2 + \frac{2}{3}$

14. $0.3x - 0.1 - 0.4x^2$ and $0.1x^2 - 0.1x + 0.6$

15. $0.2x - 0.3 + 0.5x^2$ and $-\frac{1}{10} + \frac{1}{10}x - \frac{1}{10}x^2$

16. $\quad\; -x^2 + 5x + 2$
$\underline{(+)\, 3x^2 - 7x - 2}$

17. $\quad\; -3x^2 + 2x - 4$
$\underline{(+)\quad x^2 - 4x + 7}$

18. $\quad\; -2x^4 \qquad\; + 2x - 1$
$\underline{(+)\qquad\; -x^3 - 3x + 5}$

19. $\quad\; 3x^4 \qquad - 3x + 4$
$\underline{(+)\qquad x^3 - 2x - 5}$

20. $\quad\; -3x^4 \qquad + 2x^2 - x + 5$
$\underline{(+)\qquad -2x^3 \qquad + 5x - 7}$

21. $\quad\; -5x^4 \qquad - 5x^2 + 3$
$\underline{(+)\qquad 5x^3 + 3x^2 - 5}$

22. $\quad\; 3x^3 \qquad + x - 1$
$\qquad\quad x^2 - 2x + 5$
$\underline{(+)\, 5x^3 \qquad - x}$

23. $\quad\; 5x^3 - x^2 \qquad - 3$
$\qquad\qquad\quad 5x + 9$
$\underline{(+)\quad - 3x^2 \qquad - 7}$

24. $\quad\; -\frac{1}{3}x^3 \qquad - \frac{1}{2}x + 5$
$\qquad\quad -\frac{1}{5}x^2 + \frac{1}{2}x - 1$
$\underline{(+)\, \frac{2}{3}x^3 \qquad + x - 2}$

25. $\quad\; -\frac{2}{7}x^3 + \frac{1}{6}x^2 \qquad + 2$
$\qquad\quad \frac{1}{7}x^3 \qquad + 5x - 3$
$\underline{(+)\qquad -\frac{5}{6}x^2 \qquad + 1}$

26. $\quad\; -\frac{1}{8}x^2 - \frac{1}{3}x + \frac{1}{5}$
$\qquad -x^3 + \frac{3}{8}x^2 \qquad - \frac{2}{5}$
$\underline{(+)\, -3x^3 \qquad - \frac{2}{3}x + \frac{4}{5}}$

27. $\quad\; -\frac{1}{7}x^3 \qquad\qquad + 2$
$\qquad\qquad -\frac{1}{9}x^2 - x - 3$
$\underline{(+)\, -\frac{2}{7}x^3 + \frac{2}{9}x^2 + 2x - 5}$

28. $\quad\; -2x^4 + 5x^3 - 2x^2 + 3x - 5$
$\qquad\qquad\; 8x^3 \qquad - 2x + 5$
$\qquad\; -x^4 \qquad + 3x^2 - x - 2$
$\underline{(+)\qquad\quad 6x^3 \qquad + 2x + 5}$

29. $\quad\; - 6x^3 + 2x^2 \qquad + 1$
$\qquad -x^4 + 3x^3 - 5x^2 + 3x$
$\qquad\quad - x^3 \qquad - 7x + 2$
$\underline{(+)\, -3x^4 \qquad\qquad + 3x - 1}$

30. $\quad\; - 3x^4 \qquad + 2x^2 \qquad - 5$
$\qquad x^5 + x^4 - 2x^3 + 7x^2 + 5x$
$\qquad\qquad\; 2x^4 \qquad - 2x^2 \qquad + 7$
$\underline{(+)\, 7x^5 \qquad + 2x^3 \qquad - 2x}$

⟨**B**⟩ **Subtracting Polynomials**   In Problems 31–50, subtract as indicated.

31. $(7x^2 + 2) - (3x^2 - 5)$

32. $(8x^2 - x) - (7x^2 + 3x)$

33. $(3x^2 - 2x - 1) - (4x^2 + 2x + 5)$

34. $(-3x + x^2 - 1) - (5x + 1 - 3x^2)$

35. $(-1 + 7x^2 - 2x) - (5x + 3x^2 - 7)$

36. $(7x^3 - x^2 + x - 1) - (2x^2 + 3x + 6)$

37. $(5x^2 - 2x + 5) - (3x^3 - x^2 + 5)$

38. $(3x^2 - x - 7) - (5x^3 + 5 - x^2 + 2x)$

39. $(6x^3 - 2x^2 - 3x + 1) - (-x^3 - x^2 - 5x + 7)$

40. $(x - 3x^2 + x^3 + 9) - (-8 + 7x - x^2 + x^3)$

41. $\quad\; 6x^2 - 3x + 5$
$\underline{(-)\, 3x^2 + 4x - 2}$

42. $\quad\; 7x^2 + 4x - 5$
$\underline{(-)\, 9x^2 - 2x + 5}$

43. $\quad\; 3x^2 - 2x - 1$
$\underline{(-)\; 3x^2 - 2x - 1}$

44. $\quad\; 5x^2 \qquad - 1$
$\underline{(-)\, 3x^2 - 2x + 1}$

45. $\quad\; 4x^3 \qquad - 2x + 5$
$\underline{(-)\qquad 3x^2 + 5x - 1}$

46. $\quad\; - 3x^2 + 5x - 2$
$\underline{(-)\, x^3 - 2x^2 \qquad + 5}$

47. $\quad\; 3x^3 \qquad - 2$
$\underline{(-)\qquad 2x^2 - x + 6}$

48. $\qquad\qquad x^2 - 2x + 1$
$\underline{(-)\, -3x^3 + x^2 + 5x - 2}$

49. $\quad\; -5x^3 \qquad + x - 2$
$\underline{(-)\qquad 5x^2 - 3x + 7}$

50. $\quad\; 6x^3 \qquad + 2x - 5$
$\underline{(-)\quad - 3x^2 - x}$

‹ C ›   **Finding Areas**   In Problems 51–55, find the sum of the areas of the shaded rectangles.

51.

52.
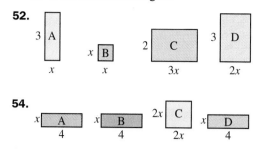

53.

54.

55.

‹ D ›   **Applications Involving Polynomials**

56. *Annual tuna consumption* Based on Census Bureau estimates, the annual consumption of canned tuna (in pounds) per person is given by the equation $C(t) = 3.87 - 0.13t$, while the consumption of turkey (in pounds) is given by the equation $T(t) = -0.15t^2 + 0.8t + 13$, where t is the number of years after 1989.

   a. How many pounds of canned tuna and how many pounds of turkey were consumed per person in 1989?

   b. Find the difference between the amount of turkey consumed and the amount of canned tuna consumed.

   c. Use your answer to part **b** to find the difference in the amount of turkey and the amount of canned tuna you would expect to be consumed in the year 2000.

57. *Annual poultry consumption* Poultry products consist of turkey and chicken. Annual chicken consumption per person (in pounds) is given by the equation $C(t) = 0.05t^3 - 0.7t^2 - t + 36$ and annual turkey consumption (in pounds) is given by the equation $T(t) = -0.15t^2 + 1.6t + 9$, where t is the number of years after 1985.

   a. Find the total annual poultry consumption per person.

   b. How much poultry was consumed per person in 1995? How much poultry was consumed in 2005?

58. *College costs* How much are you paying for tuition and fees? In a four-year public institution, the amount $T(t)$ you pay for tuition and fees (in dollars) can be approximated by $T(t) = 45t^2 + 110t + 3356$, where t is the number of years after 2000 (2000 = 0).

   a. What would you predict tuition and fees to be in 2005?

   b. The cost of books t years after 2000 can be approximated by $B(t) = 27.5t + 680$. What would be the cost of books in 2005?

   c. What polynomial would represent the cost of tuition and fees and books t years after 2000?

   d. What would you predict the cost of tuition and fees and books would be in 2005?

59. *College expenses* The three major college expenses are: tuition and fees, books, and room and board. They can be approximated, respectively, by:

$$T(t) = 45t^2 + 110t + 3356$$
$$B(t) = 27.5t + 680$$
$$R(t) = 32t^2 + 200t + 4730$$

where t is the number of years after 2000 (2000 = 0).

   a. Write a polynomial representing the total cost of tuition and fees, books, and room and board t years after 2000.

   b. What was the cost of tuition and fees, books, and room and board in 2000?

   c. What would you predict the cost of tuition and fees, books, and room and board would be in 2005?

60. *Student loans* If you are an undergraduate dependent student you can apply for a Stafford Loan. The amount of these loans can be approximated by $S(t) = 562.5t^2 + 312.5t + 2625$, where t is between 0 and 2 inclusive. If t is between 3 and 5 inclusive, then $S(t) = \$5500$.

   a. How much money can you get from a Stafford Loan the first year?

   b. What about the second year?

   c. What about the fifth year?

61. *College financial aid* Assume that you have to pay tuition and fees, books, and room and board but have your Stafford Loan to decrease expenses. (See Problems 59 and 60.) Write a polynomial that would approximate how much you would have to pay (t between 0 and 2) in your first two years if you start school in 2000.

62. *Dental services* How much do you spend in dental services? The amount can be approximated by $D(t) = -1.8t^2 + 65t + 592$ (in dollars), where t is the number of years after 2000. What about doctors and clinical services? They are more expensive and can be approximated by $C(t) = 4.75t^2 + 35t + 602$ (in dollars).

a. The total annual amount spent on dental and doctors is the sum $D(t) + C(t)$. Find this sum.

b. What was the total amount spent on dental and doctors in 2000?

c. Predict the expenditures on dental and doctor services in 2010.

Source: http://www.census.gov/., Table 121.

64. *Health expenditures* What are the annual national expenditures for health? They can be approximated by $E(t) = 0.5t^2 + 122t + 1308$ (in billion dollars), where t is the number of years after 2000. Of these, $P(t) = -1.8t^2 + 65t + 592$ are public expenditures and the rest are private.

a. What were the expenditures in 2000?

b. What were the public expenditures in 2000?

c. The private expenditures can be represented by $E(t) - P(t)$. Find this difference.

d. What were the private expenditures in 2000?

e. Predict the private expenditures for 2010.

Source: http://www.census.gov/., Table 118.

66. *Annual wages* According to the Bureau of Labor, the annual wages and salaries (in thousands of dollars) for persons under 25 years old can be approximated by $W(t) = 0.12t^3 - 0.6t^2 + t + 17$, where t is the number of years after 2000. The federal income tax paid on those wages can be approximated by $T(t) = -0.03t^3 + 0.22t^2 - 0.5t + 0.70$, where t is the number of years after 2000.

a. The wages after taxes is the difference of $W(t)$ and $T(t)$. Find this difference.

b. What are the estimated wages after taxes for 2000? For 2010?

Source: Bureau of Labor Consumer Expenditure Survey, http://www.bls.gov/.

63. *Annual expenses for medical services and medicines* The annual amount spent on medical services can be approximated by $M(t) = -t^2 + 12t + 567$ (in dollars), where t is the number of years after 2000. The amount spent on drugs and medical supplies can be approximated by $D(t) = -13t^2 + 61t + 511$ (in dollars).

a. The total amount spent on medical services and medicines is $M(t) + D(t)$. Find this sum.

b. What was the amount spent on medical services and medicines in 2000?

c. Predict the amount that will be spent on medical services and medicines in 2010.

65. *Annual wages* According to the Bureau of Labor, the annual wages and salaries (in thousands of dollars) for persons 25–34 years old can be approximated by $W(t) = 0.3t^3 - 2t^2 + 5t + 43$, where t is the number of years after 2000. The federal income tax paid on those wages can be approximated by $T(t) = t^2 - t + 3$, where t is the number of years after 2000.

a. The wages after taxes is the difference of $W(t)$ and $T(t)$. Find this difference.

b. What are the estimated wages after taxes for 2000? For 2010?

Source: Bureau of Labor Consumer Expenditure Survey, http://www.bls.gov/.

⟩⟩⟩ *Using Your Knowledge*

Business Polynomials Polynomials are also used in business and economics. For example, the revenue R may be obtained by subtracting the cost C of the merchandise from its selling price S. In symbols, this is:

$$R = S - C$$

Now the cost C of the merchandise is made up of two parts: the *variable cost* per item and the *fixed cost*. For example, if you decide to manufacture Frisbees™, you might spend \$2 per Frisbee in materials, labor, and so forth. In addition, you might have \$100 of fixed expenses. Then the cost for manufacturing x Frisbees is

$$\underbrace{\text{Cost } C \text{ of merchandise}}_{C} \text{ is } \underbrace{\overset{\text{cost per}}{\text{Frisbee}}}_{2x} \text{ and } \underbrace{\overset{\text{fixed}}{\text{expenses.}}}_{100}$$

If x Frisbees are then sold for \$3 each, the total selling price S is $3x$, and the revenue R would be:

$$R = S - C$$
$$= 3x - (2x + 100)$$
$$= 3x - 2x - 100$$
$$= x - 100$$

Thus, if the selling price S is \$3 per Frisbee, the variable costs are \$2 per Frisbee, and the fixed expenses are \$100, the revenue after selling x Frisbees is given by:

$$R = x - 100$$

In Problems 67–69, find the revenue R for the given cost C and selling price S.

67. $C = 3x + 50; S = 4x$ **68.** $C = 6x + 100; S = 8x$ **69.** $C = 7x; S = 9x$

70. In Problem 68, how many items were sold if the revenue was zero?

71. If the merchant of Problem 68 suffered a \$40 loss ($-\$40$ revenue), how many items were sold?

〉〉〉 *Write On*

72. Write the procedure you use to add polynomials.

73. Write the procedure you use to subtract polynomials.

74. Explain the difference between "subtract $x^2 + 3x - 5$ from $7x^2 - 2x + 9$" and "subtract $7x^2 - 2x + 9$ from $x^2 + 3x - 5$." What is the answer in each case?

75. List the advantages and disadvantages of adding (or subtracting) polynomials horizontally or in columns.

〉〉〉 *Concept Checker*

Fill in the blank(s) with the correct word(s), phrase, or mathematical statement.

76. $a - (b + c) =$ _____.

77. To **subtract** b from a means to find _____.

$a - b + c$ \qquad $b - a$

$a - b - c$ \qquad $a - b$

〉〉〉 *Mastery Test*

Add:

78. $3x + 3x^2 - 6$ and $-5x^2 + 10 - x$

79. $-5x + 8x^2 \quad 3$ and $-3x^2 + 4 + 8x$

Subtract:

80. $3 - 4x^2 + 5x$ from $9x^2 - 2x$

81. $9 + x^3 - 3x^2$ from $10 + 7x^2 + 5x^3$

82. Add $2x^3 + 3x - 5x^2 - 2$, $-6 + 5x - 2x^2$, and $6x^3 - 2x + 8$.

83. Find the sum of the areas of the shaded rectangles:

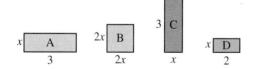

84. The number of robberies (per 100,000 population) can be approximated by $R(t) = 1.85t^2 - 19.14t + 262$, while the number of aggravated assaults is approximated by $A(t) = -0.2t^3 + 4.7t^2 - 15t + 300$, where t is the number of years after 1960.

   a. Were there more aggravated assaults or more robberies per 100,000 in 1960?

   b. Find the difference between the number of aggravated assaults and the number of robberies per 100,000.

   c. What would this difference be in the year 2000? In 2010?

〉〉〉 *Skill Checker*

Simplify:

85. $(-3x^2) \cdot (2x^3)$ **86.** $(-5x^3) \cdot (2x^4)$ **87.** $(-2x^4) \cdot (3x^5)$

88. $5(x - 3)$ **89.** $6(y - 4)$ **90.** $-3(2y - 3)$

4.6   Multiplication of Polynomials

▶ Objectives

A ▶ Multiply two monomials

B ▶ Multiply a monomial and a binomial.

C ▶ Multiply two binomials using the FOIL method.

D ▶ Solve an application involving multiplication of polynomials.

▶ To Succeed, Review How To . . .

1. Multiply expressions (pp. 309–311).
2. Use the distributive property to remove parentheses in an expression (pp. 81, 89–90).

▶ Getting Started
Deflections on a Bridge

How much does the beam bend (deflect) when a car or truck goes over the bridge? There's a formula that can tell us. For a certain beam of length L, the deflection at a distance x from one end is given by

$$(x - L)(x - 2L)$$

To multiply these two binomials, we must first learn how to do several related types of multiplication.

A ▶ Multiplying Two Monomials

We already multiplied two monomials in Section 4.1. The idea is to use the associative and commutative properties and the rules of exponents, as shown in Example 1.

EXAMPLE 1   **Multiplying two monomials**
Multiply: $(-3x^2)$ by $(2x^3)$

SOLUTION

$$(-3x^2)(2x^3) = (-3 \cdot 2)(x^2 \cdot x^3) \quad \text{Use the associative and commutative properties.}$$
$$= -6x^{2+3} \quad \text{Use the rules of exponents.}$$
$$= -6x^5$$

PROBLEM 1
Multiply: $(-4y^3)$ by $(5y^4)$

B ▶ Multiplying a Monomial and a Binomial

In Sections 1.6 and 1.7, we also multiplied a monomial and a binomial. The procedure was based on the distributive property, as shown next.

EXAMPLE 2   **Multiplying a monomial by a binomial**
Remove parentheses (simplify):

a. $5(x - 2y)$ **b.** $(x^2 + 2x)3x^4$

PROBLEM 2
Simplify:

a. $4(a - 3b)$ **b.** $(a^2 + 3a)4a^5$

Answers to PROBLEMS

1. $-20y^7$
2. a. $4a - 12b$ **b.** $4a^7 + 12a^6$

SOLUTION

a. $5(x - 2y) = 5x - 5 \cdot 2y$

$\qquad\qquad = 5x - 10y$

b. $(x^2 + 2x)3x^4 = x^2 \cdot 3x^4 + 2x \cdot 3x^4$ $\qquad$ Since $(a + b)c = ac + bc$

$\qquad\qquad = 3 \cdot x^2 \cdot x^4 + 2 \cdot 3 \cdot x \cdot x^4$

$\qquad\qquad = 3x^6 + 6x^5$

> **NOTE**
>
> You can use the commutative property first and write
>
> $$(x^2 + 2x)3x^4 = 3x^4(x^2 + 2x)$$
> $$= 3x^6 + 6x^5 \qquad \text{Same answer!}$$

C › Multiplying Two Binomials Using the FOIL Method

Another way to multiply $(x + 2)(x + 3)$ is to use the distributive property $a(b + c) = ab + ac$. Think of $x + 2$ as a, which makes x like b and 3 like c. Here's how it's done.

$$\overset{a}{}\ \overset{(b + c)}{} = \overset{a}{}\ \overset{b}{} + \overset{a}{}\ \overset{c}{}$$
$$(x + 2)(x + 3) = (x + 2)x + (x + 2)3$$
$$= x \cdot x + 2 \cdot x + x \cdot 3 + 2 \cdot 3$$
$$= x^2 + 2x + 3x + 6$$
$$= x^2 + 5x + 6$$

Similarly,

$$(x - 3)(x + 5) = (x - 3)x + (x - 3)5$$
$$= x \cdot x + (-3) \cdot x + x \cdot 5 + (-3) \cdot 5$$
$$= x^2 - 3x + 5x - 15$$
$$= x^2 + 2x - 15$$

Can you see a pattern developing? Look at the answers:

$$(x + 2)(x + 3) = x^2 + 5x + 6$$
$$(x - 3)(x + 5) = x^2 + 2x - 15$$

It seems that the *first* term in each answer (x^2) is obtained by multiplying the *first* terms in the factors (x and x). Similarly, the *last* terms (6 and -15) are obtained by multiplying the *last* terms ($2 \cdot 3$ and $-3 \cdot 5$). Here's how it works so far:

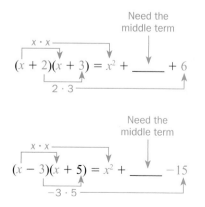

But what about the *middle* terms? In $(x + 2)(x + 3)$, the middle term is obtained by adding $3x$ and $2x$, which is the same as the result we got when we multiplied the *outer* terms (x and 3) and added the product of the inner terms (2 and x). Here's a diagram that shows how the middle term is obtained:

$$(x + 2)(x + 3) = x^2 + 3x + 2x + 6$$

$$(x - 3)(x + 5) = x^2 + 5x - 3x - 15$$

Do you see how it works now? Here is a summary of this method.

PROCEDURE

FOIL Method for Multiplying Binomials

First terms are multiplied first.

Outer terms are multiplied second.

Inner terms are multiplied third.

Last terms are multiplied last.

Of course, we call this method the **FOIL method.** We shall do one more example, step by step, to give you additional practice.

F $(x + 7)(x - 4) \rightarrow x^2$ First: $x \cdot x$

O $(x + 7)(x - 4) \rightarrow x^2 - 4x$ Outer: $-4 \cdot x$

I $(x + 7)(x - 4) \rightarrow x^2 - 4x + 7x$ Inner: $7 \cdot x$

L $(x + 7)(x - 4) = x^2 - 4x + 7x - 28$ Last: $7 \cdot (-4)$

$$= x^2 + 3x - 28$$

EXAMPLE 3   Using FOIL to multiply two binomials

Use FOIL to multiply:

a. $(x + 5)(x - 2)$ **b.** $(x - 4)(x + 3)$

SOLUTION

$$\underset{\text{(First)}}{\text{F}} \quad \underset{\text{(Outer)}}{\text{O}} \quad \underset{\text{(Inner)}}{\text{I}} \quad \underset{\text{(Last)}}{\text{L}}$$

a. $(x + 5)(x - 2) = x \cdot x - 2x + 5x - 5 \cdot 2$

$$= x^2 + 3x - 10$$

b. $(x - 4)(x + 3) = x \cdot x + 3x - 4x - 4 \cdot 3$

$$= x^2 - x - 12$$

PROBLEM 3

Use FOIL to multiply:

a. $(a + 4)(a - 3)$

b. $(a - 5)(a + 4)$

As in the case of arithmetic, we can use the ideas we've just discussed to do more complicated problems. Thus, we can use the FOIL method to multiply expressions such as $(2x + 5)$ and $(3x - 4)$. We proceed as before; just remember the properties of exponents and the FOIL sequence.

Answers to PROBLEMS

3. a. $a^2 + a - 12$

b. $a^2 - a - 20$

EXAMPLE 4   Using FOIL to multiply two binomials

Use FOIL to multiply:

a. $(2x + 5)(3x - 4)$ **b.** $(3x - 2)(5x - 1)$

SOLUTION

	(First)	(Outer)	(Inner)	(Last)
	F	O	I	L

a. $(2x + 5)(3x - 4) = \underbrace{(2x)(3x)}_{} + \underbrace{(2x)(-4)}_{} + \underbrace{5(3x)}_{} + \underbrace{(5)(-4)}_{}$

$= 6x^2 - 8x + 15x - 20$

$= 6x^2 + 7x - 20$

$$\quad\quad F \quad\quad O \quad\quad I \quad\quad L$$

b. $(3x - 2)(5x - 1) = \underbrace{(3x)(5x)}_{} + \underbrace{3x(-1)}_{} - \underbrace{2(5x)}_{} - \underbrace{2(-1)}_{}$

$= 15x^2 - 3x - 10x + 2$

$= 15x^2 - 13x + 2$

PROBLEM 4

Use FOIL to multiply:

a. $(3a + 5)(2a - 3)$

b. $(2a - 3)(4a - 1)$

Does FOIL work when the binomials to be multiplied contain more than one variable? Fortunately, yes. Again, just remember the sequence and the laws of exponents. For example, to multiply $(3x + 2y)$ by $(2x + 5y)$, we proceed as follows:

$$\quad\quad\quad F \quad\quad\quad O \quad\quad\quad I \quad\quad\quad L$$

$$(2x + 5y)(3x + 2y) = \underbrace{(2x)(3x)}_{} + \underbrace{(2x)(2y)}_{} + \underbrace{(5y)(3x)}_{} + \underbrace{(5y)(2y)}_{}$$

$$= 6x^2 + 4xy + 15xy + 10y^2$$

$$= 6x^2 + 19xy + 10y^2$$

EXAMPLE 5   Multiplying binomials involving two variables

Use FOIL to multiply:

a. $(5x + 2y)(2x + 3y)$ **b.** $(3x - y)(4x - 3y)$

SOLUTION

$$\quad\quad\quad F \quad\quad\quad O \quad\quad\quad I \quad\quad\quad L$$

a. $(5x + 2y)(2x + 3y) = \underbrace{(5x)(2x)}_{} + \underbrace{(5x)(3y)}_{} + \underbrace{(2y)(2x)}_{} + \underbrace{(2y)(3y)}_{}$

$= 10x^2 + 15xy + 4xy + 6y^2$

$= 10x^2 + 19xy + 6y^2$

$$\quad\quad\quad F \quad\quad\quad O \quad\quad\quad I \quad\quad\quad L$$

b. $(3x - y)(4x - 3y) = \underbrace{(3x)(4x)}_{} + \underbrace{(3x)(-3y)}_{} + \underbrace{(-y)(4x)}_{} + \underbrace{(-y)(-3y)}_{}$

$= 12x^2 - 9xy - 4xy + 3y^2$

$= 12x^2 - 13xy + 3y^2$

PROBLEM 5

Use FOIL to multiply:

a. $(4a + 3b)(3a + 5b)$

b. $(2a - b)(3a - 4b)$

Now one more thing. How do we multiply the expression in the *Getting Started?*

$$(x - L)(x - 2L)$$

We do it in Example 6.

Answers to PROBLEMS

4. a. $6a^2 + a - 15$

   b. $8a^2 - 14a + 3$

5. a. $12a^2 + 29ab + 15b^2$

   b. $6a^2 - 11ab + 4b^2$

EXAMPLE 6   Multiplying binomials involving two variables

Perform the indicated operation:

$$(x - L)(x - 2L)$$

SOLUTION

$$
\begin{aligned}
(x - L)(x - 2L) &= \overset{F}{\underbrace{x \cdot x}} + \overset{O}{\underbrace{(x)(-2L)}} + \overset{I}{\underbrace{(-L)(x)}} + \overset{L}{\underbrace{(-L)(-2L)}} \\
&= x^2 - \underbrace{2xL - xL} + 2L^2 \\
&= x^2 - 3xL + 2L^2
\end{aligned}
$$

PROBLEM 6

Perform the indicated operation:

$$(y - 2L)(y - 3L)$$

D › Applications Involving Multiplication of Polynomials

Suppose we wish to find out how much is spent annually on hospital care. According to the American Hospital Association, average daily room charges can be approximated by $C(t) = 160 + 14t$ (in dollars), where t is the number of years after 1990. On the other hand, the U.S. National Health Center for Health Statistics indicated that the average stay (in days) in the hospital can be approximated by $D(t) = 7 - 0.2t$, where t is the number of years after 1990. The amount spent annually on hospital care is given by

Cost per day $\times$ Number of days $= C(t) \times D(t)$

We find this product in Example 7.

EXAMPLE 7   Hospital care costs

Find: $C(t) \times D(t) = (160 + 14t)(7 - 0.2t)$

SOLUTION   We use the FOIL method:

$$
\begin{aligned}
(160 + 14t)(7 - 0.2t) &= \overset{F}{\underbrace{160 \cdot 7}} + \overset{O}{\underbrace{160 \cdot (-0.2t)}} + \overset{I}{\underbrace{14t \cdot 7}} + \overset{L}{\underbrace{14t \cdot (-0.2t)}} \\
&= 1120 - \underbrace{32t + 98t} - 2.8t^2 \\
&= 1120 + 66t - 2.8t^2
\end{aligned}
$$

Thus, the total amount spent annually on hospital care is $1120 + 66t - 2.8t^2$. Can you calculate what this amount was for 2000? What will it be for the year 2010?

PROBLEM 7

Suppose that at a certain hospital, the cost per day is $(200 + 15t)$ and the average stay (in days) is $(10 - 0.1t)$. What is the amount spent annually at this hospital?

Some students prefer a **grid method** to multiply polynomials. Thus, to do Example 4(a), $(2x + 5)(3x - 4)$, create a grid separated into four compartments. Place the term $(2x + 5)$ at the top and the term $(3x - 4)$ on the side of the grid. Multiply the rows and columns of the grid as shown. (After you get some practice, you can skip the initial step and write $6x^2$, $15x$, $-8x$, and -20 in the grid.)

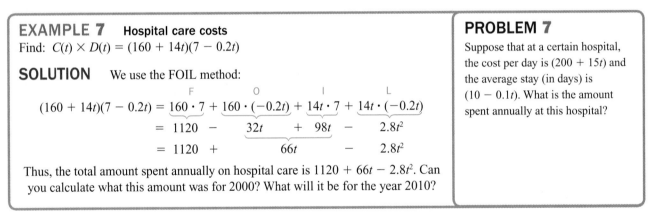

	$2x$	$+$	5
$3x$	$3x \cdot 2x$ $6x^2$		$3x \cdot 5$ $15x$
-4	$-4 \cdot 2x$ $-8x$		$-4 \cdot 5$ -20

Finish by writing the results of each of the grid boxes: $6x^2 + 15x - 8x - 20$

And combining like terms: $6x^2 + 7x - 20$

You can try using this technique in the margin problems or in the exercises!

Calculator Corner

Checking Equivalency

In the Section 4.1 *Calculator Corner,* we agreed that **two expressions are equivalent if their graphs are identical.** Thus, to check Example 1 we have to check that $(-3x^2)(2x^3) = -6x^5$. Let $Y_1 = (-3x^2)(2x^3)$ and $Y_2 = -6x^5$. Press GRAPH and the graph shown here will appear. To confirm the result numerically, press 2nd GRAPH and you get the result in the table.

X	Y₁	Y₂
0	0	0
1	-6	-6
2	-192	-192
3	-1458	-1458
4	-6144	-6144
5	-18750	-18750
6	-46656	-46656

X=0

You can check the rest of the examples except Examples 2(a), 5, and 6. Why?

› Exercises **4.6**

⟨ **A** ⟩   **Multiplying Two Monomials**   In Problems 1–6, find the product.

1. $(5x^3)(9x^2)$ **2.** $(8x^4)(9x^3)$ **3.** $(-2x)(5x^2)$ **4.** $(-3y^2)(4y^3)$

5. $(-2y^2)(-3y)$ **6.** $(-5z)(-3z)$

⟨ **B** ⟩   **Multiplying a Monomial and a Binomial**   In Problems 7–20, remove parentheses and simplify.

7. $3(x + y)$ **8.** $5(2x + y)$ **9.** $5(2x - y)$ **10.** $4(3x - 4y)$

11. $-4x(2x - 3)$ **12.** $-6x(5x - 3)$ **13.** $(x^2 + 4x)x^3$ **14.** $(x^2 + 2x)x^2$

15. $(x - x^2)4x$ **16.** $(x - 3x^2)5x$ **17.** $(x + y)3x$ **18.** $(x + 2y)5x^2$

19. $(2x - 3y)(-4y^2)$ **20.** $(3x^2 - 4y)(-5y^3)$

⟨ **C** ⟩   **Multiplying Two Binomials Using the FOIL Method**   In Problems 21–56, use the FOIL method to perform the indicated operation.

21. $(x + 1)(x + 2)$ **22.** $(y + 3)(y + 8)$ **23.** $(y + 4)(y - 9)$ **24.** $(y + 6)(y - 5)$

25. $(x - 7)(x + 2)$ **26.** $(z - 2)(z + 9)$ **27.** $(x - 3)(x - 9)$ **28.** $(x - 2)(x - 11)$

29. $(y - 3)(y - 3)$ **30.** $(y + 4)(y + 4)$ **31.** $(2x + 1)(3x + 2)$ **32.** $(4x + 3)(3x + 5)$

33. $(3y + 5)(2y - 3)$ **34.** $(4y - 1)(3y + 4)$ **35.** $(5z - 1)(2z + 9)$ **36.** $(2z - 7)(3z + 1)$

37. $(2x - 4)(3x - 11)$ **38.** $(5x - 1)(2x - 1)$ **39.** $(4z + 1)(4z + 1)$ **40.** $(3z - 2)(3z - 2)$

41. $(3x + y)(2x + 3y)$ **42.** $(4x + z)(3x + 2z)$ **43.** $(2x + 3y)(x - y)$ **44.** $(3x + 2y)(x - 5y)$

45. $(5z - y)(2z + 3y)$ **46.** $(2z - 5y)(3z + 2y)$ **47.** $(3x - 2z)(4x - z)$ **48.** $(2x - 3z)(5x - z)$

49. $(2x - 3y)(2x - 3y)$ **50.** $(3x + 5y)(3x + 5y)$ **51.** $(3 + 4x)(2 + 3x)$ **52.** $(2 + 3x)(3 + 2x)$

53. $(2 - 3x)(3 + x)$ **54.** $(3 - 2x)(2 + x)$ **55.** $(2 - 5x)(4 + 2x)$ **56.** $(3 - 5x)(2 + 3x)$

‹ D › Applications Involving Multiplication of Polynomials

57. *Area of a rectangle* The area A of a rectangle is obtained by multiplying its length L by its width W; that is, $A = LW$. Find the area of the rectangle shown in the figure.

58. *Area of a rectangle* Use the formula in Problem 57 to find the area of a rectangle of width $x - 4$ and length $x + 3$.

59. *Height of a thrown object* The height reached by an object t seconds after being thrown upward with a velocity of 96 feet per second is given by $16t(6 - t)$. Use the distributive property to simplify this expression.

60. *Resistance* The resistance R of a resistor varies with the temperature T according to the equation $R = (T + 100)(T + 20)$. Use the distributive property to simplify this expression.

61. *Gas property expression* In chemistry, when V is the volume and P is the pressure of a certain gas, we find the expression $(V_2 - V_1)(CP + PR)$, where C and R are constants. Use the distributive property to simplify this expression.

Area of a rectangle The garage shown is 40 feet by 20 feet. You want to convert it to a bigger garage with two storage areas, S_1 and S_2.

62. What is the area of the existing garage?

63. If you extend the long side by 8 feet and the short side by 5 feet, what is the area of the new garage?

64. Calculate the areas of S_1, S_2, and S_3, and write your answers in the appropriate places in the diagram.

65. Determine the total area of the new garage by adding the area of the original garage to the areas of S_1, S_2, and S_3; that is, add the answers you obtained in Problems 62 and 64.

66. Is the area of the new garage (Problem 63) the same as the answer in Problem 65?

67. If you are not sure how big you want the storage rooms, extend the long side of the garage by x feet and the short side by y feet.

　a. Find the area of S_1.

　b. Find the area of S_2.

　c. Find the area of S_3.

68. The area of the new garage is $(40 + x)(20 + y)$. Simplify this expression.

69. Add the areas of S_1, S_2, S_3, and the area of the original garage. Is the answer the same as the one you obtained in Problem 68?

70. *Blood velocity* The velocity V_r of a blood corpuscle in a vessel depends on the distance r from the center of the vessel and is given by the equation $V_r = V_m \left(1 - \frac{r^2}{R^2}\right)$, where V_m is the maximum velocity and R the radius of the vessel. Multiply this expression.

71. *Crop yield* The yield Y from a grove of Florida orange trees populated with x orange trees per acre is given by the equation $Y = x(1000 - x)$. Multiply this expression.

72. *Dividing lots* A rectangular lot is fenced and divided into two identical fenced lots as shown in the figure. If 1200 feet of fence are used, the area A of the lot is $A = w\left(600 - \frac{3}{2}w\right)$. Multiply this expression.

73. *Revenue* If x units of a product are sold at a price p the revenue R is $R = xp$. Suppose that the demand for a product is given by the equation $x = 1000 - 30p$.

   a. Write a formula for the revenue R in simplified form.

   b. What is the revenue when the price is $20 per unit?

74. *Revenue* A company manufactures and sells x jogging suits at p dollars every day. If $x = 3000 - 30p$, write a formula for the daily revenue R in simplified form and use it to find the revenue on a day in which the suits were selling for $40 each.

〉〉〉 *Using Your Knowledge*

Profitable Polynomials   Do you know how to find the profit made in a certain business? The profit P is the difference between the revenue R and the expense E of doing business. In symbols,

$$P = R - E$$

Of course, R depends on the number n of items sold and their price p (in dollars); that is,

$$R = np$$

Thus,

$$P = np - E$$

75. If $n = -3p + 60$ and $E = -5p + 100$, find P.

76. If $n = -2p + 50$ and $E = -3p + 300$, find P.

77. In Problem 75, if the price was $2, what was the profit P?

78. In Problem 76, if the price was $10, what was the profit P?

〉〉〉 *Write On*

79. Will the product of two monomials always be a monomial? Explain.

80. If you multiply a monomial and a binomial, will you ever get a trinomial? Explain.

81. Will the product of two binomials (after combining like terms) always be a trinomial? Explain.

82. Multiply:
   $$(x + 1)(x - 1) =$$
   $$(y + 2)(y - 2) =$$
   $$(z + 3)(z - 3) =$$

   What is the pattern?

〉〉〉 *Concept Checker*

Fill in the blank(s) with the correct word(s), phrase, or mathematical statement.

83. When using **FOIL** to multiply binomials, the **F** means that the _____ terms are to be multiplied.

84. When using **FOIL** to multiply binomials, the **O** means that the _____ terms are to be multiplied.

85. When using **FOIL** to multiply binomials, the **I** means that the _____ terms are to be multiplied.

86. When using **FOIL** to multiply binomials, the **L** means that the _____ terms are to be multiplied.

least	**outer**
last	**first**
inner	**final**
odd	

⟩⟩⟩ *Mastery Test*

Multiply and simplify:

87. $(-7x^4)(5x^2)$

88. $(-8a^3)(-5a^5)$

89. $(x + 7)(x - 3)$

90. $(x - 2)(x + 8)$

91. $(3x + 4)(3x - 1)$

92. $(4x + 3y)(3x + 2y)$

93. $(5x - 2y)(2x - 3y)$

94. $(x - L)(x - 3L)$

95. $6(x - 3y)$

96. $(x^3 + 5x)(-4x^5)$

⟩⟩⟩ *Skill Checker*

Find:

97. $(8x)^2$

98. $(4y)^2$

99. $(3x)^2$

100. $(-A)^2$

101. $(-3A)^2$

102. $A(-A)$

4.7   Special Products of Polynomials

▶ Objectives

Expand (simplify) binomials of the form

A ⟩ $(X + A)^2$

B ⟩ $(X - A)^2$

C ⟩ $(X + A)(X - A)$

D ⟩ Multiply a binomial by a trinomial.

E ⟩ Multiply any two polynomials.

▶ To Succeed, Review How To . . .

1. Use the FOIL method to multiply polynomials (pp. 359–362).
2. Multiply expressions (pp. 309–311).
3. Use the distributive property to simplify expressions (pp. 81, 89–90).

▶ Getting Started
Expanding Your Property

In Section 4.6, we learned how to use the distributive property and the FOIL method to multiply two binomials. In this section we shall develop patterns that will help us in multiplying certain binomial products that occur frequently. For example, do you know how to find the area of this property lot? Since the land is a square, the area is

$$(x + 10)(x + 10) = (x + 10)^2$$

The expression $(x + 10)^2$ is the square of the binomial $x + 10$. You can use the FOIL method, of course, but this type of expression is so common in algebra that we have *special products* or formulas that we use to multiply (or expand) them. We are now ready to study several of these special products. You will soon find that we've already studied the first of these special products: the FOIL method is actually special product 1!

As we implied in the *Getting Started,* there's another way to look at the FOIL method: we can format it as a special product. Do you recall the FOIL method?

$$\overset{\text{F} \qquad \text{O} \qquad \text{I} \qquad \text{L}}{(X + A)(X + B) = X^2 + XB + AX + AB}$$
$$= X^2 + (B + A)X + AB$$
$$= X^2 + (A + B)X + AB$$

Thus, we have our first special product.

PRODUCT OF TWO BINOMIALS	**Special Product 1**
	$(X + A)(X + B) = X^2 + (A + B)X + AB$ **(SP1)**

For example, $(x - 3)(x + 5)$ can be multiplied using SP1 with $X = x$, $A = -3$, and $B = 5$. Thus,

$$(x - 3)(x + 5) = x^2 + (-3 + 5)x + (-3)(5)$$
$$= x^2 + 2x - 15$$

Similarly, to expand $(2x - 5)(2x + 1)$, we can let $X = 2x$, $A = -5$, and $B = 1$ in SP1. Hence,

$$(2x - 5)(2x + 1) = (2x)^2 + (-5 + 1)2x + (-5)(1)$$
$$= 4x^2 - 8x - 5$$

Why have we gone to the trouble of reformatting FOIL? Because, as you will see, three very predictable and very handy special products can be derived from SP1.

A ⟩ Squaring Sums: $(X + A)^2$

Our second special product (SP2) deals with squaring polynomial sums. Let's see how it is developed by expanding $(X + 10)^2$. First, we start with SP1; we let $A = B$ and note that

$$(X + A)(X + A) = (X + A)^2$$

Now we let $A = 10$, so in our expansion of $(X + 10)^2$ [or $(X + A)^2$], we have

$$(X + A)(X + A) = X \cdot X + (A + A)X + A \cdot A$$

These would be B in SP1.
Note that $(A + A)X = 2AX$.

Thus,

$$(X + A)^2 = X^2 + 2AX + A^2$$

Foiled again

$(X + A)^2 = X^2 + 2AX + A^2$

Now we have our second special product, SP2:

THE SQUARE OF A BINOMIAL SUM	**Special Product 2**
	$(X + A)^2 = X^2 + 2AX + A^2$ **(SP2)** Note that $(X + A)^2 \neq X^2 + A^2$. (See the *Using Your Knowledge* in Exercises 4.7.)

Here is the pattern used in SP2:

First term		Second term		Square the first term.		Multiply the terms and double.		Square the last term.
$(X$	$+$	$A)^2$	$=$	X^2	$+$	$2AX$	$+$	A^2

If you examine this pattern carefully, you can see that SP2 is a result of the FOIL method when a binomial sum is squared. Can we obtain the same product with FOIL?

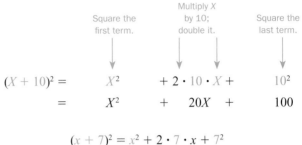

$$\begin{array}{ccccccc} & & F & O & I & & L \\ (X + A)(X + A) = & \underbrace{X \cdot X} & + & \underbrace{AX + AX} & + & \underbrace{A \cdot A} \\ = & X^2 & + & 2AX & + & A^2 \end{array}$$

The answer is yes! We are now ready to expand $(X + 10)^2$ using SP2:

$$\begin{array}{ccccc} & \text{Square the} & \text{Multiply } X \\ & \text{first term.} & \text{by 10;} & & \text{Square the} \\ & & \text{double it.} & & \text{last term.} \\ & \downarrow & \downarrow & \downarrow & \downarrow \\ (X + 10)^2 = & X^2 & + 2 \cdot 10 \cdot X + & 10^2 \\ = & X^2 & + \quad 20X \quad + & 100 \end{array}$$

Similarly,

$$(x + 7)^2 = x^2 + 2 \cdot 7 \cdot x + 7^2$$
$$= x^2 + 14x + 49$$

EXAMPLE 1 **Squaring binomial sums**

Expand the binomial sum:

a. $(x + 9)^2$ **b.** $(2x + 3)^2$ **c.** $(3x + 4y)^2$

SOLUTION

a. $(x + 9)^2 = x^2 + 2 \cdot 9 \cdot x + 9^2$ Let $A = 9$ in SP2.
$\qquad = x^2 + 18x + 81$

b. $(2x + 3)^2 = (2x)^2 + 2 \cdot 3 \cdot 2x + 3^2$ Let $X = 2x$ and $A = 3$ in SP2.
$\qquad = 4x^2 + 12x + 9$

c. $(3x + 4y)^2 = (3x)^2 + 2(4y)(3x) + (4y)^2$ Let $X = 3x$ and $A = 4y$ in SP2.
$\qquad = 9x^2 + 24xy + 16y^2$

PROBLEM 1

Expand the binomial sum:

a. $(y + 6)^2$ **b.** $(2y + 1)^2$
c. $(2x + 3y)^2$

But remember, you do not need to use SP2, you can *always* use FOIL; SP2 simply saves you time and work.

B › Squaring Differences: $(X - A)^2$

Can we also expand $(X - 10)^2$? Of course! But we first have to learn how to expand $(X - A)^2$. To do this, we simply write $-A$ instead of A in SP2 to obtain

$$(X - A)^2 = X^2 + 2(-A)X + (-A)^2$$
$$= X^2 - 2AX + A^2$$

This is the special product we need, and we call it SP3, the square of a binomial difference.

| **THE SQUARE OF A BINOMIAL DIFFERENCE** | **Special Product 3** |
| | $(X - A)^2 = X^2 - 2AX + A^2$ **(SP3)** |

Note that $(X - A)^2 \neq X^2 - A^2$, so the only difference between the square of a sum and the square of a difference is the sign preceding $2AX$. Here is the comparison:

$$(X + A)^2 = X^2 + 2AX + A^2 \hspace{3cm} \textbf{(SP2)}$$
$$(X - A)^2 = X^2 - 2AX + A^2 \hspace{3cm} \textbf{(SP3)}$$

Keeping this in mind,

$$(X - 10)^2 = X^2 - 2 \cdot 10 \cdot X + 10^2$$
$$= X^2 - 20X + 100$$

Similarly,

$$(x - 3)^2 = x^2 - 2 \cdot 3 \cdot x + 3^2$$
$$= x^2 - 6x + 9$$

EXAMPLE 2   **Squaring binomial differences**

Expand the binomial difference:

a. $(x - 5)^2$ **b.** $(3x - 2)^2$ **c.** $(2x - 3y)^2$

SOLUTION

a. $(x - 5)^2 = x^2 - 2 \cdot 5 \cdot x + 5^2$ Let $A = 5$ in SP3.
      $= x^2 - 10x + 25$

b. $(3x - 2)^2 = (3x)^2 - 2 \cdot 2 \cdot 3x + 2^2$ Let $X = 3x$ and $A = 2$ in SP3.
      $= 9x^2 - 12x + 4$

c. $(2x - 3y)^2 = (2x)^2 - 2 \cdot 3y \cdot 2x + (3y)^2$ Let $X = 2x$ and $A = 3y$ in SP3.
      $= 4x^2 - 12xy + 9y^2$

PROBLEM 2

Expand the binomial difference:

a. $(y - 4)^2$

b. $(2y - 3)^2$

c. $(3x - 2y)^2$

C ❭ Multiplying Sums and Differences: $(X + A)(X - A)$

We have one more special product, and this one is especially clever! Suppose we multiply the sum of two terms by the difference of the same two terms; that is, suppose we wish to multiply

$$(X + A)(X - A)$$

If we substitute $-A$ for B in SP1, then

$$(X + A)(X + B) = X^2 + (A + B)X + AB$$

becomes

$$(X + A)(X - A) = X^2 + (A - A)X + A(-A)$$
$$= X^2 + 0X - A^2$$
$$= X^2 - A^2$$

This gives us our last very special product, SP4.

THE PRODUCT OF THE SUM AND DIFFERENCE OF TWO TERMS

Special Product 4

$$(X + A)(X - A) = X^2 - A^2 \hspace{3cm} \textbf{(SP4)}$$

Note that $(X + A)(X - A) = (X - A)(X + A)$; so $(X - A)(X + A) = X^2 - A^2$. Thus, to multiply the sum and difference of two terms, we simply square the first term and then subtract from this the square of the last term. Checking this result using the FOIL method, we have

$$(X + A)(X - A) = X^2 \underset{0}{\underbrace{- \overset{O}{AX} + \overset{I}{AX}}} - \overset{L}{A^2}$$

$$= X^2 - A^2$$

Since the middle term is *always* zero, we have

$$(x + 3)(x - 3) = x^2 - 3^2 = x^2 - 9$$
$$(x + 6)(x - 6) = x^2 - 6^2 = x^2 - 36$$

Similarly, by the commutative property, we also have

$$(x - 3)(x + 3) = x^2 - 3^2 = x^2 - 9$$
$$(x - 6)(x + 6) = x^2 - 6^2 = x^2 - 36$$

EXAMPLE 3   Finding the product of the sum and difference of two terms

Multiply:

a. $(x + 10)(x - 10)$ **b.** $(2x + y)(2x - y)$

c. $(3x - 5y)(3x + 5y)$

SOLUTION

a. $(x + 10)(x - 10) = x^2 - 10^2$ Let $A = 10$ in SP4.
$$\qquad\qquad\qquad\quad = x^2 - 100$$

b. $(2x + y)(2x - y) = (2x)^2 - y^2$ Let $X = 2x$ and $A = y$ in SP4.
$$\qquad\qquad\qquad\quad = 4x^2 - y^2$$

c. $(3x - 5y)(3x + 5y) = (3x)^2 - (5y)^2$ Let $X = 3x$ and $A = 5y$ in SP4.
$$\qquad\qquad\qquad\qquad = 9x^2 - 25y^2$$

Note that

$$(3x - 5y)(3x + 5y) = (3x + 5y)(3x - 5y)$$

by the commutative property, so SP4 still applies.

PROBLEM 3

Multiply:

a. $(y + 9)(y - 9)$

b. $(3x + y)(3x - y)$

c. $(3x - 2y)(3x + 2y)$

D › Multiplying a Binomial by a Trinomial

Can we use the FOIL method to multiply any two polynomials? Unfortunately, no. But wait; if algebra is a generalized arithmetic, we should be able to multiply a binomial by a trinomial using the same techniques we employ to multiply, say, 23 by 342.

First let's review how we multiply 23 by 342. Here are the steps.

Step 1	Step 2	Step 3
342	342	342
$\times$ 23	$\times$ 23	$\times$ 23
1026	1026	1026
	6840	6840
		7866
$3 \times 342 = 1026$	$20 \times 342 = 6840$	$1026 + 6840 = 7866$

Now let's use this same technique to multiply two polynomials, say $(x + 5)$ and $(x^2 + x - 2)$.

Step 1	Step 2	Step 3
$x^2 + x - 2$	$x^2 + x - 2$	$x^2 + x - 2$
$x + 5$	$x + 5$	$x + 5$
$5x^2 + 5x - 10$	$5x^2 + 5x - 10$	$5x^2 + 5x - 10$
	$x^3 + x^2 - 2x$	$x^3 + x^2 - 2x$
		$x^3 + 6x^2 + 3x - 10$
$5(x^2 + x - 2)$	$x(x^2 + x - 2)$	$5x^2 + 5x - 10$
$= 5x^2 + 5x - 10$	$= x^3 + x^2 - 2x$	$x^3 + x^2 - 2x$
		$x^3 + 6x^2 + 3x - 10$

Note that in step 3, all *like* terms are placed in the same column so they can be combined.

For obvious reasons, this method is called the **vertical scheme** and can be used when one of the polynomials to be multiplied has *three or more terms*. Of course, we could have obtained the same result by using the distributive property:

$$(a + b)c = ac + bc$$

The procedure would look like this:

$$(a + b) \cdot c = a \cdot c + b \cdot c$$
$$(x + 5)(x^2 + x - 2) = x(x^2 + x - 2) + 5(x^2 + x - 2)$$
$$= x^3 + x^2 - 2x + 5x^2 + 5x - 10$$
$$= x^3 + (x^2 + 5x^2) + (-2x + 5x) - 10$$
$$= x^3 + 6x^2 + 3x - 10$$

EXAMPLE 4   **Multiplying a binomial by a trinomial**

Multiply: $(x - 3)(x^2 - 2x - 4)$

SOLUTION   Using the vertical scheme, we proceed as follows:

$$x^2 - 2x - 4$$
$$x - 3$$
$$-3x^2 + 6x + 12 \qquad \text{Multiply } x^2 - 2x - 4 \text{ by } -3.$$
$$x^3 - 2x^2 - 4x \qquad \text{Multiply } x^2 - 2x - 4 \text{ by } x.$$
$$x^3 - 5x^2 + 2x + 12 \qquad \text{Add like terms.}$$

Thus, the result is $x^3 - 5x^2 + 2x + 12$. You can also do this problem by using the distributive property $(a - b)c = ac - bc$. The procedure would look like this:

$$(a - b) \cdot c = a \cdot c - b \cdot c$$
$$(x - 3)(x^2 - 2x - 4) = x(x^2 - 2x - 4) - 3(x^2 - 2x - 4)$$
$$= x^3 - 2x^2 - 4x - 3x^2 + 6x + 12$$
$$= x^3 + (-2x^2 - 3x^2) + (-4x + 6x) + 12$$
$$= x^3 - 5x^2 + 2x + 12$$

PROBLEM 4

Multiply: $(y - 2)(y^2 - y - 3)$

Note that the same result is obtained in both cases.

E ⟩ Multiplying Two Polynomials

Here is the idea we used in Example 4.

PROCEDURE

Multiplying *Any* Two Polynomials (Term-By-Term Multiplication)

To multiply two polynomials, multiply each term of one by every term of the other and add the results.

Now that you've learned all the basic techniques used to multiply polynomials, you should be able to tackle any polynomial multiplication. To do this, you must *first* decide what special product (if any) is involved. Here are the special products we've studied.

SPECIAL PRODUCTS

$$(X + A)(X + B) = X^2 + (A + B)X + AB \qquad \textbf{(SP1 or FOIL)}$$
$$(X + A)(X + A) = (X + A)^2 = X^2 + 2AX + A^2 \qquad \textbf{(SP2)}$$
$$(X - A)(X - A) = (X - A)^2 = X^2 - 2AX + A^2 \qquad \textbf{(SP3)}$$
$$(X + A)(X - A) = X^2 - A^2 \qquad \textbf{(SP4)}$$

Of course, the FOIL method always works for the last three types of equations, but since it's more laborious, learning to recognize special products 2–4 is definitely worth the effort!

EXAMPLE 5   Using SP1 (FOIL) and the distributive property

Multiply: $3x(x + 5)(x + 6)$

SOLUTION   One way to approach this problem is to save the $3x$ multiplication until last. First, use FOIL (SP1):

$$
\begin{aligned}
3x(x + 5)(x + 6) &= 3x(x^2 + \overset{\text{F}}{6x} + \overset{\text{O}\ \ \text{I}}{5x} + \overset{\text{L}}{30}) \\
&= 3x(x^2 + 11x + 30) \\
&= 3x^3 + 33x^2 + 90x \qquad \text{Use the distributive property.}
\end{aligned}
$$

Alternatively, you can use the distributive property to multiply $(x + 5)$ by $3x$ and then proceed with FOIL, as shown here:

$$
\begin{aligned}
3x(x + 5)(x + 6) &= (3x \cdot x + 3x \cdot 5)(x + 6) & \text{Multiply } 3x(x + 5). \\
&= (3x^2 + 15x)(x + 6) & \text{Simplify.} \\
&= 3x^2 \cdot x + 3x^2 \cdot 6 + 15x \cdot x + 15x \cdot 6 & \text{Use FOIL.} \\
&= 3x^3 + 18x^2 + 15x^2 + 90x & \text{Simplify.} \\
&= 3x^3 + 33x^2 + 90x & \text{Collect like terms.}
\end{aligned}
$$

Of course, both methods produce the same result.

PROBLEM 5

Multiply: $2y(y + 2)(y + 3)$

EXAMPLE 6   Cubing a binomial

Expand: $(x + 3)^3$

SOLUTION   Recall that $(x + 3)^3 = (x + 3)(x + 3)(x + 3)$. Thus, we can square the sum $(x + 3)$ first and then use the distributive property as shown here.

PROBLEM 6

Expand: $(y + 2)^3$

Answers to PROBLEMS

5. $2y^3 + 10y^2 + 12y$ **6.** $y^3 + 6y^2 + 12y + 8$

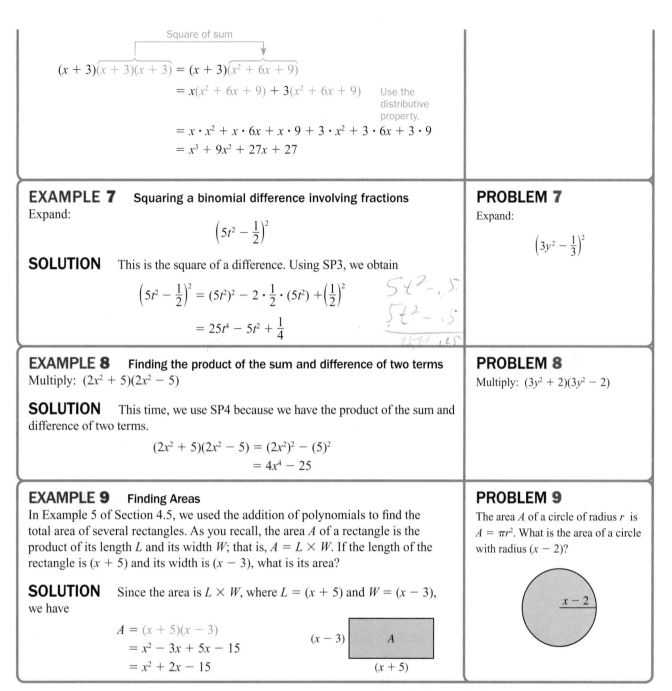

$$(x + 3)\overbrace{(x + 3)(x + 3)}^{\text{Square of sum}} = (x + 3)(x^2 + 6x + 9)$$

$$= x(x^2 + 6x + 9) + 3(x^2 + 6x + 9) \quad \text{Use the distributive property.}$$

$$= x \cdot x^2 + x \cdot 6x + x \cdot 9 + 3 \cdot x^2 + 3 \cdot 6x + 3 \cdot 9$$

$$= x^3 + 9x^2 + 27x + 27$$

EXAMPLE 7   Squaring a binomial difference involving fractions

Expand:

$$\left(5t^2 - \frac{1}{2}\right)^2$$

SOLUTION   This is the square of a difference. Using SP3, we obtain

$$\left(5t^2 - \frac{1}{2}\right)^2 = (5t^2)^2 - 2 \cdot \frac{1}{2} \cdot (5t^2) + \left(\frac{1}{2}\right)^2$$

$$= 25t^4 - 5t^2 + \frac{1}{4}$$

PROBLEM 7

Expand:

$$\left(3y^2 - \frac{1}{3}\right)^2$$

EXAMPLE 8   Finding the product of the sum and difference of two terms

Multiply: $(2x^2 + 5)(2x^2 - 5)$

SOLUTION   This time, we use SP4 because we have the product of the sum and difference of two terms.

$$(2x^2 + 5)(2x^2 - 5) = (2x^2)^2 - (5)^2$$

$$= 4x^4 - 25$$

PROBLEM 8

Multiply: $(3y^2 + 2)(3y^2 - 2)$

EXAMPLE 9   Finding Areas

In Example 5 of Section 4.5, we used the addition of polynomials to find the total area of several rectangles. As you recall, the area A of a rectangle is the product of its length L and its width W; that is, $A = L \times W$. If the length of the rectangle is $(x + 5)$ and its width is $(x - 3)$, what is its area?

SOLUTION   Since the area is $L \times W$, where $L = (x + 5)$ and $W = (x - 3)$, we have

$$A = (x + 5)(x - 3)$$

$$= x^2 - 3x + 5x - 15$$

$$= x^2 + 2x - 15$$

$(x - 3)$ $\boxed{\quad A \quad}$

$(x + 5)$

PROBLEM 9

The area A of a circle of radius r is $A = \pi r^2$. What is the area of a circle with radius $(x - 2)$?

$x - 2$

How do you know which of the special products you should use?

The trick to using these special products effectively, of course, is being able to *recognize* situations where they apply. Here are some tips.

PROCEDURE

Choosing the Appropriate Method for Multiplying Two Polynomials

1. Is the product the square of a binomial? If so, use SP2 or SP3:

$$(X + A)^2 = (X + A)(X + A) = X^2 + 2AX + A^2 \qquad \textbf{(SP2)}$$

$$(X - A)^2 = (X - A)(X - A) = X^2 - 2AX + A^2 \qquad \textbf{(SP3)}$$

Note that both answers have *three* terms.

Answers to PROBLEMS

7. $9y^4 - 2y^2 + \frac{1}{9}$ **8.** $9y^4 - 4$

9. $\pi(x - 2)^2 = \pi x^2 - 4\pi x + 4\pi$

(continued)

2. Are the two binomials in the product the sum and difference of the same two terms? If so, use SP4:

$$(X + A)(X - A) = X^2 - A^2 \qquad \textbf{(SP4)}$$

The answer has *two* terms.

3. Is the binomial product different from those in 1 and 2? If so, use FOIL. The answer will have *three* or *four* terms.

4. Is the product different from all the ones mentioned in 1, 2, and 3? If so, multiply every term of the first polynomial by every term of the second and collect like terms.

Calculator Corner

Checking Results

As we have done previously, we can use a calculator to check the results of the examples in this section. You know the procedure: let the original expression be Y_1 and the answer be Y_2. If the graphs are identical, the answer is correct. Thus, to check Example 5, let $Y_1 = 3x(x + 5)(x + 6)$ and $Y_2 = 3x^3 + 33x^2 + 90x$. Press GRAPH and you get only a partial graph, as shown here on the left. Since we are missing the bottom part of the graph, we need more values of y that are negative, so we have to fix the window we use for the graph. Press WINDOW and let Ymin $= -20$. Still, we do not see the whole graph, so try again with Ymin $= -90$. Press GRAPH and this time we see the complete graph as shown on the right.

You should now check Example 7, but use x instead of t. Can you check Example 1(c)? Why not?

A final word of advice before you do Exercises 4.7: SP1 (the FOIL method) is very important and should be thoroughly understood before you attempt the problems. This is because *all* the special products are *derived* from this formula. As a matter of fact, you can successfully complete most of the problems in Exercises 4.7 if you fully understand the result given in SP1.

Boost your grade at mathzone.com!
> Practice Problems
> NetTutor
> Self-Tests
> e-Professors
> Videos

❭ Exercises **4.7**

〈 A 〉 Squaring Sums In Problems 1–6, expand the given binomial sum.

1. $(x + 1)^2$

2. $(x + 6)^2$

3. $(2x + 1)^2$

4. $(2x + 5)^2$

5. $(3x + 2y)^2$

6. $(4x + 5y)^2$

〈 B 〉 Squaring Differences In Problems 7–16, expand the given binomial difference.

7. $(x - 1)^2$

8. $(x - 2)^2$

9. $(2x - 1)^2$

10. $(3x - 4)^2$

11. $(3x - y)^2$

12. $(4x - y)^2$

13. $(6x - 5y)^2$

14. $(4x - 3y)^2$

15. $(2x - 7y)^2$

16. $(3x - 5y)^2$

⟨ **C** ⟩   **Multiplying Sums and Differences**   In Problems 17–30, find the product.

17. $(x + 2)(x - 2)$ **18.** $(x + 1)(x - 1)$ **19.** $(x + 4)(x - 4)$

20. $(3x + y)(3x - y)$ **21.** $(3x + 2y)(3x - 2y)$ **22.** $(2x + 5y)(2x - 5y)$

23. $(x - 6)(x + 6)$ **24.** $(x - 11)(x + 11)$ **25.** $(x - 12)(x + 12)$

26. $(x - 9)(x + 9)$ **27.** $(3x - y)(3x + y)$ **28.** $(5x - 6y)(5x + 6y)$

29. $(2x - 7y)(2x + 7y)$ **30.** $(5x - 8y)(5x + 8y)$

In Problems 31–40, use the special products to multiply the given expressions.

31. $(x^2 + 2)(x^2 + 5)$ **32.** $(x^2 - 3)(x^2 + 2)$ **33.** $(x^2 + y)^2$

34. $(2x^2 + y)^2$ **35.** $(3x^2 - 2y^2)^2$ **36.** $(4x^3 - 5y^3)^2$

37. $(x^2 - 2y^2)(x^2 + 2y^2)$ **38.** $(x^2 - 3y^2)(x^2 + 3y^2)$ **39.** $(2x + 4y^2)(2x - 4y^2)$

40. $(5x^2 + 2y)(5x^2 - 2y)$

⟨ **D** ⟩   **Multiplying a Binomial by a Trinomial**   In Problems 41–52, find the product.

41. $(x + 3)(x^2 + x + 5)$ **42.** $(x + 2)(x^2 + 5x + 6)$ **43.** $(x + 4)(x^2 - x + 3)$

44. $(x + 5)(x^2 - x + 2)$ **45.** $(x + 3)(x^2 - x - 2)$ **46.** $(x + 4)(x^2 - x - 3)$

47. $(x - 2)(x^2 + 2x + 4)$ **48.** $(x - 3)(x^2 + x + 1)$ **49.** $-(x - 1)(x^2 - x + 2)$

50. $-(x - 2)(x^2 - 2x + 1)$ **51.** $-(x - 4)(x^2 - 4x - 1)$ **52.** $-(x - 3)(x^2 - 2x - 2)$

⟨ **E** ⟩   **Multiplying Two Polynomials**   In Problems 53–80, find the product.

53. $2x(x + 1)(x + 2)$ **54.** $3x(x + 2)(x + 5)$ **55.** $3x(x - 1)(x + 2)$

56. $4x(x - 2)(x + 3)$ **57.** $4x(x - 1)(x - 2)$ **58.** $2x(x - 3)(x - 1)$

59. $5x(x + 1)(x - 5)$ **60.** $6x(x + 2)(x - 4)$ **61.** $(x + 5)^3$

62. $(x + 4)^3$ **63.** $(2x + 3)^3$ **64.** $(3x + 2)^3$

65. $(2x + 3y)^3$ **66.** $(3x + 2y)^3$ **67.** $(4t^2 + 3)^2$

68. $(5t^2 + 1)^2$ **69.** $(4t^2 + 3u)^2$ **70.** $(5t^2 + u)^2$

71. $\left(3t^2 - \dfrac{1}{3}\right)^2$ **72.** $\left(4t^2 - \dfrac{1}{2}\right)^2$ **73.** $\left(3t^2 - \dfrac{1}{3}u\right)^2$

74. $\left(4t^2 - \dfrac{1}{2}u\right)^2$ **75.** $(3x^2 + 5)(3x^2 - 5)$ **76.** $(4x^2 + 3)(4x^2 - 3)$

77. $(3x^2 + 5y^2)(3x^2 - 5y^2)$ **78.** $(4x^2 + 3y^2)(4x^2 - 3y^2)$ **79.** $(4x^3 - 5y^3)(4x^3 + 5y^3)$

80. $(2x^4 - 3y^3)(2x^4 + 3y^3)$

81. Find the volume V of the cone when the height is $h = x + 4$ and the radius is $r = x + 2$.

Cone

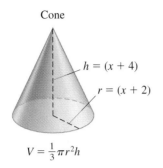

$h = (x + 4)$

$r = (x + 2)$

$V = \frac{1}{3}\pi r^2 h$

82. Find the volume V of the parallelepiped (box) when the width is $a = x$, the length is $c = x + 5$, and the height is $b = x + 4$.

Parallelepiped

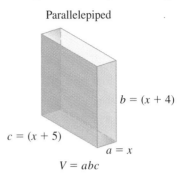

$b = (x + 4)$

$c = (x + 5)$

$a = x$

$V = abc$

Sphere

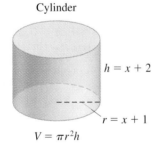

$r = x + 1$

$V = \frac{4}{3}\pi r^3$

83. Find the volume V of the sphere when the radius is $r = x + 1$.

84. Find the volume V of the hemisphere (half of the sphere) when the radius is $r = x + 2$.

85. Find the volume V of the cylinder when the height is $h = x + 2$ and the radius is $r = x + 1$.

Cylinder

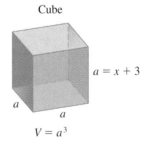

$h = x + 2$

$r = x + 1$

$V = \pi r^2 h$

86. Find the volume V of the cube when the length of the side is $a = x + 3$.

Cube

$a = x + 3$

a

a

$V = a^3$

〉〉〉 Applications

87. *Drain current* The drain current I_D for a certain gate source is modeled by the equation

$$I_D = 0.004\left(1 + \frac{V}{2}\right)^2$$

where V is the voltage at the gate source. Expand this expression.

88. *Transconductance* The equation that models a transconductance curve is

$$I_D = I_{DSS}\left[1 - \frac{V_1}{V_2}\right]^2$$

Expand this expression. (Here I_{DSS} is the current from drain to source with shorted gate, V_1 is the voltage at the gate source, and V_2 is the voltage when the gate source is off.)

89. *Heat transfer* The heat transmission between two objects of temperatures T_2 and T_1 involves the expression

$$(T_1^2 + T_2^2)(T_1^2 - T_2^2)$$

Multiply this expression.

90. *Deflection of a beam* The deflection of a certain beam involves the expression $w(l^2 - x^2)^2$. Expand this expression.

91. *Heat from convection* The heat output from a natural draught convector is given by the equation $K(t_n - t_a)^2$. Expand this expression.

92. *Pressure and volume of gases* When studying the pressure P and the volume V of a gas we deal with the expression. $(V_2 - V_1)(CP + PR)$, where C and R are constants. Multiply this expression.

93. *Relation of velocity, acceleration, and distance* The equation relating the velocity, the acceleration a and the distance s is $(v_i + v_0)(v_i - v_0) = 2as$. Multiply the expression on the left of the equation.

94. *Velocity of fluids* To find the velocity of a fluid through a pipe of radius r and outer diameter D we use the expression $k(D - 2r)(D + 2r)$, where k is a constant. Multiply this expression.

〉〉〉 *Using Your Knowledge*

Binomial Fallacies A common fallacy (mistake) when multiplying binomials is to assume that

$$(x + y)^2 = x^2 + y^2$$

Here are some arguments that should convince you this is *not* true.

95. Let $x = 1, y = 2$.

   a. What is $(x + y)^2$?

   b. What is $x^2 + y^2$?

   c. Is $(x + y)^2 = x^2 + y^2$?

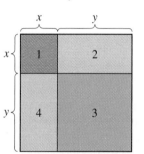

96. Let $x = 2, y = 1$.

   a. What is $(x - y)^2$?

   b. What is $x^2 - y^2$?

   c. Is $(x - y)^2 = x^2 - y^2$?

97. Look at the large square. Its area is $(x + y)^2$. The square is divided into four smaller areas numbered 1, 2, 3, and 4. What is the area of:

   a. square 1?   **b.** rectangle 2?

   c. square 3?   **d.** rectangle 4?

98. The total area of the square is $(x + y)^2$. It's also the sum of the four areas numbered 1, 2, 3, and 4. What is the sum of these four areas? (Simplify your answer.)

99. From your answer to Problem 98, what can you say about $x^2 + 2xy + y^2$ and $(x + y)^2$?

100. The sum of the areas of the squares numbered 1 and 3 is $x^2 + y^2$. Is it true that $x^2 + y^2 = (x + y)^2$?

〉〉〉 *Write On*

101. Write the procedure you use to square a binomial sum.

102. Write the procedure you use to square a binomial difference.

103. If you multiply two binomials, you usually get a trinomial. What is the exception? Explain.

104. Describe the procedure you use to multiply two polynomials.

〉〉〉 *Concept Checker*

Fill in the blank(s) with the correct word(s), phrase, or mathematical statement.

105. $(X + A)(X + B) = $ _____.

106. $(X + A)^2 = $ _____.

107. $(X - A)^2 = $ _____.

108. $(X + A)(X - A) = $ _____.

$X^2 + 2AX - A^2$ $X^2 - A^2$

$X^2 - 2AX - A^2$ $X^2 + 2AX + A^2$

$X^2 - 2AX + A^2$ $X^2 + XB + AX + AB$

$X^2 + A^2$

❯❯❯ *Mastery Test*

Expand and simplify:

109. $(x + 7)^2$

110. $(x - 2y)^2$

111. $(2x + 5y)^2$

112. $(x - 12)^2$

113. $(x^2 + 2y)^2$

114. $(x - 3y)^2$

115. $(3x - 2y)(3x + 2y)$

116. $(3x^2 + 2y)(3x^2 - 2y)$

117. $(3x + 4)(5x - 6)$

118. $(2x + 5)(5x - 2)$

119. $(7x - 2y)(3x - 4y)$

120. $(3y - 4x)(4y - 3x)$

121. $(x + 2)(x^2 + x + 1)$

122. $(x - 3)(x^2 - 3x - 1)$

123. $(x^2 - 2)(x^2 + x + 1)$

124. $(x + 3)^3$

125. $(x + 2y)^3$

126. $5x(x + 2)(x - 4)$

127. $3x(x^2 + 1)(x^2 - 1)$

128. $-4x(x^2 + 1)^2$

❯❯❯ *Skill Checker*

Simplify:

129. $\dfrac{20x^3}{10x^2}$

130. $\dfrac{28x^4}{4x^2}$

131. $\dfrac{-5x^2}{10x^2}$

132. $\dfrac{10x^2}{5x}$

4.8   Division of Polynomials

▶ Objectives

A ❯ Divide a polynomial by a monomial.

B ❯ Divide one polynomial by another polynomial.

▶ To Succeed, Review How To . . .

1. Use the distributive property to simplify expressions (pp. 80–81, 89–90).
2. Multiply polynomials (pp. 358–362).
3. Use the quotient rule of exponents (p. 311).

▶ Getting Started

Refrigerator Efficiency

In the preceding sections, we learned to add, subtract, and multiply polynomials. We are now ready for division. For example, do you know how efficient your refrigerator or your heat pump is? Their efficiency E is given by the quotient

$$E = \frac{T_1 - T_2}{T_1}$$

where T_1 and T_2 are the initial and final temperatures between which they operate. Can you do this division?

$$\frac{T_1 - T_2}{T_1}$$

Yes, you can, if you follow the steps:

$$\frac{T_1 - T_2}{T_1} = (T_1 - T_2) \div T_1$$

$$= (T_1 - T_2)\left(\frac{1}{T_1}\right) \qquad \text{Division by } T_1 \text{ is the same as multiplication by } \frac{1}{T_1}.$$

$$= T_1\left(\frac{1}{T_1}\right) - T_2\left(\frac{1}{T_1}\right) \qquad \text{Use the distributive property.}$$

$$= \frac{T_1}{T_1} - \frac{T_2}{T_1} \qquad \text{Multiply.}$$

$$= 1 - \frac{T_2}{T_1} \qquad \text{Since } \frac{T_1}{T_1} = 1.$$

In this section, we will learn how to divide a polynomial by a monomial, and then we will generalize this idea, which will enable us to divide one polynomial by another.

A ⟩ Dividing a Polynomial by a Monomial

To divide the binomial $4x^3 - 8x^2$ by the monomial $2x$, we proceed as we did in the *Getting Started*.

$$\frac{4x^3 - 8x^2}{2x} = (4x^3 - 8x^2) \div 2x$$

$$= (4x^3 - 8x^2)\left(\frac{1}{2x}\right) \qquad \text{Remember that to } \textit{divide} \text{ by } 2x, \text{ you multiply by the reciprocal } \frac{1}{2x}.$$

$$= 4x^3\left(\frac{1}{2x}\right) - 8x^2\left(\frac{1}{2x}\right) \qquad \text{Use the distributive property.}$$

$$= \frac{4x^3}{2x} - \frac{8x^2}{2x} \qquad \text{Multiply.}$$

$$= \frac{\overset{2x^2}{4x^3}}{\underset{}{2x}} - \frac{\overset{4x}{8x^2}}{\underset{}{2x}} \qquad \text{Use the quotient rule.}$$

$$= 2x^2 - 4x \qquad \text{Simplify.}$$

We then have

$$\frac{4x^3 - 8x^2}{2x} = \frac{4x^3}{2x} - \frac{8x^2}{2x} = 2x^2 - 4x$$

This example suggests the following rule.

RULE TO DIVIDE A POLYNOMIAL BY A MONOMIAL

To divide a polynomial by a monomial, divide *each* **term** in the polynomial by the *monomial*.

EXAMPLE 1   Dividing a polynomial by a monomial

Find the quotient:

a. $\dfrac{28x^4 - 14x^3}{7x^2}$

b. $\dfrac{20x^3 - 5x^2 + 10x}{10x^2}$

SOLUTION

a. $\dfrac{28x^4 - 14x^3}{7x^2} = \dfrac{28x^4}{7x^2} - \dfrac{14x^3}{7x^2} = 4x^2 - 2x$

b. $\dfrac{20x^3 - 5x^2 + 10x}{10x^2} = \dfrac{20x^3}{10x^2} - \dfrac{5x^2}{10x^2} + \dfrac{10x}{10x^2} = 2x - \dfrac{1}{2} + \dfrac{1}{x}$

PROBLEM 1

Find the quotient:

a. $\dfrac{27y^4 - 18y^3}{9y^2}$

b. $\dfrac{8y^3 - 4y^2 + 12y}{4y^2}$

Answers to PROBLEMS

1. a. $3y^2 - 2y$ **b.** $2y - 1 + \dfrac{3}{y}$

B › Dividing One Polynomial by Another Polynomial

If we wish to divide a polynomial, called the **dividend,** by another polynomial, called the **divisor,** we proceed very much as we did in long division in arithmetic. To show you that this is so, let's go through the division of 337 by 16 and $(x^2 + 3x + 3)$ (dividend) by $(x + 1)$ (divisor) side by side.

1. $16\overline{)337}$ with 2 above

Divide 33 by 16.
It goes twice.
Write 2 over the 33.

$x + 1\overline{)x^2 + 3x + 3}$ with x above

Divide x^2 by x
$\left(\dfrac{x^2}{x} = x\right)$
It goes x times.
Write x over the $3x$.

2.
$$\begin{array}{r} 2 \\ 16\overline{)337} \\ -32 \\ \hline 1 \end{array}$$

Multiply 2 by 16 and subtract the product 32 from 33 to obtain 1.

$$\begin{array}{r} x \\ x + 1\overline{)x^2 + 3x + 3} \\ (-)x^2 + x \\ \hline 0 + 2x \end{array}$$

Multiply x by $x + 1$ and subtract the product $x^2 + x$ from $x^2 + 3x$ to obtain $0 + 2x$.

3.
$$\begin{array}{r} 21 \\ 16\overline{)337} \\ -32 \\ \hline 17 \end{array}$$

Bring down the 7.
Now, divide 17 by 16. It goes once.
Write 1 after the 2.

$$\begin{array}{r} x + 2 \\ x + 1\overline{)x^2 + 3x + 3} \\ (-)x^2 + x \\ \hline 0 + 2x + 3 \end{array}$$

Bring down the 3.
Now, divide $2x$ by x
$\left(\dfrac{2x}{x} = 2\right)$
It goes 2 times.
Write $+ 2$ after the x.

4.
$$\begin{array}{r} 21 \\ 16\overline{)337} \\ -32 \\ \hline 17 \\ -16 \\ \hline 1 \end{array}$$

Multiply 1 by 16 and subtract the result from 17.
The remainder is 1.

$$\begin{array}{r} x + 2 \\ x + 1\overline{)x^2 + 3x + 3} \\ (-)x^2 + x \\ \hline 0 + 2x + 3 \\ (-)2x + 2 \\ \hline 1 \end{array}$$

Multiply 2 by $x + 1$ to obtain $2x + 2$.
Subtract this result from $2x + 3$.
The remainder is 1.

5. The answer **(quotient)** can be written as 21 R 1 (read "21 remainder 1") or as

$21 + \frac{1}{16}$, which is $21\frac{1}{16}$

The answer **(quotient)** can be written as $(x + 2)$ R 1 (read "$x + 2$ remainder 1") or as

$$x + 2 + \frac{1}{x + 1}$$

6. You can check this answer by multiplying 21 by 16 (336) and adding the remainder 1 to obtain 337, the dividend.

You can check the answer by multiplying $(x + 2)(x + 1)$, obtaining $x^2 + 3x + 2$, and then adding the remainder 1 to get $x^2 + 3x + 3$, the dividend.

EXAMPLE 2   Dividing a polynomial by a binomial

Divide: $x^2 + 2x - 17$ by $x - 3$

SOLUTION

x^2 divided by x is x.

$5x$ divided by x is 5.

$$x - 3 \overline{\smash{)}\, x^2 + 2x - 17}$$

$\begin{cases} \dfrac{x^2}{x} = x \\ x(x-3) = x^2 - 3x \end{cases}$

$(-)\, x^2 - 3x$

$5x - 17$

$\begin{cases} \dfrac{5x}{x} = 5 \\ 5(x-3) = 5x - 15 \end{cases}$

$(-)\, 5x - 15$

-2 Remainder

Thus, $(x^2 + 2x - 17) \div (x - 3) = (x + 5)\, \text{R} -2$ or

$$x + 5 + \frac{-2}{x - 3}$$

PROBLEM 2

Divide: $y^2 + y - 7$ by $y - 2$

If there are missing terms in the polynomial being divided, we insert zero coefficients, as shown in Example 3.

EXAMPLE 3   Dividing a third-degree polynomial by a binomial

Divide: $2x^3 - 2 - 4x$ by $2 + 2x$

SOLUTION   We write the polynomials in *descending* order, inserting $0x^2$ in the dividend, since the x^2 term is missing. We then have

$$2x + 2 \overline{\smash{)}\, 2x^3 + 0x^2 - 4x - 2}$$ quotient $x^2 - x - 1$

$\begin{cases} \dfrac{2x^3}{2x} = x^2 \\ \\ x^2(2x+2) = 2x^3 + 2x^2 \end{cases}$

$(-)\, 2x^3 + 2x^2$

$0 - 2x^2 - 4x$

$\begin{cases} -2x^2 \text{ divided by } 2x \text{ is } -x. \\ -x(2x+2) = -2x^2 - 2x. \end{cases}$

$(-)\, -2x^2 - 2x$

$-2x - 2$

$\begin{cases} -2x \text{ divided by } 2x \text{ is } -1. \\ -1(2x+2) = -2x - 2 \end{cases}$

$(-)\, -2x - 2$

0 There is no remainder.

Thus, $(2x^3 - 4x - 2) \div (2x + 2) = x^2 - x - 1$.

PROBLEM 3

Divide: $y^3 + y^2 - 7y - 3$ by $3 + y$

EXAMPLE 4   Dividing a fourth-degree polynomial by a binomial

Divide: $x^4 + x^3 - 3x^2 + 1$ by $x^2 - 3$

SOLUTION   We write the polynomials in *descending* order, inserting $0x$ for the missing term in the dividend. We then have

$$x^2 - 3 \overline{\smash{)}\, x^4 + x^3 - 3x^2 + 0x + 1}$$ quotient $x^2 + x$

$\begin{cases} x^4 \text{ divided by } x^2 \text{ is } x^2. \\ x^2(x^2 - 3) = x^4 - 3x^2 \end{cases}$

$(-)\, x^4 \qquad - 3x^2$

$0 + x^3 \qquad\qquad + 1$

$\begin{cases} x^3 \text{ divided by } x^2 \text{ is } x. \\ x(x^2 - 3) = x^3 - 3x \end{cases}$

$(-)\, x^3 \qquad - 3x$

$3x + 1$

We cannot divide $3x + 1$ by x^2, so we stop. The remainder is $3x + 1$.

PROBLEM 4

Divide: $y^4 + y^3 - 2y^2 + y + 1$ by $y^2 - 2$

(continued)

Answers to PROBLEMS

2. $y + 3 + \dfrac{-1}{y - 2}$ **3.** $y^2 - 2y - 1$ **4.** $y^2 + y + \dfrac{3y + 1}{y^2 - 2}$

In general, we stop the division when the degree of the remainder is less than the degree of the divisor or when the remainder is zero. Thus, $(x^4 + x^3 - 3x^2 + 1) \div (x^2 - 3) = (x^2 + x) \text{ R } (3x + 1)$. You can also write the answer as

$$x^2 + x + \frac{3x + 1}{x^2 - 3}$$

PROCEDURE

Checking the Result

1. Multiply the divisor $x^2 - 3$ by the quotient $x^2 + x$. (Use FOIL.)

2. Add the remainder $3x + 1$.

3. The result must be the dividend

$$x^4 + x^3 - 3x^2 + 1 \quad \text{(and it is!)}$$

Divisor Quotient

$$(x^2 - 3)(x^2 + x) = \quad x^4 + x^3 - 3x^2 - 3x$$
$$(+) \qquad\qquad\qquad\qquad 3x + 1$$
$$\overline{\quad x^4 + x^3 - 3x^2 \qquad + 1}$$

Calculator Corner

Checking Quotients

Now, how would you check Example 4? The original problem is $Y_1 = (x^4 + x^3 - 3x^2 + 1)/(x^2 - 3)$ and the answer is $Y_2 = x^2 + x + (3x + 1)/(x^2 - 3)$. Be very careful when entering the expressions Y_1 and Y_2. Note the parentheses! Press GRAPH. Did you get the same graph for Y_1 and Y_2? If you are not sure, let us check the numerical values by pressing 2nd GRAPH. Y_1 and Y_2 seem to have the same values, so we are probably correct. Why "probably"? Because we are not able to check that for every value of x (column 1), Y_1 and Y_2 (columns 2 and 3) have the same value. However, the lists give plausible evidence that they do.

X	Y₁	Y₂
0	-.3333	-.3333
1	0	0
2	13	13
3	13.667	13.667
4	21	21
5	30.727	30.727
6	42.576	42.576

X=0

› Exercises 4.8

⟨ **A** ⟩ **Dividing a Polynomial by a Monomial**   In Problems 1–10, divide.

1. $\dfrac{3x + 9y}{3}$

2. $\dfrac{6x + 8y}{2}$

3. $\dfrac{10x - 5y}{5}$

4. $\dfrac{24x - 12y}{6}$

5. $\dfrac{8y^3 - 32y^2 + 16y}{-4y^2}$

6. $\dfrac{9y^3 - 45y^2 + 9}{-3y^2}$

7. $10x^2 + 8x$ by x

8. $12x^2 + 18x$ by x

9. $15x^3 - 10x^2$ by $5x^2$

10. $18x^4 - 24x^2$ by $3x^2$

⟨ **B** ⟩ **Dividing One Polynomial by Another Polynomial**   In Problems 11–40, divide.

11. $x^2 + 5x + 6$ by $x + 3$

12. $x^2 + 9x + 20$ by $x + 4$

13. $y^2 + 3y - 11$ by $y + 5$

14. $y^2 + 2y - 16$ by $y + 5$

15. $2x + x^2 - 24$ by $x - 4$

16. $4x + x^2 - 21$ by $x - 3$

17. $-8 + 2x + 3x^2$ by $2 + x$

18. $-6 + x + 2x^2$ by $2 + x$

19. $2y^2 + 9y - 36$ by $7 + y$

20. $3y^2 + 13y - 32$ by $6 + y$

21. $2x^3 - 4x - 2$ by $2x + 2$

22. $3x^3 - 9x - 6$ by $3x + 3$

23. $y^4 - y^2 - 2y - 1$ by $y^2 + y + 1$

24. $y^4 - y^2 - 4y - 4$ by $y^2 + y + 2$

25. $8x^3 - 6x^2 + 5x - 9$ by $2x - 3$

26. $2x^4 - x^3 + 7x - 2$ by $2x + 3$

27. $x^3 - 8$ by $x - 2$

28. $x^3 + 64$ by $x + 4$

29. $8y^3 - 64$ by $2y - 4$

30. $27x^3 - 8$ by $3x - 2$

31. $x^4 - x^2 - 2x + 2$ by $x^2 - x - 1$

32. $y^4 - y^3 - 3y - 9$ by $y^2 - y - 3$

33. $x^5 - x^4 + 6x^2 - 5x + 3$ by $x^2 - 2x + 3$

34. $y^6 - y^5 + 6y^3 - 5y^2 + 3y$ by $y^2 - 2y + 3$

35. $m^4 - 11m^2 + 34$ by $m^2 - 3$

36. $n^3 - n^2 - 6n$ by $n^2 + 3n$

37. $\dfrac{x^3 - y^3}{x - y}$

38. $\dfrac{x^3 + 8}{x + 2}$

39. $\dfrac{x^3 + 8}{x - 2}$

40. $\dfrac{x^5 + 32}{x - 2}$

〉〉〉 **Applications**

41. *Unit costs*   The unit cost $U(x)$ of x key chains is modeled by the equation $U(x) = \frac{2x + 20}{x}$.

   a. Divide $2x + 20$ by x.

   b. What is the unit cost for one key chain?

   c. What is the unit cost for 10 key chains?

42. *Unit costs*   The unit cost $U(x)$ of x earrings is modeled by the equation $U(x) = \frac{x^2 + 10}{x}$.

   a. Divide $x^2 + 10$ by x.

   b. What is the unit cost for one earring?

   c. What is the unit cost for 10 earrings?

43. *Weight*   A box containing $2x$ books weighs $2x^2 + 4x$ pounds.

   a. How much does each book weigh?

   b. If the box has 10 books, what is the weight of each book?

Networks   Use the following information for Problems 44–46. The **Wye(Y)-delta** transform is a mathematical technique to simplify analysis of an electrical network and is used to establish the equivalence for networks of three terminals like the ones shown.

The equivalences for R_a, R_b, and R_c in the left diagram are given by the following equations and will be used in Problems 44–46.

$$R_a = \frac{R_1 R_2 + R_2 R_3 + R_3 R_1}{R_2}$$

$$R_b = \frac{R_1 R_2 + R_2 R_3 + R_3 R_1}{R_3}$$

$$R_c = \frac{R_1 R_2 + R_2 R_3 + R_3 R_1}{R_1}$$

Delta   Wye

44. Perform the indicated division by R_2 to obtain R_a.

45. Perform the indicated division by R_3 to obtain R_b.

46. Perform the indicated division by R_1 to obtain R_c.

⟩⟩⟩ *Using Your Knowledge*

Is It Profitable?   In business, the average cost per unit of a product, denoted by $\overline{C(x)}$, is defined by

$$\overline{C(x)} = \frac{C(x)}{x}$$

where $C(x)$ is a polynomial in the variable x, and x is the number of units produced.

47. If $C(x) = 3x^2 + 5x$, find $\overline{C(x)}$.

48. If $C(x) = 30 + 3x^2$, find $\overline{C(x)}$.

The average profit $\overline{P(x)}$ is

$$\overline{P(x)} = \frac{P(x)}{x}$$

where $P(x)$ is a polynomial in the variable x, and x is the number of units sold in a certain period of time.

49. If $P(x) = 50x + x^2 - 7000$ (dollars), find the average profit.

50. If in Problem 49, 100 units are sold in a period of 1 week, what is the average profit?

⟩⟩⟩ *Write On*

51. How can you check that your answer is correct when you divide one polynomial by another? Explain.

53. When you are dividing one polynomial by another, when do you stop the division process?

52. A problem in a recent test stated: "Find the quotient of $x^2 + 5x + 6$ and $x + 2$." Do you have to divide $x + 2$ by $x^2 + 5x + 6$ or do you have to divide $x^2 + 5x + 6$ by $x + 2$? Explain.

⟩⟩⟩ *Concept Checker*

Fill in the blank(s) with the correct word(s), phrase, or mathematical statement.

54. To divide a **polynomial** by a **monomial**, divide each term in the polynomial by the _____ .

55. To **check** the result of a **polynomial division, multiply** the _____ by the _____ and add the _____ .

dividend   remainder

quotient   divisor

monomial

⟩⟩⟩ *Mastery Test*

Divide:

56. $x^4 + x^3 - 2x^2 + 1$ by $x^2 - 2$

57. $2x^3 + x - 3$ by $x - 1$

58. $x^2 + 4x - 15$ by $x - 2$

59. $\dfrac{24x^4 - 18x^3}{6x^2}$

60. $\dfrac{16x^4 - 4x^2 + 8x}{8x^2}$

61. $\dfrac{-6y^3 + 12y^2 + 3}{-3y^2}$

⟩⟩⟩ *Skill Checker*

Find the LCM of the specified numbers:

62. 23 and 92

63. 20 and 18

64. 30 and 16

65. 40 and 12

66. 10, 18, and 12

67. 20, 30, and 18

68. 40, 15, and 10

❯ Collaborative Learning

How fast can you go?

How fast can you obtain information to solve a problem? Form three groups: library, the Web, and bookstore (where you can look at books, papers, and so on for free). Each group is going to research car prices. Select a car model that has been on the market for at least 5 years. Each of the groups should find:

1. The new car value and the value of a 3-year-old car of the same model

2. The estimated depreciation rate for the car

3. The estimated value of the car in 3 years

4. A graph comparing age and value of the car for the next 5 years

5. An equation of the form $C = P(1 - r)^n$ or $C = rn + b$, where n is the number of years after purchase and r is the depreciation rate

Which group finished first? Share the procedure used to obtain your information so the most efficient research method can be established.   ◪

❯ Research Questions

1. In the *Human Side of Algebra* at the beginning of this chapter, we mentioned that in the Golden Age of Greek mathematics, roughly from 300 to 200 B.C., three mathematicians "stood head and shoulders above all the others of the time." Apollonius was one. Who were the other two, and what were their contributions to mathematics?

2. Write a short paragraph about Diophantus and his contributions to mathematics.

3. Write a paper about the *Arithmetica* written by Diophantus. How many books were in this *Arithmetica?* What is its relationship to algebra?

4. It is said that "one of Diophantus' main contributions was the 'syncopation' of algebra." Explain what this means.

5. Write a short paragraph about Nicole Oresme and his mathematical achievements.

6. Expand on the discussion of the works of Nicholas Chuquet described in the *Human Side of Algebra.*

7. Write a short paper about the contributions of François Viète regarding the development of algebraic notation.

❯ Summary Chapter 4

Section	Item	Meaning	Example
4.1A	Product rule for exponents	$x^m \cdot x^n = x^{m+n}$	$x^3 \cdot x^5 = x^{3+5} = x^8$
4.1B	Quotient rule for exponents	$\dfrac{x^m}{x^n} = x^{m-n} \ (m > n, x \neq 0)$	$\dfrac{x^8}{x^3} = x^{8-3} = x^5$

(continued)

Section	Item	Meaning	Example
4.1C	Power rule for products Power rule for quotients	$(x^m y^n)^k = x^{mk} y^{nk}$ $\left(\dfrac{x}{y}\right)^m = \dfrac{x^m}{y^m} \ (y \neq 0)$	$(x^3 y^2)^4 = x^{12} y^8$ $\left(\dfrac{x^2}{y^3}\right)^4 = \dfrac{x^8}{y^{12}}$
4.2A	Zero exponent	For $x \neq 0$, $x^0 = 1$	$3^0 = 1, \ (-8)^0 = 1, \ (3x)^0 = 1$
4.2A	x^{-n}, n a positive integer	$x^{-n} = \dfrac{1}{x^n} \ (x \neq 0)$	$2^{-3} = \dfrac{1}{2^3} = \dfrac{1}{8}$
4.3A	Scientific notation	A number is in scientific notation when written in the form $M \times 10^n$, where M is a number between 1 and 10 and n is an integer.	3×10^{-3} and 2.7×10^5 are in scientific notation.
4.4A	Polynomial	An algebraic expression formed by using the operations of addition and subtraction on products of numbers and a variable raised to whole-number exponents	$x^2 + 3x - 5, \ 2x + 8 - x^3$, and $9x^7 - 3x^3 + 4x^8 - 10$ are polynomials but $\sqrt{x} - 3$, $\dfrac{x^2 - 2x + 3}{x}$, and $x^{3/2} - x$ are not.
	Terms	The parts of a polynomial separated by plus signs are the **terms.**	The terms of $x^2 - 2x + 3$ are x^2, $-2x$, and 3.
	Monomial	A polynomial with *one* term	$3x$, $7x^2$, and $-3x^{10}$ are monomials.
	Binomial	A polynomial with *two* terms	$3x + x^2$, $7x - 8$, and $x^3 - 8x^7$, are binomials.
	Trinomial	A polynomial with *three* terms	$-8 + 3x + x^2$, $7x - 8 + x^4$, and $x^3 - 8x^7 + 9$ are trinomials.
4.4B	Degree	The degree of a polynomial is the highest exponent of the variable.	The degree of $8 + 3x + x^2$ is 2, and the degree of $7x - 8 + x^4$ is 4.
4.6C	FOIL method (SP1)	To multiply two binomials such as $(x + a)(x + b)$, multiply the **F**irst terms, the **O**uter terms, the **I**nner terms, and the **L**ast terms then add.	$\quad\quad\quad\ \text{F}\quad\ \text{O}\quad\ \text{I}\quad\ \text{L}$ $(x + 2)(x + 3) = x^2 + 3x + 2x + 6$ $\quad\quad\quad\quad\quad = x^2 + 5x + 6$
4.7A	The square of a binomial sum (SP2)	$(X + A)^2 = X^2 + 2AX + A^2$	$(x + 5)^2 = x^2 + 2 \cdot 5 \cdot x + 5^2$ $\quad\quad\quad\ = x^2 + 10x + 25$
4.7B	The square of a binomial difference (SP3)	$(X - A)^2 = X^2 - 2AX + A^2$	$(x - 5)^2 = x^2 - 2 \cdot 5 \cdot x + 5^2$ $\quad\quad\quad\ = x^2 - 10x + 25$
4.7C	The product of the sum and difference of two terms (SP4)	$(X + A)(X - A) = X^2 - A^2$	$(x + 7)(x - 7) = x^2 - 7^2$ $\quad\quad\quad\quad\ = x^2 - 49$
4.8A	Dividing a polynomial by a monomial	To divide a polynomial by a monomial, divide each term in the polynomial by the monomial.	$\dfrac{35x^5 - 21x^3}{7x^2} = \dfrac{35x^5}{7x^2} - \dfrac{21x^3}{7x^2}$ $\quad\quad\quad\quad\ = 5x^3 - 3x$

Section	Item	Meaning	Example
4.8B	Dividing one polynomial by another	To divide one polynomial by another, use polynomial long division.	Divide $6x^2 + 11x - 35$ by $3x - 5$. $$\begin{array}{r} 2x + 7 \\ 3x - 5 \overline{)6x^2 + 11x - 35} \\ (-)\underline{6x^2 - 10x} \\ 21x - 35 \\ (-)\underline{21x - 35} \\ 0 \end{array}$$ Thus, $$(6x^2 + 11x - 35) \div (3x - 5)$$ is $2x + 7$.

❯Review Exercises **Chapter 4**

(If you need help with these questions, look in the section indicated in brackets.)

1. ❮**4.1A**❯ *Find the product.*

  a. **(i)** $(3a^2b)(-5ab^3)$

    (ii) $(4a^2b)(-6ab^4)$

    (iii) $(5a^2b)(-7ab^3)$

  b. **(i)** $(-2xy^2z)(-3x^2yz^4)$

    (ii) $(-3x^2yz^2)(-4xy^3z)$

    (iii) $(-4xyz)(-5xy^2z^3)$

2. ❮**4.1B**❯ *Find the quotient.*

  a. **(i)** $\dfrac{16x^6y^8}{-8xy^4}$ **b.** **(i)** $\dfrac{-8x^9y^7}{-16x^4y}$

    (ii) $\dfrac{24x^7y^6}{-4xy^3}$ **(ii)** $\dfrac{-5x^7y^8}{-10x^6y}$

    (iii) $\dfrac{-18x^8y^7}{9xy^4}$ **(iii)** $\dfrac{-3x^9y^7}{-9x^8y}$

3. ❮**4.1C**❯ *Simplify.*

  a. $(2^2)^3$ **b.** $(2^2)^2$

  c. $(3^2)^2$

4. ❮**4.1C**❯ *Simplify.*

  a. $(y^3)^2$ **b.** $(x^2)^3$

  c. $(a^4)^5$

5. ❮**4.1C**❯ *Simplify.*

  a. $(4xy^3)^2$ **b.** $(2x^2y)^3$

  c. $(3x^2y^2)^3$

6. ❮**4.1C**❯ *Simplify.*

  a. $(-2xy^3)^3$ **b.** $(-3x^2y^3)^2$

  c. $(-2x^2y^2)^3$

7. ❮**4.1C**❯ *Simplify.*

  a. $\left(\dfrac{2y^2}{x^4}\right)^2$ **b.** $\left(\dfrac{3x}{y^3}\right)^3$

  c. $\left(\dfrac{2x^2}{y^4}\right)^4$

8. ❮**4.1C**❯ *Simplify.*

  a. $(2x^4)^3(-2y^2)^2$

  b. $(3x^2)^2(-2y^3)^3$

  c. $(4x^3)^2(-2y^4)^4$

9. ❮**4.2A**❯ *Write using positive exponents and then simplify.*

  a. 2^{-3} **b.** 3^{-4}

  c. 5^{-2}

10. ❮**4.2A**❯ *Write using positive exponents.*

  a. $\left(\dfrac{1}{x}\right)^{-4}$ **b.** $\left(\dfrac{1}{y}\right)^{-3}$

  c. $\left(\dfrac{1}{z}\right)^{-5}$

11. ⟨**4.2B**⟩ *Write using negative exponents.*

    a. $\dfrac{1}{x^5}$ **b.** $\dfrac{1}{y^7}$

    c. $\dfrac{1}{z^8}$

12. ⟨**4.2C**⟩ *Multiply and simplify.*

    a. (i) $2^8 \cdot 2^{-5}$ **b. (i)** $y^{-3} \cdot y^{-5}$

       (ii) $2^6 \cdot 2^{-4}$ **(ii)** $y^{-2} \cdot y^{-3}$

       (iii) $3^6 \cdot 3^{-3}$ **(iii)** $y^{-4} \cdot y^{-2}$

13. ⟨**4.2C**⟩ *Divide and simplify.*

    a. (i) $\dfrac{x}{x^5}$ **b. (i)** $\dfrac{a^{-2}}{a^{-2}}$ **c. (i)** $\dfrac{x^{-2}}{x^{-3}}$

       (ii) $\dfrac{x}{x^7}$ **(ii)** $\dfrac{a^{-4}}{a^{-4}}$ **(ii)** $\dfrac{x^{-5}}{x^{-8}}$

       (iii) $\dfrac{x}{x^9}$ **(iii)** $\dfrac{a^{-10}}{a^{-10}}$ **(iii)** $\dfrac{x^{-7}}{x^{-9}}$

14. ⟨**4.2C**⟩ *Simplify.*

    a. $\left(\dfrac{2x^3y^4}{3x^4y^{-3}}\right)^{-2}$ **b.** $\left(\dfrac{3x^5y^{-3}}{2x^7y^4}\right)^{-3}$

    c. $\left(\dfrac{3x^{-5}y^{-4}}{2x^{-6}y^{-8}}\right)^{-2}$

15. ⟨**4.3A**⟩ *Write in scientific notation.*

    a. (i) 44,000,000

       (ii) 4,500,000

      (iii) 460,000

    b. (i) 0.0014

       (ii) 0.00015

      (iii) 0.000016

16. ⟨**4.3B**⟩ *Perform the indicated operations and write the answer in scientific notation.*

    a. (i) $(2 \times 10^2) \times (1.1 \times 10^3)$

      (ii) $(3 \times 10^2) \times (3.1 \times 10^4)$

     (iii) $(4 \times 10^2) \times (3.1 \times 10^5)$

    b. (i) $\dfrac{1.15 \times 10^{-3}}{2.3 \times 10^{-4}}$

      (ii) $\dfrac{1.38 \times 10^{-3}}{2.3 \times 10^{-4}}$

     (iii) $\dfrac{1.61 \times 10^{-3}}{2.3 \times 10^{-4}}$

17. ⟨**4.4A**⟩ *Classify as a monomial, binomial, or trinomial.*

    a. $9x^2 - 9 + 7x$ **b.** $7x^2$

    c. $3x - 1$

18. ⟨**4.4B**⟩ *Find the degree of the given polynomial.*

    a. $3x^2 - 7x + 8x^4$ **b.** $-4x + 2x^2 - 3$

    c. $8 + 3x - 4x^2$

19. ⟨**4.4C**⟩ *Write the given polynomial in descending order.*

    a. $4x^2 - 8x + 9x^4$

    b. $-3x + 4x^2 - 3$

    c. $8 + 3x - 4x^2$

20. ⟨**4.4D**⟩ *Find the value of $-16t^2 + 300$ for each value of t.*

    a. $t = 1$ **b.** $t = 3$

    c. $t = 5$

21. ⟨**4.5A**⟩ *Add the given polynomials.*

    a. $-5x + 7x^2 - 3$ and $-2x^2 - 7 + 4x$

    b. $-3x^2 + 8x - 1$ and $3 + 7x - 2x^2$

    c. $-4 + 3x^2 - 5x$ and $6x^2 - 2x + 5$

22. ⟨**4.5B**⟩ *Subtract the first polynomial from the second.*

    a. $3x - 4 + 7x^2$ from $6x^2 - 4x$

    b. $5x - 3 + 2x^2$ from $9x^2 - 2x$

    c. $6 - 2x + 5x^2$ from $2x - 5$

23. ⟨**4.6A**⟩ *Multiply.*

    a. $(-6x^2)(3x^5)$

    b. $(-8x^3)(5x^6)$

    c. $(-9x^4)(3x^7)$

24. ⟨**4.6B**⟩ *Remove parentheses (simplify).*

    a. $-2x^2(x + 2y)$

    b. $-3x^3(2x + 3y)$

    c. $-4x^3(5x + 7y)$

25. ⟨ **4.6C** ⟩ *Multiply.*
  a. $(x + 6)(x + 9)$
  b. $(x + 2)(x + 3)$
  c. $(x + 7)(x + 9)$

26. ⟨ **4.6C** ⟩ *Multiply.*
  a. $(x + 7)(x - 3)$
  b. $(x + 6)(x - 2)$
  c. $(x + 5)(x - 1)$

27. ⟨ **4.6C** ⟩ *Multiply.*
  a. $(x + 3)(x - 7)$
  b. $(x + 2)(x - 6)$
  c. $(x + 1)(x - 5)$

28. ⟨ **4.6C** ⟩ *Multiply.*
  a. $(3x - 2y)(2x - 3y)$
  b. $(5x - 3y)(4x - 3y)$
  c. $(4x - 3y)(2x - 5y)$

29. ⟨ **4.7A** ⟩ *Expand.*
  a. $(2x + 3y)^2$
  b. $(3x + 4y)^2$
  c. $(4x + 5y)^2$

30. ⟨ **4.7B** ⟩ *Expand.*
  a. $(2x - 3y)^2$
  b. $(3x - 2y)^2$
  c. $(5x - 2y)^2$

31. ⟨ **4.7C** ⟩ *Multiply.*
  a. $(3x - 5y)(3x + 5y)$
  b. $(3x - 2y)(3x + 2y)$
  c. $(3x - 4y)(3x + 4y)$

32. ⟨ **4.7D** ⟩ *Multiply.*
  a. $(x + 1)(x^2 + 3x + 2)$
  b. $(x + 2)(x^2 + 3x + 2)$
  c. $(x + 3)(x^2 + 3x + 2)$

33. ⟨ **4.7E** ⟩ *Multiply.*
  a. $3x(x + 1)(x + 2)$
  b. $4x(x + 1)(x + 2)$
  c. $5x(x + 1)(x + 2)$

34. ⟨ **4.7E** ⟩ *Expand.*
  a. $(x + 2)^3$
  b. $(x + 3)^3$
  c. $(x + 4)^3$

35. ⟨ **4.7E** ⟩ *Expand.*
  a. $\left(5x^2 - \dfrac{1}{2}\right)^2$
  b. $\left(7x^2 - \dfrac{1}{2}\right)^2$
  c. $\left(9x^2 - \dfrac{1}{2}\right)^2$

36. ⟨ **4.7E** ⟩ *Multiply.*
  a. $(3x^2 + 2)(3x^2 - 2)$
  b. $(3x^2 + 4)(3x^2 - 4)$
  c. $(2x^2 + 5)(2x^2 - 5)$

37. ⟨ **4.8A** ⟩ *Divide.*
  a. $\dfrac{18x^3 - 9x^2}{9x}$
  b. $\dfrac{20x^3 - 10x^2}{5x}$
  c. $\dfrac{24x^3 - 12x^2}{6x}$

38. ⟨ **4.8B** ⟩ *Divide.*
  a. $x^2 + 4x - 12$ by $x - 2$
  b. $x^2 + 4x - 21$ by $x - 3$
  c. $x^2 + 4x - 32$ by $x - 4$

39. ⟨ **4.8B** ⟩ *Divide.*
  a. $8x^3 - 16x - 8$ by $2 + 2x$
  b. $12x^3 - 24x - 12$ by $2 + 2x$
  c. $4x^3 - 8x - 4$ by $2 + 2x$

40. ⟨ **4.8B** ⟩ *Divide.*
  a. $2x^3 - 20x + 8$ by $x - 3$
  b. $2x^3 - 21x + 12$ by $x - 3$
  c. $3x^3 - 4x + 5$ by $x - 1$

41. ⟨ **4.8B** ⟩ *Divide.*
  a. $x^4 + x^3 - 4x^2 + 1$ by $x^2 - 4$
  b. $x^4 + x^3 - 5x^2 + 1$ by $x^2 - 5$
  c. $x^4 + x^3 - 6x^2 + 1$ by $x^2 - 6$

❯ Practice Test **Chapter 4**

(Answers on page 391)

Visit www.mhhe.com/bello to view helpful videos that provide step-by-step solutions to several of the problems below.

1. Multiply:

   a. $(2a^3b)(-6ab^3)$ **b.** $(-2x^2yz)(-6xy^3z^4)$

   c. $\dfrac{18x^5y^7}{-9xy^3}$

2. Simplify:

   a. $(2x^3y^2)^3$

   b. $(-3x^2y^3)^2$

3. Simplify:

   a. $\left(\dfrac{3}{4}\right)^3$ **b.** $\left(\dfrac{3x^3}{y^4}\right)^3$

4. Simplify and write the answer with positive exponents.

   a. $\left(\dfrac{1}{x}\right)^{-7}$ **b.** $\dfrac{x^{-6}}{x^{-6}}$ **c.** $\dfrac{x^{-6}}{x^{-7}}$

5. Simplify.

   a. $3x^2x^{-4}$ **b.** $\left(\dfrac{2x^{-3}y^4}{3x^2y^3}\right)^{-2}$

6. Write in scientific notation.

   a. 48,000,000   **b.** 0.00000037

7. Perform the indicated operations and write the answers in scientific notation.

   a. $(3 \times 10^4) \times (7.1 \times 10^6)$

   b. $\dfrac{2.84 \times 10^{-2}}{7.1 \times 10^{-3}}$

8. Classify as a monomial, binomial, or trinomial.

   a. $3x - 5$ **b.** $5x^3$ **c.** $8x^2 - 2 + 5x$

9. Write the polynomial $-3x + 7 + 8x^2$ in descending order and find its degree.

10. Find the value of $-16t^2 + 100$ when $t = 2$.

11. Add $-4x + 8x^2 - 3$ and $-5x^2 - 4 + 2x$.

12. Subtract $5x - 2 + 8x^2$ from $3x^2 - 2x$.

13. Remove parentheses (simplify): $-2x^2(x + 3y)$.

14. Multiply $(x + 8)(x - 3)$.

15. Multiply $(x + 4)(x - 6)$.

16. Multiply $(5x - 2y)(4x - 3y)$.

17. Expand $(3x + 5y)^2$.

18. Expand $(2x - 7y)^2$.

19. Multiply $(2x - 5y)(2x + 5y)$.

20. Multiply $(x + 2)(x^2 + 5x + 3)$.

21. Multiply $3x(x + 2)(x + 5)$.

22. Expand $(x + 7)^3$.

23. Expand $\left(3x^2 - \dfrac{1}{2}\right)^2$.

24. Multiply $(3x^2 + 7)(3x^2 - 7)$.

25. Divide $2x^3 - 9x + 5$ by $x - 2$.

› Answers to Practice Test **Chapter 4**

Answer		If You Missed	Review		
		Question	Section	Examples	Page
1. a. $-12a^4b^4$	**b.** $12x^3y^4z^5$	1	4.1	1, 2, 3	310–312
c. $-2x^4y^4$					
2. a. $8x^9y^6$	**b.** $9x^4y^6$	2	4.1	4, 5	312–313
3. a. $\dfrac{27}{64}$	**b.** $\dfrac{27x^9}{y^{12}}$	3	4.1	6	314
4. a. x^7 **b.** 1 **c.** x		4	4.2	1, 2, 5	321–323
5. a. $\dfrac{3}{x^2}$	**b.** $\dfrac{9x^{10}}{4y^2}$	5	4.2	4, 6	323–324
6. a. 4.8×10^7	**b.** 3.7×10^{-7}	6	4.3	1	331
7. a. 2.13×10^{11}	**b.** 4	7	4.3	3, 4	331–332
8. a. Binomial **b.** Monomial **c.** Trinomial		8	4.4	1	338
9. $8x^2 - 3x + 7;\ 2$		9	4.4	2, 3	339–340
10. 36		10	4.4	4, 5, 6	340–341
11. $3x^2 - 2x - 7$		11	4.5	1, 2	349–350
12. $-5x^2 - 7x + 2$		12	4.5	3	350–357
13. $-2x^3 - 6x^2y$		13	4.6	1, 2	358–359
14. $x^2 + 5x - 24$		14	4.6	3	360
15. $x^2 - 2x - 24$		15	4.6	3	360
16. $20x^2 - 23xy + 6y^2$		16	4.6	5	361
17. $9x^2 + 30xy + 25y^2$		17	4.7	1	368
18. $4x^2 - 28xy + 49y^2$		18	4.7	2	369
19. $4x^2 - 25y^2$		19	4.7	3	370
20. $x^3 + 7x^2 + 13x + 6$		20	4.7	4	371
21. $3x^3 + 21x^2 + 30x$		21	4.7	5	372
22. $x^3 + 21x^2 + 147x + 343$		22	4.7	6	372–373
23. $9x^4 - 3x^2 + \dfrac{1}{4}$		23	4.7	7	373
24. $9x^4 - 49$		24	4.7	8	373
25. $(2x^2 + 4x - 1)\ \text{R } 3$		25	4.8	2, 3, 4	381–382

⟩Cumulative Review **Chapters 1–4**

1. Find the additive inverse (opposite) of -7.

2. Find: $\left| -9\dfrac{9}{10} \right|$

3. Add: $-\dfrac{1}{9} + \left(-\dfrac{1}{6} \right)$

4. Subtract: $7.2 - (-6.4)$

5. Multiply: $(-2.4)(2.6)$

6. Find: $-(2^4)$

7. Divide: $-\dfrac{3}{4} \div \left(-\dfrac{7}{8} \right)$

8. Evaluate $y \div 5 \cdot x - z$ for $x = 5$, $y = 50$, $z = 3$.

9. Which property is illustrated by the following statement?
$$6 \cdot (5 \cdot 2) = (6 \cdot 5) \cdot 2$$

10. Multiply: $3(2x - 8)$

11. Combine like terms: $-8xy^3 - (-xy^3)$

12. Simplify: $5x + (x + 4) - 2(x - 3)$

13. Write in symbols: The quotient of $(m + 3n)$ and p

14. Does the number -3 satisfy the equation $12 = 15 - x$?

15. Solve for x: $5 = 5(x - 3) + 5 - 4x$

16. Solve for x: $-\dfrac{8}{7}x = -56$

17. Solve for x: $\dfrac{x}{4} - \dfrac{x}{9} = 5$

18. Solve for x: $5 - \dfrac{x}{3} = \dfrac{3(x + 1)}{7}$

19. Solve for b in the equation $S = 6a^2b$.

20. The sum of two numbers is 100. If one of the numbers is 20 more than the other, what are the numbers?

21. Dave has invested a certain amount of money in stocks and bonds. The total annual return from these investments is $615. If the stocks produce $245 more in returns than the bonds, how much money does Dave receive annually from each type of investment?

22. Train A leaves a station traveling at 30 mph. Six hours later, train B leaves the same station traveling in the same direction at 40 mph. How long does it take for train B to catch up to train A?

23. Susan purchased some municipal bonds yielding 8% annually and some certificates of deposit yielding 11% annually. If Susan's total investment amounts to $17,000 and the annual income is $1630, how much money is invested in bonds and how much is invested in certificates of deposit?

24. Graph: $-\dfrac{x}{2} + \dfrac{x}{9} \geq \dfrac{x - 9}{9}$

25. Graph the point $C(4, -4)$.

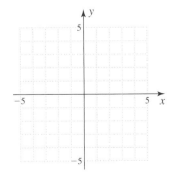

26. What are the coordinates of point A?

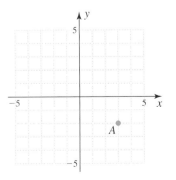

27. Determine whether the ordered pair $(-5, 1)$ is a solution of $5x + 3y = -22$.

28. Find x in the ordered pair $(x, -2)$ so that the ordered pair satisfies the equation $5x - 4y = 13$.

29. Graph: $5x + y = 5$

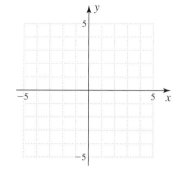

30. Graph: $3y - 15 = 0$

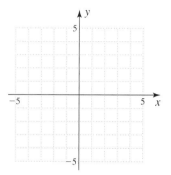

31. Find the slope of the line passing through the points $(-7, -1)$ and $(-2, 6)$.

32. What is the slope of the line $8x - 4y = 14$?

33. Find the pair of parallel lines.

  (1) $18y + 24x = 8$

  (2) $3y = 4x + 8$

  (3) $24x - 18y = 8$

34. Multiply: $(-3x^4y)(-6x^3y^2)$

35. Divide: $\dfrac{25x^2y^5}{-5xy^7}$

36. Divide: $\dfrac{x^{-6}}{x^{-9}}$

37. Multiply and simplify: $x^8 \cdot x^{-3}$

38. Simplify: $(4x^4y^{-4})^3$

39. Write in scientific notation: 8,000,000

40. Divide and express the answer in scientific notation: $(26.04 \times 10^{-5}) \div (6.2 \times 10^3)$

41. Classify as a monomial, binomial, or trinomial. $2x^2$

42. Find the degree of the polynomial: $3x^2 - 3x + 1$

43. Write the polynomial in descending order: $-3x^2 - x - 3x^3 + 4$

44. Find the value of $3x^3 + 2x^2$ when $x = 2$.

45. Add $(x + 6x^3 - 2)$ and $(-2x^3 - 1 + 8x)$.

46. Remove parentheses (simplify): $-3x(3x^2 + 4y)$

47. Find: $(3x + 2y)^2$

48. Find: $(4x - 3y)(4x + 3y)$

49. Find: $\left(5x^2 - \dfrac{1}{2}\right)^2$

50. Find: $(4x^2 + 9)(4x^2 - 9)$

51. Divide $(3x^3 - 20x^2 + 29x - 17)$ by $(x - 5)$.

52. Find an equation of the line passing through $(-2, 6)$ and with slope 4.

53. Find an equation of the line with slope -2 and y-intercept 5.

54. Graph the inequality $2x - 3y < -6$.

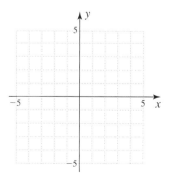

55. Graph the inequality $3x - y \geq 0$.

Chapter

5 five

▶ Factoring

The Human Side of Algebra

One of the most famous mathematicians of antiquity is Pythagoras, to whom has been ascribed the theorem that bears his name (see Section 5.7). It's believed that he was born between 580 and 569 B.C. on the Aegean island of Samos, from which he was later banned by the powerful tyrant Polycrates. When he was about 50, Pythagoras moved to Croton, a colony in southern Italy, where he founded a secret society of 300 young aristocrats called the Pythagoreans. Four subjects were studied: arithmetic, music, geometry, and astronomy. Students attending lectures were divided into two groups: *acoustici* (listeners) and *mathematici*. How did one get to join the *mathematici* in those days? One listened to the master's voice (*acoustici*) from behind a curtain for a period of 3 years! According to Burton's *History of Mathematics,* the Pythagoreans had "strange initiations, rites, and prohibitions." Among them was their refusal "to eat beans, drink wine, pick up anything that had fallen or stir a fire with an iron," but what set them apart from other sects was their philosophy that "knowledge is the greatest purification," and to them, knowledge meant mathematics. You can now gain some of this knowledge by studying Pythagoras' theorem yourself, and evaluating its greatness.

Common Factors and Grouping

▶ Objectives

A ▷ Find the greatest common factor (GCF) of numbers.

B ▷ Find the GCF of terms.

C ▷ Factor out the GCF.

D ▷ Factor a four-term expression by grouping.

▶ To Succeed, Review How To . . .

1. Write a number as a product of primes (p. 6).
2. Write a polynomial in descending order (pp. 339–340).
3. Use the distributive property (pp. 80, 89–90).

▶ Getting Started

Expansion Joints and Factoring

Have you ever noticed as you were going over a bridge that there is always a piece of metal at about the midpoint? This piece of metal is called an *expansion joint,* and it prevents the bridge from cracking when it expands or contracts. We can describe this expansion algebraically. In general, if α is the coefficient of linear expansion, L is the length of the material, and t_2 and t_1 are the high and low temperatures in degrees Celsius, then the linear expansion of a solid is

Expansion area

Summer   Winter

$$e = \alpha L t_2 - \alpha L t_1$$

The expression on the right-hand side of the equation can be written in a simpler way if we *factor* it, that is, write it as a **product** of two expressions. You will get a chance to factor this expression in Exercises 5.1, Problem 81! In this section we learn how to factor polynomials by finding a common factor and by grouping.

To *factor* an expression, we write the expression as a *product* of its factors. When two or more numbers are to be multiplied to form a product, the numbers being multiplied are the **factors.** Does this sound complicated? It isn't! Suppose we give you the numbers 3 and 5 and tell you to multiply them. You will probably write

$$3 \times 5 = 15$$

In the reverse process, we give you the product 15 and tell you to **factor** it! (Factoring is the *reverse* of multiplying.) You will then write

Product ⎯⏐   ⏐⎯ Factors

$$15 = 3 \times 5$$

Why? Because you know that multiplying 3 by 5 gives you 15. What about factoring the number 20? Here you can write

$$20 = 4 \times 5 \quad \text{or} \quad 20 = 2 \times 10$$

Note that $20 = 4 \times 5$ and $20 = 2 \times 10$ are not **completely** factored. They contain factors that are *not* prime numbers, numbers divisible only by themselves and 1. The first few prime numbers are 2, 3, 5, 7, 11, and so on. Since neither of these factorizations is *complete,* we note that in the first factorization $4 = (2 \times 2)$, so

$$20 = (2 \times 2) \times 5$$

whereas in the second factorization, $10 = (2 \times 5)$. Thus,

$$20 = 2 \times (2 \times 5)$$

In either case, the complete factorization for 20 is

$$20 = 2 \times 2 \times 5$$

From now on, when we say factor a *number,* we mean factor it *completely.* Moreover, the numerical factors are assumed to be prime numbers unless noted otherwise. Thus, we don't factor 20 as

$$20 = \frac{1}{4} \times 80$$

With these preliminaries out of the way, we can now factor some algebraic expressions. There are different factoring techniques for different situations. We will study most of them in this chapter.

A › Finding the Greatest Common Factor (GCF) of Numbers

When the numbers 15 and 20 are completely factored, we have

$$15 = 3 \cdot 5 \quad \text{and} \quad 20 = 2 \cdot 2 \cdot 5$$

As you can see, 5 is a common factor of 15 and 20. Similarly, 3 is a common factor of 18 and 30, because 3 is a factor of 18 and 3 is a factor of 30. Other common factors of 18 and 30 are 1, 2, and 6. The greatest (largest) of the common factors of 18 and 30 is 6, so 6 is the ***greatest common factor (GCF)*** of 18 and 30. Note that the GCF is the same as the GCD we already studied in Section R.1. In general, we have:

GREATEST COMMON FACTOR OF A LIST OF INTEGERS	The **greatest common factor (GCF)** of a list of integers is the **largest** common factor of the integers in the list.

EXAMPLE 1   Finding the GCF of a List of Numbers

Find the greatest common factor (GCF) of:

a. 45 and 60   **b.** 36, 60, and 108   **c.** 10, 17, and 12

SOLUTION

a. To discover the common factors, write each of the numbers as a product of primes using exponents:

$$45 = 3 \cdot 3 \cdot 5 \quad = 3^2 \cdot 5$$
$$60 = 2 \cdot 2 \cdot 3 \cdot 5 = 2^2 \cdot 3 \cdot 5$$

The common factors are 3 and 5; thus, the GCF is $3 \cdot 5 = 15$. Note that to find the GCF, we simply choose the prime factors with the *smallest* exponents and find their product.

PROBLEM 1

Find the GCF of:

a. 30 and 45

b. 45, 60, and 108

c. 20, 13, and 18

(continued)

Answers to PROBLEMS

1. a. 15   **b.** 3   **c.** 1

Here is a shortcut: Write

$$\begin{array}{c|cc} & 45 & 60 \end{array}$$

Divide each number by a prime divisor common to all numbers. The product of all divisors is the GCF.

$$\begin{array}{c|cc} 3 & 45 & 60 \\ 5 & 15 & 20 \\ & 3 & 4 \end{array}$$

$\text{GCF} = 3 \cdot 5 = 15$

b. We write each of the integers in factored form using exponents and then select the prime factors with the *smallest* exponents.

Pick the prime with the *smallest* exponent in each column.

$$\begin{aligned} 36 &= 2 \cdot 2 \cdot 3 \cdot 3 & &= 2^2 \cdot 3^2 \\ 60 &= 2 \cdot 2 \cdot 3 \cdot 5 & &= 2^2 \cdot 3 \cdot 5 \\ 108 &= 2 \cdot 2 \cdot 3 \cdot 3 \cdot 3 & &= 2^2 \cdot 3^3 \end{aligned}$$

We select 2^2 from the first column and 3 from the second. Thus, the GCF is $2^2 \cdot 3 = 12$.

Shortcut: Write

$$\begin{array}{c|ccc} & 36 & 60 & 108 \end{array}$$

Divide by a prime divisor common to all numbers:

$$\begin{array}{c|ccc} 2 & 36 & 60 & 108 \\ 2 & 18 & 30 & 54 \\ 3 & 9 & 15 & 27 \\ & 3 & 5 & 9 \end{array}$$

$\text{GCF} = 2 \cdot 2 \cdot 3 = 12$ as before.

c. First, write

$$\begin{aligned} 10 &= 2 \cdot 5 & &= 2 \cdot 5 \\ 17 &= 17 & &= 17 \\ 12 &= 2 \cdot 2 \cdot 3 &= 2^2 \cdot 3 \end{aligned}$$

There are no primes common to all three numbers, so the GCF is 1.

Shortcut: Write

$$\begin{array}{c|ccc} & 10 & 17 & 12 \end{array}$$

There is no prime divisor common to 10, 17, and 12, so the GCF is 1.

B › Finding the Greatest Common Factor (GCF) of Terms

We can also find the GCF for a list of terms. Simply write the terms in factored form using exponents, and then select the factors with the *smallest* exponents. For example, the GCF of x^5, x^3, and x^6 is x^3 because x^3 is the factor with the *smallest* exponent, 3. Similarly, to find the GCF of $a^3b^4c^2$ and $a^4b^2c^3$ write

$$a^3b^4c^2$$
$$a^4b^2c^3$$

We select the factors with the ***smallest*** exponents in each column: a^3 from the a column, b^2 from the b column, and c^2 from the c column. Thus, the GCF is $a^3 \cdot b^2 \cdot c^2$.

EXAMPLE 2 Finding the GCF of a List of Terms

Find the GCF of:

a. $24a^6$, $18a^4$, $-30a^9$ **b.** a^5b^3, b^6a^4, a^2b^5, a^8 **c.** $-x^3y$, $-xy^3$

SOLUTION

a. Write:

$$24a^6 = 2^3 \cdot 3a^6$$
$$18a^4 = 2 \cdot 3^2a^4$$
$$-30a^9 = -1 \cdot 2 \cdot 3 \cdot 5a^9$$

The common factors with the smallest exponents are 2, 3, and a^4, so the GCF $= 2 \cdot 3 \cdot a^4 = 6a^4$.

b. Rewrite b^6a^4 as a^4b^6 so we can place all terms in a column, compare exponents, and select the factor with the lowest exponent from each column:

$$a^5b^3$$
$$a^4b^6$$
$$a^2b^5$$
$$a^8$$

There are no factors containing b in a^8, so the GCF is a^2.

c. Write:

$$-x^3y = -1 \cdot 1x^3y$$
$$-xy^3 = -1 \cdot 1xy^3$$

The factors of -1 are -1 and 1, but the **greatest** (largest) common factor is 1, so the GCF $= 1 \cdot xy = xy$.

Note that in a list containing negative terms, sometimes a negative common factor is preferred. Thus, in Example 2(c) we may prefer $-xy$ as the common factor. Both answers, xy and $-xy$, are correct.

PROBLEM 2

Find the GCF of:

a. $12x^8$, $9x^5$, and $-30x^8$

b. x^4y^3, y^4x^6, x^3y^8, and y^9

C › Factoring Out the Greatest Common Factor (GCF)

We can use the ideas we have learned about GCF to write polynomials in factored form. To start, let's compare the multiplications with the factors and see whether we can discover a pattern.

Finding the Product	Finding the Factors
$4(x + y) = 4x + 4y$	$4x + 4y = 4(x + y)$
$5(a - 2b) = 5a - 10b$	$5a - 10b = 5(a - 2b)$
$2x(x + 3) = 2x^2 + 6x$	$2x^2 + 6x = 2x(x + 3)$

What do all these operations have in common? They use the *distributive property.* When multiplying, we have

$$a(b + c) = ab + ac$$

When factoring, we have

$$ab + ac = a(b + c)$$ We are factoring a monomial (a) from a binomial ($ab + ac$).

Of course, we are now more interested in the latter operation. It tells us that to factor a binomial, we must find a factor (a in this case) common to all terms. The first step in having a completely factored expression is to select the **greatest common factor, ax^n**. Here's how we do it.

GREATEST COMMON FACTOR OF A POLYNOMIAL	The term ax^n is the **greatest common factor (GCF)** of a polynomial if: 1. a is the *greatest integer* that divides each of the coefficients of the polynomial, and 2. n is the *smallest exponent* of x in all the terms of the polynomial.

Thus, to factor $6x^3 + 18x^2$, we could write

$$6x^3 + 18x^2 = 3x(2x^2 + 6x)$$

but this is not completely factored because $2x^2 + 6x$ can be factored further. Here the *greatest* integer dividing 6 and 18 is 6, and the smallest exponent of x in all terms is x^2. Thus, the complete factorization is

$$6x^3 + 18x^2 = 6x^2(x + 3)$$ Where $6x^2$ is the GCF

Note that if you write $6x^3 + 18x^2$ as the sum $2 \cdot 3 \cdot x^3 + 2 \cdot 3 \cdot 3x^2$, you have *not* factored the polynomial; you have only factored the coefficients of the terms. The complete factorization is the product $6x^2(x + 3)$.

You can check this by multiplying $6x^2(x + 3)$. Of course, it might help your accuracy and understanding if you were to write an intermediate step indicating the common factor present in each term. Thus, to factor out the GFC of $6x^3 + 18x^2$, you would write

$$6x^3 + 18x^2 = 6x^2 \cdot x + 6x^2 \cdot 3$$

$$= 6x^2(x + 3)$$ Note that since the 6 is a coefficient, it is *not* written in factored form; that is, we write $6x^2(x + 3)$ and *not* $2 \cdot 3x^2(x + 3)$.

Similarly, to factor $4x - 28$, you would write

$$4x - 28 = 4 \cdot x - 4 \cdot 7$$
$$= 4(x - 7)$$

Remember, if you write $4x - 28$ as $2 \cdot 2 \cdot x - 2 \cdot 2 \cdot 7 \cdot x$, you have factored the terms and *not* the binomial. The factorization of $4x - 28$ is the *product* $4(x - 7)$.

One more thing. When an expression such as $-3x + 12$ is to be factored, we have two possible factorizations:

$$-3(x - 4) \quad \text{or} \quad 3(-x + 4)$$

The first one, $-3(x - 4)$, is the *preferred* one, since in that case the first term of the binomial, $x - 4$, has a **positive** sign.

EXAMPLE 3   Factoring out a common factor from a binomial

Factor completely:

a. $8x + 24$ **b.** $-6y + 12$ **c.** $10x^2 - 25x^3$

PROBLEM 3

Factor completely:

a. $6a + 18$ **b.** $-9y + 27$

c. $15a^2 - 45a^3$

Answers to PROBLEMS

3. a. $6(a + 3)$ **b.** $-9(y - 3)$
   c. $15a^2(1 - 3a)$ or
      $-15a^2(3a - 1)$

SOLUTION

a. $8x + 24 = 8 \cdot x + 8 \cdot 3$

$\qquad = 8(x + 3)$ ⟶ 8 is the GCF.

b. $-6y + 12 = -6 \cdot y - 6(-2)$

$\qquad = -6(y - 2)$ ⟶ -6 is the GCF.

c. $10x^2 - 25x^3 = 5x^2 \cdot 2 - 5x^2 \cdot 5x$ ⟶ $5x^2$ is the GCF.

$\qquad = 5x^2(2 - 5x)$ or, better yet, $-5x^2(5x - 2)$

Check your results by multiplying the factors obtained, $-5x^2$ and $(5x - 2)$.

We can also factor polynomials with more than two terms, as shown next.

EXAMPLE 4   Factoring out a common factor from a polynomial

Factor completely:

a. $6x^3 + 12x^2 + 18x$ **b.** $10x^6 - 15x^5 + 20x^4 + 30x^2$

c. $2x^3 + 4x^4 + 8x^5$

SOLUTION

a. $6x^3 + 12x^2 + 18x = 6x \cdot x^2 + 6x \cdot 2x + 6x \cdot 3$ ⟶ $6x$ is the GCF.

$\qquad = 6x(x^2 + 2x + 3)$

b. $10x^6 - 15x^5 + 20x^4 + 30x^2 = 5x^2 \cdot 2x^4 - 5x^2 \cdot 3x^3 + 5x^2 \cdot 4x^2 + 5x^2 \cdot 6$

$\qquad = 5x^2(2x^4 - 3x^3 + 4x^2 + 6)$ ⟶ $5x^2$ is the GCF.

c. $2x^3 + 4x^4 + 8x^5 = 2x^3 \cdot 1 + 2x^3 \cdot 2x + 2x^3 \cdot 4x^2$

$\qquad = 2x^3(1 + 2x + 4x^2)$ ⟶ $2x^3$ is the GCF.

PROBLEM 4

Factor completely:

a. $5x^3 + 15x^2 + 20x$

b. $15x^7 - 20x^6 + 10x^5 + 25x^3$

c. $3a^3 + 9a^4 + 12a^5$

EXAMPLE 5   Factoring out a common factor that is a fraction

Factor completely: $\frac{3}{4}x^2 - \frac{1}{4}x + \frac{5}{4}$

SOLUTION   As you can see, we are not working with integers here. This is a very special situation, but this expression can still be factored. Here's how to find the common factor:

$$\frac{3}{4}x^2 - \frac{1}{4}x + \frac{5}{4} = \frac{1}{4} \cdot 3x^2 - \frac{1}{4} \cdot x + \frac{1}{4} \cdot 5$$

$$= \frac{1}{4}(3x^2 - x + 5)$$

PROBLEM 5

Factor completely: $\frac{4}{5}a^2 - \frac{1}{5}a + \frac{2}{5}$

D › Factoring by Grouping

Can we factor $x^3 + 2x^2 + 3x + 6$? It seems that there's no common factor here except 1. However, we can group and factor the first two terms and also the last two terms, and then use the distributive property. Here are the steps we use.

1. Group terms with common factors using the associative property.

$x^3 + 2x^2 + 3x + 6 = (x^3 + 2x^2) + (3x + 6)$

Same

2. Factor each resulting binomial.

$= x^2(x + 2) + 3(x + 2)$

3. Factor out the GCF, $(x + 2)$, using the distributive property.

$= (x + 2)(x^2 + 3)$

Answers to PROBLEMS

4. a. $5x(x^2 + 3x + 4)$ **b.** $5x^3(3x^4 - 4x^3 + 2x^2 + 5)$ **c.** $3a^3(1 + 3a + 4a^2)$ **5.** $\frac{1}{5}(4a^2 - a + 2)$

Thus,

$$x^3 + 2x^2 + 3x + 6 = (x + 2)(x^2 + 3)$$

Note that $x^2(x + 2) + 3(x + 2)$ can also be written as $(x^2 + 3)(x + 2)$, since $ac + bc = (a + b)c$. Hence,

$$x^3 + 2x^2 + 3x + 6 = (x^2 + 3)(x + 2)$$

Either factorization is correct. You can check this by multiplying $(x + 2)(x^2 + 3)$ or $(x^2 + 3)(x + 2)$. You will either get $x^3 + 3x + 2x^2 + 6$ or its equivalent $x^3 + 2x^2 + 3x + 6$.

EXAMPLE 6   Factor by grouping

Factor completely:

a. $3x^3 + 6x^2 + 2x + 4$ **b.** $6x^3 - 3x^2 - 4x + 2$

SOLUTION

a. We proceed by steps, as before.

1. Group terms with common factors using the associative property.

$$3x^3 + 6x^2 + 2x + 4 = (3x^3 + 6x^2) + (2x + 4)$$

Same

2. Factor each resulting binomial.

$$= 3x^2(x + 2) + 2(x + 2)$$

3. Factor out the GCF, $(x + 2)$, using the distributive property.

$$= (x + 2)(3x^2 + 2)$$

Note that if you write $3x^3 + 2x + 6x^2 + 4$ in step 1, your answer would be $(3x^2 + 2)(x + 2)$. Since by the commutative property $(3x^2 + 2)(x + 2) = (x + 2)(3x^2 + 2)$, both answers are correct. We will factor our polynomials by first writing them in *descending* order.

b. Again, we proceed by steps.

1. Group terms with common factors using the associative property.

$$6x^3 - 3x^2 - 4x + 2 = (6x^3 - 3x^2) - (4x - 2)$$

Note that $-4x + 2 = -(4x - 2)$.

Same

2. Factor each resulting binomial.

$$= 3x^2(2x - 1) - 2(2x - 1)$$

3. Factor out the GCF, $(2x - 1)$.

$$= (2x - 1)(3x^2 - 2)$$

Thus, $6x^3 - 3x^2 - 4x + 2 = (2x - 1)(3x^2 - 2)$. Note that $2x - 1$ and $3x^2 - 2$ *cannot* be factored any further, so the polynomial is completely factored.

PROBLEM 6

Factor completely:

a. $2a^3 + 6a^2 + 5a + 15$

b. $6a^3 - 2a^2 - 3a + 1$

EXAMPLE 7   Factor by grouping

Factor completely:

a. $2x^3 - 4x^2 - x + 2$ **b.** $6x^4 - 9x^2 + 4x^2 - 6$

SOLUTION

a. 1. Group terms with common factors using the associative property.

$$2x^3 - 4x^2 - x + 2 = (2x^3 - 4x^2) - (x - 2)$$

Same

2. Factor each resulting binomial.

$$= 2x^2(x - 2) - 1(x - 2)$$

3. Factor out the GCF, $(x - 2)$.

$$= (x - 2)(2x^2 - 1)$$

PROBLEM 7

Factor completely:

a. $3a^3 - 9a^2 - a + 3$

b. $6a^4 - 4a^2 + 9a^2 - 6$

Answers to PROBLEMS

6. a. $(a + 3)(2a^2 + 5)$ **b.** $(3a - 1)(2a^2 - 1)$ **7. a.** $(a - 3)(3a^2 - 1)$ **b.** $(3a^2 - 2)(2a^2 + 3)$

b. We proceed as usual.

1. Group terms with common factors using the associative property.

$$6x^4 - 9x^2 + 4x^2 - 6 = (6x^4 - 9x^2) + (4x^2 - 6)$$

2. Factor each resulting binomial.

Same

$$= 3x^2(2x^2 - 3) + 2(2x^2 - 3)$$

3. Factor out the GCF, $(2x^2 - 3)$.

$$= (2x^2 - 3)(3x^2 + 2)$$

EXAMPLE 8   Factor a polynomial with two variables by grouping

Factor completely: $6x^2 + 2xy - 9xy - 3y^2$

SOLUTION   Our steps serve us well in this situation also.

1. Group terms with common factors using the associative property.

$$6x^2 + 2xy - 9xy - 3y^2 = (6x^2 + 2xy) - (9xy + 3y^2)$$

Note that $-9xy - 3y^2 = -(9xy + 3y^2)$.

2. Factor each resulting binomial.

Same

$$= 2x(3x + y) - 3y(3x + y)$$

3. Factor out the GCF, $(3x + y)$.

$$= (3x + y)(2x - 3y)$$

PROBLEM 8

Factor completely:

$$6a^2 + 3ab - 4ab - 2b^2$$

Calculator Corner

Checking Factorization

If you believe a picture is worth a thousand words, you can use your calculator to check factorization problems. The idea is this: If in Example 6(a) you get the same picture (graph) for $3x^3 + 6x^2 + 2x + 4$ and for $(x + 2)(3x^2 + 2)$, these two polynomials must be equal; that is, $3x^3 + 6x^2 + 2x + 4 = (x + 2)(3x^2 + 2)$. Now, use your calculator to show this. Graph the polynomials by pressing Y= and entering $Y_1 = 3x^3 + 6x^2 + 2x + 4$ and $Y_2 = (x + 2)(3x^2 + 2)$. Then press GRAPH . If you get the same graph for Y_1 and Y_2, which means you see only one picture as shown in the window, you probably have the correct factorization.

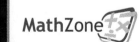

› Exercises **5.1**

‹ A › Finding the Greatest Common Factor (GCF) of Numbers   In Problems 1–6, find the GCF.

1. 20, 24   **2.**   30, 70   **3.** 16, 48, 88

4. 52, 26, 130   **5.**   8, 19, 12   **6.** 10, 41, 18

‹ B › Finding the Greatest Common Factor (GCF) of Terms   In Problems 7–20, find the GCF.

7. a^3, a^8 **8.** y^7, y^9 **9.** x^3, x^6, x^{10} **10.** b^6, b^8, b^{10}

11. $5y^6, 10y^7$ **12.** $12y^9, 24y^4$ **13.** $8x^3, 6x^7, 10x^9$ **14.** $9a^3, 6a^7, 12a^5$

Answers to PROBLEMS

8. $(2a + b)(3a - 2b)$

15. $9b^2c, 12bc^2, 15b^2c^2$ **16.** $12x^4y^3, 18x^3y^4, 6x^5y^5$ **17.** $9y^2, 6x^3, 3x^3y$ **18.** $15ab^3, 25b^3c^4, 10a^3bc$

19. $18a^4b^3z^4, 27a^5b^3z^4, 81ab^3z$ **20.** $6x^2y^2z^2, 9xy^2z^2, 15x^2y^2z^2$

⟨ **C** ⟩ **Factoring Out the Greatest Common Factor (GCF)**   In Problems 21–56, factor completely.

21. $3x + 15$ $3(x+5)$ **22.** $5x + 45$ $5(x+9)$ **23.** $9y - 18$ $9(y-2)$ **24.** $11y - 33$

25. $-5y + 20$ **26.** $-4y + 28$ **27.** $-3x - 27$ **28.** $-6x - 36$

29. $4x^2 + 32x$ **30.** $5x^3 + 20x$ **31.** $6x - 42x^2$ **32.** $7x - 14x^3$

33. $-5x^2 - 25x^4$ **34.** $-3x^3 - 18x^6$ **35.** $3x^3 + 6x^2 + 9x$

36. $8x^3 + 4x^2 - 16x$ **37.** $9y^3 - 18y^2 + 27y$ **38.** $10y^3 - 5y^2 + 10y$

39. $6x^6 + 12x^5 - 18x^4 + 30x^2$ **40.** $5x^7 - 15x^6 + 10x^3 - 20x^2$ **41.** $8y^8 + 16y^5 - 24y^4 + 8y^3$

42. $12y^9 - 4y^6 + 6y^5 + 8y^4$ **43.** $\frac{4}{7}x^3 + \frac{3}{7}x^2 - \frac{9}{7}x + \frac{3}{7}$ **44.** $\frac{2}{5}x^3 + \frac{3}{5}x^2 - \frac{2}{5}x + \frac{4}{5}$

45. $\frac{7}{8}y^9 + \frac{3}{8}y^6 - \frac{5}{8}y^4 + \frac{5}{8}y^2$ **46.** $\frac{4}{3}y^7 - \frac{1}{3}y^5 + \frac{2}{3}y^4 - \frac{5}{3}y^3$ **47.** $3(x + 4) - y(x + 4)$

$\frac{1}{3}y^3 (4y^4 - y^2 + 2y - 5)$

48. $5(y - 2) - x(y - 2)$ **49.** $x(y - 2) - (y - 2)$ **50.** $y(x + 3) - (x + 3)$

51. $c(t + s) - (t + s)$ **52.** $p(x - q) - (x - q)$ **53.** $4x^3 + 4x^4 - 12x^5$

54. $5x^6 + 10x^7 - 5x^8$ **55.** $6y^7 - 12y^9 - 6y^{11}$ **56.** $7x^9 - 7x^{13} - 14x^{15}$

⟨ **D** ⟩ **Factoring by Grouping**   In Problems 57–80, factor completely by grouping.

57. $x^3 + 2x^2 + x + 2$ **58.** $x^3 + 3x^2 + x + 3$ **59.** $y^3 - 3y^2 + y - 3$

60. $y^3 - 5y^2 + y - 5$ **61.** $4x^3 + 6x^2 + 2x + 3$ **62.** $6x^3 + 3x^2 + 2x + 1$

63. $6x^3 - 2x^2 + 3x - 1$ **64.** $6x^3 - 9x^2 + 2x - 3$ **65.** $4y^3 + 8y^2 + y + 2$

66. $2y^3 - 6y^2 - y + 3$ **67.** $2a^3 + 3a^2 + 2a + 3$ **68.** $3a^3 + 2a^2 + 3a + 2$

69. $3x^4 + 12x^2 + x^2 + 4$ **70.** $2x^4 + 2x^2 + x^2 + 1$ **71.** $6y^4 + 9y^2 + 2y^2 + 3$

72. $12y^4 + 8y^2 + 3y^2 + 2$ **73.** $4y^4 + 12y^2 + y^2 + 3$ **74.** $2y^4 + 2y^2 + y^2 + 1$

75. $3a^4 - 6a^2 - 2a^2 + 4$ **76.** $4a^4 - 12a^2 - 3a^2 + 9$ **77.** $6a - 5b + 12ad - 10bd$

78. $3x - 2y + 15xz - 10yz$ **79.** $x^2 - y - 3x^2z + 3yz$ **80.** $x^2 + 2y - 2x^2 - 4y$

〉〉〉 Applications

81. *Linear expansion* Factor $\alpha L t_2 - \alpha L t_1$, where α is the coefficient of linear expansion, L the length of the material, and t_2 and t_1 the high and low temperatures in degrees Celsius.

82. *Restoring force of spring* Factor the expression $-kx - k\ell$ (k is a constant), which represents the restoring force of a spring stretched an amount ℓ from its equilibrium position and then an additional x units.

83. *Equivalent resistance of circuits* When solving for the equivalent resistance of two circuits, we need to factor the expression $R^2 - R - R + 1$. Factor this expression by grouping.

84. *Bending moment of cantilever beam* The bending moment of a cantilever beam of length L, at x inches from its support, involves the expression $L^2 - Lx - Lx + x^2$. Factor this expression by grouping.

85. *Surface area of cylinder* The surface area A of a right circular cylinder is given by $A = 2\pi rh + 2\pi r^2$, where h is the height of the cylinder and r is its radius. Factor $2\pi rh + 2\pi r^2$.

86. *Area of trapezoid* The area A of a trapezoid is given by $A = \frac{1}{2}b_1 h + \frac{1}{2}b_2 h$, where h is the altitude of the trapezoid and b_1 and b_2 are the lengths of the bases. Factor $\frac{1}{2}b_1 h + \frac{1}{2}b_2 h$.

Formulas   In Problems 87–92 a formula is given. Factor the expression on the right side of the equation.

	Formula	Used in
87.	$Q_1 = PQ_2 - PQ_1$	Refrigeration
88.	$L = L_0 + L_0 at$	Temperature expansion
89.	$a = Vk - PV^2$	Muscle contraction
90.	$d_m = nA - A$	Optics
91.	$f_s u = fu + fv_s$	Sound
92.	$T_2 W = qT_2 - qT_1$	Energy

〉〉〉 Using Your Knowledge

Factoring Formulas   Many formulas can be simplified by factoring. Here are a few; factor the expressions given in each problem.

93. The vertical shear at any section of a cantilever beam of uniform cross section is given by

$$-w\ell + wz$$

Factor this expression.

94. The bending moment of any section of a cantilever beam of uniform cross section is given by

$$-P\ell + Px$$

Factor this expression.

95. The surface area of a square pyramid is given by

$$a^2 + 2as$$

Factor this expression.

96. The energy of a moving object is given by

$$800m - mv^2$$

Factor this expression.

97. The height of a rock thrown from the roof of a certain building is given by

$$-16t^2 + 80t + 240$$

Factor this expression. (*Hint:* -16 is a common factor.)

⟩⟩⟩ *Write On*

98. What do we mean by a factored expression, and how can you check whether the result is correct?

99. Explain the procedure you use to factor a monomial from a polynomial. Can you use the definition of GCF to factor $\frac{1}{2}x + \frac{1}{2}$? Explain.

⟩⟩⟩ *Concept Checker*

Fill in the blank(s) with the correct word(s), phrase, or mathematical statement.

100. To **factor** an expression is to write the expression as a _____ of its factors.

101. **Factoring** is the _____ of multiplying.

102. The **GCF** of a **list of integers** is the _____ **common factor** of the integers in the list.

103. ax^n is the **GCF** of a **polynomial** if n is the _____ **exponent** of x in all the terms of the polynomial.

largest	reverse
smallest	product
sum	inverse

⟩⟩⟩ *Mastery Test*

Factor:

104. $3x^3 - 6x^2 - x + 2$

105. $6x^4 + 2x^2 - 9x^2 - 3$

106. $2x^3 + 2x^2 + 3x + 3$

107. $6x^3 - 9x^2 - 2x + 3$

108. $\frac{2}{5}x^2 - \frac{3}{5}x - \frac{1}{5}$

109. $3x^6 - 6x^5 + 12x^4 + 27x^2$

110. $7x^3 + 14x^2 - 49x$

111. $12x^2 + 6xy - 10xy - 5y^2$

112. $6x + 48$

113. $-3y + 21$

114. $4x^2 - 32x^3$

115. $5x(x + b) + 6y(x + b)$

116. $3x + 7y - 12x^2 - 28xy$

117. Find the GCF of 14 and 24.

118. Find the GCF of $20x^4$, $35x^5y$, and $40xy^7$.

⟩⟩⟩ *Skill Checker*

Multiply:

119. $(x + 5)(x + 3)$

120. $(x - 5)(x + 2)$

121. $(x - 1)(x - 4)$

122. $(x + 2)(x + 1)$

123. $(x + 1)(2x - 3)$

124. $(x - 2)(x + 4)$

5.2   Factoring $x^2 + bx + c$

▶ Objective

A › Factor trinomials of the form $x^2 + bx + c$.

▶ To Succeed, Review How To . . .

1. Expand $(X + A)(X + B)$ (p. 367).
2. Multiply integers (pp. 60–61).
3. Know the definition of a prime number (p. 6).

▶ Getting Started

Supply and Demand

Why do prices go up? One reason is related to the supply of the product. Large supply, low prices; small supply, higher prices. The supply function for a certain item can be stated as

$$p^2 + 3p - 70$$

where p is the price of the product. This trinomial is factorable by using *reverse multiplication* (factoring) as we did in Section 5.1. But why do we need to know how to factor? If we know how to factor $p^2 + 3p - 70$, we can solve the equation $p^2 + 3p - 70 = 0$ by rewriting the left side in factored form as

Supply Affects Price

Small supply = Higher price

ONE PER CUSTOMER

NO LIMIT!

Large supply = Lower price

Price

Supply

$$(p + 10)(p - 7) = 0$$

Do you see the solutions $p = -10$ and $p = 7$? We will learn how to factor trinomials in this section and then use this knowledge in Sections 5.6 and 5.7 to solve equations and applications.

A › Factoring Trinomials of the Form $x^2 + bx + c$

Since factoring is the reverse of multiplying, we can use the special products that we derived from the FOIL method (Section 4.7), reverse them, and have some equally useful factoring rules. The basis for the factoring rules is special product 1 (SP1), which you also know as the FOIL method. We now rewrite this product as a factoring rule.

FACTORING RULE 1: FACTORING BY REVERSING FOIL

$$X^2 + (A + B)X + AB = (X + A)(X + B) \qquad \textbf{(F1)}$$

Thus, to factor $x^2 + bx + c$, we need to find two binomials whose product is $x^2 + bx + c$. Now suppose we wish to factor the polynomial

$$x^2 + 8x + 15$$

To do this, we use F1:

$$\underbrace{X^2} + \underbrace{(A + B)X} + \underbrace{AB}$$
$$x^2 + \quad 8x \quad + 15$$

As you can see, 15 is used instead of AB and 8 instead of $A + B$; that is, we have two numbers A and B such that $AB = 15$ and $A + B = 8$. We write the possible factors of $AB = 15$ and their sums in a table.

Factors	Sum
15, 1	16
5, 3	8

The correct numbers are $A = 5$ and $B = 3$. Now

$$X^2 + \underbrace{(A + B)}X + \underbrace{AB} \quad = (X + A)(X + B)$$
$$x^2 + \underbrace{(5 + 3)}x \quad + \underbrace{5 \cdot 3} = (x + 5)(x + 3)$$
$$x^2 + \quad 8x \quad + 15 \quad = (x + 5)(x + 3)$$

Remember, to factor $x^2 + 8x + 15$, we need two integers whose product is 15 and whose sum is 8. The integers 5 and 3 will do! So the commutative property enables us to write

$$x^2 + 8x + 15 = (x + 5)(x + 3)$$

as our answer.

What about factoring $x^2 - 8x + 15$? We still need two numbers A and B whose product is 15, but this time their sum must be -8; that is, we need $AB = 15$ and $A + B = -8$. Letting $A = -5$ and $B = -3$ will do it. Check it out:

$$(x - 5)(x - 3) = x^2 - 3x - 5x + (-5)(-3)$$
$$= x^2 - 8x + 15$$

Do you see how this works? To factor a trinomial of the form $x^2 + bx + c$, we must find two numbers A and B so that $A + B = b$ and $AB = c$. Then,

$$x^2 + bx + c = x^2 + (A + B)x + AB = (x + A)(x + B)$$

For example, to factor

$$x^2 \overset{b}{\underset{}{\ominus 3}}x \overset{c}{\underset{}{\ominus 10}}$$

we need two numbers whose product is -10 (c) and whose sum is -3 (b). Here's a table showing the possibilities for the factors and their sum.

Factors	Sum
-10, 1	-9
10, -1	9
5, -2	3
-5, 2	-3

← This is the only one in which the sum is -3.

The numbers are -5 and $+2$. Thus,

$$x^2 - 3x - 10 = (x - 5)(x + 2)$$

Note that the answer $(x + 2)(x - 5)$ is also correct by the commutative property. You can check this by multiplying $(x - 5)(x + 2)$ or $(x + 2)(x - 5)$.

Similarly, to factor $x^2 + 5x - 14$, we need two numbers whose product is -14 (so the numbers have different signs) and whose sum is 5 (so the larger one is positive). The numbers are -2 and $+7$ ($-2 \cdot 7 = -14$ and $-2 + 7 = 5$). Thus,

$$x^2 + 5x - 14 = (x - 2)(x + 7)$$

We can also write $x^2 + 5x - 14 = (x + 7)(x - 2)$.

Here are the factorizations we have obtained:

$$x^2 + 8x + 15 = (x + 5)(x + 3)$$
$$x^2 - 8x + 15 = (x - 5)(x - 3)$$
$$x^2 - 3x - 10 = (x - 5)(x + 2)$$
$$x^2 + 5x - 14 = (x - 2)(x + 7)$$

All of them involve factoring a trinomial of the form $x^2 + bx + c$; in each case, we need to find two integers whose product is b and whose sum is c. In general, here is the procedure you need:

PROCEDURE

Factoring $x^2 + bx + c$

Find *two* integers whose product is c and whose sum is b.

1. If b and c are positive, *both* integers must be positive.
2. If c is positive and b is negative, *both* integers must be negative.
3. If c is negative, *one* integer must be positive and *one* negative.

EXAMPLE 1   Factoring a trinomial with all positive terms

Factor completely: $6 + 5x + x^2$

SOLUTION   Write $6 + 5x + x^2$ in decreasing order as $x^2 + 5x + 6$. To factor $x^2 + 5x + 6$ we need two numbers with product 6 and sum 5. The possibilities are listed in the table. The two factors whose sum is 5 are 3 and 2. Thus,

Factors	Sum
6, 1	7
3, 2	5

$$x^2 + 5x + 6 = (x + 3)(x + 2)$$

CHECK $(x + 3)(x + 2) = x^2 + 2x + 3x + 3 \cdot 2$
                        $= x^2 + 5x + 6$

PROBLEM 1

Factor completely: $8 + 9x + x^2$

EXAMPLE 2   Factoring a trinomial with a negative middle term

Factor completely: $x^2 - 6x + 5$

SOLUTION   To factor $x^2 - 6x + 5$ we need two numbers with product 5 and sum -6. In order to obtain the positive product 5, both factors must be negative. There is only one possibility, so the desired numbers are -5 and -1. Thus,

Factors	Sum
$-5, -1$	-6

$$x^2 - 6x + 5 = (x - 5)(x - 1)$$

CHECK $(x - 5)(x - 1) = x^2 - 1x - 5x + (-5) \cdot (-1)$
                        $= x^2 - 6x + 5$

PROBLEM 2

Factor completely: $x^2 - 8x + 7$

Answers to PROBLEMS

1. $(x + 8)(x + 1)$

2. $(x - 7)(x - 1)$

EXAMPLE 3 Factoring a trinomial with two negative terms

Factor completely: $x^2 - 3x - 4$

SOLUTION To factor $x^2 - 3x - 4$ we need two numbers with product -4 and sum -3. To obtain the negative product -4, one number must be positive and one negative. The possibilities are shown in the table. The desired numbers are -4 and 1, so

Factors	Sum
4, −1	3
−4, 1	−3
−2, 2	0

$$x^2 - 3x - 4 = (x - 4)(x + 1)$$

CHECK $(x - 4)(x + 1) = x^2 + 1x - 4x + (-4) \cdot (1)$
$$= x^2 - 3x - 4$$

PROBLEM 3

Factor completely: $x^2 - 2x - 8$

As you recall from Section R.1, a *prime number* is a number whose only factors are itself and 1. Thus, if we want to factor the number 15 as a product of primes, we write $15 = 3 \cdot 5$. On the other hand, if we want to factor the number 17, we are unable to do it because 17 is prime. If a polynomial cannot be factored using only integers, we say that the polynomial is *prime* and call it a **prime polynomial.** Look at the next example and discover which of the two polynomials is prime.

EXAMPLE 4 Factoring a trinomial with a positive middle term and a negative last term

Factor if possible:

a. $p^2 + 3p - 70$ **b.** $p^2 + 4p - 15$

PROBLEM 4

Factor if possible:

a. $y^2 + 4y - 14$

b. $y^2 + 3y - 40$

SOLUTION

a. We need two numbers with product -70 and sum 3. Since -70 is negative, the numbers must have different signs. Moreover, since the sum of the two numbers is 3, the *larger* number must be *positive*. Here are the possibilities:

Factors	Sum
70, −1	69
35, −2	33
14, −5	9
10, −7	3

←——The only ones with sum 3

The numbers we need then are 10 and -7. Thus,

$$p^2 + 3p - 70 = (p + 10)(p - 7)$$

b. This time we need two numbers with product -15 and sum 4. Here are the possibilities:

Factors	Sum
15, −1	14
−15, 1	−14
5, −3	2
−5, 3	−2

None of the pairs of factors has a sum of 4. Thus, the trinomial $p^2 + 4p - 15$ cannot be factored using only integers. The polynomial $p^2 + 4p - 15$ is **prime.**

Answers to PROBLEMS

3. $(x - 4)(x + 2)$

4. a. $y^2 + 4y - 14$ is prime

 b. $(y + 8)(y - 5)$

EXAMPLE 5   Factoring a trinomial with two variables

Factor completely: $x^2 + 5ax + 6a^2$

SOLUTION   This problem is very similar to Example 1, except here we have two variables, x and a. The procedure, however, is the same. We need two expressions whose product is $6a^2$ and whose sum is $5a$. The possible factors are $6a$ and a, or $3a$ and $2a$. Since $3a + 2a = 5a$, the appropriate factors are $3a$ and $2a$. Thus,

$$x^2 + 5ax + 6a^2 = (x + 3a)(x + 2a)$$

CHECK $(x + 3a)(x + 2a) = x^2 + 2ax + 3ax + 3a \cdot 2a$

$$= x^2 + \quad 5ax \quad + \quad 6a^2$$

PROBLEM 5

Factor completely: $x^2 + 6ax + 8a^2$

Now, do you remember GCFs? Sometimes we have to factor out the GCF in an expression so that we can completely factor it. How? We will show you in Example 6.

EXAMPLE 6   Factoring a trinomial with a GCF

Factor completely:

a. $x^5 - 6x^4 + 9x^3$

b. $ax^6 - 3ax^5 - 18ax^4$

SOLUTION

a. The GCF of $x^5 - 6x^4 + 9x^3$ is the term with the smallest exponent, that is, x^3. Thus,

$$x^5 - 6x^4 + 9x^3 = x^3(x^2 - 6x + 9)$$

Now, we have to factor $x^2 - 6x + 9$ by finding two numbers whose product is 9 and whose sum is -6. To obtain the positive product 9, both factors must be negative; -3 and -3 will do, so

$$x^2 - 6x + 9 = (x - 3)(x - 3) = (x - 3)^2$$

Here is the complete procedure for you to follow:

$$x^5 - 6x^4 + 9x^3 = x^3(x^2 - 6x + 9) \quad \text{Factor out the GCF } x^3.$$

$$= x^3(x - 3)(x - 3) \quad \text{Factor } x^2 - 6x + 9.$$

$$= x^3(x - 3)^2 \quad \text{Write } (x - 3)(x - 3) \text{ as } (x - 3)^2.$$

You should check the result!

b. $ax^6 - 3ax^5 - 18ax^4 = ax^4(x^2 - 3x - 18) \quad \text{Factor out the GCF, } ax^4.$

$$= ax^4(x + 3)(x - 6) \quad \text{Factor } x^2 - 3x - 18 \text{ using 3 and } -6, \text{ whose product is } -18 \text{ and whose sum is } -3.$$

We leave the check for you.

PROBLEM 6

Factor completely:

a. $y^5 - 4y^4 + 4y^3$

b. $b^2y^6 - b^2y^5 - 20b^2y^4$

Boost your grade at mathzone.com!

> Practice Problems
> NetTutor
> Self-Tests
> e-Professors
> Videos

> Exercises 5.2

〈 A 〉 Factoring Trinomials of the Form $x^2 + bx + c$ In Problems 1–40, factor completely.

1. $y^2 + 6y + 8$

2. $y^2 + 10y + 21$

3. $x^2 + 7x + 10$

4. $x^2 + 13x + 22$

5. $y^2 + 3y - 10$

6. $y^2 + 5y - 24$

7. $x^2 + 5x - 14$

8. $x^2 + 5x - 36$

9. $x^2 - 6x - 7$

10. $x^2 - 7x - 8$

11. $y^2 - 5y - 14$

12. $y^2 - 4y - 12$

Answers to PROBLEMS

5. $(x + 4a)(x + 2a)$ **6. a.** $y^3(y - 2)^2$ **b.** $b^2y^4(y + 4)(y - 5)$

13. $y^2 - 3y + 2$

14. $y^2 - 11y + 30$

15. $x^2 - 5x + 4$

16. $x^2 - 12x + 27$

17. $x^2 + 3x + 4$

18. $x^2 - 5x + 6$

19. $-7 - 7x + x^2$

20. $-5 - 5x + x^2$

21. $x^2 + 3ax + 2a^2$

22. $x^2 + 12ax + 35a^2$

23. $z^2 + 6bz + 9b^2$

24. $z^2 + 4bz + 16b^2$

25. $r^2 + ar - 12a^2$

26. $r^2 + 3ar - 10a^2$

27. $x^2 + 9ax - 10a^2$

28. $x^2 + 2ax - 8a^2$

29. $-b^2 - 2by + 3y^2$

30. $-b^2 + by + 110y^2$

31. $m^2 - am - 2a^2$

32. $m^2 - 2bm - 15b^2$

33. $2t^3 + 10t^2 + 8t$

34. $3t^3 + 12t^2 + 9t$

35. $a^2x^3 + 3a^2x^2 + 2ax^2$

36. $b^3x^5 + 4b^3x^4 + 2b^3x^3$

37. $b^3x^7 + b^3x^6 - 12b^3x^5$

38. $b^5y^8 + 3b^5y^7 - 10b^5y^6$

39. $2c^5z^6 + 4c^5z^5 - 30c^5z^4$

40. $3z^8 - 12z^7y - 63z^6y$

⟩ ⟩ ⟩ Applications

In Problems 41–42, the expression

$$-5t^2 + V_0 t + h$$

represents the altitude of an object t seconds after being thrown from a height of h meters with an initial velocity of V_0 meters per second.

41. *Altitude of a thrown object*

    a. Find the expression for the altitude of an object thrown upward with an initial velocity V_0 of 5 meters per second from a building 10 meters high.

    b. Factor the expression.

    c. When the product obtained in part **b** is 0, the altitude of the object is 0 and the object is on the ground. What values of t will make the product 0?

    d. Based on the answer to part **c**, how long does it take the object to return to the ground?

43. *Descent of a rock*   The height (in feet) of a rock thrown from the roof of a building after t seconds is given by

$$-16t^2 + 32t + 240$$

    a. Factor this expression.

    b. When the product obtained in part **a** is 0, the altitude of the rock is 0 and the rock is on the ground. What values of t will make the product 0?

    c. Based on the answer to part **b**, how long does it take the rock to return to the ground?

42. *Altitude of a thrown object*

    a. Find the expression for the altitude of an object thrown upward with an initial velocity V_0 of 10 meters per second from a building 40 meters high.

    b. Factor the expression.

    c. When the product obtained in part **b** is 0, the altitude of the object is 0 and the object is on the ground. What values of t will make the product 0?

    d. Based on the answer to part **c**, how long does it take the object to return to the ground?

44. *Chlorofluorocarbon production*   Do you know what freon is? It is a gas containing chlorofluorocarbons (CFC) used in old air conditioners and linked to the depletion of the ozone layer. Their production (in thousands of tons) can be represented by the expression

$$-0.04t^2 + 2.8t + 120$$

where t is the number of years after 1960.

    a. Factor $-0.04t^2 + 2.8t + 120$ using -0.04 as the GCF.

    b. When will the product obtained in part **a** be 0?

    c. Based on the answer to part **b**, when will the production of CFCs be 0? (The Montreal Protocol called for a 50% decrease by the year 2000.)

45. *Height of a ball*   The height H of a ball thrown upward from a height of 32 feet with an initial velocity of 16 feet per second is given by the equation $H = -16t^2 + 16t + 32$.

   a. Factor $-16t^2 + 16t + 32$.

   b. What is the height of the ball after 1 second?

   c. What is the height of the ball after 2 seconds?

47. *Number of games*   A sponsor wants to organize a tournament consisting of 20 games in which each of the x teams entered plays every other team twice. To find out how many teams you need, we must solve the equation $x^2 - x - 20 = 0$.

   a. Factor $x^2 - x - 20$.

   b. The number of games played is given by $x(x - 1)$. How many games are played when five teams are entered?

49. *Spring motion*   The motion of an object suspended by a helical spring requires the solution of the equation $D^2 + 8D + 12 = 0$. Factor $D^2 + 8D + 12$.

51. *Beam deflection*   The deflection of a beam of length L requires the solution of the equation $4Lx - x^2 - 4L^2 = 0$. Factor $4Lx - x^2 - 4L^2$.

53. *Profits*   The profits p (in thousands of dollars) for a company producing x units of a product each week is given by $p = 25x - x^2 - 100$. Factor $25x - x^2 - 100$.

46. *Height of toy rocket*   The height H of a toy rocket launched from the third floor of a building 48 feet above the street with an initial velocity of 32 feet per second is given by the equation $H = -16t^2 + 32t + 48$.

   a. Factor $-16t^2 + 32t + 48$.

   b. What is the height of the ball after 2 seconds?

   c. What is the height of the ball after 3 seconds?

48. *Number of games*   If a sponsor wants a tournament consisting of 30 games in which each of the x teams entered plays every other team twice, we have to solve the equation $x^2 - x - 30 = 0$. (See Exercise 47.)

   a. Factor $x^2 - x - 30$.

   b. The number of games played is given by $x(x - 1)$. How many games are played when six teams are entered?

50. *Spring motion*   The motion of an object attached to a spring requires the solution of the equation $D^2 + 2kD + k^2 = 0$. Factor $D^2 + 2kD + k^2$.

52. *Beam length*   To find the length L of a certain beam we have to solve the equation $x^2 - 2xL + L^2 = 0$. Factor $x^2 - 2xL + L^2$.

54. *Profits*   The profits p (in thousands of dollars) for a company producing x units of a product daily is given by $p = 32x - x^2 - 60$. Factor $32x - x^2 - 60$.

〉〉〉 Write On

55. What should your first step be when factoring any trinomial?

57. Tyrone says he reworked the problem (see Problem 56) and his new answer is $(2 - x)(3 - x)$. Ana still says she got $(x - 2)(x - 3)$. Who is correct?

59. When factoring $x^2 + bx + c$ (b and c positive) as $(x + A)(x + B)$, what can you say about A and B?

61. When factoring $x^2 + bx + c$ (c negative) as $(x + A)(x + B)$, what can you say about A and B?

56. Ana says that a polynomial can be factored as $(x - 3)(x - 2)$. Tyrone insists that the answer is really $(x - 2)(x - 3)$. Who is correct?

58. Explain in your own words what a *prime polynomial* is.

60. When factoring $x^2 + bx + c$ (b negative, c positive) as $(x + A)(x + B)$, what can you say about A and B?

〉〉〉 Concept Checker

Fill in the blank(s) with the correct word(s), phrase, or mathematical statement.

62. To **factor** $x^2 + bx + c$ we need two integers whose **product** is _____ and whose **sum** is _____.

63. If in problem 62, b and c are **positive**, both integers must be _____.

64. If in problem 62, c is positive and b is negative, both integers must be _____.

65. If in problem 62, c is negative, one integer must be _____ and one _____.

b

c

positive

negative

〉〉〉 *Mastery Test*

In Problems 66–77, factor completely.

66. $8x^2 - 12xy + 2xy - 3y^2$

67. $3y^3 - 6y^2 - y + 2$

68. $3y^3 + 9y^2 + 2y + 6$

69. $2y^3 - 6y^2 - y + 3$

70. $y^2 + 5y + 6$

71. $x^2 + 7xy + 12y^2$

72. $x^2 + 5x - 14$

73. $x^2 - 7xy + 10y^2$

74. $3y^4 + 6y^5 + 9y^6$

75. $z^2 - 10z - 25$

76. $kx^3 - 2kx^2 - 15kx$

77. $2y^3 + 6y^5 + 10y^7$

〉〉〉 *Skill Checker*

78. Find: **a.** $-8 \cdot 1$ **b.** $-8 + 1$

79. Find: **a.** $-6 \cdot 2$ **b.** $-6 + 2$

80. Find: **a.** $-8 \cdot 2$ **b.** $-8 + 2$

81. Find: **a.** $-4(-1)$ **b.** $-4 + (-1)$

82. Find: **a.** $-4 \cdot 3$ **b.** $-4 + 3$

5.3

Factoring $ax^2 + bx + c, a \neq 1$

▶ Objectives

A 〉 Use the *ac* test to determine whether $ax^2 + bx + c$ is factorable.

B 〉 Factor $ax^2 + bx + c$ by grouping.

C 〉 Factor $ax^2 + bx + c$ using FOIL.

▶ To Succeed, Review How To . . .

1. Expand $(X + A)(X + B)$ (p. 367).
2. Use FOIL to expand polynomials (p. 359).
3. Multiply integers (pp. 60–61).

▶ Getting Started
When There Is Smoke We Need Water

How much water is the fire truck pumping? It depends on many factors, but one of them is the friction loss inside the hose. If the friction loss is 36 pounds per square inch, we have to know how to factor $2g^2 + g - 36$ to find the answer.

This expression is of the form $ax^2 + bx + c$, $a \neq 1$, and we can factor it two ways: by **grouping** or by using **FOIL**. How do we know whether $2g^2 + g - 36$ is even factorable? We will show you how to tell and ask you to factor $2g^2 + g - 36$ in Problem 61.

A › Use the *ac* Test to Determine if a Polynomial Is Factorable

ac TEST FOR $ax^2 + bx + c$

A trinomial of the form $ax^2 + bx + c$ is factorable if there are two integers with product *ac* and sum *b*.

Note that *a* and *c* are the first and last numbers in $ax^2 + bx + c$ (hence the name *ac* test), and *b* is the coefficient of *x*. A diagram may help you visualize this test.

ac TEST

We need two numbers whose product is *ac*.

$$ax^2 + bx + c$$

The sum of the numbers must be *b*.

Note: Before you use the *ac* test, factor the GCF and write the polynomial in descending order.

Thus, to determine whether $2g^2 + g - 36$ is factorable, we need two numbers whose product is $a \cdot c = 2 \cdot (-36) = -72$ and whose sum is $b = 1$. Since $9 \cdot (-8) = -72$ and $9 + (-8) = 1$, $2g^2 + g - 36$ is factorable.

A polynomial that cannot be factored using only factors with integer coefficients is called a **prime polynomial.**

EXAMPLE 1 **Using the *ac* test to find whether a polynomial is factorable**

Determine whether the given polynomial is factorable:

a. $6x^2 + 7x + 2$ **b.** $2x^2 + 5x + 4$

SOLUTION

a. To find out whether $6x^2 + 7x + 2$ is factorable, we use these three steps:

   1. Multiply $a = 6$ by $c = 2 (6 \cdot 2 = 12)$.

   2. Find two integers whose product is $ac = 12$ and whose sum is $b = 7$.

We need two numbers whose product is $6 \cdot 2 = 12$.

$$6x^2 + 7x + 2$$

The sum of the two numbers must be 7.

   3. A little searching will produce 4 and 3.

   CHECK $\underset{\text{Product}}{4 \times 3 = 12,}$ $\underset{\text{Sum}}{4 + 3 = 7}$

Thus, $6x^2 + 7x + 2$ is factorable.

b. Consider the trinomial $2x^2 + 5x + 4$. Here is the *ac* test for this trinomial.

   1. Multiply 2 by 4 $(2 \cdot 4 = 8)$.

   2. Find two integers whose product is 8 and whose sum is 5.

PROBLEM 1

Determine whether the polynomial is factorable:

a. $4y^2 + 3y + 2$

b. $3y^2 + 5y + 2$

(continued)

Answers to PROBLEMS

1. a. Not factorable (prime)   **b.** Factorable ($ac = 6$; factors are 3 and 2; sum $b = 3 + 2 = 5$)

3. The factors of 8 are 4 and 2, and 8 and 1. Neither pair adds up to 5 $(4 + 2 = 6, 8 + 1 = 9)$. Thus, the trinomial $2x^2 + 5x + 4$ is *not* factorable using factors containing only integer coefficients; it is a prime polynomial.

B › Factoring Trinomials of the Form $ax^2 + bx + c$ by Grouping

At this point, you should be convinced that the *ac* test really tells you whether a trinomial of the form $ax^2 + bx + c$ is factorable; however, we still don't know *how* to do the actual factorization. But we are in luck; the number *ac* still plays an important part in factoring this trinomial. In fact, the number *ac* is so important that we shall call it the **key number** in the factorization of $ax^2 + bx + c$. To get a little practice, we have found and circled the key numbers of a few trinomials.

	a	*c*	*ac*
$6x^2 + 8x + 5$	6	5	(30)
$2x^2 - 7x - 4$	2	−4	(−8)
$-3x^2 + 2x + 5$	−3	5	(−15)

As shown before, by examining the key number and the coefficient of the middle term, you can determine whether a trinomial is factorable. For example, the key number of the trinomial $6x^2 + 8x + 5$ is 30. But since there are *no* integers with sum 8 whose product is 30, this trinomial is *not* factorable. (The factors of 30 are 6 and 5, 10 and 3, 15 and 2, and 30 and 1. None of these pairs has a sum of 8.)

On the other hand, the key number for $2x^2 - 7x - 4$ is −8, and −8 has two factors (−8 and 1) whose product is −8 and whose sum is the coefficient of the middle term, −7. Thus, $2x^2 - 7x - 4$ is factorable; here are the steps:

1. Find the key number $[2 \cdot (-4) = -8]$.

$2x^2 - 7x - 4$ (−8)

2. Find the factors of the key number and use the appropriate ones to rewrite the middle term.

$2x^2 - 8x + 1x - 4$ −8, 1

          −8(1) = −8
          −8 + 1 = −7

3. Group the terms into pairs (as we did in Section 5.1).

$(2x^2 - 8x) + (1x - 4)$

4. Factor each pair.

$2x(x - 4) + 1(x - 4)$

5. Note that $(x - 4)$ is the GCF.

$(x - 4)(2x + 1)$

Thus, $2x^2 - 7x - 4 = (x - 4)(2x + 1)$. You can check to see that this is the correct factorization by multiplying $(x - 4)$ by $(2x + 1)$.

Now a word of warning: You can write the factorization of $ax^2 + bx + c$ in *two* ways. Suppose you wish to factor the trinomial $5x^2 + 7x + 2$. Here's one way:

1. Find the key number $[5 \cdot 2 = 10]$.

$5x^2 + 7x + 2$ (10)

2. Find the factors of the key number; use them to rewrite the middle term.

$5x^2 + 5x + 2x + 2$ 5, 2

3. Group the terms into pairs.

$(5x^2 + 5x) + (2x + 2)$

4. Factor each pair. $5x(x + 1) + 2(x + 1)$

5. Note that $(x + 1)$ is the GCF. $(x + 1)(5x + 2)$

Thus, $5x^2 + 7x + 2 = (x + 1)(5x + 2)$. But there's another way:

1. Find the key number $5x^2 + 7x + 2$ ⑩
      $[5 \cdot 2 = 10]$.

2. Find the factors of the key $5x^2 + 2x + 5x + 2$ 2, 5
      number and use them to
      rewrite the middle term.

3. Group the terms into pairs. $(5x^2 + 2x) + (5x + 2)$

4. Factor each pair. $x(5x + 2) + 1(5x + 2)$

5. Note that $(5x + 2)$ is the GCF. $(5x + 2)(x + 1)$

In this case, we found that

$$5x^2 + 7x + 2 = (5x + 2)(x + 1)$$

Is the correct factorization $(x + 1)(5x + 2)$ or $(5x + 2)(x + 1)$? The answer is that *both* factorizations are correct! This is because the multiplication of real numbers is commutative and the variable x, as well as the trinomials involved, also represent real numbers; thus, the *order* in which the product is written (according to the commutative property of multiplication) *makes no difference in the final answer.*

When factoring a trinomial by grouping, just remember to write first the polynomial in descending order and factor out the GCF (if any).

EXAMPLE 2   **Factoring trinomials of the form $ax^2 + bx + c$ by grouping**

Factor:

a. $4 - 3x + 6x^2$ **b.** $4x^2 - 3 - 4x$

SOLUTION

a. We first write the polynomial in descending order as $6x^2 - 3x + 4$, then proceed by steps:

    1. Find the key number $6x^2 - 3x + 4$ ㉔
        $[6 \cdot 4 = 24]$.

    2. Find the factors of the key number and use them to rewrite the middle term. Unfortunately, it's impossible to find two numbers with product 24 and sum -3. This trinomial is *not* factorable.

b. We first rewrite the polynomial (*in descending order*) as $4x^2 - 4x - 3$, and then proceed by steps.

    1. Find the key number $4x^2 - 4x - 3$ ⑫̶ (−12)
        $[4 \cdot (-3) = -12]$.

    2. Find the factors of the key $4x^2 - 6x + 2x - 3$ −6, 2
        number and use them to
        rewrite the middle term.

    3. Group the terms into pairs. $(4x^2 - 6x) + (2x - 3)$

    4. Factor each pair. $2x(2x - 3) + 1(2x - 3)$

    5. Note that $(2x - 3)$ $(2x - 3)(2x + 1)$
        is the GCF.

Thus, $4x^2 - 4x - 3 = (2x - 3)(2x + 1)$, as can easily be verified by multiplication.

PROBLEM 2

Factor:

a. $9x^2 - 2 - 3x$

b. $3 - 4x + 2x^2$

Answers to PROBLEMS

2. a. $(3x - 2)(3x + 1)$
   b. Not factorable (prime)

Factoring problems in which the third term in step 2 contains a negative number as a coefficient requires that special care be taken with the signs. For example, to factor the trinomial $4x^2 - 5x + 1$, we proceed as follows:

1. Find the key number $[4 \cdot 1 = 4]$.

$4x^2 - 5x + 1$ ④

2. Find the factors of the key number and use them to rewrite the middle term.

$4x^2 - 4x - 1x + 1$ $-4, -1$

Note that the third term has a negative coefficient, -1.

3. Group the terms into pairs. $(4x^2 - 4x) + (-1x + 1)$

4. Factor each pair. $4x(x - 1) - 1(x - 1)$ Recall that $-1(x - 1) = -x + 1$.

5. Note that $(x - 1)$ is the GCF. $(x - 1)(4x - 1)$ If the first pair has $(x - 1)$ as a factor, the second pair will also have $(x - 1)$ as a factor.

Thus, $4x^2 - 5x + 1 = (x - 1)(4x - 1)$.

EXAMPLE 3   Factoring trinomials of the form $ax^2 - bx + c$ by grouping

Factor: $5x^2 - 11x + 2$

SOLUTION

1. Find the key number $[5 \cdot 2 = 10]$.

$5x^2 - 11x + 2$ ⑩

2. Find the factors of the key number and use them to rewrite the middle term.

$5x^2 - 10x - 1x + 2$ $-10, -1$

3. Group the terms into pairs. $(5x^2 - 10x) + (-1x + 2)$

4. Factor each pair. $5x(x - 2) - 1(x - 2)$

5. Note that $(x - 2)$ is the GCF. $(x - 2)(5x - 1)$

Thus, the factorization of $5x^2 - 11x + 2$ is $(x - 2)(5x - 1)$.

PROBLEM 3

Factor: $4x^2 - 13x + 3$

So far, we have factored trinomials in one variable only. A procedure similar to the one used for factoring a trinomial of the form $ax^2 + bx + c$ can be used to factor certain trinomials in two variables. We illustrate the procedure in Example 4.

EXAMPLE 4   Factoring trinomials with two variables by grouping

Factor: $6x^2 - xy - 2y^2$

SOLUTION

1. Find the key number $[6 \cdot (-2) = -12]$.

$6x^2 - xy - 2y^2$ -12

2. Find the factors of the key number and use them to rewrite the middle term.

$6x^2 - 4xy + 3xy - 2y^2$ $-4, 3$

3. Group the terms into pairs. $(6x^2 - 4xy) + (3xy - 2y^2)$

4. Factor each pair. $2x(3x - 2y) + y(3x - 2y)$

5. Note that $(3x - 2y)$ is the GCF. $(3x - 2y)(2x + y)$

Thus, $6x^2 - xy - 2y^2 = (3x - 2y)(2x + y)$.

PROBLEM 4

Factor: $4x^2 - 4xy - 3y^2$

Answers to PROBLEMS

3. $(x - 3)(4x - 1)$
4. $(2x - 3y)(2x + y)$

C › Factoring $ax^2 + bx + c$ by FOIL (Trial and Error)

Sometimes, it's easier to factor a polynomial of the form $ax^2 + bx + c$ by using a FOIL process (*trial and error*). This is especially so when a or c is a prime number such as 2, 3, 5, 7, 11, and so on. Here's how we do this.

PROCEDURE

Factoring by FOIL (Trial and Error)

Product must be c.

$$ax^2 + bx + c = (\underline{}x + \underline{})(\underline{}x + \underline{})$$

FOIL

FOIL

Product must be a.

Now,

1. The product of the numbers in the *first* (F) blanks must be a.
2. The coefficients of the *outside* (O) products and the *inside* (I) products must add up to b.
3. The product of the numbers in the *last* (L) blanks must be c.

For example, to factor $2x^2 + 5x + 3$, we write:

$$2x^2 + 5x + 3 = (\underline{}x + \underline{})(\underline{}x + \underline{})$$

We first look for two numbers whose product is 2. These numbers are 2 and 1, or -2 and -1. We have these possibilities:

$$(2x + \underline{})(x + \underline{}) \quad \text{or} \quad (-2x + \underline{})(-x + \underline{})$$

Let's agree that we want the first coefficients inside the parentheses to be *positive*. This eliminates products involving $(-2x + \underline{})$. Now we look for numbers whose product is 3. These numbers are 3 and 1, or -3 and -1, which we substitute into the blanks, to obtain:

$$(2x + 3)(x + 1)$$
$$(2x + 1)(x + 3)$$
$$(2x - 3)(x - 1)$$
$$(2x - 1)(x - 3)$$

Since the final result must be $2x^2 + 5x + 3$, the first expression (shaded) yields the desired factorization:

$$2x^2 + 5x + 3 = (2x + 3)(x + 1)$$

You can save some time if you notice that all coefficients are positive, so the trial numbers must be positive. That leaves 2, 1 and 3, 1 as the only possibilities.

EXAMPLE 5   Factoring by FOIL (trial and error): All terms positive

Factor: $3x^2 + 7x + 2$

SOLUTION   Since we want the first coefficients in the factorization to be positive, the only two factors of 3 we consider are 3 and 1. We then look for the numbers whose product will equal 2:

$$3x^2 + 7x + 2 = (3x + \underline{})(x + \underline{})$$

These factors are 2 and 1, and the possibilities are

$$(3x + 2)(x + 1) \qquad \text{or} \qquad (3x + 1)(x + 2)$$

Since the second product, $(3x + 1)(x + 2)$, yields the correct middle term, $7x$, we have

$$3x^2 + 7x + 2 = (3x + 1)(x + 2)$$

PROBLEM 5

Factor using FOIL (trial and error):

$3x^2 + 5x + 2$

Note that the trial-and-error method is based on FOIL (Section 4.6). Thus, to *multiply* $(2x + 3)(3x + 4)$ using FOIL, we write

$$
\begin{array}{cccc}
 & \text{F} & \text{O} \quad \text{I} & \text{L} \\
(2x + 3)(3x + 4) = & 6x^2 & + 8x + 9x & + 12 \\
 = & 6x^2 & + \quad 17x & + 12 \\
 & \text{F} & \text{O + I} & \text{L} \\
 & 2 \cdot 3 & 2 \cdot 4 + 3 \cdot 3 & 3 \cdot 4
\end{array}
$$

Now to *factor* $6x^2 + 17x + 12$, we do the reverse, using trial and error. Since the factors of 6 are 6 and 1, or 3 and 2 (we won't use -6 and -1, or -3 and -2, because then the first coefficients will be negative), the possible combinations are

$$(6x + \underline{})(x + \underline{}) \qquad (3x + \underline{})(2x + \underline{})$$

Since the product of the last two numbers is 12, the possible factors are 12, 1; 6, 2; and 3, 4. The possibilities are

$$(6x + 12)(x + 1)^\star \qquad (6x + 1)(x + 12)$$
$$(6x + 6)(x + 2)^\star \qquad (6x + 2)(x + 6)^\star$$
$$(6x + 3)(x + 4)^\star \qquad (6x + 4)(x + 3)^\star$$
$$(3x + 12)(2x + 1)^\star \qquad (3x + 1)(2x + 12)^\star$$
$$(3x + 6)(2x + 2)^\star \qquad (3x + 2)(2x + 6)^\star$$
$$(3x + 3)(2x + 4)^\star \qquad (3x + 4)(2x + 3)$$

Note that in one binomial each of the starred items has a common factor, but $6x^2 + 17x + 12$ has *no* common factor other than 1. Thus, we can eliminate all starred products.

Thus, $6x^2 + 17x + 12 = (3x + 4)(2x + 3)$.

Note that if there is a common factor, we must factor it out first. Thus, to factor $12x^2 + 2x - 2$, we must *first* factor out the common factor 2, as illustrated in Example 6.

EXAMPLE 6   Factoring by FOIL (trial and error): Last term negative

Factor: $12x^2 + 2x - 2$

SOLUTION   Since 2 is a common factor, we first factor it out to obtain

$$12x^2 + 2x - 2 = 2 \cdot 6x^2 + 2 \cdot x - 2 \cdot 1$$
$$= 2(6x^2 + x - 1)$$

PROBLEM 6

Factor using FOIL (trial and error):

$18x^2 + 3x - 6$

Answers to PROBLEMS

5. $(3x + 2)(x + 1)$

6. $3(2x - 1)(3x + 2)$

Now we factor $6x^2 + x - 1$. The factors of 6 are 6, 1 or 3, 2. Thus,

$$6x^2 + x - 1 = (6x + \underline{\quad})(x + \underline{\quad})$$

or

$$6x^2 + x - 1 = (3x + \underline{\quad})(2x + \underline{\quad})$$

The product of the last two terms must be -1. The possible factors are $-1, 1$. The possibilities are

$$(6x - 1)(x + 1) \qquad (6x + 1)(x - 1)$$
$$(3x - 1)(2x + 1) \qquad (3x + 1)(2x - 1)$$

The only product that yields $6x^2 + x - 1$ is $(3x - 1)(2x + 1)$. Try it! This example shows why this method is sometimes called *trial* and error. Thus,

$$12x^2 + 2x - 2 = 2(6x^2 + x - 1)$$
$$= 2(3x - 1)(2x + 1)$$

Remember to write the common factor.

EXAMPLE 7   Factoring by FOIL (trial and error): Two variables
Factor: $6x^2 - 11xy - 10y^2$

SOLUTION   Since there are no common factors, look for the factors of 6: 6 and 1, or 3 and 2. The possibilities for our factorization are

$$(6x + \underline{\quad})(x + \underline{\quad}) \qquad \text{or} \qquad (3x + \underline{\quad})(2x + \underline{\quad})$$

The last term, $-10y^2$, has the following possible factors:

$$10y, -y \qquad -10y, y \qquad 2y, -5y \qquad -2y, 5y$$
$$-y, 10y \qquad y, -10y \qquad -5y, 2y \qquad 5y, -2y$$

Look daunting? Don't despair; just look more closely. Can you see that some trials like $(6x + 10y)(3x - y)$, which correspond to the factors $10y, -y$, or $(3x - y)(2x + 10y)$, which correspond to $-y, 10y$, can't be correct because they contain 2 as a common factor? As you can see, $6x^2 - 11xy - 10y^2$ has no common factors (other than 1), so let's try

$$(3x + 10y)(2x - y)$$

17xy is not the correct middle term. Next we try

$$(3x - 2y)(2x + 5y)$$

Again, $11xy$ is not the correct middle term, but it's close! The result is incorrect but only because of the sign of the middle term. The correct factorization is found by interchanging the signs in the binomials. That is,

$$(3x + 2y)(2x - 5y) = 6x^2 - 11xy - 10y^2$$

PROBLEM 7
Factor using FOIL (trial and error):
$6x^2 - 5xy - 6y^2$

Answers to PROBLEMS

7. $(3x + 2y)(2x - 3y)$

CAUTION

Some students and some instructors prefer the FOIL (trial-and-error) method over the grouping method. You can use either method and your answer will be the same. Which one should you use? The one you understand better or the one your instructor asks you to use!

MathZone+x

Boost your grade at mathzone.com!
> Practice   > Self-Tests
  Problems   > e-Professors
> NetTutor   > Videos

> Exercises 5.3

⟨ **A** ⟩   Use the *ac* Test to Determine if a Polynomial Is Factorable.
⟨ **B** ⟩   Factoring Trinomials of the Form $ax^2 + bx + c$ by Grouping.
⟨ **C** ⟩   Factoring $ax^2 + bx + c$ by FOIL (Trial and Error)

In Problems 1–50, determine whether the polynomial is factorable. If the polynomial is factorable, factor it. Use the FOIL method to factor if you wish.

1. $2x^2 + 5x + 3$ **2.** $2x^2 + 7x + 3$ **3.** $6x^2 + 11x + 3$

4. $6x^2 + 17x + 5$ **5.** $6x^2 + 11x + 4$ **6.** $5x^2 + 2x + 1$

7. $2x^2 + 3x - 2$ **8.** $2x^2 + x - 3$ **9.** $3x^2 + 16x - 12$

10. $6x^2 + x - 12$ **11.** $4y^2 - 11y + 6$ **12.** $3y^2 - 17y + 10$

13. $4y^2 - 8y + 6$ **14.** $3y^2 - 11y + 6$ **15.** $6y^2 - 10y - 4$

16. $12y^2 - 10y - 12$ **17.** $12y^2 - y - 6$ **18.** $3y^2 - y - 1$

19. $18y^2 - 21y - 9$ **20.** $36y^2 - 12y - 15$ **21.** $3x^2 + 2 + 7x$

22. $2x^2 + 2 + 5x$ **23.** $5x^2 + 2 + 11x$ **24.** $5x^2 + 3 + 12x$

25. $6x^2 - 5 + 15x$ **26.** $5x^2 - 8 + 6x$ **27.** $3x^2 - 2 - 5x$

28. $5x^2 - 8 - 6x$ **29.** $15x^2 - 2 + x$ **30.** $8x^2 + 15 - 14x$

31. $8x^2 + 20xy + 8y^2$ **32.** $12x^2 + 28xy + 8y^2$ **33.** $6x^2 + 7xy - 3y^2$ **34.** $3x^2 + 13xy - 10y^2$

35. $7x^2 - 10xy + 3y^2$ **36.** $6x^2 - 17xy + 5y^2$ **37.** $15x^2 - xy - 2y^2$ **38.** $5x^2 - 6xy - 8y^2$

39. $15x^2 - 2xy - 2y^2$ **40.** $4x^2 - 13xy - 3y^2$ **41.** $12r^2 + 17r - 5$ **42.** $20s^2 + 7s - 6$

43. $22t^2 - 29t - 6$ **44.** $39u^2 - 23u - 6$ **45.** $18x^2 - 21x + 6$ **46.** $12x^2 - 22x + 6$

47. $6ab^2 + 5ab + a$ **48.** $6bc^2 + 13bc + 6b$ **49.** $6x^5y + 25x^4y^2 + 4x^3y^3$ **50.** $12p^4q^3 + 11p^3q^4 + 2p^2q^5$

In Problems 51–60, first factor out -1. [*Hint:* To factor $-6x^2 + 7x + 2$, the first step will be

$$-6x^2 + 7x + 2 = -1(6x^2 - 7x - 2)$$
$$= -(6x^2 - 7x - 2)$$

then factor inside the parentheses.]

51. $-6x^2 - 7x - 2$

52. $-12y^2 - 11y - 2$

53. $-9x^2 - 3x + 2$

54. $-6y^2 - 5y + 6$

55. $-8m^2 + 10mn + 3n^2$

56. $-6s^2 + st + 2t^2$

57. $-8x^2 + 9xy - y^2$

58. $-6y^2 + 3xy + 2x^2$

59. $-x^3 - 5x^2 - 6x$

60. $-y^3 + 3y^2 - 2y$

〉〉〉 Applications

61. *Flow rate*   To find the flow g (in hundreds of gallons per minute) in 100 feet of $2\frac{1}{2}$-inch rubber-lined hose when the friction loss is 36 pounds per square inch, we need to factor the expression

$$2g^2 + g - 36$$

Factor this expression.

62. *Flow rate*   To find the flow g (in hundreds of gallons per minute) in 100 feet of $2\frac{1}{2}$-inch rubber-lined hose when the friction loss is 55 pounds per square inch, we must factor the expression

$$2g^2 + g - 55$$

Factor this expression.

63. *Equivalent resistance*   When solving for the equivalent resistance R of two electric circuits, we use the expression

$$2R^2 - 3R + 1$$

Factor this expression.

64. *Rate of ascent*   To find the time t at which an object thrown upward at 12 meters per second will be 4 meters above the ground, we must factor the expression

$$5t^2 - 12t + 4$$

Factor this expression.

65. *Box office receipts*   Have you been to the movies lately? How much was your ticket? According to the U.S. Department of Commerce, box office receipts from 1980 to 1990 can be approximated by the equation

$$B(t) \cdot P(t) = 0.004t^2 + 0.2544t + 2.72$$

where $B(t)$ is the number of people buying tickets (in thousands), $P(t)$ is the average ticket price (in dollars), and t is the number of years after 1980.

  a. Factor $0.004t^2 + 0.2544t + 2.72$.
    [*Hint:* Try $B(t) \cdot P(t) = ($ ___ $+ 1)($ ___ $+ 2.72)$.]

  b. Based on your answer to part **a**, what was the average price of a ticket in 1980? (The actual average price for 1980 was $2.72.)

  c. Based on your answer to part **a**, what was the average price of a ticket in 1985? (The actual average price for 1985 was $3.55.)

66. *Entertainment and reading expenditures*   According to the annual *Consumer Educational Survey* of the Bureau of Labor Statistics, the annual average amount of money spent for entertainment and reading can be approximated by $1300 + 4.16t + 3.12t^2$ (dollars), where t is the number of years after 1985. If you want to approximate how much is spent *weekly* on entertainment and reading, write

$$1300 + 4.16t + 3.12t^2 = 52 \cdot E(t)$$

  a. Where does the 52 come from?

  b. What does $E(t)$ represent?

  c. How much was spent weekly for entertainment and reading in 1985? (The actual amount was $25.21.)

  d. How much was spent weekly on entertainment and reading in 1995? In 2005?

67. *Stopping distance* (See page 345.)   If you are traveling at m miles per hour and want to find the speed m that will allow you to stop the car 13 feet away, you must solve the equation $5m^2 + 220m - 1225 = 0$. Factor the left side of the equation.

68. *Stopping distance*   If you are traveling at m miles per hour and want to find the speed m that will allow you to stop the car 14 feet away, you must solve the equation $5m^2 + 225m - 1250 = 0$. Factor the left side of the equation.

69. *Height of an object*   The height $H(t)$ of an object thrown downward from a height of h meters and initial velocity v_0 is given by the equation $H(t) = -5t^2 + v_0 t + h$. If an object is thrown downward at 5 meters per second from a height of 10 meters, then $H(t) = -5t^2 - 5t + 10$. Factor $-5t^2 - 5t + 10$.

70. *Height of an object*   The height $H(t)$ of an object thrown downward from a height of h meters and initial velocity v_0 is given by the equation $H(t) = -5t^2 + v_0 t + h$. If an object is thrown downward at 4 meters per second from a height of 28 meters, then $H(t) = -5t^2 - 4t + 28$. Factor $-5t^2 - 4t + 28$.

71. *Production cost*   The cost $C(x)$ of producing x units of a certain product is given by the equation $C(x) = 3x^2 - 17x + 20$. Factor $3x^2 - 17x + 20$.

72. *Production cost*   The cost $C(x)$ of producing x units of a certain product is given by the equation $C(x) = 3x^2 - 12x + 13$. Factor $5x^2 - 28x + 15$.

〉〉〉 *Using Your Knowledge*

Factoring Applications The ideas presented in this section are important in many fields. Use your knowledge to factor the given expressions.

73. To find the deflection of a beam of length L at a distance of 3 feet from its end, we must evaluate the expression

$$2L^2 - 9L + 9$$

Factor this expression.

74. The height after t seconds of an object thrown upward at 12 meters per second is

$$-5t^2 + 12t$$

To determine the time at which the object will be 7 meters above ground, we must solve the equation

$$5t^2 - 12t + 7 = 0$$

Factor $5t^2 - 12t + 7$.

75. In Problem 73, if the distance from the end is x feet, then we must use the expression

$$2L^2 - 3xL + x^2$$

Factor this expression.

〉〉〉 *Write On*

76. Mourad says that the ac key number for $2x^2 + 1 + 3x$ is 2 and hence $2x^2 + 1 + 3x$ is factorable. Bill says ac is 6. Who is correct?

77. When factoring $6x^2 + 11x + 3$ by grouping, student A writes

$$6x^2 + 11x + 3 = 6x^2 + 9x + 2x + 3$$

Student B writes

$$6x^2 + 11x + 3 = 6x^2 + 2x + 9x + 3$$

a. What will student A's answer be?

b. What will student B's answer be?

c. Who is correct, student A or student B? Explain.

78. A student gives $(3x - 1)(x - 2)$ as the answer to a factoring problem. Another student gets $(2 - x)(1 - 3x)$. Which student is correct? Explain.

〉〉〉 *Concept Checker*

Fill in the blank(s) with the correct word(s), phrase, or mathematical statement.

79. A trinomial of the form $ax^2 + bx + c$ is **factorable** if there are two integers with **product** _____ and **sum** _____.

80. When factoring $ax^2 + bx + c$ the product ac is called the _____ number.

main	b
key	c
ac	a
ad	

〉〉〉 *Mastery Test*

Factor:

81. $3x^2 - 4 - 4x$

82. $2x^2 - 11x + 5$

83. $2x^2 - xy - 6y^2$

84. $3x^2 + 5x + 2$

85. $16x^2 + 4x - 2$

86. $3x^3 + 7x^2 + 2x$

87. $3x^4 + 5x^3 - 3x^2$

88. $5x^2 - 2x + 2$

〉〉〉 *Skill Checker*

Expand:

89. $(x + 8)^2$

90. $(x - 7)^2$

91. $(3x - 2)^2$

92. $(2x + 3)^2$

93. $(2x + 3y)^2$

94. $(2x - 3y)^2$

95. $(3x + 5y)(3x - 5y)$

96. $(2x - 5y)(2x + 5y)$

97. $(x^2 + 4)(x^2 - 4)$

98. $(x^2 - 3)(x^2 + 3)$

5.4 Factoring Squares of Binomials

Objectives

A ⟩ Recognize the square of a binomial (a perfect square trinomial).

B ⟩ Factor a perfect square trinomial.

C ⟩ Factor the difference of two squares.

To Succeed, Review How To . . .

1. Expand the square of a binomial sum or difference (pp. 368–369).
2. Find the product of the sum and difference of two terms (pp. 369–370).

Getting Started

A Moment for a Crane

What is the moment (the product of a quantity and the distance from a perpendicular axis) on the crane? At x feet from its support, the moment involves the expression

$$\frac{w}{2}(x^2 - 20x + 100)$$

where w is the weight of the crane in pounds per foot. The expression $x^2 - 20x + 100$ is the result of *squaring a binomial* and is called a **perfect square trinomial** because $x^2 - 20x + 100 = (x - 10)^2$. Similarly, $x^2 + 12x + 36 = (x + 6)^2$ is the square of a binomial. We can factor these two expressions by using the special products studied in Section 4.7 in reverse. For example, $x^2 - 20x + 100$ has the same form as SP3, whereas $x^2 + 12x + 36$ looks like SP2. In this section we continue to study the *reverse* process of multiplying binomials, that of *factoring* trinomials.

A ⟩ Recognizing Squares of Binomials

We start by rewriting the products in SP2 and SP3 so you can use them for factoring.

FACTORING RULES 2 AND 3: PERFECT SQUARE TRINOMIALS

$$X^2 + 2AX + A^2 = (X + A)^2 \qquad \text{Note that } X^2 + A^2 \neq (X + A)^2 \qquad \textbf{(F2)}$$

$$X^2 - 2AX + A^2 = (X - A)^2 \qquad \text{Note that } X^2 - A^2 \neq (X - A)^2 \qquad \textbf{(F3)}$$

Note that to be the **square of a binomial** (a *perfect square trinomial*), a trinomial must satisfy three conditions:

1. The first and last terms (X^2 and A^2) must be perfect squares.

2. There must be no minus signs before A^2 or X^2.

3. The middle term is twice the product of the expressions being squared in step 1 ($2AX$) or its additive inverse ($-2AX$). Note the X and A are the terms of the binomial being squared to obtain the perfect square trinomial.

EXAMPLE 1   **Deciding whether an expression is the square of a binomial**

Determine whether the given expression is the square of a binomial:

a. $x^2 + 8x + 16$ **b.** $x^2 + 6x - 9$

c. $x^2 + 4x + 16$ **d.** $4x^2 - 12xy + 9y^2$

SOLUTION   In each case, we check the three conditions necessary for having a perfect square trinomial.

a. **1.** x^2 and $16 = 4^2$ are perfect squares.

     2. There are no minus signs before x^2 or 16.

     3. The middle term is twice the product of the expressions being squared in step 1, x and 4; that is, the middle term is $2 \cdot (x \cdot 4) = 8x$. Thus, $x^2 + 8x + 16$ *is* a perfect square trinomial (the square of a binomial).

b. **1.** x^2 and $9 = 3^2$ are perfect squares.

     2. However, there's a minus sign before the 9. Thus, $x^2 + 6x - 9$ is *not* the square of a binomial.

c. **1.** x^2 and $16 = 4^2$ are perfect squares.

     2. There are no minus signs before x^2 or 16.

     3. The middle term should be $2 \cdot (x \cdot 4) = 8x$, but instead it's $4x$. Thus, $x^2 + 4x + 16$ is *not* a perfect square trinomial.

d. **1.** $4x^2 = (2x)^2$ and $9y^2 = (3y)^2$ are perfect squares.

     2. There are no minus signs before $4x^2$ or $9y^2$.

     3. The middle term is the additive inverse of twice the product of the expressions being squared in step 1; that is, $-2 \cdot (2x \cdot 3y) = -12xy$. Thus, $4x^2 - 12xy + 9y^2$ *is* a perfect square trinomial.

PROBLEM 1

Determine whether the expression is the square of a binomial:

a. $y^2 + 9y + 9$

b. $y^2 + 4y + 4$

c. $y^2 + 8y - 16$

d. $9x^2 - 12xy + 4y^2$

Answers to PROBLEMS

1. a. No   **b.** Yes   **c.** No   **d.** Yes

B › Factoring Perfect Square Trinomials

The formulas given in F2 and F3 can be used to factor any trinomials that are perfect squares. For example, the trinomial $9x^2 + 12x + 4$ can be factored using F2 if we first notice that

1. $9x^2$ and 4 are perfect squares, since $9x^2 = (3x)^2$ and $4 = 2^2$.

2. There are no minus signs before $9x^2$ or 4.

3. $12x = 2 \cdot (2 \cdot 3x)$. (The middle term is twice the product of 2 and $3x$, the expressions being squared in step 1.)

We then write

$$\overset{X^2 \quad + \quad 2 \qquad AX \quad + A^2}{9x^2 + 12x + 4 = (3x)^2 + 2 \cdot (2 \cdot 3x) + 2^2} \qquad \text{We are letting } X = 3x, A = 2 \text{ in F2.}$$

$$= (3x + 2)^2$$

Here are some other examples of this form; study these examples carefully before you continue.

$$X^2 \quad + 2 \qquad AX \qquad + A^2$$

$$9x^2 + 6x + 1 = (3x)^2 + 2 \cdot (1 \cdot 3x) + 1^2 = (3x + 1)^2 \qquad \text{Letting } X = 3x, \\ A = 1 \text{ in F2.}$$

$$16x^2 + 24x + 9 = (4x)^2 + 2 \cdot (3 \cdot 4x) + 3^2 = (4x + 3)^2 \qquad \text{Here } X = 4x, A = 3.$$

$$4x^2 + 12xy + 9y^2 = (2x)^2 + 2 \cdot (3y \cdot 2x) + (3y)^2 = (2x + 3y)^2 \qquad \text{Here } X = 2x, A = 3y.$$

> **NOTE**
>
> The key for factoring these trinomials is to recognize that the first and last terms are *perfect squares*. (Of course, you have to check the middle term also.)

EXAMPLE 2   Factoring perfect square trinomials

Factor:

a. $x^2 + 16x + 64$ b. $25x^2 + 20x + 4$ c. $9x^2 + 12xy + 4y^2$

SOLUTION

a. We first write the trinomial in the form $X^2 + 2AX + A^2$. Thus,

$$x^2 + 16x + 64 = x^2 + 2 \cdot (8 \cdot x) + 8^2 = (x + 8)^2$$

b. $25x^2 + 20x + 4 = (5x)^2 + 2 \cdot (2 \cdot 5x) + 2^2 = (5x + 2)^2$

c. $9x^2 + 12xy + 4y^2 = (3x)^2 + 2 \cdot (2y \cdot 3x) + (2y)^2 = (3x + 2y)^2$

PROBLEM 2

Factor:

a. $y^2 + 6y + 9$

b. $4x^2 + 28xy + 49y^2$

c. $9y^2 + 12y + 4$

Of course, we use the same technique (but with F3) to factor $x^2 - 16x + 64$ or $25x^2 - 20x + 4$. Do you recall F3?

$$\underbrace{X^2}_{} - \underbrace{2AX}_{} + \underbrace{A^2}_{} = \qquad\qquad \overbrace{(X - A)^2}^{}$$
$$x^2 - 16x + 64 = x^2 - 2 \cdot (8 \cdot x) + 8^2 = (x - 8)^2$$

Similarly,

$$25x^2 - 20x + 4 = (5x)^2 - 2 \cdot (2 \cdot 5x) + 2^2 = (5x - 2)^2$$

EXAMPLE 3   Factoring perfect square trinomials

Factor:

a. $x^2 - 10x + 25$ b. $4x^2 - 12x + 9$ c. $4x^2 - 20xy + 25y^2$

SOLUTION

a. $x^2 - 10x + 25 = x^2 - 2 \cdot (5 \cdot x) + 5^2 = (x - 5)^2$

b. $4x^2 - 12x + 9 = (2x)^2 - 2 \cdot (3 \cdot 2x) + 3^2 = (2x - 3)^2$

c. $4x^2 - 20xy + 25y^2 = (2x)^2 - 2 \cdot (5y \cdot 2x) + (5y)^2 = (2x - 5y)^2$

PROBLEM 3

Factor:

a. $y^2 - 4y + 4$

b. $9y^2 - 12y + 4$

c. $9x^2 - 30xy + 25y^2$

C › Factoring the Difference of Two Squares

Can we factor $x^2 - 9$ as a product of two binomials? Note that $x^2 - 9$ has no middle term. The only special product with no middle term we've studied is the product of the sum and the difference of two terms (SP4). Here is the corresponding factoring rule:

Answers to PROBLEMS

2. a. $(y + 3)^2$ b. $(2x + 7y)^2$
   c. $(3y + 2)^2$
3. a. $(y - 2)^2$ b. $(3y - 2)^2$
   c. $(3x - 5y)^2$

FACTORING RULE 4: THE DIFFERENCE OF TWO SQUARES

$$X^2 - A^2 = (X + A)(X - A) \qquad\qquad \textbf{(F4)}$$

We can now factor binomials of the form $x^2 - 16$ and $9x^2 - 25y^2$. To do this, we proceed as follows:

$$X^2 - A^2 = (X + A)(X - A)$$

$$x^2 - 16 = (x)^2 - (4)^2 = (x + 4)(x - 4)$$ Check this by using FOIL.

and

$$X^2 - A^2 = (X + A)(X - A)$$

$$9x^2 - 25y^2 = (3x)^2 - (5y)^2 = (3x + 5y)(3x - 5y)$$

NOTE

$x^2 + A^2$ *cannot* be factored!

EXAMPLE 4   Factoring the difference of two squares

Factor:

a. $x^2 - 4$

b. $25x^2 - 9$

c. $16x^2 - 9y^2$

d. $x^4 - 16$

e. $\frac{1}{4}x^2 - \frac{1}{9}$

f. $7x^3 - 28x$

SOLUTION

$$X^2 - A^2 = (X + A)(X - A)$$

a. $x^2 - 4 = (x)^2 - (2)^2 = (x + 2)(x - 2)$

b. $25x^2 - 9 = (5x)^2 - (3)^2 = (5x + 3)(5x - 3)$

c. $16x^2 - 9y^2 = (4x)^2 - (3y)^2 = (4x + 3y)(4x - 3y)$

d. $x^4 - 16 = (x^2)^2 - (4)^2 = (x^2 + 4)(x^2 - 4)$

But $(x^2 - 4)$ itself is factorable, so

$(x^2 + 4)(x^2 - 4) = (x^2 + 4)(x + 2)(x - 2)$. Thus,

$$x^4 - 16 = \underbrace{(x^2 + 4)}(x + 2)(x - 2)$$
$$\text{Not factorable}$$

e. We start by writing $\frac{1}{4}x^2 - \frac{1}{9}$ as the difference of two squares. To obtain $\frac{1}{4}x^2$, we must square $\frac{1}{2}x$ and to obtain $\frac{1}{9}$, we must square $\frac{1}{3}$. Thus,

$$\frac{1}{4}x^2 - \frac{1}{9} = \left(\frac{1}{2}x\right)^2 - \left(\frac{1}{3}\right)^2$$

$$= \left(\frac{1}{2}x + \frac{1}{3}\right)\left(\frac{1}{2}x - \frac{1}{3}\right)$$

f. We start by finding the GCF of $7x^3 - 28x$, which is $7x$. We then write

$$7x^3 - 28x = 7x(x^2 - 4)$$
$$= 7x(x + 2)(x - 2)$$ Factor $x^2 - 4$ as $(x + 2)(x - 2)$.

PROBLEM 4

Factor:

a. $y^2 - 1$

b. $9y^2 - 25$

c. $9y^2 - 25x^2$

d. $y^4 - 81$

e. $\frac{1}{9}y^2 - \frac{1}{4}$

f. $5y^3 - 45y$

Answers to PROBLEMS

4. a. $(y + 1)(y - 1)$

b. $(3y + 5)(3y - 5)$

c. $(3y + 5x)(3y - 5x)$

d. $(y^2 + 9)(y + 3)(y - 3)$

e. $\left(\frac{1}{3}y + \frac{1}{2}\right)\left(\frac{1}{3}y - \frac{1}{2}\right)$

f. $5y(y + 3)(y - 3)$

Calculator Corner

More Factoring Checking

Can you use a calculator to check factoring? Yes, but you still have to accept the following: If the graphs of two polynomials are identical, the polynomials are identical. Consider Example 2b, $25x^2 + 20x + 4 = (5x + 2)^2$. Graph $Y_1 = 25x^2 + 20x + 4$ and $Y_2 = (5x + 2)^2$. The graphs are the same! How do you know there are two graphs? Press TRACE and ◀, then ▼. Do you see the little number at the top right of the screen? It tells you which curve is showing. The bottom of the screen shows that for both Y_1 and Y_2 when $x = 0$, $y = 4$. Graph $Y_3 = (5x + 2)^2 + 10$. Press TRACE. Are the values for Y_1, Y_2, and Y_3 the same? They shouldn't be! Do you see why? Use this technique to check other factoring problems.

› Exercises 5.4

‹ A ›   **Recognizing Squares of Binomials**   In Problems 1–10, determine whether the given expression is a perfect square trinomial (the square of a binomial).

1. $x^2 + 14x + 49$

2. $x^2 + 18x + 81$

3. $25x^2 + 10x - 1$

4. $9x^2 + 12x - 4$

5. $25x^2 + 10x + 1$

6. $9x^2 + 12x + 4$

7. $y^2 - 4y - 4$

8. $y^2 - 20y - 100$

9. $16y^2 - 40yz + 25z^2$

10. $49y^2 - 56yz + 16z^2$

‹ B ›   **Factoring Perfect Square Trinomials**   In Problems 11–34, factor completely.

11. $x^2 + 2x + 1$

12. $x^2 + 6x + 9$

13. $3x^2 + 30x + 75$

14. $2x^2 + 28x + 98$

15. $9x^2 + 6x + 1$

16. $16x^2 + 8x + 1$

17. $9x^2 + 12x + 4$

18. $25x^2 + 10x + 1$

19. $16x^2 + 40xy + 25y^2$

20. $9x^2 + 30xy + 25y^2$

21. $25x^2 + 20xy + 4y^2$

22. $36x^2 + 60xy + 25y^2$

23. $y^2 - 2y + 1$

24. $y^2 - 6y + 9$

25. $3y^2 - 24y + 48$

26. $2y^2 - 40y + 200$

27. $9x^2 - 6x + 1$

28. $4x^2 - 20x + 25$

29. $16x^2 - 56x + 49$

30. $25x^2 - 30x + 9$

31. $9x^2 - 12xy + 4y^2$

32. $16x^2 - 40xy + 25y^2$

33. $25x^2 - 10xy + y^2$

34. $49x^2 - 56xy + 16y^2$

‹ C ›   **Factoring the Difference of Two Squares**   In Problems 35–60, factor completely.

35. $x^2 - 49$

36. $x^2 - 121$

37. $9x^2 - 49$

38. $16x^2 - 81$

39. $25x^2 - 81y^2$

40. $81x^2 - 25y^2$

41. $x^4 - 1$

42. $x^4 - 256$

43. $16x^4 - 1$

44. $16x^4 - 81$

45. $\frac{1}{9}x^2 - \frac{1}{16}$

46. $\frac{1}{4}y^2 - \frac{1}{25}$

47. $\frac{1}{4}z^2 - 1$

48. $\frac{1}{9}r^2 - 1$

49. $1 - \frac{1}{4}s^2$

50. $1 - \frac{1}{9}t^2$

51. $\frac{1}{4} - \frac{1}{9}y^2$

52. $\frac{1}{9} - \frac{1}{16}u^2$

53. $\frac{1}{9} + \frac{1}{4}x^2$

54. $\frac{1}{4} + \frac{1}{25}n^2$

55. $3x^3 - 12x$

56. $4y^3 - 16y$

57. $5t^3 - 20t$

58. $7t^3 - 63t$

59. $5t - 20t^3$

60. $2s - 18s^3$

In Problems 61–74, use a variety of factoring methods to factor completely.

61. $49x^2 + 28x + 4$

62. $49y^2 + 42y + 9$

63. $x^2 - 100$

64. $x^2 - 144$

65. $x^2 + 20x + 100$

66. $x^2 + 18x + 81$

67. $9 - 16m^2$

68. $25 - 9n^2$

69. $9x^2 - 30xy + 25y^2$

70. $16x^2 - 40xy + 25y^2$

71. $z^4 - 16$

72. $16y^4 - 81$

73. $3x^3 - 75x$

74. $2y^3 - 72y$

> > > **Applications**

75. *Demand function* A business owner finds out that when x units of a product are demanded by consumers, the price per unit is a function of the demand and given by the equation $D(x) = 81 - x^2$. Factor $81 - x^2$.

76. *Demand function* A business estimates that when y units of a product are demanded by consumers, the price per units is a function of the demand and given by the equation $D(y) = 64 - y^2$. Factor $64 - y^2$.

77. *Economic supplies* The relationship between the quantity supplied S and the unit price p is given by the equation $S = C^2 - k^2p^2$, where C and k are constants. Factor $C^2 - k^2p^2$.

78. *Kinetic energy* The change in kinetic energy of a moving object of mass m with initial velocity v_1 and terminal velocity v_2 is given by the expression $\frac{1}{2}mv_1^2 - \frac{1}{2}mv_2^2$. Factor $\frac{1}{2}mv_1^2 - \frac{1}{2}mv_2^2$ completely.

79. *Parabolic distribution* The parabolic distribution of shear stress on the cross section of a certain beam is given by $\frac{3Sd^2}{2bd^3} - \frac{12Sz^2}{2bd^3}$. Factor this expression completely.

80. *Polar moment* The polar moment of inertia, J, of a hollow round shaft of inner diameter d_1 and outer diameter d is given by the expression $\frac{\pi d^4}{32} - \frac{\pi d_1^4}{32}$. Factor $\frac{\pi d^4}{32} - \frac{\pi d_1^4}{32}$ completely.

> > > **Using Your Knowledge**

How Does It Function? Many business ideas are made precise by using expressions called **functions.** These expressions are often given in unfactored form. Use your knowledge to factor the given expressions (functions).

81. When x units of an item are demanded by consumers, the price per unit is given by the **demand** function $D(x)$ (read "D of x"):

$$D(x) = x^2 - 14x + 49$$

Factor this expression.

82. When x units are supplied by sellers, the price per unit of an item is given by the **supply** function $S(x)$:

$$S(x) = x^2 + 4x + 4$$

Factor this expression.

83. When x units are produced, the cost function $C(x)$ for a certain item is given by the equation

$$C(x) = x^2 + 12x + 36$$

Factor this expression.

84. When the market price is p dollars, the supply function $S(p)$ for a certain commodity is given by the equation

$$S(p) = p^2 - 6p + 9$$

Factor this expression.

〉〉〉 *Write On*

85. The difference of two squares can be factored. Can you factor $a^2 + b^2$? Can you say that the sum of two squares can *never* be factored? Explain.

86. Can you factor $4a^2 + 16b^2$? Think of the implications for Problem 85.

87. What binomial multiplied by $(x + 2)$ gives a perfect square trinomial?

88. What binomial multiplied by $(2x - 3y)$ gives a perfect square trinomial?

〉〉〉 *Concept Checker*

Fill in the blank(s) with the correct word(s), phrase, or mathematical statement.

89. When written in **factored form**, $X^2 + 2AX + A^2 =$ _____.

90. When written in **factored form**, $X^2 - 2AX + A^2 =$ _____.

91. When written in **factored form**, $X^2 - A^2 =$ _____.

92. The expression $X^2 + A^2$ _____ be factored.

(X + A)(X − A)   **(X + A)²**

can   **(X − A)²**

cannot

〉〉〉 *Mastery Test*

Factor, if possible:

93. $x^2 - 1$

94. $9x^2 - 16$

95. $9x^2 - 25y^2$

96. $x^2 - 6x + 9$

97. $9x^2 - 24xy + 16y^2$

98. $9x^2 - 12x + 4$

99. $16x^2 + 24xy + 9y^2$

100. $x^2 + 4x + 4$

101. $9x^2 + 30x + 25$

102. $4x^2 - 20xy + 25y^2$

103. $9x^2 + 4$

104. $\frac{1}{36}x^2 - \frac{1}{49}$

105. $\frac{1}{81} - \frac{1}{4}x^2$

106. $12m^3 - 3mn^2$

107. $18x^3 - 50xy^2$

108. $9x^3 + 25xy^2$

Determine whether the expression is the square of a binomial. If it is, factor it.

109. $x^2 + 6x + 9$

110. $x^2 + 8x + 64$

111. $x^2 + 6x - 9$

112. $x^2 + 8x - 64$

113. $4x^2 - 20xy + 25y^2$

〉〉〉 *Skill Checker*

Multiply:

114. $(A + B)(A - B)$

115. $(R + r)(R - r)$

116. $(P + q)(P - q)$

Factor completely:

117. $6x^2 - 18x - 24$

118. $4x^4 + 12x^3 + 40x^2$

119. $2x^2 - 18$

120. $3x^2 - 27$

5.5   A General Factoring Strategy

▶ Objectives

A ▸ Factor the sum or difference of two cubes.

B ▸ Factor a polynomial by using the general factoring strategy.

C ▸ Factor expressions whose leading coefficient is −1.

▶ To Succeed, Review How To . . .

Factor a polynomial using the rules for reversing FOIL, perfect square trinomials, and the difference of two squares. (F1–F4) (pp. 407–409; 425–428).

▶ Getting Started
Factoring and Medicine

In an artery (see the cross section in the photo), the speed (in centimeters per second) of the blood is given by

$$CR^2 - Cr^2$$

You already know how to factor this expression; using techniques you've already learned, you would proceed as follows:

1. Factor out any common factors (in this case, C). $CR^2 - Cr^2 = C(R^2 - r^2)$

2. Look at the terms inside the parentheses. You have the difference of two square terms in the expression: $R^2 - r^2$, so you factor it. $= C(R + r)(R - r)$

3. Make sure the expression is *completely* factored. Note that $C(R + r)(R - r)$ cannot be factored further.

What we've just used here is a *strategy* for factoring polynomials—a logical way to call up any of the techniques you've studied when they fit the expression you are factoring. In this section we shall study one more type of factoring: sums or differences of cubes. We will then examine in more depth the general factoring strategy for polynomials.

A > Factoring Sums or Differences of Cubes

We've already factored the difference of two squares $X^2 - A^2$. Can we factor the difference of two cubes $X^3 - A^3$? Not only can we factor $X^3 - A^3$, we can even factor $X^3 + A^3$! Since factoring is "reverse multiplication," let's start with two multiplication problems: $(X + A)(X^2 - AX + A^2)$ and $(X - A)(X^2 + AX + A^2)$.

$$X^2 - AX + A^2$$
$$\underline{\times \qquad X + A}$$
$$\overline{AX^2 - A^2X + A^3} \quad \longleftarrow \text{Multiply } A(X^2 - AX + A^2).$$
$$\underline{X^3 - AX^2 + A^2X} \qquad\qquad \longleftarrow \text{Multiply } X(X^2 - AX + A^2).$$
$$\overline{X^3 \qquad\qquad\qquad + A^3}$$

$$X^2 + AX + A^2$$
$$\underline{\times \qquad X - A}$$
$$\overline{- AX^2 - A^2X - A^3} \quad \longleftarrow \text{Multiply } -A(X^2 + AX + A^2).$$
$$\underline{X^3 + AX^2 + A^2X} \qquad\qquad \longleftarrow \text{Multiply } X(X^2 + AX + A^2).$$
$$\overline{X^3 \qquad\qquad\qquad - A^3}$$

Thus,

Same   Different $\qquad\qquad\qquad$ Same   Different

$$X^3 + A^3 = (X + A)(X^2 - AX + A^2) \quad \text{and} \quad X^3 - A^3 = (X - A)(X^2 + AX + A^2).$$

This gives us our final factoring rules.

FACTORING RULES 5 AND 6: THE SUM AND DIFFERENCE OF TWO CUBES

$$X^3 + A^3 = (X + A)(X^2 - AX + A^2) \tag{F5}$$
$$X^3 - A^3 = (X - A)(X^2 + AX + A^2) \tag{F6}$$

NOTE

The trinomials $X^2 - AX + A^2$ and $X^2 + AX + A^2$ cannot be factored further.

EXAMPLE 1   Factoring sums and differences of cubes

Factor completely:

a. $x^3 + 27$ $\qquad\qquad$ **b.** $8x^3 + y^3$

c. $m^3 - 8n^3$ $\qquad\qquad$ **d.** $27r^3 - 8s^3$

SOLUTION

a. We rewrite $x^3 + 27$ as the sum of two cubes and then use F5:

$$x^3 + 27 = (x)^3 + (3)^3$$
$$= (x + 3)(x^2 - 3x + 3^2) \quad \text{Letting } X = x \text{ and } A = 3 \text{ in F5}$$
$$= (x + 3)(x^2 - 3x + 9)$$

b. This is also the sum of two cubes, so we write:

$$8x^3 + y^3 = (2x)^3 + (y)^3$$
$$= (2x + y)[(2x)^2 - 2xy + y^2] \quad \text{Letting } X = 2x \text{ and } A = y \text{ in F5}$$
$$= (2x + y)(4x^2 - 2xy + y^2)$$

PROBLEM 1

Factor completely:

a. $y^3 + 8$ $\qquad$ **b.** $27y^3 + 8$

c. $y^3 - 27z^3$ $\qquad$ **d.** $8a^3 - 27b^3$

(continued)

Answers to PROBLEMS

1. a. $(y + 2)(y^2 - 2y + 4)$
 b. $(3y + 2)(9y^2 - 6y + 4)$
 c. $(y - 3z)(y^2 + 3yz + 9z^2)$
 d. $(2a - 3b)(4a^2 + 6ab + 9b^2)$

c. Here we have the difference of two cubes, so we use F6. We start by writing the problem as the difference of two cubes:

$$m^3 - 8n^3 = (m)^3 - (2n)^3$$
$$= (m - 2n)[m^2 + m(2n) + (2n)^2] \quad \text{Letting } X = m \text{ and } A = 2n$$
$$= (m - 2n)(m^2 + 2mn + 4n^2) \quad \text{in F6}$$

d. We write the problem as the difference of two cubes and then use F6.

$$27r^3 - 8s^3 = (3r)^3 - (2s)^3$$
$$= (3r - 2s)[(3r)^2 + (3r)(2s) + (2s)^2] \quad \text{Letting } X = 3r \text{ and}$$
$$= (3r - 2s)(9r^2 + 6rs + 4s^2) \quad A = 2s \text{ in F6}$$

Note that you can verify all of these results by multiplying the factors in the final answer.

We can also factor $x^3 + 27$ by remembering that the result is a binomial times a trinomial, so we place parentheses accordingly.

1. To get the *binomial factor,* take the cube roots using the same sign.
2. To get the *trinomial factor,* square terms of the binomial factor and use them as the first and last terms of the trinomial. The middle term of the trinomial is the result of multiplying the two terms of the binomial factor and changing the sign.

Thus,

$$x^3 + 27$$
$$\sqrt[3]{x^3} \quad \sqrt[3]{27}$$
$$(x \ + \ 3)(x^2 - 3x + 9)$$
$$\text{square} \quad \text{square}$$
$$\text{middle term:}$$
$$\text{(change sign) } -x \cdot 3 = -3x$$

B › Using a General Factoring Strategy

We have now studied several factoring techniques. How do you know which one to use? Here is a general factoring strategy that can help you answer this question. Remember that when we say *factor,* we mean *factor completely* using integer coefficients.

PROCEDURE

A General Factoring Strategy

1. Factor out all common factors (the GCF).
2. Look at the number of terms inside the parentheses (or in the original polynomial). If there are

Four terms:	Factor by grouping.
Three terms:	Check whether the expression is a perfect square trinomial. If so, factor it. Otherwise, use the *ac* test to factor.
Two terms and *squared:*	Look for the difference of two squares ($X^2 - A^2$) and factor it. Note that $X^2 + A^2$ is not factorable.
Two terms and *cubed:*	Look for the sum of two cubes ($X^3 + A^3$) or the difference of two cubes ($X^3 - A^3$) and factor it.

3. Make sure the expression is completely factored.

You can check your results by multiplying the factors you obtain.

EXAMPLE 2   Using the general factoring strategy

Factor completely:

a. $6x^2 - 18x - 24$ **b.** $4x^4 + 12x^3 + 40x^2$

SOLUTION

a. We follow the steps in our general factoring strategy.

1. Factor out the common factor: $6x^2 - 18x - 24 = 6(x^2 - 3x - 4)$

2. $x^2 - 3x - 4$ has three terms, and it is factored by finding two numbers whose product is -4 and whose sum is -3. These numbers are 1 and -4. Thus, $x^2 - 3x - 4 = (x + 1)(x - 4)$

   We then have $6x^2 - 18x - 24 = 6(x + 1)(x - 4)$

3. This expression cannot be factored any further.

b. 1. Here the GCF is $4x^2$. Thus, $4x^4 + 12x^3 + 40x^2 = 4x^2(x^2 + 3x + 10)$

2. The trinomial $x^2 + 3x + 10$ is *not* factorable since there are no numbers whose product is 10 with a sum of 3.

3. The complete factorization is simply $4x^4 + 12x^3 + 40x^2 = 4x^2(x^2 + 3x + 10)$

PROBLEM 2

Factor completely:

a. $7a^2 - 14a - 21$

b. $5a^4 + 10a^3 + 25a^2$

EXAMPLE 3   Using the general factoring strategy with four terms

Factor completely: $3x^3 + 9x^2 + x + 3$

SOLUTION

1. There are no common factors.

2. Since the expression has four terms, we factor by grouping:

$$3x^3 + 9x^2 + x + 3 = (3x^3 + 9x^2) + (x + 3)$$
$$= 3x^2(x + 3) + 1 \cdot (x + 3)$$
$$= (x + 3)(3x^2 + 1)$$

3. This result cannot be factored any further, so the factorization is complete.

PROBLEM 3

Factor completely:

$3a^3 + 6a^2 + a + 2$

EXAMPLE 4   Using the general factoring strategy with a perfect square trinomial

The heat output from a natural draught convector is $kt_n^2 - 2kt_n t_a + kt_a^2$. ($t_n^2$ is read as "t sub n squared." The "n" is called a **subscript**.) Factor this expression.

SOLUTION   As usual, we proceed by steps.

1. The common factor is k. Hence

$$kt_n^2 - 2kt_n t_a + kt_a^2 = k(t_n^2 - 2t_n t_a + t_a^2)$$

2. $t_n^2 - 2t_n t_a + t_a^2$ is a perfect square trinomial, which factors as $(t_n - t_a)^2$. Thus,

$$kt_n^2 - 2kt_n t_a + kt_a^2 = k(t_n - t_a)^2$$

3. This expression cannot be factored further.

PROBLEM 4

Factor completely: $kt_1^2 - 2kt_1 t_2 + kt_2^2$

Answers to PROBLEMS

2. a. $7(a + 1)(a - 3)$ **b.** $5a^2(a^2 + 2a + 5)$ **3.** $(a + 2)(3a^2 + 1)$ **4.** $k(t_1 - t_2)^2$

EXAMPLE 5   Using the general factoring strategy with the difference of two squares

Factor completely: $D^4 - d^4$

PROBLEM 5

Factor completely: $m^4 - n^4$

SOLUTION

1. There are no common factors.
2. The expression has *two* squared terms separated by a minus sign, so it's the difference of *two* squares. Thus,

$$D^4 - d^4 = (D^2)^2 - (d^2)^2$$
$$= (D^2 + d^2)(D^2 - d^2)$$

3. The expression $D^2 - d^2$ is also the difference of two squares, which can be factored into $(D + d)(D - d)$. Thus,

$$D^4 - d^4 = (D^2 + d^2)(D^2 - d^2)$$
$$= (D^2 + d^2)(D + d)(D - d)$$

Note that $D^2 + d^2$, which is the *sum* of two squares, *cannot* be factored.

EXAMPLE 6   Using the general factoring strategy with sums and differences of cubes

Factor completely:

a. $8x^5 - x^2y^3$ **b.** $8x^5 + x^3y^2$

PROBLEM 6

Factor completely:

a. $27a^5 - a^2b^3$
b. $64a^5 + a^3b^2$

SOLUTION

a. We proceed as usual by steps.

1. The GCF is x^2, so we factor it out. $8x^5 - x^2y^3 = x^2(8x^3 - y^3)$

2. $8x^3 - y^3$ is the difference of two cubes with $X = 2x$ and $A = y$. $= x^2(2x - y)[(2x)^2 + (2x)y + y^2]$

3. Note that the expression cannot be factored further. $= x^2(2x - y)(4x^2 + 2xy + y^2)$

b. The GCF is x^3.

1. Factor the GCF. $8x^5 + x^3y^2 = x^3(8x^2 + y^2)$

2. $8x^2 + y^2$ is the sum of two squares and is not factorable. Thus, $8x^5 + x^3y^2 = x^3(8x^2 + y^2)$

3. Note that the expression cannot be factored further.

C › Using −1 as a Factor

In the preceding examples, we did not factor expressions in which the leading coefficient is preceded by a minus sign. Can some of these expressions be factored? The answer is yes, but we must first factor -1 from each term. Thus, to factor $-x^2 + 6x - 9$, we first write

$$-x^2 + 6x - 9 = -1 \cdot (x^2 - 6x + 9)$$
$$= -1 \cdot (x - 3)^2 \quad \text{Note that } x^2 - 6x + 9 = (x - 3)^2.$$
$$= -(x - 3)^2 \quad \text{Since } -1 \cdot a = -a, -1 \cdot (x - 3)^2 = -(x - 3)^2.$$

Answers to PROBLEMS

5. $(m^2 + n^2)(m + n)(m - n)$

6. a. $a^2(3a - b)(9a^2 + 3ab + b^2)$

   b. $a^3(64a^2 + b^2)$

EXAMPLE 7   Factoring −1 out as the GCF

Factor, if possible:

a. $-x^2 - 8x - 16$ **b.** $-4x^2 + 12xy - 9y^2$

c. $-9x^2 - 12xy + 4y^2$ **d.** $-4x^4 + 25x^2$

SOLUTION

a. We first factor out -1 to obtain

$$-x^2 - 8x - 16 = -1 \cdot (x^2 + 8x + 16)$$
$$= -1 \cdot (x + 4)^2 \quad\quad x^2 + 8x + 16 = (x + 4)^2$$
$$= -(x + 4)^2$$

b. $-4x^2 + 12xy - 9y^2 = -1 \cdot (4x^2 - 12xy + 9y^2)$
$$= -1 \cdot (2x - 3y)^2 \quad\quad 4x^2 - 12xy + 9y^2 = (2x - 3y)^2$$
$$= -(2x - 3y)^2$$

c. $-9x^2 - 12xy + 4y^2 = -1 \cdot (9x^2 + 12xy - 4y^2)$

$9x^2 = (3x)^2$ and $4y^2 = (2y)^2$, but $9x^2 + 12xy - 4y^2$ has a minus sign before the last term $4y^2$ and is *not* a perfect square trinomial, so $9x^2 + 12xy - 4y^2$ is not factorable; thus, $-9x^2 - 12xy + 4y^2 = -(9x^2 + 12xy - 4y^2)$.

d. $-4x^4 + 25x^2 = -x^2 \cdot (4x^2 - 25)$ The GCF is $-x^2$.
$$= -x^2 \cdot (2x + 5)(2x - 5) \quad\quad 4x^2 - 25 = (2x + 5)(2x - 5)$$
$$= -x^2(2x + 5)(2x - 5)$$

PROBLEM 7

Factor:

a. $-a^2 - 6a - 9$

b. $-9a^2 + 12ab - 4b^2$

c. $-4a^2 - 12ab + 9b^2$

d. $-4a^4 + 9a^2$

MathZone+x

Boost your grade at mathzone.com!
> Practice   > Self-Tests
   Problems   > e-Professors
> NetTutor   > Videos

❯ Exercises 5.5

❮ **A** ❯   **Factoring Sums or Differences of Cubes**   In Problems 1–10, factor completely.

1. $x^3 + 8$ **2.** $z^3 + 8$ **3.** $8m^3 - 27$

4. $8y^3 - 27x^3$ **5.** $27m^3 - 8n^3$ **6.** $27x^2 - x^5$

7. $64s^3 - s^6$ **8.** $t^7 - 8t^4$ **9.** $27x^4 + 8x^7$

10. $8y^8 + 27y^5$

❮ **B** ❯   **Using a General Factoring Strategy**   In Problems 11–46, factor completely.

11. $3x^2 - 3x - 18$ **12.** $4x^2 - 12x - 16$ **13.** $5x^2 + 11x + 2$ **14.** $6x^2 + 19x + 10$

15. $3x^3 + 6x^2 + 21x$ **16.** $6x^3 + 18x^2 + 12x$ **17.** $2x^4 - 4x^3 - 10x^2$ **18.** $3x^4 - 12x^3 - 9x^2$

19. $4x^4 + 12x^3 + 18x^2$ **20.** $5x^4 + 25x^3 + 30x^2$ **21.** $3x^3 + 6x^2 + x + 2$ **22.** $2x^3 + 8x^2 + x + 4$

23. $3x^3 + 3x^2 + 2x + 2$ **24.** $4x^3 + 8x^2 + 3x + 6$ **25.** $2x^3 + 2x^2 - x - 1$ **26.** $3x^3 + 6x^2 - x - 2$

Answers to PROBLEMS

7. a. $-(a + 3)^2$ **b.** $-(3a - 2b)^2$ **c.** $-(4a^2 + 12ab - 9b^2)$ **d.** $-a^2(2a + 3)(2a - 3)$

Web IT   go to **mhhe.com/bello** for more lessons

Web IT go to mhhe.com/bello *for more lessons*

27. $3x^2 + 24x + 48$ **28.** $2x^2 + 12x + 18$ **29.** $kx^2 + 4kx + 4k$ **30.** $kx^2 + 10kx + 25k$

31. $4x^2 - 24x + 36$ **32.** $5x^2 - 20x + 20$ **33.** $kx^2 - 12kx + 36k$ **34.** $kx^2 - 10kx + 25k$

35. $3x^3 + 12x^2 + 12x$ **36.** $2x^3 + 16x^2 + 32x$ **37.** $18x^3 + 12x^2 + 2x$ **38.** $12x^3 + 12x^2 + 3x$

39. $12x^4 - 36x^3 + 27x^2$ **40.** $18x^4 - 24x^3 + 8x^2$ **41.** $x^4 - 1$ **42.** $x^4 - 16$

43. $x^4 - y^4$ **44.** $x^4 - z^4$ **45.** $x^4 - 16y^4$ **46.** $x^4 - 81y^4$

⟨ **C** ⟩   **Using −1 as a Factor**   In Problems 47–70, factor.

47. $-x^2 - 6x - 9$ **48.** $-x^2 - 10x - 25$ **49.** $-x^2 - 4x - 4$

50. $-x^2 - 12x - 36$ **51.** $-4x^2 - 4xy - y^2$ **52.** $-9x^2 - 6xy - y^2$

53. $-9x^2 - 12xy - 4y^2$ **54.** $-4x^2 - 12xy - 9y^2$ **55.** $-4z^2 + 12zy - 9y^2$

56. $-9x^2 + 12xy - 4y^2$ **57.** $-18x^3 - 24x^2y - 8xy^2$ **58.** $-12x^3 - 36x^2y - 27xy^2$

59. $-18x^3 - 60x^2y - 50xy^2$ **60.** $-12x^3 - 60x^2y - 75xy^2$ **61.** $-x^3 + x$

62. $-x^3 + 9x$ **63.** $-x^4 + 4x^2$ **64.** $-x^4 + 16x^2$

65. $-4x^4 + 9x^2$ **66.** $-9x^4 + 4x^2$ **67.** $-2x^4 + 16x$

68. $-24x^4 + 3x$ **69.** $-16x^5 - 2x^2$ **70.** $-3x^5 + 24x^2$

❯❯❯   **Applications**

71. *Supply function*   When x units are supplied by sellers, the price per unit of an item is given by the equation $S(x) = x^3 + 216$. Factor $x^3 + 216$.

72. *Supply function*   When x units are supplied by sellers, the price per unit of an item is given by the equation $S(x) = x^3 + 343$. Factor $x^3 + 343$.

73. *Production cost*   When x units of a certain item are produced, the cost is given by the equation $C(x) = 8x^3 - 1$. Factor $8x^3 - 1$.

74. *Production cost*   When x units of a certain item are produced, the cost is given by the equation $C(x) = 8x^3 - 125$. Factor $8x^3 - 125$.

❯❯❯   **Using Your Knowledge**

Factoring Engineering Problems   Many of the ideas presented in this section are used by engineers and technicians. Use your knowledge to factor the given expressions.

75. The bend allowance needed to bend a piece of metal of thickness t through an angle A when the inside radius of the bend is R_1 is given by the expression

$$\frac{2\pi A}{360}R_1 + \frac{2\pi A}{360}Kt$$

where K is a constant. Factor this expression.

76. The change in kinetic energy of a moving object of mass m with initial velocity v_1 and terminal velocity v_2 is given by the expression

$$\frac{1}{2}mv_1^2 - \frac{1}{2}mv_2^2$$

Factor this expression.

77. The parabolic distribution of shear stress on the cross section of a certain beam is given by the expression

$$\frac{3Sd^2}{2bd^3} - \frac{12Sz^2}{2bd^3}$$

Factor this expression.

78. The polar moment of inertia J of a hollow round shaft of inner diameter d_1 and outer diameter d is given by the expression

$$\frac{\pi d^4}{32} - \frac{\pi d_1^4}{32}$$

Factor this expression.

⟩⟩⟩ *Write On*

79. Write the procedure you use to factor the sum of two cubes.

80. Write the procedure you use to factor the difference of two cubes.

81. A student factored $x^4 - 7x^2 - 18$ as $(x^2 + 2)(x^2 - 9)$. The student did not get full credit on the answer. Why?

⟩⟩⟩ *Concept Checker*

Fill in the blank(s) with the correct word(s), phrase, or mathematical expression.

82. In **factored form** $X^3 + A^3 = $ _____

83. In **factored form** $X^3 - A^3 = $ _____

84. The **first** step in the **general factoring strategy** is to _____ out all **common** factors.

85. The **second** step in the **general factoring strategy** is to _____ at the **number** of terms.

$(X - A)(X^2 + AX + A^2)$ **factor**

$(X - A)(X^2 - AX + A^2)$ $(X + A)(X^2 + AX + A^2)$

ignore   **look**

$(X + A)(X^2 - AX + A^2)$

⟩⟩⟩ *Mastery Test*

Factor completely:

86. $8x^2 - 16x - 24$

87. $5x^4 - 10x^3 + 20x^2$

88. $3x^3 + 12x^2 + x + 4$

89. $6x^2 - x - 35$

90. $2x^4 + 7x^3 - 15x^2$

91. $27t^3 - 64$

92. $kt_n^2 + 2kt_n t_a + kt_a^2$

93. $x^4 - 81$

94. $-z^2 - 10z - 25$

95. $-9x^2 - 30xy - 25y^2$

96. $-9x^2 + 30xy - 25y^2$

97. $64y^3 + 27x^3$

98. $-9z^4 + 4z^2$

99. $-x^5 - x^2y^3$

⟩⟩⟩ *Skill Checker*

Use the *ac* test to factor (if possible).

100. $10x^2 + 11x + 6$

101. $10x^2 + 13x - 3$

102. $3x^2 - 5x - 1$

103. $2x^2 - 5x - 3$

104. $2x^2 - 3x - 2$

5.6 Solving Quadratic Equations by Factoring

▶ **Objective**

A ❯ Solve quadratic equations by factoring.

▶ **To Succeed, Review How To . . .**

1. Factor an expression of the form $ax^2 + bx + c$ (pp. 407–411).
2. Solve a linear equation (pp. 132–136).

▶ **Getting Started**

Quadratics and Gravity

Let's suppose the girl throws the ball with an initial velocity of 4 meters per second. If she releases the ball 1 meter above the ground, the *height* of the ball after t seconds is

$$-5t^2 + 4t + 1$$

To find out how long it takes the ball to hit the ground, we set this expression equal to zero. The reason for this is that when the ball *is* on the ground, the height is zero. Thus, we write

Remember, the girl's hand is 1 meter *above* the ground.

$$-5t^2 \ + \ 4t \ + \ 1 \ = \ 0$$

This is the height after t seconds. This is the height when the ball hits the ground.

The equation is a *quadratic equation,* an equation in which the greatest exponent of the variable is 2 and which can be written as $at^2 + bt + c = 0, a \neq 0.$ We learn how to solve these equations next.

A ❯ Solving Quadratic Equations by Factoring

How can we solve the equation given in the *Getting Started?* For starters, we make the leading coefficient in $-5t^2 + 4t + 1 = 0$ positive by multiplying each side of the equation by -1 to obtain

$$5t^2 - 4t - 1 = 0 \qquad -1(-5t^2 + 4t + 1) = 5t^2 - 4t - 1 \text{ and } -1 \cdot 0 = 0$$

Since $a = 5$ and $c = 1$, the ac number is -5 and we write

$$5t^2 \underbrace{- 5t + 1t}_{-4t} - 1 = 0 \qquad \text{Write } -4t \text{ as } -5t + 1t.$$

$$5t(t - 1) + 1(t - 1) = 0 \qquad \text{The GCF is } (t - 1).$$

$$(t - 1)(5t + 1) = 0$$

(You can also use reverse FOIL or trial and error.) At this point, we note that the product of two expressions $(t - 1)$ and $(5t + 1)$ gives us a result of zero. What does this mean?

We know that if we have two numbers and at least one of them is zero, then their product is zero. For example,

$$-5 \cdot 0 = 0 \qquad \tfrac{3}{2} \cdot 0 = 0 \qquad x \cdot 0 = 0$$

$$0 \cdot 8 = 0 \qquad 0 \cdot x = 0 \qquad 0 \cdot 0 = 0$$

As you can see, in all these cases at least one of the factors is zero. In general, it can be shown that if the product of the two factors is zero, at least one of the factors *must be* zero. We shall call this idea the **zero product property.**

ZERO PRODUCT PROPERTY	If $A \cdot B = 0$, then $A = 0$ or $B = 0$ (or both A and B are equal to 0).

Now, let's go back to our original equation. We can think of $(t - 1)$ as A and $(5t + 1)$ as B. Then our equation

$$(t - 1)(5t + 1) = 0$$

becomes

$$A \cdot B = 0$$

By the zero product property, if $A \cdot B = 0$, then

$$A = 0 \quad \text{or} \quad B = 0$$

Thus,

$$t - 1 = 0 \qquad \text{or} \qquad 5t + 1 = 0$$
$$t = 1 \qquad\qquad\qquad 5t = -1 \qquad \text{We added 1 in the first equation and subtracted 1 in the second.}$$
$$t = 1 \qquad\qquad\qquad t = -\tfrac{1}{5}$$

> **NOTE**
>
> When solving quadratic equations, you usually get two answers, but sometimes in application problems one of the answers must be discarded. It is always a good idea to check that the answers you get apply to the original conditions of the problem.

Thus, the ball reaches the ground after 1 second or after $-\tfrac{1}{5}$ second. The second answer is negative, which is impossible because we start timing when $t = 0$, so we can see that the ball thrown by the girl at 4 meters per second will reach the ground after $t = 1$ second. You can check this by letting $t = 1$ in the original equation:

$$-5t^2 + 4t + 1 \overset{?}{=} 0$$
$$-5(1)^2 + 4(1) + 1 \overset{?}{=} 0$$
$$-5 + 4 + 1 \overset{?}{=} 0$$
$$0 = 0$$

Similarly, if we want to find how long a ball thrown from level ground at 10 meters per second takes to return to the ground, we need to solve the equation

$$5t^2 - 10t = 0$$

Factoring, we obtain

$$5t(t - 2) = 0$$

By the zero product property,

$$5t = 0 \qquad \text{or} \qquad t - 2 = 0$$
$$t = 0 \qquad\qquad\qquad t = 2 \qquad \text{If } 5t = 0, \text{ then } t = 0.$$

Thus, the ball returns to the ground after 2 seconds. (The other possible answer, $t = 0$, indicates that the ball was on the ground when $t = 0$, which is true.)

In the preceding discussion, we solved equations in which the highest exponent of the variable was 2. These equations are called *quadratic equations*. A quadratic equation is an equation in which the greatest exponent of the variable is 2. Moreover, to solve these quadratic equations, one of the sides of the equation must equal zero. These two ideas can be summarized as follows.

QUADRATIC EQUATION IN STANDARD FORM	**If a, b, and c are real numbers ($a \neq 0$),** $$ax^2 + bx + c = 0$$ is a **quadratic equation in standard form.**

To solve a quadratic equation by the method of factoring, the equation must be in *standard form.*

EXAMPLE 1   Solving a quadratic equation by factoring

Solve:

a. $3x^2 + 11x - 4 = 0$

b. $6x^2 - x - 2 = 0$

SOLUTION

a. The equation is in standard form. To solve this equation, we must first factor the left-hand side. Here's how we do it.

$$3x^2 + \underline{11x} - 4 = 0$$
The key number is -12 ($= 3(-4)$); then determine $12(-1) = -12$ and $12 + (-1) = 11$.

$$3x^2 + \underline{12x - 1x} - 4 = 0$$
Rewrite the middle term, $11x$.

$$3x(x + 4) - 1(x + 4) = 0$$
Factor each pair.

$$(x + 4)(3x - 1) = 0$$
Factor out the GCF, $(x + 4)$.

$$x + 4 = 0 \quad \text{or} \quad 3x - 1 = 0$$
Use the zero product property.

$$x = -4 \qquad\qquad 3x = 1$$
Solve each equation.

$$x = -4 \qquad\qquad x = \frac{1}{3}$$

Thus, the possible solutions are $x = -4$ and $x = \frac{1}{3}$. To verify that -4 is a correct solution, we substitute -4 in the original equation to obtain

$$3(-4)^2 + 11(-4) - 4 = 3(16) - 44 - 4$$
$$= 48 - 44 - 4 = 0$$

We leave it to you to verify that $\frac{1}{3}$ is also a solution.

b. As before, we must factor the left-hand side.

$$6x^2 - \underline{x} - 2 = 0$$
The key number is -12.

$$6x^2 - \underline{4x + 3x} - 2 = 0$$
Rewrite the middle term, $-x$.

$$2x(3x - 2) + 1(3x - 2) = 0$$
Factor each pair.

$$(3x - 2)(2x + 1) = 0$$
Factor out the GCF, $(3x - 2)$.

$$3x - 2 = 0 \quad \text{or} \quad 2x + 1 = 0$$
Use the zero product property.

$$3x = 2 \qquad\qquad 2x = -1$$
Solve each equation.

$$x = \frac{2}{3} \qquad\qquad x = -\frac{1}{2}$$

Thus, the solutions are $\frac{2}{3}$ and $-\frac{1}{2}$. You can check this by substituting these values in the original equation.

PROBLEM 1

Solve:

a. $3x^2 + 8x - 3 = 0$

b. $6x^2 - 7x - 3 = 0$

Answers to PROBLEMS

1. a. $x = -3$ and $x = \frac{1}{3}$ **b.** $x = \frac{3}{2}$ and $x = -\frac{1}{3}$

EXAMPLE 2   Solving a quadratic equation *not* in standard form

Solve: $10x^2 + 13x = 3$

SOLUTION   This equation is not in standard form, so we can't solve it as written. However, if we subtract 3 from each side of the equation, we have

$$10x^2 + 13x - 3 = 0$$

which is in standard form. We can now solve by factoring using trial and error or the *ac* test. To use the *ac* test, we write

$10x^2 + \underline{13x} - 3 = 0$	The key number is -30.
$10x^2 + \underline{15x - 2x} - 3 = 0$	Rewrite the middle term, 13x.
$5x(2x + 3) - 1(2x + 3) = 0$	Factor each pair.
$(2x + 3)(5x - 1) = 0$	Factor out the GCF, $(2x + 3)$.
$2x + 3 = 0 \qquad$ or $\qquad 5x - 1 = 0$	Use the zero product property.
$2x = -3 \qquad\qquad 5x = 1$	Solve each equation.
$x = -\dfrac{3}{2} \qquad\qquad x = \dfrac{1}{5}$	

Thus, the solutions of the equation are

$$x = -\frac{3}{2} \quad \text{and} \quad x = \frac{1}{5}$$

Check this!

PROBLEM 2

Solve: $6x^2 + 13x = 5$

Sometimes we need to simplify before we write an equation in standard form. For example, to solve the equation

$$(3x + 1)(x - 1) = 3(x + 1) - 2$$

we need to remove parentheses by multiplying the factors involved and write the equation in standard form. Here's how we do it:

$(3x + 1)(x - 1) = 3(x + 1) - 2$	Given
$3x^2 - 2x - 1 = 3x + 3 - 2$	Multiply $(3x + 1)(x - 1)$ and $3(x + 1)$.
$3x^2 - 2x - 1 = 3x + 1$	Simplify on the right.
$3x^2 - 5x - 2 = 0$	Subtract 3x and 1 from each side.

Now we factor using the *ac* test. (We could also use trial and error.)

$3x^2 \underline{- 6x + 1x} - 2 = 0$	The key number is -6.
$3x(x - 2) + 1(x - 2) = 0$	Factor each pair.
$(x - 2)(3x + 1) = 0$	Factor out the GCF, $(x - 2)$.
$x - 2 = 0 \qquad$ or $\qquad 3x + 1 = 0$	Use the zero product property.
$x = 2 \qquad\qquad 3x = -1$	
$x = 2 \qquad$ or $\qquad x = -\dfrac{1}{3}$	Solve each equation.

Remember to check this by substituting $x = 2$ and then $x = -\frac{1}{3}$ in the original equation.

EXAMPLE 3 Solving a quadratic equation by simplifying first

Solve: $(2x + 1)(x - 2) = 2(x - 1) + 3$

SOLUTION

$2x^2 - 3x - 2 = 2x - 2 + 3$	Multiply.
$2x^2 - 3x - 2 = 2x + 1$	Simplify.
$2x^2 - 5x - 2 = 1$	Subtract 2x.
$2x^2 - 5x - 3 = 0$	Subtract 1.
$2x^2 - 6x + 1x - 3 = 0$	Rewrite the middle term, $-5x$.
$2x(x - 3) + 1(x - 3) = 0$	Factor each pair.
$(x - 3)(2x + 1) = 0$	Factor out the GCF, $(x - 3)$.
$x - 3 = 0$ or $2x + 1 = 0$	Use the zero product property.
$x = 3$ $x = -\dfrac{1}{2}$	Solve each equation.

Remember to check these answers.

PROBLEM 3

Solve: $(4n + 1)(n - 2) = 4(n + 1) - 3$

Finally, we can always use the general factoring strategy developed in Section 5.5 to solve certain equations. We illustrate this possibility in Example 4.

EXAMPLE 4 Solving a quadratic equation by simplifying first

Solve: $(4x - 1)(x - 1) = 2(x + 2) - 3x - 4$

SOLUTION

$4x^2 - 5x + 1 = 2x + 4 - 3x - 4$	Multiply.
$4x^2 - 5x + 1 = -x$	Simplify.
$4x^2 - 4x + 1 = 0$	Add x.
$4x^2 - 2x - 2x + 1 = 0$	Rewrite the middle term.
$2x(2x - 1) - 1(2x - 1) = 0$	Factor each pair.
$(2x - 1)(2x - 1) = 0$	Factor out the GCF, $(2x - 1)$.
$2x - 1 = 0$ or $2x - 1 = 0$	Use the zero product property.
$x = \dfrac{1}{2}$ $x = \dfrac{1}{2}$	Solve the equations.

Don't forget to check the answer!

PROBLEM 4

Solve:

$(3m + 1)(3m + 2) = 5(m + 1) - 2m - 4$

Note that in Example 4 there is really only *one* solution, $x = \frac{1}{2}$. A lot of work could be avoided if you notice that the expression $4x^2 - 4x + 1 = 0$ is a perfect square trinomial (F3) that can be factored as $(2x - 1)^2$ or, equivalently, $(2x - 1)(2x - 1)$. To avoid extra work, follow the general factoring strategy from Section 5.5.

In Example 5, we will solve several quadratic equations **not** in standard form. To do so, we make the right-hand side of the equation 0 and use the general factoring strategy to factor the left-hand side.

EXAMPLE 5 Solving quadratic equations using the general factoring strategy

Solve:

a. $9z^2 - 16 = 0$ **b.** $y(3y + 7) = -2$ **c.** $m^2 = 3m$

PROBLEM 5

Solve:

a. $16x^2 - 9 = 0$

b. $y(2y + 3) = -1$

c. $n^2 = 4n$

Answers to PROBLEMS

3. $n = -\dfrac{1}{4}$ and $n = 3$ **4.** $m = -\dfrac{1}{3}$ **5. a.** $x = \dfrac{3}{4}$ and $x = -\dfrac{3}{4}$ **b.** $y = -\dfrac{1}{2}$ and $y = -1$ **c.** $n = 0$ and $n = 4$

SOLUTION

a. The right-hand side is 0, so we use the general factoring strategy to factor $9z^2 - 16$:

$$9z^2 - 16 = 0 \qquad \text{Given}$$
$$(3z + 4)(3z - 4) = 0 \qquad \text{Factor the difference of squares.}$$
$$3z + 4 = 0 \quad \text{or} \quad 3z - 4 = 0 \qquad \text{Use the zero product property.}$$
$$z = -\frac{4}{3} \qquad\qquad z = \frac{4}{3} \qquad \text{Solve the equations.}$$

Check that the solutions are $-\frac{4}{3}$ and $\frac{4}{3}$ by substituting each number in the original equation.

b. We start by adding 2 to both sides of the equation so that the right-hand side is 0

$$y(3y + 7) + 2 = -2 + 2 \qquad \text{Add 2.}$$
$$3y^2 + 7y + 2 = 0 \qquad \text{Simplify.}$$
$$(3y + 1)(y + 2) = 0 \qquad \text{Factor.}$$
$$3y + 1 = 0 \quad \text{or} \quad y + 2 = 0 \qquad \text{Use the zero product property.}$$
$$y = -\frac{1}{3} \qquad\qquad y = -2 \qquad \text{Solve.}$$

Check to make sure the solutions are $-\frac{1}{3}$ and -2.

c. Start by subtracting $3m$, then factor the result:

$$m^2 = 3m \qquad \text{Given}$$
$$m^2 - 3m = 0 \qquad \text{Subtract } 3m.$$
$$m(m - 3) = 0 \qquad \text{Factor.}$$
$$m = 0 \quad \text{or} \quad m - 3 = 0 \qquad \text{Use the zero product property.}$$
$$m = 0 \qquad\qquad m = 3 \qquad \text{Solve.}$$

Check the solutions 0 and 3 in the original equation.

The zero product property can be applied to solve certain **nonquadratic** equations, as long as one side of the equation is 0 and the other side can be written as a product. Example 6 illustrates such a situation.

EXAMPLE 6   **Extending the zero product property to solve other types of equations**

Solve: $(v - 2)(v^2 - v - 12) = 0$

SOLUTION   The right-hand side of the equation is 0 as needed. Avoid the temptation of multiplying the expressions on the left-hand side! Remember, what we need is a *product* of factors so we can use the zero product property. With this in mind, factor $v^2 - v - 12$ as shown:

$$(v - 2)(v^2 - v - 12) = 0 \qquad \text{Given}$$
$$(v - 2)(v - 4)(v + 3) = 0 \qquad \text{Factor } v^2 - v - 12.$$
$$v - 2 = 0 \quad \text{or} \quad v - 4 = 0 \quad \text{or} \quad v + 3 = 0 \qquad \text{Use the zero product property.}$$
$$v = 2 \qquad\qquad v = 4 \qquad\qquad v = -3 \qquad \text{Solve.}$$

The solutions are 2, 4, and -3. Check that this is the case by substituting each number in the original equation.

PROBLEM 6

Solve: $(m - 3)(m^2 - m - 2) = 0$

Answers to PROBLEMS

6. $m = 3$ and $m = 2$ and $m = -1$

Before you attempt the exercises, we remind you of the steps used to solve quadratic equations by factoring:

PROCEDURE TO SOLVE QUADRATICS BY FACTORING

1. Perform the necessary operations on both sides of the equation so that the right-hand side is 0.
2. Use the general factoring strategy to factor the left side of the equation, if necessary.
3. Use the zero product property and make each factor on the left equal 0.
4. Solve each of the resulting equations.
5. Check the results by substituting the solutions obtained in step 4 in the original equation.

Calculator Corner

Solving Quadratics

Let's look at Example 3, where we have to solve $(2x + 1)(x - 2) = 2(x - 1) + 3$. As we mentioned, *you have to know the algebra* to write the equation in the standard form, $2x^2 - 5x - 3 = 0$. Now graph $Y = 2x^2 - 5x - 3$ (see Window 1). Where are the points where $Y = 0$? They are on the horizontal axis (the x-axis). For any point on the x-axis, its y-value is $y = 0$. The graph appears to have x-values 3 and $-\frac{1}{2}$ at the two points where the graph crosses the x-axis ($y = 0$). You can use your TRACE and ZOOM keys to confirm this.

Some calculators have a "zero" feature that tells you when $y = 0$. On a TI-83 Plus, enter 2nd TRACE 2 to activate the zero feature. The calculator asks you to select a left bound—that is, a point on the curve to the left of where the curve crosses the horizontal axis. Pick one near $-\frac{1}{2}$ using your ◆ and ◆ keys to move the cursor. Press ENTER. Select a right bound—that is, a point on the curve to the right of where the curve crosses the horizontal axis, and press ENTER again. The calculator then asks you to guess. Use the calculator's guess by pressing ENTER. The zero is given as $-.5$ (see Window 1). Do the same to find the other zero, $x = 3$.

Use these techniques to solve some of the other examples in this section and some of the problems in Exercises 5.6.

Zero
X=-.5   Y=0
Window 1

Boost your grade at mathzone.com!

> Practice Problems   > Self-Tests
> NetTutor   > e-Professors
                             > Videos

⟩ Exercises **5.6**

⟨ **A** ⟩ **Solving Quadratic Equations by Factoring**   In Problems 1–20, solve the given equation.

1. $2x^2 + 7x + 3 = 0$ **2.** $2x^2 + 5x + 3 = 0$ **3.** $2x^2 + x - 3 = 0$ **4.** $6x^2 + x - 12 = 0$

5. $3y^2 - 11y + 6 = 0$ **6.** $4y^2 - 11y + 6 = 0$ **7.** $3y^2 - 2y - 1 = 0$ **8.** $12y^2 - y - 6 = 0$

9. $6x^2 + 11x = -4$ **10.** $5x^2 + 6x = -1$ **11.** $3x^2 - 5x = 2$ **12.** $12x^2 - x = 6$

13. $5x^2 + 6x = 8$ **14.** $6x^2 + 13x = 5$ **15.** $5x^2 - 13x = -8$ **16.** $3x^2 + 5x = -2$

17. $3y^2 = 17y - 10$ **18.** $3y^2 = 2y + 1$ **19.** $2y^2 = -5y - 2$ **20.** $5y^2 = -6y + 8$

In Problems 21–36, solve (you may use the factoring rules for perfect square trinomials, F2 and F3).

21. $9x^2 + 6x + 1 = 0$ **22.** $x^2 + 14x + 49 = 0$ **23.** $y^2 - 8y = -16$

24. $y^2 - 20y = -100$ **25.** $9x^2 + 12x = -4$ **26.** $25x^2 + 10x = -1$

27. $4y^2 - 20y = -25$ **28.** $16y^2 - 56y = -49$ **29.** $x^2 = -10x - 25$

30. $x^2 = -16x - 64$ **31.** $(2x - 1)(x - 3) = 3x - 5$ **32.** $(3x + 1)(x - 2) = x + 7$

33. $(2x + 3)(x + 4) = 2(x - 1) + 4$ **34.** $(5x - 2)(x + 2) = 3(x + 1) - 7$ **35.** $(2x - 1)(x - 1) = x - 1$

36. $(3x - 2)(3x - 1) = 1 - 3x$

In Problems 37–60, solve.

37. $4x^2 - 1 = 0$ **38.** $9x^2 - 1 = 0$ **39.** $4y^2 - 25 = 0$

40. $25y^2 - 9 = 0$ **41.** $z^2 = 9$ **42.** $z^2 = 25$

43. $25x^2 = 49$ **44.** $9x^2 = 64$ **45.** $m^2 = 5m$

46. $m^2 = 9m$ **47.** $2n^2 = 10n$ **48.** $3n^2 = 12n$

49. $y(y + 11) = -24$ **50.** $y(y + 15) = -56$ **51.** $y(y - 16) = -63$

52. $y(y - 19) = -88$ **53.** $(v - 2)(v^2 + 3v + 2) = 0$ **54.** $(v - 1)(v^2 + 5v + 6) = 0$

55. $(m^2 - 3m + 2)(m - 4) = 0$ **56.** $(m^2 - 4m + 3)(m - 6) = 0$ **57.** $(n^2 - 3n - 4)(n + 2) = 0$

58. $(n^2 - 4n - 5)(n + 8) = 0$ **59.** $(x^2 + 2x - 3)(x - 1) = 0$ **60.** $(x^2 + 3x - 4)(x - 1) = 0$

〉〉〉 Applications

The next four problems are from Section 5.3. In Section 5.3 we asked you to *factor* the equations. Here, we go one step further: *Solve* the equations!

61. *Stopping distance* (See page 345) If you are traveling at m miles per hour and want to find the speed m that will allow you to stop the car 13 feet away, you must solve the equation $5m^2 + 220m - 1225 = 0$.

   a. Solve the equation.

   b. Since m represents the velocity of the car, can you use both answers? Which is the correct answer?

62. *Stopping distance* If you are traveling at m miles per hour and want to find the speed m that will allow you to stop the car 14 feet away, you must solve the equation $5m^2 + 225m - 1250 = 0$.

   a. Solve the equation.

   b. Since m represents the velocity of the car, can you use both answers? Which answer is correct?

63. *Height of an object* The height $H(t)$ after t seconds of an object thrown downward from a height of h meters and initial velocity v_0 is modeled by the equation $H(t) = -5t^2 + v_0t + h$. If an object is thrown downward at 5 meters per second from a height of 10 meters, then $H(t) = -5t^2 - 5t + 10$.

   a. How long would it take for the object to reach the ground $[H(t) = 0]$?

   b. The equation $-5t^2 - 5t + 10 = 0$ has two answers. What are the solutions? Since t represents the number of seconds, can you use both answers? Which answer is correct?

64. *Height of an object* The height $H(t)$ of an object thrown downward from a height of h meters and initial velocity v_0 is modeled by the equation $H(t) = -5t^2 + v_0t + h$. If an object is thrown downward at 4 meters per second from a height of 28 meters, then $H(t) = -5t^2 - 4t + 28$.

   a. How long would it take for the object to reach the ground $[H(t) = 0]$?

   b. The equation $-5t^2 - 4t + 28 = 0$ has two answers. What are the solutions? Since t represents the number of seconds, can you use both answers? Which answer is correct?

〉〉〉 *Using Your Knowledge*

The "Function" of Quadratics   The ideas you've studied in this section are useful for solving problems in business.

65. Solve the equation in Problem 81 of the *Using Your Knowledge* in Exercises 5.4 when $D(x) = 0$.

66. Solve the equation in Problem 84 of Exercises 5.4 when $S(p) = 0$.

〉〉〉 *Write On*

67. What is a quadratic equation and what procedure do you use to solve these equations?

68. Write the steps you use to solve $x(x - 1) = 0$ and then write the difference between this procedure and the one you would use to solve $3x(x - 1) = 0$.

69. The equation $x^2 + 2x + 1 = 0$ can be solved by factoring. How many solutions does it have? The original equation is equivalent to $(x + 1)^2 = 0$. Why do you think -1 is called a *double root* for the equation?

〉〉〉 *Concept Checker*

Fill in the blank(s) with the correct word(s), phrase, or mathematical statement.

70. An equation of the form $ax^2 + bx + c = 0, a \neq 0$, is called a _____ equation.

71. The zero product property states that if $A \cdot B = 0$ then _____ or _____.

linear	**B ≠ 0**
A = 0	**B = 0**
A ≠ 0	**quadratic**

〉〉〉 *Mastery Test*

Solve:

72. $10x^2 - 13x = 3$

73. $(3x - 2)(x - 1) = 2(x + 3) + 2$

74. $(9x - 2)(x - 1) = 2(x + 1) - 7x - 1$

75. $5x^2 + 9x - 2 = 0$

76. $3x^2 - 2x - 5 = 0$

77. $x(x - 1) = 0$

78. $2x(x + 3) = 0$

79. $(x - 4)(x^2 + 4x - 5) = 0$

80. $25y^2 - 36 = 0$

81. $m(3m + 5) = -2$

82. $n^2 = 11n$

〉〉〉 *Skill Checker*

Simplify:

83. $H^2 + (3 + H)^2$

84. $H^2 + (6 + H)^2$

Expand:

85. $(H - 9)(H + 3)$

86. $(H - 8)(H + 4)$

5.7   Applications of Quadratics

▶ Objectives

Use the RSTUV procedure to solve:

A〉 Consecutive integer problems

B〉 Area and perimeter problems

C〉 Problems involving the Pythagorean Theorem

D〉 Motion problems (braking distance)

▶ To Succeed, Review How To . . .

1. Solve integer problems (p. 145).
2. Use FOIL to expand polynomials (p. 359).
3. Multiply integers (pp. 60–61).

▶ Getting Started

Stop That Car!

The diagram shows the distance needed to stop a car traveling at the indicated speeds. At speed m (in miles per hour), that stopping distance $D(m)$ in feet is given by

$$D(m) = 0.05m^2 + 2.2m + 0.75$$

If you are one car length (13 feet) behind a stopped car, how fast can you be traveling and still be able to stop before hitting the car?

Auto stopping distance	
Speed (mi/hr)	Distance (feet)
55	273
65	355
75	447

In this case, the actual distance is 13 feet, so we have to solve

$$
\begin{aligned}
D(m) = 0.05m^2 + 2.2m + 0.75 &= 13 \\
5m^2 + 220m + 75 &= 1300 \qquad \text{Multiply by 100 (clear the decimals).} \\
5m^2 + 220m - 1225 &= 0 \qquad \text{Subtract 1300 (write in standard form).} \\
m^2 + 44m - 245 &= 0 \qquad \text{Divide by 5.} \\
(m - 5)(m + 49) &= 0 \qquad \text{Factor.} \\
m = 5 \quad \text{or} \quad m &= -49 \qquad \text{Use the zero product property and solve.}
\end{aligned}
$$

So, you can be going 5 miles per hour and stop in 13 feet. If your speed is over 5 miles per hour, you will hit the car!

A ⟩ Consecutive Integer Problems

Do you remember the integer problems of Chapter 2? Here's the terminology we need to solve a problem involving integers and quadratic equations.

Terminology	Notation	Examples
Two consecutive integers	$n, n + 1$	$3, 4; -6, -5$
Three consecutive integers	$n, n + 1, n + 2$	$7, 8, 9; -4, -3, -2$
Two consecutive even integers	$n, n + 2$	$8, 10; -6, -4$
Two consecutive odd integers	$n, n + 2$	$13, 15; -21, -19$

As before, we solve this type problem using the RSTUV method.

EXAMPLE 1   A consecutive integer problem

The product of two consecutive even integers is 10 more than 7 times the larger of the two integers. Find the integers.

SOLUTION

1. **Read the problem.** We are asked to find two consecutive even integers.
2. **Select the unknown.** Let n and $n + 2$ be the integers ($n + 2$ being the larger).
3. **Think of a plan.** We first translate the problem:

$$\underbrace{\text{The product of two consecutive integers}}_{n(n+2)} \quad \underbrace{\text{is}}_{=} \quad \underbrace{10}_{10} \quad \underbrace{\text{more than}}_{+} \quad \underbrace{\begin{array}{c}7 \text{ times}\\ \text{the larger.}\end{array}}_{7(n+2)}$$

4. **Use algebra to solve the equation.**

$$\begin{array}{ll} n^2 + 2n = 10 + 7n + 14 & \text{Use the distributive property.} \\ n^2 + 2n = 24 + 7n & \text{Simplify.} \\ n^2 + 2n - 24 - 7n = 0 & \text{Subtract } 24 + 7n. \\ n^2 - 5n - 24 = 0 & \text{Simplify.} \\ (n - 8)(n + 3) = 0 & \text{Factor } (-8 \cdot 3 = -24, -8 + 3 = -5). \\ n - 8 = 0 \quad \text{or} \quad n + 3 = 0 & \text{Use the zero product property.} \\ n = 8 \qquad\qquad n = -3 & \text{Solve each equation.} \end{array}$$

If $n = 8$ is the first integer, the second is $n + 2 = 8 + 2 = 10$. The solution $n = -3$ is not acceptable because -3 is *not* even. Thus, there is only one pair of integers that satisfies the problem: 8 and 10.

5. **Verify the solution.**

$$\underbrace{\text{The product of two consecutive integers}}_{8 \cdot 10} \quad \underbrace{\text{is}}_{=} \quad \underbrace{10}_{10} \quad \underbrace{\text{more than}}_{+} \quad \underbrace{\begin{array}{c}7 \text{ times}\\ \text{the larger.}\end{array}}_{7 \cdot 10} \quad \text{True.}$$

PROBLEM 1

The product of two consecutive odd integers is 10 more than 5 times the smaller of the two integers. Find the integers.

B ⟩ Area and Perimeter

Quadratic equations are also used in geometry. As you recall, if L and W are the length and width of a rectangle, then the perimeter P is $P = 2L + 2W$, and the area A is $A = LW$. Let's use these ideas in the next example.

Answers to PROBLEMS

1. 5 and 7

EXAMPLE 2   Finding the dimensions of a room

A rectangular room is 4 feet longer than it is wide. The area of the room is numerically equal to its perimeter plus 92. What are the dimensions of the room?

SOLUTION

1. Read the problem. We are asked to find the dimensions of the room; we also know it is a rectangle.

2. Select the unknown. Let W be the width. Since the room is 4 feet longer than it is wide, the length is $W + 4$.

3. Think of a plan. Two measurements are involved: the area and the perimeter. Let's start with a picture:

The area of the room is

$$W(W + 4)$$

The perimeter of the room is

$$2(W + 4) + 2W = 4W + 8$$

$L = W + 4$

Look at the wording of the problem. We can restate it as follows:

The area of the room	is equal to	its perimeter plus 92.
$W(W + 4)$	$=$	$(4W + 8) + 92$

4. Use algebra to solve the equation.

$W^2 + 4W = 4W + 100$	Simplify both sides.
$W^2 - 100 = 0$	Subtract $4W + 100$.
$(W + 10)(W - 10) = 0$	Factor.
$W + 10 = 0$ or $W - 10 = 0$	Use the zero product property.
$W = -10$ $W = 10$	Solve each equation.

Since a room can't have a negative width, discard the -10, so the width is 10 feet and the length is 4 feet longer or 14 feet. Thus, the dimensions of the room are 10 feet by 14 feet.

5. Verify the solution. The area of the room is $10 \cdot 14 = 140$ and the perimeter is $2 \cdot 10 + 2 \cdot 14 = 20 + 28 = 48$. The area (140) must exceed the perimeter (48) by 92. Does $140 = 48 + 92$? Yes, so our answer is correct.

PROBLEM 2

What are the dimensions of the room in Example 2 if the area exceeds the perimeter by 56?

EXAMPLE 3   Finding the dimensions of a monitor

Denise's monitor is 3 inches wider than it is high and has 130 square inches of viewing area.

a. Find the dimensions of her monitor.

b. Denise wants a Trinitron® monitor, which is 4 inches wider than it is high and has 62 more square inches of viewing area than her current monitor. What are the dimensions of the Trinitron monitor?

$H + 3$

130 in.²

H

PROBLEM 3

a. What are the dimensions of Denise's current monitor if it is 3 inches wider than it is high and has 108 square inches of viewing area?

b. What are the dimensions of a Cinema monitor, which is 6 inches wider than it is high and has double the viewing area of the monitor in part **a**?

(continued)

Answers to PROBLEMS

2. 8 ft by 12 ft

3. a. 9 in. by 12 in.

   b. 12 in. by 18 in.

SOLUTION

a. 1. Read the problem. We want the dimensions of the monitor.

2. Select the unknown. Let H be the height of the monitor.

3. Think of a plan. Draw a picture. H is the height. The width is 3 inches more than the height; that is, the width is $H + 3$. The area of the rectangle is the height H times the width, $H + 3$. It is also 130 square inches. Thus,

$$H(H + 3) = 130$$

4. Use algebra to solve the equation.

$$H(H + 3) = 130$$
$$H^2 + 3H = 130 \qquad \text{Simplify.}$$
$$H^2 + 3H - 130 = 0 \qquad \text{Subtract 130.}$$
$$(H + 13)(H - 10) = 0 \qquad \text{Factor.}$$
$$H + 13 = 0 \quad \text{or} \quad H - 10 = 0 \qquad \text{Use the zero product property.}$$
$$H = -13 \quad \text{or} \quad H = 10 \qquad \text{Solve.}$$

Discard $H = -13$. (Why?) Thus, the dimensions of the monitor are 10 inches by 13 inches.

b. What are the Trinitron dimensions? Let the height be H, so the width is $H + 4$. The area of the Trinitron screen is $H(H + 4)$, which is 62 square inches more than her old screen; that is,

$$H(H + 4) = 130 + 62 = 192$$

4. Use algebra to solve the equation.

$$H(H + 4) = 192$$
$$H^2 + 4H = 192 \qquad \text{Simplify.}$$
$$H^2 + 4H - 192 = 0 \qquad \text{Subtract 192.}$$
$$(H - 12)(H + 16) = 0 \qquad \text{Use the zero product property.}$$
$$H = 12 \quad \text{or} \quad H = -16 \qquad \text{Solve.}$$

Discard $H = -16$. The dimensions of the Trinitron are 12 inches by 16 inches.

5. Verify the solution!

C ⟩ The Pythagorean Theorem

Suppose you were to buy a 20-inch television set. What does that 20-inch measurement represent? It's the diagonal length of the rectangular screen. The relationship between the length L, the height H, and the diagonal measurement d of the screen can be found by using the Pythagorean Theorem, which is stated here.

PYTHAGOREAN THEOREM

If the longest side of a right triangle (a triangle with a 90° angle) is of length c and the other two sides are of length a and b, respectively, then

$$a^2 + b^2 = c^2$$

Note that the longest side of the triangle (opposite the 90° angle) is called the **hypotenuse**, and the two shorter sides are called the **legs** of the triangle.

In the case of the television set, $H^2 + L^2 = 20^2$. If the screen is 16 inches long ($L = 16$), can we find its height H? Substituting $L = 16$ into $H^2 + L^2 = 20^2$, we obtain

$$H^2 + 16^2 = 20^2$$

$$H^2 + 256 = 400 \qquad \text{Simplify.}$$

$$H^2 - 144 = 0 \qquad \text{Subtract 400.}$$

$$(H - 12)(H + 12) = 0 \qquad \text{Factor.}$$

$$H - 12 = 0 \quad \text{or} \quad H + 12 = 0 \qquad \text{Use the zero product property.}$$

$$H = 12 \qquad\qquad H = -12 \qquad \text{Solve each equation.}$$

Since H represents the height of the screen, we discard -12 as an answer. (The height cannot be negative.) Thus, the height of the screen is 12 inches.

EXAMPLE 4 Calculating a computer monitor's dimensions

The screen of a rectangular computer monitor is 3 inches wider than it is high, and its diagonal is 6 inches longer than its height. What are the dimensions of the monitor?

SOLUTION

1. Read the problem. We are asked to find the dimensions, which involves finding the width, height, and diagonal of the screen.

2. Select the unknown. Since all measurements are given in terms of the height, let H be the height.

3. Think of a plan. It's a good idea to start with a picture so we can see the relationships among the measurements. Note that the width is $3 + H$ (3 inches wider than the height) and the diagonal is $6 + H$ (6 inches longer than the height). We enter this information in a diagram, as shown here.

4. Use the Pythagorean Theorem to describe the relationship among the measurements and then solve the resulting equation. According to the Pythagorean Theorem:

$$H^2 + (3 + H)^2 = (6 + H)^2$$

$$H^2 + 9 + 6H + H^2 = 36 + 12H + H^2 \qquad \begin{array}{l}\text{Expand } (3 + H)^2 \\ \text{and } (6 + H)^2.\end{array}$$

$$H^2 + 9 + 6H + H^2 - 36 - 12H - H^2 = 0 \qquad \begin{array}{l}\text{Subtract} \\ 36 + 12H + H^2.\end{array}$$

$$H^2 - 6H - 27 = 0 \qquad \text{Simplify.}$$

$$(H - 9)(H + 3) = 0 \qquad \begin{array}{l}\text{Factor } (-9 \cdot 3 = -27; \\ -9 + 3 = -6).\end{array}$$

$$H - 9 = 0 \quad \text{or} \quad H + 3 = 0 \qquad \begin{array}{l}\text{Use the zero} \\ \text{product property.}\end{array}$$

$$H = 9 \qquad\qquad H = -3 \qquad \text{Solve each equation.}$$

Leg a Hypotenuse c

90°

Leg b

$3 + H$

$6 + H$

H

PROBLEM 4

A rectangular book is 2 inches higher than it is wide. If the diagonal is 4 inches longer than the width of the book, what are the dimensions of the cover?

(continued)

Since H is the height, we discard -3, so the height of the monitor is 9 inches, the width is 3 more inches, or 12 inches, and the diagonal is 6 more inches than the height, or $9 + 6 = 15$ inches.

5. Verify the solution!   Looking at the diagram and using the Pythagorean Theorem, we see that $9^2 + 12^2$ must be 15^2, that is,

$$9^2 + 12^2 = 15^2$$

Since $81 + 144 = 225$ is a true statement, our dimensions are correct.

By the way, television sets claiming to be 25 inches or 27 inches (meaning the length of the screen measured diagonally is 25 or 27 inches) hardly ever measure 25 or 27 inches. This is easy to confirm using the Pythagorean Theorem. For example, a 27-inch Panasonic® has a screen that is 16 inches high and 21 inches long. Can it really be 27 inches diagonally? If this were the case, $16^2 + 21^2$ would equal 27^2. Is this true? Measure a couple of TV or computer screens and see whether the manufacturers' claims are true!

D › Motion Problems: Braking Distance

In the *Getting Started,* we gave a formula for the stopping distance for a car traveling at the indicated speeds. That distance involves the **braking distance *b*,** the distance it takes to stop a car *after* the brakes are applied. This distance is given by

$$b = 0.06v^2$$

where v is the speed of the car when the brakes are applied. It also involves the reaction distance

$$r = 1.5tv$$

where t is the driver's reaction time (in seconds) and v is the speed of the car in miles per hour.

EXAMPLE 5   Finding the speed of a car based on braking distance
A car traveled 96 feet *after* the brakes were applied. How fast was the car going when the brakes were applied?

SOLUTION

1. Read the problem. We want to find how fast the car was going.

2. Select the unknown. Let v represent the velocity.

3. Think of a plan. The braking distance b is 96.

The formula for b is $b = 0.06v^2$.

Thus, $0.06v^2 = 96$.

4. Use algebra to solve the equation.

$$0.06v^2 = 96$$
$$6v^2 = 9600 \quad \text{Multiply by 100.}$$
$$6v^2 - 9600 = 0 \quad \text{Subtract 9600.}$$
$$6(v^2 - 1600) = 0 \quad \text{Factor out the GCF.}$$
$$6(v + 40)(v - 40) = 0 \quad \text{Factor.}$$
$$v + 40 = 0 \quad \text{or} \quad v - 40 = 0 \quad \text{Solve.}$$
$$v = -40 \quad \text{or} \quad v = 40$$

Discard the -40 because the velocity was positive, so the velocity of the car was 40 miles per hour.

PROBLEM 5

A car traveled 150 feet *after* the brakes were applied. How fast was the car going when the brakes were applied?

Answers to PROBLEMS

5. 50 mi/hr

5. Verify the solution. Substitute $v = 40$ in $b = 0.06v^2$:

$$b = 0.06(40)^2 = 0.06(1600) = 96$$

so the answer is correct.

Before you attempt to solve the Exercises, practice translating!

TRANSLATE THIS

1. The product of two consecutive even integers is 20 less than 10 times the larger of the two integers.

2. The product of two consecutive odd integers is 10 more than 5 times the smaller of the two integers.

3. A rectangular room is 5 feet longer than it is wide. What is the area A of the room in terms of the width W?

4. The viewing area of a TV is $A = WH$, where W is the width and H is the height of the rectangular screen. If the height H is 20 inches more than the width, what is the area A?

5. If the width W of a rectangle is 20 inches less than its height H, what is the area of the rectangle?

6. The screen on a rectangular computer monitor is 4 inches wider than it is high. If the diagonal is 8 more inches than its height H, write an equation relating the length of the sides of the screen in terms of H.

The third step in the RSTUV procedure is to **TRANSLATE** the information into an equation. In Problems 1–10 **TRANSLATE** the sentence and match the correct translation with one of the equations A–O

A. $A = (H - 20)H$

B. $A = W(W + 5)$

C. $A = \frac{1}{2}(h + 2)h$

D. $n(n + 2) = 5n + 10$

E. $(H + 4)^2 = H^2 + (H + 8)^2$

F. $A = (H - 20) + H$

G. $F = 0.06v^2$

H. $(H + 4)^2 + H^2 = (H + 8)^2$

I. $A = (h + 2)h$

J. $L^2 + (L - 4)^2 = (L + 4)^2$

K. $n(n + 2) = 10(n + 2) - 20$

L. $L^2 + (L + 4)^2 = (L - 4)^2$

M. $n^2 + (n + 2)^2 = (n + 4)^2$

N. $n(n + 1) = 10(n + 1) - 20$

O. $A = W(W + 20)$

7. The braking distance b for a car traveling at v miles per hour is the product of 0.06 and the square of the velocity. If a car travels F feet after the brakes were applied, what is the equation for the braking distance F?

8. Write the area A of a triangle whose base is 2 inches longer than its height h.

9. The hypotenuse of a right triangle is 4 inches longer than the length L of the longer side. If the remaining side is 4 inches shorter than the length L, write an equation relating the sides of the triangle.

10. The sides of a right triangle are consecutive even integers. Write an equation relating the length of the sides.

MathZone

Boost your grade at mathzone.com!
> Practice Problems   > Self-Tests
> NetTutor   > e-Professors
           > Videos

› Exercises 5.7

‹ A ›   **Consecutive Integer Problems**   In Problems 1–6, use the RSTUV method to solve the integer problems.

1. The product of two consecutive integers is 8 less than 10 times the smaller of the two integers. Find the integers.

2. The product of two consecutive even integers is 4 more than 5 times the smaller of the two integers. Find the integers.

3. The product of two consecutive odd integers is 7 more than their sum. What are the integers?

4. Find three consecutive even integers such that the square of the largest exceeds the sum of the squares of the other two by 12.

5. Find two consecutive even integers such that the square of the larger integer is 4 less than 10 times the smaller integer.

6. The sum of two consecutive integers is 29 less than their product. Find the integers.

⟨ **B** ⟩   **Area and Perimeter**   Use the information in the table and the RSTUV method to solve Problems 7–14.

Name	Geometric Shapes	Name	Geometric Shapes
Triangle Area $= \frac{1}{2}bh$	h b	**Rectangle** Area $= LW$	W L
Trapezoid Area $= \frac{1}{2}h(b_1 + b_2)$	b_1 h b_2	**Circle** Area $= \pi r^2$	r
Parallelogram Area $= Lh$	h L W		

7. The area A of a triangle is 40 square inches. Find its dimensions if the base b is 2 inches more than its height h.

8. The area A of a trapezoid is 105 square centimeters. If the height h is the same length as the smaller side b_1 and 1 inch less than the longer side b_2, find the height h.

9. The area A of a parallelogram is 150 square inches. If the length of the parallelogram is 5 inches more than its height, what are the dimensions?

10. The area of a rectangle is 96 square centimeters. If the width of the rectangle is 4 centimeters less than its length, what are the dimensions of the rectangle?

11. The area of a circle is 49π square units. If the radius r is $x + 3$ units, find x and then find the radius of the circle.

12. A rectangular room is 5 feet longer than it is wide. The area of the room numerically exceeds its perimeter by 100. What are the dimensions of the room?

13. The biggest lasagna ever made had an area of 250 square feet. If it was made in the shape of a rectangle 45 feet longer than wide, what were its dimensions? (It was made in Dublin and weighed 3609 pounds, 10 ounces.)

14. The biggest strawberry shortcake needed 360 square feet of strawberry topping. If this area was numerically 225 more than 3 times its length, what were the dimensions of this rectangular shortcake?

⟨ **C** ⟩   **The Pythagorean Theorem**   In Problems 15–20, use the RSTUV procedure to solve the problems.

15. The biggest television ever built is the 289 Sony® Jumbo Tron. (This means the screen measured 289 feet diagonally.) If the length of the screen was 150 feet, how high was the TV.

16. A television screen is 2 inches longer than it is high. If the diagonal length of the screen is 2 inches more than its length, what is the diagonal measurement of this screen?

17. The hypotenuse of a right triangle is 4 inches longer than the shortest side and 2 inches longer than the remaining side. Find the dimensions of the triangle.

18. The hypotenuse of a right triangle is 16 inches longer than the shortest side and 2 inches longer than the remaining side. Find the dimensions of the triangle.

19. One of the sides of a right triangle is 3 inches longer than the shortest side. If the hypotenuse is 3 inches longer than the longer side, what are the dimensions of the triangle?

20. The sides of a right triangle are consecutive even integers. Find their lengths.

⟨ **D** ⟩   **Motion Problems: Braking Distance**   In Problems 21–26, solve the motion problems. Do not round answers resulting in decimals. (*Hint:* For Problems 21–24, use the formula from Example 5.)

21. A car traveled 54 feet after the driver applied the brakes. How fast was the car going when the brakes were applied?

22. A car traveled 216 feet after the brakes were applied. How fast was the car going when the brakes were applied?

23. A car left a 73.5-foot skid mark on the road.
  a. How fast was the car going when the brakes were applied and the skid mark was made?
  b. If the speed limit was 30 miles per hour, was the car speeding?

24. A police officer measured a skid mark to be 150 feet long. The driver said he was going under the speed limit, which was 45 miles per hour. Was the driver correct? How long would the skid mark have to be if he was indeed going 45 miles per hour when the brakes were applied?

25. Pedro reacts very quickly. In fact, his reaction time is 0.4 second. When driving on a highway, Pedro saw a danger signal ahead and tried to stop. If his car traveled 120 feet before stopping, how fast was he going when he saw the sign? Use

$$d = 1.5tv + 0.06v^2$$

for the stopping distance, where d is in feet, t is in seconds, and v is in miles per hour.

26. Vilmos' reaction time is 0.3 seconds. He tried to avoid hitting a student walking on the road.

   a. If his car traveled 33 feet before stopping, how fast was he going?

   b. How many feet would he travel at that speed, using the formula given in the *Getting Started?*

⟩⟩⟩   Applications

27. *Games in a tournament*   The total number of games G played in a tournament in which each of the t teams entered plays every other team twice is modeled by the equation $G = t^2 - t$. Find the number of teams t entered in a 20 game tournament.

28. *Games in a tournament*   If in Problem 27 the total number of games ($G = t^2 - t$) played is 30, how many teams were entered in the tournament?

29. *Handshakes*   Before a meeting of the Big 10 countries, the attending delegates shook every other delegate's hand only once. If 28 handshakes were exchanged and the number of possible handshakes is modeled by the equation $N = \frac{1}{2}(d^2 - d)$, where d is the number of delegates attending, how many delegates were at the meeting?

30. *Handshakes*   A bipartisan committee of c congressmen exchanges 55 handshakes. If the number of possible handshakes is modeled by the equation $N = \frac{1}{2}(c^2 - c)$ and every person shook every other person's hand only once, how many congressmen were in the committee?

31. *Cost of tutoring*   The tutorial center spent $1300 tutoring x students. If the cost of tutoring x students is given by $(x^2 + 10x + 100)$ dollars, how many students were tutored?

32. *Cost of tutoring*   The tutorial center will offer individualized tutoring for x students at a price $P = (x^2 + 25x)$ dollars a month. How many students can be tutored when the price P is $350 per student?

⟩⟩⟩   Using Your Knowledge

Motion Problems   In Problems 33–36, use the fact that the height $H(t)$ of an object thrown downward from a height of h meters and initial velocity V_0 is modeled by the equation

$$H(t) = -5t^2 - V_0 t + h$$

33. An object is thrown downward at 5 meters per second from a height of 10 meters. How long does it take the object to hit the ground?

34. An object is thrown downward from a height of 28 meters with an initial velocity of 4 meters per second. How long does it take the object to hit the ground?

35. An object is thrown downward from a building 15 meters high at 10 meters per second. How long does it take the object to hit the ground?

36. How long does it take a package thrown downward from a plane at 10 meters per second to hit the ground 175 meters below?

⟩⟩⟩   Write On

37. In the *Getting Started,* the stopping distance is given by the equation

$$D(m) = 0.05m^2 + 2.2m + 0.75$$

In Problem 25, it is

$$d = 1.5tv + 0.06v^2$$

If you are traveling at 20 miles per hour, which formula can you use to evaluate your stopping distance? Why?

38. What additional information do you need in Problem 37 to use the second formula?

39. Name at least three factors that would influence the stopping distance of a car. Explain.

40. Suppose you are driving at 20 miles per hour. What reaction time will give the same stopping distance for both formulas?

⟩⟩⟩   Concept Checker

Fill in the blank(s) with the correct word(s), phrase, or mathematical

41. According to the Pythagorean theorem, if a, b, and c are the lengths of the sides of a right triangle (with c the hypotenuse), then _____.

$$a^2 = b^2 + c^2 \quad \textbf{crashing}$$

$$a^2 + b^2 = c^2 \quad \textbf{braking}$$

42. The distance it takes to stop a car after the brakes are applied is called the _____ distance.

> > >   *Mastery Test*

43. The product of two consecutive odd integers is the difference between 23 and the sum of the two integers. What are the integers?

44. A rectangular room is 2 feet longer than it is wide. Its area numerically exceeds 10 times its longer side by 28. What are the dimensions of the room?

45. The hypotenuse of a right triangle is 8 inches longer than the shortest side and 1 inch longer than the remaining side. What are the dimensions of the triangle?

46. The product of the length of the sides of a right triangle is 325 less than the length of the hypotenuse squared. If the longer side is 5 units longer than the shorter side, what are the dimensions of the triangle?

47. The distance b (in feet) it takes to stop a car *after* the brakes are applied is given by the equation

$$b = 0.06v^2$$

A car traveled 150 feet after the driver applied the brakes. How fast was the car going when the brakes were applied?

> > >   *Skill Checker*

48. Write $\frac{3}{8}$ with a denominator of 16.

49. Write $\frac{5}{6}$ with a denominator of 18.

In Problems 50–54, factor:

50. $6x - 12y$

51. $18x - 36y$

52. $x^2 + 2x - 15$

53. $x^2 + 5x - 6$

54. $x^2 - 9$

> **Collaborative Learning**

Shortcut patterns for solving quadratic equations

In this chapter we have used several methods to solve quadratic equations. Now we are going to develop a shortcut to solve these quadratic equations by letting you discover some applicable patterns.

Form three groups of students. Each of the groups will solve the equations assigned to them and enter the required information.

Recall that to factor $x^2 + bx + c = 0$, we need two numbers whose product is c and whose sum is b.

Group 1	Numbers required to factor	Solutions
$x^2 + 3x + 2 = 0$	2, 1	$-2, -1$
$x^2 - 7x + 12 = 0$		
$x^2 + x - 2 = 0$		
$x^2 - x - 2 = 0$		
Group 2		
$x^2 + 5x + 6 = 0$	2, 3	$-2, -3$
$x^2 - 5x - 6 = 0$		
$x^2 + x - 6 = 0$		
$x^2 - x - 6 = 0$		
Group 3		
$x^2 + 7x + 12 = 0$	3, 4	$-3, -4$
$x^2 - 3x + 2 = 0$		
$x^2 + x - 12 = 0$		
$x^2 - x - 12 = 0$		

Based on the patterns you see, have each of the groups complete the following conjecture (guess):

The factorization of $x^2 + bx + c = 0$ requires two factors F_1 and F_2 whose product is _____ and whose sum is _____.

The solutions of $x^2 + bx + c = 0$ are _____ and _____.

See whether all groups agree.

Is there a pattern that can be used to solve $ax^2 + bx + c = 0$ when $a \neq 1$? First, recall that to factor $ax^2 + bx + c$ we need two factors whose product is ac and whose sum is b. Here are some examples and some work for all groups.

	ac	Factors	Solutions
$4x^2 - 4x + 1 = 0$	4	$-2, -2$	$\frac{1}{2}$
$3x^2 - 5x - 2 = 0$	-6	$-6, 1$	$2, -\frac{1}{3}$
$10x^2 + 13x - 3 = 0$	-30	$15, -2$	$-\frac{3}{2}, \frac{1}{5}$

Based on the patterns you see, have each of the groups complete the following conjecture (guess):

The factorization of $ax^2 + bx + c = 0$ requires two factors F_1 and F_2 whose product is _____ and whose sum is _____.

The solutions of $ax^2 + bx + c = 0$ are _____ and _____.

See whether all groups agree.

Research Questions

1. Proposition 4 in Book II of Euclid's *Elements* states:

   If a straight line be cut at random, the square on the whole is equal to the squares on the segments and twice the rectangles contained by the segments.

   This statement corresponds to one of the rules of factoring we've studied in this chapter. Which rule is it, and how does it relate?

2. Proposition 5 of Book II of the *Elements* also corresponds to one of the factoring formulas we studied. Which one is it, and what does the proposition say?

3. Many scholars ascribe the Pythagorean Theorem to Pythagoras. However, other versions of the theorem exist:

   a. The ancient Chinese proof

   b. Bhaskara's proof

   c. Euclid's proof

   d. Garfield's proof

   e. Pappus's generalization

   Select three of these versions and write a paper giving details, if possible, telling where they appeared, who authored them, and what they said.

4. There are different versions regarding Pythagoras's death. Write a short paper detailing the circumstances of his death, where it occurred, and how.

5. Write a report about Pythagorean triples.

6. In *The Human Side of Algebra,* we mentioned that the Pythagoreans studied arithmetic, music, geometry, and astronomy. Write a report about the Pythagoreans' theory of music.

7. Write a report about the Pythagoreans' theory of astronomy.

› Summary **Chapter 5**

Section	Item	Meaning	Example
5.1A	Greatest common factor (GCF) of numbers	The GCF of a list of integers is the largest common factor of the integers in the list.	The GCF of 45 and 75 is 15.
5.1B	Greatest common factor (GCF) of variable terms	The GCF of a list of terms is the product of each of the variables raised to the lowest exponent to which they occur.	The GCF of x^2y^3, x^4y^2, and x^5y^4 is x^2y^2.
5.1C	Greatest common factor (GCF) of a polynomial	ax^n is the GCF of a polynomial if a divides each of the coefficients and n is the smallest exponent of x in the polynomial.	$3x^2$ is the GCF of $6x^4 - 9x^3 + 3x^2$.
5.2A	Factoring Rule 1: Reversing FOIL (F1)	$X^2 + (A + B)X + AB = (X + A)(X + B)$	$x^2 + 8x + 15 = (x + 5)(x + 3)$
	Factoring a trinomial of the form $x^2 + bx + c$	To factor $x^2 + bx + c$, find two numbers whose product is c and whose sum is b.	To factor $x^2 + 7x + 10$, find two numbers whose product is 10 and whose sum is 7. The numbers are 5 and 2. Thus, $x^2 + 7x + 10 = (x + 5)(x + 2)$.
5.3A	ac test	$ax^2 + bx + c$ is factorable if there are two integers with product ac and sum b.	$3x^2 + 8x + 5$ is factorable. (There are two integers whose product is 15 and whose sum is 8: 5 and 3.) $2x^2 + x + 3$ is not factorable. (There are no integers whose product is 6 and whose sum is 1.)
5.4A, B	Factoring squares of binomials (F2 and F3)	$X^2 + 2AX + A^2 = (X + A)^2$ $X^2 - 2AX + A^2 = (X - A)^2$	$x^2 + 10x + 25 = (x + 5)^2$ $x^2 - 10x + 25 = (x - 5)^2$
5.4C	Factoring the difference of two squares (F4)	$X^2 - A^2 = (X + A)(X - A)$	$x^2 - 36 = (x + 6)(x - 6)$
5.5A	Factoring the sum (F5) or difference (F6) of cubes	$X^3 + A^3 = (X + A)(X^2 - AX + A^2)$ $X^3 - A^3 = (X - A)(X^2 + AX + A^2)$	$x^3 + 64 = (x + 4)(x^2 - 4x + 16)$ $x^3 - 64 = (x - 4)(x^2 + 4x + 16)$
5.5B	General factoring strategy	1. Factor out the GCF. 2. Look at the number of terms inside the parentheses or in the original polynomial. 　　*Four terms:* Grouping 　　*Three terms:* Perfect square trinomial or *ac* test 　　*Two terms:* Difference of two squares, sum of two cubes, difference of two cubes 3. Make sure the expression is completely factored.	

Section	Item	Meaning	Example
5.6A	Quadratic equation	An equation in which the greatest exponent of the variable is 2. If a, b, and c are real numbers, $ax^2 + bx + c = 0$ is in standard form.	$x^2 + 3x - 7 = 0$ is a quadratic equation in standard form.
	Zero product property	If $A \cdot B = 0$, then $A = 0$ or $B = 0$.	If $(x + 1)(x + 2) = 0$, then $x + 1 = 0$ or $x + 2 = 0$.
5.7C	Pythagorean Theorem	If the longest side of a right triangle (a triangle with a 90° angle) is of length c and the two other sides are of length a and b, then $a^2 + b^2 = c^2$.	If the length of leg a is 3 inches and the length of leg b is 4 inches, then the length h of the hypotenuse is $3^2 + 4^2 = h^2$ Thus, $9 + 16 = h^2$ $25 = h^2$ $5 = h$
5.7D	Braking distance $b = 0.06v^2$	If a car is moving v miles per hour, b is the distance (in feet) needed to stop the car *after* the brakes are applied.	The braking distance b for a car moving at 20 miles per hour is $b = 0.06(20^2) = 24$ feet.

⟩ Review Exercises **Chapter 5**

(If you need help with these exercises, look in the section indicated in brackets.)

1. ⟨ **5.1A** ⟩ *Find the GCF of:*
   a. 60 and 90

   b. 12 and 18

   c. 27, 80, and 17

3. ⟨ **5.1C** ⟩ *Factor.*
   a. $20x^3 - 55x^5$

   b. $14x^4 - 35x^6$

   c. $16x^7 - 40x^9$

5. ⟨ **5.1D** ⟩ *Factor.*
   a. $3x^3 - 21x^2 - x + 7$

   b. $3x^3 + 18x^2 + x + 6$

   c. $4x^3 - 8x^2y + x - 2y$

2. ⟨ **5.1B** ⟩ *Find the GCF of:*
   a. $24x^7, 18x^5, -30x^{10}$

   b. $18x^8, 12x^9, -20x^{10}$

   c. $x^6y^4, y^7x^5, x^3y^6, x^9$

4. ⟨ **5.1C** ⟩ *Factor.*
   a. $\frac{3}{7}x^6 - \frac{5}{7}x^5 + \frac{2}{7}x^4 - \frac{1}{7}x^2$

   b. $\frac{4}{9}x^7 - \frac{2}{9}x^6 + \frac{2}{9}x^5 - \frac{1}{9}x^3$

   c. $\frac{3}{8}x^9 - \frac{7}{8}x^8 + \frac{3}{8}x^7 - \frac{1}{8}x^5$

6. ⟨ **5.2A** ⟩ *Factor.*
   a. $x^2 + 8x + 7$

   b. $x^2 - 8x - 9$

   c. $x^2 + 6x + 5$

7. ⟨ **5.2A** ⟩ *Factor.*

   a. $x^2 - 7x + 10$

   b. $x^2 - 9x + 14$

   c. $x^2 + 2x - 8$

9. ⟨ **5.3B** ⟩ *Factor.*

   a. $6x^2 - 17xy + 5y^2$

   b. $6x^2 - 7xy + 2y^2$

   c. $6x^2 - 11xy + 4y^2$

11. ⟨ **5.4B** ⟩ *Factor.*

   a. $9x^2 + 12xy + 4y^2$

   b. $9x^2 + 30xy + 25y^2$

   c. $9x^2 + 24xy + 16y^2$

13. ⟨ **5.4B** ⟩ *Factor.*

   a. $4x^2 - 12xy + 9y^2$

   b. $4x^2 - 20xy + 25y^2$

   c. $4x^2 - 28xy + 49y^2$

15. ⟨ **5.4C** ⟩ *Factor.*

   a. $16x^2 - 81y^2$

   b. $25x^2 - 64y^2$

   c. $9x^2 - 100y^2$

17. ⟨ **5.5A** ⟩ *Factor.*

   a. $8y^3 - 27x^3$

   b. $64y^3 - 125x^3$

   c. $8m^3 - 125n^3$

19. ⟨ **5.5B** ⟩ *Factor.*

   a. $2x^3 - 2x^2 - 4x$

   b. $3x^3 - 6x^2 - 9x$

   c. $4x^3 - 12x^2 - 16x$

21. ⟨ **5.5B** ⟩ *Factor.*

   a. $9kx^2 + 12kx + 4k$

   b. $9kx^2 + 30kx + 25k$

   c. $4kx^2 + 20kx + 25k$

23. ⟨ **5.5C** ⟩ *Factor.*

   a. $-x^3 - y^3$

   b. $-8m^3 - 27n^3$

   c. $-64n^3 - m^3$

25. ⟨ **5.5C** ⟩ *Factor.*

   a. $-4x^2 - 12xy + 9y^2$

   b. $-25x^2 - 30xy + 9y^2$

   c. $-16x^2 - 24xy + 9y^2$

8. ⟨ **5.3B** ⟩ *Factor.*

   a. $6x^2 - 6 + 5x$

   b. $6x^2 - 1 + x$

   c. $6x^2 - 5 + 13x$

10. ⟨ **5.4B** ⟩ *Factor.*

   a. $x^2 + 4x + 4$

   b. $x^2 + 10x + 25$

   c. $x^2 + 8x + 16$

12. ⟨ **5.4B** ⟩ *Factor.*

   a. $x^2 - 4x + 4$

   b. $x^2 - 6x + 9$

   c. $x^2 - 12x + 36$

14. ⟨ **5.4C** ⟩ *Factor.*

   a. $x^2 - 36$

   b. $x^2 - 49$

   c. $x^2 - 81$

16. ⟨ **5.5A** ⟩ *Factor.*

   a. $m^3 + 125$

   b. $n^3 + 64$

   c. $y^3 + 8$

18. ⟨ **5.5B** ⟩ *Factor.*

   a. $3x^3 - 6x^2 + 27x$

   b. $3x^3 - 6x^2 + 30x$

   c. $4x^3 - 8x^2 + 32x$

20. ⟨ **5.5B** ⟩ *Factor.*

   a. $2x^3 + 8x^2 + x + 4$

   b. $2x^3 + 10x^2 + x + 5$

   c. $2x^3 + 12x^2 + x + 6$

22. ⟨ **5.5C** ⟩ *Factor.*

   a. $-3x^4 + 27x^2$

   b. $-4x^4 + 64x^2$

   c. $-5x^4 + 20x^2$

24. ⟨ **5.5C** ⟩ *Factor.*

   a. $-y^3 + x^3$

   b. $-8m^3 + 27n^3$

   c. $-64t^3 + 125s^3$

26. ⟨ **5.6A** ⟩ *Solve.*

   a. $x^2 - 4x - 5 = 0$

   b. $x^2 - 5x - 6 = 0$

   c. $x^2 - 6x - 7 = 0$

27. ⟨**5.6A**⟩ *Solve.*

    a. $2x^2 + x = 10$

    b. $2x^2 + 3x = 5$

    c. $2x^2 + x = 3$

29. ⟨**5.7A**⟩ *Find the integers if the product of two consecutive even integers is:*

    a. 4 more than 5 times the smaller of the integers

    b. 4 more than twice the smaller of the integers

    c. 10 more than 11 times the smaller of the integers

28. ⟨**5.6A**⟩ *Solve.*

    a. $(3x + 1)(x - 2) = 2(x - 1) - 4$

    b. $(2x + 1)(x - 4) = 6(x - 4) - 1$

    c. $(2x + 1)(x - 1) = 3(x + 2) - 1$

30. ⟨**5.7C**⟩ *The hypotenuse of a right triangle is 6 inches longer than the longest side of the triangle and 12 inches longer than the remaining side. What are the dimensions of the triangle?*

> Practice Test **Chapter 5**

(Answers on page 465)

Visit www.mhhe.com/bello to view helpful videos that provide step-by-step solutions to several of the problems below.

1. Find the GCF of 40 and 60.

2. Find the GCF of $18x^2y^4$ and $30x^3y^5$.

3. Factor $10x^3 - 35x^5$.

4. Factor $\frac{4}{5}x^6 - \frac{3}{5}x^5 + \frac{2}{5}x^4 - \frac{1}{5}x^2$.

5. Factor $2x^3 + 6x^2y + x + 3y$.

6. Factor $x^2 - 8x + 12$.

7. Factor $6x^2 - 3 + 7x$.

8. Factor $6x^2 - 11xy + 3y^2$.

9. Factor $4x^2 + 12xy + 9y^2$.

10. Factor $x^2 - 14x + 49$.

11. Factor $9x^2 - 12xy + 4y^2$.

12. Factor $x^2 - 100$.

13. Factor $16x^2 - 25y^2$.

14. Factor $125t^3 + 27s^3$.

15. Factor $8y^3 - 125x^3$.

16. Factor $3x^3 - 6x^2 + 24x$.

17. Factor $2x^3 - 8x^2 - 10x$.

18. Factor $2x^3 + 6x^2 + x + 3$.

19. Factor $4kx^2 + 12kx + 9k$.

20. Factor $-9x^4 + 36x^2$.

21. Factor $-9x^2 - 24xy - 16y^2$.

22. Solve $x^2 - 3x - 10 = 0$.

23. Solve $2x^2 - x = 15$.

24. Solve $(2x - 3)(x - 4) = 2(x - 1) - 1$.

25. Solve $y(2y + 7) = -3$.

26. The product of two consecutive odd integers is 13 more than 10 times the larger of the two integers. Find the integers.

27. The product of two consecutive integers is 14 less than 10 times the smaller of the two integers. What are the integers?

28. The area of a rectangle is numerically 44 more than its perimeter. If the length of the rectangle is 8 inches more than its width, what are the dimensions of the rectangle?

29. A rectangular 10-inch television screen (measured diagonally) is 2 inches wider than it is high. What are the dimensions of the screen?

30. A car traveled 24 feet after the brakes were applied. How fast was the car going when the brakes were applied? ($b = 0.06v^2$)

> Answers to Practice Test **Chapter 5**

Answer	If You Missed	Review		
	Question	Section	Examples	Page
1. 20	1	5.1	1	397–398
2. $6x^2y^4$	2	5.1	2	399
3. $5x^3(2 - 7x^2)$	3	5.1	3, 4	400–401
4. $\frac{1}{5}x^2(4x^4 - 3x^3 + 2x^2 - 1)$	4	5.1	5	401
5. $(x + 3y)(2x^2 + 1)$	5	5.1	6–8	402–403
6. $(x - 2)(x - 6)$	6	5.2	1–4	409–410
7. $(3x - 1)(2x + 3)$	7	5.3	2, 3, 5, 6	417–418, 420–421
8. $(3x - y)(2x - 3y)$	8	5.3	4, 7	418, 421
9. $(2x + 3y)^2$	9	5.4	2	427
10. $(x - 7)^2$	10	5.4	3a, b	427
11. $(3x - 2y)^2$	11	5.4	3c	427
12. $(x + 10)(x - 10)$	12	5.4	4a, b	428
13. $(4x + 5y)(4x - 5y)$	13	5.4	4c	428
14. $(5t + 3s)(25t^2 - 15st + 9s^2)$	14	5.5	1a, b	433
15. $(2y - 5x)(4y^2 + 10xy + 25x^2)$	15	5.5	1c, d	434
16. $3x(x^2 - 2x + 8)$	16	5.5	2	435
17. $2x(x + 1)(x - 5)$	17	5.5	2	435
18. $(x + 3)(2x^2 + 1)$	18	5.5	3	435
19. $k(2x + 3)^2$	19	5.5	4	435
20. $-9x^2(x + 2)(x - 2)$	20	5.5	7	437
21. $-(3x + 4y)^2$	21	5.5	7	437
22. $x = 5$ or $x = -2$	22	5.6	1	442
23. $x = 3$ or $x = -\frac{5}{2}$	23	5.6	2	443
24. $x = 5$ or $x = \frac{3}{2}$	24	5.6	3, 4	444
25. $y = -3$ or $y = -\frac{1}{2}$	25	5.6	5	444–445
26. 11 and 13 or -3 and -1	26	5.7	1	450
27. 7 and 8 or 2 and 3	27	5.7	1	450
28. 6 in. by 14 in.	28	5.7	2, 3	451–452
29. 6 in. by 8 in.	29	5.7	4	453–454
30. 20 mi/hr	30	5.7	5	454–455

> Cumulative Review Chapters 1–5

1. Add: $-\frac{3}{8} + \left(-\frac{1}{6}\right)$

2. Subtract: $6.6 - (-9.8)$

3. Multiply: $(-5.5)(5.7)$

4. Find: $-(5^2)$

5. Divide: $-\frac{5}{6} \div \left(-\frac{5}{18}\right)$

6. Evaluate $y \div 5 \cdot x - z$ for $x = 6$, $y = 60$, $z = 3$.

7. Which property is illustrated by the following statement?
$7 \cdot (6 \cdot 4) = (7 \cdot 6) \cdot 4$

8. Combine like terms: $-8xy^4 - (-9xy^4)$

9. Simplify: $2x - (x + 3) - 2(x + 4)$

10. Write in symbols: The quotient of $(d + 5e)$ and f

11. Solve for x: $5 = 3(x - 1) + 5 - 2x$

12. Solve for x: $\frac{x}{7} - \frac{x}{9} = 2$

13. Solve for x: $8 - \frac{x}{3} = \frac{6(x + 1)}{7}$

14. The sum of two numbers is 105. If one of the numbers is 25 more than the other, what are the numbers?

15. Train A leaves a station traveling at 50 miles per hour. Two hours later, train B leaves the same station traveling in the same direction at 60 miles per hour. How long does it take for train B to catch up to train A?

16. Susan purchased some municipal bonds yielding 11% annually and some certificates of deposit yielding 12% annually. If Susan's total investment amounts to $6000 and the annual income is $680, how much money is invested in bonds and how much is invested in certificates of deposit?

17. Graph: $-\frac{x}{6} + \frac{x}{2} \geq \frac{x - 2}{2}$

18. Graph the point $C(3, -3)$.

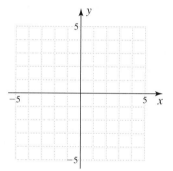

19. Determine whether the ordered pair $(-3, -3)$ is a solution of $5x - y = -18$.

20. Find x in the ordered pair $(x, 3)$ so that the ordered pair satisfies the equation $2x - 3y = -5$.

21. Graph: $x + y = 4$

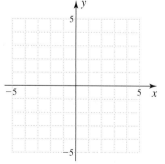

22. Graph: $4y - 20 = 0$

23. Find the slope of the line passing through the points $(7, 7)$ and $(-4, 5)$.

24. What is the slope of the line $6x - 2y = 15$?

25. Find the pair of parallel lines.

    (1) $4y = x - 7$

    (2) $7x - 28y = -7$

    (3) $28y + 7x = -7$

26. Find an equation of the line passsing through $(-5, 3)$ and with slope 6.

27. Find the equation of the line with slope -3 and y intercept 4.

28. Graph $4x - 3y < -12$.

29. Graph $2x - y \geq 0$.

30. Simplify: $\dfrac{20x^6y^3}{-5x^4y^7}$

31. Simplify: $\dfrac{x^{-7}}{x^{-9}}$

32. Multiply and simplify: $x \cdot x^{-5}$

33. Simplify: $(3x^3y^{-3})^{-3}$

34. Write in scientific notation: 0.000048

35. Divide and express the answer in scientific notation: $(5.98 \times 10^{-3}) \div (1.3 \times 10^2)$

36. Find the degree of the polynomial: $6x^2 + x + 2$.

37. Find the value of $x^3 + 2x^2 + 1$ when $x = -2$.

38. Add $(-3x^2 - 2x^3 - 6)$ and $(5x^3 - 7 - 4x^2)$.

39. Simplify: $-4x^4(8x^2 + 4y)$

40. Expand: $(3x + 2y)^2$

41. Multiply: $(5x - 7y)(5x + 7y)$

42. Expand: $\left(2x^2 - \dfrac{1}{5}\right)^2$

43. Multiply: $(5x^2 + 9)(5x^2 - 9)$

44. Divide: $(2x^3 + x^2 - 5x - 9)$ by $(x - 2)$.

45. Factor completely: $12x^6 - 14x^9$

46. Factor completely: $\dfrac{4}{5}x^7 - \dfrac{3}{5}x^6 + \dfrac{4}{5}x^5 - \dfrac{2}{5}x^3$

47. Factor completely: $x^2 - 12x + 27$

48. Factor completely: $20x^2 - 23xy + 6y^2$

49. Factor completely: $25x^2 - 49y^2$

50. Factor completely: $-5x^4 + 80x^2$

51. Factor completely: $3x^3 - 6x^2 - 9x$

52. Factor completely: $2x^2 + 5x + 6x + 15$

53. Factor completely: $9kx^2 + 6kx + k$

54. Solve for x: $4x^2 + 17x = 15$

Section

Chapter

six

▷ **Rational Expressions**

The Human Side of Algebra

The concept of whole number is one of the oldest in mathematics. The concept of rational numbers (so named because they are *ratios* of whole numbers) developed much later because nonliterate tribes had no need for such a concept. Rational numbers evolved over a long period of time, stimulated by the need for certain types of measurement. For example, take a rod of length 1 unit and cut it into two equal pieces. What is the length of each piece? One-half, of course. If the same rod is cut into four equal pieces, then each piece is of length $\frac{1}{4}$. Two of these pieces will have length $\frac{2}{4}$, which tells us that we should have $\frac{2}{4} = \frac{1}{2}$.

It was ideas such as these that led to the development of the arithmetic of the rational numbers.

During the Bronze Age, Egyptian hieroglyphic inscriptions show the reciprocals of integers by using an elongated oval sign. Thus, $\frac{1}{8}$ and $\frac{1}{20}$ were respectively written as

In this chapter, we generalize the concept of a rational number to that of a *rational expression*—that is, the quotient of two polynomials.

6.1   Building and Reducing Rational Expressions

▶ Objectives

A ⟩ Determine the values that make a rational expression undefined.

B ⟩ Build fractions.

C ⟩ Reduce (simplify) a rational expression to lowest terms.

▶ To Succeed, Review How To . . .

1. Factor polynomials (pp. 407–409, 416–421, 425–428, 433–437).
2. Write a fraction with specified denominator (pp. 3–4).
3. Simplify fractions (pp. 4–6).

▶ Getting Started

Recycling Waste and Rational Expressions

We've already mentioned that algebra is generalized arithmetic. In arithmetic we study the natural numbers, the whole numbers, the integers, and the rational numbers. In algebra we've studied expressions and polynomials, and we have also discussed how they follow rules similar to those used with the real numbers. We will find that rational expressions in algebra follow the same rules as rational numbers in arithmetic.

Waste Generated

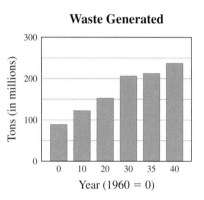

Year (1960 = 0)

In arithmetic, a **rational number** is a number that can be written in the form $\frac{a}{b}$, where a and b are integers and b is not zero. As usual, a is called the *numerator* and b, the *denominator*. A similar approach is used in algebra.

In algebra, an expression of the form $\frac{A}{B}$, where A and B are *polynomials* and B is not zero, is called an **algebraic fraction,** or a **rational expression.** So if $G(t) = -0.001t^3 + 0.06t^2 + 2.6t + 88.6$ is a polynomial representing the amount of waste generated in the United States (in millions of tons), $R(t) = 0.06t^2 - 0.59t + 6.4$ is a polynomial representing the amount of waste recovered (in millions of tons), and t is the number of years after 1960, then

Waste Recovered

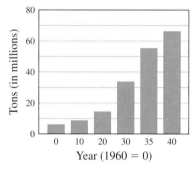

Year (1960 = 0)

$$\frac{R(t)}{G(t)}$$

is a *rational expression* representing the *fraction* of the waste recovered in those years. Thus, in 1960 (when $t = 0$), the fraction is

$$\frac{R(0)}{G(0)} = \frac{0.06(0)^2 - 0.59(0) + 6.4}{-0.001(0)^3 + 0.06(0)^2 + 2.6(0) + 88.6} = \frac{6.4}{88.6}$$

which is about 7%. What percent of the waste will be recovered in the year 2010 when $t = 2010 - 1960$?

In this section we shall learn how to determine when rational expressions are undefined and how to write them with a given denominator and then simplify them.

A ⟩ Undefined Values of Rational Expressions

DEFINITION OF RATIONAL EXPRESSION	An **algebraic fraction** or a **rational expression** is an expression of the form $\frac{A}{B}$ where A and B are *polynomials* and $B \neq 0$.

The expressions

$$\frac{8}{x}, \qquad \frac{x^2 + 2x + 3}{x + 5}, \qquad \frac{y}{y - 1}, \qquad \text{and} \qquad \frac{x^2 + 3x + 9}{x^2 - 4x + 4}$$

are rational expressions. Of course, since we don't want the denominators of these expressions to be zero, we must place some restrictions on these denominators. Let's see what these restrictions must be.

For $\frac{8}{x}$, x cannot be 0 ($x \neq 0$) because then we would have $\frac{8}{0}$, which is not defined. For

$$\frac{x^2 + 2x + 3}{x + 5}, \quad x \neq -5 \qquad \text{If } x \text{ is } -5, \frac{x^2 + 2x + 3}{x + 5} = \frac{25 - 10 + 3}{0},$$
which is undefined.

For

$$\frac{y}{y - 1}, \quad y \neq 1 \qquad \text{If } y \text{ is } 1, \frac{y}{y - 1} = \frac{1}{0}, \text{ which is undefined.}$$

and for

$$\frac{x^2 + 3x + 9}{x^2 - 4x + 4} = \frac{x^2 + 3x + 9}{(x - 2)^2}, \quad x \neq 2 \qquad \text{If } x \text{ is } 2, \frac{x^2 + 3x + 9}{x^2 - 4x + 4} = \frac{4 + 6 + 9}{4 - 8 + 4} = \frac{19}{0}$$
—again, undefined.

To avoid stating repeatedly that the denominators of rational expressions must not be zero, we make the following rule.

> **RULE**
>
> **Avoiding Zero Denominators**
> The variables in a rational expression must not be replaced by numbers that make the denominator zero.

To find the value or values that make the denominator zero: (1) set the denominator equal to zero, (2) solve the resulting equation for the variable.

For example, we found that

$$\frac{x^2 + 2x + 3}{x + 5}$$

is undefined for $x = -5$ by letting $x + 5 = 0$ and solving for x, to obtain $x = -5$. Similarly, we can find the values that make

$$\frac{1}{x^2 + 3x - 4}$$

undefined by letting

$$x^2 + 3x - 4 = 0$$
$$(x + 4)(x - 1) = 0 \qquad \text{Factor.}$$
$$x + 4 = 0 \quad \text{or} \quad x - 1 = 0 \qquad \text{Solve each equation.}$$
$$x = -4 \qquad\qquad x = 1$$

Thus,

$$\frac{1}{x^2 + 3x - 4}$$

is undefined when $x = -4$ or $x = 1$. (Check this out!)

EXAMPLE 1   Finding values that make rational expressions undefined

Find the values for which the rational expression is undefined:

a. $\dfrac{n}{2n - 3}$ **b.** $\dfrac{x + 1}{x^2 + 5x - 6}$ **c.** $\dfrac{m + 2}{m^2 + 2}$

SOLUTION

a. We set the denominator equal to zero (that makes the expression undefined) and solve:

$$2n - 3 = 0$$
$$2n = 3 \quad \text{Add 3.}$$
$$n = \frac{3}{2} \quad \text{Divide by 2.}$$

Thus,

$$\frac{n}{2n - 3}$$

is undefined for $n = \frac{3}{2}$.

b. We proceed as before and set the denominator equal to zero:

$$x^2 + 5x - 6 = 0$$
$$(x + 6)(x - 1) = 0 \quad \text{Factor.}$$
$$x + 6 = 0 \quad \text{or} \quad x - 1 = 0 \quad \text{Solve each equation.}$$
$$x = -6 \qquad\qquad x = 1$$

Thus,

$$\frac{x + 1}{x^2 + 5x - 6}$$

is undefined for $x = -6$ or $x = 1$.

c. Setting the denominator equal to zero, we have

$$m^2 + 2 = 0$$

so that $m^2 = -2$. But the square of any real number is not negative, so there are no real numbers m for which the denominator is zero. (Note that m^2 is greater than or equal to zero, so $m^2 + 2$ is always positive.) There are *no values* for which

$$\frac{m + 2}{m^2 + 2}$$

is undefined.

PROBLEM 1

Find values for which the rational expression is undefined:

a. $\dfrac{m}{2m + 3}$ **b.** $\dfrac{y + 1}{y^2 + 4y - 5}$

c. $\dfrac{n + 3}{n^2 + 3}$

B › Building Fractions

Now that we know that we must avoid zero denominators, recall what we did with fractions in arithmetic. First, we learned how to recognize which ones are equal, and then we used this idea to reduce or build fractions. Let's talk about equality first.

Answers to PROBLEMS

1. a. $m = -\dfrac{3}{2}$ **b.** $y = -5$ or
$y = 1$ **c.** No values

What does this picture tell you (if you write the value of the coins using fractions)? It states that

$$\frac{1}{2} = \frac{2}{4}$$

One half-dollar   equals   two quarters

As a matter of fact, the following is also true:

$$\frac{1}{2} = \frac{2}{4} = \frac{3}{6} = \frac{4}{8} = \frac{x}{2x} = \frac{x^2}{2x^2}$$

and so on. Do you see the pattern? Here is another way to write it:

$$\frac{1}{2} = \frac{1 \cdot 2}{2 \cdot 2} = \frac{2}{4} \qquad \text{Note that } \frac{2}{2} = 1.$$

$$\frac{1}{2} = \frac{1 \cdot 3}{2 \cdot 3} = \frac{3}{6} \qquad \text{Here, } \frac{3}{3} = 1.$$

$$\frac{1}{2} = \frac{1 \cdot 4}{2 \cdot 4} = \frac{4}{8}$$

$$\frac{1}{2} = \frac{1 \cdot x}{2 \cdot x} = \frac{x}{2x}$$

$$\frac{1}{2} = \frac{1 \cdot x^2}{2 \cdot x^2} = \frac{x^2}{2x^2}$$

We can always obtain a rational expression that is equivalent to another rational expression $\frac{A}{B}$ by multiplying the numerator and denominator of the given rational expression by the same nonzero number or expression, C. Here is this rule stated in symbols.

RULE

Fundamental Rule of Rational Expressions

$$\frac{A}{B} = \frac{A \cdot C}{B \cdot C} \qquad (B \neq 0, \, C \neq 0)$$

Note that

$$\frac{A}{B} = \frac{A \cdot C}{B \cdot C} \qquad \text{because} \qquad \frac{C}{C} = \mathbf{1}$$

and multiplying by **1** does not change the value of the expression.

This idea is very important for adding or subtracting rational expressions (Section 6.3). For example, to add $\frac{1}{2}$ and $\frac{3}{4}$, we write

$$\frac{1}{2} = \frac{2}{4} \qquad \text{Note that } \frac{1}{2} = \frac{1 \cdot 2}{2 \cdot 2} = \frac{2}{4}.$$

$$\frac{+\frac{3}{4} = \frac{3}{4}}{\frac{5}{4}}$$

Here we've used the fundamental rule of rational expressions to write the $\frac{1}{2}$ as an equivalent fraction with a denominator of 4, namely, $\frac{2}{4}$. Now, suppose we wish to write $\frac{3}{8}$ with a denominator of 16. How do we do this? First, we write the problem as

$$\frac{3}{8} = \frac{?}{16} \qquad \text{Note that } 16 = 8 \cdot 2.$$

Multiply by 2.

and notice that to get 16 as the denominator, we need to multiply the 8 by 2. Of course, we must do the same to the numerator 3; we obtain

Multiply by 2.

By the fundamental rule of rational expressions, if we multiply the denominator by 2, we must multiply the numerator by 2.

$$\frac{3}{8} = \frac{6}{16}$$ Note that $\frac{3 \cdot 2}{8 \cdot 2} = \frac{6}{16}$.

Similarly, to write

$$\frac{5x}{3y}$$

with a denominator of $6y^3$, we first write the new equivalent expression

$$\frac{?}{6y^3}$$

with the old denominator $3y$ factored out:

$$\frac{5x}{3y} = \frac{?}{6y^3} = \frac{?}{3y(2y^2)}$$ Write $6y^3$ as $3y(2y^2)$.

Multiply by $2y^2$.

Since the multiplier is $2y^2$, we have

Multiply by $2y^2$.

$$\frac{5x}{3y} = \frac{5x(2y^2)}{3y(2y^2)} = \frac{10xy^2}{6y^3}$$

Thus,

$$\frac{5x}{3y} = \frac{10xy^2}{6y^3}$$ We multiplied the denominator by $2y^2$, so we have to multiply the numerator by $2y^2$.

EXAMPLE 2   Rewriting rational expressions with a specified denominator

Write:

a. $\frac{5}{6}$ with a denominator of 18   **b.** $\frac{2x}{9y^2}$ with a denominator of $18y^3$

c. $\frac{3x}{x-1}$ with a denominator of $x^2 + 2x - 3$

SOLUTION

a. $\frac{5}{6} = \frac{?}{18}$

Multiply by 3.
Multiply by 3.

$$\frac{5}{6} = \frac{15}{18}$$ Note that $\frac{5 \cdot 3}{6 \cdot 3} = \frac{15}{18}$.

PROBLEM 2

Write:

a. $\frac{4}{7}$ with a denominator of 14

b. $\frac{2y}{7x^2}$ with a denominator of $14x^3$

c. $\frac{2x}{x+2}$ with a denominator of $x^2 + x - 2$

---

b. Since $18y^3 = 9y^2(2y)$,

$$\frac{2x}{9y^2} = \frac{?}{9y^2(2y)}$$

Multiply by $2y$.

Multiply by $2y$.

$$\frac{2x}{9y^2} = \frac{2x(2y)}{9y^2(2y)} = \frac{4xy}{18y^3}$$

c. We first note that $x^2 + 2x - 3 = (x - 1)(x + 3)$. Thus,

$$\frac{3x}{x - 1} = \frac{?}{(x - 1)(x + 3)}$$

Multiply by $(x + 3)$.

Multiply by $(x + 3)$.

$$\frac{3x}{x - 1} = \frac{3x(x + 3)}{(x - 1)(x + 3)} = \frac{3x^2 + 9x}{x^2 + 2x - 3}$$

C ⟩ Reducing Rational Expressions

Now that we know how to build up rational expressions, we are now ready to use the reverse process—that is, to **reduce** them. In Example 2(a), we wrote $\frac{5}{6}$ with a denominator of 18; that is, we found out that

$$\frac{5}{6} = \frac{5 \cdot 3}{6 \cdot 3} = \frac{15}{18}$$

Of course, you will probably agree that $\frac{5}{6}$ is written in a "simpler" form than $\frac{15}{18}$. Certainly you will eventually agree—though it is hard to see at first glance—that

$$\frac{5}{3} = \frac{5(x + 3)(x^2 - 4)}{3(x + 2)(x^2 + x - 6)}$$

and that $\frac{5}{3}$ is the "simpler" of the two expressions. In algebra, the process of removing all factors common to the numerator and denominator is called **reducing to lowest terms**, or **simplifying** the expression. How do we reduce the rational expression

$$\frac{5(x + 3)(x^2 - 4)}{3(x + 2)(x^2 + x - 6)}$$

to lowest terms? We proceed by steps.

PROCEDURE

Reducing Rational Expressions to Lowest Terms (Simplifying)

1. Write the numerator and denominator of the expression in factored form.
2. Find the factors that are common to the numerator and denominator.
3. Replace the quotient of the common factors by the number 1, since $\frac{a}{a} = 1$.
4. Rewrite the expression in simplified form.

Let's use this procedure to reduce

$$\frac{5(x + 3)(x^2 - 4)}{3(x + 2)(x^2 + x - 6)}$$

to lowest terms:

1. Write the numerator and denominator in factored form.

$$\frac{5(x + 3)(x + 2)(x - 2)}{3(x + 2)(x + 3)(x - 2)}$$

2. Find the factors that are common to the numerator and denominator (we rearranged them so the common factors are in columns).

$$\frac{5(x + 2)(x + 3)(x - 2)}{3(x + 2)(x + 3)(x - 2)}$$

3. Replace the quotient of the common factors by the number 1. Remember, $\frac{a}{a} = 1$.

$$\frac{5\overset{1}{(x + 2)}\overset{1}{(x + 3)}\overset{1}{(x - 2)}}{3\underset{}{(x + 2)(x + 3)(x - 2)}}$$

4. Rewrite the expression in simplified form.

$$\frac{5}{3}$$

The whole procedure can be written as

$$\frac{5(x + 3)(x^2 - 4)}{3(x + 2)(x^2 + x - 6)} = \frac{5\overset{1}{(x + 3)}\overset{1}{(x + 2)}\overset{1}{(x - 2)}}{3(x + 2)(x + 3)(x - 2)} = \frac{5}{3}$$

EXAMPLE 3   Reducing rational expressions to lowest terms

Simplify:

a. $\dfrac{-5x^2y}{15xy^3}$

b. $\dfrac{-8(x^2 - y^2)}{-4(x + y)}$

SOLUTION

a. We use our four-step procedure.

1. Write in factored form.

$$\frac{-5x^2y}{15xy^3} = \frac{(-1) \cdot 5 \cdot x \cdot x \cdot y}{3 \cdot 5 \cdot x \cdot y \cdot y \cdot y}$$

2. Find the common factors.

$$= \frac{5 \cdot x \cdot y(-1) \cdot x}{5 \cdot x \cdot y \cdot 3 \cdot y \cdot y}$$

3. Replace the quotient of the common factors by 1. Remember, $\frac{a}{a} = 1$.

$$= \frac{1 \cdot (-1) \cdot x}{3 \cdot y \cdot y}$$

4. Rewrite in simplified form.

$$= \frac{-x}{3y^2}$$

The whole process can be written as

$$\frac{-5x^2y}{15xy^3} = -\frac{\overset{-1x}{\cancel{5x^2y}}}{\underset{3\ y^2}{\cancel{15xy^3}}} = \frac{-x}{3y^2}$$

b. We again use our four-step procedure.

1. Write the fractions in factored form.

$$\frac{-8(x^2 - y^2)}{-4(x + y)} = \frac{(-1) \cdot 2 \cdot 2 \cdot 2(x + y)(x - y)}{(-1) \cdot 2 \cdot 2(x + y)}$$

2. Find the common factors.

$$= \frac{(-1) \cdot 2 \cdot 2 \cdot (x + y)(x - y) \cdot 2}{(-1) \cdot 2 \cdot 2 \cdot (x + y)}$$

PROBLEM 3

Simplify:

a. $\dfrac{-3x^2y}{6xy^4}$

b. $\dfrac{-6(m^2 - n^2)}{-3(m - n)}$

---

3. Replace the quotient of $= 1 \cdot (x - y) \cdot 2$
the common factors by 1.

4. Rewrite in simplified form. $= 2(x - y)$

The abbreviated form is

$$\frac{-8(x^2 - y^2)}{-4(x + y)} = \frac{-\overset{2}{\cancel{8}}(\cancel{x + y})(x - y)}{-\underset{1}{\cancel{4}}(\cancel{x + y})} = 2(x - y)$$

You may have noticed that in Example 3(a) we wrote the answer as

$$\frac{-x}{3y^2}$$

It could be argued that since

$$\frac{-5}{15} = -\frac{1}{3}$$

the answer should be

$$-\frac{x}{3y^2}$$

However, to avoid confusion, we agree to write

$$-\frac{x}{3y^2} \quad \text{as} \quad \frac{-x}{3y^2}$$

with the *negative* sign in the numerator. In general, since a fraction has **three** possible signs (numerator, denominator, and the sign of the fraction itself), we use the following conventions.

FORMS OF A FRACTION

$-\dfrac{a}{b}$ is written as $\dfrac{-a}{b}$ $\dfrac{-a}{-b}$ is written as $\dfrac{a}{b}$

$\dfrac{a}{-b}$ is written as $\dfrac{-a}{b}$ $-\dfrac{-a}{b}$ is written as $\dfrac{a}{b}$

$-\dfrac{-a}{-b}$ is written as $\dfrac{-a}{b}$ $-\dfrac{a}{-b}$ is written as $\dfrac{a}{b}$

STANDARD FORM OF A FRACTION

The forms

$$\frac{a}{b} \quad \text{and} \quad \frac{-a}{b}$$

are called the **standard forms** of a fraction.

The forms $\frac{a}{b}$ and $\frac{-a}{b}$ are the preferred forms to write answers involving fractions, as we shall see in Example 4.

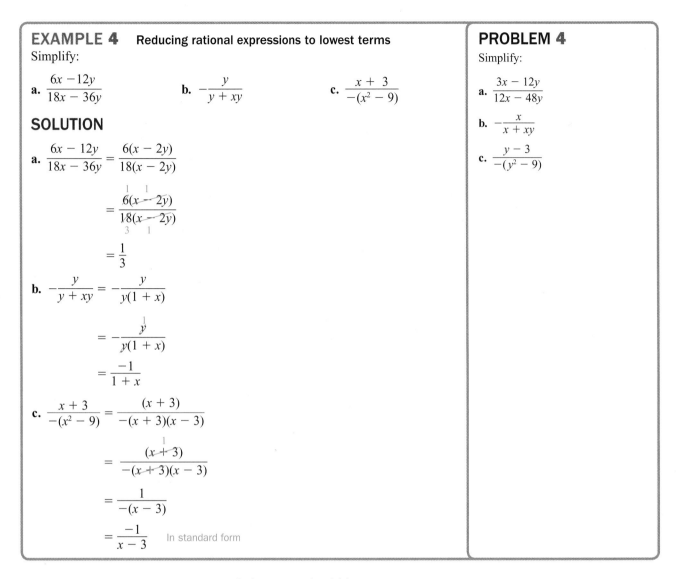

EXAMPLE 4   Reducing rational expressions to lowest terms

Simplify:

a. $\dfrac{6x - 12y}{18x - 36y}$ **b.** $-\dfrac{y}{y + xy}$ **c.** $\dfrac{x + 3}{-(x^2 - 9)}$

SOLUTION

a. $\dfrac{6x - 12y}{18x - 36y} = \dfrac{6(x - 2y)}{18(x - 2y)}$

$= \dfrac{\overset{1}{6}(x - 2y)}{\underset{3}{18}(x - 2y)}$

$= \dfrac{1}{3}$

b. $-\dfrac{y}{y + xy} = -\dfrac{y}{y(1 + x)}$

$= -\dfrac{\overset{1}{y}}{y(1 + x)}$

$= \dfrac{-1}{1 + x}$

c. $\dfrac{x + 3}{-(x^2 - 9)} = \dfrac{(x + 3)}{-(x + 3)(x - 3)}$

$= \dfrac{\overset{1}{(x + 3)}}{-(x + 3)(x - 3)}$

$= \dfrac{1}{-(x - 3)}$

$= \dfrac{-1}{x - 3}$ In standard form

PROBLEM 4

Simplify:

a. $\dfrac{3x - 12y}{12x - 48y}$

b. $-\dfrac{x}{x + xy}$

c. $\dfrac{y - 3}{-(y^2 - 9)}$

Is there a way in which

$$\frac{-1}{x - 3}$$

can be simplified further? The answer is yes. See whether you can see the reasons behind each step:

$$\frac{-1}{x - 3} = \frac{-1}{-(3 - x)} \qquad -(3 - x) = -3 + x = x - 3$$

$$= \frac{1}{3 - x}$$

Thus,

$$\frac{-1}{x - 3} = \frac{1}{3 - x}$$

Note that $\frac{-1}{x-3}$ involves finding the additive inverse of 1 in the numerator and subtracting 3 from x in the denominator, whereas $\frac{1}{3-x}$ only involves subtracting x from 3 in the denominator. Less work!

And why, you might ask, is this answer simpler than the other? Because

$$\frac{-1}{x - 3}$$

is a quotient involving a subtraction ($x - 3$) and an additive inverse (-1), but

$$\frac{1}{3 - x}$$

is a quotient that involves only a subtraction ($3 - x$). Be aware of these simplifications when writing your answers!

Here is a situation that occurs very frequently in algebra. Can you reduce this expression?

$$\frac{a - b}{b - a}$$

Look at the following steps. Note that the original denominator, $b - a$, can be written as $-(a - b)$ since $-(a - b) = -a + b = b - a$. Thus, $b - a = -(a - b)$.

$$\frac{a - b}{b - a} = \frac{\overset{1}{\cancel{(a - b)}}}{-\cancel{(a - b)}} \qquad \text{Write } b - a \text{ as } -(a - b) \text{ in the denominator.}$$

$$= \frac{1}{-1}$$

$$= -1$$

QUOTIENT OF ADDITIVE INVERSES	For any real numbers a and b, where $b - a \neq 0$, $$\frac{a - b}{b - a} = -1$$

This means that a fraction whose numerator and denominator are additive inverses (such as $a - b$ and $b - a$) equals -1. We use this idea in Example 5.

EXAMPLE 5   Reducing rational expressions to lowest terms

Reduce to lowest terms:

a. $\dfrac{x^2 - 16}{4 + x}$ b. $-\dfrac{x^2 - 16}{x + 4}$ c. $\dfrac{x^2 + 2x - 15}{3 - x}$

SOLUTION

a. $\dfrac{x^2 - 16}{4 + x} = \dfrac{(x + 4)(x - 4)}{\cancel{x + 4}}$ $4 + x = x + 4$ by the commutative property.

$= x - 4$

b. $-\dfrac{x^2 - 16}{x + 4} = -\dfrac{(x + 4)(x - 4)}{(x + 4)}$

$= -\dfrac{\overset{1}{\cancel{(x + 4)}}(x - 4)}{\cancel{(x + 4)}}$ Note that $\frac{x + 4}{x + 4} = 1$.

$= -(x - 4)$

$= -x + 4$

$= 4 - x$

c. $\dfrac{x^2 + 2x - 15}{3 - x} = \dfrac{(x + 5)(x - 3)}{3 - x}$

$= \dfrac{\overset{-1}{\cancel{(x - 3)}}(x + 5)}{\cancel{(3 - x)}}$ Note that $\frac{x - 3}{3 - x} = -1$.

$= -(x + 5)$

PROBLEM 5

Reduce to lowest terms:

a. $\dfrac{y^2 - 9}{3 + y}$

b. $\dfrac{y^2 - 9}{y + 3}$

c. $\dfrac{y^2 + y - 12}{3 - y}$

(continued)

Answers to PROBLEMS

5. a. $y - 3$ b. $y - 3$

c. $-(y + 4)$

We have used the fact that

$$\frac{a - b}{b - a} = -1$$

in the second step to simplify

$$\frac{(x - 3)}{(3 - x)}$$

Note that the final answer is written as $-(x + 5)$ instead of $-x - 5$, since $-(x + 5)$ is considered to be simpler.

Calculator Corner

Checking Quotients

How can you check that your answers are correct? If you have a calculator, you can do it two ways:

1. *By looking at the graph*   If the graph of the original problem and the graph of the answer coincide, the answer is probably correct.
   Thus, to check the results of Example 5(c), enter

   $$Y_1 = \frac{(x^2 + 2x - 15)}{(3 - x)} \quad \text{and} \quad Y_2 = -(x + 5)$$

   Note the parentheses in the numerator and denominator of Y_1, and remember that the "−" in $-(x + 5)$ has to be entered with the (−) key.
2. *By looking at the numerical values of Y_1 and Y_2* (Press 2nd GRAPH.) The graph and the numerical values are as shown.
   One more thing: When you are looking at the table of values, what happens when $x = 3$? Why?

X	Y1	Y2
0	-5	-5
1	-6	-6
2	-7	-7
3	ERROR	-8
4	-9	-9
5	-10	-10
6	-11	-11
X=3		

MathZone

› Exercises **6.1**

‹ **A** › **Undefined Values of Rational Expressions**   In Problems 1–20, find the value(s) for which the rational expression is undefined.

1. $\dfrac{x}{x - 7}$

2. $\dfrac{3y}{2y + 9}$

3. $\dfrac{y - 5}{2y + 8}$

4. $\dfrac{y + 4}{3y - 9}$

5. $\dfrac{x + 9}{x^2 - 9}$

6. $\dfrac{y + 4}{y^2 - 16}$

7. $\dfrac{y + 3}{y^2 + 5}$

8. $\dfrac{y + 4}{y^2 + 1}$

9. $\dfrac{2y + 9}{y^2 - 6y + 8}$

10. $\dfrac{3x + 5}{x^2 - 7x + 6}$

11. $\dfrac{x^2 - 4}{x^3 + 8}$

12. $\dfrac{y^2 - 9}{y^3 + 1}$

13. $\dfrac{x^2 + 4}{x^3 - 27}$

14. $\dfrac{x^2 + 16}{x^3 - 8}$

15. $\dfrac{x - 1}{x^2 + 6x + 9}$

16. $\dfrac{x-4}{x^2+8x+16}$

17. $\dfrac{y-3}{y^2-6y-16}$

18. $\dfrac{y-5}{y^2-10y+25}$

19. $\dfrac{x+4}{x^3+2x^2+x}$

20. $\dfrac{x+5}{x^3+4x^2+4x}$

‹ **B** › **Building Fractions**   In Problems 21–32, write the given fraction as an equivalent one with the indicated denominator.

21. $\dfrac{3}{7}$ with a denominator of 21

22. $\dfrac{5}{9}$ with a denominator of 36

23. $-\dfrac{8}{11}$ with a denominator of 22

24. $\dfrac{-5}{17}$ with a denominator of 51

25. $\dfrac{5x}{6y^2}$ with a denominator of $24y^3$

26. $\dfrac{7y}{5x^3}$ with a denominator of $10x^4$

27. $\dfrac{-3x}{7y}$ with a denominator of $21y^4$

28. $\dfrac{-4y}{7x^2}$ with a denominator of $28x^3$

29. $\dfrac{4x}{x+1}$ with a denominator of x^2-x-2

30. $\dfrac{5y}{y-1}$ with a denominator of y^2+2y-3

31. $\dfrac{-5x}{x+3}$ with a denominator of x^2+x-6

32. $\dfrac{-3y}{y-4}$ with a denominator of y^2-2y-8

‹ **C** › **Reducing Rational Expressions**   In Problems 33–70, reduce to lowest terms (simplify).

33. $\dfrac{7x^3y}{14xy^4}$

34. $\dfrac{24xy^3}{6x^3y}$

35. $\dfrac{-9xy^5}{3x^2y}$

36. $\dfrac{-24x^3y^2}{48xy^4}$

37. $\dfrac{-6x^2y}{-12x^3y^4}$

38. $\dfrac{-9xy^4}{-18x^5y}$

39. $\dfrac{-25x^3y^2}{-5x^2y^4}$

40. $\dfrac{-30x^2y^2}{-6x^3y^5}$

41. $\dfrac{6(x^2-y^2)}{18(x+y)}$

42. $\dfrac{12(x^2-y^2)}{48(x+y)}$

43. $\dfrac{-9(x^2-y^2)}{3(x+y)}$

44. $\dfrac{-12(x^2-y^2)}{3(x-y)}$

45. $\dfrac{-6(x+y)}{24(x^2-y^2)}$

46. $\dfrac{-8(x+3)}{40(x^2-9)}$

47. $\dfrac{-5(x-2)}{-10(x^2-4)}$

48. $\dfrac{-12(x-2)}{-60(x^2-4)}$

49. $\dfrac{-3(x-y)}{-3(x^2-y^2)}$

50. $\dfrac{-10(x+y)}{-10(x^2-y^2)}$

51. $\dfrac{4x-4y}{8x-8y}$

52. $\dfrac{6x+6y}{2x+2y}$

53. $\dfrac{4x-8y}{12x-24y}$

54. $\dfrac{15y-45x}{5y-15x}$

55. $-\dfrac{6}{6+12y}$

56. $-\dfrac{4}{8+12x}$

57. $-\dfrac{x}{x+2xy}$

58. $-\dfrac{y}{2y+6xy}$

59. $-\dfrac{6y}{6xy+12y}$

60. $-\dfrac{4x}{8xy+16x}$

61. $\dfrac{3x-2y}{2y-3x}$

62. $\dfrac{5y-2x}{2x-5y}$

63. $\dfrac{x^2+4x-5}{1-x}$

64. $\dfrac{x^2-2x-15}{5-x}$

65. $\dfrac{x^2-6x+8}{4-x}$

66. $\dfrac{x^2-8x+15}{3-x}$

67. $\dfrac{2-x}{x^2+4x-12}$

68. $\dfrac{3-x}{x^2+3x-18}$

69. $-\dfrac{3-x}{x^2-5x+6}$

70. $-\dfrac{4-x}{x^2-3x-4}$

$\rangle\rangle\rangle$ *Applications*

71. *Annual advertising expenditures*   How much is spent annually on advertising? The total amount is given by the polynomial $S(t) = -0.3t^2 + 10t + 50$ (millions), where t is the number of years after 1980. The amounts spent on national and local advertisement (in millions) are given by the respective polynomials

$$N(t) = -0.13t^2 + 5t + 30$$

and

$$L(t) = -0.17t^2 + 5t + 20$$

a. What was the total amount spent on advertising in 1980 ($t = 0$)? In 2000? In 2010?

b. What amount was spent on national advertising in 1980? In 2000? In 2010?

c. What amount was spent on local advertising in 1980? In 2000? In 2010?

d. What does the rational expression

$$\frac{N(t)}{S(t)}$$

represent?

73. *Expenditures for television advertising*   The amount spent annually on television advertising can be approximated by the polynomial $T(t) = -0.07t^2 + 2t + 11$ (millions), where t is the number of years after 1980. Use the information in Problem 71 to find what percent of the total amount spent annually on advertising would be

a. Spent on television in the year 2010.

b. Spent on local advertising in the year 2010.

c. What does the rational fraction

$$\frac{T(t)}{S(t)}$$

represent?

72. *Annual advertising expenditures*   Use the information given in Problem 71 to answer these questions:

a. What percent of the total amount spent in 1980 was for national advertising? In 2000?

b. What percent of the total amount spent in 1980 was for local advertising? In 2000?

c. What rational expression represents the percent spent for national advertising?

d. What percent of all advertising would be spent on national advertising in the year 2010?

74. *Television spot advertising*   The estimated fees for spot advertising on television can be approximated by the polynomial $C(t) = 0.04t^3 - t^2 + 6t + 2.5$ (billions), where t is the number of years after 1980. Of this amount, automotive spot advertising fees can be approximated by the polynomial $A(t) = 0.02t^3 - 0.5t^2 + 3t + 0.3$ (billions).

a. What was the total amount spent on TV spot advertising in 1980?

b. What was the amount spent on automotive TV spot advertising in 1980?

c. What was the total amount spent on TV spot advertising in 1990? In 2000? What will it be in 2010?

d. What was the amount spent on automotive TV spot advertising in 1990? In 2000? What will it be in 2010?

e. What does the rational fraction

$$\frac{A(t)}{C(t)}$$

represent?

Use the table for Problems 75–77. The table shows the Local Online Ad Spending (in millions) for New York, Los Angeles, and Chicago.

Projected U.S. Local Online Ad Spending, 2003–2005						
Spending Rank		**DMA**	**Local Online Ad Spending ($ millions)**			**Change (%)**
2004	2005		2003	2004	2005	2004–2005
1	1	New York City	$137.15	$184.09	$240.49	30.6
2	2	Los Angeles	$ 85.90	$106.67	$141.32	32.5
3	3	Chicago	$ 59.29	$ 76.38	$100.58	31.7
		Total	$282.34	$367.14	$582.39	

Source: http://www.clickz.com.

75. *New York online ad spending* The polynomial $N(t) = 0.83t^3 - 5.5t^2 + 55t$ (millions), where t is the number of years after 2000, represents the estimated local online ad spending for New York.

  a. What is the estimated spending (in millions) in 2003, 2004, and 2005?

  b. Compare the estimated spending values with the New York values for 2003, 2004, and 2005 shown in the table.

  c. Now, we are ready to do some more predictions. Based on $N(t)$, what do you predict will be the local online ad spending for New York in 2010?

  d. The total amount spent on local online ads can be approximated by $T(t) = 13.45t^3 - 96.4t^2 + 262t$, so we can find a rational expression representing the **fraction** of the local online ad spending for New York. Find that rational expression.

  e. Use the rational expression found in part **d** to predict what percent of the total local online ads will be spent by New York in 2010.

76. *L.A. online ad spending* The polynomial $L(t) = 1.7t^3 - 13.7t^2 + 54t$ (millions), where t is the number of years after 2000, represents the estimated local online ad spending for Los Angeles.

  a. What is the estimated spending (in millions) in 2003, 2004, and 2005?

  b. Compare the estimated spending values with the Los Angeles values for 2003, 2004, and 2005 shown in the table.

  c. Now, we are ready to do some more predictions. Based on $L(t)$, what do you predict is the local online ad spending for Los Angeles in 2010?

  d. The total amount spent on local online ads can be approximated by $T(t) = 13.45t^3 - 96.4t^2 + 262t$ so we can find a rational expression representing the **fraction** of the local online ad spending for Los Angeles. Find that rational expression.

  e. Use the rational expression found in part **d** to predict what percent of the total local online ads will be spent by Los Angeles in 2010.

77. *Chicago online ad spending* The polynomial $C(t) = 8.5t^3 - 68t^2 + 148t$ (millions), where t is the number of years after 2000, represents the estimated local online ad spending for Chicago.

  a. What is the estimated spending (in millions) in 2003, 2004, and 2005?

  b. Compare the estimated spending values with the Chicago values for 2003, 2004, and 2005 shown in the table.

  c. Based on $C(t)$ what do you predict will be the local online ad spending for Chicago in 2010?

  d. The total amount spent on local online ads can be approximated by $T(t) = 13.45t^3 - 96.4t^2 + 262t$ so we can find a rational expression representing the **fraction** of the local online ad spending for Chicago. Find that rational expression.

  e. Use the rational expression found in part **d** to predict what percent of the total local online ads will be spent by Chicago in 2010.

U.S. population As of October 2006 the U.S. population reached 300 million. What happens after that? We can project the U.S. population by using the polynomial $P(t) = 2.8t + 281$ (million), where t is the number of years after 2000. The Census Bureau uses the projections shown in the graph.

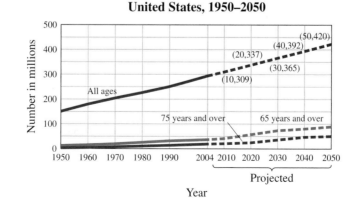

**Total Population and Older Population:
United States, 1950–2050**

78. Use $P(t) = 2.8t + 281$ (million), where t is the number of years after 2000 to find the projected population in the year:

   a. 2010   **b.** 2020

   c. 2030   **d.** 2040

   e. 2050

   f. Are the results close to those shown on the graph?

80. The polynomial $T(t) = 0.13t + 20$ (million), where t is the number of years after 2000, projects the population of persons 20–24 (T).

   a. Write the rational expression $\frac{T(t)}{P(t)}$.

   b. Use the rational expression obtained in part **a** to project, to the nearest percent, the percent of the population 20–24 in the year 2000 and in the year 2050.

   c. Is the 20–24 population increasing or decreasing?

79. The polynomial $S(t) = 1.1t + 33$ (million), where t is the number of years after 2000, projects the population over age 65 (65+).

   a. Write the rational expression $\frac{S(t)}{P(t)}$.

   b. Use the rational expression obtained in part **a** to project, to the nearest percent, the percent of the 65+ population in the year 2000 and in the year 2050.

   c. Is the 65+ population increasing or decreasing?

〉〉〉 *Using Your Knowledge*

Ratios   There is an important relationship between fractions and *ratios*. In general, a **ratio** is a way of comparing two or more numbers. For example, if there are 10 workers in an office, 3 women and 7 men, the ratio of women to men is 3 to 7 or

$$\frac{3}{7} \quad \longleftarrow \text{Number of women} \\ \longleftarrow \text{Number of men}$$

On the other hand, if there are 6 men and 4 women in the office, the **reduced ratio** of women to men is

$$\frac{4}{6} = \frac{2}{3} \quad \longleftarrow \text{Number of women} \\ \longleftarrow \text{Number of men}$$

Use your knowledge to solve the following problems.

81. A class is composed of 40 men and 60 women. Find the reduced ratio of men to women.

82. Do you know the teacher-to-student ratio in your school? Suppose your school has 10,000 students and 500 teachers.

   a. Find the reduced teacher-to-student ratio.

   b. If the school wishes to maintain a $\frac{1}{20}$ ratio and the enrollment increases to 12,000 students, how many teachers are needed?

83. The transmission ratio in your automobile is defined by

$$\text{Transmission ratio} = \frac{\text{Engine speed}}{\text{Drive shaft speed}}$$

   a. If the engine is running at 2000 revolutions per minute and the drive shaft speed is 500 revolutions per minute, what is the reduced transmission ratio?

   b. If the transmission ratio of a car is 5 to 1, and the drive shaft speed is 500 revolutions per minute, what is the engine speed?

We will examine ratios more closely in Section 6.6.

〉〉〉 *Write On*

84. Write the procedure you use to determine the values for which the rational expression

$$\frac{P(x)}{Q(x)}$$

   is undefined.

85. Consider the rational expression

$$\frac{1}{x^2 + a}$$

   a. Is this expression always defined when a is positive? Explain. (*Hint:* Let $a = 1, 2$, and so on.)

   b. Is this expression always defined when a is negative? Explain. (*Hint:* Let $a = -1, -2$, and so on.)

86. Write the procedure you use to reduce a fraction to lowest terms.

87. If

$$\frac{P(x)}{Q(x)}$$

is equal to -1, what is the relationship between $P(x)$ and $Q(x)$?

〉〉〉   *Concept Checker*

Fill in the blank(s) with the correct word(s), phrase, or mathematical statement.

88. The **variables** in a rational expression **must not** be replaced by numbers that **make the** denominator _____.

89. The **Fundamental Rule of Rational Expressions** states that $\frac{A}{B} =$ _____.

90. The **standard form** of the fraction $-\frac{a}{b}$ is _____.

91. The **standard form** of the fraction $\frac{a}{-b}$ is _____.

92. The **standard form** of the fraction $-\frac{-a}{-b}$ is _____.

93. The **quotient** of **additive inverses** $\frac{a-b}{b-a} =$ _____.

1

−1

0

$$\frac{A \cdot C}{B \cdot C}$$

$$\frac{-a}{b}$$

〉〉〉   *Mastery Test*

Reduce to lowest terms (simplify):

94. $\dfrac{x^2 - 9}{3 + x}$

95. $\dfrac{x^2 - 9}{x + 3}$

96. $\dfrac{x^2 - 3x - 10}{5 - x}$

97. $\dfrac{10x - 15y}{4x - 6y}$

98. $-\dfrac{x}{xy + x}$

99. $\dfrac{x + 4}{-(x^2 - 16)}$

100. $\dfrac{-3xy^2}{12x^2y}$

101. $\dfrac{-6(x^2 - y^2)}{-3(x - y)}$

102. Write $\frac{7}{8}$ with a denominator of 16.

103. Write $\frac{3x}{8y^2}$ with a denominator of $24y^3$.

104. Write $\frac{4x}{x + 2}$ with a denominator of $x^2 - x - 6$.

105. Find the values for which $\frac{x^2 + 1}{x^2 - 4}$ is undefined.

106. Find the values for which $\frac{x + 4}{x^2 - 6x + 8}$ is undefined.

〉〉〉   *Skill Checker*

Multiply:

107. $\dfrac{3}{2} \cdot \dfrac{4}{9}$

108. $\dfrac{3}{5} \cdot \dfrac{10}{9}$

Factor:

109. $x^2 + 2x - 3$

110. $x^2 + 7x + 12$

111. $x^2 - 7x + 10$

112. $x^2 + 3x - 4$

6.2 Multiplication and Division of Rational Expressions

Objectives

A > Multiply two rational expressions.

B > Divide one rational expression by another.

To Succeed, Review How To . . .

1. Multiply, divide, and reduce fractions (pp. 4–5, 10–12).
2. Factor trinomials (pp. 407–409, 416–421, 425–427).
3. Factor the difference of two squares (pp. 427–428).

Getting Started

Gearing for Multiplication

How fast can the last (smallest) gear in this compound gear train go? It depends on the speed of the first gear, and the number of teeth in all the gears! The formula that tells us the number of revolutions per minute (rpm) the last gear can turn is

$$\text{rpm} = \frac{T_1}{t_1} \cdot \frac{T_2}{t_2} \cdot R$$

where T_1 and T_2 are the numbers of teeth in the driving gears, t_1 and t_2 are the numbers of teeth in the driven gears, and R is the number of revolutions per minute the first driving gear is turning. Many useful formulas require that we know how to multiply and divide rational expressions, and we shall learn how to do so in this section.

A > Multiplying Rational Expressions

Can you simplify this expression?

$$\frac{T_1}{t_1} \cdot \frac{T_2}{t_2} \cdot R$$

Of course you can, if you remember how to multiply fractions in arithmetic.* As you recall, in arithmetic the product of two fractions is another fraction whose numerator is the product of the original numerators and whose denominator is the product of the original denominators. Here's how we state this rule in symbols.

> **RULE**
>
> **Multiplying Rational Expressions**
>
> $$\frac{A}{B} \cdot \frac{C}{D} = \frac{AC}{BD}, \qquad B \neq 0, D \neq 0$$

Thus, the formula in the *Getting Started* can be simplified to

$$\text{rpm} = \frac{T_1 T_2 R}{t_1 t_2}$$

---

*Multiplication and division of arithmetic fractions is covered in Section R.2.

If we assume that $R = 12$ and then count the teeth in the gears, we get $T_1 = 48$, $T_2 = 40$, $t_1 = 24$, and $t_2 = 24$. To find the revolutions per minute, we write

$$\text{rpm} = \frac{48}{24} \cdot \frac{40}{24} \cdot \frac{12}{1}$$

Then we have the following:

1. Reduce each fraction.

$$\frac{\overset{2}{48}}{\underset{1}{24}} \cdot \frac{\overset{5}{40}}{\underset{3}{24}} \cdot \frac{12}{1} = \frac{2}{1} \cdot \frac{5}{3} \cdot \frac{12}{1}$$

2. Multiply the numerators.

$$\frac{120}{1 \cdot 3 \cdot 1}$$

3. Multiply the denominators.

$$\frac{120}{3}$$

4. Reduce the answer.

$$40$$

Thus, the speed of the final gear is 40 revolutions per minute. Here is what we have done.

> **PROCEDURE**
>
> **Multiplying Rational Expressions**
> **1.** Reduce each expression if possible.
> **2.** Multiply the numerators to obtain the new numerator.
> **3.** Multiply the denominators to obtain the new denominator.
> **4.** Reduce the answer if possible.

Note that you could also write

$$\frac{2}{1} \cdot \frac{5}{\underset{1}{3}} \cdot \frac{\overset{4}{12}}{1}$$

and obtain $2 \cdot 5 \cdot 4 = 40$ as before.

EXAMPLE 1   **Multiplying rational expressions**

Multiply:

a. $\dfrac{x}{6} \cdot \dfrac{7}{y}$

b. $\dfrac{3x^2}{2} \cdot \dfrac{4y}{9x}$

SOLUTION

a. $\dfrac{x}{6} \cdot \dfrac{7}{y} = \dfrac{7x}{6y}$ ⟵—— Multiply numerators.
⟵—— Multiply denominators.

b. $\dfrac{3x^2}{2} \cdot \dfrac{4y}{9x} = \dfrac{12x^2y}{18x}$ ⟵—— Multiply numerators.
⟵—— Multiply denominators.

$$= \frac{\overset{2 \cdot x}{12x^2y}}{\underset{3 \cdot 1}{18x}}$$ Reduce the answer.

$$= \frac{2xy}{3}$$

PROBLEM 1

Multiply:

a. $\dfrac{m}{4} \cdot \dfrac{5}{n}$

b. $\dfrac{5y^3}{2} \cdot \dfrac{4x}{15y}$

Answers to PROBLEMS

1. a. $\dfrac{5m}{4n}$ **b.** $\dfrac{2xy^2}{3}$

The procedure used to find the product in Example 1(b) can be shortened if we do some reduction beforehand. Thus, we can write

$$\frac{\overset{1x}{\cancel{3x^2}}}{\underset{1}{\cancel{2}}} \cdot \frac{\overset{2}{\cancel{4y}}}{\underset{3}{\cancel{9x}}} = \frac{2xy}{3}$$

We use this idea in Example 2.

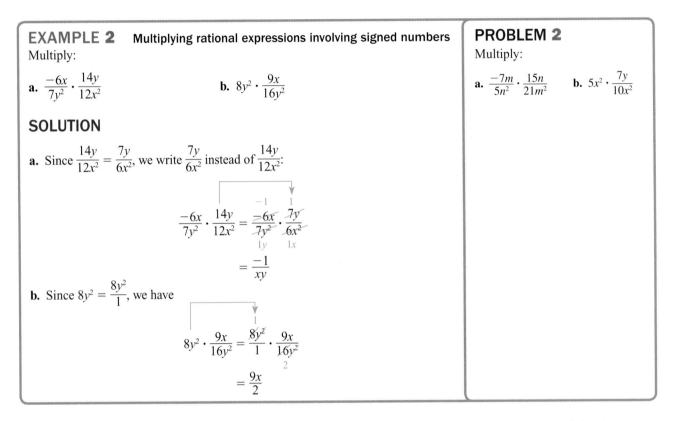

EXAMPLE 2   Multiplying rational expressions involving signed numbers

Multiply:

a. $\dfrac{-6x}{7y^2} \cdot \dfrac{14y}{12x^2}$ **b.** $8y^2 \cdot \dfrac{9x}{16y^2}$

SOLUTION

a. Since $\dfrac{14y}{12x^2} = \dfrac{7y}{6x^2}$, we write $\dfrac{7y}{6x^2}$ instead of $\dfrac{14y}{12x^2}$:

$$\frac{-6x}{7y^2} \cdot \frac{14y}{12x^2} = \frac{-6x}{7y^2} \cdot \frac{7y}{6x^2}$$

$$= \frac{-1}{xy}$$

b. Since $8y^2 = \dfrac{8y^2}{1}$, we have

$$8y^2 \cdot \frac{9x}{16y^2} = \frac{8y^2}{1} \cdot \frac{9x}{16y^2}$$

$$= \frac{9x}{2}$$

PROBLEM 2

Multiply:

a. $\dfrac{-7m}{5n^2} \cdot \dfrac{15n}{21m^2}$ **b.** $5x^2 \cdot \dfrac{7y}{10x^2}$

Can all problems be done as in these examples? Yes, but note the following.

> **NOTE**
>
> When the numerators and denominators involved are binomials or trinomials, it isn't easy to do the reductions we just did in the examples *unless* the numerators and denominators involved are *factored*.

Thus, to multiply

$$\frac{x^2 + 2x - 3}{x^2 + 7x + 12} \cdot \frac{x + 4}{x + 5}$$

we *first* factor and then multiply. The result is,

$$\frac{x^2 + 2x - 3}{x^2 + 7x + 12} \cdot \frac{x + 4}{x + 5} = \frac{(x - 1)(x + 3)}{(x + 4)(x + 3)} \cdot \frac{(x + 4)}{(x + 5)}$$

$$= \frac{x - 1}{x + 5}$$

Answers to PROBLEMS

2. a. $-\dfrac{1}{mn}$ **b.** $\dfrac{7y}{2}$

Thus, when multiplying fractions involving trinomials, we factor the trinomials and reduce the answer if possible. But note the following.

> ### NOTE
>
> Only **factors** can be divided out (canceled), never terms. Thus,
>
> $$\frac{x\overset{1}{\cancel{y}}}{\cancel{y}} = x$$
>
> but
>
> $$\frac{x + y}{y}$$
>
> cannot be reduced (simplified) further.

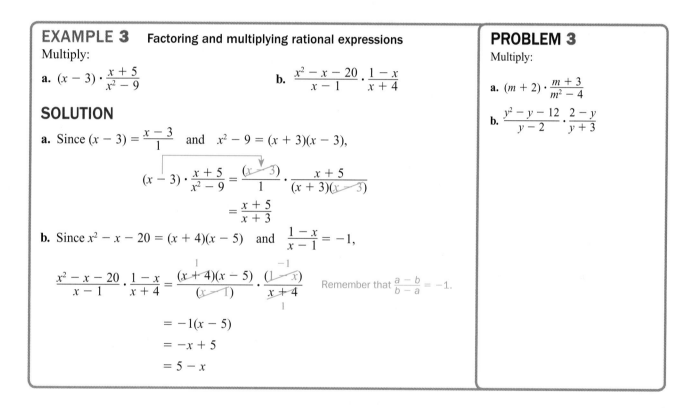

EXAMPLE 3   Factoring and multiplying rational expressions

Multiply:

a. $(x - 3) \cdot \dfrac{x + 5}{x^2 - 9}$ **b.** $\dfrac{x^2 - x - 20}{x - 1} \cdot \dfrac{1 - x}{x + 4}$

SOLUTION

a. Since $(x - 3) = \dfrac{x - 3}{1}$ and $x^2 - 9 = (x + 3)(x - 3)$,

$$(x - 3) \cdot \frac{x + 5}{x^2 - 9} = \frac{(\cancel{x - 3})}{1} \cdot \frac{x + 5}{(x + 3)(\cancel{x - 3})}$$

$$= \frac{x + 5}{x + 3}$$

b. Since $x^2 - x - 20 = (x + 4)(x - 5)$ and $\dfrac{1 - x}{x - 1} = -1$,

$$\frac{x^2 - x - 20}{x - 1} \cdot \frac{1 - x}{x + 4} = \frac{(x + 4)(x - 5)}{(\cancel{x - 1})} \cdot \frac{\overset{-1}{(\cancel{1 - x})}}{x + 4}$$

Remember that $\dfrac{a - b}{b - a} = -1$.

$$= -1(x - 5)$$
$$= -x + 5$$
$$= 5 - x$$

PROBLEM 3

Multiply:

a. $(m + 2) \cdot \dfrac{m + 3}{m^2 - 4}$

b. $\dfrac{y^2 - y - 12}{y - 2} \cdot \dfrac{2 - y}{y + 3}$

B ❯ Dividing Rational Expressions

What about dividing rational expressions? We are in luck! The division of rational expressions uses the same rule as in arithmetic.

> ### RULE
>
> **Dividing Rational Expressions**
>
> $$\frac{A}{B} \div \frac{C}{D} = \frac{A}{B} \cdot \frac{D}{C} = \frac{AD}{BC}, \qquad B, C, \text{ and } D \neq 0$$

Answers to PROBLEMS

3. a. $\dfrac{m + 3}{m - 2}$ **b.** $4 - y$

Thus, to divide $\frac{A}{B}$ by $\frac{C}{D}$, we simply **invert** $\frac{C}{D}$ (interchange the numerator and denominator) and multiply. That is, to divide $\frac{A}{B}$ by $\frac{C}{D}$, we multiply $\frac{A}{B}$ by the **reciprocal** (inverse) of $\frac{C}{D}$. For example, to divide

$$\frac{x+4}{x-5} \div \frac{x^2+3x-4}{x^2-7x+10}$$

we use the given rule and write

$$\frac{x+4}{x-5} \div \frac{x^2+3x-4}{x^2-7x+10} = \frac{x+4}{x-5} \cdot \frac{x^2-7x+10}{x^2+3x-4}$$

$$= \frac{x+4}{x-5} \cdot \frac{(x-5)(x-2)}{(x+4)(x-1)} \qquad \text{First factor numerator and denominator.}$$

$$= \frac{x-2}{x-1} \qquad \text{Reduce.}$$

Here is another example.

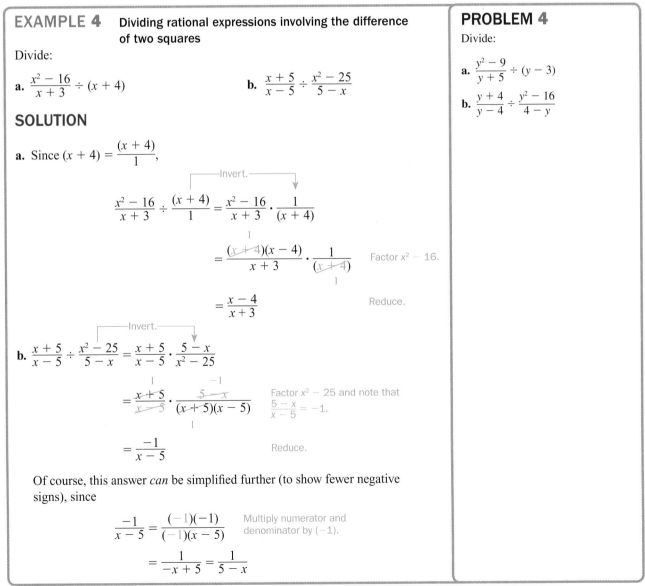

EXAMPLE 4   Dividing rational expressions involving the difference of two squares

Divide:

a. $\dfrac{x^2-16}{x+3} \div (x+4)$

b. $\dfrac{x+5}{x-5} \div \dfrac{x^2-25}{5-x}$

SOLUTION

a. Since $(x+4) = \dfrac{(x+4)}{1}$,

$$\frac{x^2-16}{x+3} \div \frac{(x+4)}{1} = \frac{x^2-16}{x+3} \cdot \frac{1}{(x+4)}$$

$$= \frac{(x+4)(x-4)}{x+3} \cdot \frac{1}{(x+4)} \qquad \text{Factor } x^2 - 16.$$

$$= \frac{x-4}{x+3} \qquad \text{Reduce.}$$

b. $\dfrac{x+5}{x-5} \div \dfrac{x^2-25}{5-x} = \dfrac{x+5}{x-5} \cdot \dfrac{5-x}{x^2-25}$

$$= \frac{x+5}{x-5} \cdot \frac{5-x}{(x+5)(x-5)} \qquad \begin{array}{l}\text{Factor } x^2 - 25 \text{ and note that}\\ \frac{5-x}{x-5} = -1.\end{array}$$

$$= \frac{-1}{x-5} \qquad \text{Reduce.}$$

Of course, this answer *can* be simplified further (to show fewer negative signs), since

$$\frac{-1}{x-5} = \frac{(-1)(-1)}{(-1)(x-5)} \qquad \begin{array}{l}\text{Multiply numerator and}\\ \text{denominator by } (-1).\end{array}$$

$$= \frac{1}{-x+5} = \frac{1}{5-x}$$

PROBLEM 4

Divide:

a. $\dfrac{y^2-9}{y+5} \div (y-3)$

b. $\dfrac{y+4}{y-4} \div \dfrac{y^2-16}{4-y}$

Answers to PROBLEMS

4. a. $\dfrac{y+3}{y+5}$ **b.** $\dfrac{-1}{y-4} = \dfrac{1}{4-y}$

Note: $\dfrac{1}{4-y}$ is preferred!

Here's another example.

EXAMPLE 5   Factoring and dividing rational expressions

Divide:

a. $\dfrac{x^2 + 5x + 4}{x^2 - 2x - 3} \div \dfrac{x^2 - 4}{x^2 - 6x + 8}$

b. $\dfrac{x^2 - 1}{x^2 + x - 6} \div \dfrac{x^2 - 4x + 3}{x^2 - 4}$

SOLUTION

Invert.

a. $\dfrac{x^2 + 5x + 4}{x^2 - 2x - 3} \div \dfrac{x^2 - 4}{x^2 - 6x + 8} = \dfrac{x^2 + 5x + 4}{x^2 - 2x - 3} \cdot \dfrac{x^2 - 6x + 8}{x^2 - 4}$

$= \dfrac{(x + 4)(x + 1)}{(x - 3)(x + 1)} \cdot \dfrac{(x - 4)(x - 2)}{(x + 2)(x - 2)}$ Factor.

$= \dfrac{(x + 4)(x - 4)}{(x - 3)(x + 2)}$

$= \dfrac{x^2 - 16}{x^2 - x - 6}$

Invert.

b. $\dfrac{x^2 - 1}{x^2 + x - 6} \div \dfrac{x^2 - 4x + 3}{x^2 - 4} = \dfrac{x^2 - 1}{x^2 + x - 6} \cdot \dfrac{x^2 - 4}{x^2 - 4x + 3}$

$= \dfrac{(x + 1)(x - 1)}{(x + 3)(x - 2)} \cdot \dfrac{(x + 2)(x - 2)}{(x - 1)(x - 3)}$ Factor.

$= \dfrac{(x + 1)(x + 2)}{(x + 3)(x - 3)}$

$= \dfrac{x^2 + 3x + 2}{x^2 - 9}$

PROBLEM 5

Divide:

a. $\dfrac{y^2 - 3y + 2}{y^2 - 4y + 3} \div \dfrac{y^2 - 49}{y^2 - 5y - 14}$

b. $\dfrac{y^2 - 1}{y^2 - y - 6} \div \dfrac{y^2 - 3y + 2}{y^2 - 9}$

A final word of warning! Be very careful when you reduce fractions.

CAUTION

You may cancel factors, but you must not cancel terms.

$$\dfrac{\cancel{a}\,b}{\cancel{a}\,c} = \dfrac{b}{c} \quad \text{Yes!}$$

$$\dfrac{\cancel{a} + b}{\cancel{a} + c} \quad \text{No!}$$

Thus,

$$\dfrac{\cancel{x^2} + 5x + 6}{\cancel{x^2} + 8x + 15} = \dfrac{5x + 6}{8x + 15} \quad \text{is wrong!}$$

Note that x^2 is a **term,** *not* a factor. The correct way is to *factor* first and then cancel. Thus,

$$\dfrac{x^2 + 5x + 6}{x^2 + 8x + 15} = \dfrac{(x + 3)(x + 2)}{(x + 3)(x + 5)} = \dfrac{x + 2}{x + 5}$$

Of course,

$$\dfrac{x + 2}{x + 5}$$

cannot be reduced further. To write

$$\frac{x + 2}{x + 5} = \frac{2}{5} \quad \text{is wrong!}$$

Again, x is a term, not a factor. (If you had

$$\frac{2x}{5x} = \frac{2}{5}$$

that would be correct. In the expressions $2x$ and $5x$, x is a *factor* that may be canceled.)
Why is

$$\frac{x + 2}{x + 5} \neq \frac{2}{5}?$$

Try it when x is 4.

$$\frac{x + 2}{x + 5} = \frac{4 + 2}{4 + 5} = \frac{6}{9} = \frac{2}{3}$$

Thus, the answer *cannot* be $\frac{2}{5}$!

> **Exercises 6.2**

MathZone+

Boost your grade at mathzone.com!
> Practice Problems
> NetTutor
> Self-Tests
> e-Professors
> Videos

〈 **A** 〉 **Multiplying Rational Expressions** In Problems 1–30, multiply and simplify.

1. $\dfrac{x}{3} \cdot \dfrac{8}{y}$

2. $\dfrac{-x}{4} \cdot \dfrac{7}{y}$

3. $\dfrac{-6x^2}{7} \cdot \dfrac{14y}{9x}$

4. $\dfrac{-5x^3}{6y^2} \cdot \dfrac{18y}{-10x}$

5. $7x^2 \cdot \dfrac{3y}{14x^2}$

6. $11y^2 \cdot \dfrac{4x}{33y}$

7. $\dfrac{-4y}{7x^2} \cdot 14x^3$

8. $\dfrac{-3y^3}{8x^2} \cdot -16y$

9. $(x - 7) \cdot \dfrac{x + 1}{x^2 - 49}$

10. $3(x + 1) \cdot \dfrac{x + 2}{x^2 - 1}$

11. $-2(x + 2) \cdot \dfrac{x - 1}{x^2 - 4}$

12. $-3(x - 1) \cdot \dfrac{x - 2}{x^2 - 1}$

13. $\dfrac{3}{x - 5} \cdot \dfrac{x^2 - 25}{x + 1}$

14. $\dfrac{1}{x - 2} \cdot \dfrac{x^2 - 4}{x - 1}$

15. $\dfrac{x^2 - x - 6}{x - 2} \cdot \dfrac{2 - x}{x - 3}$

16. $\dfrac{x^2 + 3x - 4}{3 - x} \cdot \dfrac{x - 3}{x + 4}$

17. $\dfrac{x - 1}{3 - x} \cdot \dfrac{x + 3}{1 - x}$

18. $\dfrac{2x - 1}{5 - 3x} \cdot \dfrac{3x - 5}{1 - 2x}$

19. $\dfrac{3(x - 5)}{14(4 - x)} \cdot \dfrac{7(x - 4)}{6(5 - x)}$

20. $\dfrac{7(1 - x)}{10(x - 5)} \cdot \dfrac{5(5 - x)}{14(x - 1)}$

21. $\dfrac{6x^3}{x^2 - 16} \cdot \dfrac{x^2 - 5x + 4}{3x^2}$

22. $\dfrac{3a^4}{a^2 - 4} \cdot \dfrac{a^2 - a - 2}{9a^3}$

23. $\dfrac{y^2 + 2y - 3}{y - 5} \cdot \dfrac{y^2 - 3y - 10}{y^2 + 5y - 6}$

24. $\dfrac{f^2 + 2f - 8}{f^2 + 7f + 12} \cdot \dfrac{f^2 + 2f - 3}{f^2 - 3f + 2}$

25. $\dfrac{2y^2 + y - 3}{6 - 11y - 10y^2} \cdot \dfrac{5y^3 - 2y^2}{3y^2 - 5y + 2}$

26. $\dfrac{3x^2 - x - 2}{2 - x - 6x^2} \cdot \dfrac{2x^4 - x^3}{3x^2 - 2x - 1}$

27. $\dfrac{15x^2 - x - 2}{2x^2 + 5x - 18} \cdot \dfrac{2x^2 + x - 36}{3x^2 - 11x - 4}$

28. $\dfrac{6x^2 + x - 1}{3x^2 + 5x + 2} \cdot \dfrac{3x^2 - x - 2}{2x^2 - x - 1}$

29. $\dfrac{27y^3 + 8}{6y^2 + 19y + 10} \cdot \dfrac{4y^2 - 25}{9y^2 - 6y + 4}$

30. $\dfrac{8y^3 + 27}{10y^2 + 19y + 6} \cdot \dfrac{25y^2 - 4}{4y^2 - 6y + 9}$

‹ **B** › **Dividing Rational Expressions**   In Problems 31–64, divide and simplify.

31. $\dfrac{x^2 - 1}{x + 2} \div (x + 1)$

32. $\dfrac{x^2 - 4}{x - 3} \div (x + 2)$

33. $\dfrac{x^2 - 25}{x - 3} \div 5(x + 5)$

34. $\dfrac{x^2 - 16}{8(x - 3)} \div 4(x + 4)$

35. $(x + 3) \div \dfrac{x^2 - 9}{x + 4}$

36. $4(x - 4) \div \dfrac{8(x^2 - 16)}{5}$

37. $\dfrac{-3}{x - 4} \div \dfrac{6(x + 3)}{5(x^2 - 16)}$

38. $\dfrac{-6}{x - 2} \div \dfrac{3(x - 1)}{7(x^2 - 4)}$

39. $\dfrac{-4(x + 1)}{3(x + 2)} \div \dfrac{-8(x^2 - 1)}{6(x^2 - 4)}$

40. $\dfrac{-10(x^2 - 1)}{6(x^2 - 4)} \div \dfrac{5(x + 1)}{-3(x + 2)}$

41. $\dfrac{x + 3}{x - 3} \div \dfrac{x^2 - 1}{3 - x}$

42. $\dfrac{4 - x}{x + 1} \div \dfrac{x - 4}{x^2 - 1}$

43. $\dfrac{x^2 - 4}{7(x^2 - 9)} \div \dfrac{x + 2}{14(x + 3)}$

44. $\dfrac{x^2 - 25}{3(x^2 - 1)} \div \dfrac{5 - x}{6(x + 1)}$

45. $\dfrac{3(x^2 - 36)}{14(5 - x)} \div \dfrac{6(6 - x)}{7(x^2 - 25)}$

46. $\dfrac{6(x^2 - 1)}{35(x^2 - 4)} \div \dfrac{12(1 - x)}{7(2 - x)}$

47. $\dfrac{x + 2}{x - 1} \div \dfrac{x^2 + 5x + 6}{x^2 - 4x + 4}$

48. $\dfrac{x - 3}{x + 2} \div \dfrac{x^2 - 4x + 3}{x^2 - x - 6}$

49. $\dfrac{x - 5}{x + 3} \div \dfrac{5(x - 5)}{x^2 + 9x + 18}$

50. $\dfrac{x - 3}{x + 4} \div \dfrac{2(x - 3)}{x^2 + 2x - 8}$

51. $\dfrac{x^2 + 2x - 3}{x - 5} \div \dfrac{x^2 + 6x + 9}{x^2 - 2x - 15}$

52. $\dfrac{x^2 - 3x + 2}{x^2 - 5x + 6} \div \dfrac{x^2 - 5x + 4}{x^2 - 7x + 12}$

53. $\dfrac{x^2 - 1}{x^2 + 3x - 10} \div \dfrac{x^2 - 3x - 4}{x^2 - 25}$

54. $\dfrac{x^2 - 4x - 21}{x^2 - 10x + 25} \div \dfrac{x^2 + 2x - 3}{x^2 - 6x + 5}$

55. $\dfrac{x^2 + 3x - 4}{x^2 + 7x + 12} \div \dfrac{x^2 + x - 2}{x^2 + 5x + 6}$

56. $\dfrac{x^2 + x - 2}{x^2 + 6x - 7} \div \dfrac{x^2 - 3x - 10}{x^2 + 5x - 14}$

57. $\dfrac{x^2 - y^2}{x^2 - 2xy} \div \dfrac{x^2 + xy - 2y^2}{x^2 - 4y^2}$

58. $\dfrac{x^2 + xy - 2y^2}{x^2 - 4y^2} \div \dfrac{x^2 - y^2}{x^2 - 2xy}$

59. $\dfrac{x^2 + 2xy - 3y^2}{y^2 - 7y + 10} \div \dfrac{x^2 + 5xy - 6y^2}{y^2 - 3y - 10}$

60. $\dfrac{x^2 + 2xy - 8y^2}{x^2 + 7xy + 12y^2} \div \dfrac{x^2 - 3xy + 2y^2}{x^2 + 2xy - 3y^2}$

61. $\dfrac{2x^2 - x - 28}{3x^2 - x - 2} \div \dfrac{4x^2 + 16x + 7}{3x^2 + 11x + 6}$

62. $\dfrac{15x^2 - x - 2}{2x^2 + 5x - 18} \div \dfrac{3x^2 - 11x - 4}{2x^2 + x - 36}$

63. $\dfrac{(a^3 - 27)(a^2 - 9)}{(a - 3)^2(a + 3)^3} \div \dfrac{a^2 + 3a + 9}{a^2 + 3a}$

64. $\dfrac{(y^3 - 8)(y^2 - 4)}{(y + 2)^2(y - 2)^3} \div \dfrac{y^2 + 2y + 4}{y^2 - 2y}$

> > >   *Applications*

65. *Price and demand* If the price for x units of a product is given by the expression

$$\frac{3x + 9}{4}$$

and the demand for these is given by the expression

$$\frac{600}{x^2 + 3x}$$

find the product of the price and the demand.

66. *Price and demand* If the price for x units of a product is given by the expression

$$\frac{4x + 8}{5}$$

and the demand is given by the expression

$$\frac{200}{x^2 + 2x}$$

what is the product of the price and the demand?

67. *Price and demand* If the price P for x units of a product is given by the expression

$$\frac{5x + 10}{2}$$

and the demand D is given by the expression

$$\frac{400}{x^2 + 2x}$$

find PD, the product of the price and the demand.

68. *Current, voltage, and resistance* In a simple electrical circuit, the current I is the quotient of the voltage E and the resistance R. If the resistance changes with the time t according to

$$R = \frac{t^2 + 9}{t^2 + 6t + 9}$$

and the voltage changes according to the formula

$$E = \frac{4t}{t + 3}$$

find the current I.

69. *Current, voltage, and resistance* If in Problem 68 the resistance is given by the equation

$$R = \frac{t^2 + 5}{t^2 + 4t + 4}$$

and the voltage is given by the equation

$$E = \frac{5t}{t + 2}$$

find the current I.

70. *Current, voltage, and resistance* In a simple electric circuit, the current is modeled by the equation $I = \frac{E}{R}$, where E is the voltage and R is the resistance. If

$$R = \frac{t^2 + 4}{t^2 + 4t + 4}$$

and

$$E = \frac{3t}{t + 2}$$

find I.

> > >   *Using Your Knowledge*

Resistance, Molecules, and Reordering

71. In the study of parallel resistors, the expression

$$R \cdot \frac{R_T}{R - R_T}$$

occurs, where R is a known resistance and R_T a required one. Perform the multiplication.

72. The molecular model predicts that the pressure of a gas is given by

$$\frac{2}{3} \cdot \frac{mv^2}{2} \cdot \frac{N}{v}$$

where m, v, and N represent the mass, velocity, and total number of molecules, respectively. Perform the multiplication.

73. Suppose a store orders 3000 items each year. If it orders x units at a time, the number N of reorders is

$$N = \frac{3000}{x}$$

If there is a fixed \$20 reorder fee and a \$3 charge per item, the cost of each order is

$$C = 20 + 3x$$

The yearly reorder cost C_R is then given by

$$C_R = N \cdot C$$

Find C_R.

〉〉〉 Write On

74. Write the procedure you use to multiply two rational expressions.

75. Write the procedure you use to divide one rational expression by another.

76. Explain why you cannot "cancel" the x's in

$$\frac{x+5}{x+2}$$

to obtain an answer of $\frac{5}{2}$ but you can cancel the x's in $\frac{5x}{2x}$.

77. Explain what the statement "you can cancel factors but you cannot cancel terms" means and give examples.

〉〉〉 Concept Checker

Fill in the blank(s) with the correct word(s), phrase, or mathematical statement.

78. $\frac{A}{B} \cdot \frac{C}{D} = \underline{\hspace{1cm}}.$

$$\frac{AD}{BC} \qquad \frac{AC}{BD}$$

79. $\frac{A}{B} \div \frac{C}{D} = \underline{\hspace{1cm}}.$

$$\frac{AD}{AC} \qquad \frac{BC}{AD}$$

〉〉〉 Mastery Test

Divide and simplify:

80. $\frac{x^2 - 3x + 2}{x^2 - 4x + 3} \div \frac{x^2 - 49}{x^2 + 5x - 14}$

81. $\frac{x^2 - 1}{x^2 - x - 6} \div \frac{x^2 - 3x + 2}{x^2 - 9}$

82. $\frac{x^2 - 25}{x + 1} \div (x + 5)$

83. $\frac{x + 6}{x - 6} \div \frac{x^2 - 36}{6 - x}$

84. $\frac{x^3 + 8}{x - 2} \div \frac{x^2 - 2x + 4}{x^2 - 4}$

85. $\frac{x^3 - 27}{x + 3} \div \frac{x^2 + 3x + 9}{x^2 - 9}$

Multiply and simplify:

86. $(x - 4) \cdot \frac{x + 8}{x^2 - 16}$

87. $\frac{x^2 + x - 6}{x - 3} \cdot \frac{3 - x}{x + 3}$

88. $\frac{-3x}{4y^2} \cdot \frac{18y}{12x^2}$

89. $\frac{6x}{11y^2} \cdot 22y^2$

90. $\frac{x^3 - 64}{x + 1} \cdot \frac{x^2 - 1}{x^2 + 4x + 16}$

91. $\frac{2x^2 - x - 28}{3x^2 - x - 2} \cdot \frac{3x^2 + 11x + 6}{4x^2 + 16x + 7}$

〉〉〉 Skill Checker

Add:

92. $\frac{7}{12} + \frac{1}{12}$

93. $\frac{7}{8} + \frac{2}{5}$

94. $\frac{7}{12} + \frac{1}{18}$

Subtract:

95. $\frac{7}{8} - \frac{2}{5}$

96. $\frac{7}{12} - \frac{1}{18}$

97. $\frac{5}{2} - \frac{1}{6}$

98. $\frac{8}{3} - \frac{3}{4}$

6.3 Addition and Subtraction of Rational Expressions

▶ Objectives

A ❯ Add and subtract rational expressions with the same denominator.

B ❯ Add and subtract rational expressions with different denominators.

C ❯ Solve applications involving rational expressions.

▶ To Succeed, Review How To . . .

1. Find the LCD of two or more fractions (pp. 13–15).
2. Add and subtract fractions (pp. 12–18).

▶ Getting Started
Tennis, Anyone?

The racket hits the ball with such tremendous force that the ball is distorted. Can we find out how much force? The answer is

$$\frac{mv}{t} - \frac{mv_0}{t}$$

where

$$m = \text{mass of ball}$$
$$v = \text{velocity of racket}$$
$$v_0 = \text{"initial" velocity of racket}$$
$$t = \text{time of contact}$$

Since the expressions involved have the same denominator, subtracting them is easy. As in arithmetic, we simply subtract the numerators and keep the same denominator. Thus,

$$\frac{mv}{t} - \frac{mv_0}{t} = \frac{mv - mv_0}{t}$$ ◀— Subtract numerators.
◀— Keep the denominator.

In this section we shall learn how to add and subtract rational expressions.

A ❯ Adding and Subtracting Rational Expressions with the Same Denominator

As you recall from Section R.2, $\frac{1}{5} + \frac{2}{5} = \frac{3}{5}$, $\frac{1}{7} + \frac{4}{7} = \frac{5}{7}$, and $\frac{1}{11} + \frac{8}{11} = \frac{9}{11}$. The same procedure works for rational expressions. For example,

$$\frac{3}{x} + \frac{5}{x} = \frac{3+5}{x} = \frac{8}{x}$$ ◀— Add numerators.
◀— Keep the denominator.

Similarly,

$$\frac{5}{x+1} + \frac{2}{x+1} = \frac{5+2}{x+1} = \frac{7}{x+1}$$ ◀— Add numerators.
◀— Keep the denominator.

and

$$\frac{5}{7(x-1)} + \frac{2}{7(x-1)} = \frac{5+2}{7(x-1)} = \frac{\overset{1}{7}}{\underset{1}{7(x-1)}} = \frac{1}{x-1}$$

For subtraction,

$$\frac{8}{x+5} - \frac{2}{x+5} = \frac{8-2}{x+5} = \frac{6}{x+5}$$ ← Subtract numerators.
                                                ← Keep the denominator.

and

$$\frac{8}{9(x-3)} - \frac{2}{9(x-3)} = \frac{8-2}{9(x-3)} = \frac{\overset{2}{6}}{\underset{3}{9}(x-3)} = \frac{2}{3(x-3)}$$

EXAMPLE 1   Adding and subtracting rational expressions:
                 Same denominator

Add or Subtract:

a. $\dfrac{8}{3(x-2)} + \dfrac{1}{3(x-2)}$ **b.** $\dfrac{7}{5(x+4)} - \dfrac{2}{5(x+4)}$

SOLUTION

a. $\dfrac{8}{3(x-2)} + \dfrac{1}{3(x-2)} = \dfrac{8+1}{3(x-2)} = \dfrac{\overset{3}{9}}{\underset{1}{3}(x-2)} = \dfrac{3}{x-2}$ Remember to
reduce the answer.

b. $\dfrac{7}{5(x+4)} - \dfrac{2}{5(x+4)} = \dfrac{7-2}{5(x+4)} = \dfrac{\overset{1}{5}}{\underset{1}{5}(x+4)} = \dfrac{1}{x+4}$

PROBLEM 1

Add or Subtract:

a. $\dfrac{4}{5(y-1)} + \dfrac{1}{5(y-1)}$

b. $\dfrac{9}{7(y+2)} - \dfrac{2}{7(y+2)}$

B › Adding and Subtracting Rational Expressions with Different Denominators

Not all rational expressions have the same denominator. To add or subtract rational expressions with different denominators, we again rely on our experiences in arithmetic. Let's practice adding rational numbers before we try rational expressions.

EXAMPLE 2   Adding fractions: Different denominators

Add: $\dfrac{7}{12} + \dfrac{5}{18}$

SOLUTION   We first must find a common denominator—that is, a *multiple* of 12 and 18. Of course, it's more convenient to use the smallest one available. In general, the **lowest common denominator (LCD)** of two fractions is the **smallest** number that is a multiple of *both* denominators. To find the LCD, we can use successive divisions as in Section R.2.

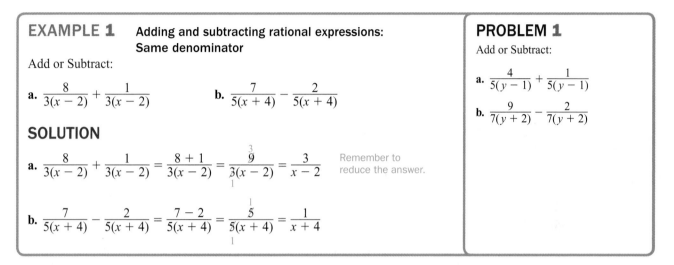

The LCD is $2 \times 3 \times 2 \times 3 = 36$. Better yet, we can also factor both numbers and write each factor in a column to obtain

⎰Pick the number with
⎱the highest exponent.

$$12 = 2 \cdot 2 \cdot 3 = 2^2 \cdot 3^1$$

$$18 = 2 \cdot 3 \cdot 3 = 2^1 \cdot 3^2$$ Note that all the 2's and all the 3's
are written in separate columns.

PROBLEM 2

Add: $\dfrac{5}{12} + \dfrac{7}{18}$

(continued)

Note that since we need a number that is a multiple of 12 and 18, we select the factors raised to the *highest* power in each column—that is, 2^2 and 3^2. The product of these factors is the LCD. Thus, the LCD of 12 and 18 is $2^2 \cdot 3^2 = 4 \cdot 9 = 36$, as before. We then write each fraction with a denominator of 36 and add.

$$\frac{7}{12} = \frac{7 \cdot 3}{12 \cdot 3} = \frac{21}{36}$$ We multiply the denominator 12 by 3 (to get 36), so we do the same to the numerator.

$$\frac{5}{18} = \frac{5 \cdot 2}{18 \cdot 2} = \frac{10}{36}$$ Here we multiply the denominator 18 by 2 to get 36, so we do the same to the numerator.

$$\frac{7}{12} + \frac{5}{18} = \frac{21}{36} + \frac{10}{36} = \frac{31}{36}$$ Can you see how this one is done?

The procedure can also be written as

$$\frac{7}{12} = \frac{21}{36}$$
$$+ \frac{5}{18} = \frac{10}{36}$$
$$\overline{\phantom{+\frac{5}{18}=} \frac{31}{36}}$$

NOTE

Make sure you know how to do this type of problem before you go on. The idea in adding and subtracting rational expressions is the same as the idea used with rational numbers. In fact, problems $1-20$ in Exercises 6.3 have two parts: one with rational numbers and one with rational expressions. If you know how to do one, you should be able to do the other one.

Now, let's practice subtraction of fractions.

EXAMPLE 3   **Subtracting fractions: Different denominators**

Subtract: $\dfrac{11}{15} - \dfrac{5}{18}$

SOLUTION   We can use successive divisions to find the LCD, writing

$$3 \underline{| 15 \quad 18}$$
$$ 5 \quad 6$$

The LCD is $3 \times 5 \times 6 = 90$. Better yet, we can factor the denominators and write them as follows:

$$15 = 3 \cdot 5 \quad = \quad 3 \cdot 5$$ Note that all the 2's, all the 3's, and
$$18 = 2 \cdot 3 \cdot 3 = 2 \cdot 3^2$$ all the 5's are in separate columns.

As before, the LCD is

$$2 \cdot 3^2 \cdot 5 = 2 \cdot 9 \cdot 5 = 90$$

Then we write $\frac{11}{15}$ and $\frac{5}{18}$ as equivalent fractions with a denominator of 90.

$$\frac{11}{15} = \frac{11 \cdot 6}{15 \cdot 6} = \frac{66}{90}$$ Multiply the numerator and denominator by 6.

$$\frac{5}{18} = \frac{5 \cdot 5}{18 \cdot 5} = \frac{25}{90}$$ Multiply the numerator and denominator by 5.

PROBLEM 3

Subtract: $\dfrac{13}{15} - \dfrac{7}{12}$

Answers to PROBLEMS

3. $\dfrac{17}{60}$

and then subtract

$$\frac{11}{15} - \frac{5}{18} = \frac{66}{90} - \frac{25}{90}$$

$$= \frac{66 - 25}{90}$$

$$= \frac{41}{90}$$

Of course, it's possible that the denominators involved have no common factors. In this case, the LCD is the *product* of the denominators. Thus, to add $\frac{3}{5}$ and $\frac{4}{7}$, we use $5 \cdot 7 = 35$ as the LCD and write

$$\frac{3}{5} = \frac{3 \cdot 7}{5 \cdot 7} = \frac{21}{35}$$ Multiply the numerator and denominator by 7.

$$\frac{4}{7} = \frac{4 \cdot 5}{7 \cdot 5} = \frac{20}{35}$$ Multiply the numerator and denominator by 5.

Thus,

$$\frac{3}{5} + \frac{4}{7} = \frac{21}{35} + \frac{20}{35} = \frac{41}{35}$$

Similarly, the expression

$$\frac{4}{x} + \frac{5}{3}$$

has $3x$ as the LCD. We then write $\frac{4}{x}$ and $\frac{5}{3}$ as equivalent fractions with $3x$ as the denominator and add:

$$\frac{4}{x} = \frac{4 \cdot 3}{x \cdot 3} = \frac{12}{3x}$$ Multiply the numerator and denominator by 3.

$$\frac{5}{3} = \frac{5 \cdot x}{3 \cdot x} = \frac{5x}{3x}$$ Multiply the numerator and denominator by x.

Thus,

$$\frac{4}{x} + \frac{5}{3} = \frac{12}{3x} + \frac{5x}{3x}$$

$$= \frac{12 + 5x}{3x}$$

Are you ready for a more complicated problem? First, let's state a generalized procedure for adding and subtracting fractions with different denominators.

PROCEDURE

Adding (or Subtracting) Fractions with Different Denominators

1. Find the LCD.
2. Write all fractions as equivalent ones with the LCD as the denominator.
3. Add (or subtract) numerators and keep denominators.
4. Reduce if possible.

Let's use these steps to add

$$\frac{x + 1}{x^2 + x - 2} + \frac{x + 3}{x^2 - 1}$$

1. We first find the LCD of the denominators. To do this, we factor the denominators.

$$x^2 + x - 2 = (x + 2)(x - 1)$$
$$x^2 - 1 = \quad (x - 1)(x + 1)$$
$$(x + 2)(x - 1)(x + 1) \quad \text{The LCD.}$$

2. We then write

$$\frac{x + 1}{x^2 + x - 2} \quad \text{and} \quad \frac{x + 3}{x^2 - 1}$$

as equivalent fractions with $(x + 2)(x - 1)(x + 1)$ as denominator.

$$\frac{x + 1}{x^2 + x - 2} = \frac{x + 1}{(x + 2)(x - 1)} = \frac{(x + 1)(x + 1)}{(x + 2)(x - 1)(x + 1)}$$

$$\frac{x + 3}{x^2 - 1} = \frac{x + 3}{(x + 1)(x - 1)} = \frac{(x + 3)(x + 2)}{(x + 1)(x - 1)(x + 2)}$$

$$= \frac{(x + 3)(x + 2)}{(x + 2)(x - 1)(x + 1)}$$

3. Add the numerators and keep the denominator.

$$\frac{x + 1}{x^2 + x - 2} + \frac{x + 3}{x^2 - 1} = \frac{(x + 1)(x + 1)}{(x + 2)(x - 1)(x + 1)} + \frac{(x + 3)(x + 2)}{(x + 2)(x - 1)(x + 1)}$$

$$= \frac{(x^2 + 2x + 1) + (x^2 + 5x + 6)}{(x + 2)(x - 1)(x + 1)}$$

$$= \frac{2x^2 + 7x + 7}{(x + 2)(x - 1)(x + 1)} \quad \text{Note that the denominator is left as an indicated product.}$$

4. The answer is not reducible, since there are no factors common to the numerator and denominator.

We use this procedure to add and subtract rational expressions in Example 4.

EXAMPLE 4   **Adding and subtracting rational expressions: Different denominators**

Add or Subtract:

a. $\dfrac{7}{8} + \dfrac{2}{x}$

b. $\dfrac{2}{x - 1} - \dfrac{1}{x + 2}$

SOLUTION

a. Since 8 and x don't have any common factors, the LCD is $8x$. We write $\frac{7}{8}$ and $\frac{2}{x}$ as equivalent fractions with $8x$ as denominator and add.

$$\frac{7}{8} = \frac{7 \cdot x}{8 \cdot x} = \frac{7x}{8x}$$

$$\frac{2}{x} = \frac{2 \cdot 8}{x \cdot 8} = \frac{16}{8x}$$

Thus,

$$\frac{7}{8} + \frac{2}{x} = \frac{7x}{8x} + \frac{16}{8x}$$

$$= \frac{7x + 16}{8x}$$

Note that when the denominators A and B do not have any common factors, the common denominator is AB.

PROBLEM 4

Add or Subtract:

a. $\dfrac{3}{5} + \dfrac{2}{y}$

b. $\dfrac{2}{y - 2} - \dfrac{1}{y + 1}$

Answers to PROBLEMS

4. a. $\dfrac{3y + 10}{5y}$ **b.** $\dfrac{y + 4}{(y - 2)(y + 1)}$

b. Since $(x - 1)$ and $(x + 2)$ don't have any common factors, the LCD of

$$\frac{2}{x - 1} \quad \text{and} \quad \frac{1}{x + 2}$$

is $(x - 1)(x + 2)$. We then write

$$\frac{2}{x - 1} \quad \text{and} \quad \frac{1}{x + 2}$$

as equivalent fractions with $(x - 1)(x + 2)$ as the denominator.

$$\frac{2}{x - 1} = \frac{2 \cdot (x + 2)}{(x - 1)(x + 2)} \qquad \text{Multiply numerator and denominator by } (x + 2).$$

$$\frac{1}{x + 2} = \frac{1 \cdot (x - 1)}{(x + 2)(x - 1)} = \frac{(x - 1)}{(x - 1)(x + 2)} \qquad \text{Multiply numerator and denominator by } (x - 1).$$

Hence,

$$\frac{2}{x - 1} - \frac{1}{x + 2} = \frac{2 \cdot (x + 2)}{(x - 1)(x + 2)} - \frac{(x - 1)}{(x - 1)(x + 2)}$$

$$= \frac{2(x + 2) - (x - 1)}{(x - 1)(x + 2)} \qquad \leftarrow \text{Subtract numerators.} \\ \leftarrow \text{Keep denominator.}$$

$$= \frac{2x + 4 - x + 1}{(x - 1)(x + 2)} \qquad \text{Remember that} \\ -(x - 1) = -x + 1.$$

$$= \frac{x + 5}{(x - 1)(x + 2)} \qquad \leftarrow \text{Simplify numerator.} \\ \leftarrow \text{Keep denominator.}$$

EXAMPLE 5   Subtracting rational expressions: Different denominators

Subtract:

$$\frac{x - 2}{x^2 - x - 6} - \frac{x + 3}{x^2 - 9}$$

SOLUTION   We use the four-step procedure.

1. To find the LCD, we factor the denominators to obtain

$$x^2 - x - 6 = \quad (x - 3)(x + 2)$$
$$x^2 - 9 = (x + 3)(x - 3)$$

$$(x + 3)(x - 3)(x + 2) \quad \text{The LCD.}$$

2. We write each fraction as an equivalent one with the LCD as the denominator. Hence,

$$\frac{x - 2}{x^2 - x - 6} = \frac{(x - 2)(x + 3)}{(x - 3)(x + 2)(x + 3)}$$

$$\frac{x + 3}{x^2 - 9} = \frac{(x + 3)(x + 2)}{(x + 3)(x - 3)(x + 2)}$$

PROBLEM 5

Subtract:

$$\frac{x - 3}{(x + 1)(x - 2)} - \frac{x + 3}{x^2 - 4}$$

(continued)

Answers to PROBLEMS

5. $\dfrac{-5x - 9}{(x + 1)(x + 2)(x - 2)}$

3. $\dfrac{x-2}{x^2-x-6} - \dfrac{x+3}{x^2-9} = \dfrac{(x-2)(x+3)}{(x+3)(x-3)(x+2)} - \dfrac{(x+3)(x+2)}{(x+3)(x-3)(x+2)}$

$= \dfrac{(x^2+x-6)-(x^2+5x+6)}{(x+3)(x-3)(x+2)}$

$= \dfrac{x^2+x-6-x^2-5x-6}{(x+3)(x-3)(x+2)}$ Note that $-(x^2+5x+6) = -x^2-5x-6$.

$= \dfrac{-4x-12}{(x+3)(x-3)(x+2)}$

$= \dfrac{-4(x+3)}{(x+3)(x-3)(x+2)}$ Factor the numerator and keep the denominator.

4. Reduce.

$$\frac{-4\cancel{(x+3)}}{\cancel{(x+3)}(x-3)(x+2)} = \frac{-4}{(x-3)(x+2)}$$

C › Applications Involving Rational Expressions

EXAMPLE 6   Graduate engineering students
Do you want to be an engineer? The chart shows the total number (T) and number of female (F) graduate science/engineering students (in thousands). These numbers can be approximated by the polynomials

$$T(t) = -0.4t^2 + 10t + 360$$
$$F(t) = -0.1t^2 + 6t + 125$$

where t is the number of years after 1985.

a. What fraction of the total number is female?

b. What fraction of the total number was female in 1985?

c. What percent of the total was female in 1985? (Answer to the nearest percent.)

d. What fraction of the total would you predict to be female in 2005?

e. Is the percent of female engineering graduate students increasing?

SOLUTION

a. The fraction of the total number [$T(t)$] that is female [$F(t)$] is

$$\frac{F(t)}{T(t)} = \frac{-0.1t^2 + 6t + 125}{-0.4t^2 + 10t + 360}$$

b. Since t is the number of years after 1985, $t = 0$ in 1985. Substituting $t = 0$ in

$$\frac{F(t)}{T(t)} = \frac{-0.1t^2 + 6t + 125}{-0.4t^2 + 10t + 360}$$

we obtain

$$\frac{F(0)}{T(0)} = \frac{-0.1(0)^2 + 6(0) + 125}{-0.4(0)^2 + 10(0) + 360} = \frac{125}{360} = \frac{25}{72}$$

c. To convert $\frac{25}{72}$ to a percent, divide 25 by 72 and multiply by 100, obtaining 34.72 or 35% (to the nearest percent).

PROBLEM 6

a. What is the ratio of total students to female students?

b. According to part **e** of Example 6, the percent of female graduates is increasing, but these figures are only a model. Can this increasing trend continue indefinitely? *Hint:* What would be the theoretical percent of female students in 2020 ($t = 35$)? In 2021?

Graduates in Science/Eng.

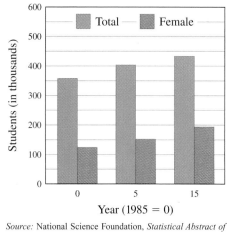

Students (in thousands) / Year (1985 = 0)

Total   Female

Source: National Science Foundation, *Statistical Abstract of the United States,* Table 784.

Answers to PROBLEMS

6. a. $\dfrac{-0.4t^2 + 10t + 360}{-0.1t^2 + 6t + 125}$ **b.** No. In 2020 the percent of female students would be 97%, and in 2021 it would be 105%.

d. In 2005, $t = 2005 - 1985 = 20$. Substituting 20 in part **a** yields

$$\frac{F(20)}{T(20)} = \frac{-0.1(20)^2 + 6(20) + 125}{-0.4(20)^2 + 10(20) + 360} = \frac{205}{400} = \frac{41}{80} \quad \text{or about} \quad 51\%$$

e. Yes, it is increasing from 35% in 1985 to 51% in 2005.

And now, a last word before you go on to the exercise set. At this point, you can see that there are great similarities between algebra and arithmetic. In fact, in this very section we use the arithmetic addition of fractions as a model to do the algebraic addition of rational expressions. To show that these similarities are very strong and also to give you more practice, Problems 1–20 in the exercise set consist of two similar problems, an arithmetic one and an algebraic one. Use the practice and experience gained in working one to do the other.

MathZone

› Exercises **6.3**

‹ **A** › Adding and Subtracting Rational Expressions with the Same Denominator
‹ **B** › Adding and Subtracting Rational Expressions with Different Denominators

In Problems 1–40, perform the indicated operations.

1. a. $\frac{2}{7} + \frac{3}{7}$

   b. $\frac{3}{x} + \frac{8}{x}$

2. a. $\frac{5}{9} + \frac{2}{9}$

   b. $\frac{9}{x-1} + \frac{2}{x-1}$

3. a. $\frac{8}{9} - \frac{2}{9}$

   b. $\frac{6}{x} - \frac{2}{x}$

4. a. $\frac{4}{7} - \frac{2}{7}$

   b. $\frac{6}{x+4} - \frac{2}{x+4}$

5. a. $\frac{6}{7} + \frac{8}{7}$

   b. $\frac{3}{2x} + \frac{7}{2x}$

6. a. $\frac{1}{9} + \frac{2}{9}$

   b. $\frac{3}{8(x-2)} + \frac{1}{8(x-2)}$

7. a. $\frac{8}{3} - \frac{2}{3}$

   b. $\frac{11}{3(x+1)} - \frac{9}{3(x+1)}$

8. a. $\frac{3}{8} - \frac{1}{8}$

   b. $\frac{7}{15(x-1)} - \frac{2}{15(x-1)}$

9. a. $\frac{8}{9} + \frac{4}{9}$

   b. $\frac{7x}{4(x+1)} + \frac{3x}{4(x+1)}$

10. a. $\frac{15}{14} + \frac{3}{14}$

   b. $\frac{29x}{15(x-3)} + \frac{4x}{15(x-3)}$

11. a. $\frac{3}{4} - \frac{1}{3}$

   b. $\frac{7}{x} - \frac{3}{8}$

12. a. $\frac{5}{7} - \frac{2}{5}$

   b. $\frac{x}{3} - \frac{7}{x}$

13. a. $\frac{1}{5} + \frac{1}{7}$

   b. $\frac{4}{x} + \frac{x}{9}$

14. a. $\frac{1}{3} + \frac{1}{9}$

   b. $\frac{5}{x} + \frac{6}{3x}$

15. a. $\frac{2}{5} - \frac{4}{15}$

   b. $\frac{4}{7(x-1)} - \frac{3}{14(x-1)}$

16. a. $\frac{9}{2} - \frac{5}{8}$

   b. $\frac{8}{9(x+3)} - \frac{5}{36(x+3)}$

17. a. $\frac{4}{7} + \frac{3}{8}$

   b. $\frac{3}{x+1} + \frac{5}{x-2}$

18. a. $\frac{2}{9} + \frac{4}{5}$

   b. $\frac{2x}{x+2} + \frac{3x}{x-4}$

19. a. $\frac{7}{8} - \frac{1}{3}$

   b. $\frac{6}{x-2} - \frac{3}{x+1}$

20. a. $\frac{6}{7} - \frac{2}{3}$

   b. $\frac{4x}{x+1} - \frac{4x}{x+2}$

21. $\frac{x+1}{x^2+3x-4} + \frac{x+2}{x^2-16}$

22. $\dfrac{x-2}{x^2-9}+\dfrac{x+1}{x^2-x-12}$

23. $\dfrac{3x}{x^2+3x-10}+\dfrac{2x}{x^2+x-6}$

24. $\dfrac{x+3}{x^2-x-2}+\dfrac{x-1}{x^2+2x+1}$

25. $\dfrac{1}{x^2-y^2}+\dfrac{5}{(x+y)^2}$

26. $\dfrac{3x}{(x+y)^2}+\dfrac{5x}{x-y}$

27. $\dfrac{2}{x-5}-\dfrac{3x}{x^2-25}$

28. $\dfrac{x+3}{x^2-x-2}-\dfrac{x-1}{x^2+2x+1}$

29. $\dfrac{x-1}{x^2+3x+2}-\dfrac{x+7}{x^2+5x+6}$

30. $\dfrac{2}{x^2+3xy+2y^2}-\dfrac{1}{x^2-xy-2y^2}$

31. $\dfrac{y}{y^2-1}+\dfrac{y}{y+1}$

32. $\dfrac{3y}{y^2-4}-\dfrac{y}{y+2}$

33. $\dfrac{3y+1}{y^2-16}-\dfrac{2y-1}{y-4}$

34. $\dfrac{2y+1}{y^2-4}-\dfrac{3y-1}{y+2}$

35. $\dfrac{x+1}{x^2-x-2}+\dfrac{x-1}{x^2+2x+1}$

36. $\dfrac{y+3}{y^2+y-6}+\dfrac{y-2}{y^2+3y-10}$

37. $\dfrac{a}{a-w}-\dfrac{w}{a+w}-\dfrac{a^2+w^2}{a^2-w^2}$

38. $\dfrac{x}{x-y}+\dfrac{y}{x+y}-\dfrac{x^2+y^2}{x^2-y^2}$

39. $\dfrac{1}{a^3+8}+\dfrac{a+1}{a^2-2a+4}$

40. $\dfrac{c}{c^3-1}+\dfrac{2}{c^2+c+1}$

⟨ C ⟩ Applications Involving Rational Expressions

41. *Odds in favor of an event* If the odds in favor of an event are f to u, the probability p of the event happening is $p=\dfrac{f}{f+u}$ and the probability of the event **not** happening is $q=\dfrac{u}{f+u}$.

a. Write $p+q$ as a single rational expression.

b. Simplify the fraction.

43. *Noise analysis* When analyzing the noise in an electronic device, we need to add the expressions $\dfrac{1}{g_m}+\dfrac{8}{g_m^2}$, where g_m is the maximum noise conductance. Write $\dfrac{1}{g_m}+\dfrac{8}{g_m^2}$ as a single rational expression in reduced form.

42. *Measuring noise* The noise measure M of a system is given by the equation $M=\dfrac{GF}{G-1}-\dfrac{G}{G-1}$, where F is the noise figure and G is the associated gain of the device. Perform the subtraction and write M as a single rational expression in reduced form.

Thin lenses The equation used to make thin lenses states that

$$\frac{1}{f}=\left(\frac{n}{n_m}-1\right)\left[\frac{1}{R_1}-\frac{1}{R_2}+\frac{(n-1)d}{nR_1R_2}\right]$$

where f is the distance from the lens, R_1 and R_2 are positive or negative depending on the convexity of the lens, n is the refractive index of the lens material, and n_m is the refractive index of the medium around the lens. Now, don't panic, you don't have to know all that! You only have to know how to add or subtract rational expressions as in Problems 44–45.

44. Write $\left(\dfrac{n}{n_m}-1\right)$ as a single fraction in reduced form.

45. Write $\dfrac{1}{R_1}-\dfrac{1}{R_2}+\dfrac{(n-1)d}{nR_1R_2}$ as a single fraction in reduced form.

46. *Second law of motion* When working with Kepler's second law of motion, we have the expression

$$\frac{GM}{(1-\varepsilon)a} - \frac{GM}{(1+\varepsilon)a}$$

where G is the gravitational constant, M the mass of the object, a the object's semimajor axis, and ε is the specific

orbital energy. Perform the subtraction and write the answer as a single fraction in reduced form.

47. *Moment of a beam* The moment M of a cantilever beam of length L, x units from the end is given by the expression

$$-\frac{w_0 x^3}{6L} + \frac{w_0 Lx}{2} - \frac{w_0 L^2}{3}$$

Write this expression as a single rational expression in reduced form.

48. *Deflection of a beam* The deflection d of the beam of Problem 47 involves the expression

$$\frac{-x^4}{24L} + \frac{Lx^2}{4} - \frac{L^2 x}{3}$$

Write this expression as a single rational expression in reduced form.

49. *Planetary Motion* In astronomy, planetary motion is given by the expression.

$$\frac{p^2}{2mr^2} - \frac{gmM}{r}$$

Write this expression as a single rational expression in reduced form.

50. *Pendulum* The motion of a pendulum is given by the expression

$$\frac{P_1^2 + P_2^2}{2(h_1 + h_2)} + \frac{P_1^2 - P_2^2}{2(h_1 - h_2)}$$

Write this expression as a single rational expression in reduced form.

⟩⟩⟩ *Using Your Knowledge*

Continuing the Study of Fractions In Chapter 1, we mentioned that some numbers cannot be written as the ratio of two integers. These numbers are called **irrational numbers.** We can approximate irrational numbers by using a type of fraction called a **continued fraction.** Here's how we do it. From a table of square roots, or a calculator, we find that

$$\sqrt{2} \approx 1.4142 \qquad \approx \text{means "approximately equal."}$$

Can we find some continued fraction to approximate $\sqrt{2}$?

51. Try $1 + \frac{1}{2}$ (write it as a decimal).

52. Try $1 + \dfrac{1}{2 + \frac{1}{2}}$ (write it as a decimal).

53. Try $1 + \dfrac{1}{2 + \dfrac{1}{2 + \frac{1}{2}}}$ (write it as a decimal).

54. Look at the pattern for the approximation of $\sqrt{2}$ given in Problems 51–53. What do you think the next approximation (when written as a continued fraction) will be?

55. How close is the approximation for $\sqrt{2}$ in Problem 53 to the value $\sqrt{2} \approx 1.4142$?

⟩⟩⟩   *Write On*

56. Write the procedure you use to find the LCD of two rational expressions.

57. Write the procedure you use to find the sum of two rational expressions:

    a. with the same denominator.

    b. with different denominators.

58. Write the procedure you use to find the difference of two rational expressions:

    a. with the same denominator.

    b. with different denominators.

⟩⟩⟩   *Concept Checker*

Fill in the blank(s) with the correct word(s), phrase, or mathematical statement.

59. The **first step** in **adding** or **subtracting** fractions with **different denominators** is to **find** the _____ of the fractions.

60. The **second step** in **adding** or **subtracting** fractions with **different denominators** is to **write** all fractions as **equivalent fractions** with the _____ as the denominator.

GCF

LCD

⟩⟩⟩   *Mastery Test*

61. The fraction of the waste recovered in the United States can be approximated by

$$\frac{R(t)}{G(t)}$$

Of this,

$$\frac{P(t)}{G(t)}$$

is paper and paperboard. If

$$P(t) = 0.02t^2 - 0.25t + 6$$

$$G(t) = 0.04t^2 + 2.34t + 90$$

$$R(t) = 0.04t^2 - 0.59t + 7.42$$

and t represents the number of years after 1960, find the fraction of the waste recovered that is *not* paper and paperboard.

Perform the indicated operations.

62. $\dfrac{x-3}{x^2-x-2} - \dfrac{x+3}{x^2-4}$

63. $\dfrac{5}{x-1} - \dfrac{3}{x+3}$

64. $\dfrac{4}{5} + \dfrac{3}{x}$

65. $\dfrac{4}{5(x-2)} + \dfrac{6}{5(x-2)}$

66. $\dfrac{11}{3(x+2)} - \dfrac{2}{3(x+2)}$

67. $\dfrac{x+3}{x^2-x-2} - \dfrac{x-3}{x^2-4}$

⟩⟩⟩   *Skill Checker*

Perform the indicated operations:

68. $2 + \dfrac{2}{9}$

69. $1 \div \dfrac{20}{9}$

70. $1 \div \dfrac{30}{7}$

71. $12x\left(\dfrac{2}{x} + \dfrac{3}{2x}\right)$

72. $12x\left(\dfrac{4}{3x} - \dfrac{1}{4x}\right)$

73. $x^2\left(1 - \dfrac{1}{x^2}\right)$

74. $x^2\left(1 + \dfrac{1}{x}\right)$

6.4 Complex Fractions

⊳ Objective

A ⟩ Simplify a complex fraction using one of two methods.

⊳ To Succeed, Review How To . . .

Add, subtract, multiply, and divide fractions (pp. 10–18).

⊳ Getting Started
Planetary Models and Complex Fractions

We've already learned how to do the four fundamental operations using rational expressions. In some instances, we want to find the quotient of two expressions that contain fractions in the numerator or the denominator or in both. For example, the model of the planets in our solar system shown here is similar to the model designed by the seventeenth-century Dutch mathematician and astronomer Christian Huygens. The gears used in the model were especially difficult to design since they had to make each of the planets revolve around the sun at different rates. For example, Saturn goes around the sun in

$$29 + \frac{1}{2 + \frac{2}{9}} \text{ yr}$$

The expression

$$\frac{1}{2 + \frac{2}{9}}$$

is a *complex fraction*. In this section we shall learn how to simplify complex fractions.

A ⟩ Simplifying Complex Fractions

Before we simplify *complex fractions,* we need a formal definition.

COMPLEX FRACTION	A **complex fraction** is a fraction that has one or more fractions in its numerator, denominator, or both.

Complex fractions can be simplified in either of two ways:

> **PROCEDURE**
>
> **Simplifying Complex Fractions**
> 1. Multiply numerator and denominator by the LCD of the fractions involved, or
> 2. Perform the operations indicated in the numerator and denominator of the complex fraction, and then divide the simplified numerator by the simplified denominator.

We illustrate these two methods by simplifying the complex fraction

$$\frac{1}{2 + \frac{2}{9}}$$

Method 1. Multiply numerator and denominator by the LCD of the fractions involved (in our case, by 9).

$$\frac{1}{2 + \frac{2}{9}} = \frac{9 \cdot 1}{9\left(2 + \frac{2}{9}\right)}$$ Note that $9\left(2 + \frac{2}{9}\right) = 9 \cdot 2 + 9 \cdot \frac{2}{9} = 18 + 2$.

$$= \frac{9}{18 + 2}$$

$$= \frac{9}{20}$$

Method 2. Perform the operations indicated in the numerator and denominator of the complex fraction and then divide the numerator by the denominator.

$$\frac{1}{2 + \frac{2}{9}} = \frac{1}{\frac{18}{9} + \frac{2}{9}} = \frac{1}{\frac{20}{9}}$$ Add $2 + \frac{2}{9}$.

$$= 1 \div \frac{20}{9}$$ Write $\frac{1}{\frac{20}{9}}$ as $1 \div \frac{20}{9}$.

$$= 1 \cdot \frac{9}{20}$$ Multiply by the reciprocal of $\frac{20}{9}$.

$$= \frac{9}{20}$$

Either procedure also works for more complicated rational expressions. In Example 1 we simplify a complex fraction using both methods. Compare the results and see which method you prefer!

EXAMPLE 1   Simplifying complex fractions: Both methods

Simplify:

$$\frac{\frac{1}{a} + \frac{2}{b}}{\frac{3}{a} - \frac{1}{b}}$$

SOLUTION

Method 1. The LCD of the fractions involved is ab, so we multiply the numerator and denominator by ab to obtain

$$\frac{ab \cdot \left(\frac{1}{a} + \frac{2}{b}\right)}{ab \cdot \left(\frac{3}{a} - \frac{1}{b}\right)} = \frac{ab \cdot \frac{1}{a} + ab \cdot \frac{2}{b}}{ab \cdot \frac{3}{a} - ab \cdot \frac{1}{b}}$$ Note that $ab \cdot \frac{1}{a} = b$, $ab \cdot \frac{2}{b} = 2a$,

$ab \cdot \frac{3}{a} = 3b$, and $ab \cdot \frac{1}{b} = a$.

$$= \frac{b + 2a}{3a - a}$$

PROBLEM 1

Simplify:

$$\frac{\frac{2}{a} - \frac{3}{b}}{\frac{1}{a} + \frac{2}{b}}$$

Answers to PROBLEMS

1. $\frac{2b - 3a}{b + 2a}$

Method 2. Add the fractions in the numerator and subtract the fractions in the denominator. In both cases, the LCD of the fractions is ab. Hence,

$$\frac{\frac{1}{a}+\frac{2}{b}}{\frac{3}{a}-\frac{1}{b}} = \frac{\frac{b}{ab}+\frac{2a}{ab}}{\frac{3b}{ab}-\frac{a}{ab}} \qquad \text{Write the fractions with their LCD.}$$

$$= \frac{\frac{b+2a}{ab}}{\frac{3b-a}{ab}} \qquad \text{Add in the numerator, subtract in the denominator.}$$

$$= \frac{b+2a}{ab}\cdot\frac{ab}{3b-a} \qquad \text{Multiply by the reciprocal of } \frac{3b-a}{ab}.$$

$$= \frac{b+2a}{3b-a} \qquad \text{Simplify.}$$

EXAMPLE 2   Simplifying complex fractions: Method 1

Simplify:

$$\frac{\frac{2}{x}+\frac{3}{2x}}{\frac{4}{3x}-\frac{1}{4x}}$$

SOLUTION   We must first find the LCD of x, $2x$, $3x$, and $4x$. Now

$$
\begin{array}{ll}
x = & \quad\quad x \\
2x = 2 & \quad\quad \cdot x \\
3x = & \;3 \quad \cdot x \\
4x = 2^2 & \quad\quad \cdot x
\end{array}
\qquad \text{Write the factors in columns.}
$$

The LCD is $2^2 \cdot 3 \cdot x = 12x$. Multiplying numerator and denominator by $12x$, we have

$$\frac{\frac{2}{x}+\frac{3}{2x}}{\frac{4}{3x}-\frac{1}{4x}} = \frac{12x\cdot\left(\frac{2}{x}+\frac{3}{2x}\right)}{12x\cdot\left(\frac{4}{3x}-\frac{1}{4x}\right)}$$

$$= \frac{12x\cdot\frac{2}{x}+12x\cdot\frac{3}{2x}}{12x\cdot\frac{4}{3x}-12x\cdot\frac{1}{4x}} \qquad \text{Use the distributive property.}$$

$$= \frac{12\cdot 2+6\cdot 3}{4\cdot 4-3\cdot 1} \qquad \text{Simplify.}$$

$$= \frac{24+18}{16-3}$$

$$= \frac{42}{13}$$

EXAMPLE 3   Simplifying complex fractions: Method 1

Simplify:

$$\frac{1-\frac{1}{x^2}}{1+\frac{1}{x}}$$

PROBLEM 2

Simplify:

$$\frac{\frac{1}{4x}+\frac{2}{3x}}{\frac{3}{2x}-\frac{1}{x}}$$

PROBLEM 3

Simplify:

$$\frac{1-\frac{1}{x^2}}{1-\frac{1}{x}}$$

(continued)

SOLUTION   Here the LCD of the fraction involved is x^2. Thus,

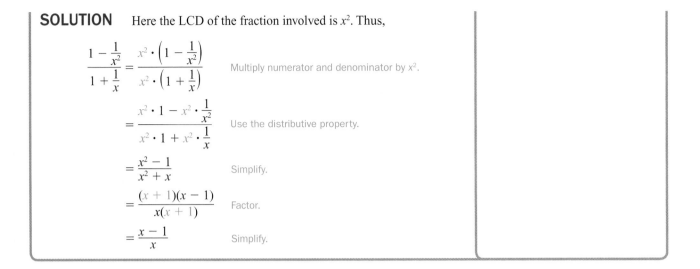

$$\frac{1-\dfrac{1}{x^2}}{1+\dfrac{1}{x}} = \frac{x^2 \cdot \left(1-\dfrac{1}{x^2}\right)}{x^2 \cdot \left(1+\dfrac{1}{x}\right)}$$ Multiply numerator and denominator by x^2.

$$= \frac{x^2 \cdot 1 - x^2 \cdot \dfrac{1}{x^2}}{x^2 \cdot 1 + x^2 \cdot \dfrac{1}{x}}$$ Use the distributive property.

$$= \frac{x^2 - 1}{x^2 + x}$$ Simplify.

$$= \frac{(x+1)(x-1)}{x(x+1)}$$ Factor.

$$= \frac{x-1}{x}$$ Simplify.

> Web IT   go to mhhe.com/**bello** *for more lessons*

MathZone+x

Boost your grade at mathzone.com!
> Practice   > Self-Tests
  Problems   > e-Professors
> NetTutor   > Videos

> **Exercises 6.4**

< **A** >   **Simplifying Complex Fractions**   In Problems 1–24, simplify.

1. $\dfrac{\dfrac{1}{2}}{2+\dfrac{1}{2}}$

2. $\dfrac{\dfrac{1}{4}}{3+\dfrac{1}{4}}$

3. $\dfrac{\dfrac{1}{2}}{2-\dfrac{1}{2}}$

4. $\dfrac{\dfrac{1}{4}}{3-\dfrac{1}{4}}$

5. $\dfrac{a-\dfrac{a}{b}}{1+\dfrac{a}{b}}$

6. $\dfrac{1-\dfrac{1}{a}}{1+\dfrac{1}{a}}$

7. $\dfrac{\dfrac{1}{a}+\dfrac{1}{b}}{\dfrac{1}{a}-\dfrac{1}{b}}$

8. $\dfrac{\dfrac{2}{a}+\dfrac{1}{b}}{\dfrac{2}{a}-\dfrac{1}{b}}$

9. $\dfrac{\dfrac{1}{2a}+\dfrac{1}{3b}}{\dfrac{4}{a}-\dfrac{3}{4b}}$

10. $\dfrac{\dfrac{1}{2a}+\dfrac{1}{4b}}{\dfrac{2}{a}-\dfrac{3}{5b}}$

11. $\dfrac{\dfrac{1}{3}+\dfrac{3}{4}}{\dfrac{3}{8}-\dfrac{1}{6}}$

12. $\dfrac{\dfrac{1}{5}+\dfrac{3}{2}}{\dfrac{5}{8}-\dfrac{3}{10}}$

13. $\dfrac{2+\dfrac{1}{x}}{4-\dfrac{1}{x^2}}$

14. $\dfrac{3+\dfrac{1}{x}}{9-\dfrac{1}{x^2}}$

15. $\dfrac{2+\dfrac{2}{x}}{1+\dfrac{1}{x}}$

16. $\dfrac{5+\dfrac{5}{x^2}}{1+\dfrac{1}{x^2}}$

17. $\dfrac{\dfrac{1}{y}+\dfrac{1}{x}}{\dfrac{x}{y}-\dfrac{y}{x}}$

18. $\dfrac{\dfrac{1}{x}+\dfrac{1}{y}}{\dfrac{y}{x}-\dfrac{x}{y}}$

19. $\dfrac{x-2-\dfrac{8}{x}}{x-3-\dfrac{4}{x}}$

20. $\dfrac{x-2-\dfrac{15}{x}}{x-3-\dfrac{10}{x}}$

21. $\dfrac{\dfrac{1}{x+5}}{\dfrac{4}{x^2-25}}$

22. $\dfrac{\dfrac{1}{x-3}}{\dfrac{2}{x^2-9}}$

23. $\dfrac{\dfrac{1}{x^2-16}}{\dfrac{2}{x+4}}$

24. $\dfrac{\dfrac{3}{x^2-64}}{\dfrac{4}{x+8}}$

> > >   *Using Your Knowledge*

Around the Sun in Complex Fractions   In the *Getting Started* of this section, we mentioned that Saturn takes

$$29 + \frac{1}{2+\dfrac{2}{9}} \text{ yr}$$

to orbit the sun. Since we have shown that

$$\frac{1}{2 + \frac{2}{9}} = \frac{9}{20}$$

we know that Saturn takes $29 + \frac{9}{20} = 29\frac{9}{20}$ years to orbit the sun.

In Problems 25–30, use your knowledge to simplify the number of years it takes the following planets to orbit the sun.

25. Mercury, $\dfrac{1}{4 + \frac{1}{6}}$ yr

26. Venus, $\dfrac{1}{1 + \frac{2}{3}}$ yr

In Exercises 27–30, write your answer as a mixed number.

27. Jupiter, $11 + \dfrac{1}{1 + \frac{7}{43}}$ yr

28. Mars, $1 + \dfrac{1}{1 + \frac{3}{22}}$ yr

29. Uranus, $84 + \dfrac{1}{90 + \frac{10}{11}}$

30. Neptune, $164 + \dfrac{1}{1 + \frac{21}{79}}$

As you may know, there are three new "dwarf" planets: Ceres, Eris, and the newly reclassified Pluto. How long does it take (in years) these planets to orbit the sun? Write your answer as a mixed number.

We will see in Problems 31–33.

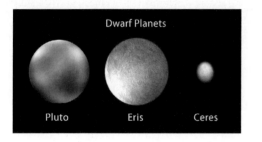

Dwarf Planets

Pluto　　　Eris　　　Ceres

31. Ceres, $4 + \dfrac{1}{1 + \frac{2}{3}}$

32. Eris, $556 + \dfrac{1}{1 + \frac{3}{7}}$

33. Pluto, $248 + \dfrac{1}{1 + \frac{23}{27}}$

34. Look at the results of Problems 25–33.
  a. Which planet takes the longest time to orbit the sun?
  b. Which planet takes the shortest time to orbit the sun?

〉〉〉 *Write On*

35. What is a complex fraction?

36. List the advantages and disadvantages of Method 1 when simplifying a complex fraction.

37. List the advantages and disadvantages of Method 2 when simplifying a complex fraction.

38. Which method do you prefer to simplify complex fractions? Why?

39. How do you know which method to use when simplifying complex fractions?

〉〉〉 *Concept Checker*

Fill in the blank(s) with the correct word(s), phrase, or mathematical statement.

40. A _____ **fraction** is a **fraction** that has **one or more fractions** in its **numerator, denominator,** or **both.**

41. The **first** step to **simplify** a **complex fraction** is to **multiply its numerator** and **denominator** by the _____ of all the fractions involved.

simplified　　GCF

complex　　LCD

❯❯❯ *Mastery Test*

Simplify:

42. $\dfrac{\dfrac{2}{a} - \dfrac{3}{b}}{\dfrac{1}{a} + \dfrac{2}{b}}$

43. $\dfrac{\dfrac{1}{4x} + \dfrac{2}{3x}}{\dfrac{3}{2x} - \dfrac{1}{x}}$

44. $\dfrac{1 - \dfrac{1}{x^2}}{1 - \dfrac{1}{x}}$

45. $\dfrac{w + 2 - \dfrac{18}{w - 5}}{w - 1 - \dfrac{12}{w - 5}}$

46. $\dfrac{\dfrac{8x}{3x + 1} - \dfrac{3x - 1}{x}}{\dfrac{4x}{3x + 1} - \dfrac{2x - 1}{x}}$

47. $\dfrac{\dfrac{3}{m - 4} - \dfrac{16}{m - 3}}{\dfrac{2}{m - 3} - \dfrac{15}{m + 5}}$

❯❯❯ *Skill Checker*

Solve:

48. $17x = 408$

49. $19w = 2356$

50. $18L = 2232$

51. $9x + 24 = x$

52. $10x + 36 = x$

53. $5x = 4x + 3$

54. $6x = 5x + 5$

6.5 Solving Equations Containing Rational Expressions

▶ Objectives

A❯ Solve equations that contain rational expressions.

B❯ Solve a fractional equation for a specified variable.

▶ To Succeed, Review How To . . .

1. Solve linear equations (pp. 132–136).
2. Find the LCD of two or more rational expressions (pp. 497–502).

▶ Getting Started

Salute One of the Largest Flags

This American flag (one of the largest ever) was displayed in J. L. Hudson's store in Detroit. By law, the ratio of length to width of the American flag should be $\frac{19}{10}$. If the length of this flag was 235 feet, what should its width be to conform with the law? To solve this problem, we let W be the width of the flag and set up the equation

$$\frac{19}{10} = \frac{235}{W} \quad \begin{matrix} \leftarrow \text{Length} \\ \leftarrow \text{Width} \end{matrix}$$

This equation is an example of a *fractional equation*. A **fractional equation** is an equation that contains one or more rational expressions. To solve this equation, we must clear the denominators involved. We did this in Chapter 2 (Section 2.3) by multiplying each term by the LCD. Since the LCD of $\frac{19}{10}$ and $\frac{235}{W}$ is $10W$, we have

$$10W \cdot \frac{19}{10} = \frac{235}{W} \cdot 10W \qquad \text{Multiply by the LCD.}$$

$$19W = 2350 \qquad \text{Simplify.}$$

$$W = \frac{2350}{19} \qquad \text{Divide by 19.}$$

$$W \approx 124 \qquad \text{Approximate the answer.}$$

$$\begin{array}{r} 123.6 \\ 19\overline{)2350.0} \\ \underline{19} \\ 45 \\ \underline{38} \\ 70 \\ \underline{57} \\ 130 \\ \underline{114} \\ 16 \end{array}$$

(By the way, the flag was only 104 feet long and weighed 1500 pounds. It was *not* an official flag.) In this section we shall learn how to solve fractional equations.

A ⟩ Solving Fractional Equations

The first step in solving fractional equations is to multiply each side of the equation by the **least common multiple (LCM)** of the denominators present. This is equivalent to multiplying *each term* by the LCD because if you have the equation

$$\frac{a}{b} + \frac{c}{d} = \frac{e}{f}$$

then multiplying each side by L, the LCM of the denominators, gives

$$L \cdot \left(\frac{a}{b} + \frac{c}{d} \right) = L \cdot \frac{e}{f}$$

or, using the distributive property,

$$L \cdot \frac{a}{b} + L \cdot \frac{c}{d} = L \cdot \frac{e}{f}$$

Thus, we have the following result.

LEAST COMMON MULTIPLE

Multiplying each side of the equation

$$\frac{a}{b} + \frac{c}{d} = \frac{e}{f} \text{ (where } b, d, \text{ and } f \neq 0)$$

by L is equivalent to multiplying each *term* by L, that is,

$$L \left(\frac{a}{b} + \frac{c}{d} \right) = L \left(\frac{e}{f} \right) \text{ is equivalent to}$$

$$L \cdot \frac{a}{b} + L \cdot \frac{c}{d} = L \cdot \frac{e}{f}.$$

The procedure for solving fractional equations is similar to the one used to solve linear equations. You may want to review the procedure for solving linear equations (see pp. 132–136) before continuing with this section.

EXAMPLE 1 Solving fractional equations

Solve:

$$\frac{3}{4} + \frac{2}{x} = \frac{1}{12}$$

SOLUTION The LCD of $\frac{3}{4}$, $\frac{2}{x}$, and $\frac{1}{12}$ is $12x$.

1. Clear the fractions; the LCD is $12x$. $12x \cdot \frac{3}{4} + 12x \cdot \frac{2}{x} = 12x \cdot \frac{1}{12}$

2. Simplify. $9x + 24 = x$
3. Subtract 24 from each side. $9x = x - 24$
4. Subtract x from each side. $8x = -24$
5. Divide each side by 8. $x = -3$

The answer is -3.

6. Here is the check:

$$\frac{3}{4} + \frac{2}{x} \stackrel{?}{=} \frac{1}{12}$$

$$\begin{array}{c|c} \frac{3}{4} + \frac{2}{-3} & \frac{1}{12} \\ \hline \frac{3 \cdot 3}{4 \cdot 3} + \frac{2 \cdot 4}{-3 \cdot 4} & \\ \frac{9}{12} - \frac{8}{12} & \\ \frac{1}{12} & \end{array}$$

PROBLEM 1

Solve:

$$\frac{4}{5} + \frac{1}{x} = \frac{3}{10}$$

In some cases, the denominators involved may be more complicated. Nevertheless, the procedure used to solve the equation remains the same. Thus, we can also use the six-step procedure to solve

$$\frac{2x}{x - 1} + 3 = \frac{4x}{x - 1}$$

as shown in Example 2.

EXAMPLE 2 Solving fractional equations

Solve:

$$\frac{2x}{x - 1} + 3 = \frac{4x}{x - 1}$$

SOLUTION Since $x - 1$ is the only denominator, it must be the LCD. We then proceed by steps.

1. Clear the fractions; the LCD is $(x - 1)$. $(x - 1) \cdot \frac{2x}{x - 1} + 3(x - 1) = (x - 1) \cdot \frac{4x}{x - 1}$

2. Simplify. $2x + 3x - 3 = 4x$
                                           $5x - 3 = 4x$
3. Add 3. $5x = 4x + 3$
4. Subtract $4x$. $x = 3$
5. Division is not necessary.
6. You can easily check this by substitution.

Thus, the answer is $x = 3$, as can easily be verified by substituting 3 for x in the original equation.

PROBLEM 2

Solve:

$$\frac{8}{x - 1} - 4 = \frac{2x}{x - 1}$$

Answers to PROBLEMS

1. $x = -2$ **2.** $x = 2$

So far, the denominators used in the examples have *not* been factorable. In the cases where they are, it's very important that we factor them *before* we find the LCD. For instance, to solve the equation

$$\frac{x}{x^2-16} + \frac{4}{x-4} = \frac{1}{x+4}$$

we first note that

$$x^2 - 16 = (x+4)(x-4)$$

We then write

as

$$\frac{x}{x^2-16} + \frac{4}{x-4} = \frac{1}{x+4}$$ The denominator x^2-16 has been factored as $(x+4)(x-4)$.

$$\frac{x}{(x+4)(x-4)} + \frac{4}{x-4} = \frac{1}{x+4}$$

The solution to this equation is given in the next example.

EXAMPLE 3   **Solving fractional equations**

Solve:

$$\frac{x}{x^2-16} + \frac{4}{x-4} = \frac{1}{x+4}$$

SOLUTION   Since $x^2 - 16 = (x+4)(x-4)$, we write the equation with the $x^2 - 16$ factored as

$$\frac{x}{(x+4)(x-4)} + \frac{4}{x-4} = \frac{1}{x+4}$$

1. Clear the fractions; the LCD is $(x+4)(x-4)$.

$$(x+4)(x-4) \cdot \frac{x}{(x+4)(x-4)} + (x+4)(x-4) \cdot \frac{4}{x-4}$$
$$= (x+4)(x-4) \cdot \frac{1}{x+4}$$

2. Simplify.
$$x + 4(x+4) = x - 4$$
$$x + 4x + 16 = x - 4$$
$$5x + 16 = x - 4$$

3. Subtract 16. $5x = x - 20$

4. Subtract x. $4x = -20$

5. Divide by 4. $x = -5$

Thus, the solution is $x = -5$.

6. You can easily check this by substituting -5 for x in the original equation.

PROBLEM 3

Solve:

$$\frac{x}{x^2-16} + \frac{4}{x-4} = \frac{1}{x+4}$$

By now, you've probably noticed that we always recommend checking the possible or proposed solution by *direct substitution* into the original equation. This is done to avoid **extraneous solutions,** that is, possible or proposed solutions that do not satisfy the original equation when substituted for the variable.

Answers to PROBLEMS

3. $x = -5$

EXAMPLE 4   Solving fractional equations: No-solution case

Solve:

$$\frac{x}{x+4} - \frac{2}{5} = \frac{-4}{x+4}$$

SOLUTION   Here the denominators are 5 and $(x+4)$.

1. Clear the fractions; the LCD is $5(x+4)$.

$$5(x+4) \cdot \frac{x}{x+4} - \frac{2}{5} \cdot 5(x+4) = \frac{-4}{x+4} \cdot 5(x+4)$$

2. Simplify.

$$5x - 2x - 8 = -20$$
$$3x - 8 = -20$$

3. Add 8.

$$3x = -12$$

4. The variable is already isolated, so step 4 is not necessary.

5. Divide by 3.

$$x = -4$$

Thus, the solution seems to be $x = -4$.

6. But now let's do the check. If we substitute -4 for x in the original equation, we have

$$\frac{-4}{-4+4} - \frac{2}{5} = \frac{-4}{-4+4}$$

or

$$\frac{-4}{0} - \frac{2}{5} = \frac{-4}{0}$$

Division by zero is not defined.

Two of the terms are not defined. Thus, this equation has no solution.

PROBLEM 4

Solve:

$$\frac{x}{x-2} + \frac{3}{4} = \frac{2}{x-2}$$

NOTE

Remember, no matter how careful you are when you get an answer, call it a *possible* or a *proposed* answer. A given number is *not* an answer until you show that, when the number is substituted for the variable in the original equation, the result is a true statement.

Finally, we must point out that the equations resulting when clearing denominators are not *always* linear equations—that is, equations that can be written in the form $ax + b = c$ $(a \neq 0)$. For example, to solve the equation

$$\frac{x^2}{x+2} = \frac{4}{x+2}$$

we first multiply by the LCD $(x+2)$ to obtain

$$(x+2) \cdot \frac{x^2}{x+2} = (x+2) \cdot \frac{4}{x+2}$$

or

$$x^2 = 4$$

Answers to PROBLEMS

4. No solution

In this equation, the variable x has a 2 as an exponent; thus, it is a **quadratic** equation and can be solved when written in standard form—that is, by writing the equation as

$x^2 - 4 = 0$ Recall that a quadratic equation is an equation that can be written in standard form as $ax^2 + bx + c = 0$ ($a \neq 0$).

$$(x + 2)(x - 2) = 0 \quad \text{Factor.}$$

$$x + 2 = 0 \qquad \text{or} \qquad x - 2 = 0 \quad \text{Use the zero product property.}$$

$$x = -2 \qquad \text{or} \qquad x = 2 \quad \text{Solve each equation.}$$

Thus, $x = 2$ is a solution since

$$\frac{2^2}{2 + 2} = \frac{4}{2 + 2}$$

However, for $x = -2$,

$$\frac{x^2}{x + 2} = \frac{2^2}{-2 + 2} = \frac{4}{0}$$

and the denominator $x + 2$ becomes 0. Thus, $x = -2$ is *not* a solution; -2 is called an *extraneous* root. The only solution is $x = 2$.

EXAMPLE 5   **Solving fractional equations: Extraneous roots case**

Solve:

$$1 + \frac{3}{x - 2} = \frac{12}{x^2 - 4}$$

SOLUTION   Since $x^2 - 4 = (x + 2)(x - 2)$, the LCD is $(x + 2)(x - 2)$. We then write the equation with the denominator $x^2 - 4$ in factored form and multiply each term by the LCD as before. Here are the steps.

$$(x + 2)(x - 2) \cdot 1 + (x + 2)(x - 2) \cdot \frac{3}{x - 2} \quad \text{The LCD is } (x + 2)(x - 2).$$

$$= (x + 2)(x - 2) \cdot \frac{12}{(x + 2)(x - 2)}$$

$$(x^2 - 4) + 3(x + 2) = 12 \quad \text{Simplify.}$$

$$x^2 - 4 + 3x + 6 = 12$$

$$x^2 + 3x + 2 = 12$$

$$x^2 + 3x - 10 = 0 \quad \text{Subtract 12 from both sides to write in standard form.}$$

$$(x + 5)(x - 2) = 0 \quad \text{Factor.}$$

$$x + 5 = 0 \quad \text{or} \quad x - 2 = 0 \quad \text{Use the zero product property.}$$

$$x = -5 \qquad\qquad x = 2 \quad \text{Solve each equation.}$$

Since $x = 2$ makes the denominator $x - 2$ equal to zero, the only possible solution is $x = -5$. This solution can be verified in the original equation.

PROBLEM 5

Solve:

$$1 - \frac{4}{x^2 - 1} = \frac{-2}{x - 1}$$

B ⟩ Solving Fractional Equations for a Specified Variable

In Section 2.6, we solved a formula for a specified variable. We can also solve fractional equations for a specified variable. Here S_n is the sum of an arithmetic sequence:

$$S_n = \frac{n(a_1 + a_n)}{2}$$

Answers to PROBLEMS

5. $x = -3$

(By the way, don't be intimidated by subscripts such as n in S_n, which is read "S sub n"; they are simply used to distinguish one variable from another—for example, a_1 is different from a_n because the subscripts are different.) So, to solve for n in this equation, we proceed as follows:

$$S_n = \frac{n(a_1 + a_n)}{2}$$

Given
Since the only denominator is 2, the LCD is 2.

1. Clear any fractions; the LCD is 2:

$$2 \cdot S_n = 2 \cdot \frac{n(a_1 + a_n)}{2}$$

2. Simplify.

$$2S_n = n(a_1 + a_n)$$

3. Since we want n by itself, divide by $(a_1 + a_n)$:

$$\frac{2S_n}{a_1 + a_n} = n$$

Thus, the solution is $n = \dfrac{2S_n}{a_1 + a_n}$.

EXAMPLE 6   **Solving for a specified variable**

The sum S_n of a geometric sequence is

$$S_n = \frac{a_1(1 - r^n)}{1 - r}$$

Solve for a_1.

SOLUTION   This time, the only denominator is $1 - r$, so it must be the LCD. As before, we proceed by steps.

$$S_n = \frac{a_1(1 - r^n)}{1 - r}$$ Given

1. Clear any fractions; the LCD is $1 - r$. $(1 - r)S_n = (1 - r) \cdot \dfrac{a_1(1 - r^n)}{1 - r}$

2. Simplify. $(1 - r)S_n = a_1(1 - r^n)$

3. Divide both sides by $1 - r^n$ to isolate a_1. $\dfrac{(1 - r)S_n}{1 - r^n} = a_1$

Thus, the solution is $a_1 = \dfrac{(1 - r)S_n}{1 - r^n}$.

PROBLEM 6

Solve for a in

$$S = \frac{a(r^n - 1)}{r - 1}$$

You will have more opportunities to practice solving for a specified variable in the *Using Your Knowledge*.

> Exercises **6.5**

⟨ **A** ⟩ **Solving Fractional Equations**   In Problems 1–50, solve (if possible).

1. $\dfrac{x}{4} = \dfrac{3}{2}$

2. $\dfrac{x}{8} = \dfrac{-7}{4}$

3. $\dfrac{3}{x} = \dfrac{3}{4}$

4. $\dfrac{6}{x} = \dfrac{-2}{7}$

5. $-\dfrac{8}{3} = \dfrac{16}{x}$

6. $-\dfrac{5}{6} = \dfrac{10}{x}$

7. $\dfrac{4}{3} = \dfrac{x}{9}$

8. $-\dfrac{3}{7} = \dfrac{x}{14}$

9. $\dfrac{2}{5} + \dfrac{3}{x} = \dfrac{23}{20}$

Answers to PROBLEMS

6. $a = \dfrac{S(r - 1)}{r^n - 1}$

10. $\dfrac{6}{7} + \dfrac{2}{x} = \dfrac{3}{21}$

11. $\dfrac{3}{x} - \dfrac{2}{7} = \dfrac{11}{35}$

12. $\dfrac{4}{x} - \dfrac{2}{9} = \dfrac{22}{63}$

13. $\dfrac{3}{5} + \dfrac{7x}{10} = 2$

14. $\dfrac{2}{7} + \dfrac{4x}{21} = \dfrac{2}{3}$

15. $\dfrac{3x}{4} - \dfrac{1}{5} = \dfrac{13}{10}$

16. $\dfrac{2x}{3} - \dfrac{1}{4} = -1$

17. $\dfrac{3}{x+2} = \dfrac{4}{x-1}$

18. $\dfrac{2}{x-2} = \dfrac{5}{x+1}$

19. $\dfrac{-1}{x+1} = \dfrac{3}{x+5}$

20. $\dfrac{2}{x-1} = \dfrac{-3}{x+9}$

21. $\dfrac{3x}{x-3} + 2 = \dfrac{5x}{x-3}$

22. $\dfrac{2x}{x-2} + 18 = \dfrac{8x}{x-2}$

23. $\dfrac{5x}{x+1} - 6 = \dfrac{3x}{x+1}$

24. $\dfrac{5x}{x+1} - 2 = \dfrac{2x}{x+1}$

25. $\dfrac{x}{x^2-25} + \dfrac{5}{x-5} = \dfrac{1}{x+5}$

26. $\dfrac{x}{x^2-64} + \dfrac{8}{x-8} = \dfrac{1}{x+8}$

27. $\dfrac{x}{x^2-49} + \dfrac{7}{x-7} = \dfrac{1}{x+7}$

28. $\dfrac{x}{x^2-1} + \dfrac{1}{x-1} = \dfrac{1}{x+1}$

29. $\dfrac{x}{x+3} + \dfrac{3}{4} = \dfrac{-3}{x+3}$

30. $\dfrac{1}{5} + \dfrac{x}{x-2} = \dfrac{2}{x-2}$

31. $\dfrac{x}{x-4} - \dfrac{2}{7} = \dfrac{4}{x-4}$

32. $\dfrac{x}{x-8} - \dfrac{1}{5} = \dfrac{8}{x-8}$

33. $1 + \dfrac{2}{x-1} = \dfrac{4}{x^2-1}$

34. $1 + \dfrac{2}{x-3} = \dfrac{5}{x^2-9}$

35. $2 - \dfrac{6}{x^2-1} = \dfrac{-3}{x-1}$

36. $2 - \dfrac{4}{x^2-4} = \dfrac{-1}{x-2}$

37. $\dfrac{4}{x-3} - \dfrac{2}{x-1} = \dfrac{2}{x+2}$

38. $\dfrac{5}{x+1} - \dfrac{1}{x+2} = \dfrac{13}{x+5}$

39. $\dfrac{2x}{x^2-1} + \dfrac{4}{x-1} = \dfrac{1}{x-1}$

40. $\dfrac{3x-2}{x^2-4} + \dfrac{4}{x+2} = \dfrac{1}{x-2}$

41. $\dfrac{2z+7}{z-3} + 1 = \dfrac{z+5}{z-6} + 2$

42. $\dfrac{4x-2}{x+4} - 3 = \dfrac{2-5x}{x-2} + 6$

43. $\dfrac{2y-5}{2} - \dfrac{1}{y-1} = y+1$

44. $\dfrac{4}{x+1} - \dfrac{3}{x} = \dfrac{1}{x-2}$

45. $\dfrac{5}{2v-1} + \dfrac{2}{v} = \dfrac{18v}{4v^2-1}$

46. $\dfrac{3y}{y^2-9} - \dfrac{4}{y+3} = \dfrac{6-y}{y^2+3y}$

47. $\dfrac{z+7}{z-1} - \dfrac{z+3}{z-2} = \dfrac{3}{z-6}$

48. $\dfrac{x+1}{x-3} - \dfrac{x+5}{x-2} = \dfrac{-3}{x-1}$

49. $\dfrac{2}{x^2-4x+3} - \dfrac{5}{x^2-x-6} =$

$\dfrac{x-7}{(x-1)(x-3)(x+2)}$

50. $\dfrac{y-5}{y^2-4} - \dfrac{1}{y^2+2y-8} =$

$\dfrac{y^2}{(y^2-4)(y+4)}$

⟨ **B** ⟩   **Solving Fractional Equations for a Specified Variable**

Use the following information for Problems 51–53. Suppose you inflate a balloon for a party. How can you make it smaller? One way is to squeeze it. This will increase the pressure P of the air inside the balloon and decrease its volume V. The law stating the relationship between the pressure P and the volume V is called Boyle's law after Robert Boyle, an Irish physicist and chemist.

The law simply states that for a fixed amount of gas the product of the pressure P and the volume V is a constant, that is, $PV = $ constant provided the temperature is not changed.

If you start heating the balloon, the pressure will change!

This means that if you have two different balloons with two different pressures and volumes, the product of their respective pressures and volumes is the same constant. Thus,

$$P_1 V_1 = P_2 V_2$$

51. Solve for V_2.

52. Suppose you have a 2-liter (2-L) volume balloon so $V_1 = 2$ at the regular atmospheric pressure $P_1 = 1$ [1 atmosphere (atm) = 14.7 pounds per square inch]. If you go on a submarine with your balloon and the pressure increases to $P_2 = 2$ atmospheres, what is the volume V_2 of the balloon? (*Hint:* Use the results of Problem 51.)

53. Now, suppose you go on the space shuttle with your 2-liter balloon ($V_1 = 2$ L and $P_1 = 1$ atm) and the pressure on the space shuttle is only half as much as on earth ($P_2 = \frac{1}{2}$ atmosphere). What is the volume V_2 of your balloon?

Use the following information for Problems 54–56. How can your make your balloon smaller without squeezing it, that is, leaving the pressure constant? You can freeze it! The law stating the relationship between the volume of a gas and its temperature is called Charles' law after the French chemist, physicist, and mathematician Jacques Charles. The law simply states that for a fixed pressure, the quotient of the volume V of a gas and its temperature T (in Kelvins) is a constant. Thus,

$$\frac{V_1}{T_1} = \frac{V_2}{T_2}$$

54. Solve for V_2.

55. Consider a balloon with a volume $V_1 = 21$ liters at $T_1 = 273$ Kelvins. What will be the new volume V_2 if the temperature increases to $T_2 = 299$ Kelvins?

56. Solve for T_2.

⟩⟩⟩   *Using Your Knowledge*

Looking for Variables in the Right Places   Use your knowledge of fractional equations to solve the given problem for the indicated variable.

57. The area A of a trapezoid is

$$A = \frac{h(b_1 + b_2)}{2}$$

Solve for h.

58. When studying an electric circuit, we have to work with the equation

$$\frac{1}{R} = \frac{1}{R_1} + \frac{1}{R_2}$$

Solve for R.

59. In refrigeration we find the formula

$$\frac{Q_1}{Q_2 - Q_1} = P$$

Solve for Q_1.

60. When studying the expansion of metals, we find the formula

$$\frac{L}{1 + at} = L_0$$

Solve for t.

61. Manufacturers of camera lenses use the formula

$$\frac{1}{f} = \frac{1}{a} + \frac{1}{b}$$

Solve for f.

〉〉〉 *Write On*

62. Explain the difference between adding two rational expressions such as

$$\frac{1}{x} + \frac{1}{2}$$

and solving an equation such as

$$\frac{1}{x} + \frac{1}{2} = 1$$

63. Write the procedure you use to solve a fractional equation.

64. In Section 6.1, we used the fundamental rule of rational expressions to multiply the numerators and denominators of *fractions* so that the resulting fractions have the LCD as their denominator. In this section we multiplied each *term* of an equation by the LCD. Explain the difference.

〉〉〉 *Concept Checker*

Fill in the blank(s) with the correct word(s), phrase, or mathematical statement.

65. **Multiplying** each side of $\frac{a}{b} + \frac{c}{d} = \frac{e}{f}$ (where b, d, and $f \neq 0$) by the **least common multiple** L of the denominators is **equivalent** to multiplying each _____ in the equation by L.

66. A given number n is **not** a solution (it is only a ***proposed*** solution) of an equation until you show that **when n is substituted for the variable** in the **original** equation, the result is a _____ statement.

false	**letter**
true	**term**

〉〉〉 *Mastery Test*

Solve:

67. $1 - \dfrac{4}{x^2 - 1} = \dfrac{-2}{x - 1}$

68. $\dfrac{x}{x - 2} + \dfrac{3}{4} = \dfrac{2}{x - 2}$

69. $\dfrac{x}{x^2 - 9} + \dfrac{3}{x - 3} = \dfrac{1}{x + 3}$

70. $\dfrac{8}{x - 1} - 4 = \dfrac{2x}{x - 1}$

71. $\dfrac{4}{5} + \dfrac{1}{x} = \dfrac{3}{10}$

72. $\dfrac{1}{x^2 + 2x - 3} + \dfrac{1}{x^2 - 9} = \dfrac{1}{(x - 1)(x^2 - 9)}$

Solve for the specified variable:

73. $C = \dfrac{5}{9}(F - 32); F$

74. $A = P(1 + r); r$

〉〉〉 *Skill Checker*

Solve:

75. $4(x + 3) = 45$

76. $\dfrac{d}{3} + \dfrac{d}{4} = 1$

77. $\dfrac{h}{4} + \dfrac{h}{6} = 1$

78. $60(R - 5) = 40(R + 5)$

79. $30(R - 5) = 10(R + 15)$

80. $\dfrac{n}{90} = \dfrac{1}{15}$

6.6 Ratio, Proportion, and Applications

▶ Objectives

A ⟩ Solve proportions.

B ⟩ Solve applications involving ratios and proportions.

▶ To Succeed, Review How To . . .

1. Find the LCD of two or more fractions (pp. 13–15).
2. Solve linear equations (pp. 132–136).
3. Use the RSTUV method to solve word problems (p. 143–144).

▶ Getting Started

Coffee, Ratio, and Proportion

The manufacturer of this jar of instant coffee claims that 4 ounces of its instant coffee is equivalent to 1 pound (16 ounces) of regular coffee. In mathematics, we say that the ratio of instant coffee used to regular coffee used is 4 to 16. The ratio 4 to 16 can be written as the fraction $\frac{4}{16}$ or in symbols as 4:16.

In this section we shall learn how to use ratios to solve proportions and to solve different applications involving these proportions.

A ⟩ Solving Proportions

DEFINITION OF A RATIO	A **ratio** is a quotient of two numbers. There are *three* ways in which the ratio of a number *a* to another number *b* can be written: **1.** *a* to *b* **2.** *a:b* **3.** $\frac{a}{b}$

The equality of two ratios is called a **proportion.**

Let's look more closely at the coffee jar in the *Getting Started*. The label on the back claims that 4 ounces of instant coffee will make 60 cups of coffee. Thus, the ratio of ounces to cups, when written as a fraction, is

$$\frac{4}{60} = \frac{1}{15} \qquad \begin{matrix} \leftarrow \text{Ounces} \\ \leftarrow \text{Cups} \end{matrix}$$

The fraction $\frac{1}{15}$ is called the *reduced ratio* of ounces of coffee to cups of coffee. It tells us that 1 ounce of coffee will make 15 cups of coffee. Now suppose you want to make 90 cups of coffee. How many ounces do you need? The ratio of ounces to cups is $\frac{1}{15}$, and we need to know how many ounces will make 90 cups. Let *n* be the number of ounces needed. Then

$$\frac{1}{15} = \frac{n}{90} \qquad \begin{matrix} \leftarrow \text{Ounces} \\ \leftarrow \text{Cups} \end{matrix}$$

Note that in both fractions the numerator indicates the number of ounces and the denominator indicates the number of cups. The equation $\frac{1}{15} = \frac{n}{90}$ is an equality between two ratios. In mathematics, an equality between ratios is called a **proportion.** Thus, $\frac{1}{15} = \frac{n}{90}$ is a proportion. To solve this proportion, which is simply a fractional equation, we proceed as before. First, since the LCM of 15 and 90 is 90:

1. Multiply by the LCM. $90 \cdot \frac{1}{15} = 90 \cdot \frac{n}{90}$

2. Simplify. $6 = n$

Thus, we need 6 ounces of instant coffee to make 90 cups.
       We use the same ideas in Example 1.

EXAMPLE 1   Writing ratios and solving proportions

A car travels 140 miles on 8 gallons of gas.

a. What is the reduced ratio of miles to gallons? (Note that another way to state this ratio is to use miles *per* gallon.)

b. How many gallons will be needed to travel 210 miles?

SOLUTION

a. The ratio of miles to gallons is

$$\frac{140}{8} = \frac{35}{2} \quad \begin{array}{l} \leftarrow \text{Miles} \\ \leftarrow \text{Gallons} \end{array}$$

b. Let g be the gallons needed. The ratio of miles to gallons is $\frac{35}{2}$; it is also $\frac{210}{g}$. Thus,

$$\frac{210}{g} = \frac{35}{2}$$

Multiplying by $2g$, the LCD, we have

$$2g \cdot \frac{210}{g} = 2g \cdot \frac{35}{2}$$
$$420 = 35g$$
$$g = \frac{420}{35} = 12$$

Hence 12 gallons of gas will be needed to travel 210 miles.

PROBLEM 1

A car travels 150 miles on 9 gallons of gas. How many gallons does it need to travel 900 miles?

Proportions are so common that we use a shortcut method to solve them. The method depends on the fact that if two fractions $\frac{a}{b}$ and $\frac{c}{d}$ are equivalent, their **cross products** are also equal. Here is the rule.

RULE

Cross Products

If $\frac{a}{b} \times \frac{c}{d}$, then $ad = bc$ $(b \neq 0 \text{ and } d \neq 0)$

Note that if $\quad \frac{a}{b} = \frac{c}{d}$

then $\qquad bd \cdot \frac{a}{b} = bd \cdot \frac{c}{d}$

$\qquad\qquad ad = bc$

Thus, since $\frac{1}{2} = \frac{2}{4}$, $1 \cdot 4 = 2 \cdot 2$, and since $\frac{3}{9} = \frac{1}{3}$, $3 \cdot 3 = 9 \cdot 1$. To solve the proportion

$$\frac{210}{g} = \frac{35}{2}$$

of Example 1(b), we use cross products and write:

$$210 \cdot 2 = 35g$$

$$\frac{210 \cdot 2}{35} = g \qquad \text{Dividing by 35 yields } g, \text{ as before.}$$

$$12 = g$$

> **NOTE**
>
> This technique avoids having to find the LCD first, but it applies only when you have *one* term on each side of the equation!

B > Applications Involving Ratios and Proportions

Ratios and proportions can be used to solve work problems. For example, suppose a worker can finish a certain job in 3 days, whereas another worker can do it in 4 days. How many days would it take both workers working together to complete the job? Before we solve this problem, let's see how ratios play a part in the problem itself.

Since the first worker can do the job in 3 days, she does $\frac{1}{3}$ of the job in 1 day.

The second worker can do the job in 4 days, so he does $\frac{1}{4}$ of the job in 1 day.

Working together they do the job in d days, and in 1 day, they do $\frac{1}{d}$ of the job.

Here's what happens in 1 day:

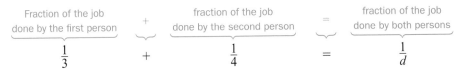

Fraction of the job done by the first person	+	fraction of the job done by the second person	=	fraction of the job done by both persons
$\frac{1}{3}$	$+$	$\frac{1}{4}$	$=$	$\frac{1}{d}$

This is a fractional equation that can be solved by multiplying each term by the LCD, $3 \cdot 4 \cdot d = 12d$. Note that we can't "cross multiply" here because we have three terms! Thus, we do it like this:

$$\frac{1}{3} + \frac{1}{4} = \frac{1}{d} \qquad \text{Given}$$

$$12d \cdot \frac{1}{3} + 12d \cdot \frac{1}{4} = 12d \cdot \frac{1}{d} \qquad \text{Multiply each term by the LCD, } 12d.$$

$$4d + 3d = 12 \qquad \text{Simplify.}$$

$$7d = 12 \qquad \text{Combine like terms.}$$

$$d = \frac{12}{7} = 1\frac{5}{7} \qquad \text{Divide by 7.}$$

Thus, if they work together they can complete the job in $1\frac{5}{7}$ days.

EXAMPLE 2   Work problems and ratios

A computer can do a job in 4 hours. Computer sharing is arranged with another computer that can finish the job in 6 hours. How long would it take for both computers operating simultaneously to finish the job?

SOLUTION   We use the RSTUV method.

1. Read the problem. We are asked to find the time it will take both computers working together to complete the job.

2. Select the unknown. Let h be the number of hours it takes to complete the job when both computers are operating simultaneously.

3. Think of a plan. Translate the problem:

The first computer does $\frac{1}{4}$ of the job in 1 hour.

The second computer does $\frac{1}{6}$ of the job in 1 hour.

When both work together, they do $\frac{1}{h}$ of the job in 1 hour.

The sum of the fractions of the job done in 1 hour by each computer, $\frac{1}{4} + \frac{1}{6}$, must equal the fraction of the job done each hour when they are operating simultaneously, $\frac{1}{h}$. Thus,

$$\frac{1}{4} + \frac{1}{6} = \frac{1}{h}$$

4. Use algebra to solve the problem. To solve this equation, we proceed as usual.

$$12h \cdot \frac{1}{4} + 12h \cdot \frac{1}{6} = 12h \cdot \frac{1}{h} \qquad \text{Multiply by the LCD, } 12h.$$

$$3h + 2h = 12 \qquad \text{Simplify.}$$

$$5h = 12$$

$$h = \frac{12}{5} = 2\frac{2}{5} \qquad \text{Divide by 5.}$$

$$= 2.4 \text{ hr}$$

Thus, both computers working together take 2.4 hours to do the job.

5. Verify the solution. We leave the verification to you.

PROBLEM 2

A worker can finish a report in 5 hours. Another worker can do it in 8 hours. How many hours will it take to finish the report if both workers work on it?

Other types of problems often require ratios and proportions for their solution—for example, distance, rate, and time problems. We've already mentioned that the formula relating these three variables is

$$D = RT$$

Distance ⎯⎯⎯ Time, Rate

We use this formula to solve a motion problem in Example 3.

EXAMPLE 3 Boat speed and proportions

Suppose you are cruising down a river one sunny afternoon in your powerboat. Before you know it, you've gone 60 miles. Now, it's time to get back. Perhaps you can make it back in the same time? Wrong! This time you cover only 40 miles in the same time. What happened? You traveled 60 miles downstream in the same time it took to travel 40 miles upstream! Oh yes, the current—it was flowing at 5 miles per hour. What was the speed of your boat in still water?

SOLUTION We use the RSTUV method.

1. Read the problem. You are asked to find the speed of your boat in still water.

2. Select the unknown. Let R be the speed of the boat in still water.

speed downstream: $R + 5$ Current helps.

speed upstream: $R - 5$ Current hinders.

3. Think of a plan. Translate the problem. The time taken downstream and upstream must be the same:

$$T_{up} = T_{down}$$

Since $D = RT$,

$$T = \frac{D}{R}$$

$$\underbrace{T_{up}} = \underbrace{T_{down}}$$

$$\frac{60}{R + 5} = \frac{40}{R - 5}$$

4. Use the cross-product rule to solve the problem.

$$60(R - 5) = 40(R + 5)$$ Cross multiply.

$$60R - 300 = 40R + 200$$ Simplify.

$$60R = 40R + 500$$ Add 300.

$$20R = 500$$ Subtract 40R.

$$R = 25$$ Divide by 20.

Thus, the speed of the boat in still water is 25 miles per hour.

5. Verify the solution. Substitute 25 for R in

$$\frac{60}{R + 5} = \frac{40}{R - 5}$$

$$\frac{60}{25 + 5} = \frac{40}{25 - 5}$$

$$2 = 2$$ A true statement

PROBLEM 3

A freight train travels 120 miles in the same time a passenger train covers 140 miles. If the passenger train is 5 miles per hour faster, what is the speed of the freight train?

Now, let's use proportion to look at the infamous 1994 Major League baseball season. Remember the 1994 season? It was cut short by a work stoppage (strike); instead of the usual 162 games, each team played only between 110 and 115 games. The abbreviated season deprived several players of an opportunity to break some of baseball's most coveted records. For example, San Francisco's Matt Williams hit 43 home runs in 112 games that year. If he could have maintained this pace for 162 games, how many homers would he have hit? Using proportions, if we assume that he would have hit h home runs in 162 games, this translates as

$$\frac{\text{Homers}}{\text{Games}} \longrightarrow \frac{h}{162} = \frac{43}{112}$$

$$112h = 43 \cdot 162$$ Cross multiply.

$$h = \frac{43 \cdot 162}{112} = 62 \text{ (to the nearest whole number)}$$ Divide by 112.

And this *would* have broken Roger Maris's record of 61 home runs in one season!

Answers to PROBLEMS

3. 30 mi/hr

EXAMPLE 4   Baseball records and proportions

Do you know what a "walk" or a "base on balls" means? This is baseball lingo for when a batter advances to first base after four pitches that are balls; sometimes he's walked intentionally. In the 1994 baseball season, Frank Thomas of the Chicago White Sox walked 121 times in 113 games! The record for "walks" is held by Babe Ruth, who had 170 walks in 152 games in the 1923 season. If Thomas could have maintained his 1994 ratio of walks for 152 games, would he have broken Ruth's record?

SOLUTION   As usual, we use our RSTUV method.

1. Read the problem. We are asked to find out whether the ratio of Thomas's walks would exceed the ratio set by Ruth—that is, would it be more than 170 walks in 152 games?

2. Select the unknown. Let w be the number of walks Thomas would receive in 152 games. His *hypothetical* ratio of walks to games is then $\frac{w}{152}$.

3. Think of a plan. We know that he had 121 walks in 113 games, so his *actual* ratio of walks to games was $\frac{121}{113}$. We can now construct a proportion using this information:

$$\frac{w}{152} = \frac{121}{113}$$

4. Use the cross-product rule to solve the problem.

$113w = 121 \cdot 152$ Cross multiply.

$w = \dfrac{121 \cdot 152}{113}$ Divide by 113.

$w = 163$ (to the nearest whole number)   Simplify.

Ruth's record stands! Since 163 is less than 170, Thomas wouldn't have broken Ruth's record had he continued to walk at the same rate as in the first 113 games. (Some purists say that a modern baseball season consists of 162 games. Would he have broken the record in 162 games? See Problem 4.)

5. Verify the solution. We leave the verification to you.

PROBLEM 4

Would Thomas have broken Ruth's record in 162 games?

In everyday life many objects are *similar* but *not* the same; that is, they have the same shape but are not necessarily the same size. Thus, a penny and a nickel are similar, and two computer or television screens may be similar. In geometry, figures that have exactly the same shape but not necessarily the same size are called **similar figures.** Look at the two similar triangles:

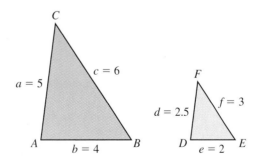

Side *a* corresponds to side *d*.
Side *b* corresponds to side *e*.
Side *c* corresponds to side *f*.

To show that corresponding sides are in proportion, we write the ratios of the corresponding sides:

$$\frac{a}{d} = \frac{5}{2.5} = 2, \qquad \frac{b}{e} = \frac{4}{2} = 2, \qquad \text{and} \qquad \frac{c}{f} = \frac{6}{3} = 2$$

As you can see, corresponding sides are proportional. We use this idea to solve Example 5.

Answers to PROBLEMS

4. Yes. In 162 games, Thomas would have had 173 walks. However, the record would be for 162 games, not for 152 games!

EXAMPLE 5   Similar triangles

Two similar triangles measured in centimeters (cm) are shown. Find f for the triangle on the right:

PROBLEM 5

Find d in the diagram.

SOLUTION   Since the triangles are similar, the corresponding sides must be proportional. Thus,

$$\frac{f}{3} = \frac{4}{6}$$

Since $\frac{4}{6} = \frac{2}{3}$, we have

$$\frac{f}{3} = \frac{2}{3}$$

$$3f = 6 \quad \text{Cross multiply.}$$

Solving this equation for f, we get

$$f = \frac{6}{3} = 2 \text{ cm}$$

Thus, f is 2 centimeters long.

> ## Exercises 6.6

Web IT *go to* **mhhe.com/bello** *for more lessons*

⟨ **A** ⟩   **Solving Proportions**   In Problems 1–13, solve the proportion problems.

1. Do you think your gas station sells a lot of gas? The greatest number of gallons sold through a single pump is claimed by the Downtown Service Station in New Zealand. It sold 7400 Imperial gallons in 24 hours. At that rate, how many Imperial gallons would it sell in 30 hours?

2. Do you like lemons? Bob Blackmore ate three whole lemons in 24 seconds. At that rate, how many lemons could he eat in 1 minute (60 seconds)?

3. Michael Cisneros ate 25 tortillas in 15 minutes. How many could he eat in 1 hour at the same rate?

4. If your lawn is yellowing, it may need "essential minor elements" administered at the rate of 2.5 pounds per 100 square feet of lawn. If your lawn covers 150 square feet, how many pounds of essential minor elements do you need?

5. If you have roses, they may need Ironite® administered at the rate of 4 pounds per 100 square feet. If your rose garden is 15 feet by 10 feet, how many pounds of Ironite do you need?

6. The directions for a certain fertilizer recommend that 35 pounds of fertilizer be applied per 1000 square feet of lawn.

   a. If your lawn covers 1400 square feet, how many pounds of fertilizer do you need?

   b. If fertilizer comes in 40-pound bags, how many bags do you need?

Answers to PROBLEMS

5. $d = \frac{8}{3} = 2\frac{2}{3}$ cm

7. You can fertilize your flowers by using 6-6-6 fertilizer at the rate of 2 pounds per 100 square feet of flower bed. If your flower bed is 12 feet by 15 feet, how many pounds of fertilizer do you need?

8. Petunias can be planted in a mixture consisting of two parts peat moss to five parts of potting soil. If you have a 20-pound bag of potting soil, how many pounds of peat moss do you need to make a mixture to plant your petunias?

9. According to the U.S. Bureau of the Census, there were 400 homeless persons "visible on the street" for every 1200 homeless in shelters in Dallas.

    a. If the homeless persons visible on the street grew to 750, how many would you expect to find in shelters?

    b. If each shelter houses 50 homeless persons, how many shelters would be needed when there are 600 homeless persons visible on the street?

10. In Minneapolis, there were 30 homeless persons visible on the street for every 1050 homeless in shelters. If the number of homeless visible on the street grew to 75, how many would you expect to find in shelters?

11. What is the ratio of "homeless in shelters" to "homeless visible on the street" in

    a. Dallas? (See Problem 9.)

    b. Minneapolis? (See Problem 10.)

    c. Where is the ratio of "homeless in shelters" to "homeless visible on the street" greater, Dallas or Minneapolis? Why do you think this is so?

12. According to the U.S. Bureau of the Census, about 40,000 persons living in Alabama were born in a foreign country. If Alabama had 4 million people, how many persons born in a foreign country would you expect if the population of Alabama increased to 5 million people?

13. In the year 2005, the U.S. population was about 295 million people, 36 million of which were foreign born. If the population in 2007 increased to 300 million people, how many foreign-born persons would you expect to be living in the United States? Answer to the nearest million.

⟨ B ⟩ Applications Involving Ratios and Proportions

14. Mr. Gerry Harley, of England, shaved 130 men in 60 minutes. If another barber takes 5 hours to shave the 130 men, how long would it take both men working together to shave the 130 men?

15. Mr. J. Moir riveted 11,209 rivets in 9 hours (a world record). If it takes another man 12 hours to do this job, how long would it take both men to rivet the 11,209 rivets?

16. It takes a printer 3 hours to print a certain document. If a faster printer can print the document in 2 hours, how long would it take both printers working together to print the document?

17. A computer can send a company's e-mail in 5 minutes. A faster computer does it in 3 minutes. How long would it take both computers operating simultaneously to send out the company's e-mail?

18. Two secretaries working together typed the company's annual report in 4 hours. If one of them could type the report by herself in 6 hours, how long would it take the other secretary working alone to type the report?

19. Two fax machines can together send all the office mailings in 2 hours. When one of the machines broke down, it took 3 hours to send all the mailings. If the number of faxes sent was the same in both cases, how many hours would it take the remaining fax machine to send all the mailings?

20. The train *Trés Grande Vitesse* covers the 264 miles from Paris to Lyons in the same time a regular train covers 124 miles. If the *Trés Grande Vitesse* is 70 miles per hour faster, how fast is it?

21. The strongest current on the East Coast of the United States is at St. Johns River in Pablo Creek, Florida, where the current reaches a speed of 6 miles per hour. A motorboat can travel 4 miles downstream on Pablo Creek in the same time it takes to go 16 miles upstream. (No, it is *not* wrong! The St. Johns flows *up*stream.) What is the speed of the boat in still water?

22. The strongest current in the United States occurs at Pt. Kootzhahoo in Chatam, Alaska. If a boat that travels at 10 miles per hour in still water takes the same time to travel 2 miles upstream in this area as it takes to travel 18 miles downstream, what is the speed of the current?

23. One of the fastest point-to-point trains in the world is the *New Tokaido* from Osaka to Okayama. This train covers 450 miles in the same time a regular train covers 180 miles. If the *Tokaido* is 60 miles per hour faster than the regular train, how fast is it?

24. The world's strongest current, reaching 18 miles per hour, is the Saltstraumen in Norway. A motorboat can travel 48 miles downstream in the Saltstraumen in the same time it takes to go 12 miles upstream. What is the speed of the boat in still water?

25. In the abbreviated 1994 baseball season, Ken Griffey of the Seattle Mariners hit 40 home runs in 111 games.

    a. At this rate, how many home runs would he have hit during a regular season (162 games)? Answer to the nearest whole number.

    b. Would he have broken Roger Maris's record of 61 home runs in one season?

26. Frank Thomas gained 291 total bases in 113 games in the 1994 baseball season.

    a. At this rate, how many games would he need to reach 457 total bases, Babe Ruth's Major League record? Answer to the nearest whole number.

    b. A regular season consists of 162 games. Would Thomas have broken Ruth's record?

27. Frank Thomas scored 106 runs in 113 games in the 1994 baseball season. At this rate, how many games would he need to score 177 runs (Babe Ruth's record)? Answer to the nearest whole number. Could he have broken the record in a regular baseball season consisting of 162 games?

28. The modern record for the most singles in a season belongs to Lloyd James Waner, who hit 198 singles in 150 games in 1927. To the nearest whole number, how many singles would a player have to hit in a 162-game season to equal Waner's record?

In Problems 29–41, the pairs of triangles are similar. Find the lengths of the indicated unknown sides.

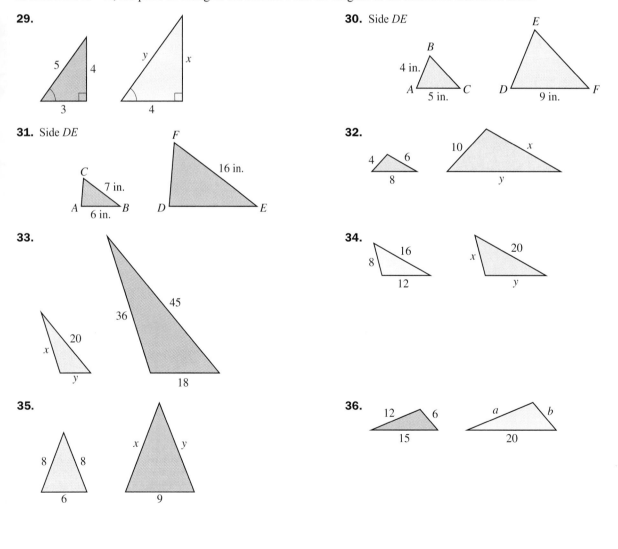

29.

30. Side *DE*

31. Side *DE*

32.

33.

34.

35.

36.

37.

38.

39.

40.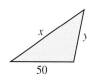

41.

> > > **Applications**

42. *Film processing*   A film processing department can process nine rolls of film in 2 hours. At that rate, how long will it take them to process 20 rolls of film?

43. *Blueprints*   If a blueprint uses a scale of $\frac{1}{2}$ inch = 3 feet, find the scaled-down dimensions of a room that is 9 feet × 12 feet.

44. *Travel*   Latrice and Bob want to drive from Miramar, Florida, to Pittsburgh. Pennsylania, a distance of approximately 1000 miles. Each day they plan to drive 7 hours and cover 425 miles.

   a. At that rate, how many driving hours should the trip take?

   b. At that rate, how many days should the trip take?

45. *Cycling*   Marquel is an avid bicyclist and averages 16 miles per hour when he rides. He likes to ride 80 miles per day. He plans to take a 510-mile, round trip, between New Orleans and Panama City.

   a. At that rate, how many hours of riding will it take Marquel to complete the trip?

   b. At that rate, how many days should it take Marquel to complete the trip?

46. *Batting average*   Early in the season a softball player has a 0.250 batting average (5 hits in 20 times at bat). How many consecutive hits would she need to bring her average to 0.348?

> > > **Using Your Knowledge**

A Matter of Proportion   Proportions are used in many areas other than mathematics. The following problems are typical.

47. In a certain experiment, a physicist stretched a spring 3 inches by applying a 7-pound force to it. How many pounds of force would be required to make the spring stretch 8 inches?

48. Suppose a photographer wishes to enlarge a 2-inch by 3-inch picture so that the longer side is 10 inches. How wide does she have to make it?

49. In carpentry the pitch of a rafter is the ratio of the rise to the run of the rafter. What is the rise of a rafter having a pitch of $\frac{2}{5}$ if the run is 15 feet?

50. In an automobile the rear axle ratio is the ratio of the number of teeth in the ring gear to the number of teeth in the pinion gear. A car has a 3-to-1 rear axle ratio, and the ring gear has 60 teeth. How many teeth does the pinion gear have?

51. A zoologist took 250 fish from a lake, tagged them, and released them. A few days later 53 fish were taken from the lake, and 5 of them were found to be tagged. Approximately how many fish were originally in the lake?

〉〉〉 *Write On*

52. What is a proportion?

53. Write the procedure you use to solve a proportion.

54. We can solve proportions by using their cross products. Can you solve the equation

$$\frac{x}{2} + \frac{x}{4} = 3$$

by using cross products? Explain.

55. Find some examples of similar figures and then write your own definition of similar figures.

〉〉〉 *Concept Checker*

Fill in the blank(s) with the correct word(s), phrase, or mathematical statement.

56. A _____ is a **quotient of two numbers.**

57. Using the **cross-product** rules, if $\frac{a}{b} = \frac{c}{d}$ (where $b \neq 0$ and $d \neq 0$), then _____.

58. If R and T are the **rate** and **time**, respectively, the **distance** D is given by _____.

59. Figures that have **exactly the same shape** but **not** necessarily the **same size** are called _____ figures.

$ac = bd$	$\frac{R}{T}$
$ad = bc$	**equal**
ratio	**similar**
RT	

〉〉〉 *Mastery Test*

60. A freight train travels 90 miles in the same time a passenger train travels 105 miles. If the passenger train is 5 miles per hour faster, what is the speed of the freight train?

61. A typist can finish a report in 5 hours. Another typist can do it in 8 hours. How many hours will it take to finish the report if both work on it?

62. A car travels 150 miles on 9 gallons of gas. How many gallons does the car need to travel 800 miles?

63. A baseball player has 160 singles in 120 games. If he continues at this rate, how many singles will he hit in 150 games?

64. Find $\frac{d}{2}$ for the triangle in Example 5.

〉〉〉 *Skill Checker*

65. Solve: $600 = 120k$

66. Solve: $300 = 15k$

67. Solve: $60 = \frac{k}{4}$

68. Solve: $23 = \frac{k}{5}$

6.7   Direct and Inverse Variation

▶ Objectives

A ⟩ Find and solve equations of direct variation given values of the variables.

B ⟩ Find and solve equations of inverse variation given values of the variables.

C ⟩ Solve applications involving variation.

▶ To Succeed, Review How To . . .

Solve linear equations (pp. 132–136).

▶ Getting Started
Don't Forget the Tip!

Jasmine is a server at CDB restaurant. Aside from her tips, she gets $2.88/hour. In 1 hour, she earns $2.88; in 2 hr, she earns $5.76; in 3 hr, she earns $8.64, and so on. We can form the set of ordered pairs (1, 2.88), (2, 5.76), (3, 8.64) using the number of hours she works as the first coordinate and the amount she earns as the second coordinate. Note that the ratio of second coordinates to first coordinates is the same number:

$$\frac{2.88}{1} = 2.88, \quad \frac{5.76}{2} = 2.88, \quad \frac{8.64}{3} = 2.88,$$

and so on.

When the ratio of ordered pairs of numbers is constant, we say that there is a **direct variation.** In this case, the earnings E **vary directly** (or are **directly proportional**) to the number of hours h, that is,

$$\frac{E}{h} = 2.88 \quad \text{(a constant)} \quad \text{or} \quad E = 2.88h$$

Note that in the graph of $E = 2.88h$ or the more familiar $y = 2.88x$, the constant $k = 2.88$ is the **slope** of the line.

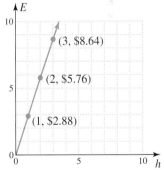

A ⟩ Direct Variation

In general, if we have a situation in which a variable **varies directly** or **is directly proportional** to another variable, we make the following definition.

> **DEFINITION**
>
> **y varies directly as x (or y is proportional to x)** if there is a constant k such that $y = kx$. The constant k is the **constant of variation or proportionality.**

In real life, we can find k by experimenting. Suppose you are dieting and want to eat some McDonald's® fries, but only 100 calories worth. The fries come in small, medium, or large sizes with 210, 450, or 540 calories, respectively. You could use **estimation** and eat approximately half of a small order $\left(\frac{210}{2} \approx 105 \text{ calories} \right)$, but you can estimate better! The numbers of fries in the small, medium, and large sizes are 45, 97, and 121; thus, the calories per fry are:

$$\frac{210}{45} \approx 4.7 \text{ (small)}, \quad \frac{450}{97} \approx 4.6 \text{ (medium)}, \quad \frac{540}{121} \approx 4.5 \text{ (large)}$$

Source: Author (counted and ate!).

EXAMPLE 1   Using direct variation

The number C of calories in an order of McDonald's french fries varies directly as the number n of fries in the order. Suppose you buy a large order of fries (121 fries, 540 calories).

a. Write an equation of variation.

b. Find k.

c. If you want to eat 100 calories worth, how many fries would you eat?

SOLUTION

a. If C varies directly as n, then $C = kn$.

b. A large order of fries has $C = 540$ calories and $n = 121$ fries. Thus,

$$C = kn \quad \text{becomes} \quad 540 = k(121) \quad \text{or} \quad k = \frac{540}{121} \approx 4.5$$

c. If you want to eat $C = 100$ calories worth,

$$C = kn = 4.5n \quad \text{becomes} \quad 100 = 4.5n$$

Solving for n by dividing both sides by 4.5, we get

$$n = \frac{100}{4.5} \approx 22$$

So, if you eat about 22 fries, you will have consumed about 100 calories. Note that this is indeed about half of a small order!

PROBLEM 1

The number C of calories in an order of Burger King® fries also varies directly as the number n of fries in the order. An order of king-size fries has 590 calories and 101 fries.

a. Write an equation of variation.

b. Find k.

c. If you want to eat 100 calories worth, how many fries would you eat?

Now that we have eaten some calories, let's see how we can spend (burn) some calories. Read on.

EXAMPLE 2   Using direct variation

If you weigh about 160 pounds and you jog (5 miles per hour) or ride a bicycle (12 miles per hour), the number C of calories used is proportional to the time t (in minutes).

a. Find an equation of variation.

b. If jogging for 15 minutes uses 150 calories, find k.

c. How many calories would you use if you jog for 20 minutes?

d. To lose a pound, you have to use about 3500 calories. How many minutes do you have to jog to lose 1 pound?

PROBLEM 2

If you weigh 120 pounds and you jog (5 miles per hour) or ride a bicycle (12 miles per hour), the number C of calories used is proportional to the time t (in minutes).

a. Find an equation of variation.

b. If jogging for 15 minutes uses 105 calories, find k.

Answers to PROBLEMS

1. a. $C = kn$ **b.** $k = \frac{590}{101} \approx 5.8$ **c.** $\frac{100}{5.8} \approx 17$ (about 17 fries)   **2. a.** $C = kt$ **b.** $C = 7t$ **c.** 140 calories   **d.** 500 minutes

SOLUTION

a. Since the number C of calories used is proportional to the time t (in minutes), the equation of variation is $C = kt$.

b. If jogging for 15 minutes ($t = 15$) uses $C = 150$ calories, then

$C = kt$ becomes $150 = k(15)$

Dividing both sides by 15 $10 = k$

Note that now $C = kt$ becomes $C = 10t$

c. We want to know how many calories C you would use if you jog for $t = 20$ minutes. Substitute 20 for t in $C = 10t$

obtaining $C = 10(20) = 200$

Thus, you will use 200 calories if you jog for 20 minutes.

d. To lose 1 pound, you need to use $C = 3500$ calories.

$C = 10t$ becomes $3500 = 10t$

Dividing both sides by 10 $350 = t$

Thus, you need 350 minutes of jogging to lose 1 pound! By the way, if you jog for 350 minutes at 5 miles per hour, you will be 29 miles away!

c. How many calories would you use if you jog for 20 minutes?

d. To lose a pound, you have to use about 3500 calories. How many minutes do you have to jog to lose 1 pound?

B 〉 Inverse Variation

Sometimes, as one quantity increases, a related quantity decreases proportionately. For example, the more time we spend practicing a task, the less time it will take us to do the task. In this case, we say that the quantities **vary inversely** as each other.

DEFINITION	y **varies inversely as x** (or is **inversely proportional** to x) if there is a constant k such that $$y = \frac{k}{x}$$

EXAMPLE 3   Inverse proportion problem

The rate of speed v at which a car travels is inversely proportional to the time t it takes to travel a given distance.

a. Write the equation of variation.

b. If a car travels at 60 miles per hour for 3 hours, what is k, and what does it represent?

SOLUTION

a. The equation is

$$v = \frac{k}{t}$$

b. We know that $v = 60$ when $t = 3$. Thus,

$$60 = \frac{k}{3}$$

$$k = 180$$

In this case, k represents the distance traveled, and the new equation of variation is

$$v = \frac{180}{t}$$

PROBLEM 3

Suppose a car travels at 55 miles per hour for 2 hours.

a. Find k and explain what it represents.

b. Write the new equation of variation.

Answers to PROBLEMS

3. a. $k = 110$ and represents the distance traveled.   **b.** $v = \frac{110}{t}$

EXAMPLE 4   Boom boxes and inverse proportion

Have you ever heard one of those loud "boom" boxes or a car sound system that makes your stomach tremble? The loudness L of sound is inversely proportional to the square of the distance d that you are from the source.

a. Write an equation of variation.

b. The loudness of rap music coming from a boom box 5 feet away is 100 decibels (dB). Find k.

c. If you move to 10 feet away from the boom box, how loud is the sound?

SOLUTION

a. The equation is

$$L = \frac{k}{d^2}$$

b. We know that $L = 100$ for $d = 5$, so that

$$100 = \frac{k}{5^2} = \frac{k}{25}$$

Multiplying both sides by 25, we find that $k = 2500$, and the new equation of variation is

$$L = \frac{2500}{d^2}$$

c. When $d = 10$,

$$L = \frac{2500}{10^2} = 25 \text{ dB}$$

PROBLEM 4

Suppose the loudness is 80 decibels at 5 feet.

a. Find k.

b. Write the new equation of variation.

c. If you move 10 feet away, how loud is the sound now?

> **Answers to PROBLEMS**
>
> 4. a. $k = 2000$
>   b. $L = \frac{2000}{d^2}$
>   c. 20 dB
> 5. $g = \frac{3}{2}A$

C ⟩ Applications Involving Variation

EXAMPLE 5   Using direct variation

Figure 6.1 shows the number of gallons of water, g (in millions), produced by an inch of snow in different cities. Note that the larger the area of the city, the more gallons of water are produced, so g is directly proportional to A, the area of the city (in square miles).

a. Write an equation of variation.

b. If the area of St. Louis is about 62 square miles, what is k?

c. Find the amount of water produced by 1 inch of snow falling in Anchorage, Alaska, with an area of 1700 square miles.

SOLUTION

a. Since g is directly proportional to A, $g = kA$.

b. From Figure 6.1 we can see that $g = 100$ (in millions) is the number of gallons of water produced by 1 inch of snow in St. Louis. Since it is given that $A = 62$, $g = kA$ becomes

$$100 = k \cdot 62 \quad \text{or} \quad k = \frac{100}{62} = \frac{50}{31}$$

c. For Anchorage, $A = 1700$, thus, $g = \frac{50}{31} \cdot 1700 \approx 2742$ million gallons of water.

PROBLEM 5

If the area of the city of Boston is about 40 square miles and 60 (million) gallons of water are produced by 1 inch of snow, find an equation of variation for Boston.

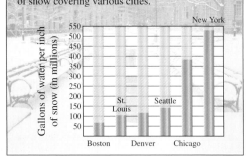

Snow brings needed water
Here's how much water is in each inch of snow covering various cities.

>**Figure 6.1**

Source: Data from USA Today, 1994.

⟩ Exercises **6.7**

⟨ A ⟩ Direct Variation   In Problems 1–7, solve the direct variation problems.

1. Do you know how much your skin weighs? Your weight W (in pounds) is proportional to the weight S of your skin, which can be obtained by dividing W by 16.

   a. Write an equation of variation.

   b. What is k?

   c. If you weigh 160 pounds, what is the weight of your skin?

2. How much blood do you have? Your weight W (in kilograms) is directly proportional to your blood volume V (in liters) and can be obtained by dividing W by 12.

   a. Write an equation of variation.

   b. What is k?

   c. If you weigh 64 kilograms (about 140 pounds), what is your blood volume?

3. Your weight W (in pounds) is directly proportional to your basal metabolic rate R (number of calories you burn when at rest) and can be obtained by dividing W by 10.

   a. Write an equation of variation.

   b. What is k?

   c. If you weigh 160 pounds, what is the value of your basal metabolic rate?

4. The amount I of annual interest received on a savings account is directly proportional to the amount m of money you have in the account.

   a. Write an equation of variation.

   b. If \$480 produced \$26.40 in interest, what is k?

   c. How much annual interest would you receive if the account had \$750?

5. The number R of revolutions a record makes as it is being played varies directly with the time t that it is on the turntable.

   a. Write an equation of variation.

   b. A record that lasted $2\frac{1}{2}$ minutes made 112.5 revolutions. What is k?

   c. If a record makes 108 revolutions, how long does it take to play the entire record?

6. The distance d an automobile travels after the brakes have been applied varies directly as the square of its speed s.

   a. Write an equation of variation.

   b. If the stopping distance for a car going 30 miles per hour is 54 feet, what is k?

   c. What is the stopping distance for a car going 60 miles per hour ?

7. The weight of a person varies directly as the cube of the person's height h (in inches). The **threshold weight** T (in pounds) for a person is defined as the "crucial weight, above which the mortality risk for the person rises astronomically."

   a. Write an equation of variation relating T and h.

   b. If $T = 196$ when $h = 70$, find k to five decimal places.

   c. To the nearest pound, what is the threshold weight for a person 75 inches tall?

⟨ B ⟩ Inverse Variation   In Problems 8–15, solve the inverse variation problems.

8. To remain popular, the number S of new songs a rock band needs to produce each year is inversely proportional to the number y of years the band has been in the business.

   a. Write an equation of variation.

   b. If, after 3 years in the business, the band needs 50 new songs, how many songs will it need after 5 years?

9. When a camera lens is focused at infinity, the f-stop on the lens varies inversely with the diameter d of the aperture (opening).

   a. Write an equation of variation.

   b. If the f-stop on a camera is 8 when the aperture is $\frac{1}{2}$ inch, what is k?

   c. Find the f-stop when the aperture is $\frac{1}{4}$ inch.

10. Boyle's law states that if the temperature is held constant, then the pressure P of an enclosed gas varies inversely as the volume V. If the pressure of the gas is 24 pounds per square inch when the volume is 18 cubic inches, what is the pressure if the gas is compressed to 12 cubic inches?

11. For the gas of Problem 10, if the pressure is 24 pounds per square inch when the volume is 18 cubic inches, what is the volume if the pressure is increased to 40 pounds per square inch?

12. The weight W of an object varies inversely as the square of its distance d from the center of the Earth.

    a. Write an equation of variation.

    b. An astronaut weighs 121 pounds on the surface of the Earth. If the radius of the Earth is 3960 miles, find the value of k for this astronaut. (Do not multiply out your answer.)

    c. What will this astronaut weigh when she is 880 miles above the surface of the Earth?

14. The price P of oil varies inversely with the supply S (in million barrels per day). In the year 2000, the price of one barrel of oil was \$26.00 and OPEC production was 24 million barrels per day.

    a. Write an equation of variation.

    b. What is k?

    c. If OPEC plans to increase production to 28 million barrels per day, what would the price of one barrel be? Answer to the nearest cent.

13. One of the manuscript pages of this book had about 600 words and was typed using a 12-point font. Suppose the average number w of words that can be printed on a manuscript page is inversely proportional to the font size s.

    a. Write an equation of variation.

    b. What is k?

    c. How many words could be typed on the page if a 10-point font was used?

15. According to the National Center for Health Statistics, the number b of births (per 1000 women) is inversely proportional to the age a of the woman. The number b of births (per 1000 women) for 27-year-olds is 110.

    a. Write an equation of variation.

    b. What is k?

    c. What would you expect the number b of births (per 1000 women) to be for 33-year-old women?

⟨ C ⟩ Applications Involving Variation

16. *Miles per gallon*   The number of miles m you can drive in your car is directly proportional to the amount of fuel g in your gas tank.

    a. Write an equation of variation.

    b. The greatest distance yet driven without refueling on a single fill in a standard vehicle is 1691.6 miles. If the twin tanks used to do this carried a total of 38.2 gallons of fuel, what is k (round to two decimal places)?

    c. How many miles per gallon is this?

18. *Responses to radio call-in contest*   Have you called in on a radio contest lately? According to Don Burley, a radio talk-show host in Kansas City, the listener response to a radio call-in contest is directly proportional to the size of the prize.

    a. If 40 listeners call when the prize is \$100, write an equation of variation using N for the number of listeners and P for the prize in dollars.

    b. How many calls would you expect for a \$5000 prize?

20. *Atmospheric carbon dioxide concentration*   The concentration of carbon dioxide (CO_2) in the atmosphere has been increasing due to automobile emissions, electricity generation, and deforestation. In 1965, CO_2 concentration was 319.9 parts per million (ppm), and 23 years later, it had increased to 351.3 ppm. The *increase I*, of carbon dioxide concentration in the atmosphere is directly proportional to the number n of years elapsed since 1965.

    a. Write an equation of variation for I.

    b. Find k (round to two decimal places).

    c. What would you predict the CO_2 concentration to be in the year 2000 (round to two decimal places)?

22. *Water depth and temperature*   At depths of more than 1000 meters (a kilometer), water temperature T (in degrees Celsius) in the Pacific Ocean varies inversely as the water depth d (in meters). If the water temperature at 4000 meters is 1°C, what would it be at 8000 meters?

17. *Distance and speed of car*   The distance d (in miles) traveled by a car is directly proportional to the average speed s (in miles per hour) of the car, even when driving in reverse!

    a. Write an equation of variation.

    b. The highest average speed attained in any nonstop reverse drive of more than 500 miles is 28.41 miles per hour. If the distance traveled was 501 miles, find k (round to two decimal places).

    c. What does k represent in this situation?

19. *Cricket chirps and temperature*   The number C of chirps a cricket makes each minute is directly proportional to 37 less than the temperature F in degrees Fahrenheit.

    a. If a cricket chirps 80 times when the temperature is 57°F, what is the equation of variation?

    b. How many chirps per minute would the cricket make when the temperature is 90°F?

21. *Percent increase of college graduates*   The *cumulative increase I* in the *percent* of college graduates in the United States after 1930 among persons 25 years and older is proportional to the square of the number of years after 1930. In 1940, the cumulative increase was about 5%.

    a. Write an equation of variation for I if n is the number of years elapsed since 1930.

    b. Find k.

    c. What would you predict the cumulative percent increase to be in the year 2000?

23. *Blood alcohol concentration*   You might think that the BAC is directly proportional to how many beers you drink in an hour. Strangely enough, for both males and females of a specific weight, the BAC is directly proportional to $(N - 1)$, which is one less than the number of beers consumed during the last hour.

   a. Write an equation of variation.

   b. For a 150-pound man, the average BAC after 3 beers is 0.052. Find k.

   c. What is the BAC after 5 beers?

   d. How many beers can the man drink before going over the 0.08 limit?

24. *Blood alcohol concentration*   For a 130-pound female, the average BAC is also directly proportional to one less than the number of beers consumed during the last hour $(N - 1)$.

   a. Write an equation of variation.

   b. For a 130-pound female, the average BAC after 3 beers is 0.066. Find k.

   c. What is the BAC after 5 beers?

   d. How many beers can the woman drink before going over the 0.08 limit?

25. *BAC and weight*   If you drank 3 beers in the last hour, your blood alcohol content (BAC) is inversely proportional to your weight W. For a 130-pound male, the BAC after drinking 3 beers is 0.06.

   a. Write an equation of variation.

   b. What is the BAC of a 260-pound male after drinking 3 beers?

   c. In most states, you are legally drunk if your BAC is 0.08 or higher. What is the weight of a male whose BAC is exactly 0.08 after drinking 3 beers in the last hour?

   d. What would your BAC be if you weigh more than the male in part **c**?

26. *BAC and weight*   If you drank 3 beers in the last hour, your blood alcohol content (BAC) is inversely proportional to your weight W. For a 130-pound female, the BAC after drinking 3 beers is 0.066.

   a. Write an equation of variation.

   b. What is the BAC of a 260-pound female after drinking those 3 beers?

   c. In most states, you are legally drunk if your BAC is 0.08 or higher. What is the weight of a female whose BAC is exactly 0.08 after drinking 3 beers in the last hour?

   d. What would your BAC be if you weigh more than the female in part **c**?

〉〉〉   *Using Your Knowledge*

27. Do you know what your Sun Protection Factor (SPF) is? The SPF of a sunscreen indicates the time period you can stay in the sun without burning, based on your complexion. The time T you are in the sun is directly proportional to your SPF (S). For example, suppose you can stay in the sun for 15 minutes without burning. Then, $T = 15S$ and we assume that your S is 1. If you want to stay in the sun for 30 minutes, $T = 30 = 15S$ and now you need a sunscreen with an SPF of 2, written as

SPF2. Thus, you can stay out twice as much time in the sun $(2 \times 15 = 30)$ without burning.

   a. Write an equation of variation relating the time T and the SPF S.

   b. What SPF do you need if you want to stay out for an hour without burning?

〉〉〉   *Write On*

28. Write in your own words what it means for two variables to be directly proportional. Give examples of variables that are directly proportional.

29. Write in your own words what it means for two variables to be inversely proportional. Give examples of variables that are inversely proportional.

30. Explain in your own words the type of relationship (direct or inverse) that should exist between the blood alcohol concentration (BAC) and:

   a. The weight of the person

   b. The number of beers the person has had in the last hour

   c. The gender of the person

31. In Problems 23 and 24, we stated that for a specific weight the BAC is directly proportional to one less than the number of beers consumed in the last hour $(N - 1)$. Why do you think it is not directly proportional to the number N of beers consumed in the last hour? (You can do a Web search to explore this further.)

32. In the definition of direct variation, there is a constant of proportionality, k. In your own words, what is this k and what does it represent?

〉〉〉   *Concept Checker*

Fill in the blank(s) with the correct word(s), phrase, or mathematical statement.

33. y **varies** _____ **as** x **if there is a constant** k **such that** $y = kx$.

inversely

34. y **varies** _____ **as** x **if there is a constant** k **such that** $y = \frac{k}{x}$.

directly

〉 〉 〉 *Mastery Test*

35. The number of calories C in an order of KFC Colonel's Crispy Strips varies directly as the number n of strips in the order. Suppose you buy an order of the Strips (3 strips, 300 calories).

    a. Write an equation of variation.

    b. Find k.

    c. If you want to eat 200 calories, how many strips would you eat?

36. If you weigh about 160 pounds and ride a bicycle at 12 miles per hour, the number C of calories used is proportional to the time t (in minutes) you ride.

    a. Find an equation of variation.

    b. If bicycling for 15 minutes burns about 105 calories, find k.

    c. How many calories would you burn if you bicycled for 20 minutes?

    d. To lose a pound, you have to burn about 3500 calories. How many minutes do you have to bicycle to lose 1 pound?

37. The rate of speed v at which a train travels is inversely proportional to the time t it takes the train to travel a given distance.

    a. Write the equation of variation.

    b. If a train travels at 30 miles per hour for 4 hours, find k.

    c. What does the constant k represent in this situation?

38. If you are in the front row of a rock concert (10 feet away) the loudness L of the music is inversely proportional to the square of the distance d you are from the source.

    a. Write an equation of variation.

    b. If the loudness of the music in the front row (10 feet away) is 110 decibels, find k.

    c. If you move 25 feet away from the music, how loud is the sound?

〉 〉 〉 *Skill Checker*

In Problems 39–42, graph the equation.

39. $x + 2y = 4$

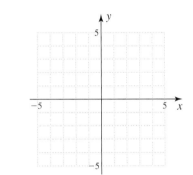

40. $2y - x = 0$

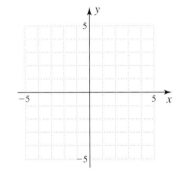

41. $y - 2x = 4$

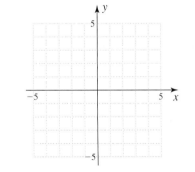

42. $2y - 4x = 4$

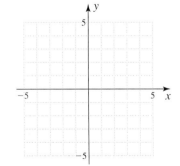

In Problems 43–44, determine whether the lines are parallel.

43. $x + y = 4$
      $2x - y = -1$

44. $x + 2y = 4$
      $2x + 4y = 6$

〉Collaborative Learning

BAC Meters

In this section we have stated that the blood alcohol concentration (BAC) is directly proportional to one less than the number of beers consumed during the last hour ($N - 1$). No, there will be no beer drinking, but we want to corroborate this information.

1. Form several groups and find websites that will automatically tell you the BAC based on the number of beers consumed during the last hour. (Try an online search for "BAC meter.") How do the BAC values compare with those given for the 150-pound man and 130-pound woman mentioned in Problems 23 and 24?

2. Do the meters you found corroborate the information in Problems 23 and 24?

3. What are the factors considered by the BAC meter your group is using?

4. Compare the results you get with different online BAC meters. Are the results different? Why do you think that is?

〉Collaborative Learning

Golden Rectangles

A **golden rectangle** is described as one of the most pleasing shapes to the human eye. Fill in the blanks in the table.

Item	Length	Width	Ratio of length to width	Golden rectangle?
3 × 5 card				
8.5 × 11 sheet of paper				
Desktop				
Textbook				
Teacher's desk				
ID card				
Anything else of interest . . .				

Source: Oswego (NY) City School District Regents Exam Prep Center.

1. Which is the ratio of length to width in a golden rectangle?

2. Form several groups. Each group select a topic:

   The golden rectangle in architecture

   The golden section in art

   The golden section in music

   Magical properties of the golden section

   Make a report to the rest of the group about your findings.

〉Research Questions

In this chapter we have written rational numbers either as fractions ($\frac{2}{7}$) or decimals (0.285714285714285714 . . .). The ancient Egyptians used a number system based on *unit fractions*—fractions with a 1 in the numerator. This idea let them represent numbers such as $\frac{1}{7}$ easily enough; other numbers such as $\frac{2}{7}$ were represented as sums of unit fractions (e.g., $\frac{2}{7} = \frac{1}{4} + \frac{1}{28}$). Further, the same fraction could not be used twice (so $\frac{2}{7} = \frac{1}{7} + \frac{1}{7}$ is not allowed). We call a formula representing a sum of distinct unit fractions an *Egyptian fraction*. (*Source*: David Eppstein, ICS, University of California, Irvine.) Do a web search for Egyptian fractions to answer the following questions.

(continued)

1. Write $\frac{3}{7}$ as an Egyptian fraction.

2. There are many algorithms (procedures) that will show how to write a fraction as an Egyptian fraction. Name three of these algorithms.

3. One of the algorithms uses *continued fractions*. Describe what a *continued fraction* is and how they are used to write fractions as Egyptian fractions.

4. The Indian mathematician Aryabhata (d. 550 A.D.) used a continued fraction to solve a linear *indeterminate equation*. What is an *indeterminate equation*?

5. Write a short paragraph about Aryabhata and his contributions to mathematics.

6. Can you write $\frac{5}{6}$ as an Egyptian fraction? Find two sites that will do it for you.

7. Use the sites you found in question 6 to write $\frac{5}{11}$ as an Egyptian fraction. Do you get the same answer at both sites?

❭Summary Chapter 6

Section	Item	Meaning	Example
6.1	Rational number	A number that can be written as $\frac{a}{b}$, a and b integers and b not 0	$\frac{3}{4}$, $-\frac{6}{5}$, 0, -8, and $1\frac{1}{3}$ are rational numbers.
	Rational expression	An expression of the form $\frac{A}{B}$, where A and B are polynomials, $B \neq 0$	$\frac{x^2 + 3}{x - 1}$ and $\frac{y^2 + y - 1}{3y^3 + 7y + 8}$ are rational expressions.
6.1B	Fundamental rule of rational expressions	$\dfrac{A}{B} = \dfrac{A \cdot C}{B \cdot C}$ $B \neq 0, C \neq 0$	$\dfrac{3}{4} = \dfrac{3 \cdot 5}{4 \cdot 5}$ and $\dfrac{x}{x^2 + 2} = \dfrac{5 \cdot x}{5 \cdot (x^2 + 2)}$
6.1C	Reducing a rational expression	The process of removing a common factor from the numerator and denominator of a rational expression	$\dfrac{8}{6} = \dfrac{2 \cdot 4}{2 \cdot 3} = \dfrac{4}{3}$ is reduced.
	Standard form of a fraction	$\frac{a}{b}$ and $\frac{-a}{b}$ ($b \neq 0$) are the standard forms of a fraction.	$-\frac{-a}{-b}$ is written as $\frac{-a}{b}$ and $\frac{a}{-b}$ is written as $\frac{-a}{b}$.
6.2A	Multiplication of rational expressions	$\dfrac{A}{B} \cdot \dfrac{C}{D} = \dfrac{A \cdot C}{B \cdot D}$ $B \neq 0, D \neq 0$	$\dfrac{3}{4} \cdot \dfrac{7}{5} = \dfrac{3 \cdot 7}{4 \cdot 5} = \dfrac{21}{20}$
6.2B	Division of rational expressions	$\dfrac{A}{B} \div \dfrac{C}{D} = \dfrac{A \cdot D}{B \cdot C}$ $B \neq 0, C \neq 0, D \neq 0$	$\dfrac{3}{7} \div \dfrac{5}{4} = \dfrac{3 \cdot 4}{7 \cdot 5} = \dfrac{12}{35}$
6.3B	Addition of rational expressions with different denominators	$\dfrac{A}{B} + \dfrac{C}{D} = \dfrac{AD + BC}{BD}$ $B \neq 0, D \neq 0$	$\dfrac{1}{6} + \dfrac{1}{5} = \dfrac{5 + 6}{6 \cdot 5} = \dfrac{11}{30}$ $\dfrac{x}{6} + \dfrac{x}{5} = \dfrac{11x}{30}$
	Subtraction of rational expressions with different denominators	$\dfrac{A}{B} - \dfrac{C}{D} = \dfrac{AD - BC}{BD}$ $B \neq 0, D \neq 0$	$\dfrac{1}{5} - \dfrac{1}{6} = \dfrac{6 - 5}{5 \cdot 6} = \dfrac{1}{30}$ $\dfrac{x}{5} - \dfrac{x}{6} = \dfrac{x}{30}$

Section	Item	Meaning	Example
6.4	Complex fraction	A fraction that has fractions in the numerator, denominator, or both	$\dfrac{\frac{x}{x+2}}{x^2+x+1}$ is a complex fraction.
	Simplifying complex fractions	You can simplify a complex fraction by: 1. Multiplying numerator and denominator by the LCD of all fractions involved or 2. Performing the indicated operations in the numerator and denominator and then dividing the numerator by the denominator	Simplify $\dfrac{\frac{1}{x}+1}{\frac{1}{x}-1}$ Method 1: $\dfrac{x\left(\frac{1}{x}+1\right)}{x\left(\frac{1}{x}-1\right)}=\dfrac{1+x}{1-x}$ Method 2: $\dfrac{\frac{1}{x}+\frac{x}{x}}{\frac{1}{x}-\frac{x}{x}}=\dfrac{\frac{1+x}{x}}{\frac{1-x}{x}}=\dfrac{1+x}{x}\cdot\dfrac{x}{1-x}$ $=\dfrac{1+x}{1-x}$
6.5	Fractional equation	An equation containing one or more rational expressions	$\frac{x}{2}+\frac{x}{3}=1$ is a fractional equation.
6.6A	Ratio	A quotient of two numbers	3 to 4, 3:4, and $\frac{3}{4}$ are ratios.
	Proportion	An equality between ratios	3:4 as x:6, or $\frac{3}{4}=\frac{x}{6}$
	Cross products	If $\frac{a}{b}=\frac{c}{d}$ (where $b\neq 0$ and $d\neq 0$), then $ad=bc$. ad and bc are the cross products.	If $\frac{3}{4}=\frac{x}{6}$, then $3\cdot 6=4\cdot x$. $3\cdot 6$ and $4\cdot x$ are the cross products.
6.6B	Similar figures	Figures that have the same shape but not necessarily the same size	□ and □ are similar figures.
6.7A	Direct variation	y varies directly as x if $y=kx$. We also say y is proportional to x, where k is the constant of proportionality.	The cost C of bagels is proportional to the number n you buy. $C=kn$ and k is the price of one bagel.
6.7B	Inverse variation	y varies inversely as x if $y=\frac{k}{x}$, where k is the constant of proportionality.	The acceleration a of an object is inversely proportional to its mass m: $a=\frac{k}{m}$.

❭ Review Exercises Chapter 6

(If you need help with these exercises, look in the section indicated in brackets.)

1. ❬ **6.1B** ❭ *Write the given fraction with the indicated denominator.*

  a. $\frac{5x}{8y}$ with a denominator of $16y^2$

  b. $\frac{3x}{4y^2}$ with a denominator of $16y^3$

  c. $\frac{2y}{3x^3}$ with a denominator of $15x^5$

2. ❬ **6.1C** ❭ *Reduce the given fraction to lowest terms.*

  a. $\dfrac{-9(x^2-y^2)}{3(x+y)}$

  b. $\dfrac{-10(x^2-y^2)}{5(x+y)}$

  c. $\dfrac{-16(x^2-y^2)}{-4(x+y)}$

3. ⟨ **6.1A, C** ⟩ *Determine the values for which the given fraction is undefined, then simplify the expression.*

a. $\dfrac{-x}{x^2 + x}$

b. $\dfrac{-x}{x^2 - x}$

c. $\dfrac{-x}{x - x^2}$

4. ⟨ **6.1C** ⟩ *Reduce to lowest terms.*

a. $\dfrac{x^2 - 3x - 18}{6 - x}$

b. $\dfrac{x^2 - 2x - 8}{4 - x}$

c. $\dfrac{x^2 - 2x - 15}{5 - x}$

5. ⟨ **6.2A** ⟩ *Multiply.*

a. $\dfrac{3y^2}{7} \cdot \dfrac{14x}{9y}$

b. $\dfrac{7y^2}{5} \cdot \dfrac{15x}{14y}$

c. $\dfrac{6y^3}{7} \cdot \dfrac{28x}{3y}$

6. ⟨ **6.2A** ⟩ *Multiply.*

a. $(x - 3) \cdot \dfrac{x + 2}{x^2 - 9}$

b. $(x - 5) \cdot \dfrac{x + 1}{x^2 - 25}$

c. $(x - 4) \cdot \dfrac{x + 5}{x^2 - 16}$

7. ⟨ **6.2B** ⟩ *Divide.*

a. $\dfrac{x^2 - 9}{x + 2} \div (x + 3)$

b. $\dfrac{x^2 - 16}{x + 1} \div (x + 4)$

c. $\dfrac{x^2 - 25}{x + 4} \div (x + 5)$

8. ⟨ **6.2B** ⟩ *Divide.*

a. $\dfrac{x + 5}{x - 5} \div \dfrac{x^2 - 25}{5 - x}$

b. $\dfrac{x + 1}{x - 1} \div \dfrac{x^2 - 1}{1 - x}$

c. $\dfrac{x + 2}{x - 2} \div \dfrac{x^2 - 4}{2 - x}$

9. ⟨ **6.3A** ⟩ *Add.*

a. $\dfrac{3}{2(x - 1)} + \dfrac{1}{2(x - 1)}$

b. $\dfrac{5}{6(x - 2)} + \dfrac{7}{6(x - 2)}$

c. $\dfrac{3}{4(x + 1)} + \dfrac{1}{4(x + 1)}$

10. ⟨ **6.3A** ⟩ *Subtract.*

a. $\dfrac{7}{2(x + 1)} - \dfrac{3}{2(x + 1)}$

b. $\dfrac{11}{5(x + 2)} - \dfrac{1}{5(x + 2)}$

c. $\dfrac{17}{7(x + 3)} - \dfrac{3}{7(x + 3)}$

11. ⟨ **6.3B** ⟩ *Add.*

a. $\dfrac{2}{x + 2} + \dfrac{1}{x - 2}$

b. $\dfrac{3}{x + 1} + \dfrac{1}{x - 1}$

c. $\dfrac{4}{x + 3} + \dfrac{1}{x - 3}$

12. ⟨ **6.3B** ⟩ *Subtract.*

a. $\dfrac{x - 1}{x^2 + 3x + 2} - \dfrac{x + 7}{x^2 + 5x + 6}$

b. $\dfrac{x + 3}{x^2 - x - 2} - \dfrac{x - 1}{x^2 + 2x + 1}$

c. $\dfrac{x - 1}{x^2 + 3x + 2} - \dfrac{x + 1}{x^2 + x - 2}$

13. ⟨ **6.4A** ⟩ *Simplify.*

a. $\dfrac{\dfrac{3}{2x} - \dfrac{1}{x}}{\dfrac{2}{3x} + \dfrac{3}{4x}}$

b. $\dfrac{\dfrac{3}{2x} - \dfrac{1}{3x}}{\dfrac{2}{x} + \dfrac{1}{4x}}$

c. $\dfrac{\dfrac{3}{2x} - \dfrac{1}{x}}{\dfrac{3}{4x} + \dfrac{4}{3x}}$

14. ⟨ **6.5A** ⟩ *Solve.*

a. $\dfrac{2x}{x - 1} + 3 = \dfrac{4x}{x - 1}$

b. $\dfrac{6x}{x - 5} + 7 = \dfrac{8x}{x - 5}$

c. $\dfrac{5x}{x - 4} + 6 = \dfrac{7x}{x - 4}$

15. ⟨ **6.5A** ⟩ *Solve.*

a. $\dfrac{x}{x^2 - 4} + \dfrac{2}{x - 2} = \dfrac{x - 3}{x^2 - x - 6}$

b. $\dfrac{x}{x^2 - 16} + \dfrac{4}{x - 4} = \dfrac{x - 5}{x^2 - x - 20}$

c. $\dfrac{x}{x^2 - 25} + \dfrac{5}{x - 5} = \dfrac{x - 6}{x^2 - x - 30}$

16. ⟨ **6.5A** ⟩ *Solve.*

a. $\dfrac{x}{x + 6} - \dfrac{1}{7} = \dfrac{-6}{x + 6}$

b. $\dfrac{x}{x + 7} - \dfrac{1}{8} = \dfrac{-7}{x + 7}$

c. $\dfrac{x}{x + 8} - \dfrac{1}{9} = \dfrac{-8}{x + 8}$

17. ⟨ **6.5A** ⟩ *Solve.*

a. $3 + \dfrac{5}{x - 4} = \dfrac{50}{x^2 - 16}$

b. $4 + \dfrac{6}{x - 5} = \dfrac{84}{x^2 - 25}$

c. $5 + \dfrac{7}{x - 6} = \dfrac{126}{x^2 - 36}$

18. ⟨ **6.5B** ⟩ *Solve for the indicated variable.*

a. $A = \dfrac{a_1(1 - b)}{1 - b^n}; a_1$

b. $B = \dfrac{b_1(1 - c^n)}{1 - c}; b_1$

c. $C = \dfrac{c_1(1 - d)^n}{d + 1}; c_1$

19. ⟨ **6.6A** ⟩ *A car travels 160 miles on 7 gallons of gas.*

a. How many gallons will it need to travel 240 miles?

b. Repeat the problem where the car travels 180 miles on 9 gallons of gas and we wish to go 270 miles.

c. Repeat the problem where the car travels 200 miles on 12 gallons and we wish to go 300 miles.

20. ⟨ **6.6A** ⟩ *Solve using cross products.*

a. $\dfrac{x + 3}{6} = \dfrac{7}{2}$

b. $\dfrac{x + 4}{8} = \dfrac{9}{5}$

c. $\dfrac{x + 5}{2} = \dfrac{6}{5}$

21. ⟨ **6.6B** ⟩ *A person can do a job in 6 hours. Another person can do it in 8 hours.*

a. How long would it take to do the job if both of them work together?

b. Repeat the problem where the first person takes 10 hours and the second person takes 8 hours.

c. Repeat the problem where the first person takes 9 hours and the second person takes 6 hours.

22. ⟨ **6.6B** ⟩ *A boat can travel 10 miles against a current in the same time it takes to travel 30 miles with the current. What is the speed of the boat in still water if the current flows at:*

a. 2 miles per hour?

b. 4 miles per hour?

c. 6 miles per hour?

23. ⟨ **6.6B** ⟩ *A baseball player has 30 home runs in 120 games. At that rate, how many home runs will he have in:*

a. 128 games?

b. 140 games?

c. 160 games?

24. ⟨ **6.6B** ⟩ *A company wants to produce 1500 items in 1 year (12 months). To attain this goal, how many items should be produced by the end of:*

a. September (the 9th month)?

b. October (the 10th month)?

c. November (the 11th month)?

25. ⟨ **6.6B** ⟩ *Find the unknown in the given similar triangle.*

a. Find x. **b.** Find b. **c.** Find s.

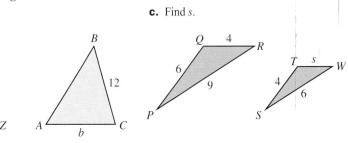

26. ⟨ **6.7A** ⟩

a. If you walk for m minutes, the number C of calories used is proportional to the time you walk. If 30 calories are used when you walk for 12 minutes, write an equation of variation and find the number of calories used when you walk for an hour.

b. If you row for m minutes, the number C of calories used is proportional to the time you row. If 140 calories are used when you row for 20 minutes, write an equation of variation and find the number of calories used when you row for an hour.

c. If you play tennis for m minutes, the number C of calories used is proportional to the time you play. If 180 calories are used when you play for 180 minutes, write an equation of variation and find the number of calories used when you play for 45 minutes

27. ⟨ **6.7B** ⟩

a. The amount F of force you exert on a wrench handle to loosen a rusty bolt varies inversely with the length L of the handle. If $k = 30$, write an equation of variation and find the force needed when the handle is 6 inches long.

b. What force is needed when the handle is 10 inches long?

c. What force is needed when the handle is 15 inches long?

> Practice Test **Chapter 6**

(Answers on page 548)

Visit www.mhhe.com/bello to view helpful videos that provide step-by-step solutions to several of the problems below.

1. Write $\frac{3x}{7y}$ with a denominator of $21y^3$.

2. Reduce $\dfrac{-6(x^2 - y^2)}{3(x - y)}$ to lowest terms.

3. Determine the values for which the expression $\dfrac{-x}{x + x^2}$ is undefined then simplify.

4. Reduce to lowest terms $\dfrac{x^2 + 2x - 8}{2 - x}$.

In Problems 5–12, perform the indicated operations and simplify.

5. Multiply $\dfrac{2y^2}{7} \cdot \dfrac{21x}{4y}$.

6. Multiply $(x - 2) \cdot \dfrac{x + 3}{x^2 - 4}$.

7. Divide $\dfrac{x^2 - 4}{x + 5} \div (x - 2)$.

8. Divide $\dfrac{x + 3}{x - 3} \div \dfrac{x^2 - 9}{3 - x}$.

9. Add $\dfrac{5}{2(x - 2)} + \dfrac{1}{2(x - 2)}$.

10. Subtract $\dfrac{7}{3(x + 1)} - \dfrac{1}{3(x + 1)}$.

11. Add $\dfrac{2}{x + 1} + \dfrac{1}{x - 1}$.

12. Subtract $\dfrac{x + 1}{x^2 + x - 2} - \dfrac{x + 2}{x^2 - 1}$.

13. Simplify $\dfrac{\dfrac{1}{x} - \dfrac{2}{3x}}{\dfrac{3}{4x} + \dfrac{1}{2x}}$.

14. Solve $\dfrac{3x}{x - 2} + 4 = \dfrac{5x}{x - 2}$.

15. Solve $\dfrac{x}{x^2 - 9} + \dfrac{3}{x - 3} = \dfrac{1}{x + 3}$.

16. Solve $\dfrac{x}{x + 5} - \dfrac{1}{6} = \dfrac{-5}{x + 5}$.

17. Solve $2 + \dfrac{4}{x - 3} = \dfrac{24}{x^2 - 9}$.

18. Solve for d_1 in

$$D = \frac{d_1(1 - d^n)}{(1 + d)^n}.$$

19. Solve $\dfrac{x + 5}{7} = \dfrac{11}{6}$.

20. A car travels 150 miles on 9 gallons of gas. How many gallons will it need to travel 400 miles?

21. A woman can paint a house in 5 hours. Another one can do it in 8 hours. How long would it take to paint the house if both women work together?

22. A boat can travel 10 miles against a current in the same time it takes to travel 30 miles with the current. If the speed of the current is 8 miles per hour, what is the speed of the boat in still water?

23. A baseball player has 20 singles in 80 games. At that rate, how many singles will he have in 160 games?

24. A conversion van company wants to finish 150 vans in 1 year (12 months). To attain this goal, how many vans should be finished by the end of April (the fourth month)?

25. Find the unknown in the given similar triangles.

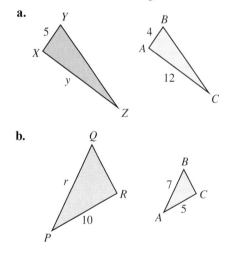

a.

b.

26. Have you raked leaves lately? If you rake leaves for m minutes, the number C of calories used is proportional to the time you rake. If 60 calories are used when you rake for 30 minutes, write an equation of variation and find the number of calories used when you rake for $2\frac{1}{2}$ hours.

27. The maximum weight W that can be supported by a 2-by-4-inch piece of pinewood varies inversely with its length L. If the maximum weight W that can be supported by a 10-foot-long 2-by-4 piece of pine is 500 pounds, find an equation of variation and the maximum weight W that can be supported by a 25-foot length of 2-by-4 pine.

Answer	If You Missed		Review	
	Question	Section	Examples	Page
1. $\dfrac{9xy^2}{21y^3}$	1	6.1	2	474–475
2. $-2(x + y)$ or $-2x - 2y$	2	6.1	3	476–477
3. $\dfrac{-1}{1 + x}$; undefined for $x = 0$ and $x = -1$	3	6.1	4, 1	478, 472
4. $-(x + 4)$ or $-x - 4$	4	6.1	5	479–480
5. $\dfrac{3xy}{2}$	5	6.2	1, 2	487–488
6. $\dfrac{x + 3}{x + 2}$	6	6.2	3	489
7. $\dfrac{x + 2}{x + 5}$	7	6.2	4	490
8. $\dfrac{-1}{x - 3}$ or $\dfrac{1}{3 - x}$	8	6.2	4, 5	490–491
9. $\dfrac{3}{x - 2}$	9	6.3	1a	497
10. $\dfrac{2}{x + 1}$	10	6.3	1b	497
11. $\dfrac{3x - 1}{x^2 - 1} = \dfrac{3x - 1}{(x + 1)(x - 1)}$	11	6.3	2, 4a	497–498, 500
12. $\dfrac{-2x - 3}{(x + 2)(x + 1)(x - 1)}$	12	6.3	3, 4b, 5	498–499, 500–501, 501–502
13. $\dfrac{4}{15}$	13	6.4	1, 2	508–509
14. $x = 4$	14	6.5	1, 2	514
15. $x = -4$	15	6.5	3	515
16. No solution	16	6.5	4	516
17. $x = -5$	17	6.5	5	517
18. $d_1 = \dfrac{D(1 + d)^n}{(1 - d^n)}$	18	6.5	6	518
19. $x = \dfrac{47}{6}$	19	6.6	1	523
20. 24	20	6.6	1	523
21. $3\dfrac{1}{13}$ hr	21	6.6	2	525
22. 16 mi/hr	22	6.6	3	526
23. 40	23	6.6	4	527
24. 50	24	6.6	4	527
25. **a.** $y = 15$ **b.** $r = 14$	25	6.6	5	528
26. $C = 2m$; 300	26	6.7	1, 2	534–535
27. $W = \dfrac{5000}{L}$; 200 lb	27	6.7	3, 4	535–536

› Cumulative Review **Chapters 1–6**

1. Add: $-\dfrac{1}{6} + \left(-\dfrac{2}{5}\right)$

2. Subtract: $-5.6 - (-8.3)$

3. Find: $(-5)^2$

4. Divide: $-\dfrac{7}{8} \div \left(-\dfrac{7}{16}\right)$

5. Evaluate $y \div 5 \cdot x - z$ for $x = 6$, $y = 60$, $z = 3$.

6. Simplify: $x + 4(x - 3) + (x - 2)$

7. Write in symbols: The quotient of $(a - b)$ and c.

8. Solve for x: $5 = 3(x - 1) + 4 - 2x$

9. Solve for x: $\dfrac{x}{7} - \dfrac{x}{9} = 2$

10. The sum of two numbers is 95. If one of the numbers is 35 more than the other, what are the numbers?

11. Susan purchased some municipal bonds yielding 8% annually and some certificates of deposit yielding 10% annually. If Susan's total investment amounts to $12,000 and the annual income is $1020, how much money is invested in bonds and how much is invested in certificates of deposit?

12. Graph: $-\dfrac{x}{2} + \dfrac{x}{4} \geq \dfrac{x - 4}{4}$

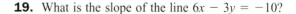

13. Graph the point $C(-1, -3)$.

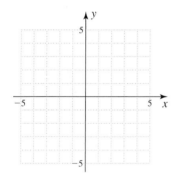

14. Determine whether the ordered pair $(-3, -3)$ is a solution of $5x - y = -12$.

15. Find x in the ordered pair $(x, -2)$ so that the ordered pair satisfies the equation $3x - y = 11$.

16. Graph: $2x + y = 2$

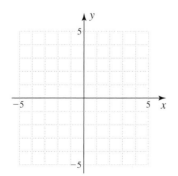

17. Graph: $3y + 6 = 0$

18. Find the slope of the line passing through the points $(-7, -9)$ and $(9, -2)$.

19. What is the slope of the line $6x - 3y = -10$?

20. Find the pair of parallel lines.

(1) $15y + 20x = 8$

(2) $20x - 15y = 8$

(3) $3y = 4x + 8$

21. Find an equation of the line passing through $(-4, 3)$ and with slope 6.

22. Find an equation of the line with slope -2 and y-intercept 5.

23. Graph $3x - 4y < -12$.

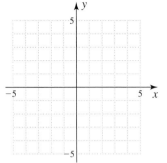

24. Graph $3x - y \geq 0$.

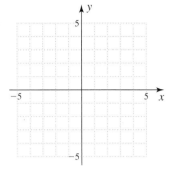

25. Simplify: $\dfrac{x^{-8}}{x^{-9}}$

26. Multiply and simplify: $x^5 \cdot x^{-7}$

27. Simplify: $(3x^3y^{-2})^{-4}$

28. Write in scientific notation: 0.00036

29. Divide and express the answer in scientific notation: $(14.04 \times 10^{-3}) \div (7.8 \times 10^2)$

30. Find the value of $x^2 + 3x - 2$ when $x = -2$.

31. Add $(2x^4 - 6x^6 - 7)$ and $(-6x^6 + 3 + 7x^4)$.

32. Find: $\left(5x^2 - \dfrac{1}{5}\right)^2$

33. Multiply: $(9x^2 + 2)(9x^2 - 2)$

34. Divide $(3x^3 + 11x^2 + 18)$ by $(x + 4)$.

35. Factor completely: $6x^2 - 9x^4$

36. Factor completely: $\dfrac{2}{5}x^7 - \dfrac{4}{5}x^6 + \dfrac{4}{5}x^5 - \dfrac{1}{5}x^3$

37. Factor completely: $x^2 - 15x + 56$

38. Factor completely: $15x^2 - 37xy + 20y^2$

39. Factor completely: $16x^2 - 9y^2$

40. Factor completely: $-4x^4 + 4x^2$

41. Factor completely: $3x^3 - 6x^2 - 9x$

42. Factor completely: $4x^2 + 4x + 3x + 3$

43. Factor completely: $16kx^2 - 8kx + k$

44. Solve for x: $2x^2 + x = 15$

45. Write $\dfrac{7x}{6y}$ with a denominator of $12y^3$.

46. Reduce to lowest terms: $\dfrac{-8(x^2 - y^2)}{4(x - y)}$

47. Reduce to lowest terms: $\dfrac{x^2 + 4x - 5}{1 - x}$

48. Multiply: $(x - 7) \cdot \dfrac{x + 3}{x^2 - 49}$

49. Divide: $\dfrac{x + 3}{x - 3} \div \dfrac{x^2 - 9}{3 - x}$

50. Add: $\dfrac{3}{2(x + 8)} + \dfrac{9}{2(x + 8)}$

51. Subtract: $\dfrac{x + 5}{x^2 + x - 30} - \dfrac{x + 6}{x^2 - 25}$

52. Simplify: $\dfrac{\dfrac{2}{x} + \dfrac{3}{2x}}{\dfrac{1}{3x} - \dfrac{1}{4x}}$

53. Solve for x: $\dfrac{3x}{x - 2} + 3 = \dfrac{4x}{x - 2}$

54. Solve for x: $\dfrac{x}{x^2 - 49} + \dfrac{7}{x - 7} = \dfrac{1}{x + 7}$

55. Solve for x: $\dfrac{x}{x + 4} - \dfrac{1}{5} = \dfrac{-4}{x + 4}$

56. Solve for x: $1 + \dfrac{2}{x - 3} = \dfrac{12}{x^2 - 9}$

57. A car travels 120 miles on 6 gallons of gas. How many gallons will it need to travel 460 miles?

58. Solve for x: $\dfrac{x + 8}{9} = \dfrac{2}{11}$

59. Maria can paint a kitchen in 6 hours, and James can paint the same kitchen in 7 hours. How long would it take for both working together to paint the kitchen?

60. An enclosed gas exerts a pressure P on the walls of a container. This pressure is directly proportional to the temperature T of the gas. If the pressure is 5 pounds per square inch when the temperature is 250° F. find k.

61. If the temperature of a gas is held constant, the pressure P varies inversely as the volume V. A pressure of 1960 pounds per square inch is exerted by 7 cubic feet of air in a cylinder fitted with a piston. Find k.

Chapter

7

seven

▶ Solving Systems of Linear Equations and Inequalities

The Human Side of Algebra

The first evidence of a systematic method of solving systems of linear equations is provided in the *Nine Chapters of the Mathematical Arts,* the oldest arithmetic textbook in existence. The method for solving a system of three equations with three unknowns occurs in the 18 problems of the eighth chapter, entitled "The Way of Calculating by Arrays." Unfortunately, the original copies of the *Nine Chapters* were destroyed in 213 B.C. However, the Chinese mathematician Liu Hui wrote a commentary on the *Nine Chapters* in A.D. 263, and information concerning the original work comes to us through this commentary.

In modern times, when a large number of equations or inequalities has to be solved, a method called the **simplex method** is used. This method, based on the **simplex algorithm,** was developed in the 1940s by George B. Dantzig. It was first used by the Allies of World War II to solve logistics problems dealing with obtaining, maintaining, and transporting military equipment and personnel.

7.1 Solving Systems of Equations by Graphing

▶ Objectives

A ❭ Solve a system of two equations in two variables by graphing.

B ❭ Determine whether a system of equations is consistent, inconsistent, or dependent.

C ❭ Solve an application involving systems of equations.

▶ To Succeed, Review How To . . .

1. Graph the equation of a line (pp. 209–211, 235–240).
2. Determine whether an ordered pair is a solution of an equation (pp. 221–222).

▶ Getting Started

Supply, Demand, and Intersections

Can you tell from the graph when the energy supply and the demand were about the same? This happened where the line representing the supply and the line representing the demand intersect, or around 1980. When the demand and the supply are *equal,* prices reach *equilibrium.* If x is the year and y the number of millions of barrels of oil per day, the point (x, y) at which the demand is the same as the supply—that is, the point at which the graphs *intersect*—is (1980, 85). We can graph a pair of linear equations and find a point of intersection if it exists. This point of intersection is an ordered pair of numbers such as (1980, 85) and is a **solution** of both equations. This means that when you substitute 1980 for x and 85 for y in the original equations, the results are true statements. In this section we learn how to find the solution of a system of two equations in two variables by using the graphical method, which involves graphing the equations and finding their point of intersection, if it exists.

A ❭ Solving a System by Graphing

The solution of a linear equation in one variable, as studied in Chapter 2, was a single number. Thus if we solve *two* linear equations in two variables simultaneously, we expect to get two numbers. We write this solution as an ordered pair. For example, the solution of

$$x + 2y = 4$$
$$2y - x = 0$$

is (2, 1). This can be checked by letting $x = 2$ and $y = 1$ in both equations:

$$x + 2y = 4 \qquad\qquad 2y - x = 0$$
$$2 + 2(1) = 4 \qquad\qquad 2(1) - 2 = 0$$
$$2 + 2 = 4 \qquad\qquad 2 - 2 = 0$$
$$4 = 4 \qquad\qquad\qquad 0 = 0$$

Clearly, a true statement results in both cases. We call a system of two linear equations a **system of simultaneous equations.** To solve one of these systems, we need to find (if possible) all ordered pairs of numbers that satisfy **both** equations. Thus, to *solve* the system

$$x + 2y = 4$$
$$2y - x = 0$$

This system is called a system of *simultaneous* equations because we have to find a solution that satisfies both equations.

we graph each of the equations in the same coordinate axis in the usual way. To graph $x + 2y = 4$, we find the intercepts using the following table:

x	y
0	2
4	0

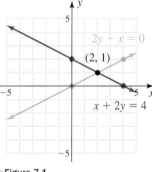

>Figure 7.1

The intercepts (0, 2) and (4, 0), as well as the completed graph, are shown in red in Figure 7.1.

The equation $2y - x = 0$ is graphed similarly using the following table:

x	y
0	0

Note that we need another point in the table. We can pick any x we want and then find y. If we pick $x = 4$, then $2y - 4 = 0$, or $y = 2$, giving us the point (4, 2). The graph of the equation $2y - x = 0$ is shown in blue. The lines intersect at (2, 1), which is the solution of the system of equations. (Recall that we checked that this point satisfies the equations by substituting 2 for x and 1 for y in each equation.)

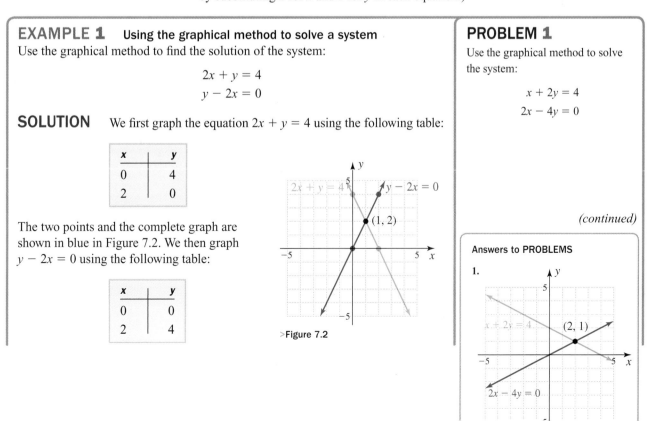

EXAMPLE 1   Using the graphical method to solve a system

Use the graphical method to find the solution of the system:

$$2x + y = 4$$
$$y - 2x = 0$$

SOLUTION   We first graph the equation $2x + y = 4$ using the following table:

x	y
0	4
2	0

The two points and the complete graph are shown in blue in Figure 7.2. We then graph $y - 2x = 0$ using the following table:

x	y
0	0
2	4

>Figure 7.2

PROBLEM 1

Use the graphical method to solve the system:

$$x + 2y = 4$$
$$2x - 4y = 0$$

(continued)

Answers to PROBLEMS

1.

The graph of $y - 2x = 0$ is shown in red. The lines intersect at $(1, 2)$. Note that $(1, 2)$ is part of both lines, so $(1, 2)$ is the **solution** of the system of equations. We can check this by letting $x = 1$ and $y = 2$ in

$$2x + y = 4$$
$$y - 2x = 0$$

thus obtaining the true statements

$$2(1) + 2 = 4$$
$$2 - 2(1) = 0$$

EXAMPLE 2   Solving an inconsistent system

Use the graphical method to find the solution of the system:

$$y - 2x = 4$$
$$2y - 4x = 12$$

SOLUTION   We first graph the equation $y - 2x = 4$ using the following table:

x	y
0	4
−2	0

The two points, as well as the completed graph, are shown in blue in Figure 7.3. We then graph $2y - 4x = 12$ using the following table:

x	y
0	6
−3	0

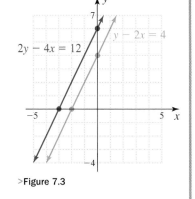

>Figure 7.3

The graph of $2y - 4x = 12$ is shown in red. The two lines appear to be parallel; they do not intersect. If we examine the equations more carefully, we see that by dividing both sides of the second equation by 2, we get $y - 2x = 6$. Thus, one equation says $y - 2x = 4$, and the other says that $y - 2x = 6$. Hence, both equations cannot be true at the same time, and their graphs cannot intersect.

To confirm this, note that $y - 2x = 6$ is equivalent to $y = 2x + 6$ and $y - 2x = 4$ is equivalent to $y = 2x + 4$. Since $y = 2x + 6$ and $y = 2x + 4$ both have slope 2 but different y-intercepts, their graphs are **parallel lines.** Thus, there is *no solution* for this system, since the two lines do not have any points in common; the system is said to be **inconsistent.**

PROBLEM 2

Use the graphical method to solve the system:

$$y - 3x = 3$$
$$2y - 6x = 12$$

Answers to PROBLEMS

2. No solution

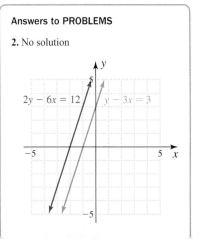

EXAMPLE 3 Solving a dependent system

Use the graphical method to solve the system:

$$2x + y = 4$$
$$2y + 4x = 8$$

SOLUTION We use the table

x	y
0	4
2	0

to graph $2x + y = 4$, which is shown in color in Figure 7.4. To graph $2y + 4x = 8$, we first let $x = 0$ to obtain $2y = 8$ or $y = 4$. For $y = 0$, $4x = 8$ or $x = 2$. Thus, the two points in our second table will be

x	y
0	4
2	0

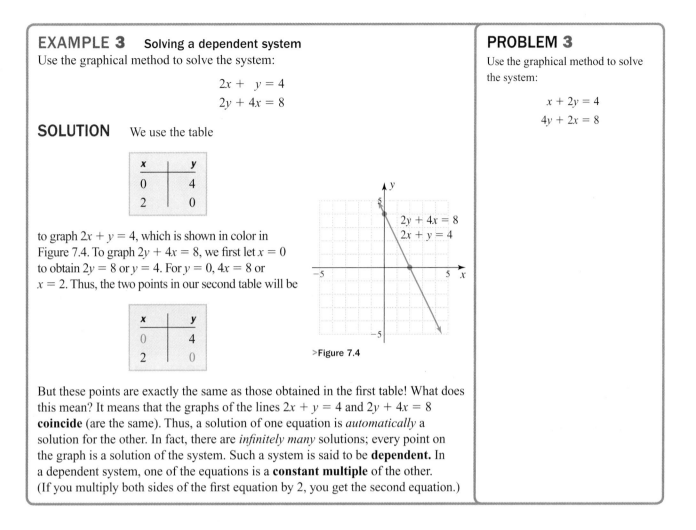

>Figure 7.4

But these points are exactly the same as those obtained in the first table! What does this mean? It means that the graphs of the lines $2x + y = 4$ and $2y + 4x = 8$ **coincide** (are the same). Thus, a solution of one equation is *automatically* a solution for the other. In fact, there are *infinitely many* solutions; every point on the graph is a solution of the system. Such a system is said to be **dependent.** In a dependent system, one of the equations is a **constant multiple** of the other. (If you multiply both sides of the first equation by 2, you get the second equation.)

PROBLEM 3

Use the graphical method to solve the system:

$$x + 2y = 4$$
$$4y + 2x = 8$$

B › Finding Consistent, Inconsistent, and Dependent Systems of Equations

As you can see from the examples we've given, a system of equations can have exactly *one* solution (when the lines *intersect,* as in Figure 7.5), *no* solution (when the lines are *parallel,* as in Figure 7.6), or *infinitely many* solutions (when the graphs of the two lines are *identical,* as in Figure 7.7). These examples illustrate the three possible solutions to a system of simultaneous equations.

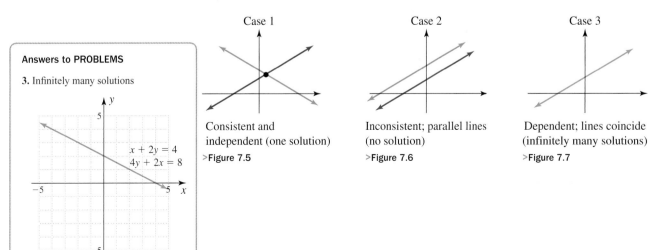

Case 1
Consistent and independent (one solution)
>Figure 7.5

Case 2
Inconsistent; parallel lines (no solution)
>Figure 7.6

Case 3
Dependent; lines coincide (infinitely many solutions)
>Figure 7.7

Answers to PROBLEMS

3. Infinitely many solutions

> **POSSIBLE SOLUTIONS TO A SYSTEM OF SIMULTANEOUS EQUATIONS**
> 1. **Consistent and independent systems:** The graphs of the equations intersect at *one* point, whose coordinates give the solution of the system.
> 2. **Inconsistent systems:** The graphs of the equations are *parallel* lines; there is *no* solution for the system.
> 3. **Dependent systems:** The graphs of the equations *coincide* (are the same). There are *infinitely many* solutions for the system.

The following table will help you further.

Type of Lines	Slopes	y-Intercept	Number of Solutions	Type of System
Intersecting	Different	Same or different	One	Consistent
Parallel	Same	Different	None	Inconsistent
Coinciding	Same	Same	Infinite	Dependent

EXAMPLE 4   Classifying a system by graphing

Use the graphical method to solve the given system of equations. Classify each system as consistent (one solution), inconsistent (no solution), or dependent (infinitely many solutions).

a. $x + y = 4$
  $2y - x = -1$

b. $x + 2y = 4$
  $2x + 4y = 6$

c. $x + 2y = 4$
  $4y + 2x = 8$

SOLUTION

a. The respective tables for $x + y = 4$ and $2y - x = -1$ are

The graphs of these two lines are shown in Figure 7.8. As you can see, the solution is (3, 1). (Check this!) The system is *consistent*.

>Figure 7.8

PROBLEM 4

Use the graphical method to solve the given system. Classify each system as consistent (one solution), inconsistent (no solution), or dependent (infinitely many solutions).

a. $x + y = 4$
  $2y - x = 2$

b. $2x + y = 4$
  $2y + 4x = 6$

c. $2x + y = 4$
  $2y + 4x = 8$

Answers to PROBLEMS

4. a. Consistent; solution (2, 2)   **b.** Inconsistent; no solution   **c.** Dependent; infinitely many solutions

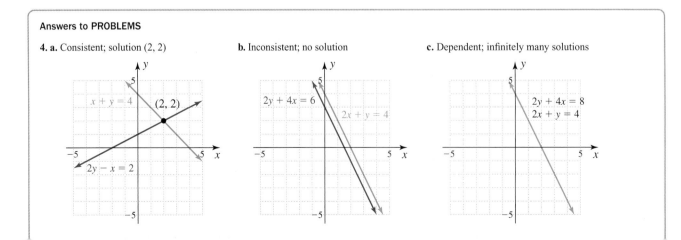

b. The respective tables for $x + 2y = 4$ and $2x + 4y = 6$ are

x	y
0	2
4	0

x	y
0	$\frac{3}{2}$
3	0

The graphs of the two lines are shown in Figure 7.9. There is no solution because the lines are parallel. To see this, we solve $2x + 4y = 6$ and $x + 2y = 4$ for y to obtain

$$y = -\frac{1}{2}x + \frac{3}{2} \qquad \text{and} \qquad y = -\frac{1}{2}x + 2$$

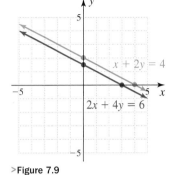

> Figure 7.9

These equations represent two lines with the same slope and different y-intercepts. Thus, the lines are parallel, there is no solution, and the system is *inconsistent.*

c. The respective tables for $x + 2y = 4$ and $4y + 2x = 8$ are

x	y
0	2
4	0

x	y
0	2
4	0

and they are identical. So, there are infinitely many solutions because the lines coincide (see Figure 7.10). The system is *dependent,* and the solutions are all the points on the graph. For example, $(0, 2)$, $(4, 0)$, and $(2, 1)$ are solutions.

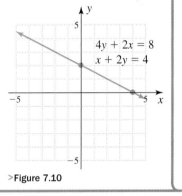

> Figure 7.10

Note that in a *dependent* system, one equation is a constant *multiple* of the other. Thus,

$$x + 2y = 4 \qquad \text{and} \qquad 4y + 2x = 8$$

are a dependent system because $4y + 2x = 8$ is a constant *multiple* of $x + 2y = 4$. Note that

$$2(x + 2y) = 2(4)$$

becomes

$$2x + 4y = 8 \qquad \text{or} \qquad 4y + 2x = 8$$

which is the second equation.

A HELPFUL HINT

If you want to know what *type* of solutions you are going to have, solve both equations for y to obtain the system

$$y = m_1x + b_1$$
$$y = m_2x + b_2$$

When $m_1 \neq m_2$, the lines intersect (there is **one** solution).
When $m_1 = m_2$ and $b_1 = b_2$, there is only one line (**infinitely many** solutions).
When $m_1 = m_2$ and $b_1 \neq b_2$, the lines are parallel (there is **no** solution).

C › Applications Involving Systems of Equations

Most of the problems that we've discussed use x- and y-values ranging from -10 to 10. This is not the case when working real-life applications! For example, the prices for connecting to the Internet using America Online® (AOL) and CompuServe® (CS) are identical: $9.95 per month plus $2.95 for every hour over 5 hours. If you are planning on using more than 5 hours, however, both companies have a special plan.

AOL: $19.95 for 20 hours plus $2.95 per hour after 20 hours*

CS: $24.95 for 20 hours plus $1.95 per hour after 20 hours

For convenience, let's round the numbers as follows:

AOL: $20 for 20 hours plus $3 per hour after 20 hours

CS: $25 for 20 hours plus $2 per hour after 20 hours

EXAMPLE 5   Graphing and analyzing systems of equations

Make a graph of both prices to see which is a better deal.

SOLUTION   Let's start by making a table where h represents the number of hours used and p the monthly price. Keep in mind that when the number of hours is 20 or less the price is fixed: $20 for AOL and $25 for CS.

Price for AOL		
h	**p**	
0	20	
5	20	
10	20	
15	20	
20	20	
25	35	$20 + (25 - 20)3 = 35$
30	50	$20 + (30 - 20)3 = 50$
35	65	$20 + (35 - 20)3 = 65$

Price for CS		
h	**p**	
0	25	
5	25	
10	25	
15	25	
20	25	
25	35	$25 + (25 - 20)2 = 35$
30	45	$25 + (30 - 20)2 = 45$
35	55	$25 + (35 - 20)2 = 55$

To graph these two functions, we let h run from 0 to 50 and p run from 0 to 100 in increments of 5. After 20 hours, the price p for h hours with AOL is

$$p = 20 + (h - 20) \cdot 3 = 20 + 3h - 60 = 3h - 40$$

The price for CS is

$$p = 25 + (h - 20) \cdot 2 = 25 + 2h - 40 = 2h - 15$$

The graph is shown in Figure 7.11. As you can see, if you plan on using 20 hours or less, the price is fixed for both AOL and CS. From 20 to 25 hours, AOL has a lower price, but if you are using more than 25 hours, the price for CS is lower. At exactly 25 hours, the price is the same for both, $35. Note that the complete graph is in quadrant I, since neither the number of hours h nor the price p is negative.

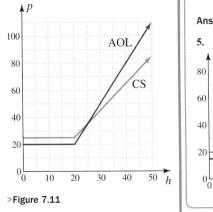

>**Figure 7.11**

PROBLEM 5

One Internet provider charges $20 for 20 hours plus $1.50 per hour after 20 hours. Another Internet provider charges $15 for 20 hours plus $2.00 per hour after 20 hours. Make a graph of both prices to see which is a better deal.

Answers to PROBLEMS

5.

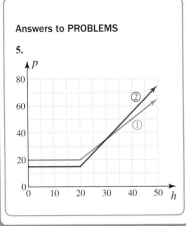

*As of this writing, AOL has a new plan costing $23.90 per month with no limit on the number of hours used, but you can opt for using AOL for free. CompuServe has been acquired by AOL.

Calculator Corner

Solving Systems of Equations

If you have a calculator, you can solve a linear system of equations very easily. However, you must know how to solve an equation for a specified variable. Thus, in Example 1, you can solve $2x + y = 4$ for y to obtain $y = -2x + 4$, and graph $Y_1 = -2x + 4$. Next, solve $y - 2x = 0$ for y to obtain $y = 2x$ and graph $Y_2 = 2x$. To find the intersection, use a decimal window and the trace and zoom features of your calculator, or better yet, if you have an intersection feature, press `2nd` `TRACE` 5 and follow the prompts. (The graph is shown in Window 1.)

Example 2 is done similarly. Solve $y - 2x = 4$ for y to obtain $y = 2x + 4$. Next, solve $2y - 4x = 12$ for y to get $y = 2x + 6$. You have the equations $y = 2x + 4$ and $y = 2x + 6$, but you don't need a graph to know that there's no solution! Since the lines have the same slope, algebra tells you that the lines are parallel! (Sometimes algebra is better than your calculator.) To verify this, graph $y = 2x + 4$ and $y = 2x + 6$. The results are shown in Window 2.

To solve Example 3, solve $2x + y = 4$ for y to obtain $y = -2x + 4$. Next, solve $2y + 4x = 8$ for y to get $y = -2x + 4$. Again, you don't need a calculator to see that the lines are the same! Their graph appears in Window 3.

Window 1   Window 2   Window 3

› Exercises **7.1**

‹ **A** ›   **Solving a System by Graphing**
‹ **B** ›   **Finding Consistent, Inconsistent, and Dependent Systems of Equations**

In Problems 1–30, solve by graphing. Label each system as consistent (write the solution), inconsistent (no solution), or dependent (infinitely many solutions).

1. $x + y = 4$
    $x - y = -2$

2. $x + y = 3$
    $x - y = -5$

3. $x + 2y = 0$
    $x - \; y = -3$

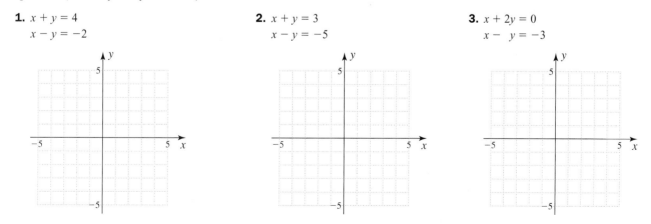

Web IT go to **mhhe.com/bello** for more lessons

4. $y + 2x = -3$
$y - x = 3$

5. $3x - 2y = 6$
$6x - 4y = 12$

6. $2x + y = -2$
$8x + 4y = 8$

7. $3x - y = -3$
$y - 3x = 3$

8. $4x - 2y = 8$
$y - 2x = -4$

9. $2x - y = -2$
$y = 2x + 4$

10. $2x + y = -2$
$y = -2x + 4$

11. $y = -2$
$2y = x - 2$

12. $3y = 6 - x$
$y = 3$

13. $x = 3$
$y = 2x - 4$

14. $y = -x + 2$
$x = -1$

15. $x + y = 3$
$2x - y = 0$

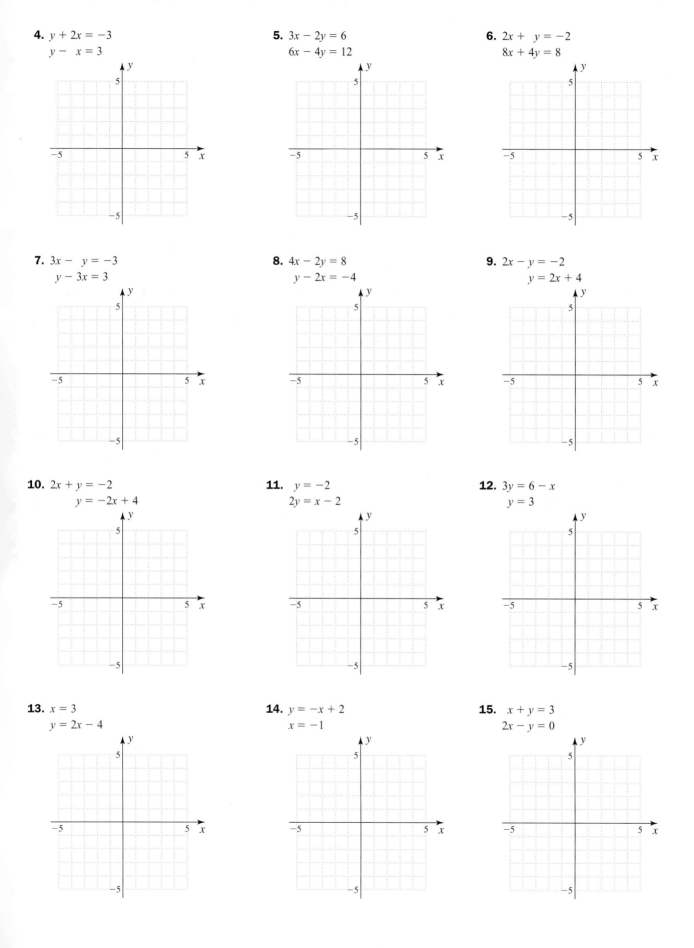

16. $x + y = 5$
    $x - 4y = 0$

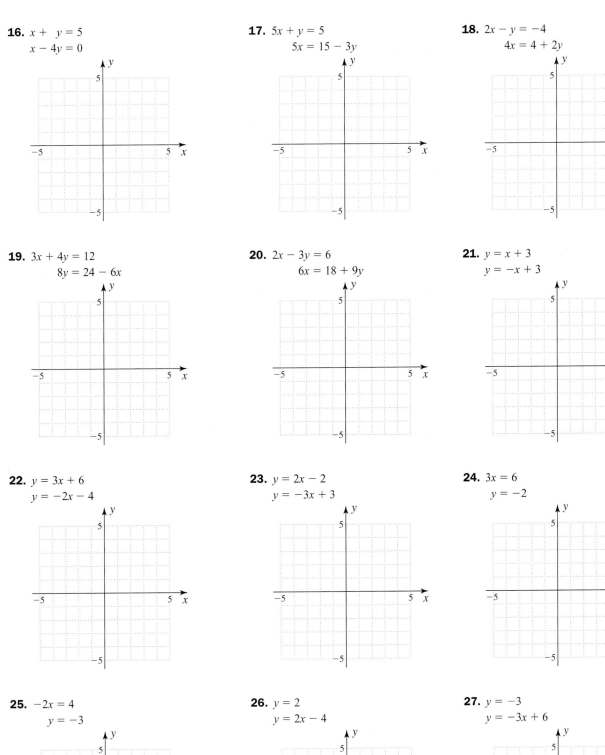

17. $5x + y = 5$
    $5x = 15 - 3y$

18. $2x - y = -4$
    $4x = 4 + 2y$

19. $3x + 4y = 12$
    $8y = 24 - 6x$

20. $2x - 3y = 6$
    $6x = 18 + 9y$

21. $y = x + 3$
    $y = -x + 3$

22. $y = 3x + 6$
    $y = -2x - 4$

23. $y = 2x - 2$
    $y = -3x + 3$

24. $3x = 6$
    $y = -2$

25. $-2x = 4$
    $y = -3$

26. $y = 2$
    $y = 2x - 4$

27. $y = -3$
    $y = -3x + 6$

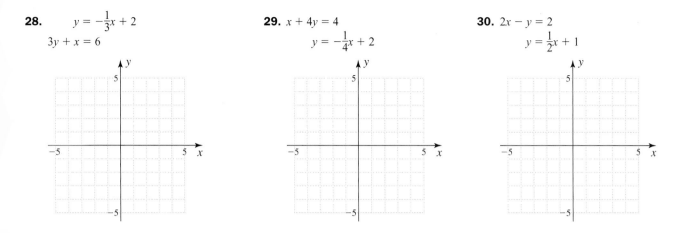

28. $y = -\frac{1}{3}x + 2$
$3y + x = 6$

29. $x + 4y = 4$
$y = -\frac{1}{4}x + 2$

30. $2x - y = 2$
$y = \frac{1}{2}x + 1$

⟨ **C** ⟩ **Applications Involving Systems of Equations** In Problems 31–35, use the following information. You want to watch 10 movies at home each month. You have two options:

OPTION 1 Get cable service. The cost is $20 for the installation fee and $35 per month.

OPTION 2 Buy a DVD player and rent movies. The cost is $200 for a DVD player and $25 a month for movie rental fees.

31. *Cost of cable service*

a. If C is the cost of installing cable service plus the monthly fee for m months, write an equation for C in terms of m.

b. Complete the following table where C is the cost of cable service for m months:

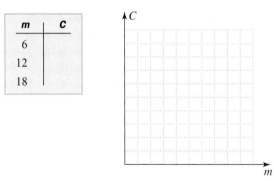

m	C
6	
12	
18	

c. Graph the information obtained in parts **a** and **b**. (*Hint:* Let m run from 0 to 30 and C run from 0 to 1000 in increments of 100.)

33. *Graphical comparison* Make a graph of the information obtained in Problems 31 and 32 on the same coordinate axis.

32. *Cost of renting movies*

a. If C is the total cost of buying the DVD player plus renting the movies for m months, write an equation for C in terms of m.

b. Complete the following table where C is the cost of buying a DVD player and renting movies for m months:

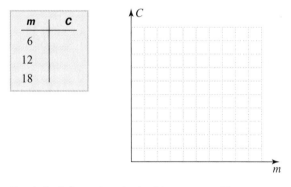

m	C
6	
12	
18	

c. Graph the information obtained in parts **a** and **b**.

34. *Cable service* Based on the graph for Problem 33, when is the cable service cheaper?

35. *When is renting movies cheaper?* Based on the graph for Problem 33, when is the DVD player and rental option cheaper?

36. *Lower wages, higher tips*   At Grady's restaurant, servers earn $80 a week plus tips, which amount to $5 per table.

   a. Write an equation for the weekly wages W based on serving t tables.

   b. Complete the following table where W is the wages and t is the number of tables served:

   c. Graph the information obtained in parts **a** and **b**.

38. *Graphical comparison*   Graph the information from Problems 36 and 37 on the same coordinate axis. Based on the graph, answer the following questions.

   a. When does a server at Grady's make more money than a server at El Centro?

   b. When does a server at El Centro make more money than a server at Grady's?

40. *Better buy?*   Based on the graphs obtained in Problem 39, which plan would you buy, A or B? Explain.

37. *Higher wages, lower tips*   At El Centro restaurant, servers earn $100 a week, but the average tip per table is only $3.

   a. Write an equation for the weekly wages W based on serving t tables.

   b. Complete the following table where W is the wages and t is the number of tables served:

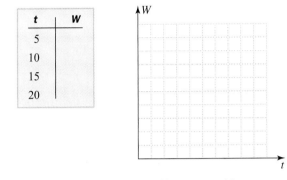

   c. Graph the information obtained in parts **a** and **b**.

39. *Cell phone plans*   How much do you pay a month? At the present time, two companies have cell phone plans that cost $19.95 per month. However, plan A costs $0.60 per minute of airtime during peak hours, while plan B costs $0.45 per minute of airtime during peak hours. Plan A offers a free phone with its plan, while plan B's phone costs $45. For comparison purposes, since the monthly cost is the same for both plans, the cost C is based on the price of the phone plus the number m of minutes of airtime used.

   a. Write an equation for the cost C of plan A.

   b. Write an equation for the cost C of plan B.

   c. Using the same coordinate axis, make a graph for the costs of plans A and B. (*Hint:* Let m and C run from 0 to 500.)

41. *Catering at school*   Here are the actual catering prices for Jefferson City Public Schools.

Jefferson City Public Schools		
Basket Lunch	**Deli Buffet**	**Hot Buffet**
$5.00 per person	$5.50 per person	$6.00 per person

Suppose x students buy Basket Lunches costing $5 each and y students buy the Hot Buffet costing $6 each.

a. Write an equation for the cost C_B of the x Basket Lunches.

b. Write an equation for the cost C_H of the y Hot Buffets.

c. Write an expression that represents the total number of lunches purchased. If this total is 50, write an equation for the total number of lunches purchased.

d. The total bill for the 50 lunches is $5x + 6y$. If this total is $270, write an equation for the total bill.

e. Graph the equations $x + y = 50$ (the total number of lunches purchased) and $5x + 6y = 270$ (the total cost) on the same coordinate axis and find the number of Basket Lunches and Hot Buffet lunches purchased.

Lunches at Jefferson City

42. *Breakfast at restaurant*   Here are some breakfast prices at a restaurant.

Breakfast	
Continental I	$7.99 ≈ $8.00
Continental II	$8.99 ≈ $9.00

Suppose x guests buy the Continental I breakfast costing $8 each and y guests buy the Continental II costing $9 each.

a. Write an equation for the cost C_I of the x Continental I breakfasts.

b. Write an equation for the cost C_{II} of the y Continental II breakfasts.

c. Write an expression that represents the total number of breakfasts purchased. If this total is 26, write an equation for the total number of lunches purchased.

d. The total bill for the 26 breakfasts is $8x + 9y$. If this total is $216, write an equation for the total bill.

e. Graph the equations $x + y = 26$ (the total number of breakfasts purchased) and $8x + 9y = 216$ (the total cost) on the same coordinate axis and find the number of Continental I and Continental II breakfasts purchased.

Breakfast at Restaurant

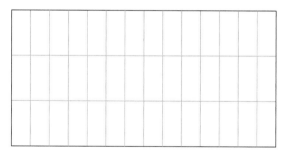

How do we find the y? Remember that $x + y = 26$ and from the graph we can see that $x = 18$. Thus, $x + y = 26$ becomes $18 + y = 26$, or $y = 8$.

43. *Breakfast at College* Here are some breakfast prices at College.

Breakfast

Breakfast croissant	$6
Cheddar cheese omelet	$7
Egg and sausage burritos	$7.25

Suppose x guests buy the breakfast croissant costing $6 each and y guests buy the cheddar cheese omelet costing $7 each.

a. Write an equation for the cost C_B of the x breakfast croissants.

b. Write an equation for the cost C_C of the y breakfast omelets.

c. Write an expression that represents the total number of breakfasts purchased. If this total is 20, write an equation for the total number of breakfasts purchased.

d. The total bill for the 20 breakfasts is $6x + 7y$. If this total is $126, write an equation for the total bill.

e. Graph the equations $x + y = 20$ (the total number of breakfasts purchased) and $6x + 7y = 126$ (the total cost) on the same coordinate axis and find the number of croissant and omelet breakfasts purchased.

Breakfast at College

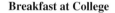

To find y after you know from the graph that $x = 14$, substitute the 14 in the equation $x + y = 20$ obtaining $14 + y = 20$, which means that $y = 6$.

44. *Saturated fats* According to *Restaurant Confidential* by Jacobson and Hurley "all burgers are not created equal. You can't rely on calorie-counting guides that don't make the distinction among burgers served at fast-food establishments, family restaurants and dinner houses." What's the problem?

The saturated fats! If x is the saturated fat content (in grams) in a McDonald's Quarter Pounder and y is the amount of saturated fat in a family-style restaurant, when you eat two Quarter Pounders ($2x$ saturated fat grams) and only one of the family-style burgers (y saturated fat grams), the amount of saturated fat is **30 grams,** 8 over the recommended daily allowance for a female between the ages of 19 and 50. If you add the saturated fats in the McDonald's (x grams) and the family-style restaurant (y grams) the result $x + y$ is exactly **22 grams,** the recommended amount of saturated fat you should have the whole day, even if you ate nothing else!

a. Write an equation for the amount of saturated fats when you eat one McDonald's and one family-style restaurant burger.

b. Write an equation for the amount of saturated fats P_Q contained in two Quarter Pounders.

c. Write an equation for the amount of saturated fats P_F contained in one family-style restaurant burger.

d. The total grams you eat when consuming two Quarter Pounders and one family-style restaurant burger is $2x + y$, or **30** grams. Write an equation for the total grams of fat in the meal.

e. We have not said how many grams of saturated fats there are in each burger, but we can find out! Graph $x + y = 22$ and $2x + y = 30$ on the same coordinate axis and find x (saturated fat grams in the McDonald's) and y (saturated fat grams in the family-style restaurant).

Saturated Fat in Burgers

Note that when $x = 8$, $x + y = 22$ becomes $8 + y = 22$, that is, $y = 14$.

45. *Comparing McDonald's and Burger King*   If x is the saturated fat content (in grams) in a McDonald's Quarter Pounder and y is the amount of saturated fat in a Burger King Whopper Jr, when you eat two Quarter Pounders ($2x$ saturated fat grams) and only one Whopper Jr. burger (y saturated fat grams), the amount of saturated fat is **24 grams.** If you add the saturated fats in the McDonald's (x grams) and the Whopper Jr. (y grams) the result $x + y$ is only **16 grams,** under the recommended amount of saturated fats you should have the whole day.

   a. Write an equation for the amount of saturated fats when you eat one McDonald's and one Whopper Jr. burger.

   b. Write an equation for the amount of saturated fats P_Q contained in two Quarter Pounders.

   c. Write an equation for the amount of saturated fats P_W contained in one Whopper Jr. burger.

   d. The total grams you eat when consuming two Quarter Pounders and one Whopper Jr. burger is $2x + y$, or 24 grams. Write an equation for the total grams of fat in the meal.

   e. We have not said how many grams of saturated fats are in each burger. Graph $x + y = 16$ and $2x + y = 24$ on the same coordinate axis and find x (saturated fat grams in the McDonald's) and y (saturated fat grams in the Whopper Jr.) and find out!

McDonald's vs. Burger King

If you find from the graph that $x = 8$ and you have $x + y = 16$, we can substitute 8 for x, obtaining $8 + y = 16$, which means $y = 8$.

46. *Comparing McDonald's and Wendy's*   Any burgers with less saturated fat than McDonald's? Let's try Wendy's Jr. Bacon Cheeseburger. If x is the saturated fat content (in grams) in a McDonald's Quarter Pounder and y is the amount of saturated fat in Wendy's Jr. Bacon Cheeseburger, when you eat two Quarter Pounders ($2x$ saturated fat grams) and only one Wendy's burger (y saturated fat grams), the amount of saturated fat is **23 grams.** If you add the saturated fats in the McDonald's (x grams) and the Wendy's burger (y grams), the result $x + y$ is only **15 grams,** under the recommended amount of saturated fats you should have in a day.

   a. Write an equation for the amount of saturated fats when you eat one McDonald's and one Wendy's Jr. Bacon Cheeseburger.

   b. Write an equation for the amount of saturated fats P_Q contained in two Quarter Pounders.

   c. Write an equation for the amount of saturated fats P_B contained in one Bacon Cheeseburger.

   d. The total grams you eat when consuming two Quarter Pounders and one Bacon Cheeseburger is $2x + y$, or 23 grams. Write an equation for the total grams of fat in the meal.

   e. We have not said how many grams of saturated fats are in each burger. Graph $x + y = 15$ and $2x + y = 23$ on the same coordinate axis and find x (saturated fat grams in the McDonald's) and y (saturated fat grams in the Bacon Cheeseburger) and find out!

McDonald's vs. Wendy's

By the way, after you find $x = 8$, how do you find y? Since we know that $x + y = 15$, substitute 8 for x in $x + y = 15$, obtaining $8 + y = 15$ or $y = 7$.

〉〉〉 *Using Your Knowledge*

Art Books for Sale!   The table shows prices for art books and the number of books *supplied* in three different years.

Demand Function		Supply Function	
Price ($)	Quantity (hundreds)	Price ($)	Quantity (hundreds)
$28	17	$26	12
$35	15	$35	15
$42	13	$44	18

Note that *fewer* books are *demanded* by consumers as the price of the books goes up. (Consumers are not willing to pay that much for the book.)

47. Graph the points corresponding to the demand function in the coordinate system shown.

48. Draw a line passing through the points. The line is an approximation to the demand function.

On the other hand, book sellers *supply fewer* books when prices go down and more books when prices go up.

49. Graph the points corresponding to the supply function on the same coordinate system.

50. Draw a line passing through the points. The line is an approximation to the supply function.

51. What is the point of intersection of the two lines?

52. At the point of intersection, what is the price?

53. At the point of intersection, what is the quantity of books sold?

〉〉〉 *Write On*

54. What does the solution of a system of linear equations represent?

55. Define a consistent, an inconsistent, and a dependent system of equations.

56. How can you tell graphically whether a system is consistent, inconsistent, or dependent?

Suppose you have a system of equations and you solve both equations for *y* to obtain:

$$y = m_1x + b_1$$
$$y = m_2x + b_2$$

What can you say about the graph of the system when

57. $m_1 = m_2$ and $b_1 \neq b_2$? How many solutions do you have? Explain.

58. $m_1 = m_2$ and $b_1 = b_2$? How many solutions do you have? Explain.

59. $m_1 \neq m_2$? How many solutions do you have? Explain.

〉〉〉 *Concept Checker*

Fill in the blank(s) with the correct word(s), phrase, or mathematical statement.

60. The ordered pair **(a, b)** that is the point of **intersection** of a **system of equations** consisting of two lines in the plane is the _____ of the system.

61. In a **consistent** and **independent** system of equations, the **graph of the equations** intersect at _____ point.

62. In an **inconsistent** system of equations, the **graph of the equations** are _____ lines.

63. In an **inconsistent** system of equations, there is _____ solution for the system.

64. In a **dependent** system of equations, the graph of the equations _____.

65. A **dependent** system of equations has _____ many solutions.

finite	**solution**
intersect	**no**
many	**infinitely**
perpendicular	**parallel**
one	**coincide**

〉〉〉 *Mastery Test*

Use the graphical method to find the solution of the system of equations (if it exists).

66. $x + 2y = 4$
      $2x - 4y = 0$

67. $y - 3x = 3$
      $2y = 6x + 12$

68. $x + 2y = 4$
      $4y = -2x + 8$

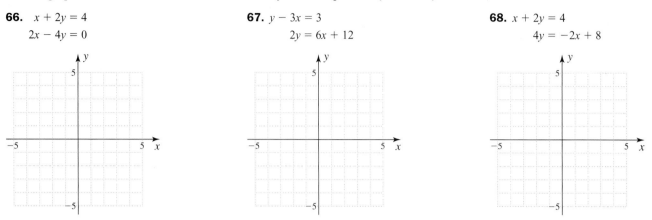

Graph and classify each system as consistent, inconsistent, or dependent; if the system is consistent, find the solution.

69. $x + y = 4$
      $2x - y = 2$

70. The monthly cost of Internet provider A is $10 for the first 5 hours and $3 for each hour after 5. Provider B's cost is $15 for the first 5 hours and $2 for each hour after 5. Make a graph of the cost C for providers A and B when h hours of airtime are used. (*Hint:* Let h run from 0 to 25 and C run from 0 to 60.)

71. $2x + y = 4$
      $2y + 4x = 6$

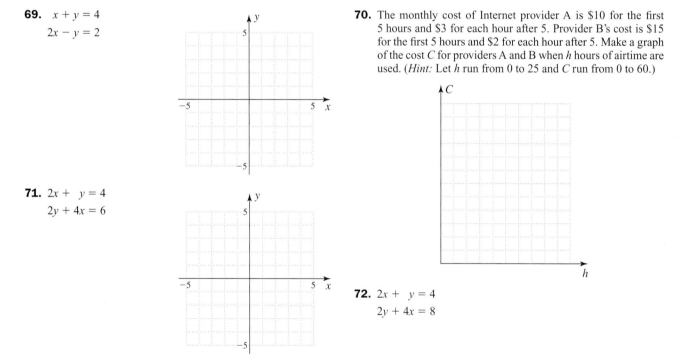

72. $2x + y = 4$
      $2y + 4x = 8$

〉〉〉 *Skill Checker*

Solve.

73. $3x + 72 = 2.5x + 74$

74. $5x + 20 = 3.5x + 26$

Determine whether the given point is a solution of the equation.

75. $(3, 5)$; $2x + y = 11$

76. $(-1, 4)$; $2x - y = -6$

77. $(-1, 2)$; $2x - y = 0$

78. $(-2, 6)$; $3x - y = 0$

7.2   Solving Systems of Equations by Substitution

▶ Objectives

Use the substitution
method to:

A ▶ Solve a system of
equations in two
variables.

B ▶ Determine whether a
system of equations
is consistent,
inconsistent, or
dependent.

C ▶ Solve applications
involving systems of
equations.

▶ To Succeed, Review How To . . .

1. Solve linear equations (pp. 132–136).

2. Determine whether an ordered pair satisfies an equation (pp. 221–222).

▶ Getting Started

Supply, Demand, and Substitution

Does this graph look familiar? As we noted in the preceding section's *Getting Started,*
the supply and the demand were about the same in the year 1980. However, this solution is only an *approximate* one because the *x*-scale representing the years is numbered
at 5-year intervals, and it's hard to pinpoint the exact point at which the lines intersect.
Suppose we have the equations for the supply and the demand between 1975 and 1985.
These equations are

Supply: $y = 2.5x + 72$ x = the number of years elapsed after 1975

Demand: $y = 3x + 70$

Can we now tell *exactly* where the lines meet? Not graphically! For one thing, if we let
$x = 0$ in the first equation, we obtain $y = 2.5 \cdot 0 + 72$, or $y = 72$. Thus, we either need a
piece of graph paper with 72 units, or we have to make each division on the graph paper
10 units, thereby *losing* accuracy. But there's a way out. We can use an *algebraic* method
rather than a *graphical* one. Since we are looking for the point at which the supply (y)
is the same as the demand (y), we may *substitute* the expression for y in the demand
equation—that is, $3x + 70$—into the supply equation. Thus, we have

$$\overbrace{\text{Demand } (y)}^{} = \overbrace{\text{Supply } (y)}^{}$$
$$3x + 70 = 2.5x + 72$$
$$3x = 2.5x + 2 \qquad \text{Subtract 70.}$$
$$0.5x = 2 \qquad \text{Subtract 2.5x.}$$
$$\frac{0.5x}{0.5} = \frac{2}{0.5} \qquad \text{Divide by 0.5.}$$
$$x = 4 \qquad \text{Simplify.}$$

Thus, 4 years after 1975 (or in 1979), the supply equaled the demand. At this time the demand was

$$y = 3(4) + 70 = 12 + 70 = 82 \text{ (million barrels)}$$

In this section we learn to solve equations by the ***substitution method.*** This method is recommended for solving systems in which *one* equation is solved, or can be easily solved, for one of the variables.

A ⟩ Using the Substitution Method to Solve a System of Equations

Here is a summary of the substitution method, which we just used in the *Getting Started.*

PROCEDURE

Solving a System of Equations by the Substitution Method
1. Solve one of the equations for x or y.
2. Substitute the resulting expression into the **other** equation. (Now you have an equation in one variable.)
3. Solve the new equation for the variable.
4. Substitute the value of that variable into one of the original equations and solve this equation to get the value for the second variable.
5. Check the solution by substituting the numerical values of the variables in both equations.

The idea is to solve one of the equations for a variable and substitute the result in the **other** equation. It does *not* matter which equation or which variable you pick, the final answer will be the same. However, when possible solve for the variable with a coefficient of one.

EXAMPLE 1   Solving a system by substitution

Solve the system:

$$x + y = 8$$
$$2x - 3y = -9$$

SOLUTION   We use the five-step procedure.

1. Solve one of the equations for x or y
   (we solve the first equation for y, $y = 8 - x$
   since y has a coefficient of 1).
2. Substitute $8 - x$ for y in
   $2x - 3y = -9$. $2x - 3(8 - x) = -9$

   Note that you must substitute $y = 8 - x$ into the **second** equation. Substituting in the first equation gives

   $$x + (8 - x) = 8$$
   $$8 = 8$$

   which is, of course, true.

PROBLEM 1

Solve the system:

$$x + y = 5$$
$$2x - 4y = -8$$

---

3. Solve the new equation for the variable.

$$2x - 3(8 - x) = -9$$
$$2x - 24 + 3x = -9 \quad \text{Simplify.}$$
$$5x - 24 = -9 \quad \text{Combine like terms.}$$
$$5x = 15 \quad \text{Add 24 to both sides.}$$
$$x = 3 \quad \text{Divide by 5.}$$

4. Substitute the value of the variable $x = 3$ into one of the original equations. (We substitute in the equation $x + y = 8$.) Then solve for the second variable. Our solution is the ordered pair $(3, 5)$.

$$3 + y = 8$$
$$y = 5$$

5. CHECK   When $x = 3$ and $y = 5$,

$$x + y = 8$$

becomes

$$3 + 5 = 8$$
$$8 = 8$$

which is true. Then the second equation

$$2x - 3y = -9$$

becomes

$$2(3) - 3(5) = -9$$
$$6 - 15 = -9$$
$$-9 = -9$$

which is also true. Thus, our solution $(3, 5)$ is correct.

EXAMPLE 2   Solving an inconsistent system by substitution

Solve the system:

$$x + 2y = 4$$
$$2x = -4y + 6$$

SOLUTION   We use the five-step procedure.

1. Solve one of the equations for one of the variables (we solve the first equation for x since x has a coefficient of 1). $x = 4 - 2y$

2. Substitute $x = 4 - 2y$ into $2x = -4y + 6$.

$$2(4 - 2y) = -4y + 6$$
$$8 - 4y = -4y + 6 \quad \text{Simplify.}$$
$$8 - 4y + 4y = -4y + 4y + 6 \quad \text{Add 4y.}$$
$$8 = 6$$

3. There is no equation to solve. The result, $8 = 6$, is never true. It is a contradiction. Since our procedure is correct, we conclude that the given system has *no solution;* it is *inconsistent.*

4. We do not need step 4.

5. CHECK   Note that if you divide the second equation by 2, you get $x = -2y + 3$ or $x + 2y = 3$ which *contradicts* the first equation, $x + 2y = 4$.

PROBLEM 2

Solve the system:

$$x - 3y = 6$$
$$2x - 6y = 8$$

Answers to PROBLEMS

2. No solution; the system is inconsistent.

EXAMPLE 3   Solving a dependent system by substitution

Solve the system:

$$x + 2y = 4$$
$$4y + 2x = 8$$

SOLUTION   As before, we use the five-step procedure.

1. Solve the first equation for x. $x = 4 - 2y$
2. Substitute $x = 4 - 2y$ into $4y + 2x = 8$. $4y + 2(4 - 2y) = 8$
3. There is no equation to solve. Note that $4y + 8 - 4y = 8$ Simplify.
   in this case we have obtained the true $8 = 8$
   statement $8 = 8$, regardless of the value
   we assign to either x or to y.
4. We do not need step 4 because the equations are *dependent;* that is, there are
   infinitely many solutions.
5. **CHECK**   If we let $x = 0$ in the equation $x + 2y = 4$, we obtain $2y = 4$, or
   $y = 2$. Similarly, if we let $x = 0$ in the equation $4y + 2x = 8$, we obtain
   $4y = 8$, or $y = 2$, so $(0, 2)$ is a solution of both equations. It can also be
   shown that $x = 2$, $y = 1$ satisfies both equations. Therefore, $(2, 1)$ is another
   solution, and so on. Note that if you divide the second equation by 2 and
   rearrange, you get $x + 2y = 4$, which is identical to the first equation. Thus, any
   solution of the first equation is also a solution of the second equation; that is,
   the solution consists of all points satisfying $x + 2y = 4$.

PROBLEM 3

Solve the system:

$$x - 3y = 6$$
$$6y - 2x = -12$$

EXAMPLE 4   Simplifying and solving a system by substitution

Solve the system:

$$-2x = -y + 2$$
$$6 - 3x + y = -4x + 5$$

SOLUTION   The second equation has x's and constants on both sides, so we
first simplify it by adding $4x$ and subtracting 6 from both sides to obtain

$$6 - 3x + y + 4x - 6 = -4x + 5 + 4x - 6$$
$$x + y = -1$$

We now have the equivalent system

$$-2x = -y + 2$$
$$x + y = -1$$

Solving the second equation for x, we get $x = -y - 1$. Substituting $-y - 1$ for x
in the first equation, we have

$$-2(-y - 1) = -y + 2$$
$$2y + 2 = -y + 2$$
$$3y = 0 \quad \text{Add } y, \text{ subtract 2.}$$
$$y = 0 \quad \text{Divide by 3.}$$

Since $-2x = -y + 2$ and $y = 0$, we have

$$-2x = 0 + 2$$
$$x = -1$$

Thus, the system is *consistent* and its solution is $(-1, 0)$. You can verify this by
substituting -1 for x and 0 for y in the two original equations.

PROBLEM 4

Solve the system:

$$-3x = -y + 6$$
$$6 - 3x + y = -5x + 2$$

Answers to PROBLEMS

3. Infinitely many solutions. The equations are dependent. Some solutions are $(0, -2)$, $(6, 0)$, and $(3, -1)$ **4.** $(-2, 0)$; consistent system

If a system has equations that contain fractions, we clear the fractions by multiplying each side by the LCD (Remember? LCD is the lowest common denominator), and then we solve the resulting system, as shown in Example 5.

EXAMPLE 5   Solving a system involving fractions

Solve the system:

$$2x + \frac{y}{4} = -1$$

$$\frac{x}{4} + \frac{3y}{8} = \frac{5}{4}$$

SOLUTION   Multiply both sides of the first equation by 4, and both sides of the second equation by 8 (the LCM of 4 and 8) to obtain

$$4\left(2x + \frac{y}{4}\right) = (4)(-1) \qquad \text{or equivalently} \qquad 8x + y = -4$$

$$8\left(\frac{x}{4} + \frac{3y}{8}\right) = 8\left(\frac{5}{4}\right) \qquad \text{or equivalently} \qquad 2x + 3y = 10$$

Solving the first equation for y, we have $y = -8x - 4$. Now we substitute $-8x - 4$ for y in $2x + 3y = 10$:

$$2x + 3(-8x - 4) = 10$$
$$2x - 24x - 12 = 10 \qquad \text{Simplify.}$$
$$-22x = 22 \qquad \text{Simplify and add 12.}$$
$$x = -1 \qquad \text{Divide by } -22.$$

Substituting -1 for x in $2x + \frac{y}{4} = -1$, we get $2(-1) + \frac{y}{4} = -1$ or $y = 4$. Thus, the system is *consistent* and its solution is $(-1, 4)$. Verify this!

PROBLEM 5

Solve the system:

$$2x + \frac{y}{3} = -1$$

$$\frac{x}{4} + \frac{y}{6} = \frac{1}{4}$$

B > Consistent, Inconsistent, and Dependent Systems

When we use the substitution method, one of three things can occur:

1. The equations are *consistent;* there is only *one* solution (x, y).
2. The equations are *inconsistent;* we get a contradictory (false) statement, and there will be *no* solution.
3. The equations are *dependent;* we get a statement that is true for all values of the remaining variable, and there will be *infinitely many* solutions.

Keep this in mind when you do the exercise set, and be very careful with your arithmetic!

C > Applications Involving Systems of Equations

Remember the prices for connecting to the Internet via America Online (AOL) and CompuServe* (CS)? The cost for each of the plans is as follows:

AOL: $20 for 20 hours plus $3 for each hour after 20 hours
CS: $25 for 20 hours plus $2 for each hour after 20 hours

---

*Note that CompuServe has been aquired by AOL.

This means that the price p for h hours of service when $h \geq 20$ is

$$\text{AOL:} \quad p \; = \; \underbrace{20}_{\$20} \; + \; \underbrace{3(h - 20)}_{\substack{\$3 \text{ for each} \\ \text{hour after } 20}}$$

$$\text{CS:} \quad p \; = \; \underbrace{25}_{\$25} \; + \; \underbrace{2(h - 20)}_{\substack{\$2 \text{ for each} \\ \text{hour after } 20}}$$

EXAMPLE 6   Substitution and Internet prices

Based on the information above, when is the price for both services the same?

SOLUTION   To find when the price p is the same for both services, we substitute $p = 20 + 3(h - 20)$ into the second equation to obtain

$$20 + 3(h - 20) = 25 + 2(h - 20)$$

$$20 + 3h - 60 = 25 + 2h - 40 \qquad \text{Use the distributive property.}$$

$$3h - 40 = 2h - 15 \qquad \text{Simplify.}$$

$$3h = 2h + 25 \qquad \text{Add 40 to both sides.}$$

$$h = 25 \qquad \text{Subtract } h \text{ from both sides.}$$

Thus, if you use 25 hours, the price is the same for AOL and CS.

CHECK   The price for 25 hours of AOL service is

$$p = 20 + 3(25 - 20) = 20 + 15 = \$35$$

The price for 25 hours of CS service is

$$p = 25 + 2(25 - 20) = 25 + 10 = \$35$$

Thus, when using 25 hours, the price is the same for both services: $35.

PROBLEM 6

Quickie Internet charges $25 for 10 hours plus $2.50 for each hour after 10. When is this price the same as the AOL price, $20 + 3(h - 20)$?

> Exercises **7.2**

⟨ **A** ⟩   Using the Substitution Method to Solve a System of Equations

⟨ **B** ⟩   Consistent, Inconsistent, and Dependent Systems

In Problems 1–32, use the substitution method to find the solution, if possible. Label each system as consistent (one solution), inconsistent (no solution), or dependent (infinitely many solutions). If the system is consistent, give the solution.

1. $\quad y = 2x - 4$
$\quad -2x = y - 4$

2. $\quad y = 2x + 2$
$\quad -x = y + 1$

3. $\quad x + y = 5$
$\quad 3x + y = 9$

4. $\quad x + y = 5$
$\quad 3x + y = 3$

5. $\quad y - 4 = 2x$
$\quad y = 2x + 2$

6. $\quad y + 5 = 4x$
$\quad y = 4x + 7$

7. $\quad x = 8 - 2y$
$\quad x + 2y = 4$

8. $\quad x = 4 - 2y$
$\quad x - 2y = 0$

9. $x + 2y = 4$
$\quad x = -2y + 4$

Answers to PROBLEMS

6. When 80 hours are used

10. $x + 3y = 6$
     $x = -3y + 6$

11. $x = 2y + 1$
     $y = 2x + 1$

12. $y = 3x + 2$
     $x = 3y + 2$

13. $2x - y = -4$
     $4x = 4 + 2y$

14. $5x + y = 5$
     $5x = 15 - 3y$

15. $x = 5 - y$
     $0 = x - 4y$

16. $x = 3 - y$
     $0 = 2x - y$

17. $x + 1 = y + 3$
     $x - 3 = 3y - 7$

18. $x - 1 = 2y + 12$
     $x + 6 = 3 - 6y$

19. $2y = -x + 4$
     $8 + x - 4y = -2y + 4$

20. $y - 1 = 2x + 1$
     $3x + y + 2 = 5x + 6$

21. $3x + y - 5 = 7x + 2$
     $y + 3 = 4x - 2$

22. $4x + 2y + 1 = 4 + 3x + 5$
     $x - 3 = 5 - 2y$

23. $4x - 2y - 1 = 3x - 1$
     $x + 2 = 6 - 2y$

24. $8 + y - 4x = -2x + 4$
     $2x + 3 = -y + 7$

25. $\dfrac{x}{6} + \dfrac{y}{2} = 1$

     $5x - 2y = 13$

26. $\dfrac{x}{8} - \dfrac{5y}{8} = 1$

     $-7x + 8y = 25$

27. $3x - y = 12$

     $-\dfrac{x}{2} + \dfrac{y}{6} = -2$

28. $x - 3y = -4$

     $-\dfrac{x}{6} + \dfrac{y}{2} = \dfrac{2}{3}$

29. $\dfrac{y}{4} + x = \dfrac{3}{8}$

     $y = 8 - 4x$

30. $y = 1 - 5x$

     $\dfrac{y}{5} + x = \dfrac{3}{10}$

31. $3x + \dfrac{y}{3} = 5$

     $\dfrac{x}{2} - \dfrac{2y}{3} = 3$

32. $3y + \dfrac{x}{3} = 5$

     $\dfrac{y}{2} - \dfrac{2x}{3} = 3$

⟨ C ⟩ Applications Involving Systems of Equations

33. *Internet service costs* The Information Network charges a $20 fee for 15 hours of Internet service plus $3 for each additional hour while InterServe Communications charges $20 for 15 hours plus $2 for each additional hour.

  a. Write an equation for the price p when you use h hours of Internet service with The Information Network.

  b. Write an equation for the price p when you use h hours of Internet service with InterServe Communications.

  c. When is the price p for both services the same?

34. *Internet service costs* TST On Ramp charges $10 for 10 hours of Internet service plus $2 for each additional hour.

  a. Write an equation for the price p when you use h hours of Internet service with TST On Ramp.

  b. When is the price of TST the same as that for InterServe Communications (see Problem 33)? (*Hint:* The algebra won't tell you—try a graph!)

35. *Cell phone costs* Phone Company A has a plan costing $20 per month plus 60¢ for each minute m of airtime, while Company B charges $50 per month plus 40¢ for each minute m of airtime. When is the cost for both companies the same?

36. *Cell phone charges* Sometimes phone companies charge an activation fee to "turn on" your cell phone. One company charges $50 for the activation fee, $40 for your cell phone, and 60¢ per minute m of airtime. Another company charges $100 for your cell phone and 40¢ a minute of airtime. When is the cost for both companies the same?

37. *Wages and tips* Le Bon Ton restaurant pays its servers $50 a week plus tips, which average $10 per table. Le Magnifique pays $100 per week but tips average only $5 per table. How many tables t have to be served so that the weekly income of a server is the same at both restaurants?

38. *Fitness center costs* The Premier Fitness Center has a $200 initiation fee plus $25 per month. Bodies by Jacques has an initial charge of $500 but charges only $20 per month. At the end of which month is the cost the same?

39. *Cable company costs* One cable company charges $35 for the initial installation plus $20 per month. Another company charges $20 for the initial installation plus $35 per month. At the end of which month is the cost the same for both companies?

40. *Plumber rates* A plumber charges $20 an hour plus $60 for the house call. Another plumber charges $25 an hour, but the house call is only $50. What is the least number of hours for which the costs for both plumbers are the same?

41. *Temperature conversions* The formula for converting degrees Celsius *C* to degrees Fahrenheit *F* is

$$F = \frac{9}{5} C + 32$$

When is the temperature in degrees Fahrenheit the same as that in degrees Celsius?

42. *Temperature conversions* The formula for converting degrees Fahrenheit *F* to degrees Celsius *C* is

$$C = \frac{5}{9} (F - 32)$$

When is the temperature in degrees Celsius the same as that in degrees Fahrenheit?

43. *Supply and demand* The supply *y* of a certain item is given by the equation $y = 2x + 8$, where *x* is the number of days elapsed. If the demand is given by the equation $y = 4x$, how many days will the supply equal the demand?

44. *Supply and demand* The supply of a certain item is given by the equation $y = 3x + 8$, where *x* is the number of days elapsed. If the demand is given by the equation $y = 4x$, in how many days will the supply equal the demand?

45. *Supply and demand* A company has 10 units of a certain item and can manufacture 5 items each day; thus, the supply is given by the equation $y = 5x + 10$. If the demand for the item is $y = 7x$, where *x* is the number of days elapsed, in how many days will the demand equal the supply?

46. *Supply and demand* Clonker Manufacturing has 12 clonkers in stock. The company manufactures 3 more clonkers each day. If the clonker demand is 7 each day, in how many days will the supply equal the demand?

47. *College expenses* According to the College Board, parents will pay *x* dollars for tuition and room and board at public colleges this year. They are also likely to spend *y* dollars on textbooks, supplies, transportation, and "other."

  a. The total cost for tuition, room and board (*x*) plus textbooks, supplies, transportation, and "other" (*y*) comes to $15,127. Write an equation for this fact.

  b. The tuition, room and board *x* is $9127 more than the textbooks, supplies and transportation, and "other" *y*. Write an equation for this fact.

  c. Use the substitution method to solve the systems of two equations obtained in parts **a** and **b** to find *x* (cost of tuition and room and board) and *y* (cost of textbooks, supplies, transportation, and "other").

Source: www.msnbc.msn.com.

48. *Cost of computers, books, and supplies* The "Average Cost of Attendance" page at the University of Florida estimates that the expenses for computers *C* plus the expenses for books and supplies *B* will amount to $1820.

  a. Write an equation for this fact.

  b. The books and supplies *B* are only $20 more than the computers *C*. Write an equation for this fact.

  c. Use the substitution method to solve the system of equations obtained in parts **a** and **b** to find the estimated cost of computers *C* and the estimated cost of books and supplies *B*.

Source: www.sfa.ufl.edu.

49. *Transportation and personal/miscellaneous expenses* At California State University at Long Beach the expenses for transportation *T* plus personal/miscellaneous expenses *PM* amount to $3222.

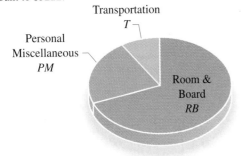

Transportation

Personal Miscellaneous *PM*

T

Room & Board *RB*

  a. Write an equation for this fact.

  b. The difference between *PM* and *T* is $1350. Write an equation for this fact.

  c. Use the substitution method to solve the system of equations obtained in parts **a** and **b** to find the transportation expenses *T* and the personal/miscellaneous expenses *PM*.

Source: www.csulb.edu.

50. *Textbook costs* A report released by the Student Public Interest Group indicates that the average cost *F* of the 22 most frequently assigned textbooks is $65.72 more than the cost *L* of less expensive alternatives.

  a. Write an equation for the average cost *F*.

  b. The cost *L* of those less expensive counterparts is about $\frac{1}{2}$ the cost *F* of the most frequently assigned textbooks. Write an equation for the cost *L* of the less expensive counterparts.

  c. Use the substitution method to solve the systems of two equations obtained in parts **a** and **b** to find the average cost *F* of the 22 most frequently assigned textbooks and the cost *L* of the less expensive counterparts.

Source: www.alternet.org.

〉〉〉 *Using Your Knowledge*

Regeneration, Resistors, and Revenue   The ideas presented in this section are important in many fields. Use your knowledge to solve the following problems.

51. The total inductance of the inductors L_1 and L_2 in an oscillator must be 400 microhenrys. Thus,

$$L_1 = -L_2 + 400$$

To provide the correct regeneration for the oscillator circuit requires

$$\frac{L_2}{L_1} = 4, \quad \text{that is,} \quad L_2 = 4L_1$$

Solve the system

$$L_1 = -L_2 + 400$$

$$L_2 = 4L_1$$

by substitution.

52. The equations for the resistors in a voltage divider must be such that

$$R_1 = 3R_2$$

$$R_1 + R_2 = 400$$

Solve for R_1 and R_2 using the substitution method.

53. The total revenue R for a certain manufacturer is given by the equation

$$R = 5x$$

the total cost C is given by the equation

$$C = 4x + 500$$

where x represents the number of units produced and sold.

a. Use your knowledge of the substitution method to write the equation that will result when

$$R = C$$

b. The point at which $R = C$ is called the *break-even point*. Find the number of units the manufacturer must produce and sell in order to break even.

〉〉〉 *Write On*

54. If you are solving the system

$$2x + y = -7$$

$$3x - 2y = 7$$

which variable would you solve for in the first step of the five-step procedure given in the text?

56. In Example 5, we multiplied both sides of the first equation by 4 and both sides of the second equation by 8. Would you get the same answer if you multiplied both sides of the first equation by 8 and both sides of the second equation by 4? Why is this approach not a good idea?

55. When solving a system of equations using the substitution method, how can you tell whether the system is

a. consistent?

b. inconsistent?

c. dependent?

〉〉〉 *Concept Checker*

Fill in the blank(s) with the correct word(s), phrase, or mathematical statement

57. When a system of equations is **consistent**, there is _____ **solution** to the system.

58. The **solution** of a **consistent** system of equations is an _____ pair of numbers.

59. An **inconsistent** system of equations has _____ solutions.

60. A **dependent** system of equations has _____ solutions.

many	**ordered**
one	**finite**
no	**infinitely many**

⟩⟩⟩ *Mastery Test*

Solve and label the system as consistent, inconsistent, or dependent. If the system is consistent, write the solution.

61. $x - 3y = 6$
$\quad\ 2x - 6y = 8$

62. $x - 3y = 6$
$\quad\ 6y - 2x = -12$

63. $x + y = 5$
$\quad\ 2x - 3y = -5$

64. $2y + x = 3$
$\quad\ x - 3y = 0$

65. $3x - 3y + 1 = 5 + 2x$

$\qquad x - 3y = 4$

66. $\qquad 5x + y = 5$

$\quad\ 5x + y - 10 = 5 - 2y$

67. $\dfrac{x}{2} - \dfrac{y}{4} = -1$

$\qquad 2x = 2 + y$

68. $x + \dfrac{y}{5} = 1$

$\qquad \dfrac{x}{3} + \dfrac{y}{5} = 1$

69. A store is selling a Sony® DSS system for $300. The basic monthly charge is $50. An RCA® system is selling for $500 with a $30 monthly charge. What is the least number of months for which the prices of both systems are the same?

⟩⟩⟩ *Skill Checker*

Solve:

70. $-1.50b = -6$

71. $-0.5x = -4$

72. $-0.2y = -6$

73. $9x = -9$

74. $11y = -11$

7.3   Solving Systems of Equations by Elimination

▶ Objectives

A ⟩ Solve a system of equations in two variables by elimination.

B ⟩ Determine whether a system is consistent, inconsistent, or dependent.

C ⟩ Solve applications involving systems of equations.

▶ To Succeed, Review How To . . .

1. Use the RSTUV method for solving word problems (pp. 143–144).
2. Solve linear equations (pp. 132–136).

▶ Getting Started
Using Elimination When Buying Coffee

We've studied two methods for solving systems of equations. The *graphical method* gives us a visual model of the system and allows us to find approximate solutions. The *substitution method* gives exact solutions but is best used when either of the given equations has at least one coefficient of 1 or -1. If graphing or substitution is not desired or feasible, there is another method we can use: the *elimination method,* sometimes called the **addition** or **subtraction method.**

The man in the photo is selling coffee, ground to order. A customer wants 10 pounds of a mixture of coffee A costing $6 per pound and coffee B costing $4.50 per pound. If the price for the purchase is $54, how many pounds of each of the coffees will be in the mixture?

To solve this problem, we use an idea that we've already learned: solving a system of equations. In this problem we want a precise answer, so we use the elimination method, which we will learn next.

A ⟩ Solving Systems of Equations by Elimination

To solve the coffee problem in the *Getting Started,* we first need to organize the information. The information can be summarized in a table like this:

	Price	Pounds	Total Price
Coffee A	6.00	a	$6a$
Coffee B	4.50	b	$4.50b$
Totals		$a + b$	$6a + 4.50b$

Since the customer bought a total of 10 pounds of coffee, we know that

$$a + b = 10$$

Also, since the purchase came to $54, we know that

$$6a + 4.50b = 54$$

So, the system of equations we need to solve is

$$a + b = 10$$
$$6a + 4.50b = 54$$

To solve this system, we shall use the **elimination method,** which consists of replacing the given system by equivalent systems until we get a system with an obvious solution. To do this, we first write the equations in the form $Ax + By = C$. Recall that an *equivalent* system is one that has the *same* solution as the given one. For example, the equation

$$A = B$$

is equivalent to the equation

$$kA = kB \quad (k \neq 0)$$

This means that you can multiply both sides of the equation $A = B$ by the same nonzero expression k and obtain the equivalent equation $kA = kB$.

Also, the system

$$A = B$$
$$C = D$$

is equivalent to the system

$$A = B$$
$$k_1A + k_2C = k_1B + k_2D \quad (k_1 \text{ and } k_2 \text{ not both 0})$$

We can check this by multiplying both sides of $A = B$ by k_1 and both sides of $C = D$ by k_2 and adding. That is the theory, but how do you do it?

Let's return to the coffee problem. We multiply the first equation in the given system by -6; we then get the equivalent system

$$
\begin{array}{ll}
-6a - 6b = -60 & \text{Multiply by } -6 \text{ because we} \\
\underline{(+)\ 6a + 4.50b = 54} & \text{want the coefficients of } a \text{ to be} \\
0 - 1.50b = -6 & \text{opposites (like } -6 \text{ and 6).}
\end{array}
$$

Add the equations.

$$-1.50b = -6$$
$$b = 4 \qquad \text{Divide by } -1.50.$$
$$a + 4 = 10 \qquad \text{Substitute 4 for } b \text{ in } a + b = 10.$$
$$a = 6 \qquad \text{Solve for } a.$$

Thus, the mixture contains 6 pounds of coffee A and 4 pounds of coffee B. This answer can be verified. If the customer bought 6 pounds of coffee A and 4 of coffee B, she did indeed buy 10 pounds. Her price for coffee A was $6 \cdot 6 = \$36$ and for coffee B, $4 \cdot 4.50 = \$18$. Thus, the entire cost was $\$36 + \$18 = \$54$, as stated.

What have we done here? Well, this technique depends on the fact that one (or both) of the equations in a system can be multiplied by a nonzero number to obtain two equivalent equations with opposite coefficients of x (or y). Here is the idea.

ELIMINATION METHOD   One or both of the equations in a system of simultaneous equations can be multiplied (or divided) by any nonzero number to obtain an *equivalent* system in which the coefficients of the x's (or of the y's) are opposites, thus, *eliminating x or y* when the equations are added.

EXAMPLE 1   Solving a consistent system by elimination

Solve the system:

$$2x + y = 1$$
$$3x - 2y = -9$$

SOLUTION   Remember the idea: we multiply one or both of the equations by a number or numbers that will cause either the coefficients of x or the coefficients of y to be opposites. We can do this by multiplying the first equation by 2:

$2x + y = 1$ Multiply by 2. $\rightarrow$ $4x + 2y = 2$

$3x - 2y = -9$ Leave as is. $\rightarrow$ $3x - 2y = -9$

Add the equations. $\overline{7x + 0 = -7}$

$7x = -7$

Divide by 7. $x = -1$

Substitute -1 for x in $2x + y = 1$. $2(-1) + y = 1$

$-2 + y = 1$

Add 2. $y = 3$

Thus, the solution of the system is $(-1, 3)$.

CHECK   When $x = -1$ and $y = 3$, $2x + y = 1$ becomes

$$2(-1) + 3 = 1$$
$$-2 + 3 = 1$$
$$1 = 1$$

a true statement, and $3x - 2y = -9$ becomes

$$3(-1) - 2(3) = -9$$
$$-3 - 6 = -9$$
$$-9 = -9$$

which is also true.

PROBLEM 1

Solve the system:

$$3x + y = 1$$
$$3x - 4y = 11$$

Answers to PROBLEMS

1. $(1, -2)$

B › Determining Whether a System Is Consistent, Inconsistent, or Dependent

There are *three* possibilities when solving simultaneous linear equations.

1. *Consistent* and *independent* equations have *one* solution.
2. *Inconsistent* equations have *no* solution. You can recognize them when you get a contradiction (a false statement) in your work, (See Example 2. In Example 2, we get $0 = -12$, a contradiction.)
3. *Dependent* equations have *infinitely many* solutions. You can recognize them when you get a true statement such as $0 = 0$ (See Example 3). Remember that any solution of one of these equations is a solution of the other.

Pay close attention to the position of the variables in the equations. All the equations **except** those solved by substitution should be written in the form

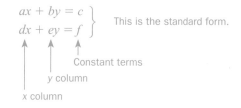

$$\left. \begin{array}{l} ax + by = c \\ dx + ey = f \end{array} \right\}$$ This is the standard form.

Constant terms
y column
x column

If the equations are not in this form, and you are not using the substitution method, rewrite them using this form. It helps to keep things straight!

Note that as it is mentioned in (2) above, not all systems have solutions. How do we find out if they don't? Let's look at the next example, which shows a **contradiction** for a system that has **no** solution.

EXAMPLE 2   Solving an inconsistent system by elimination

Solve the system:

$$2x + 3y = 3$$
$$4x + 6y = -6$$

SOLUTION   In this case, we try to eliminate the variable x by multiplying the first equation by -2.

$$2x + 3y = 3 \quad \xrightarrow{\text{Multiply by } -2.} \quad -4x - 6y = -6$$

$$4x + 6y = -6 \quad \xrightarrow{\text{Leave as is.}} \quad \underline{4x + 6y = -6}$$

$$0 + \ 0 = -12 \quad \text{Add.}$$

$$0 = -12$$

Of course, this is a contradiction, so there is no solution; the system is *inconsistent*.

PROBLEM 2

Solve the system:

$$3x + 2y = 1$$
$$6x + 4y = 12$$

EXAMPLE 3   Solving a dependent system by elimination

Solve the system:

$$2x - 4y = 6$$
$$-x + 2y = -3$$

PROBLEM 3

Solve the system:

$$3x - 6y = 9$$
$$-x + 2y = -3$$

Answers to PROBLEMS

2. No solution; inconsistent
3. Infinitely many solutions; dependent

(continued)

SOLUTION   Here we try to eliminate the variable x by multiplying the second equation by 2. We obtain

$$2x - 4y = 6 \xrightarrow{\text{Leave as is.}} 2x - 4y = 6$$

$$-x + 2y = -3 \xrightarrow{\text{Multiply by 2.}} \underline{-2x + 4y = -6}$$

$$0 + \ 0 = 0 \qquad \text{Add.}$$

$$0 = 0$$

Lo and behold, we've eliminated both variables! However, notice that if we had multiplied the second equation in the original system by -2, we would have obtained

$$2x - 4y = 6 \xrightarrow{\text{Leave as is.}} 2x - 4y = 6$$

$$-x + 2y = -3 \xrightarrow{\text{Multiply by } -2.} 2x - 4y = 6$$

This means that the first equation is a constant multiple of the second one; that is, they are equivalent equations. When a system of equations consists of two equivalent equations, the system is said to be *dependent,* and any solution of one equation is a solution of the other. Because of this, the system has *infinitely many* solutions. For example, if we let x be 0 in the first equation, then $y = -\frac{3}{2}$, and $(0, -\frac{3}{2})$ is a solution of the system. Similarly, if we let y be 0 in the first equation, then $x = 3$, and we obtain the solution $(3, 0)$. Many other solutions are possible; try to find some of them.

Finally, in some cases, we cannot multiply just one of the equations by an integer that will cause the coefficients of one of the variables to be opposites. For example, to solve the system

$$2x + 3y = 3$$
$$5x + 2y = 13$$

we must multiply **both** equations by integers chosen so that the coefficients of one of the variables will be opposites. We can do this using either of the following methods.

METHOD 1

Solving by elimination: x or y?
To eliminate x, multiply the first equation by 5 and the second one by -2 to obtain an equivalent system.

$$2x + 3y = 3 \xrightarrow{\text{Multiply by 5.}} 10x + 15y = 15$$

$$5x + 2y = 13 \xrightarrow{\text{Multiply by } -2.} \underline{-10x - \ 4y = -26}$$

Add. $\qquad\qquad\qquad\qquad\qquad\qquad\quad 0 + 11y = -11$

$$11y = -11$$

Divide by 11. $\qquad\qquad\qquad\qquad\qquad\qquad\qquad y = -1$

Substitute -1 for y in $2x + 3y = 3$. $\qquad 2x + 3\,(-1) = 3$

Simplify. $\qquad\qquad\qquad\qquad\qquad\qquad 2x - 3 = 3$

Add 3. $\qquad\qquad\qquad\qquad\qquad\qquad\qquad 2x = 6$

Divide by 2. $\qquad\qquad\qquad\qquad\qquad\qquad\qquad x = 3$

Thus, the solution of the system is $(3, -1)$. This time we eliminated the x and solved for y. Alternatively, we can eliminate the y first, as shown next.

METHOD 2

This time, we eliminate the y:

$$2x + 3y = 3 \quad \xrightarrow{\text{Multiply by } -2.} \quad -4x - 6y = -6$$

$$5x + 2y = 13 \quad \xrightarrow{\text{Multiply by } 3.} \quad \underline{15x + 6y = 39}$$

Add. $\qquad\qquad\qquad\qquad\qquad\qquad 11x + 0 = 33$

$$11x = 33$$

Divide by 11. $\qquad\qquad\qquad\qquad\qquad x = 3$

Substitute 3 for x in $2x + 3y = 3$. $\qquad 2(3) + 3y = 3$

Simplify. $\qquad\qquad\qquad\qquad\qquad 6 + 3y = 3$

Subtract 6. $\qquad\qquad\qquad\qquad\qquad 3y = -3$

Divide by 3. $\qquad\qquad\qquad\qquad\qquad y = -1$

Thus, the solution is $(3, -1)$, as before.

CHECK: When $x = 3$ and $y = -1$, $2x + 3y = 3$ becomes

$$2(3) + 3(-1) = 3$$
$$6 - 3 \quad = 3$$
$$3 \quad = 3$$

which is true. Substituting $x = 3$ and $y = -1$ in $5x + 2y = 13$ yields

$$5(3) + 2(-1) = 13$$
$$15 - 2 \quad = 13$$

which is also true. Thus, $(3, -1)$ is the solution!

EXAMPLE 4   Writing in standard form and solving by elimination

Solve the system:

$$5y + 2x = 9$$
$$2y = 8 - 3x$$

SOLUTION   We first write the system in standard form—that is, the x's first, then the y's, and then the constants on the other side of the equation. The result is the equivalent system

$$2x + 5y = 9$$
$$3x + 2y = 8$$

Now we multiply the first equation by 3 and the second one by -2 so that, upon addition, the x's will be eliminated.

$$2x + 5y = 9 \quad \xrightarrow{\text{Multiply by } 3.} \quad 6x + 15y = \quad 27$$

$$3x + 2y = 8 \quad \xrightarrow{\text{Multiply by } -2.} \quad \underline{-6x - 4y = -16}$$

Add. $\qquad\qquad\qquad\qquad\qquad\qquad 0 + 11y = 11$

$$11y = 11$$

Divide by 11. $\qquad\qquad\qquad\qquad\qquad y = 1$

Substitute 1 for y in $2x + 5y = 9$. $\qquad 2x + 5(1) = 9$

Simplify. $\qquad\qquad\qquad\qquad\qquad 2x + 5 = 9$

Subtract 5. $\qquad\qquad\qquad\qquad\qquad 2x = 4$

Divide by 2. $\qquad\qquad\qquad\qquad\qquad x = 2$

Thus, the solution is $(2, 1)$ and the system is *consistent*. You should verify this by substituting $x = 2$ and $y = 1$ in the original equation to make sure they satisfy both equations.

PROBLEM 4

Solve the system:

$$5x + 4y = 6$$
$$3y = 4x - 11$$

Answers to **PROBLEMS**

4. $(2, -1)$

C › Applications Involving Systems of Equations

EXAMPLE 5   Cell phone plans

Do you have a cell phone? The time the phone is used, called airtime, is usually charged by the minute at two different rates: peak and off-peak. Suppose your plan charges $0.60 for each peak-time minute and $0.45 for each off-peak minute. If your airtime cost $54 and you've used 100 minutes of airtime, how many minutes p of peak time and how many minutes n of off-peak time did you use?

PROBLEM 5

Big Cell Phone has a plan that charges $0.50 for each peak-time minute and $0.40 for each off-peak minute. If airtime cost $44 and 100 minutes were used, how many minutes p of peak time and how many minutes n of off-peak time were used?

SOLUTION   To solve this problem, we need two equations involving the two unknowns, p and n. We know that the charges amount to $54 and that 100 minutes of airtime were used. How can we accumulate $54 of charges?

Since peak time costs $0.60 per minute, peak times cost $0.60p$.

Since off-peak time costs $0.45 per minute, off-peak times cost $0.45n$.

The total cost is $54, so we add peak and off-peak costs:

$$0.60p + 0.45n = 54$$

Also, the total number of minutes is 100. Thus, $p + n = 100$. To try to *eliminate n,* we multiply both sides of the second equation by -0.45 and then add:

$$0.60p + 0.45n = 54 \quad \xrightarrow{\text{Leave as is.}} \quad 0.60p + 0.45n = 54$$
$$p + n = 100 \quad \xrightarrow{\text{Multiply by } -0.45} \quad \underline{-0.45p - 0.45n = -45}$$
$$\text{Add.} \qquad 0.15p = 9$$
$$\text{Divide both sides by } 0.15. \qquad p = \frac{9}{0.15} = 60$$

Thus, $p = 60$ minutes of peak time were used and the rest of the 100 minutes used—that is, $100 - 60 = 40$—were off-peak minutes. This means that $n = 40$. You can check that $p = 60$ and $n = 40$ by substituting in the original equations.

› Exercises **7.3**

⟨ **A** ⟩   Solving Systems of Equations by Elimination
⟨ **B** ⟩   Determining Whether a System is Consistent, Inconsistent, or Dependent

In Problems 1–30, use the elimination method to solve each system. If the system is not consistent, state whether it is inconsistent or dependent.

1. $x + y = 3$
$x - y = -1$

2. $x + y = 5$
$x - y = 1$

3. $x + 3y = 6$
$x - 3y = -6$

4. $x + 2y = 4$
$x - 2y = 8$

5. $2x + y = 4$
$4x + 2y = 0$

6. $3x + 5y = 2$
$6x + 10y = 5$

7. $2x + 3y = 6$
$4x + 6y = 2$

8. $3x - 5y = 4$
$-6x + 10y = 0$

Answers to PROBLEMS

5. 40 minutes of peak and 60 minutes of off-peak

9. $x - 5y = 15$
$x + 5y = 5$

10. $-3x + 2y = 1$
$2x + y = 4$

11. $x + 2y = 2$
$2x + 3y = -10$

12. $3x - 2y = -1$
$x + 7y = -8$

13. $3x - 4y = 10$
$5x + 2y = 34$

14. $5x - 4y = 6$
$3x + 2y = 8$

15. $11x - 3y = 25$
$5x + 8y = 2$

16. $12x + 8y = 8$
$7x - 5y = 24$

17. $2x + 3y = 21$
$3x = y + 4$

18. $2x - 3y = 16$
$x = y + 7$

19. $x = 1 + 2y$
$-y = x + 5$

20. $3y = 1 - 2x$
$3x = -4y - 1$

21. $\frac{x}{4} + \frac{y}{3} = 4$
$\frac{x}{2} - \frac{y}{6} = 3$

(*Hint:* Multiply by the LCD first.)

22. $\frac{x}{5} + \frac{y}{6} = 5$
$\frac{2x}{5} + \frac{y}{3} = -2$

(*Hint:* Multiply by the LCD first.)

23. $\frac{1}{4}x - \frac{1}{3}y = -\frac{5}{12}$
$\frac{1}{5}x + \frac{2}{5}y = 1$

(*Hint:* Multiply by the LCD first.)

24. $\frac{x}{2} + \frac{y}{2} = \frac{5}{2}$
$\frac{x}{2} - \frac{y}{3} = \frac{5}{2}$

(*Hint:* Multiply by the LCD first.)

25. $\frac{x}{8} + \frac{y}{8} = 1$
$\frac{x}{2} - \frac{y}{2} = -1$

26. $\frac{x}{5} + \frac{y}{5} = 1$
$\frac{x}{4} - \frac{y}{4} = \frac{1}{4}$

27. $\frac{x}{2} - \frac{y}{3} = 1$
$\frac{x}{2} + \frac{y}{2} = \frac{7}{2}$

28. $\frac{x}{3} + \frac{y}{2} = \frac{7}{3}$
$\frac{x}{3} - \frac{y}{2} = 0$

29. $\frac{2x}{9} - \frac{y}{2} = -1$
$x - \frac{9y}{4} = -\frac{9}{2}$

30. $-\frac{8x}{49} + \frac{5y}{49} = -1$
$\frac{2x}{3} - \frac{5y}{12} = \frac{49}{12}$

⟨ **C** ⟩ **Applications Involving Systems of Equations**

31. *Coffee blends* The Holiday House blends Costa Rican coffee that sells for $8 a pound and Indian Mysore coffee that sells for $9 a pound to make 1-pound bags of its Gourmet Blend coffee, which sells for $8.20 a pound. How much Costa Rican and how much Indian coffee should go into each pound of the Gourmet Blend?

32. *Coffee blends* The Holiday House also blends Colombian Swiss Decaffeinated coffee that sells for $11 a pound and High Mountain coffee that sells for $9 a pound to make its Lower Caffeine coffee, 1-pound bags of which sell for $10 a pound. How much Colombian and how much High Mountain should go into each pound of the Lower Caffeine mixture?

33. *Tea blends* Oolong tea that sells for $19 per pound is blended with regular tea that sells for $4 per pound to produce 50 pounds of tea that sells for $7 per pound. How much Oolong and how much regular should go into the mixture?

34. *Metal alloys* If the price of copper is 65¢ per pound and the price of zinc is 30¢ per pound, how many pounds of copper and zinc should be mixed to make 70 pounds of brass, which sells for 45¢ per pound?

Use the following information for Problems 35–36. If you have high blood pressure, heart disease, or if you just want to maintain your health you should monitor your intake of **sodium** (< 2400 milligrams mg per day), **total fat** (< 65 grams per day), and **calories** (< 2000 per day). How can we do that? Let us concentrate on daily intakes.
Source: www.pamf.org.

35. *Daily sodium intake* Suppose you go to lunch at McDonald's and dinner at Burger King. You eat **one** Burger King Fire-Grilled Chicken Salad containing c grams of sodium and **two** McDonald's Double Quarter Pounders with cheese with

m milligrams of sodium each. You just ate 3100 milligrams of sodium which is over the daily recommended limit! Your sodium consumption for the day is $c + 2m = 3100$.

The next day you change your routine. Now you eat **two** Grilled Chicken Salads from Burger King at c milligrams of sodium each and just **one** Quarter Pounder with m milligrams of sodium. You just consumed 4205 milligrams of sodium. Which is still over the daily recommended limit. Your sodium consumption for that day is $2c + m = 4205$.

Solve the system:

$$2c + m = 4205$$
$$c + 2m = 3100$$

and find c (milligrams of sodium in the Grilled Chicken Salad) and m (milligrams of sodium in the burger).

Chicken Salad

Double Quarter Pounder with Cheese

36. *Daily sodium intake* Suppose you decide to eat **two** Shrimp Garden Salads with vinaigrette dressing with s milligrams of sodium in each and **one** McDonald hamburger with h milligrams of sodium. You are still over the daily recommended amount of sodium intake at 4070 milligrams! Your sodium consumption is $2s + h = 4070$.

Next you limit yourself to **one** Shrimp Garden Salad with s milligrams of sodium and **one** hamburger with h milligrams. Congratulations, you are finally under: you had 2300 milligrams of sodium!

Your sodium consumption is $s + h = 2300$

Solve the system:

$$s + h = 2300$$
$$2s + h = 4070$$

and find s (milligrams of sodium in the salad) and h (milligrams of sodium in the burger).

37. *Daily fat intake* The recommended maximum **total** fat intake is **65** grams per day. To be under that, eat chicken and salads: three Wendy's Ultimate Chicken Grilled Sandwiches at c grams per sandwich and **one** Wendy's Taco Supreme Salad containing t total grams of fat, yielding $3c + t$ or 52 grams of total fat—under the goal! The next day, try **one** Ultimate Chicken Grilled Sandwich and **two** Taco Salads with $c + 2t$ or 69 grams of total fat. We are a little over the desired 65 grams.

Solve the system:

$$3c + t = 52$$
$$c + 2t = 69$$

and find c (grams of total fat on Wendy's Ultimate Chicken Grill) and t (grams of total fat on the Wendy' Taco Salad).

Source: www.cspinet.org.

38. *Daily fat intake* Let's stick to chicken: two orders of McDonald's McNuggetts (6 pieces per order) have m grams of total fat each and **one** McDonald's California Cobb Salad with Grilled Chicken contains c grams of total fat. So far, we consumed 50 grams or $2m + c$ total fat. **One** order of the McNuggets and **one** salad contain $m + c$ or 35 total fat calories.

Solve the system:

$$2m + c = 50$$
$$m + c = 35$$

and find m (total fat grams in the McNuggets) and c (total fat grams in the salad).

Source: www.cspinet.org.

39. *Caloric intake* The recommended caloric intake is less than 2000 calories per day. The difference in calories between **one** Burger King Dutch Apple pie (a calories) and **one** Burger King Chili (c calories) is $a - c = 150$ calories. The Chili (c calories) and the Apple Pie (a calories) yield a mere $c + a$ or 530 calories.

Solve the system:

$$a - c = 150$$
$$c + a = 530$$

to find the calories c in the Chili and a in the Dutch Apple pie.

Source: www.cspinet.org.

40. *Caloric intake* Suppose we order the Wendy's Ultimate Chicken Grill Sandwich, Salad with Low Fat Honey Mustard, and Iced Tea at u calories instead of the Classic Triple with Cheese, Great Biggie Fries, and Biggie Cola at a whopping c calories. The difference is $c - u$ or 1240 calories. Even if you eat **four** Wendy's Ultimate Chicken Grill Sandwiches the caloric difference is $4u - c$ or 290.

Solve the system:

$$c - u = 1240$$
$$4u - c = 290$$

to find the calories u in the Wendy's Ultimate Chicken Grill Sandwich, Salad with Low Fat Honey Mustard, and Iced Tea and calories c in the Classic Triple with Cheese, Great Biggie Fries, and Biggie Cola.

Source: www.cspinet.org.

❭❭❭ Using Your Knowledge

Tweedledee and Tweedledum   Have you ever read *Alice in Wonderland?* Do you know who the author is? It's Lewis Carroll, of course. Although better known as the author of *Alice in Wonderland,* Lewis Carroll was also an accomplished mathematician and logician. Certain parts of his second book, *Through the Looking Glass,* reflect his interest in mathematics. In this book, one of the characters, Tweedledee, is talking to Tweedledum. Here is the conversation.

Tweedledee: The sum of your weight and twice mine is 361 pounds.

Tweedledum: Contrariwise, the sum of your weight and twice mine is 360 pounds.

41. If Tweedledee weighs x pounds and Tweedledum weighs y pounds, find their weights using the ideas of this section.

❭❭❭ Write On

42. When solving a system of equations by elimination, how would you recognize if the pair of equations is:

    a. consistent?   **b.** inconsistent?   **c.** dependent?

43. Explain why the system

$$2x + 5y = 9$$
$$3x + 2y = 8$$

is easier to solve by elimination rather than substitution.

44. Write the procedure you use to solve a system of equations by the elimination method.

❭❭❭ Concept Checker

Fill in the blank(s) with the correct word(s), phrase, or mathematical statement.

45. When solving a system of equations using the **elimination** method, the idea is to **multiply** the equations by a number that will yield a(n) _____ **system** in which the **coefficients** of the x's or the y's are **opposites.**

46. When solving a system of equations using the **elimination** method **the system** can be one of _____ types.

one	two
opposite	**three**
equivalent	

❭❭❭ Mastery Test

Solve the systems; if the system is not consistent, state whether it is inconsistent or dependent.

47. $2x + 5y = 9$
    $4x - 3y = 11$

48. $5x - 4y = 7$
    $4x + 2y = 16$

49. $2x - 5y = 5$
    $2x - y = 4 + x$

50. $3x - 2y = 6$
    $-6x = -4y - 12$

51. $\dfrac{x}{6} - \dfrac{y}{2} = 1$
    $-\dfrac{x}{4} + \dfrac{3y}{4} = -\dfrac{3}{4}$

52. A 10-pound bag of coffee sells for $114 and contains a mixture of coffee A, which costs $12 a pound, and coffee B, which costs $10 a pound. How many pounds of each of the coffees does the bag contain?

❭❭❭ Skill Checker

Write an expression corresponding to the given sentence.

53. The sum of the numbers of nickels (n) and dimes (d) equals 300.

54. The difference of h and w is 922.

55. The product of 4 and $(x - y)$ is 48.

56. The quotient of x and y is 80.

57. The number m is 3 less than the number n.

58. The number m is 5 more than the number n.

7.4 Coin, General, Motion, and Investment Problems

▶ Objectives

Solve word problems:

A ▶ Involving coins.

B ▶ Of a general nature.

C ▶ Using the distance formula *D = RT*.

D ▶ Involving the interest formula *I = PR*.

▶ To Succeed, Review How To . . .

1. Use the RSTUV method to solve word problems (pp. 143–144).
2. Solve a system of two equations with two unknowns (pp. 552–555, 570–573, 579–580).

▶ Getting Started
Money Problems

Patty's upset; she needs *help!* Why? Because *she* hasn't learned about systems of equations, but we have, so we can help her! In the preceding sections we studied systems of equations. We now use that knowledge to solve word problems involving two variables.

PEANUTS reprinted by permission of United Feature Syndicate, Inc.

Before we tackle Patty's problem, let's get down to nickels, dimes, and quarters! Suppose you are down to your last nickel: You have 5¢.

Follow the pattern:

If you have 2 nickels, you have 5 · 2 = 10 cents.   5 · 1
If you have 3 nickels, you have 5 · 3 = 15 cents.   5 · 2
If you have *n* nickels, you have 5 · *n* = 5*n* cents.   5 · 3
    5 · *n*

The same thing can be done with dimes.

Follow the pattern:

If you have 1 dime, you have 10 · 1 = 10 cents.   10 · 1
If you have 2 dimes, you have 10 · 2 = 20 cents.   10 · 2
If you have *n* dimes, you have 10 · *n* = 10*n* cents.   10 · *n*

We can construct a table that will help us summarize the information:

	Value (cents)	×	How Many	=	Total Value
Nickels	5		*n*		5*n*
Dimes	10		*d*		10*d*
Quarters	25		*q*		25*q*
Half-dollars	50		*h*		50*h*

In this section we shall use information like this and systems of equations to solve word problems.

A › Solving Coin and Money Problems

Now we are ready to help poor Patty! As usual, we use the RSTUV method. If you've forgotten how that goes, this is a good time to review it (see p. 143).

EXAMPLE 1   Patty's coin problem
Read the cartoon in the *Getting Started* again for the details of Patty's problem.

SOLUTION

1. Read the problem. Patty is asked how many dimes and quarters the man has.

2. Select the unknowns. Let d be the number of dimes the man has and q the number of quarters.

3. Think of a plan. We translate each of the sentences in the cartoon:

a. "A man has 20 coins consisting of dimes and quarters."

$$20 = d + q$$

b. The next sentence seems hard to translate. So, let's look at the easy part first—how much money he has now. Since he has d dimes, the table in the *Getting Started* tells us that he has $10d$ (cents). He also has q quarters, which are worth $25q$ (cents). Thus, he has

$$(10d + 25q) \text{ cents}$$

What would happen if the dimes were quarters and the quarters were dimes? We simply would change the amount the coins are worth, and he would have

$$(25d + 10q) \text{ cents}$$

Now let's translate the sentence:

If the dimes were quarters   he would   90¢ more than
and the quarters were dimes,   have   he has now.
‾‾‾‾‾‾‾‾‾‾‾‾‾‾‾‾‾‾‾‾‾‾‾‾‾   ‾‾‾‾‾‾‾   ‾‾‾‾‾‾‾‾‾‾‾
$25d + 10q$ = $(10d + 25q) + 90$

If we put the information from parts **a** and **b** together, we have the following system of equations:

$$d + q = 20$$
$$25d + 10q = 10d + 25q + 90$$

Now we need to write this system in standard form—that is, with all the variables and constants in the proper columns. We do this by subtracting $10d$ and $25q$ from both sides of the second equation to obtain

$$d + q = 20$$
$$15d - 15q = 90$$

We then divide each term in the second equation by 15 to get

$$d + q = 20$$
$$d - q = 6$$

PROBLEM 1
Pedro has 20 coins consisting of pennies and nickels. If the pennies were nickels and the nickels were pennies, he would have 16 cents less. How many pennies and how many nickels does Pedro have?

(continued)

4. Use the elimination method to solve the problem. To eliminate q, we simply add the two equations:

$$d + q = 20$$
$$\underline{d - q = 6}$$
$$2d = 26 \qquad \text{Add.}$$

Number of dimes: $d = 13$ \qquad Divide by 2.

$$13 + q = 20 \qquad \text{Substitute 13 for } d \text{ in } d + q = 20.$$

Number of quarters: $q = 7$

Thus, the man has 13 dimes ($1.30) and 7 quarters ($1.75), a total of $3.05.

5. Verify the solution. If the dimes were quarters and the quarters were dimes, the man would have 13 quarters ($3.25) and 7 dimes ($0.70), a total of $3.95, which is indeed $0.90 more than the $3.05 he now has. Patty, you got your help!

Let's solve another coin problem.

EXAMPLE 2   Jack's coin problem

Jack has $3 in nickels and dimes. He has twice as many nickels as he has dimes. How many nickels and how many dimes does he have?

SOLUTION   As usual, we use the RSTUV method.

1. Read the problem. We are asked to find the numbers of nickels and dimes.

2. Select the unknowns. Let n be the number of nickels and d the number of dimes.

3. Think of a plan. If we translate the problem and use the table in the *Getting Started,* Jack has $3 (300 cents) in nickels and dimes:

$$300 = 5n + 10d$$

He has twice as many nickels as he has dimes:

$$n = 2d$$

We then have the system

$$5n + 10d = 300$$
$$n = 2d$$

4. Use the substitution method to solve the problem. This time it's easy to use the substitution method.

$$5n + 10d = 300 \xrightarrow{\text{Letting } n = 2d.} 5(2d) + 10d = 300$$

Simplify. $10d + 10d = 300$

Combine like terms. $20d = 300$

Divide by 20. $d = 15$

Substitute 15 for d in $n = 2d$. $n = 2(15) = 30$

Thus, Jack has 15 dimes ($1.50) and 30 nickels ($1.50).

5. Verify the solution. Since Jack has $3 ($1.50 + $1.50) and he does have twice as many nickels as dimes, the answer is correct.

PROBLEM 2

Jill has $1.50 in nickels and dimes. She has twice as many dimes as she has nickels. How many nickels and how many dimes does she have?

B › Solving General Problems

We can use systems of equations to solve many problems. Here is an interesting one.

EXAMPLE 3   A heavy marriage

The greatest weight difference recorded for a married couple is 922 pounds (Mills Darden of North Carolina and his wife Mary). Their combined weight is 1118 pounds. What is the weight of each of the Dardens? (He is the heavy one.)

SOLUTION

1. **Read the problem.** We are asked to find the weight of each of the Dardens.
2. **Select the unknowns.** Let h be the weight of Mills and w be the weight of Mary.
3. **Think of a plan.** We translate the problem. The weight difference is 922 pounds:

$$h - w = 922$$

Their combined weight is 1118 pounds:

$$h + w = 1118$$

We then have the system

$$h - w = 922$$
$$h + w = 1118$$

4. **Use the elimination method to solve the problem.** Using the elimination method, we have

$$
\begin{array}{ll}
h - w = 922 & \\
\underline{h + w = 1118} & \\
\quad 2h = 2040 & \text{Add.} \\
\quad\ h = 1020 & \text{Divide by 2.} \\
1020 + w = 1118 & \text{Substitute 1020 for } h \text{ in } h + w = 1118. \\
\quad\ w = 98 & \text{Subtract 1020.}
\end{array}
$$

Thus, Mary weighs 98 pounds and Mills weighs 1020 pounds.

5. **Verify the solution.** $h - w$ becomes $1020 - 98 = 922$, which is true. Moreover, $1020 + 98 = 1118$ which is also true. You can also verify this in the *Guinness Book of Records*.

PROBLEM 3

When a couple of astronauts stood on a scale together before a mission, their combined weight was 320 pounds. The difference in their weights was 60 pounds, and the woman was lighter than the man. What was the weight of each?

C 〉 Solving Motion Problems

Remember the motion problems we solved in Section 2.5? They can also be done using two variables. The procedure is about the same. We write the given information in a chart labeled $R \times T = D$ and then use our RSTUV method, as demonstrated in Example 4.

EXAMPLE 4   Currents and boating

The world's strongest current is the Saltstraumen in Norway. The current is so strong that a boat that travels 48 miles downstream (with the current) in 1 hour takes 4 hours to go the same 48 miles upstream (against the current). How fast is the current flowing?

SOLUTION

1. **Read the problem.** We are asked to find the speed of the current. Note that the speed of the boat downstream has two components: the boat speed and the current speed.

PROBLEM 4

A plane travels 800 miles against a storm in 4 hours. Giving up, the pilot turns around and flies back 800 miles to the airport in only 2 hours with the aid of the tailwind. Find the speed of the wind and the speed of the plane in still air.

(continued)

Answers to PROBLEMS

3. Man: 190 lb; woman: 130 lb
4. Plane speed: 300 mi/hr;
   wind speed: 100 mi/hr

2. Select the unknowns. Let x be the speed of the boat in still water and y be the speed of the current. Then $(x + y)$ is the speed of the boat going downstream; $(x - y)$ is the speed of the boat going upstream.

3. Think of a plan. We enter this information in a chart:

	R	×	T	=	D	
Downstream:	$x + y$		1		48	⟶ $x + y = 48$
Upstream:	$x - y$		4		48	⟶ $4(x - y) = 48$

4. Use the elimination method to solve the problem. Our system of equations can be simplified as follows:

$$x + y = 48 \xrightarrow{\text{Leave as is.}} x + y = 48$$
$$4(x - y) = 48 \xrightarrow{\text{Divide by 4.}} \underline{x - y = 12}$$

Add. $2x = 60$

Divide by 2. $x = 30$

Substitute 30 for x in $x + y = 48$. $30 + y = 48$

Subtract 30. $y = 18$

Thus, the speed of the boat in still water is $x = 30$ miles per hour, and the speed of the current is 18 miles per hour.

5. Verify the solution. We leave the verification to you.

D ⟩ Solving Investment Problems

The investment problems we solved in Section 2.5 can also be worked using two variables. These problems use the formula $I = PR$ to find the *annual* interest I on a principal P at a rate R. The procedure is similar to that used to solve distance problems and uses the same strategy: use a table to enter the information, obtain a system of two equations and two unknowns, and solve the system. We show this strategy next.

EXAMPLE 5   Clayton's credit card problem

Clayton owes a total of $10,900 on two credit cards with annual interest rates of 12% and 18%, respectively. If he pays a total of $1590 in interest for the year, how much does he owe on each card?

SOLUTION

1. Read the problem. We are asked to find the amount owed on *each* card.

2. Select the unknowns. Let x be the amount Clayton owes on the first card and y be the amount he owes on the second card.

3. Think of a plan. We make a table similar to the one in Example 4 but using the heading $P \times R = I$.

12%

THRIFTEBANK

4221 7032 7124 3004
4221
2004   08/08   PASS
CLAYTON D SIZTION   PORT

18%

GLOBE BANK

4221 0732 2174 0377
4221
2004   08/08   PASS
CLAYTON D SIZTION   PORT

	P	×	R	=	I
Card A	x		0.12		$0.12x$
Card B	y		0.18		$0.18y$

Since the total amount owed is $10,900: $x + y = 10,900$

Since the total interest paid is $1590: $0.12x + 0.18y = 1590$

PROBLEM 5

Dorothy owes $11,000 on two credit cards with annual interest rates of 9% and 12%. If she pays $1140 in interest for the year, how much does she owe on each card?

Thus, we have to solve the system

$$x + y = 10{,}900$$
$$0.12x + 0.18y = 1590$$

4. Use the substitution method to solve the problem. We solve for x in the first equation to obtain $x = 10{,}900 - y$. Now we substitute $x = 10{,}900 - y$ in

$$0.12x + 0.18y = 1590$$
$$0.12(10{,}900 - y) + 0.18y = 1590$$
$$1308 - 0.12y + 0.18y = 1590 \qquad \text{Simplify.}$$
$$0.06y = 282 \qquad \text{Subtract 1308 and combine } y\text{'s.}$$
$$y = 4700 \qquad \text{Divide by 0.06.}$$

Since

$$x = 10{,}900 - y$$
$$x = 10{,}900 - 4700$$
$$x = 6200$$

Thus, Clayton owes \$6200 on card A and \$4700 on card B.

5. Verify the solution. Since $0.12 \cdot \$6200 + 0.18 \cdot \$4700 = \$744 + \$846 = \$1590$ (the amount Clayton paid in interest), the amounts of \$6200 and \$4700 are correct.

Before you attempt to solve the Exercises, practice translating!

TRANSLATE THIS

The third step in the RSTUV procedure is to **TRANSLATE** the information into an equation. In Problems 1–10 **TRANSLATE** the sentence and match the correct translation with one of the equations **A–O.**

1. If you have C cents consisting of n nickels and d dimes, what is the equation relating C, n, and d?

2. If you have C cents consisting of d dimes and q quarters, what is the equation relating C, d, and q?

3. A person is h inches tall and another person is i inches tall. If the difference in their heights is 10 inches and the sum of their heights is 110 inches, what system of equations represents these facts?

4. If s is the speed of a boat in still water and c is the speed of the current, find an equation representing the distance D traveled downstream in T hours.

5. If s is the speed of a boat in still water and c is the speed of the current, find an equation representing the distance D traveled upstream in T hours.

A. $h - i = 10; h + i = 110$
B. $D = (s - c)T$
C. $C = 10n + 5d$
D. $C = 10d + 25q$
E. $C = 25d + 10q$
F. $I = 0.10x$
$ + 0.08(10{,}000 - x)$
G. $m = RT$
H. $I = 0.10x + 0.08y$
I. $I = 0.10x + 0.08$
$ (x - 10{,}000)$
J. $m = (R + W)T$
K. $C = 5n + 10d$
L. $10 - h = i; h + i = 110$
M. $m = (W - R)T$
N. $D = (s + c)T$
O. $m = (R - W)T$

6. A person invests x dollars at 10% and y dollars at 8%. If the interest from both investments is I, find an equation for I.

7. A person invests \$10,000, x dollars at 10% and the rest at 8%. Write an equation for the interest I earned on the \$10,000.

8. A plane travels m miles in T hours flying at a rate of R miles per hour. Write an equation relating m, T, and R.

9. A plane flies for T hours at R miles per hour with a tail wind of W miles per hour. If the plane travels m miles, write an equation relating T, R, W, and m.

10. A plane flies for T hours at R miles per hour with a head wind of W miles per hour. If the plane travels m miles, write an equation relating T, R, W, and m.

› Exercises **7.4**

〈**A**〉 **Solving Coin and Money Problems**
〈**B**〉 **Solving General Problems**

In Problems 1–6, solve the money problems.

1. Mida has $2.25 in nickels and dimes. She has four times as many dimes as nickels. How many dimes and how many nickels does she have?

2. Dora has $5.50 in nickels and quarters. She has twice as many quarters as she has nickels. How many of each coin does she have?

3. Mongo has 20 coins consisting of nickels and dimes. If the nickels were dimes and the dimes were nickels, he would have 50¢ more than he now has. How many nickels and how many dimes does he have?

4. Desi has 10 coins consisting of pennies and nickels. Strangely enough, if the nickels were pennies and the pennies were nickels, she would have the same amount of money as she now has. How many pennies and nickels does she have?

5. Don had $26 in his pocket. If he had only $1 bills and $5 bills, and he had a total of 10 bills, how many of each of the bills did he have?

6. A person went to the bank to deposit $300. The money was in $10 and $20 bills, 25 bills in all. How many of each did the person have?

In Problems 7–14, find the solution.

7. The sum of two numbers is 102. Their difference is 16. What are the numbers?

8. The difference between two numbers is 28. Their sum is 82. What are the numbers?

9. The sum of two integers is 126. If one of the integers is 5 times the other, what are the integers?

10. The difference between two integers is 245. If one of the integers is 8 times the other, find the integers.

11. The difference between two numbers is 16. One of the numbers exceeds the other by 4. What are the numbers?

12. The sum of two numbers is 116. One of the numbers is 50 less than the other. What are the numbers?

13. Longs Peak is 145 feet higher than Pikes Peak. If you were to put these two peaks on top of each other, you would still be 637 feet short of reaching the elevation of Mount Everest, 29,002 feet. Find the elevations of Longs Peak and Pikes Peak.

14. Two brothers had a total of $7500 in separate bank accounts. One of the brothers complained, and the other brother took $250 and put it in the complaining brother's account. They now had the same amount of money! How much did each of the brothers have in the bank before the transfer?

〈**C**〉 **Solving Motion Problems** In Problems 15–20, solve the motion problems.

15. A plane flying from city A to city B at 300 miles per hour arrives $\frac{1}{2}$ hour later than scheduled. If the plane had flown at 350 miles per hour, it would have made the scheduled time. How far apart are cities A and B?

16. A plane flies 540 miles with a tailwind in $2\frac{1}{4}$ hours. The plane makes the return trip against the same wind and takes 3 hours. Find the speed of the plane in still air and the speed of the wind.

17. A motorboat runs 45 miles downstream in $2\frac{1}{2}$ hours and 39 miles upstream in $3\frac{1}{4}$ hours. Find the speed of the boat in still water and the speed of the current.

18. A small plane travels 520 miles with the wind in 3 hours, 20 minutes ($3\frac{1}{3}$ hours), the same time that it takes to travel 460 miles against the wind. What is the plane's speed in still air?

19. If Bill drives from his home to his office at 40 miles per hour, he arrives 5 minutes early. If he drives at 30 miles per hour, he arrives 5 minutes late. How far is it from his home to his office?

20. An unidentified plane approaching the U.S. coast is sighted on radar and determined to be 380 miles away and heading straight toward the coast at 600 miles per hour. Five minutes ($\frac{1}{12}$ hour) later, a U.S. jet, flying at 720 miles per hour, scrambles from the coastline to meet the plane. How far from the coast does the interceptor meet the plane?

〈**D**〉 **Solving Investment Problems** In Problems 21–23, solve the investment problems.

21. Fred invested $20,000, part at 6% and the rest at 8%. Find the amount invested at each rate if the annual income from the two investments is $1500.

22. Maria invested $25,000, part at 7.5% and the rest at 6%. If the annual interest from the two investments amounted to $1620, how much money was invested at each rate?

23. Dominic has a savings account that pays 5% annual interest and some certificates of deposit that pay 7% annually. His total interest from the two investments is $1100 and the total amount invested is $18,000. How much money does he have in the savings account?

〉 〉 〉 *Applications*

24. *Hurricane damages* Two of the costliest hurricanes in U.S. history were Andrew (1992) and Hugo (1989), in that order; together they caused $33.5 billion in damages. If the difference in damages caused by Andrew and Hugo was $19.5 billion, how much damage did each of them cause?

Source: University of Colorado Natural Hazards Center.

25. *Higher education enrollments* In a recent year, the total enrollment in public and private institutions of higher education was 15 million students. If there were 8.4 million more students enrolled in public institutions than in private, how many students were enrolled in public and how many were enrolled in private institutions?

Source: U.S. Department of Education.

26. *Education expenditures* In a recent year, the education expenditures for public and private institutions amounted to $187 billion. If $49 billion more was spent in public institutions than in private, what were the education expenditures for public and for private institutions?

Source: U.S. Department of Education.

27. *Premium cable services* The total number of subscribers for Home Box Office and Showtime in a recent year was 28,700,000. If Home Box Office had 7300 more subscribers than Showtime, how many subscribers did each of the services have?

Source: National Cable Television Association.

28. *Automobile and home accidents* In a recent year, the cost of motor vehicle and home accidents reached $241.7 billion. If motor vehicle accidents caused losses that were $85.1 billion more than those caused by home accidents, what were the losses in each category?

Source: National Safety Council Accident Facts.

Refer to the table for Problems 29–31. The table shows the four most intense hurricanes to hit the U.S. mainland.

Rank	Hurricane	Location	Year	Category	Damage (in billions)
1.	Katrina	La./Miss.	2005	3	K
2.	Andrew	Fla./La.	1992	5	A
3.	Charley	Fla.	2004	4	C
4.	Wilma	Fla.	2005	3	W

Hurricane Katrina

29. *Hurricane damage* The combined damage from Katrina K and Andrew A reached $122.5 billion. The difference in the damage they caused was $69.5 billion. Find the damage (in billions) caused by Katrina and Andrew.

30. *Hurricane damage* The most intense hurricane recorded in the Atlantic basin was Wilma in October 2005. The combined damage of Wilma and Charley amounted to $29.4 billion, but the dollar difference in damages was a mere $0.6 billion. Find the damage (in billions) caused by Wilma W and Charley C.

31. *Hurricane wind speeds* As you can see from the table, Katrina was a category 3 hurricane (winds 111–130 miles per hour) and Andrew a category 5 (winds of more than 155 miles per hour). At landfall, Andrew's winds were 25 miles per hour stronger than Katrina's and 10 miles more than the 155 miles per hour needed to make it a category 5. Find the wind speeds in mph for Katrina and Andrew.

32. *Hurricane barometric pressure* Hurricane Katrina was stronger than Andrew at landfall. How can that be? Here is the explanation: Barometric pressure is the most accurate representation of a storm's power. The lower the barometric pressure, the more intense the storm is and Katrina's barometric pressure K was 0.12 inches lower than Andrew's A. As a matter of fact, Andrew's barometric pressure was only 0.06 inches above the 27.17 needed to make it a category 5 hurricane. Find the barometric pressure K for Katrina and A for Andrew.

Most Intense U.S. Hurricanes at Landfall

Hurricane	Year	Barometric Pressure (inches)	Wind Speed (mph)
"Labor Day" (Fla. Keys)	1935	26.35	160
Camille	1969	26.84	190
Katrina	2005	K	140
Andrew	1992	A	165

Source: National Weather Service

〉 〉 〉 *Write On*

33. Make up a problem involving coins, and write a solution for it using the RSTUV procedure.

34. Make up a problem whose solution involves the distance formula $D = RT$, and write a solution for it using the RSTUV procedure.

35. Make up a problem whose solution involves the interest formula $I = PR$, and write a solution for it using the RSTUV procedure.

36. The problems in Examples 1–4 have precisely the information you need to solve them. In real life, however, irrelevant information is often present. This type of information is called a **red herring.** Find some problems with red herrings and point them out.

〉 〉 〉 *Concept Checker*

Fill in the blank(s) with the correct word(s), phrase, or mathematical statement.

37. If you have d **dimes** and q **quarters,** you have _____ cents.

38. If you have n **nickels** and d **dimes,** you have _____ cents.

39. The **distance** D traveled by an object moving at a **rate** R for a **time** T equals _____.

40. If P is the **principal** and R the **rate,** the **interest** I is _____.

RD	**5n + 10d**
RT	**PR**
10n + 5d	**IR**
10d + 25q	

〉 〉 〉 *Mastery Test*

41. Jill has \$2 in nickels and dimes. She has twice as many nickels as she has dimes. How many nickels and how many dimes does she have?

42. At birth, the Stimson twins weighed a total of 35 ounces. If their weight difference was 3 ounces, what was the weight of each of the twins?

43. A plane travels 1200 miles with a tailwind in 3 hours. It takes 4 hours to travel the same distance against the wind. Find the speed of the wind and the speed of the plane in still air.

44. Harper makes two investments totaling \$10,000. The first investment pays 8% annually, and the second investment pays 5%. If the annual return from both investments is \$600, how much has Harper invested at each rate?

45. In the 2004 presidential election, the 121 million people who voted gave the Republican ticket of Bush and Cheney 3 million more votes than the Democratic ticket of Kerrey and Edwards. How many votes (in millions) did each ticket get?

Graph the inequalities.

46. $x + 2y < 4$

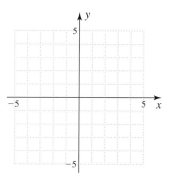

47. $x - 2y > 6$

48. $x > y$

49. $x \leq 2y$

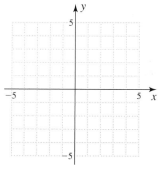

7.5   Systems of Linear Inequalities

▶ Objective

A ⟩ Solve a system of linear inequalities by graphing.

▶ To Succeed, Review How To . . .

Graph a linear inequality (pp. 275–278).

▶ Getting Started
Inequalities and Hospital Stays

According to the American Hospital Association, between 1980 and 2005 (inclusive), the average length of a hospital stay was $y = 7.74 - 0.09x$ (days), where x is the number of years after 1980, as shown in the first graph.

The graph of $y > 7.74 - 0.09x$ represents those longer-than-average stays. If we graph those longer-than-average stays *between* 1980 ($x = 0$) and 2005 ($x = 25$), we get the shaded area *above* the line $y = 7.74 - 0.09x$ and *between* $x = 0$ and $x = 25$. Since the line $y = 7.74 - 0.09x$ is *not* part of the graph, the line itself is shown dashed. The region satisfying the inequalities

$$y > 7.74 - 0.09x, \quad x > 0, \quad \text{and} \quad x < 25$$

is shown shaded in the second graph.

In this section we learn how to solve linear inequalities graphically by finding the set of points that satisfy *all* the inequalities in the system.

A ⟩ Solving a System of Linear Inequalities by Graphing

It turns out that we can use the procedure we studied in Section 3.7 to solve a system of linear inequalities.

PROCEDURE

Solving a System of Inequalities

Graph each inequality on the *same* set of axes using the following steps:

1. Graph the line that is the boundary of the region. If the inequality involves ≤ or ≥, draw a **solid** line; if it involves < or >, draw a **dashed** line.

2. Use any point (a, b) *not* on the line as a test point. Substitute the values of a and b for x and y in the inequality. If a *true* statement results, shade the side of the line containing the test point. If a *false* statement results, shade the other side.

The **solution set** is the set of points that satisfies *all* the inequalities in the system.

EXAMPLE 1 Solving systems of inequalities involving horizontal and vertical lines by graphing

Graph the solution of the system: $x \leq 0$
$$y \geq 2$$

SOLUTION Since $x = 0$ is a vertical line corresponding to the y-axis, $x \leq 0$ consists of the graph of the line $x = 0$ and all points to the *left*, as shown in Figure 7.12. The condition $y \geq 2$ defines all points on the line $y = 2$ and *above*, as shown in Figure 7.13.

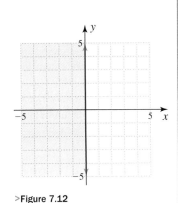

>Figure 7.12

PROBLEM 1

Graph the solution of the system:
$$x \leq 2$$
$$y \geq 0$$

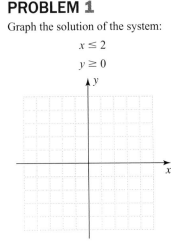

Answer on page 599

The solution set is the set satisfying *both* conditions, that is, where $x \le 0$ *and* $y \ge 2$. This set is the darker area in Figure 7.14.

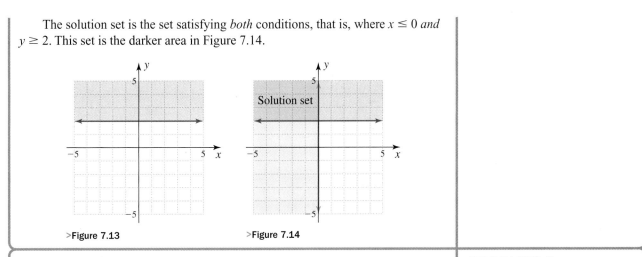

>Figure 7.13   >Figure 7.14

EXAMPLE 2   Solving systems of inequalities using a test point

Graph the solution of the system: $x + 2y \le 5$
$$x - y < 2$$

SOLUTION   First we graph the lines $x + 2y = 5$ and $x - y = 2$. Using $(0, 0)$ as a test point, $x + 2y \le 5$ becomes $0 + 2 \cdot 0 \le 5$, a true statement. So we shade the region containing $(0, 0)$: the points *on or below* the line $x + 2y = 5$. (See Figure 7.15.) The inequality $x - y < 2$ is also satisfied by the test point $(0, 0)$, so we shade the points *above* the line $x - y = 2$. This line is drawn dashed to indicate that the points on it do *not* satisfy the inequality $x - y < 2$. (See Figure 7.16.) The solution set of the system is shown in Figure 7.17 by the darker region and the portion of the solid line forming one boundary of the region.

>Figure 7.15

>Figure 7.16

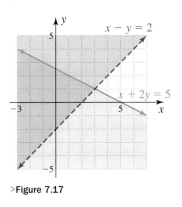

>Figure 7.17

PROBLEM 2

Graph the solution of the system:
$$2x + y \le 5$$
$$x - y < 3$$

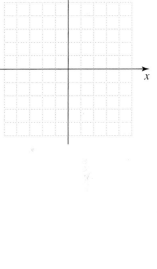

Answers to PROBLEMS

1.

2.

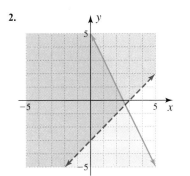

EXAMPLE 3   Solving systems of inequalities using a test point that does not satisfy both inequalities

Graph the solution of the system: $y + x \geq 2$
$$y - x \leq 2$$

SOLUTION   Graph the lines $y + x = 2$ and $y - x = 2$. Since the test point $(0, 0)$ does *not* satisfy the inequality $y + x \geq 2$, we shade the region that does *not* contain $(0, 0)$: the points *on or above* the line $y + x = 2$. (See Figure 7.18.) The test point $(0, 0)$ *does* satisfy the inequality $y - x \leq 2$, so we shade the points *on or below* the line $y - x = 2$. (See Figure 7.19.) The solution set of the system is the darker region in Figure 7.20 and includes parts of both lines.

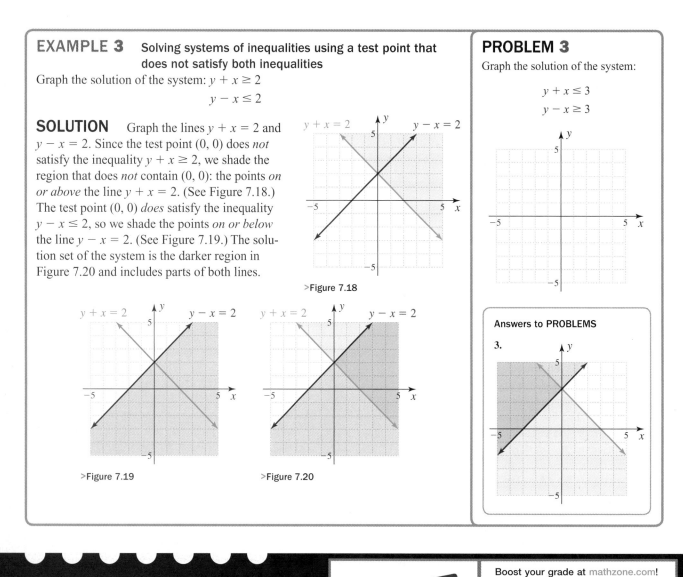

>Figure 7.18

>Figure 7.19

>Figure 7.20

PROBLEM 3

Graph the solution of the system:

$$y + x \leq 3$$
$$y - x \geq 3$$

Answers to PROBLEMS

3.

Boost your grade at mathzone.com!

> Practice Problems   > Self-Tests
> NetTutor   > e-Professors
                             > Videos

MathZone

❭ Exercises 7.5

〈A〉 Solving a System of Linear Inequalities by Graphing   In Problems 1–15, graph the solution set of the system of inequalities.

1. $x \geq 0$ and $y \leq 2$

2. $x > 1$ and $y < 3$

3. $x < -1$ and $y > -2$

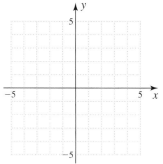

4. $x - y \geq 2$
$x + y \leq 6$

5. $x + 2y \leq 3$
$x < y$

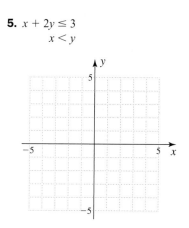

6. $5x - y > -1$
$-x + 2y \leq 6$

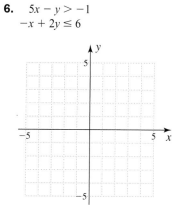

7. $4x - y > -1$
$-2x - y \leq -3$

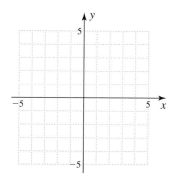

8. $2x - 3y < 6$
$4x - 3y > 12$

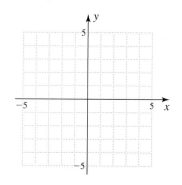

9. $-2x + y > 3$
$5x - y \leq -10$

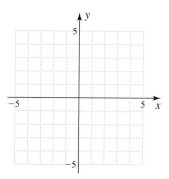

10. $2x - 5y \leq 10$
$3x + 2y < 6$

11. $2x - 3y < 5$
$x \geq y$

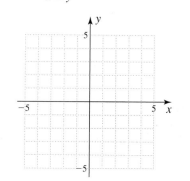

12. $x \leq 2y$
$x + y < 4$

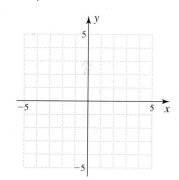

13. $x + 3y \leq 6$
$x > y$

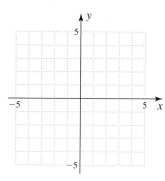

14. $2x - y < 2$
$x \leq y$

15. $3x + y > 6$
$x \leq y$

〉 〉 〉 Applications

16. *Sugar and protein* A McDonald's Filet-O-Fish (FOF) has 8 grams of sugar, while a Burger King Tender Grilled Chicken (TGC) sandwich has 9 grams of sugar. You should limit your sugar intake to less than 40 grams per day. The RDA (Recommended Dietary Allowance) for protein is at least 50 grams per day: the FOF has 14 grams of protein and the TGC has 37 grams.

To satisfy your RDA's by eating x FOF and y TGC you have to satisfy the system:

$$8x + 9y < 40$$
$$14x + 37y \geq 50$$

Graph the system and give at least two ordered pairs that satisfy the inequalities.

17. *Carbohydrates and protein* A McDonald's Filet-O-Fish (FOF) sandwich has 42 grams of carbohydrates and a Burger King Tender Grilled Chicken (TGC) sandwich has 53 grams of carbohydrates. Their protein contents 14 grams and 37 grams, respectively. To satisfy the RDA (Recommended Dietary Allowance) of carbohydrates (<300) and protein (≥ 50) when eating x FOF and y TGC you have to satisfy the system of inequalities:

$$42x + 53y < 300$$
$$14x + 37y \geq 50$$

Graph the system and give at least two ordered pairs that satisfy the inequalities.

18. *Carbohydrates and sugar* A runner wants to maintain the RDA (Recommanded Dietary Allowance) of carbs (< 300) by eating x McDonald's Filet-O-Fish (FOF) and y Burger King Tender Grilled Chicken (TGC). At the same time sugar levels must be kept at more than the minimum RDA of 40 grams. A system of equations satisfying these two conditions is:

$$42x + 53y < 300$$
$$8x + 9y \geq 40$$

Graph the system and give at least two ordered pairs that satisfy the inequalities.

〉 〉 〉 Using Your Knowledge

Inequalities in Exercise   The target zone used to gauge your effort when performing aerobic exercises is determined by your pulse rate p and your age a. In this *Using Your Knowledge* the target zone is the solution set of the inequalities in Problems 19–21. What is your target zone? You have to do Problems 19–21 to find out! Graph the given inequalities on the set of axes provided.

19. $p \geq -\dfrac{2a}{3} + 150$

20. $p \leq -a + 190$

21. $10 \leq a \leq 70$

〉 〉 〉 Write On

22. Write the steps you would take to graph the inequality

$$ax + by > c \quad (a \neq 0)$$

23. Describe the solution set of the inequality $x \geq k$.

24. Describe the solution set of the inequality $y < k$.

⟩⟩⟩ *Concept Checker*

Fill in the blank(s) with the correct item word(s), phrase, or mathematical statement.

25. When graphing a **linear inequality** involving ≤ or ≥ the **graph** is a _____ line.

26. When graphing a **linear inequality** involving < or > the **graph** is a _____ line.

direct   **solid**

dashed

⟩⟩⟩ *Mastery Test*

Graph the solution set for each system.

27. $x > 2$
    $y < 3$

28. $x + y > 4$
    $x - y \leq 2$

29. $3x - y < -1$
    $x + 2y \leq 2$

30. $2x - 3y \geq 6$
    $-x + 2y < -4$

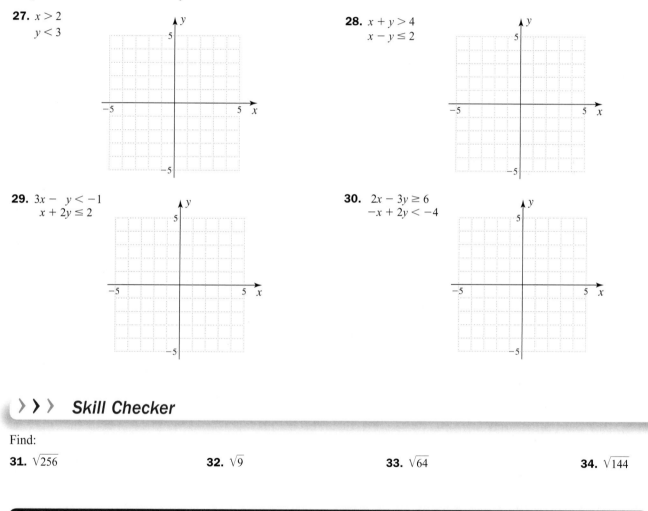

⟩⟩⟩ *Skill Checker*

Find:

31. $\sqrt{256}$

32. $\sqrt{9}$

33. $\sqrt{64}$

34. $\sqrt{144}$

⟩**Collaborative Learning**

Women and Men in the Workforce

The figure on the next page shows the percentages of women (W) and men (M) in the workforce since 1955. Can you tell from the graph in which year the percentages will be the same? It will happen after 2005! If t is the number of years after 1955, W the percentage of women in the workforce, and M the percentage of men, the graphs will *intersect* at a point (a, b). The coordinates of the point of intersection (if there is one) will be the common solution of both equations. If the percentages W and M are approximated by

$$W = 0.55t + 34.4 \text{ (percent)}$$
$$M = -0.20t + 83.9 \text{ (percent)}$$

<CONTINUED>

you can find the year after 1955 when the same percentages of women and men are in the workforce by graphing M and W and locating the point of intersection.

Form three teams of students: **Team 1,** *The Estimators;* **Team 2,** *The Graphicals;* and **Team 3,** *The Substituters.*

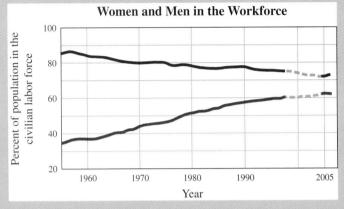

Source: U.S. Bureau of Labor Statistics

Here are the assignments:

Team 1 Extend the lines for the percentage of men (M), the percentage of women (W), and the horizontal axis.

1. In what year do the lines seem to intersect?

2. What would be the percentage of men and women in the labor force in that year?

Team 2 Get some graph paper with a 20-by-20 grid, each grid representing 1 unit. Graph the lines

$$M = -0.20t + 83.9 \text{ (percent)}$$
$$W = 0.55t + 34.4 \text{ (percent)}$$

where t is the number of years after 1955.

1. Where do the lines intersect?

2. What does the point of intersection represent?

Team 3 The point at which the lines intersect will satisfy both equations; consequently at that point $M = W$. Substitute $-0.20t + 83.9$ for M and $0.55t + 34.4$ for W and solve for t.

1. What value did you get?

2. What is the point of intersection for the two lines, and what does it represent?

Compare the results of teams 1, 2, and 3. Do they agree? Why or why not? Which is the most accurate method for obtaining an answer for this problem?

> **Research Questions**

1. Write a report on the content of the *Nine Chapters of the Mathematical Arts*.

2. Write a paragraph detailing how the copies of the *Nine Chapters* were destroyed.

3. Write a short biography on the Chinese mathematician Liu Hui.

4. Write a report explaining the relationship between systems of linear equations and inequalities and the simplex method.

5. Write a report on the simplex algorithm and its developer.

> Summary **Chapter 7**

Section	Item	Meaning	Example
7.1A	System of simultaneous equations	A set of equations that may have a common solution	$x + y = 2$ $x - y = 4$ is a system of equations.
	Inconsistent system	A system with no solution	$x + y = 2$ $x + y = 3$ is an inconsistent system.
	Dependent system	A system in which both equations are equivalent	$x + y = 2$ $2x + 2y = 4$ is a dependent system.
7.2A	Substitution method	A method used to solve systems of equations by solving one equation for one variable and substituting this result in the other equation	To solve the system $x + y = 2$ $2x + 3y = 6$ solve the first equation for x: $x = 2 - y$ and substitute in the other equation: $2(2 - y) + 3y = 6$ $4 - 2y + 3y = 6$ $y = 2$ $x = 2 - 2 = 0$
7.3A	Elimination method	A method used to solve systems of equations by multiplying by numbers that will cause the coefficients of one of the variables to be opposites	To solve the system $x + 2y = 5$ $x - y = -1$ multiply the second equation by 2 and add to the first equation.
7.4	RSTUV method	A method for solving word problems consisting of **R**eading, **S**electing the variables, **T**hinking of a plan to solve the problem, **U**sing algebra to solve, and **V**erifying the answer	
7.5A	Solution set of a system of inequalities	The set of points that satisfy all inequalities in the system	The solution set of the system $x + y \geq 2$ $-x + y \leq -1$ are the shaded points in the figure.

> Review Exercises Chapter 7

(If you need help with these exercises, look in the section indicated in brackets.)

1. ⟨**7.1A, B**⟩ *Use the graphical method to solve the system (if possible).*

   a. $2x + y = 4$
   $y - 2x = 0$

   b. $x + y = 4$
   $y - x = 0$

   c. $x + y = 4$
   $y - 3x = 0$

2. ⟨**7.1A, B**⟩ *Use the graphical method to solve the system (if possible).*

   a. $y - 3x = 3$
   $2y - 6x = 12$

   b. $y - 2x = 2$
   $2y - 4x = 8$

   c. $y - 3x = 6$
   $2y - 6x = 6$

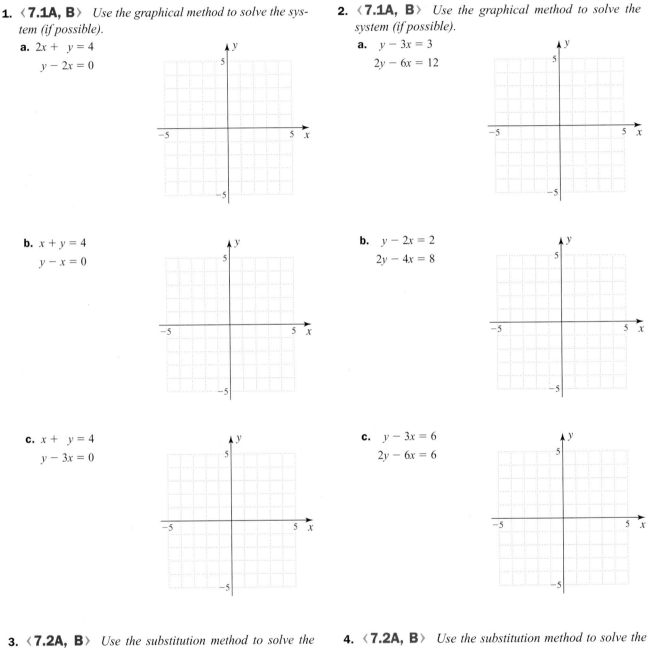

3. ⟨**7.2A, B**⟩ *Use the substitution method to solve the system (if possible).*

   a. $x + 4y = 5$
   $2x + 8y = 15$

   b. $x + 3y = 6$
   $3x + 9y = 12$

   c. $x + 4y = 5$
   $2x + 13y = 15$

4. ⟨**7.2A, B**⟩ *Use the substitution method to solve the system (if possible).*

   a. $x + 4y = 5$
   $2x + 8y = 10$

   b. $x + 3y = 6$
   $3x + 9y = 18$

   c. $x + 4y = 5$
   $-2x - 8y = -10$

5. ⟨**7.3A**⟩ *Use the elimination method to solve the system (if possible).*

  a. $3x + 2y = 1$
     $2x + \ y = 0$

  b. $3x + 2y = 4$
     $2x + \ y = 3$

  c. $3x + 2y = -7$
     $2x + \ y = -4$

6. ⟨**7.3B**⟩ *Use the elimination method to solve the system (if possible).*

  a. $\quad 2x - 3y = 6$
     $-4x + 6y = -2$

  b. $\quad 3x - 2y = 8$
     $-9x + 6y = -4$

  c. $\quad 3x - 5y = 6$
     $-3x + 5y = -12$

7. ⟨**7.3B**⟩ *Use the elimination method to solve the system (if possible).*

  a. $3y + 2x = 1$
     $6y + 4x = 2$

  b. $2y + 3x = 1$
     $6x + 4y = 2$

  c. $3y + 4x = -11$
     $8x + 6y = -22$

8. ⟨**7.4A**⟩ *Desi has $3 in nickels and dimes. How many nickels and how many dimes does she have if:*

  a. she has the same number of nickels and dimes?

  b. she has 4 times as many nickels as she has dimes?

  c. she has 10 times as many nickels as she has dimes?

9. ⟨**7.4B**⟩ *The sum of two numbers is 180. What are the numbers if:*

  a. their difference is 40?

  b. their difference is 60?

  c. their difference is 80?

10. ⟨**7.4C**⟩ *A plane flew 2400 miles with a tailwind in 3 hours. What was the plane's speed in still air if the return trip took*

  a. 8 hours?

  b. 10 hours?

  c. 12 hours?

11. ⟨**7.4D**⟩ *An investor bought some municipal bonds yielding 5% annually and some certificates of deposit yielding 10% annually. If the total investment amounts to $20,000, how much money is invested in bonds and how much in certificates of deposit (CDs) if the annual interest is*

  a. $1750?

  b. $1150?

  c. $1500?

12. ⟨**7.5**⟩ *Graph the solution set of the system.*

  a. $x > 4$
     $y < -1$

  b. $x + y > 3$
     $x - y < 4$

  c. $2x + \ y \le 4$
     $x - 2y > 2$

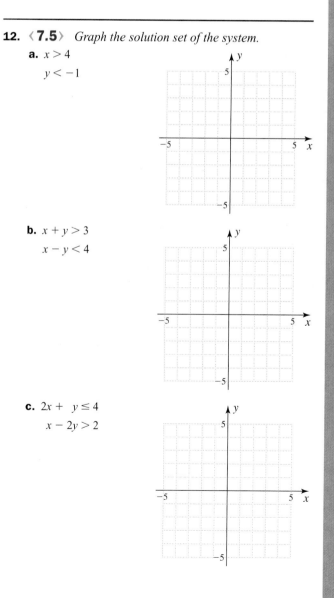

> Practice Test **Chapter 7**

(Answers on page 609)

Visit www.mhhe.com/bello to view helpful videos that provide step-by-step solutions to several of the problems below.

1. Use the graphical method to solve the system.

$$x + 2y = 4$$
$$2y - x = 0$$

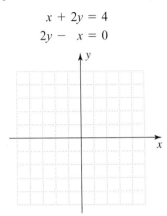

2. Use the graphical method to solve the system.

$$y - 2x = 2$$
$$2y - 4x = 8$$

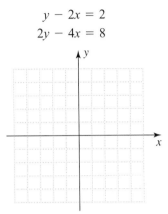

3. Use the substitution method to solve the system (if possible).

$$x + 3y = 6$$
$$2x + 6y = 8$$

4. Use the substitution method to solve the system (if possible).

$$x + 3y = 6$$
$$3x + 9y = 18$$

5. Use the elimination method to solve the system (if possible).

$$2x + 3y = -8$$
$$3x + y = -5$$

6. Use the elimination method to solve the system (if possible).

$$3x - 2y = 6$$
$$-6x + 4y = -2$$

7. Use the elimination method to solve the system (if possible).

$$2y + 3x = -12$$
$$6x + 4y = -24$$

8. Eva has $2 in nickels and dimes. She has twice as many dimes as nickels. How many nickels and how many dimes does she have?

9. The sum of two numbers is 140. Their difference is 90. What are the numbers?

10. A plane flies 600 miles with a tailwind in 2 hours. It takes the same plane 3 hours to fly the 600 miles when flying against the wind. What is the plane's speed in still air?

11. Herbert invests $10,000, part at 5% and part at 6%. How much money is invested at each rate if his annual interest is $568?

12. Graph the solution set of the system.

$$x - 2y > 4$$
$$2x - y \le 6$$

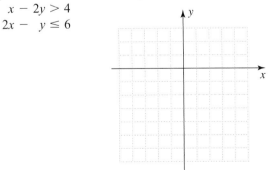

❭ Answers to Practice Test **Chapter 7**

Answer	If You Missed Question	Section	Review Examples	Page
1. The solution is (2, 1).	1	7.1	1	553–554

Answer	If You Missed Question	Section	Review Examples	Page
2. The lines are parallel; there is no solution (inconsistent).	2	7.1	2	554

Answer	If You Missed Question	Section	Review Examples	Page
3. No solution (inconsistent)	3	7.2	2	571
4. Dependent (infinitely many solutions)	4	7.2	3	572
5. $(-1, -2)$	5	7.3	1	580
6. No solution (inconsistent)	6	7.3	2	581
7. Dependent (infinitely many solutions)	7	7.3	3	581–582
8. 8 nickels; 16 dimes	8	7.4	1, 2	589–590
9. 115 and 25	9	7.4	3	591
10. 250 mi/hr	10	7.4	4	591–592
11. $3200 at 5%; $6800 at 6%	11	7.4	5	592–593
12.	12	7.5	1, 2, 3	598–600

❯Cumulative Review Chapters 1–7

1. Add: $-\dfrac{3}{7} + \left(-\dfrac{1}{6}\right)$

2. Find: $(-4)^4$

3. Divide: $-\dfrac{1}{4} \div \left(-\dfrac{1}{8}\right)$

4. Evaluate $y \div 2 \cdot x - z$ for $x = 6$, $y = 24$, $z = 3$.

5. Simplify: $2x - (x + 4) - 2(x + 3)$

6. Write in symbols: The quotient of $(x - 5y)$ and z

7. Solve for x: $3 = 3(x - 2) + 3 - 2x$

8. Solve for x: $\dfrac{x}{4} - \dfrac{x}{9} = 5$

9. Graph: $-\dfrac{x}{3} + \dfrac{x}{8} \geq \dfrac{x - 8}{8}$

10. Graph the point $C(1, -2)$.

11. Determine whether the ordered pair $(-1, -1)$ is a solution of $4x + 3y = -1$.

12. Find x in the ordered pair $(x, 2)$ so that the ordered pair satisfies the equation $3x - 2y = -13$.

13. Graph: $x + y = 1$

14. Graph: $2y + 8 = 0$

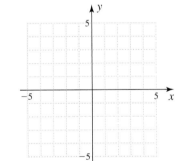

15. Find the slope of the line passing through the points $(-2, -3)$ and $(5, -4)$.

16. What is the slope of the line $12x - 4y = 14$?

17. Find the pair of parallel lines.

　(1) $2y = -x + 6$

　(2) $-4x - 8y = 6$

　(3) $8y - 4x = 6$

18. Simplify: $(2x^4y^{-3})^{-3}$

19. Write in scientific notation: 0.000035

20. Divide and express the answer in scientific notation: $(5.46 \times 10^{-3}) \div (2.6 \times 10^4)$

21. Find (expand): $\left(6x^2 - \dfrac{1}{4}\right)^2$

22. Divide $(2x^3 - x^2 - 2x + 4)$ by $(x - 1)$.

23. Factor completely: $x^2 - 11x + 30$

24. Factor completely: $15x^2 - 29xy + 12y^2$

25. Factor completely: $81x^2 - 64y^2$

26. Factor completely: $-3x^4 + 12x^2$

27. Factor completely: $4x^3 - 4x^2 - 8x$

28. Factor completely: $4x^2 + 3x + 4x + 3$

29. Factor completely: $25kx^2 - 30kx + 9k$

30. Solve for x: $3x^2 + 4x = 15$

31. Write $\dfrac{4x}{3y}$ with a denominator of $15y^2$.

32. Reduce to lowest terms: $\dfrac{-9(x^2 - y^2)}{3(x - y)}$

33. Reduce to lowest terms: $\dfrac{x^2 - 4x - 12}{6 - x}$

34. Multiply: $(x - 5) \cdot \dfrac{x - 2}{x^2 - 25}$

35. Divide: $\dfrac{x + 4}{x - 4} \div \dfrac{x^2 - 16}{4 - x}$

36. Add: $\dfrac{9}{2(x - 9)} + \dfrac{3}{2(x - 9)}$

37. Subtract: $\dfrac{x + 2}{x^2 + x - 6} - \dfrac{x + 3}{x^2 - 4}$

38. Simplify: $\dfrac{\dfrac{4}{x} - \dfrac{3}{4x}}{\dfrac{2}{3x} - \dfrac{1}{2x}}$

39. Solve for x: $\dfrac{5x}{x - 5} - 2 = \dfrac{x}{x - 5}$

40. Solve for x: $\dfrac{x}{x^2 - 4} + \dfrac{2}{x - 2} = \dfrac{1}{x + 2}$

41. Solve for x: $\dfrac{x}{x + 4} - \dfrac{1}{5} = \dfrac{-4}{x + 4}$

42. Solve for x: $2 + \dfrac{8}{x - 2} = \dfrac{32}{x^2 - 4}$

43. A van travels 60 miles on 4 gallons of gas. How many gallons will it need to travel 195 miles?

44. Solve for x: $\dfrac{x + 1}{4} = \dfrac{9}{5}$

45. Sandra can paint a kitchen in 5 hours, and Roger can paint the same kitchen in 4 hours. How long would it take for both working together to paint the kitchen?

46. Find an equation of the line that passes through the point $(-6, -2)$ and has slope $m = -4$.

47. Find an equation of the line having slope 5 and y-intercept 2.

48. Graph: $4x - 3y > -12$

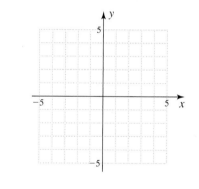

49. Graph: $-y \le -3x + 6$

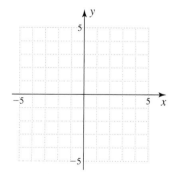

50. An enclosed gas exerts a pressure P on the walls of the container. This pressure is directly proportional to the temperature T of the gas. If the pressure is 7 pounds per square inch when the temperature is $350°F$, find k.

51. If the temperature of a gas is held constant, the pressure P varies inversely as the volume V. A pressure of 1560 pounds per square inch is exerted by 4 cubic feet of air in a cylinder fitted with a piston. Find k.

52. Graph the system and find the solution (if possible):

$$x + 2y = 6$$
$$2y - x = -2$$

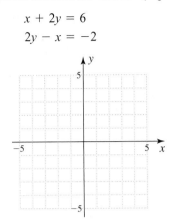

53. Graph the system and find the solution (if possible):

$$y - x = -1$$
$$2y - 2x = -4$$

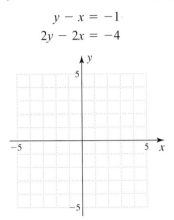

54. Solve by substitution (if possible):

$$x - 4y = 5$$
$$-3x + 12y = -17$$

55. Solve by substitution (if possible):

$$x - 4y = -13$$
$$3x - 12y = -39$$

56. Solve the system (if possible):

$$3x + 4y = 7$$
$$2x + 3y = 5$$

57. Solve the system (if possible):

$$4x - 3y = 1$$
$$4x - 3y = -4$$

58. Solve the system (if possible):

$$2y - x = 0$$
$$-2x + 4y = 0$$

59. Kaye has $4.20 in nickels and dimes. She has three times as many dimes as nickels. How many nickels and how many dimes does she have?

60. The sum of two numbers is 165. Their difference is 95. What are the numbers?

Chapter

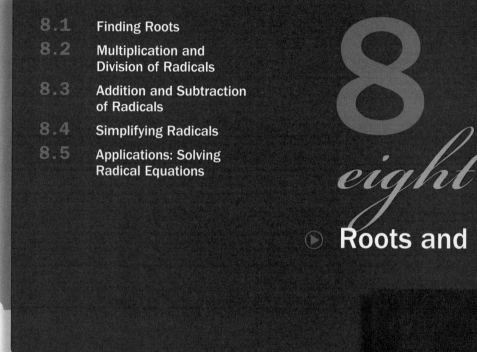

8
eight

▶ Roots and Radicals

The Human Side of Algebra

Beginning in the tenth century and extending into the fourteenth century, Chinese mathematics shifted to arithmetical algebra. A Chinese mathematician discovered the relation between extracting roots and the array of binomial coefficients in Pascal's triangle. This discovery and iterated multiplications were used to extend root extraction and to solve equations of higher degree than cubic, an effort that reached its apex with the work of four prominent thirteenth-century Chinese algebraists.

The square root sign $\sqrt{}$ can be traced back to Christoff Rudolff (1499–1545), who wrote it as $\sqrt{}$ with only two strokes. Rudolff thought that $\sqrt{}$ resembled the small letter r, the first letter in the word *radix,* which means root. One way of computing square roots was developed by the English mathematician Isaac Newton (1642–1727), and the process is aptly named Newton's method. With the advent of calculators and computers, Newton's method, as well as the square root tables popular a few years ago, are seldom used anymore.

Sir Isaac Newton

8.1   Finding Roots

▶ Objectives

A ⟩ Find the square root of a number.

B ⟩ Square a radical expression.

C ⟩ Classify the square root of a number and approximate it with a calculator.

D ⟩ Find higher roots of numbers.

E ⟩ Solve applications involving square roots.

▶ To Succeed, Review How To . . .

1. Find the square of a number (p. 61).
2. Raise a number to a power (pp. 61, 62).

▶ Getting Started

Square Roots and Round Wheels

The first bicycles were built in 1839 by Kirkpatrick Macmillan of Scotland. These bicycles were heavy and unstable, so in 1870, James Starley of Great Britain set out to reduce their weight. The result, shown in the photo, was a lighter bicycle but one that was likely to tip over when going around corners. As a matter of fact, even now, the greatest speed s (in miles per hour) at which a cyclist can safely take a corner of radius r (in feet) is given by the equation

$$s = 4\sqrt{r}$$ Read "s equals 4 times the square root of r."

The concept of a *square root* is related to the concept of squaring a number. The number 36, for example, is the square of 6 because $6^2 = 36$. Six, on the other hand, is the square root of 36—that is, $\sqrt{36} = 6$. Similarly,

$$\sqrt{16} = 4 \qquad \text{because} \qquad 4^2 = 16$$

$$\sqrt{\frac{4}{9}} = \frac{2}{3} \qquad \text{because} \qquad \left(\frac{2}{3}\right)^2 = \left(\frac{4}{9}\right)$$

In this section we learn how to find square roots.

A ⟩ Finding Square Roots

To find $\sqrt{a}$, we have to find the number b so that $b^2 = a$; that is,

$$\sqrt{a} = b \qquad \text{means that} \qquad b^2 = a$$

Note that since $b^2 = a$, a must be **nonnegative.**

For example,

$$\sqrt{4} = 2 \qquad \text{because} \qquad 2^2 = 4$$
$$\sqrt{9} = 3 \qquad \text{because} \qquad 3^2 = 9$$
$$\sqrt{\frac{1}{4}} = \frac{1}{2} \qquad \text{because} \qquad \left(\frac{1}{2}\right)^2 = \frac{1}{4}$$

The symbol $\sqrt{}$ is called a **radical sign,** and it indicates the **positive** square root of a number (except $\sqrt{0} = 0$). The expression under the radical sign is called the **radicand.** An algebraic expression containing a radical is called a **radical expression.** We can summarize our discussion as follows.

SQUARE ROOT

If a is a positive real number,

$\sqrt{a} = b$ is the **positive** square root of a so that $b^2 = a$.

$-\sqrt{a} = b$ is the **negative** square root of a so that $b^2 = a$.

If $a = 0$, $\sqrt{a} = 0$ and $\sqrt{0} = 0$ since $0^2 = 0$.

Note that the symbol $-\sqrt{}$ represents the *negative* square root. Thus, the square root of 4 is $\sqrt{4} = 2$, but the *negative* square root of 4 is $-\sqrt{4} = -2$. When we write $\sqrt{a}$ we mean the **positive** or **principal** square root of a.

EXAMPLE 1   Finding square roots
Find:

a. $\sqrt{121}$ **b.** $-\sqrt{100}$ **c.** $\sqrt{\dfrac{25}{36}}$

SOLUTION

a. $\sqrt{121} = 11$ Since $11^2 = 121$

b. $-\sqrt{100} = -10$ Since $10^2 = 100$

c. $\sqrt{\dfrac{25}{36}} = \dfrac{5}{6}$ Since $\left(\dfrac{5}{6}\right)^2 = \dfrac{25}{36}$

PROBLEM 1
Find:

a. $\sqrt{169}$ **b.** $-\sqrt{64}$

c. $\sqrt{\dfrac{81}{25}}$

B › Squaring Radical Expressions

Since $b = \sqrt{a}$ means that $b^2 = a$,

$$\sqrt{a} \cdot \sqrt{a} = (\sqrt{a})^2 = a$$

Similarly, since $b = -\sqrt{a}$ means that $b^2 = a$,

$$(-\sqrt{a}) \cdot (-\sqrt{a}) = (-\sqrt{a})^2 = a$$

Thus, we have the following rule.

RULE

Squaring a Square Root

When the square root of a nonnegative real number a is squared, the result is that positive real number; that is, $(\sqrt{a})^2 = a$ and $(-\sqrt{a})^2 = a$. Also, $(\sqrt{0})^2 = 0$.

Answers to PROBLEMS

1. a. 13   **b.** -8 **c.** $\dfrac{9}{5}$

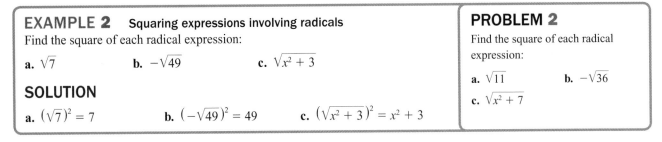

EXAMPLE 2   Squaring expressions involving radicals
Find the square of each radical expression:

a. $\sqrt{7}$ **b.** $-\sqrt{49}$ **c.** $\sqrt{x^2 + 3}$

SOLUTION

a. $(\sqrt{7})^2 = 7$ **b.** $(-\sqrt{49})^2 = 49$ **c.** $(\sqrt{x^2 + 3})^2 = x^2 + 3$

PROBLEM 2

Find the square of each radical expression:

a. $\sqrt{11}$ **b.** $-\sqrt{36}$

c. $\sqrt{x^2 + 7}$

C ⟩ Classifying and Approximating Square Roots

If a number has a rational number for its square root, it's called a **perfect square.** Thus, 1, 4, 9, 16, $\frac{4}{25}$, $\frac{9}{4}$, and $\frac{81}{16}$ are examples of perfect squares. If a number is *not* a perfect square, its square root is *irrational.* Thus, $\sqrt{2}, \sqrt{3}, \sqrt{5}, \sqrt{\frac{3}{4}}$, and $\sqrt{\frac{5}{6}}$ are irrational. As you recall, an irrational number *cannot* be written as the ratio of two integers. Its decimal representation is nonrepeating and nonterminating. A decimal approximation for such a number can be obtained by using the [2nd] [x²] keys on your calculator. Using the symbol ≈ (which means "is approximately equal to"), we have

$$\sqrt{2} \approx 1.4142136$$
$$\sqrt{3} \approx 1.7320508$$
$$\sqrt{5} \approx 2.236068$$

Of course, not all real numbers have a *real-number* square root. As you recall, $\sqrt{a} = b$ is *equivalent* to $b^2 = a$. Since b^2 is nonnegative, a has to be nonnegative.

SQUARE ROOT OF A NEGATIVE NUMBER

If a is **negative**, $\sqrt{a}$ is not a real number.

For example, $\sqrt{-7}, \sqrt{-4}$, and $\sqrt{-\frac{4}{5}}$ are *not* real numbers. Try some possibilities. Suppose you say $\sqrt{-4} = -2$. By the definition of square root $(-2)^2$ must be -4. But $(-2)^2 = 4$ (not -4), so $\sqrt{-4}$ is *not* a real number b because you can't find a number b such that $b^2 = -4$.

EXAMPLE 3   Classifying and approximating real numbers
Classify the given numbers as rational, irrational, or not a real number. Approximate the irrational numbers using a calculator.

a. $\sqrt{\frac{81}{25}}$ **b.** $-\sqrt{64}$ **c.** $\sqrt{13}$ **d.** $\sqrt{-64}$

SOLUTION

a. $\frac{81}{25} = \left(\frac{9}{5}\right)^2$ is a perfect square, so $\sqrt{\frac{81}{25}} = \frac{9}{5}$ is a rational number.

b. $64 = 8^2$ is a perfect square, so $-\sqrt{64} = -8$ is a rational number.

c. 13 is *not* a perfect square, so $\sqrt{13} \approx 3.6055513$ is irrational.

d. $\sqrt{-64}$ is *not* a real number, since there is no number that we can square and get -64 for an answer. (Try it!)

PROBLEM 3

Classify as rational, irrational, or not a real number. Approximate the irrational numbers using a calculator.

a. $\sqrt{\frac{36}{25}}$ **b.** $-\sqrt{81}$

c. $\sqrt{19}$ **d.** $\sqrt{-9}$

Answers to PROBLEMS

2. a. 11   **b.** 36   **c.** $x^2 + 7$

3. a. $\frac{6}{5}$; Rational   **b.** -9; Rational

   c. Irrational. $\sqrt{19} \approx 4.3588989$

   d. Not real

D › Finding Higher Roots of Numbers

We've already seen that, by definition,

$$\sqrt{a} = b \qquad \text{means that} \qquad b^2 = a$$

Similarly,

$$\sqrt[3]{a} = b \qquad \text{means that} \qquad b^3 = a$$

$$\sqrt[4]{a} = b \qquad \text{means that} \qquad b^4 = a$$

Because of these relationships, finding the square root of a number can be regarded as the **inverse** of finding the square of the number. Similarly, the inverse of finding the **cube** or **fourth power** of a number a is to find $\sqrt[3]{a}$, the **cube root** of a, or $\sqrt[4]{a}$, the **fourth root** of a. In general, $\sqrt[n]{a}$ is called the **nth root of a** and the number n is the **index** or **order** of the radical. Note that we could write $\sqrt[2]{a}$ for the square root, but the simpler notation $\sqrt{a}$ is customary because the square root is the most widely performed operation.

EXAMPLE 4   **Finding higher roots of real numbers**

Find each root if possible:

a. $\sqrt[4]{16}$ **b.** $\sqrt[4]{-81}$ **c.** $\sqrt[3]{8}$

d. $\sqrt[3]{-27}$ **e.** $-\sqrt[4]{16}$

SOLUTION

a. We need to find a number whose fourth power is 16, that is, $(?)^4 = 16$. Since $2^4 = 16$, $\sqrt[4]{16} = 2$.

b. We need to find a number whose fourth power is -81. There is no such real number, so $\sqrt[4]{-81}$ is *not* a real number.

c. We need to find a number whose cube is 8. Since $2^3 = 8$, $\sqrt[3]{8} = 2$.

d. We need to find a number whose cube is -27. Since $(-3)^3 = -27$, $\sqrt[3]{-27} = -3$.

e. We need to find a number whose fourth power is 16. Since $2^4 = 16$, $-\sqrt[4]{16} = -2$.

PROBLEM 4

Find each root if possible:

a. $\sqrt[4]{81}$

b. $\sqrt[4]{-16}$

c. $\sqrt[3]{-64}$

d. $\sqrt[3]{-216}$

e. $-\sqrt[4]{625}$

E › Applications Involving Square Roots

EXAMPLE 5   **Simplifying square roots**

The time t (in seconds) it takes an object dropped from a distance d (in feet) to reach the ground is given by the equation

$$t = \sqrt{\frac{d}{16}}$$

The highest regularly performed dive used to be at La Quebrada in Acapulco. How long does it take the divers to reach the water 100 feet below?

SOLUTION   Using

$$t = \sqrt{\frac{d}{16}} \qquad \text{and} \qquad d = 100$$

we have

$$t = \sqrt{\frac{100}{16}} = \frac{10}{4} = 2.5 \text{ sec}$$

By the way, the water is only 12 feet deep, but the height of the dive has been given as high as 118 feet and as low as 87.5 feet!

100 ft

PROBLEM 5

How long does it take the diver to travel 81 feet?

Answers to PROBLEMS

4. a. 3   **b.** Not real   **c.** −4   **d.** −6   **e.** −5   **5.** 2.25 sec

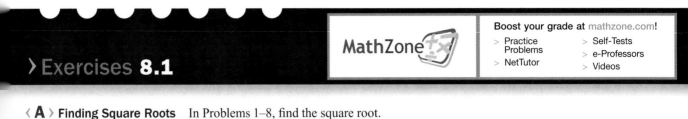

> Exercises **8.1**

⟨ **A** ⟩ **Finding Square Roots**   In Problems 1–8, find the square root.

1. $\sqrt{25}$

2. $\sqrt{36}$

3. $-\sqrt{9}$

4. $-\sqrt{81}$

5. $\sqrt{\dfrac{16}{9}}$

6. $\sqrt{\dfrac{25}{4}}$

7. $-\sqrt{\dfrac{4}{81}}$

8. $-\sqrt{\dfrac{9}{100}}$

In Problems 9–12, find all the square roots (positive and negative) of each number.

9. $\dfrac{25}{81}$

10. $\dfrac{36}{49}$

11. $\dfrac{49}{100}$

12. $\dfrac{1}{9}$

⟨ **B** ⟩ **Squaring Radical Expressions**   In Problems 13–20, find the square of each radical expression.

13. $\sqrt{5}$

14. $\sqrt{8}$

15. $-\sqrt{11}$

16. $-\sqrt{13}$

17. $\sqrt{x^2 + 1}$

18. $-\sqrt{a^2 + 2}$

19. $-\sqrt{3y^2 + 7}$

20. $\sqrt{8z^2 + 1}$

⟨ **C** ⟩ **Classifying and Approximating Square Roots**   In Problems 21–32, find and classify the given numbers as rational, irrational, or not a real number. Use a calculator to approximate the irrational numbers. *Note:* The number of decimals in your calculator display may be different.

21. $\sqrt{36}$

22. $-\sqrt{4}$

23. $\sqrt{-100}$

24. $\sqrt{-9}$

25. $-\sqrt{64}$

26. $\sqrt{\dfrac{9}{49}}$

27. $\sqrt{\dfrac{16}{9}}$

28. $\sqrt{-\dfrac{4}{81}}$

29. $-\sqrt{6}$

30. $\sqrt{7}$

31. $-\sqrt{2}$

32. $-\sqrt{12}$

⟨ **D** ⟩ **Finding Higher Roots of Numbers**   In Problems 33–40, find each root if possible.

33. $\sqrt[4]{81}$

34. $\sqrt[4]{-256}$

35. $-\sqrt[4]{81}$

36. $\sqrt[3]{27}$

37. $-\sqrt[3]{64}$

38. $-\sqrt[3]{-64}$

39. $-\sqrt[3]{-125}$

40. $-\sqrt[3]{-8}$

⟨ **E** ⟩ **Applications Involving Square Roots**

41. *Ribbon Falls*   Ribbon Falls in California is the tallest continuous waterfall in the United States, with a drop of about 1600 feet. How long does it take the water to travel from the top of the waterfall to the bottom? (*Hint:* See Example 5.) (Actually, the waterfall is dry from late July to early April.)

42. *Jump from airship*   Have you heard about the *Hindenburg* airship? On June 22, 1936, Colonel Harry A. Froboess of Switzerland jumped almost 400 feet from the *Hindenburg* into the Bodensee. How long did his jump take? (*Hint:* See Example 5.)

43. *Highest dive into airbag*   The time t (in seconds) it takes an object dropped from a distance d (in meters) to reach the ground is given by the equation

$$t = \sqrt{\dfrac{d}{5}}$$

The highest reported dive into an airbag is 100 meters from the top of Vegas World Hotel and Casino by stuntman Dan Koko. About how long did it take Koko to reach the airbag below?

44. *Time of highest dive*   Colonel Froboess of Problem 42 jumped about 125 meters from the *Hindenburg* to the Bodensee. Using the formula in Problem 43, how long did his jump take? Is your answer consistent with that of Problem 42?

45. *Pythagorean theorem*   The Greek mathematician Pythagoras discovered a theorem that states that the lengths a, b, and c of the sides of a right triangle (a triangle with a 90° angle) are related by the formula $a^2 + b^2 = c^2$. (See the diagram.) This means that $c = \sqrt{a^2 + b^2}$. If the shorter sides of a right triangle are 4 inches and 3 inches, find the length of the longest side c (called the hypotenuse).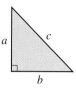

46. *Pythagorean theorem*   Use the Pythagorean theorem stated in Problem 45 to find how high on the side of a building a 13-foot ladder will reach if the base of the ladder is 5 feet from the wall on which the ladder is resting.

47. *View from in-flight airplane*   How far can you see from an airplane in flight? It depends on how high the airplane is. As a matter of fact, your view V_m (in miles) is modeled by the equation $V_m = 1.22\sqrt{a}$, where a is the altitude of the plane in feet. What is your view when your plane is cruising at an altitude of 40,000 feet?

48. *Dimensions of a square*   The area of a square is 144 square inches. How long is each side of the square?

49. *Dimensions of square room*   A square room has an area of 121 square feet. What are the dimensions of the room?

50. *Dimensions of a quilt*   A square quilt has an area of 169 square units. What are its dimensions?

Curving without skidding.   Use the following information for Problems 51 and 52. The maximum speed v (in miles per hour) a car can travel on a concrete highway curve without skidding is modeled by the equation $v = \sqrt{9r}$, where r is the radius of the curve in feet.

51. If the radius on the first curve in the photo is 100 feet, what is the maximum speed v a car can travel on this curve without skidding?

52. The radius on the second curve in the photo is only 64 feet. What is the maximum speed v a car can travel on this curve without skidding?

Velocity   Use the following information for Problems 53 and 54. If an object is dropped from a height of h feet its velocity v (in feet per second) when it hits the ground is given by the equation $v = \sqrt{2gh}$, where g is the acceleration due to gravity, 32 feet per second squared (32 ft/sec²).

53. Gumercindo was leaning over the railing of a balcony 81 feet above the street when his Blackberry slipped out of his hand. Find the Blackberry's velocity when it hit the ground.

54. Kanisha was riding the Kinda Ka roller coaster and after it went over the top at 441 feet high, her iPod slipped out of her pocket and fell to the ground. What was the velocity of the iPod when it hit the ground?

〉〉〉 *Using Your Knowledge*

Interpolation   Suppose you want to approximate $\sqrt{18}$ *without* using your calculator. Since $\sqrt{16} = 4$ and $\sqrt{25} = 5$, you have $\sqrt{16} = 4 < \sqrt{18} < \sqrt{25} = 5$. This means that $\sqrt{18}$ is between 4 and 5. To find a better approximation of $\sqrt{18}$, you can use a method that mathematicians call **interpolation.** Don't be scared by this name! The process is *really* simple. If you want to find an approximation for $\sqrt{18}$, follow the steps in the diagram; $\sqrt{20}$ and $\sqrt{22}$ are also approximated.

$$
\begin{array}{ccc}
\begin{array}{l} 18 - 16 = 2 \\ \left[\begin{array}{l} \sqrt{16} = 4 \\ \sqrt{18} \approx 4 + \frac{2}{9} \end{array}\right] \\ \sqrt{25} = 5 \\ 25 - 16 = 9 \end{array}
&
\begin{array}{l} 20 - 16 = 4 \\ \left[\begin{array}{l} \sqrt{16} = 4 \\ \sqrt{20} \approx 4 + \frac{4}{9} \end{array}\right] \\ \sqrt{25} = 5 \\ 25 - 16 = 9 \end{array}
&
\begin{array}{l} 22 - 16 = 6 \\ \left[\begin{array}{l} \sqrt{16} = 4 \\ \sqrt{22} \approx 4 + \frac{6}{9} \end{array}\right] \\ \sqrt{25} = 5 \\ 25 - 16 = 9 \end{array}
\end{array}
$$

As you can see,

$$\sqrt{18} \approx 4\frac{2}{9} \qquad \sqrt{20} \approx 4\frac{4}{9} \qquad \text{and} \qquad \sqrt{22} \approx 4\frac{6}{9} = 4\frac{2}{3}$$

Do you see a pattern? What is $\sqrt{24}$? Using a calculator, we obtain

$$\sqrt{18} \approx 4.2, \qquad \sqrt{20} \approx 4.5, \quad \text{and} \quad \sqrt{22} \approx 4.7$$

$$\sqrt{18} \approx 4\frac{2}{9} \approx 4.2, \quad \sqrt{20} \approx 4\frac{4}{9} \approx 4.4, \quad \sqrt{22} \approx 4\frac{6}{9} \approx 4.7$$

As you can see, we can get very close approximations!

Use this knowledge to approximate the following roots. Give the answer as a mixed number.

55. a. $\sqrt{26}$ **b.** $\sqrt{28}$ **c.** $\sqrt{30}$ **56. a.** $\sqrt{67}$ **b.** $\sqrt{70}$ **c.** $\sqrt{73}$

〉〉〉 **Write On**

57. If a is a *nonnegative* number, how many square roots does a have? Explain.

58. If a is *negative*, how many square roots does a have? Explain.

59. If $a = 0$, how many square roots does a have?

60. How many real-number cube roots does a positive number have?

61. How many real-number cube roots does a negative number have?

62. How many cube roots does zero have?

63. Suppose you want to find $\sqrt{a}$. What assumption must you make about the variable a?

64. If you are given a whole number less than 200, how would you determine whether the square root of the number is rational or irrational without a calculator? Explain.

〉〉〉 **Concept Checker**

Fill in the blank(s) with the correct word(s), phrase, or mathematical statement.

65. $\sqrt{a} = b$ is the **positive** square root of a so that _____.

66. $-\sqrt{a} = b$ is the **negative** square root of a so that _____.

67. $\sqrt{0} =$ _____.

68. For $a \geq 0$, $(\sqrt{a})^2 =$ _____.

69. For $a \geq 0$, $(-\sqrt{a})^2 =$ _____.

70. If a is **negative**, $\sqrt{a}$ is _____ a **real number.**

not defined	**a**
$a^2 = b$	**not**
$b^2 = a$	**always**
$-a$	**0**

⟩⟩⟩ *Mastery Test*

Find:

71. $\sqrt{\dfrac{16}{49}}$

72. $-\sqrt{\dfrac{36}{25}}$

73. $\sqrt{144}$

Find the square of each radical expression:

74. $-\sqrt{28}$

75. $\sqrt{17}$

76. $-\sqrt{x^2 + 9}$

Find and classify as rational, irrational, or not a real number. Approximate the irrational numbers using a calculator:

77. $-\sqrt{\dfrac{49}{121}}$

78. $\sqrt{-100}$

79. $\sqrt{15}$

Find if possible:

80. $\sqrt[4]{1296}$

81. $\sqrt[3]{-125}$

82. $\sqrt[4]{-1}$

83. $-\sqrt[3]{-27}$

84. The time t in seconds it takes an object dropped from a distance d (in feet) to reach the ground is given by the equation $t = \sqrt{\dfrac{d}{16}}$. How long does it take an object dropped from a distance of 81 feet to reach the ground?

⟩⟩⟩ *Skill Checker*

Find $\sqrt{b^2 - 4ac}$ given the specified value for each variable.

85. $a = 1, b = -4, c = 3$ **86.** $a = 1, b = -1, c = -2$ **87.** $a = 2, b = 5, c = -3$ **88.** $a = 6, b = -7, c = -3$

8.2

Multiplication and Division of Radicals

▶ Objectives

A ⟩ Multiply and simplify radicals using the product rule.

B ⟩ Divide and simplify radicals using the quotient rule.

C ⟩ Simplify radicals involving variables.

D ⟩ Simplify higher roots.

▶ To Succeed, Review How To . . .

1. Write a number as a product of primes (p. 6).
2. Write a number using specified powers as factors (pp. 15, 16).

▶ Getting Started
Skidding Is No Accident!

Have you seen your local police measuring skid marks at the scene of an accident? The speed s (in miles per hour) a car was traveling if it skidded d feet on a dry concrete road is given by $s = \sqrt{24d}$. If a car left a 50-foot skid mark at the scene of an accident and the speed limit was 30 miles per hour, was the driver speeding? To find the answer, we need to use the formula $s = \sqrt{24d}$. Letting $d = 50, s = \sqrt{24 \cdot 50} = \sqrt{1200}$. How can we simplify $\sqrt{1200}$? If we use factors that are perfect squares, we get

$$\sqrt{1200} = \sqrt{100 \cdot 4 \cdot 3}$$
$$= \sqrt{100} \cdot \sqrt{4} \cdot \sqrt{3}$$
$$= 10 \cdot 2 \cdot \sqrt{3}$$
$$= 20\sqrt{3}$$

which is about 35 miles per hour. So the driver *was* exceeding the speed limit! In solving this problem, we have made the assumption that $\sqrt{a \cdot b} = \sqrt{a} \cdot \sqrt{b}$. Is this assumption always true? In this section you will learn that this is indeed the case!

A ⟩ Using the Product Rule for Radicals

Is it true that $\sqrt{a \cdot b} = \sqrt{a} \cdot \sqrt{b}$? Let's check some examples.

$$\sqrt{4 \cdot 9} = \sqrt{36} = 6, \quad \sqrt{4} \cdot \sqrt{9} = 2 \cdot 3 = 6, \quad \text{thus,} \quad \sqrt{4 \cdot 9} = \sqrt{4} \cdot \sqrt{9}$$

Similarly,

$$\sqrt{9 \cdot 16} = \sqrt{144} = 12, \quad \sqrt{9} \cdot \sqrt{16} = 3 \cdot 4 = 12, \quad \text{thus,} \quad \sqrt{9 \cdot 16} = \sqrt{9} \cdot \sqrt{16}$$

In general, we have the following rule

PRODUCT RULE FOR RADICALS

If a and b are nonnegative numbers,

$$\sqrt{a \cdot b} = \sqrt{a} \cdot \sqrt{b}$$

The product rule can be used to *simplify* radicals—that is, to make sure that no perfect squares remain under the radical sign. Thus, $\sqrt{15}$ is simplified, but $\sqrt{18}$ is not because $\sqrt{18} = \sqrt{9 \cdot 2}$, and $\sqrt{9 \cdot 2} = \sqrt{9} \cdot \sqrt{2} = 3\sqrt{2}$ using the product rule.

EXAMPLE 1   Simplifying expressions using the product rule

Simplify:

a. $\sqrt{75}$ **b.** $\sqrt{48}$

SOLUTION   To use the product rule, you have to remember the square roots of a few perfect square numbers such as 4, 9, 16, 25, 36, 49, 64, and 81 and try to write the number under the radical using one of these numbers as a factor.

a. Since $\sqrt{75} = \sqrt{25 \cdot 3}$, we write $\sqrt{75} = \sqrt{25 \cdot 3} = \sqrt{25} \cdot \sqrt{3} = 5\sqrt{3}$.

b. Similarly, $\sqrt{48} = \sqrt{16 \cdot 3}$. Thus, $\sqrt{48} = 4\sqrt{3}$.

PROBLEM 1

Simplify:

a. $\sqrt{72}$ **b.** $\sqrt{98}$

We can also use the product rule to actually multiply radicals. Thus,

$$\sqrt{3} \cdot \sqrt{5} = \sqrt{3 \cdot 5} = \sqrt{15} \quad \text{and} \quad \sqrt{8} \cdot \sqrt{2} = \sqrt{16} = 4$$

EXAMPLE 2   Multiplying expressions using the product rule

Multiply:

a. $\sqrt{6} \cdot \sqrt{5}$ **b.** $\sqrt{20} \cdot \sqrt{5}$ **c.** $\sqrt{5} \cdot \sqrt{x}, x > 0$

PROBLEM 2

Multiply:

a. $\sqrt{5} \cdot \sqrt{7}$ **b.** $\sqrt{40} \cdot \sqrt{10}$

c. $\sqrt{6} \cdot \sqrt{x}, x > 0$

Answers to PROBLEMS

1. **a.** $6\sqrt{2}$ **b.** $7\sqrt{2}$

2. **a.** $\sqrt{35}$ **b.** 20   **c.** $\sqrt{6x}$

SOLUTION

a. $\sqrt{6} \cdot \sqrt{5} = \sqrt{30}$ **b.** $\sqrt{20} \cdot \sqrt{5} = \sqrt{100} = 10$

c. $\sqrt{5} \cdot \sqrt{x} = \sqrt{5x}$

B > Using the Quotient Rule for Radicals

We already know that

$$\sqrt{\frac{81}{16}} = \frac{9}{4}$$

It's also true that

$$\sqrt{\frac{81}{16}} = \frac{\sqrt{81}}{\sqrt{16}} = \frac{9}{4}$$

This rule can be stated as follows.

QUOTIENT RULE FOR RADICALS

If a and b are positive numbers,

$$\sqrt{\frac{a}{b}} = \frac{\sqrt{a}}{\sqrt{b}}$$

EXAMPLE 3   Simplifying expressions using the quotient rule

Simplify:

a. $\sqrt{\frac{5}{9}}$ **b.** $\frac{\sqrt{18}}{\sqrt{6}}$ **c.** $\frac{14\sqrt{20}}{7\sqrt{10}}$

SOLUTION

a. $\sqrt{\frac{5}{9}} = \frac{\sqrt{5}}{\sqrt{9}} = \frac{\sqrt{5}}{3}$ **b.** $\frac{\sqrt{18}}{\sqrt{6}} = \sqrt{\frac{18}{6}} = \sqrt{3}$

c. $\frac{14\sqrt{20}}{7\sqrt{10}} = \frac{2\sqrt{20}}{\sqrt{10}} = 2\sqrt{\frac{20}{10}} = 2\sqrt{2}$

PROBLEM 3

Simplify:

a. $\sqrt{\frac{7}{16}}$ **b.** $\frac{\sqrt{30}}{\sqrt{6}}$

c. $\frac{16\sqrt{30}}{8\sqrt{10}}$

C > Simplifying Radicals Involving Variables

We can simplify radicals involving variables as long as the expressions under the radical sign are defined. This means that these expressions must *not* be negative. What about the answers? Note that $\sqrt{2^2} = 2$ and $\sqrt{(-2)^2} = \sqrt{4} = 2$. Thus, the square root of a squared nonzero number is always positive. We use absolute values to indicate this.

ABSOLUTE VALUE OF A RADICAL

For any real number a,

$$\sqrt{a^2} = |a|$$

Of course, in examples and problems where the variables are assumed to be positive, the absolute-value bars are not necessary.

EXAMPLE 4   Simplifying radical expressions involving variables

Simplify (assume all variables represent positive real numbers):

a. $\sqrt{49x^2}$ **b.** $\sqrt{81n^4}$ **c.** $\sqrt{50y^8}$ **d.** $\sqrt{x^{11}}$

SOLUTION

a. $\sqrt{49x^2} = \sqrt{49} \cdot \sqrt{x^2}$ Use the product rule.

$\quad\quad\quad = 7x$

b. $\sqrt{81n^4} = \sqrt{81} \cdot \sqrt{n^4}$ Use the product rule.

$\quad\quad\quad = 9n^2$ Since $(n^2)^2 = n^4$, $\sqrt{n^4} = n^2$.

c. $\sqrt{50y^8} = \sqrt{50} \cdot \sqrt{y^8}$ Use the product rule.

$\quad\quad\quad = \sqrt{25 \cdot 2} \cdot \sqrt{y^8}$ Factor 50 using a perfect square.

$\quad\quad\quad = 5\sqrt{2} \cdot y^4$ Since $\sqrt{25} = 5$ and $\sqrt{y^8} = y^4$

$\quad\quad\quad = 5y^4\sqrt{2}$

d. $\sqrt{x^{11}} = \sqrt{x^{10} \cdot x}$ Since $x^{10} \cdot x = x^{11}$

$\quad\quad\quad = x^5\sqrt{x}$ Since $\sqrt{x^{10}} = x^5$

PROBLEM 4

Simplify (all variables are positive):

a. $\sqrt{81y^2}$ **b.** $\sqrt{9n^4}$

c. $\sqrt{72x^{10}}$ **d.** $\sqrt{y^{13}}$

D ⟩ Simplifying Higher Roots

The product and quotient rules can be generalized for higher roots as shown in the box.

> **PROPERTIES OF RADICALS**
>
> For all real numbers where the indicated roots exist,
>
> $$\sqrt[n]{a \cdot b} = \sqrt[n]{a} \cdot \sqrt[n]{b} \quad \text{and} \quad \sqrt[n]{\frac{a}{b}} = \frac{\sqrt[n]{a}}{\sqrt[n]{b}}$$

The idea when simplifying *cube* roots is to find factors that are *perfect cubes*. Similarly, to simplify *fourth roots,* we look for factors that are *perfect fourth powers*. This means that to simplify $\sqrt[4]{48}$, we need to find a perfect fourth-power factor of 48. Since $48 = 16 \cdot 3 = 2^4 \cdot 3$,

$$\sqrt[4]{48} = \sqrt[4]{2^4 \cdot 3} = 2 \cdot \sqrt[4]{3} = 2\sqrt[4]{3}$$

EXAMPLE 5   Simplifying radical expressions involving higher roots

Simplify:

a. $\sqrt[3]{54}$ **b.** $\sqrt[4]{162}$ **c.** $\sqrt[3]{\frac{27}{8}}$

SOLUTION

a. We are looking for a *cube* root, so we need to find a *perfect cube* factor of 54. Since $2^3 = 8$ is *not* a factor of 54, we try $3^3 = 27$:

$$\sqrt[3]{54} = \sqrt[3]{27 \cdot 2} = \sqrt[3]{3^3 \cdot 2} = 3\sqrt[3]{2}$$

PROBLEM 5

Simplify:

a. $\sqrt[3]{16}$ **b.** $\sqrt[4]{112}$ **c.** $\sqrt[3]{\frac{64}{27}}$

Answers to PROBLEMS

4. a. $9y$ **b.** $3n^2$ **c.** $6x^5\sqrt{2}$

d. $y^6\sqrt{y}$

5. a. $2\sqrt[3]{2}$ **b.** $2\sqrt[4]{7}$ **c.** $\frac{4}{3}$

b. This time we need a *perfect fourth* factor of 162. Since $2^4 = 16$ is *not* a factor of 162, we try $3^4 = 81$:

$$\sqrt[4]{162} = \sqrt[4]{81 \cdot 2} = \sqrt[4]{3^4 \cdot 2} = 3\sqrt[4]{2}$$

c. This time we are looking for *perfect cube* factors of 27 and 8. Now $3^3 = 27$ and $2^3 = 8$, so

$$\sqrt[3]{\frac{27}{8}} = \frac{\sqrt[3]{3^3}}{\sqrt[3]{2^3}} = \frac{3}{2}$$

NOTE

You can find $\sqrt[3]{\frac{27}{8}}$ *without* using the quotient rule. Since $3^3 = 27$ and $2^3 = 8$, you can write

$$\sqrt[3]{\frac{27}{8}} = \sqrt[3]{\frac{3^3}{2^3}} = \sqrt[3]{\left(\frac{3}{2}\right)^3} = \frac{3}{2}$$

Calculator Corner

Finding Roots of Numbers

You can find the root of a number using your calculator. Start by graphing $y = \sqrt{x}$. To find $\sqrt{25}$, use TRACE to move the cursor slowly through the x-values—$x = 1$, $x = 2$, $x = 3$, and so on until you get to $x = 25$. The y-value for $x = 25$ is $y = \sqrt{25} = 5$. (See Window 1.) Did you notice as you traced to the right on the curve $y = \sqrt{x}$ that the y-values for $\sqrt{1}$, $\sqrt{2}$, $\sqrt{3}$, and so on were displayed? In fact, what you have here is a built-in square root table! How would you obtain $\sqrt{8}$ as displayed in Window 2?

X=25   Y=5
Window 1

X=8   Y=2.8284271
Window 2

Finally, there are some other things to learn from the graph in Window 1. Can you obtain $\sqrt{-2}$? Why not? Can you find fourth roots by graphing $y = \sqrt[4]{x}$? Note that the calculator doesn't have a fourth-root button. If you know that $\sqrt[4]{x} = x^{1/4}$, then you can graph $y = x^{1/4}$ and use a similar procedure to find the fourth root of a number.

Now look at Window 3, where $\sqrt[4]{81}$ is shown. What is the value for $\sqrt[4]{81}$? Can you get $\sqrt[4]{-1}$? Why not? Can you now discover how to obtain $\sqrt[3]{8}$ and $\sqrt[3]{-8}$ with your calculator? Do you get the same answer as you would doing it with paper and pencil?

X=81   Y=3
Window 3

> **Exercises 8.2**

⟨ **A** ⟩ **Using the Product Rule for Radicals**   In Problems 1–20, simplify.

1. $\sqrt{45}$ **2.** $\sqrt{128}$ **3.** $\sqrt{125}$ **4.** $\sqrt{175}$ **5.** $\sqrt{180}$

6. $\sqrt{162}$ **7.** $\sqrt{200}$ **8.** $\sqrt{245}$ **9.** $\sqrt{384}$ **10.** $\sqrt{486}$

11. $\sqrt{125}$ **12.** $\sqrt{80}$ **13.** $\sqrt{600}$ **14.** $\sqrt{324}$ **15.** $\sqrt{361}$

16. $\sqrt{648}$ **17.** $\sqrt{700}$ **18.** $\sqrt{726}$ **19.** $\sqrt{432}$ **20.** $\sqrt{507}$

In Problems 21–30, find the product. Assume all variables represent positive numbers.

21. $\sqrt{3} \cdot \sqrt{5}$ 　　**22.** $\sqrt{5} \cdot \sqrt{11}$ 　　**23.** $\sqrt{27} \cdot \sqrt{3}$ 　　**24.** $\sqrt{2} \cdot \sqrt{32}$ 　　**25.** $\sqrt{7} \cdot \sqrt{7}$

26. $\sqrt{15} \cdot \sqrt{15}$ 　　**27.** $\sqrt{3} \cdot \sqrt{x}$ 　　**28.** $\sqrt{5} \cdot \sqrt{y}$ 　　**29.** $\sqrt{2a} \cdot \sqrt{18a}$ 　　**30.** $\sqrt{3b} \cdot \sqrt{12b}$

‹ **B** › **Using the Quotient Rule for Radicals** In Problems 31–40, find the quotient. Assume all variables represent positive numbers.

31. $\sqrt{\dfrac{2}{25}}$ 　　**32.** $\sqrt{\dfrac{5}{81}}$ 　　**33.** $\dfrac{\sqrt{20}}{\sqrt{5}}$ 　　**34.** $\dfrac{\sqrt{27}}{\sqrt{9}}$ 　　**35.** $\dfrac{\sqrt{27}}{\sqrt{3}}$

36. $\dfrac{\sqrt{72}}{\sqrt{2}}$ 　　**37.** $\dfrac{15\sqrt{30}}{3\sqrt{10}}$ 　　**38.** $\dfrac{52\sqrt{28}}{26\sqrt{4}}$ 　　**39.** $\dfrac{18\sqrt{40}}{3\sqrt{10}}$ 　　**40.** $\dfrac{22\sqrt{40}}{11\sqrt{10}}$

‹ **C** › **Simplifying Radicals Involving Variables** In Problems 41–50, simplify. Assume all variables represent positive numbers.

41. $\sqrt{100a^2}$ 　　**42.** $\sqrt{16b^2}$ 　　**43.** $\sqrt{49a^4}$ 　　**44.** $\sqrt{64x^8}$

45. $-\sqrt{32a^6}$ 　　**46.** $-\sqrt{18b^{10}}$ 　　**47.** $\sqrt{m^{13}}$ 　　**48.** $\sqrt{n^{17}}$

49. $-\sqrt{27m^{11}}$ 　　**50.** $-\sqrt{50n^7}$

‹ **D** › **Simplifying Higher Roots** In Problems 51–60, simplify.

51. $\sqrt[3]{40}$ 　　**52.** $\sqrt[3]{108}$ 　　**53.** $\sqrt[3]{-16}$ 　　**54.** $\sqrt[3]{-48}$

55. $\sqrt[3]{\dfrac{8}{27}}$ 　　**56.** $\sqrt[3]{\dfrac{27}{64}}$ 　　**57.** $\sqrt[4]{48}$ 　　**58.** $\sqrt[4]{243}$

59. $\sqrt[3]{\dfrac{64}{27}}$ 　　**60.** $\dfrac{\sqrt[4]{16}}{\sqrt[4]{81}}$

〉〉〉 **Applications**

Horizon Use the following information for Problems 61–63. Because of the Earth curvature, the distance you can see in each direction is bounded by a circle called the **horizon**. The size of this circle depends on how high you are, the greater your height *h* (in feet), the larger the circle. The distance *d* (in miles) you can see when your height is *h* is given by the equation:

$$d = \frac{\sqrt{3h}}{\sqrt{2}}$$

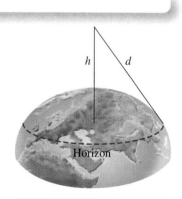

61. Suppose you are on the second floor of a building that is 54 feet high. How far in each direction can you see from the top of this building? (*Hint:* Use the quotient rule of radicals to simplify.)

62. Suppose you are at the top of a tower that is 486 feet tall. How far in each direction can you see from the top of this tower?

63. The Taipei 101, at this time the tallest building in the world, is 1670 feet tall. How far can you see from the top of the Taipei Tower? Answer to the nearest mile.

〉〉〉 *Write On*

64. Explain why $\sqrt{(-2)^2}$ is not -2.

65. Explain why $\sqrt[3]{(-2)^3} = -2$.

66. Which of the following is in simplified form? Explain.

    a. $-\sqrt{18}$ **b.** $\sqrt{41}$ **c.** $\sqrt{144}$

67. Assume that p is a prime number.

    a. Is $\sqrt{p}$ rational or irrational?

    b. Is $\sqrt{p}$ in simplified form? Explain.

〉〉〉 *Concept Checker*

Fill in the blank(s) with the correct word(s), phrase, or mathematical statement.

68. If a and b are **nonnegative** numbers, $\sqrt{a \cdot b} = $ _____.

69. If a and b are **positive** numbers, $\sqrt{\dfrac{a}{b}} = $ _____.

70. For **any** real number a, $\sqrt{a^2} = $ _____.

71. For **all real** numbers where the roots exist, $\sqrt[n]{a \cdot b} = $ _____.

a $\dfrac{\sqrt{a}}{\sqrt{b}}$

$|a|$

$\sqrt[n]{a} \cdot \sqrt[n]{b}$

$\sqrt{a} \cdot \sqrt{b}$

〉〉〉 *Mastery Test*

Simplify (assume all variables represent positive numbers):

72. $\sqrt[4]{80}$ **73.** $\sqrt[3]{135}$ **74.** $\sqrt[4]{16x^4}$

75. $\sqrt[3]{\dfrac{27}{64}}$ **76.** $\dfrac{\sqrt[4]{81}}{\sqrt[4]{16}}$ **77.** $\sqrt{100x^6}$

78. $\sqrt{32x^7}$ **79.** $\dfrac{\sqrt{28}}{\sqrt{14}}$ **80.** $\dfrac{\sqrt{72}}{\sqrt{2}}$

81. $\sqrt{\dfrac{4}{9}}$ **82.** $\dfrac{\sqrt{17}}{\sqrt{36}}$ **83.** $\sqrt{72} \cdot \sqrt{2}$

84. $\sqrt{18} \cdot \sqrt{3x^2}$

〉〉〉 *Skill Checker*

Combine like terms:

85. $5x + 7x$ **86.** $8x^2 - 3x^2$ **87.** $9x^3 + 7x^3 - 2x^3$

8.3   Addition and Subtraction of Radicals

▶ Objectives

A ▷ Add and subtract like radicals.

B ▷ Use the distributive property to simplify radicals.

C ▷ Rationalize the denominator in an expression.

▶ To Succeed, Review How To . . .

1. Combine like terms (pp. 86–89).
2. Simplify radicals (pp. 621–625).
3. Use the distributive property (pp. 80–81, 89–90).

▶ Getting Started
A Broken Pattern

In the preceding section we learned the product rule for radicals: $\sqrt{a \cdot b} = \sqrt{a} \cdot \sqrt{b}$, which we know means that the square root of a product is the product of the square roots. Similarly, the square root of a quotient is the quotient of the square roots; that is,

$$\sqrt{\frac{a}{b}} = \frac{\sqrt{a}}{\sqrt{b}}$$

Is the square root of a sum or difference the sum or difference of the square roots? Let's look at an example. Is $\sqrt{9 + 16}$ the same as $\sqrt{9} + \sqrt{16}$? First, $\sqrt{9 + 16} = \sqrt{25} = 5$. But $\sqrt{9} + \sqrt{16} = 3 + 4 = 7$. Thus,

$$\sqrt{9 + 16} \neq \sqrt{9} + \sqrt{16} \quad \text{because} \quad 5 \neq 3 + 4$$

In general,
$$\sqrt{a + b} \neq \sqrt{a} + \sqrt{b}$$

So, what can be done with sums and differences involving radicals? We will learn that next.

A ❭ Adding and Subtracting Radicals

Expressions involving radicals can be handled using simple arithmetic rules. For example, like terms can be combined,

$$3x + 7x = 10x$$

and
$$3\sqrt{6} + 7\sqrt{6} = 10\sqrt{6}$$

Similarly,

$$9x - 2x = 7x$$

and
$$9\sqrt{3} - 2\sqrt{3} = 7\sqrt{3}$$

Test your understanding of this idea in Example 1.

EXAMPLE 1   Adding and subtracting expressions involving radicals
Simplify:

a. $6\sqrt{7} + 9\sqrt{7}$ **b.** $8\sqrt{5} - 2\sqrt{5}$

SOLUTION

a. $6\sqrt{7} + 9\sqrt{7} = 15\sqrt{7}$ (just like $6x + 9x = 15x$)

b. $8\sqrt{5} - 2\sqrt{5} = 6\sqrt{5}$ (just like $8x - 2x = 6x$)

PROBLEM 1
Simplify:

a. $8\sqrt{5} + 2\sqrt{5}$ **b.** $7\sqrt{6} - 2\sqrt{6}$

Of course, you may have to simplify before combining **like radical terms**—that is, terms in which the *radical factors* are *exactly* the same. (For example, $4\sqrt{3}$ and $5\sqrt{3}$ are like radical terms.) Here is a problem that may seem difficult:

$$\sqrt{48} + \sqrt{75}$$

In this case, $\sqrt{48}$ and $\sqrt{75}$ are *not* like terms, and only like terms can be combined. However,

$$\sqrt{48} = \sqrt{16 \cdot 3} = \sqrt{16} \cdot \sqrt{3} = 4\sqrt{3}$$
$$\sqrt{75} = \sqrt{25 \cdot 3} = \sqrt{25} \cdot \sqrt{3} = 5\sqrt{3}$$

Now,

$$\sqrt{48} + \sqrt{75} = 4\sqrt{3} + 5\sqrt{3}$$
$$= 9\sqrt{3}$$

EXAMPLE 2   Adding and subtracting expressions involving radicals
Simplify:

a. $\sqrt{80} + \sqrt{20}$ **b.** $\sqrt{75} + \sqrt{12} - \sqrt{147}$

SOLUTION

a.
$$\sqrt{80} = \sqrt{16 \cdot 5} = \sqrt{16} \cdot \sqrt{5} = 4\sqrt{5}$$
$$\sqrt{20} = \sqrt{4 \cdot 5} = \sqrt{4} \cdot \sqrt{5} = 2\sqrt{5}$$
$$\sqrt{80} + \sqrt{20} = 4\sqrt{5} + 2\sqrt{5} = 6\sqrt{5}$$

b.
$$\sqrt{75} = \sqrt{25 \cdot 3} = \sqrt{25} \cdot \sqrt{3} = 5\sqrt{3}$$
$$\sqrt{12} = \sqrt{4 \cdot 3} = \sqrt{4} \cdot \sqrt{3} = 2\sqrt{3}$$
$$\sqrt{147} = \sqrt{49 \cdot 3} = \sqrt{49} \cdot \sqrt{3} = 7\sqrt{3}$$
$$\sqrt{75} + \sqrt{12} - \sqrt{147} = 5\sqrt{3} + 2\sqrt{3} - 7\sqrt{3} = 0$$

PROBLEM 2
Simplify:

a. $\sqrt{150} + \sqrt{24}$

b. $\sqrt{20} + \sqrt{80} - \sqrt{45}$

B › Using the Distributive Property to Simplify Expressions

Now let's see how we can use the distributive property to simplify an expression.

Answers to PROBLEMS

1. a. $10\sqrt{5}$ **b.** $5\sqrt{6}$

2. a. $7\sqrt{6}$ **b.** $3\sqrt{5}$

EXAMPLE 3   Multiplying expressions involving radicals

Simplify:

a. $\sqrt{5}\left(\sqrt{40} - \sqrt{2}\right)$ **b.** $\sqrt{2}\left(\sqrt{2} - \sqrt{3}\right)$

SOLUTION

a. Using the distributive property,

$$\sqrt{5}\left(\sqrt{40} - \sqrt{2}\right) = \sqrt{5}\sqrt{40} - \sqrt{5}\sqrt{2}$$

$$= \sqrt{200} - \sqrt{10} \qquad \text{Since } \sqrt{5}\sqrt{40} = \sqrt{200}$$
$$\text{and } \sqrt{5}\sqrt{2} = \sqrt{10}$$

$$= \sqrt{100 \cdot 2} - \sqrt{10} \qquad \text{Since } \sqrt{100 \cdot 2} = \sqrt{100} \cdot \sqrt{2} = 10\sqrt{2}$$

$$= 10\sqrt{2} - \sqrt{10}$$

b. $\sqrt{2}\left(\sqrt{2} - \sqrt{3}\right) = \sqrt{2}\sqrt{2} - \sqrt{2}\sqrt{3}$ Use the distributive property.

$$= 2 - \sqrt{6} \qquad \text{Since } \sqrt{2}\sqrt{2} = 2$$

PROBLEM 3

Simplify:

a. $\sqrt{3}\left(\sqrt{45} - \sqrt{2}\right)$

b. $\sqrt{5}\left(\sqrt{5} - \sqrt{3}\right)$

C ⟩ Rationalizing Denominators

In Chapter 9 the solution of some quadratic equations will be of the form

$$\sqrt{\frac{9}{5}}$$

If we use the quotient rule for radicals, we obtain

$$\sqrt{\frac{9}{5}} = \frac{\sqrt{9}}{\sqrt{5}} = \frac{3}{\sqrt{5}}$$

The expression $\frac{3}{\sqrt{5}}$ contains the square root of a nonperfect square, which is an irrational number. To simplify $\frac{3}{\sqrt{5}}$, we **rationalize the denominator.** This means we remove all radicals from the denominator.

> ### PROCEDURE
>
> **Rationalizing Denominators**
>
> **Method 1.** Multiply *both* the *numerator* and *denominator* of the fraction by the **square root** in the denominator; or
>
> **Method 2.** Multiply numerator and denominator by the square root of a number that makes the denominator the square root of a perfect square.

Thus, to rationalize the denominator in $\frac{3}{\sqrt{5}}$, we multiply the numerator and denominator by $\sqrt{5}$ to obtain

$$\frac{3}{\sqrt{5}} = \frac{3 \cdot \sqrt{5}}{\sqrt{5} \cdot \sqrt{5}}$$

$$= \frac{3\sqrt{5}}{5} \qquad \text{Since } \sqrt{5} \cdot \sqrt{5} = 5$$

Note that the idea in rationalizing

$$\sqrt{\frac{a}{b}} = \frac{\sqrt{a}}{\sqrt{b}}$$

is to make the denominator $\sqrt{b}$ a square root of a perfect square. Multiplying numerator and denominator by $\sqrt{b}$ *always* works, but you can save time if you find a factor smaller than $\sqrt{b}$ that will make the denominator a square root of a perfect square. Thus, when rationalizing $\frac{\sqrt{3}}{\sqrt{8}}$, you could *first* multiply numerator and denominator by $\sqrt{8}$. However, it's *better* to multiply numerator and denominator by $\sqrt{2}$, as shown in Example 4.

EXAMPLE 4   Rationalizing denominators: Two methods

Write with a rationalized denominator: $\sqrt{\dfrac{3}{8}}$

SOLUTION

Method 1. Use the quotient rule and then multiply numerator and denominator by $\sqrt{8}$.

$$\sqrt{\frac{3}{8}} = \frac{\sqrt{3}}{\sqrt{8}} \qquad \text{Use the quotient rule.}$$

$$= \frac{\sqrt{3} \cdot \sqrt{8}}{\sqrt{8} \cdot \sqrt{8}} \qquad \text{Multiply numerator and denominator by } \sqrt{8}.$$

$$= \frac{\sqrt{24}}{8} \qquad \text{Since } \sqrt{3} \cdot \sqrt{8} = \sqrt{24} \text{ and } \sqrt{8} \cdot \sqrt{8} = 8$$

$$= \frac{\sqrt{4 \cdot 6}}{8} \qquad \text{Since } 24 = 4 \cdot 6 \text{ and 4 is a perfect square}$$

$$= \frac{2 \cdot \sqrt{6}}{8} \qquad \text{Since } \sqrt{4} = 2$$

$$= \frac{\sqrt{6}}{4} \qquad \text{Divide numerator and denominator by 2.}$$

Method 2. If you noticed that multiplying numerator and denominator by $\sqrt{2}$ yields $\sqrt{16} = 4$ in the denominator, you could have obtained

$$\sqrt{\frac{3}{8}} = \frac{\sqrt{3}}{\sqrt{8}} \qquad \text{Use the quotient rule.}$$

$$= \frac{\sqrt{3} \cdot \sqrt{2}}{\sqrt{8} \cdot \sqrt{2}} \qquad \text{Multiply by } \sqrt{2} \text{ so the denominator is } \sqrt{16} = 4.$$

$$= \frac{\sqrt{6}}{\sqrt{16}} \qquad \text{Since } \sqrt{3} \cdot \sqrt{2} = \sqrt{6}$$

$$= \frac{\sqrt{6}}{4} \qquad \text{Since } \sqrt{16} = 4$$

You get the same answer as in Method 1 by multiplying by $\sqrt{2}$. As you can see fewer steps were involved when simplifying! *You can save time by looking for factors that make the denominator the square root of a perfect square.*

EXAMPLE 5   Rationalizing denominators by making them perfect squares

Write with a rationalized denominator: $\sqrt{\dfrac{x^2}{32}} \quad (x > 0)$

SOLUTION   Since

$$\sqrt{\frac{x^2}{32}} = \frac{\sqrt{x^2}}{\sqrt{32}}$$

we could multiply numerator and denominator by $\sqrt{32}$. However, multiplying by $\sqrt{2}$ gives a denominator of $\sqrt{64} = 8$. Thus,

$$\sqrt{\frac{x^2}{32}} = \frac{\sqrt{x^2} \cdot \sqrt{2}}{\sqrt{32} \cdot \sqrt{2}}$$

$$= \frac{x\sqrt{2}}{\sqrt{64}}$$

$$= \frac{x\sqrt{2}}{8}$$

PROBLEM 4

Write with a rationalized denominator: $\sqrt{\dfrac{5}{12}}$

PROBLEM 5

Write with a rationalized denominator:

$$\sqrt{\frac{x^2}{50}} \quad (x > 0)$$

Answers to PROBLEMS

4. $\dfrac{\sqrt{15}}{6}$ 5. $\dfrac{x\sqrt{2}}{10}$

> Exercises **8.3**

⟨ **A** ⟩   **Adding and Subtracting Radicals**   In Problems 1–16, perform the indicated operations and simplify.

1. $6\sqrt{7} + 4\sqrt{7}$

2. $4\sqrt{11} + 9\sqrt{11}$

3. $9\sqrt{13} - 4\sqrt{13}$

4. $6\sqrt{10} - 2\sqrt{10}$

5. $\sqrt{32} + \sqrt{50} - \sqrt{72}$

6. $\sqrt{12} + \sqrt{27} - \sqrt{75}$

7. $\sqrt{162} + \sqrt{50} - \sqrt{200}$

8. $\sqrt{48} + \sqrt{75} - \sqrt{363}$

9. $9\sqrt{48} - 5\sqrt{27} + 3\sqrt{12}$

10. $3\sqrt{32} - 5\sqrt{8} + 4\sqrt{50}$

11. $5\sqrt{7} - 3\sqrt{28} - 2\sqrt{63}$

12. $3\sqrt{28} - 6\sqrt{7} - 2\sqrt{175}$

13. $-5\sqrt{3} + 8\sqrt{75} - 2\sqrt{27}$

14. $-6\sqrt{99} + 6\sqrt{44} - \sqrt{176}$

15. $-3\sqrt{45} + \sqrt{20} - \sqrt{5}$

16. $-5\sqrt{27} + \sqrt{12} - 5\sqrt{48}$

⟨ **B** ⟩   **Using the Distributive Property to Simplify Expressions**   In Problems 17–30, simplify.

17. $\sqrt{10}(\sqrt{20} - \sqrt{3})$

18. $\sqrt{10}(\sqrt{30} - \sqrt{2})$

19. $\sqrt{6}(\sqrt{14} + \sqrt{5})$

20. $\sqrt{14}(\sqrt{18} + \sqrt{3})$

21. $\sqrt{3}(\sqrt{3} - \sqrt{2})$

22. $\sqrt{6}(\sqrt{6} - \sqrt{5})$

23. $\sqrt{5}(\sqrt{2} + \sqrt{5})$

24. $\sqrt{3}(\sqrt{2} + \sqrt{3})$

25. $\sqrt{6}(\sqrt{2} - \sqrt{3})$

26. $\sqrt{5}(\sqrt{15} - \sqrt{27})$

27. $2(\sqrt{2} - 5)$

28. $\sqrt{5}(3 - \sqrt{5})$

29. $\sqrt{2}(\sqrt{6} - 3)$

30. $\sqrt{3}(\sqrt{6} - 4)$

⟨ **C** ⟩   **Rationalizing Denominators**   In Problems 31–50, rationalize the denominator. Assume all variables are positive real numbers.

31. $\dfrac{3}{\sqrt{6}}$

32. $\dfrac{6}{\sqrt{7}}$

33. $\dfrac{-10}{\sqrt{5}}$

34. $\dfrac{-9}{\sqrt{3}}$

35. $\dfrac{\sqrt{8}}{\sqrt{2}}$

36. $\dfrac{\sqrt{48}}{\sqrt{3}}$

37. $\dfrac{-\sqrt{2}}{\sqrt{5}}$

38. $\dfrac{-\sqrt{3}}{\sqrt{7}}$

39. $\dfrac{\sqrt{2}}{\sqrt{8}}$

40. $\dfrac{\sqrt{3}}{\sqrt{12}}$

41. $\dfrac{\sqrt{x^2}}{\sqrt{18}}$

42. $\dfrac{\sqrt{a^4}}{\sqrt{32}}$

43. $\dfrac{\sqrt{a^2}}{\sqrt{b}}$

44. $\dfrac{\sqrt{x^4}}{\sqrt{y}}$

45. $\sqrt{\dfrac{3}{10}}$

46. $\sqrt{\dfrac{2}{27}}$

47. $\sqrt{\dfrac{x^2}{32}}$

48. $\sqrt{\dfrac{x}{18}}$

49. $\sqrt{\dfrac{x^4}{20}}$

50. $\sqrt{\dfrac{x^6}{72}}$

⟩⟩⟩   **Applications**

Meteorology   Use the following information for Problems 51 and 52. How long will a storm last? According to meteorologists the time t (in hours) a storm will last is based on the diameter d (in miles) of the storm and given by the equation:

$$t = \sqrt{\dfrac{d^3}{216}}$$

51. How long would a storm 6 miles in diameter last?

52. *Meteorology*   A storm 3 miles in diameter is threatening a baseball game, which must be resumed within an hour or the game will have to be postponed.

a. How long would the storm last? (Give the answer in rationalized form).

b. Will the game be resumed or postponed?

54. *Cone*   The radius r of a cone is given by the equation $r = \sqrt{\frac{3V}{\pi h}}$, where V is the volume of the cone and h is its height. Write an expression for r with a rationalized denominator.

53. *Sphere*   The radius r of a sphere is given by the equation $r = \sqrt{\frac{S}{4\pi}}$, where S is the surface area of the sphere. Write an expression for r with a rationalized denominator.

Surface area

Radius

〉〉〉   *Using Your Knowledge*

"Radical" Shortcuts   Suppose you want to rationalize the denominator in the expression $\sqrt{\frac{3}{32}}$. Using the quotient rule, we can write

$$\sqrt{\frac{3}{32}} = \frac{\sqrt{3}}{\sqrt{32}} = \frac{\sqrt{3} \cdot \sqrt{32}}{\sqrt{32} \cdot \sqrt{32}} = \frac{\sqrt{96}}{32}$$

$$= \frac{\sqrt{16 \cdot 6}}{32} = \frac{4 \cdot \sqrt{6}}{32} = \frac{\sqrt{6}}{8}$$

A shorter way is as follows:

$$\sqrt{\frac{3}{32}} = \sqrt{\frac{3 \cdot 2}{32 \cdot 2}} = \sqrt{\frac{6}{64}} = \frac{\sqrt{6}}{8}$$

55. Use this shorter procedure to do Problems 45–50.

〉〉〉   *Write On*

56. In Problems 31–50, we specified that *all* variables should be positive real numbers. Specify which variables *have to be* positive real numbers and in which problems this must occur.

58. Write an explanation of what is meant by like radicals.

57. In Example 4, we rationalized the denominator in the expression

$$\sqrt{\frac{3}{8}} = \frac{\sqrt{3}}{\sqrt{8}}$$

by *first* using the quotient rule and writing

$$\sqrt{\frac{3}{8}} = \frac{\sqrt{3}}{\sqrt{8}}$$

and then multiplying numerator and denominator by $\sqrt{2}$. How can you do the example without first using the quotient rule?

⟩⟩⟩ Concept Checker

Fill in the blank(s) with the correct word(s), phrase, or mathematical statement.

59. **Like radical terms** are terms in which the radical **factors** are _____ the same.	**almost** **radicals**
60. To **rationalize** the denominator means to **remove** all _____ from the **denominator**.	**numbers** **exactly**

⟩⟩⟩ Mastery Test

Simplify:

61. $8\sqrt{3} + 5\sqrt{3}$

62. $7\sqrt{5} - 2\sqrt{5}$

63. $\sqrt{18} + 3\sqrt{2}$

64. $\sqrt{32} - 3\sqrt{2}$

65. $7\sqrt{18} + 5\sqrt{2} - 7\sqrt{8}$

66. $3(\sqrt{5} - 2)$

67. $\sqrt{3}(5 - \sqrt{3})$

68. $\sqrt{18}(\sqrt{2} - 2)$

Rationalize the denominator, assuming all variables are positive real numbers:

69. $\dfrac{3}{\sqrt{7}}$

70. $\sqrt{\dfrac{x^6}{10}}$

71. $\dfrac{\sqrt{x}}{\sqrt{2}}$

72. $\dfrac{\sqrt{x^2}}{\sqrt{20}}$

⟩⟩⟩ Skill Checker

Multiply:

73. $(x + 3)(x - 3)$

74. $(a + b)(a - b)$

75. $\dfrac{6x + 12}{3}$

76. $\dfrac{4x - 8}{2}$

8.4   Simplifying Radicals

▶ Objectives

A ⟩ Simplify a radical expression involving products, quotients, sums, or differences.

B ⟩ Use the conjugate of a number to rationalize the denominator of an expression.

C ⟩ Reduce a fraction involving a radical by factoring.

▶ To Succeed, Review How To . . .

1. Find the root of an expression (pp. 614–615).
2. Add, subtract, multiply, and divide radicals (pp. 622–623, 628–629).
3. Rationalize the denominator in an expression (pp. 630–631).
4. Expand binomials of the form $(x \pm y)^2$ (pp. 366–369).

▶ Getting Started

Breaking the Sound Barrier

How fast can this plane travel? The answer is classified information, but it exceeds twice the speed of sound (747 miles per hour). It is said that the plane's speed is more than Mach 2. The formula for calculating the Mach number is

$$M = \sqrt{\dfrac{2}{\gamma}} \quad \sqrt{\dfrac{P_2 - P_1}{P_1}}$$

where P_1 and P_2 are air pressures and γ is a constant. This expression can be simplified by multiplying the radical expressions and then rationalizing the denominator (we discuss how to do this in the *Using Your Knowledge*). In this section you will learn how to simplify more complicated radical expressions such as the one on page 634 by using the techniques we've discussed in the last three sections. What did we do in those sections? We found the roots of algebraic expressions, multiplied and divided radicals, added and subtracted radicals, and rationalized denominators. You will see that the procedure used to *completely* simplify a radical involves steps in which you will perform these tasks in precisely the order in which the topics were studied.

A ⟩ Simplifying Radical Expressions

In the preceding sections you were asked to "simplify" expressions involving radicals. To make this idea more precise and to help you simplify radical expressions, we use the following rules.

RULES

Simplifying Radical Expressions

1. Whenever possible, write the *rational-number* representation of a radical expression. For example, write

$$\sqrt{81} \text{ as } 9, \quad \sqrt{\frac{4}{9}} \text{ as } \frac{2}{3} \quad \text{and} \quad \sqrt[3]{\frac{1}{8}} \text{ as } \frac{1}{2}$$

2. Use the product rule $\sqrt{x} \cdot \sqrt{y} = \sqrt{xy}$ to write indicated products as a single radical. For example, write

$$\sqrt{6} \text{ instead of } \sqrt{2} \cdot \sqrt{3} \quad \text{and} \quad \sqrt{2ab} \text{ instead of } \sqrt{2a} \cdot \sqrt{b}$$

3. Use the quotient rule

$$\frac{\sqrt{x}}{\sqrt{y}} = \sqrt{\frac{x}{y}}$$

to write indicated quotients as a single radical. For example, write

$$\frac{\sqrt{6}}{\sqrt{2}} \text{ as } \sqrt{3} \quad \text{and} \quad \frac{\sqrt[3]{10}}{\sqrt[3]{5}} \text{ as } \sqrt[3]{2}$$

4. If a radicand has a perfect square as a factor, write the radical expression as the product of the square root of the perfect square and the radical of the other factor. A similar statement applies to cubes and higher roots. For example, write

$$\sqrt{18} = \sqrt{9 \cdot 2} \text{ as } 3\sqrt{2} \quad \text{and} \quad \sqrt[3]{54} = \sqrt[3]{27 \cdot 2} \text{ as } 3\sqrt[3]{2}$$

5. Combine like radicals whenever possible. For example,

$$2\sqrt{5} + 8\sqrt{5} = (2 + 8)\sqrt{5} = 10\sqrt{5}$$
$$9\sqrt{11} - 2\sqrt{11} = (9 - 2)\sqrt{11} = 7\sqrt{11}$$

6. Rationalize the denominator of algebraic expressions. For example,

$$\frac{3}{\sqrt{2}} = \frac{3 \cdot \sqrt{2}}{\sqrt{2} \cdot \sqrt{2}} = \frac{3\sqrt{2}}{2}$$

Now let's use these rules.

EXAMPLE 1   Simplifying radicals: Sums and differences

Simplify:

a. $\sqrt{9 + 16} - \sqrt{4}$ **b.** $9\sqrt{6} - \sqrt{2} \cdot \sqrt{3}$ **c.** $\dfrac{\sqrt{6x^3}}{\sqrt{2x^2}},\quad x > 0$

SOLUTION   In each case, a rule applies.

a. Since $\sqrt{9 + 16} = \sqrt{25} = 5$ and $\sqrt{4} = 2$,   *Use Rules 1 and 5.*

$\sqrt{9 + 16} - \sqrt{4} = 5 - 2 = 3$

b. $9\sqrt{6} - \sqrt{2} \cdot \sqrt{3} = 9\sqrt{6} - \sqrt{6} = (9 - 1)\sqrt{6} = 8\sqrt{6}$ *Use Rules 2 and 5.*

c. $\dfrac{\sqrt{6x^3}}{\sqrt{2x^2}} = \sqrt{\dfrac{6x^3}{2x^2}} = \sqrt{3x}$ *Use Rule 3.*

PROBLEM 1

Simplify:

a. $\sqrt{60 + 4} - \sqrt{9}$

b. $8\sqrt{10} - \sqrt{2}\sqrt{5}$

c. $\dfrac{\sqrt{10x^3}}{\sqrt{5x^2}},\quad x > 0$

EXAMPLE 2   Simplifying quotients

Simplify:

a. $\dfrac{\sqrt[3]{256}}{\sqrt[3]{2}}$ **b.** $\dfrac{9}{\sqrt[3]{4}}$

SOLUTION   A rule applies in each case.

a. $\dfrac{\sqrt[3]{256}}{\sqrt[3]{2}} = \sqrt[3]{128}$ *Use Rule 3.*

$\quad\quad = \sqrt[3]{64 \cdot 2}$ *Since $64 \cdot 2 = 128$ and 64 is a perfect cube*

$\quad\quad = 4\sqrt[3]{2}$ *Use Rule 4.*

b. We have to make the denominator the cube root of a perfect cube. To do this, we multiply numerator and denominator by $\sqrt[3]{2}$. Note that the denominator will be $\sqrt[3]{4} \cdot \sqrt[3]{2} = \sqrt[3]{8} = 2$. Thus, we have

$$\frac{9}{\sqrt[3]{4}} = \frac{9 \cdot \sqrt[3]{2}}{\sqrt[3]{4} \cdot \sqrt[3]{2}}$$

$$= \frac{9 \cdot \sqrt[3]{2}}{\sqrt[3]{8}}$$

$$= \frac{9\sqrt[3]{2}}{2}$$

PROBLEM 2

Simplify:

a. $\dfrac{\sqrt[3]{108}}{\sqrt[3]{2}}$ **b.** $\dfrac{7}{\sqrt[3]{9}}$

Do you recall the FOIL method? It and the special products can also be used to simplify expressions involving radicals. We do this next.

EXAMPLE 3   Using FOIL to simplify products

Simplify:

a. $(\sqrt{2} + 5\sqrt{3})(\sqrt{2} - 4\sqrt{3})$ **b.** $(\sqrt{3} + 2\sqrt{5})(\sqrt{3} - 2\sqrt{5})$

SOLUTION   Using the FOIL method, we have

a. $(\sqrt{2} + 5\sqrt{3})(\sqrt{2} - 4\sqrt{3})$

$$= \underset{F}{\underbrace{\sqrt{2} \cdot \sqrt{2}}} + \underset{O}{\underbrace{\sqrt{2}(-4\sqrt{3})}} + \underset{I}{\underbrace{5\sqrt{3}(\sqrt{2})}} + \underset{L}{\underbrace{(5\sqrt{3})(-4\sqrt{3})}}$$

$\quad = \quad\quad 2 \quad - \quad\quad 4\sqrt{6} \quad + \quad\quad 5\sqrt{6} \quad - \quad\quad 20 \cdot 3$ *Use the product rule.*

$\quad = \quad\quad 2 \quad + \quad\quad\quad\quad \sqrt{6} \quad\quad\quad - \quad\quad 60$ *Combine radicals.*

$\quad = -58 + \sqrt{6}$ *Since $2 - 60 = -58$*

PROBLEM 3

Simplify:

a. $(\sqrt{3} + 5\sqrt{2})(\sqrt{3} - 4\sqrt{2})$

b. $(2 + 3\sqrt{5})(2 - 3\sqrt{5})$

Answers to PROBLEMS

1. a. 5   **b.** $7\sqrt{10}$ **c.** $\sqrt{2x}$

2. a. $3\sqrt[3]{2}$ **b.** $\dfrac{7\sqrt[3]{3}}{3}$

3. a. $-37 + \sqrt{6}$ **b.** -41

b. Using the special products formula SP4 (p. 367), we have

$$(X + A)(X - A) = X^2 - A^2$$
$$(\sqrt{3} + 2\sqrt{5})(\sqrt{3} - 2\sqrt{5}) = (\sqrt{3})^2 - (2\sqrt{5})^2$$
$$= 3 - \left[(2)^2(\sqrt{5})^2\right]$$
$$= 3 - (4)(5)$$
$$= 3 - 20$$
$$= -17$$

B > Using Conjugates to Rationalize Denominators

In Example 3(b), the product of the sum $\sqrt{3} + 2\sqrt{5}$ and the difference $\sqrt{3} - 2\sqrt{5}$ is the rational number -17. This is no coincidence! The expressions $\sqrt{3} + 2\sqrt{5}$ and $\sqrt{3} - 2\sqrt{5}$ are **conjugates** of each other. In general, the expressions $a\sqrt{b} + c\sqrt{d}$ and $a\sqrt{b} - c\sqrt{d}$ are conjugates of each other. Their product is obtained by using the special products formula SP4 (p. 369) and always results in a rational number. Here is one way of using conjugates to simplify radical expressions.

> **PROCEDURE**
>
> **Using Conjugates to Simplify Radical Expressions**
>
> To simplify an algebraic expression with two terms in the denominator, at least one of which is a square root, multiply both numerator and denominator by the conjugate of the denominator.

EXAMPLE 4   Using conjugates to rationalize denominators

Simplify:

a. $\dfrac{7}{\sqrt{5} + 1}$

b. $\dfrac{3}{\sqrt{5} - \sqrt{3}}$

SOLUTION

a. The denominator $\sqrt{5} + 1$ has two terms, one of which is a radical. To simplify

$$\frac{7}{\sqrt{5} + 1}$$

multiply numerator and denominator by the conjugate of the denominator, which is $\sqrt{5} - 1$. [*Note:* $(\sqrt{5} + 1)(\sqrt{5} - 1) = (\sqrt{5})^2 - (1)^2$.] Thus,

$$\frac{7}{\sqrt{5} + 1} = \frac{7 \cdot (\sqrt{5} - 1)}{(\sqrt{5} + 1)(\sqrt{5} - 1)}$$ Multiply the numerator and denominator by $\sqrt{5} - 1$.

$$= \frac{7\sqrt{5} - 7}{(\sqrt{5})^2 - (1)^2}$$ Use the distributive property and SP4.

$$= \frac{7\sqrt{5} - 7}{5 - 1}$$ Since $(\sqrt{5})^2 = 5$

$$= \frac{7\sqrt{5} - 7}{4}$$

PROBLEM 4

Simplify:

a. $\dfrac{5}{\sqrt{7} + 1}$

b. $\dfrac{5}{\sqrt{6} - \sqrt{2}}$

(continued)

Answers to PROBLEMS

4. a. $\dfrac{5\sqrt{7} - 5}{6}$ **b.** $\dfrac{5\sqrt{6} + 5\sqrt{2}}{4}$

b. This time we multiply the numerator and denominator of the fraction by the conjugate of $\sqrt{5} - \sqrt{3}$, which is $\sqrt{5} + \sqrt{3}$.

$$\frac{3}{\sqrt{5} - \sqrt{3}} = \frac{3(\sqrt{5} + \sqrt{3})}{(\sqrt{5} - \sqrt{3})(\sqrt{5} + \sqrt{3})}$$ Multiply the numerator and denominator by $\sqrt{5} + \sqrt{3}$.

$$= \frac{3\sqrt{5} + 3\sqrt{3}}{(\sqrt{5})^2 - (\sqrt{3})^2}$$ Use the distributive property and SP4.

$$= \frac{3\sqrt{5} + 3\sqrt{3}}{5 - 3}$$ Since $(\sqrt{5})^2 = 5$ and $(\sqrt{3})^2 = 3$

$$= \frac{3\sqrt{5} + 3\sqrt{3}}{2}$$

C › Reducing Fractions Involving Radicals by Factoring

In the next chapter we will encounter solutions of quadratic equations written as

$$\frac{8 + \sqrt{20}}{4}$$

To simplify these expressions, first note that $\sqrt{20} = \sqrt{4 \cdot 5} = 2\sqrt{5}$. Thus,

$$\frac{8 + \sqrt{20}}{4} = \frac{8 + 2\sqrt{5}}{4}$$ Since $\sqrt{20} = 2\sqrt{5}$

$$= \frac{2 \cdot (4 + \sqrt{5})}{2 \cdot 2}$$ Factor the numerator and denominator.

$$= \frac{4 + \sqrt{5}}{2}$$ Divide by 2.

EXAMPLE 5   **Reducing fractions by factoring first**

Simplify:

a. $\dfrac{-4 + \sqrt{8}}{2}$ **b.** $\dfrac{-8 + \sqrt{28}}{4}$

SOLUTION

a. Since $\sqrt{8} = \sqrt{4 \cdot 2} = 2\sqrt{2}$, we have

$$\frac{-4 + \sqrt{8}}{2} = \frac{-4 + 2\sqrt{2}}{2}$$

$$= \frac{2 \cdot (-2 + \sqrt{2})}{2}$$ Factor the numerator and denominator.

$$= -2 + \sqrt{2}$$ Divide numerator and denominator by 2.

b. Since $\sqrt{28} = \sqrt{4 \cdot 7} = 2\sqrt{7}$, we have

$$\frac{-8 + \sqrt{28}}{4} = \frac{-8 + 2\sqrt{7}}{4}$$

$$= \frac{2(-4 + \sqrt{7})}{2 \cdot 2}$$ Factor the numerator and denominator.

$$= \frac{-4 + \sqrt{7}}{2}$$ Divide numerator and denominator by 2.

PROBLEM 5

Simplify:

a. $\dfrac{-9 + \sqrt{18}}{3}$

b. $\dfrac{-12 + \sqrt{24}}{6}$

Answers to PROBLEMS

5. a. $-3 + \sqrt{2}$ **b.** $\dfrac{-6 + \sqrt{6}}{3}$

› Exercises 8.4

‹ A › Simplifying Radical Expressions   In Problems 1–36, simplify. Assume all variables represent positive real numbers.

1. $\sqrt{36} + \sqrt{100}$

2. $\sqrt{9} + \sqrt{25}$

3. $\sqrt{144 + 25}$

4. $\sqrt{24^2 + 10^2}$

5. $\sqrt{4} - \sqrt{36}$

6. $\sqrt{64} - \sqrt{121}$

7. $\sqrt{13^2 - 12^2}$

8. $\sqrt{17^2 - 8^2}$

9. $15\sqrt{10} + \sqrt{90}$

10. $8\sqrt{7} + \sqrt{28}$

11. $14\sqrt{11} - \sqrt{44}$

12. $3\sqrt{13} - \sqrt{52}$

13. $\sqrt[3]{54} - \sqrt[3]{8}$

14. $\sqrt[3]{81} - \sqrt[3]{16}$

15. $5\sqrt[3]{16} - 3\sqrt[3]{54}$

16. $\sqrt[3]{250} - \sqrt[3]{128}$

17. $\sqrt{\dfrac{9x^2}{x}}$

18. $\sqrt{\dfrac{16x^5}{x^2}}$

19. $\sqrt{\dfrac{81y^7}{16y^5}}$

20. $\sqrt{\dfrac{4x^3y^4}{3z}}$

21. $\sqrt{\dfrac{64a^4b^6}{3ab^4}}$

22. $\sqrt{\dfrac{25a^5b^6}{7a^2b^4c^2}}$

23. $\sqrt[3]{\dfrac{8b^6c^{10}}{27bc}}$

24. $\sqrt[3]{\dfrac{64ab^4}{125a^4b}}$

25. $\dfrac{\sqrt[3]{500}}{\sqrt[3]{2}}$

26. $\dfrac{\sqrt[3]{243}}{\sqrt[3]{3}}$

27. $\dfrac{6}{\sqrt[3]{9}}$

28. $\dfrac{7}{\sqrt[3]{2}}$

29. $(\sqrt{3} + 6\sqrt{5})(\sqrt{3} - 4\sqrt{5})$

30. $(\sqrt{5} + 6\sqrt{2})(\sqrt{5} - 3\sqrt{2})$

31. $(\sqrt{2} + 3\sqrt{3})(\sqrt{2} + 3\sqrt{3})$

32. $(3\sqrt{2} + \sqrt{3})(3\sqrt{2} + \sqrt{3})$

33. $(5\sqrt{2} - 3\sqrt{3})(5\sqrt{2} - \sqrt{3})$

34. $(7\sqrt{5} - 4\sqrt{2})(7\sqrt{5} - 4\sqrt{2})$

35. $(\sqrt{13} + 2\sqrt{2})(\sqrt{13} - 2\sqrt{2})$

36. $(\sqrt{17} + 3\sqrt{5})(\sqrt{17} - 3\sqrt{5})$

‹ B › Using Conjugates to Rationalize Denominators   In Problems 37–50, rationalize the denominator.

37. $\dfrac{3}{\sqrt{2} + 1}$

38. $\dfrac{5}{\sqrt{5} + 1}$

39. $\dfrac{4}{\sqrt{7} - 1}$

40. $\dfrac{6}{\sqrt{7} - 2}$

41. $\dfrac{\sqrt{2}}{2 + \sqrt{3}}$

42. $\dfrac{\sqrt{3}}{3 + \sqrt{2}}$

43. $\dfrac{\sqrt{5}}{2 - \sqrt{3}}$

44. $\dfrac{\sqrt{6}}{3 - \sqrt{5}}$

45. $\dfrac{\sqrt{5}}{\sqrt{2} + \sqrt{3}}$

46. $\dfrac{\sqrt{2}}{\sqrt{5} + \sqrt{3}}$

47. $\dfrac{\sqrt{3}}{\sqrt{5} - \sqrt{2}}$

48. $\dfrac{6}{\sqrt{6} - \sqrt{2}}$

49. $\dfrac{\sqrt{3} + \sqrt{2}}{\sqrt{3} - \sqrt{2}}$

50. $\dfrac{\sqrt{5} - \sqrt{2}}{\sqrt{5} + \sqrt{2}}$

‹ C › Reducing Fractions Involving Radicals by Factoring   In Problems 51–64, reduce the fraction.

51. $\dfrac{-8 + \sqrt{16}}{2}$

52. $\dfrac{-4 + \sqrt{36}}{4}$

53. $\dfrac{-6 - \sqrt{4}}{6}$

54. $\dfrac{-8 - \sqrt{16}}{8}$

55. $\dfrac{2 + 2\sqrt{3}}{6}$

56. $\dfrac{6 + 2\sqrt{7}}{8}$

57. $\dfrac{-2 + 2\sqrt{23}}{4}$

58. $\dfrac{-6 - 2\sqrt{6}}{4}$

59. $\dfrac{-6 + 3\sqrt{10}}{9}$

60. $\dfrac{-20 + 5\sqrt{10}}{15}$

61. $\dfrac{-8 + \sqrt{28}}{6}$

62. $\dfrac{15 - \sqrt{189}}{6}$

63. $\dfrac{-9 + \sqrt{243}}{6}$

64. $\dfrac{-6 + \sqrt{180}}{9}$

〉〉〉 Applications

The **golden mean** or **golden ratio,** is the number $\varphi = \dfrac{\sqrt{5}+1}{2}$. Here is an amazing property of the golden mean: **the golden mean φ equals its reciprocal $\left(\dfrac{1}{\varphi}\right)$ plus 1, that is, $\varphi = \dfrac{1}{\varphi} + 1$.** We will prove this, step by step, in Problems 65–69.

65. Find the reciprocal $\dfrac{1}{\varphi}$ of $\dfrac{\sqrt{5}+1}{2}$.

66. From Problem 65, $\dfrac{1}{\varphi} = \dfrac{2}{\sqrt{5}+1}$. Rationalize the denominator in $\dfrac{2}{\sqrt{5}+1}$.

67. From Problem 66, $\dfrac{1}{\varphi} = \dfrac{\sqrt{5}-1}{2}$. Find $\dfrac{\sqrt{5}-1}{2} + 1$.

68. From Problem 67, $\dfrac{\sqrt{5}-1}{2} + 1 = \dfrac{\sqrt{5}+1}{2}$. Rewrite $\dfrac{\sqrt{5}-1}{2} + 1 = \dfrac{\sqrt{5}+1}{2}$ substituting $\varphi = \dfrac{\sqrt{5}+1}{2}$ and $\dfrac{1}{\varphi} = \dfrac{\sqrt{5}-1}{2}$.

69. Write in words: $\varphi = \dfrac{1}{\varphi} + 1$. Use "The golden mean" for φ. Is that the statement we wanted to show?

In Problems 70–74, we will prove that "If we have a number x so that its reciprocal plus 1 is the number x," then that number must be

$$\varphi = \frac{\sqrt{5}+1}{2}$$

70. Let x be a real number. Translate: $\dfrac{1}{x} + 1 = x$.

71. Add $\dfrac{1}{x} + 1$

72. From Problem 71, $\dfrac{1}{x} + 1 = \dfrac{x+1}{x}$, substitute this result in $\dfrac{1}{x} + 1 = x$.

73. **a.** Use "cross products" to rewrite $\dfrac{x+1}{x} = x$ as an equivalent equation.

b. Rewrite the equation obtained in **a** with all the terms on the right and 0 on the left.

74. Verify that a solution of $x^2 - x - 1 = 0$ is $x = \dfrac{\sqrt{5}+1}{2}$. Suggestion: Find x^2, then find $-x$, then find $x^2 - x - 1$. What do you get?

〉〉〉 Using Your Knowledge

Simplifying Mach Numbers The Mach number M mentioned in the *Getting Started* is given by the equation

$$\sqrt{\frac{2}{\gamma}} \; \sqrt{\frac{P_2 - P_1}{P_1}}$$

75. Write this expression as a single radical.

76. Rationalize the denominator of the expression obtained in Problem 75.

❯❯❯ *Write On*

77. Write in your own words the procedure you use to simplify any expression containing radicals.

78. Explain the difference between rationalizing the denominator in an algebraic expression whose denominator has only one term involving a radical and one whose denominator has two terms, at least one of which involves a radical.

79. Suppose you wish to rationalize the denominator in the expression

$$\frac{1}{\sqrt{2} + 1}$$

and you decide to multiply numerator and denominator by $\sqrt{2} + 1$. Would you obtain a rational denominator? What should you multiply by?

❯❯❯ *Concept Checker*

Fill in the blank(s) with the correct word(s), phrase, or mathematical statement.

80. The **conjugate** of $a\sqrt{b} + c\sqrt{d}$ is _____.

81. $\left(a\sqrt{b} + c\sqrt{d}\right)\left(a\sqrt{b} - c\sqrt{d}\right) = $ _____.

$a\sqrt{b} + c\sqrt{d}$ $a\sqrt{b} - c\sqrt{d}$

$a^2 b - c^2 d$ $a^2 b + c^2 d$

❯❯❯ *Mastery Test*

Simplify:

82. $\dfrac{4 + \sqrt{36}}{8}$

83. $\dfrac{-4 + \sqrt{28}}{2}$

84. $\dfrac{-6 - \sqrt{72}}{2}$

85. $\dfrac{3}{\sqrt{2} + 2}$

86. $\dfrac{\sqrt{2} + \sqrt{3}}{\sqrt{5} - \sqrt{2}}$

87. $\dfrac{3}{\sqrt[3]{2}}$

88. $\dfrac{5}{\sqrt[3]{9}}$

89. $\dfrac{\sqrt[3]{256}}{\sqrt[3]{2}}$

90. $\dfrac{\sqrt{12x^4}}{\sqrt{3x^2}}$

91. $\dfrac{\sqrt[3]{16a^5}}{\sqrt[3]{2a^3}}$

92. $\left(\sqrt{7} + \sqrt{3}\right)\left(\sqrt{7} - \sqrt{3}\right)$

93. $\left(\sqrt{5} + 2\sqrt{3}\right)\left(\sqrt{5} - 3\sqrt{3}\right)$

94. $\left(\sqrt{2} + \sqrt{3}\right)\left(3\sqrt{2} - 5\sqrt{3}\right)$

❯❯❯ *Skill Checker*

Find the square of each radical expression:

95. $\sqrt{x - 1}$

96. $\sqrt{x + 7}$

97. $2\sqrt{x}$

98. $-3\sqrt{y}$

99. $\sqrt{x^2 + 2x + 1}$

100. $\sqrt{x^2 - 2x + 7}$

Factor completely:

101. $x^2 - 3x$

102. $x^2 + 4x$

103. $x^2 - 3x + 2$

104. $x^2 + 4x + 3$

8.5   Applications: Solving Radical Equations

▶ Objectives

A ▷ Solve equations with one square root term containing the variable.

B ▷ Solve equations with two square root terms containing the variable.

C ▷ Solving applications involving radical equations.

▶ To Succeed, Review How To . . .

1. Square a radical expression (pp. 615–616).
2. Square a binomial (pp. 367–369).
3. Solve quadratic equations by factoring (pp. 440–446).

▶ Getting Started

Your Weight and Your Life

Has your doctor said that you are a little bit overweight? What does that mean? Can it be quantified? The "threshold weight" T (in pounds) for a man between 40 and 49 years of age is defined as "the crucial weight above which the mortality risk rises astronomically." In plain language, this means that if you get too fat, you are almost certainly going to die as a result! The formula linking T and the height h in inches is

$$12.3 \sqrt[3]{T} = h$$

Can you solve for T in this equation? To start, we divide both sides of the equation by 12.3 so that the radical term $\sqrt[3]{T}$ is *isolated* (by itself) on one side of the equation. We obtain

$$\sqrt[3]{T} = \frac{h}{12.3}$$

Now we can cube each side of the equation to get

$$(\sqrt[3]{T})^3 = \left(\frac{h}{12.3}\right)^3$$

or

$$T = \left(\frac{h}{12.3}\right)^3$$

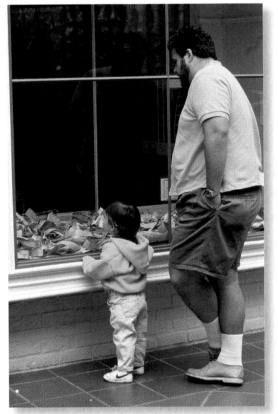

Thus, if a 40-year-old man is 73.8 inches tall, his threshold weight should be

$$T = \left(\frac{73.8}{12.3}\right)^3 = 6^3 = 216 \text{ lb}$$

In this section we shall use a new technique to solve equations involving radicals—raising both sides of an equation to a power. We did just that when we *cubed* both sides of the equation

$$\sqrt[3]{T} = \frac{h}{12.3}$$

so we could solve for T. We will use this method of solving radical equations in this section.

A ⟩ Solving Equations with One Square Root Term Containing the Variable

The equations $12.3\sqrt[3]{T} = h$, $\sqrt{x + 1} = 2$ and $\sqrt{x + 1} - x = -1$ are examples of **radical equations**. A **radical equation** has variables in one or more radicals.

The properties of equality studied in Chapter 2 are not enough to solve equations such as $\sqrt{x + 1} = 2$. A new property that we need to solve these types of equations is stated as follows.

RAISING BOTH SIDES OF AN EQUATION TO A POWER

If both sides of the equation $A = B$ are *squared,* all solutions of $A = B$ are *among* the solutions of the new equation $A^2 = B^2$.

Note that this property can yield a new equation that has *more* solutions than the original equation. For example, the equation $x = 3$ has one solution, 3. If we square both sides of the equation $x = 3$, we have

$$x^2 = 3^2$$
$$x^2 = 9$$

which has *two* solutions, $x = 3$ and $x = -3$. The -3 *does not* satisfy the original equation $x = 3$, and it is called an **extraneous** solution. Because of this, we must check our answers carefully when we solve equations with radicals by substituting the answers in the *original* equation and discarding any extraneous solutions.

EXAMPLE 1 Solving equations in which the radical expression is isolated

Solve:

a. $\sqrt{x + 3} = 4$

b. $\sqrt{x + 3} = x + 3$

SOLUTION

a. We shall proceed in steps.

1. $\sqrt{x + 3} = 4$ The square root term is isolated.

2. $\left(\sqrt{x + 3}\right)^2 = 4^2$ Square each side.

3. $x + 3 = 16$ Simplify.

4. $x = 13$ Subtract 3.

5. Now we check this answer in the original equation:

$$\sqrt{13 + 3} \stackrel{?}{=} 4$$ Substitute $x = 13$ in the original equation.

$$\sqrt{16} = 4$$ A true statement

Thus, $x = 13$ is the only solution of $\sqrt{x + 3} = 4$.

PROBLEM 1

Solve:

a. $\sqrt{x + 1} = 5$

b. $\sqrt{x + 1} = x + 1$

(continued)

Answers to PROBLEMS

1. a. 24   **b.** -1; 0

b. We again proceed in steps.

1. $\sqrt{x + 3} = x + 3$ The square root term is isolated.

2. $(\sqrt{x + 3})^2 = (x + 3)^2$ Square each side.

3. $x + 3 = x^2 + 6x + 9$ Simplify.

4. $0 = x^2 + 5x + 6$ Subtract $x + 3$.

    $0 = (x + 3)(x + 2)$ Factor.

    $x + 3 = 0$ or $x + 2 = 0$ Set each factor equal to zero.

    $x = -3$ $x = -2$ Solve each equation.

Thus, the proposed solutions are -3 and -2.

5. We check these proposed solutions in the original equation.

If $x = -3$,

$$\sqrt{-3 + 3} \stackrel{?}{=} -3 + 3$$
$$\sqrt{0} = 0 \qquad \text{True.}$$

If $x = -2$,

$$\sqrt{-2 + 3} \stackrel{?}{=} -2 + 3$$
$$\sqrt{1} = 1 \qquad \text{True.}$$

Thus, -3 and -2 are the solutions of $\sqrt{x + 3} = x + 3$.

EXAMPLE 2   Solving equations by first isolating the radical expression

Solve: $\sqrt{x + 1} - x = -1$

PROBLEM 2

Solve: $\sqrt{x + 3} - x = -3$

SOLUTION   First we must isolate the radical term $\sqrt{x + 1}$ by adding x to both sides of the equation.

$\sqrt{x + 1} - x = -1$ Given

1. $\sqrt{x + 1} = x - 1$ Add x.

2. $(\sqrt{x + 1})^2 = (x - 1)^2$ Square each side.

3. $x + 1 = x^2 - 2x + 1$ Simplify.

4. $0 = x^2 - 3x$ Subtract $x + 1$.

    $0 = x(x - 3)$ Factor.

    $x = 0$ or $x - 3 = 0$ Set each factor equal to zero.

    $x = 0$ $x = 3$ Solve each equation.

Thus, the proposed solutions are 0 and 3.

5. The check is as follows:

If $x = 0$,

$$\sqrt{0 + 1} - 0 \stackrel{?}{=} -1$$
$$\sqrt{1} - 0 \stackrel{?}{=} -1 \qquad \text{False.}$$

If $x = 3$,

$$\sqrt{3 + 1} - 3 \stackrel{?}{=} -1$$
$$\sqrt{4} - 3 = -1 \qquad \text{True.}$$

Thus, the equation $\sqrt{x + 1} - x = -1$ has *one* solution, 3.

Answers to PROBLEMS

2. 6

We can now generalize the steps we've been using to solve equations that have one square root term containing the variable.

> **PROCEDURE**
>
> **Solving Radical Equations**
> 1. Isolate the square root terms containing the variable.
> 2. Square both sides of the equation.
> 3. Simplify and repeat steps 1 and 2 if there is a square root term containing the variable.
> 4. Solve the resulting linear or quadratic equation.
> 5. Check all proposed solutions in the original equation.

B › Solving Equations with Two Square Root Terms Containing the Variable

The equations $\sqrt{y+4} = \sqrt{2y+3}$ and $\sqrt{y+7} - 3\sqrt{2y-3} = 0$ are different from the equations we have just solved because they have *two* square root terms containing the variable. However, we can still solve them using the five-step procedure.

EXAMPLE 3   Solving equations using the five-step procedure

Solve:

a. $\sqrt{y+4} = \sqrt{2y+3}$

b. $\sqrt{y+7} - 3\sqrt{2y-3} = 0$

SOLUTION

a. Since the square root terms containing the variable are isolated, we first square each side of the equation and then solve for y.

1. $\sqrt{y+4} = \sqrt{2y+3}$ The radicals are isolated.

2. $\left(\sqrt{y+4}\right)^2 = \left(\sqrt{2y+3}\right)^2$ Square both sides.

3. $y + 4 = 2y + 3$ Simplify.

4. $4 = y + 3$ Subtract y.
            $1 = y$ Subtract 3.

Thus, the proposed solution is 1.

5. Let's check this: If $y = 1$,

$$\sqrt{1+4} \stackrel{?}{=} \sqrt{2 \cdot 1 + 3}$$
$$\sqrt{5} = \sqrt{5}$$ True.

Thus, the solution of $\sqrt{y+4} = \sqrt{2y+3}$ is 1.

PROBLEM 3

Solve:

a. $\sqrt{x+4} = \sqrt{2x+1}$

b. $\sqrt{x+3} - 3\sqrt{2x-11} = 0$

(continued)

Answers to PROBLEMS

3. a. 3   **b.** 6

b. We start by isolating the square root terms containing the variable by adding $3\sqrt{2y-3}$ to both sides.

$$\sqrt{y+7} - 3\sqrt{2y-3} = 0 \qquad \text{Given}$$

1. $\qquad\qquad \sqrt{y+7} = 3\sqrt{2y-3} \qquad \text{Add } 3\sqrt{2y-3}.$

2. $\qquad\qquad (\sqrt{y+7})^2 = (3\sqrt{2y-3})^2 \qquad \text{Square both sides.}$

3. $\qquad\qquad y+7 = 3^2(2y-3) \qquad \text{Simplify.}$

4. $\qquad\qquad y+7 = 9(2y-3) \qquad \text{Since } 3^2 = 9$

$\qquad\qquad\qquad y+7 = 18y-27 \qquad \text{Simplify.}$

$\qquad\qquad\qquad 7 = 17y-27 \qquad \text{Subtract } y.$

$\qquad\qquad\qquad 34 = 17y \qquad \text{Add 27.}$

$\qquad\qquad\qquad 2 = y \qquad \text{Divide by 17.}$

5. And our check: If $y = 2$,

$$\sqrt{2+7} - 3\sqrt{2 \cdot 2 - 3} \stackrel{?}{=} 0$$

$$\sqrt{9} - 3\sqrt{1} \stackrel{?}{=} 0$$

$$3 - 3 = 0 \qquad \text{True.}$$

Thus, the solution of $\sqrt{y+7} - 3\sqrt{2y-3} = 0$ is 2.

C › Applications Involving Radical Equations

Let's see how square roots can be used in a real-world application.

EXAMPLE 4 Cell phones and radicals

If you have a cell phone, how long are your calls? According to the Cellular Telecommunications Industry Association, the average length L of a local call in 2000 was 2.56 minutes, and can be approximated by $L = \sqrt{t+5}$, where L is the length of the call in minutes t years after 2000. In what year would you expect the average length of a call to be 3 minutes?

SOLUTION   To predict when the length of a call will be 3 minutes, we have to find t when $L = 3$. Thus, we have to solve the equation $\sqrt{t+5} = 3$. We will use our five-step procedure.

1. $\quad \sqrt{t+5} = 3 \qquad$ The radicals are isolated.

2. $(\sqrt{t+5})^2 = 3^2 \qquad$ Square both sides.

3. $\quad t+5 = 9 \qquad$ Simplify.

4. $\qquad t = 4 \qquad$ Subtract 5 from both sides.

5. We check this: If $t = 4$,

$$\sqrt{4+5} = 3 \qquad \text{True.}$$

Thus, 4 years after 2000—that is, in 2004—the average length of a cellular call was 3 minutes.

A recent study by Bridgeratings.com indicated that the average length of cell phone calls in 2005 and 2006 was 3.09 and 3.29 minutes, respectively, and increasing!

PROBLEM 4

According to the information in Example 4 when would you expect the average length of a call to be 4 min?

Answers to PROBLEMS

4. 11 years after 2000, that is, in the year 2011.

› Exercises 8.5

〈 A 〉 Solving Equations with One Square Root Term Containing the Variable   In Problems 1–20, solve the given equations.

1. $\sqrt{x} = 4$

2. $\sqrt{x} = -3$

3. $\sqrt{x-1} = -2$

4. $\sqrt{x+1} = 2$

5. $\sqrt{y} - 2 = 0$

6. $\sqrt{y} + 3 = 0$

7. $\sqrt{y+1} - 3 = 0$

8. $\sqrt{y-1} + 2 = 0$

9. $\sqrt{x+1} = x - 5$

10. $\sqrt{x+14} = x + 2$

11. $\sqrt{x+4} = x + 2$

12. $\sqrt{x+9} = x - 3$

13. $\sqrt{x-1} - x = -3$

14. $\sqrt{x-2} - x = -4$

15. $y - 10 - \sqrt{5y} = 0$

16. $y - 3 - \sqrt{4y} = 0$

17. $\sqrt{y} + 20 = y$

18. $\sqrt{y} + 12 = y$

19. $4\sqrt{y} = y + 3$

20. $6\sqrt{y} = y + 5$

〈 B 〉 Solving Equations with Two Square Root Terms Containing the Variable   In Problems 21–30, solve the given equation.

21. $\sqrt{y+3} = \sqrt{2y-3}$

22. $\sqrt{y+7} = \sqrt{3y+3}$

23. $\sqrt{3x+1} = \sqrt{2x+6}$

24. $\sqrt{4x-3} = \sqrt{3x-2}$

25. $2\sqrt{x+5} = \sqrt{8x+4}$

26. $3\sqrt{x+2} = 2\sqrt{x+7}$

27. $\sqrt{4x-1} - \sqrt{x+10} = 0$

28. $\sqrt{3x+6} - \sqrt{5x+4} = 0$

29. $\sqrt{3y-2} - \sqrt{2y+3} = 0$

30. $\sqrt{5y+7} - \sqrt{3y+11} = 0$

〈 C 〉 Applications Involving Radical Equations

31. *Radius of a sphere*   The radius r of a sphere is given by the equation

$$r = \sqrt{\frac{S}{4\pi}}$$

where S is the surface area. If the radius of a sphere is 2 feet, what is its surface area? Use $\pi \approx 3.14$.

32. *Radius of a cone*   The radius r of a cone is given by the equation

$$r = \sqrt{\frac{3V}{\pi h}}$$

where V is the volume of the cone and h is its height. If a 10-centimeter-high ice cream cone has a radius of 2 centimeters, what is the volume of the ice cream in the cone? Use $\pi \approx 3.14$ and round to one decimal place.

33. *Time to fall*   The time t (in seconds) it takes a body to fall d feet is given by the equation

$$t = \sqrt{\frac{d}{16}}$$

How far would a body fall in 3 seconds?

34. *Velocity of a falling body*   After traveling d feet, the velocity v (in feet per second) of a falling body starting from rest is given by the equation $v = \sqrt{64d}$. If a body that started from rest is traveling at 44 feet per second, how far has it fallen?

35. *Length of pendulum cycle*   A pendulum of length L feet takes

$$t = 2\pi\sqrt{\frac{L}{32}} \text{ (seconds)}$$

to go through a complete cycle. If a pendulum takes 2 seconds to go through a complete cycle, how long is the pendulum? Use $\pi \approx \frac{22}{7}$ and round to two decimal places.

36. *Vehicle emissions*   According to Environmental Protection Agency (EPA) figures, the estimated amount A of particulate matter (in short tons) emitted by transportation sources (cars, buses, and so on) is modeled by the equation $A = \sqrt{2 + 0.2y}$, where y is the number of years after 1984. In what year would the amount of particulate matter emitted by transportation sources reach 2 short tons?

37. *Sulfur oxide emissions*   According to EPA figures, the amount A of sulfur oxides (in short tons) emitted by transportation sources is modeled by the equation $A = \sqrt{1 + 0.04y}$, where y is the number of years after 1989. In what year would the amount of sulfur oxides emitted by transportation sources reach 1.2 short tons?

38. *Distance seen from tall buildings*   The maximum distance d in kilometers you can see from a tall building is modeled by the equation $d = 110\sqrt{h}$, where h is the height of the building in kilometers. If the maximum distance you can see from the tallest building in the world is 77 kilometers, how high is the building? Write your answer in decimal form.

39. *Speed* The speed v (in miles per hour) a car can travel in a concrete highway curve without skidding is modeled by the equation $v = \sqrt{9r}$, where r is the radius of the curve in feet. What is the radius r if the speed limit is 30 miles per hour?

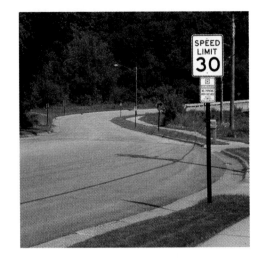

40. *Radius of a curve* Using the equation in Problem 39, find the radius r of the curve when the speed limit is 45 miles per hour.

41. *Terminal velocity* If air resistance is neglected, the terminal velocity v of a body (in meters per second) depends on the height h of the body above the ground and is given by the equation $v = \sqrt{20h + v_0}$, where $v_0 = 5\text{m/sec}$ is the initial velocity of the object. If the terminal velocity v of an object is 25 meters per second, how far has the object fallen?

42. *Velocity* If the velocity v of an object (in feet per second) is modeled by the equation $v = \sqrt{64h + v_0}$, where v_0 is the initial velocity and h is the height (in feet), how far has an object with an initial velocity of 16 feet per second fallen when the velocity of the object is 20 feet per second?

> > > ## Using Your Knowledge

Working with Higher Roots Step 2 in the five-step procedure for solving radical equations directs us to "square each side of the equation." Thus, to solve $\sqrt{x} = 2$, we square each side of the equation to obtain 4 as our solution. If we have an equation of the form $\sqrt[3]{x} = 2$, we can **cube** each side of the equation to obtain $2^3 = 8$. You can check that 8 is the correct solution by substituting 8 for x in $\sqrt[3]{x} = 2$ to obtain $\sqrt[3]{8} = 2$, a true statement.

Use this knowledge to solve the following equations.

43. $\sqrt[3]{x} = 3$ **44.** $\sqrt[3]{x} = -4$ **45.** $\sqrt[3]{x + 1} = 2$ **46.** $\sqrt[3]{x - 1} = -2$

Generalize the idea used in Problems 43–46 to solve the following problems.

47. $\sqrt[4]{x} = 2$ **48.** $\sqrt[4]{x + 1} = 1$ **49.** $\sqrt[4]{x - 1} = -2$ **50.** $\sqrt[4]{x - 1} = 2$

> > > ## Write On

51. Consider the equation $\sqrt{x} + 3 = 0$. What should the first step be in solving this equation? If you follow the rest of the steps in the procedure to solve radical equations, you should conclude that this equation has no real-number solution. Can you write an explanation of why this is so *after the first step* in the procedure?

52. Consider the equation $\sqrt{x + 1} = -\sqrt{x + 2}$. Write an explanation of why this equation has no real-number solutions; then follow the procedure given in the text to prove that this is the case.

53. Write your explanation of a "proposed" solution for solving equations involving radicals. Why do you think they are called "proposed" solutions?

54. Why is it necessary to check proposed solutions in the original equation when solving equations involving radicals?

> > > ## Concept Checker

Fill in the blank(s) with the correct word(s), phrase, or mathematical statement.

55. If **both** sides of the equation $A = B$ are **squared**, all **solutions of** $A = B$ are _____ the **solutions** of the new equation $A^2 = B^2$.

56. An _____ **solution** is a **possible (proposed)** solution that **does not** satisfy the **original** equation.

equal to   extraneous

among   extra

⟩⟩⟩ *Mastery Test*

57. The total monthly cost C (in millions of dollars) of running daily flights between two cities is modeled by the equation $C = \sqrt{0.3p + 1}$, where p is the number of passengers in thousands. If the monthly cost C for a certain month was $2 million, what was the number of passengers for the month?

Solve if possible:

58. $\sqrt{y + 6} = \sqrt{2y + 3}$

59. $\sqrt{3y + 10} = \sqrt{y + 14}$

60. $\sqrt{4x - 3} - \sqrt{x + 3} = 0$

61. $\sqrt{5x + 1} - \sqrt{x + 9} = 0$

62. $\sqrt{x + 1} - x = -5$

63. $\sqrt{x + 2} - x = -4$

64. $\sqrt{x + 2} = 3$

65. $\sqrt{x - 1} = -2$

66. $\sqrt{x - 3} = x - 3$

67. $\sqrt{x - 2} = x - 2$

⟩⟩⟩ *Skill Checker*

Find:

68. $\sqrt{169}$

69. $\sqrt{49}$

70. $\sqrt{\dfrac{81}{16}}$

71. $\sqrt{\dfrac{5}{16}}$

72. $\sqrt{\dfrac{7}{4}}$

⟩ Collaborative Learning

The Golden Ratio

Form two groups of students. Each group should find a different picture of the Mona Lisa and compute the ratio of length to width of the frame. Are the answers about the same for all groups?

Each group should compare their answer to the golden ratio, $\dfrac{\sqrt{5} + 1}{2} \approx 1.618$. Are the answers close? Let us now construct a golden rectangle:

Group 1   Start with a 1-by-1 square. What is the ratio of length to width?

Group 2   Add another 1-by-1 square to the right of the original square. What is the ratio of length to width?

Group 1   Add a 2-by-2 square under the previous rectangle. What is the ratio of length to width?

Group 2   Add a 3-by-3 square to the right of the previous rectangle. What is the ratio of length to width?

Group 1   Add a 5-by-5 square under the previous rectangle. What is the ratio of length to width?

Are the ratios approximating the golden ratio?

Now, let us take some measurements. Group 1 will take the smaller measurements (x) and Group 2 the larger measurements (y). Record all measurements to the nearest tenth of a centimeter.

Group 1 (x) Smaller Measurement	Group 2 (y) Larger Measurement
Height of belly button from the floor	Total height
Belly button to top of head	Belly button height from floor
Chin to top of head	Belly button to chin

Look at the ratio of each individual measurement y to x in each row. Are the ratios close to the golden ratio? Compute the average of the individual measurements in each row (for example, the average of the heights of belly buttons in row 1 and the average of total heights in row 1). Look at the ratios of the averages of the y (larger) measurements to the averages of the x (smaller) measurements (for example, average of total heights to average of heights of belly buttons). Are they now closer to the golden ratio?

1. Write a paragraph about Christoff Rudolff, the inventor of the square root sign, and indicate where the square root sign was first used.

2. A Chinese mathematician discovered the relationship between extracting roots and the array of binomial coefficients in Pascal's triangle. Write a paragraph about this mathematician.

3. Name the four Chinese algebraists who discovered the relationship between root extraction and the coefficients of Pascal's triangle and who then extended the idea to solve higher-than-cubic equations.

4. Write a paragraph about Newton's method for finding square roots and illustrate its use by finding the square root of 11, for example.

> Summary Chapter 8

Section	Item	Meaning	Example
8.1A	$\sqrt{a}$ $-\sqrt{a}$	$\sqrt{a} = b$ is equivalent to $b^2 = a$. $-\sqrt{a} = b$ is equivalent to $b^2 = a$.	$\sqrt{4} = 2$ because $2^2 = 4$. $-\sqrt{4} = -2$ because $(-2)^2 = 4$.
8.1C	$\sqrt{a}$ if a is negative	If a is negative, $\sqrt{a}$ is not a real number.	$\sqrt{-16}$ and $\sqrt{-7}$ are not real numbers.
8.1D	$\sqrt[n]{a}$	The nth root of a	$\sqrt[3]{8} = 2$, $\sqrt[4]{81} = 3$
8.2A	Product rule for radicals	$\sqrt{a \cdot b} = \sqrt{a} \cdot \sqrt{b}$	$\sqrt{16 \cdot 9} = \sqrt{16} \cdot \sqrt{9} = 4 \cdot 3 = 12$ $\sqrt{18} = \sqrt{9 \cdot 2} = \sqrt{9} \cdot \sqrt{2} = 3\sqrt{2}$
8.2B	Quotient rule for radicals	$\sqrt{\dfrac{a}{b}} = \dfrac{\sqrt{a}}{\sqrt{b}}$	$\sqrt{\dfrac{9}{4}} = \dfrac{\sqrt{9}}{\sqrt{4}} = \dfrac{3}{2}$
8.2C	$\sqrt{a^2} = |a|$	The square root of a real number a is the absolute value of a.	$\sqrt{5^2} = |5|$, $\sqrt{(-3)^2} = |-3| = 3$
8.2D	Properties of radicals	For real numbers where the roots exist, $\sqrt[n]{a \cdot b} = \sqrt[n]{a} \ \sqrt[n]{b}$ $\sqrt[n]{\dfrac{a}{b}} = \dfrac{\sqrt[n]{a}}{\sqrt[n]{b}}$	$\sqrt[4]{80} = \sqrt[4]{2^4 \cdot 5} = 2\sqrt[4]{5}$ $\sqrt[3]{\dfrac{8}{27}} = \dfrac{\sqrt[3]{8}}{\sqrt[3]{27}} = \dfrac{\sqrt[3]{2^3}}{\sqrt[3]{3^3}} = \dfrac{2}{3}$
8.3C	Rationalizing the denominator	Multiply the numerator and denominator of the fraction by the square root in the denominator.	$\dfrac{1}{\sqrt{3}} = \dfrac{1 \cdot \sqrt{3}}{\sqrt{3} \cdot \sqrt{3}} = \dfrac{\sqrt{3}}{3}$
8.4B	Conjugate	$a + b$ and $a - b$ are conjugates.	To rationalize the denominator of $\dfrac{1}{\sqrt{5} - \sqrt{3}}$, multiply the numerator and denominator of $\dfrac{1}{\sqrt{5} - \sqrt{3}}$ by the conjugate of $\sqrt{5} - \sqrt{3}$, which is $\sqrt{5} + \sqrt{3}$.
8.5A, B	Raising both sides of an equation to a power	If both sides of the equation $A = B$ are squared, all solutions of $A = B$ are among the solutions of the new equation $A^2 = B^2$.	If both sides of the equation $\sqrt{x} = 3$ are squared, all solutions of $\sqrt{x} = 3$ are among the solutions of $(\sqrt{x})^2 = 3^2$, that is, $x = 9$.

〉Review Exercises Chapter 8

(If you need help with these exercises, look in the sections indicated in brackets.)

1. ⟨ **8.1A, C** ⟩ *Find the root if possible.*

    a. $\sqrt{81}$ b. $\sqrt{-64}$

    c. $\sqrt{\dfrac{36}{25}}$

2. ⟨ **8.1A, C** ⟩ *Find the root if possible.*

    a. $-\sqrt{36}$ b. $-\sqrt{\dfrac{64}{25}}$

    c. $\sqrt{-\dfrac{9}{4}}$

3. ⟨ **8.1B** ⟩ *Find the square of each radical expression.*

    a. $\sqrt{8}$ b. $\sqrt{25}$ c. $\sqrt{17}$

4. ⟨ **8.1B** ⟩ *Find the square of each radical expression.*

    a. $-\sqrt{36}$ b. $-\sqrt{17}$ c. $-\sqrt{64}$

5. ⟨ **8.1B** ⟩ *Find the square of each radical expression.*

    a. $\sqrt{x^2 + 1}$ b. $\sqrt{x^2 + 4}$

    c. $-\sqrt{x^2 + 5}$

6. ⟨ **8.1C** ⟩ *Find and classify each number as rational, irrational, or not a real number. Approximate the irrational numbers using a calculator.*

    a. $\sqrt{11}$ b. $-\sqrt{25}$

    c. $\sqrt{-9}$

7. ⟨ **8.1C** ⟩ *Classify each number as rational, irrational, or not a real number.*

    a. $\sqrt{\dfrac{9}{4}}$ b. $-\sqrt{\dfrac{9}{4}}$

    c. $\sqrt{-\dfrac{9}{4}}$

8. ⟨ **8.1D** ⟩ *Find each root if possible.*

    a. $\sqrt[3]{64}$ b. $\sqrt[3]{-8}$ c. $-\sqrt[4]{81}$

9. ⟨ **8.1D** ⟩ *Find each root if possible.*

    a. $\sqrt[4]{16}$ b. $-\sqrt[4]{16}$ c. $\sqrt[4]{-16}$

10. ⟨ **8.1E** ⟩ *If an object is dropped from a distance d (in feet), it takes*

$$t = \sqrt{\dfrac{d}{16}}$$

   seconds to reach the ground. How long does it take an object to reach the ground if it is dropped from:

    a. 121 feet   b. 144 feet   c. 169 feet

11. ⟨ **8.2A** ⟩ *Simplify.*

    a. $\sqrt{32}$ b. $\sqrt{48}$ c. $\sqrt{196}$

12. ⟨ **8.2A** ⟩ *Multiply.*

    a. $\sqrt{3} \cdot \sqrt{7}$ b. $\sqrt{12} \cdot \sqrt{3}$ c. $\sqrt{5}\,\sqrt{y}, y > 0$

13. ⟨ **8.2B** ⟩ *Simplify.*

    a. $\sqrt{\dfrac{3}{16}}$ b. $\sqrt{\dfrac{5}{36}}$ c. $\sqrt{\dfrac{9}{4}}$

14. ⟨ **8.2B** ⟩ *Simplify.*

    a. $\dfrac{\sqrt{8}}{\sqrt{2}}$ b. $\dfrac{\sqrt{21}}{\sqrt{3}}$ c. $\dfrac{6\sqrt{50}}{2\sqrt{10}}$

15. ⟨ **8.2C** ⟩ *Simplify. Assume all variables represent positive real numbers.*

    a. $\sqrt{36x^2}$ b. $\sqrt{100y^4}$ c. $\sqrt{81n^8}$

16. ⟨ **8.2C** ⟩ *Simplify. Assume all variables represent positive real numbers.*

    a. $\sqrt{72y^{10}}$ b. $\sqrt{147z^8}$ c. $\sqrt{48x^{12}}$

17. ⟨ **8.2C** ⟩ *Simplify. Assume all variables represent positive real numbers.*

    a. $\sqrt{y^{15}}$ b. $\sqrt{y^{13}}$ c. $\sqrt{50n^7}$

18. ⟨ **8.2D** ⟩ *Simplify.*

    a. $\sqrt[3]{24}$ b. $\sqrt[3]{\dfrac{8}{27}}$ c. $\sqrt[3]{-\dfrac{125}{64}}$

19. ⟨ **8.2D** ⟩ *Simplify.*

    a. $\sqrt[4]{81}$ b. $\sqrt[4]{48}$ c. $\sqrt[4]{80}$

20. ⟨ **8.3A** ⟩ *Add and simplify.*

    a. $7\sqrt{3} + 8\sqrt{3}$ b. $\sqrt{32} + 5\sqrt{2}$ c. $\sqrt{12} + \sqrt{48}$

21. ⟨ **8.3A** ⟩ *Subtract and simplify.*

    a. $9\sqrt{11} - 6\sqrt{11}$ **b.** $\sqrt{50} - 4\sqrt{2}$

    c. $\sqrt{108} - \sqrt{75}$

22. ⟨ **8.3B** ⟩ *Simplify.*

    a. $\sqrt{3}\left(\sqrt{20} - \sqrt{2}\right)$ **b.** $\sqrt{5}\left(\sqrt{5} - \sqrt{3}\right)$

    c. $\sqrt{7}\left(\sqrt{7} - \sqrt{98}\right)$

23. ⟨ **8.3C** ⟩ *Write with a rationalized denominator.*

    a. $\sqrt{\dfrac{5}{8}}$ **b.** $\sqrt{\dfrac{x^2}{50}}, \quad x > 0$ **c.** $\sqrt{\dfrac{y^2}{27}}, \quad y > 0$

24. ⟨ **8.4A** ⟩ *Simplify.*

    a. $\sqrt{32 + 4} - \sqrt{9}$ **b.** $\sqrt{18 + 7} - \sqrt{4}$

    c. $\sqrt{60 + 4} - \sqrt{16}$

25. ⟨ **8.4A** ⟩ *Simplify.*

    a. $8\sqrt{15} - \sqrt{3} \cdot \sqrt{5}$ **b.** $7\sqrt{6} - \sqrt{2} \cdot \sqrt{3}$

    c. $9\sqrt{14} - 2\sqrt{7} \cdot \sqrt{2}$

26. ⟨ **8.4A** ⟩ *Simplify.*

    a. $\dfrac{\sqrt[3]{162}}{\sqrt[3]{2}}$ **b.** $\dfrac{\sqrt[3]{135}}{\sqrt[3]{5}}$ **c.** $\dfrac{\sqrt[3]{192}}{\sqrt[3]{24}}$

27. ⟨ **8.4A** ⟩ *Simplify by rationalizing the denominator.*

    a. $\dfrac{7}{\sqrt[3]{4}}$ **b.** $\dfrac{5}{\sqrt[3]{9}}$ **c.** $\dfrac{9}{\sqrt[3]{25}}$

28. ⟨ **8.4A** ⟩ *Multiply and simplify.*

    a. $\left(\sqrt{3} + 3\sqrt{2}\right)\left(\sqrt{3} - 5\sqrt{2}\right)$

    b. $\left(\sqrt{7} + 3\sqrt{5}\right)\left(\sqrt{7} - 2\sqrt{5}\right)$

29. ⟨ **8.4A** ⟩ *Multiply and simplify.*

    a. $\left(\sqrt{7} + 2\sqrt{3}\right)\left(\sqrt{7} - 2\sqrt{3}\right)$

    b. $\left(\sqrt{11} + 3\sqrt{5}\right)\left(\sqrt{11} - 3\sqrt{5}\right)$

30. ⟨ **8.4B** ⟩ *Simplify by rationalizing the denominator.*

    a. $\dfrac{3}{\sqrt{3} + 1}$ **b.** $\dfrac{5}{\sqrt{2} - 1}$

31. ⟨ **8.4B** ⟩ *Simplify by rationalizing the denominator.*

    a. $\dfrac{7}{\sqrt{3} - \sqrt{2}}$ **b.** $\dfrac{2}{\sqrt{5} - \sqrt{2}}$

32. ⟨ **8.4C** ⟩ *Simplify.*

    a. $\dfrac{-8 + \sqrt{8}}{2}$ **b.** $\dfrac{-16 + \sqrt{12}}{4}$

33. ⟨ **8.5A** ⟩ *Solve.*

    a. $\sqrt{x + 2} = 3$ **b.** $\sqrt{x - 2} = -2$

34. ⟨ **8.5A** ⟩ *Solve.*

    a. $\sqrt{x + 5} = x - 1$ **b.** $\sqrt{x + 10} = x - 2$

35. ⟨ **8.5A** ⟩ *Solve.*

    a. $\sqrt{x + 4} - x = -2$ **b.** $\sqrt{x + 2} - x = -4$

36. ⟨ **8.5B** ⟩ *Solve.*

    a. $\sqrt{y + 5} = \sqrt{3y - 3}$

    b. $\sqrt{y + 5} = \sqrt{2y + 5}$

37. ⟨ **8.5B** ⟩ *Solve.*

    a. $\sqrt{y + 8} - 3\sqrt{2y - 1} = 0$

    b. $\sqrt{y + 9} - 3\sqrt{2y + 1} = 0$

38. ⟨ **8.5C** ⟩ *The total daily cost C (thousand dollars) of producing a certain product is given by the equation $C = \sqrt{0.2x + 1}$, where x is the number of items in hundreds. How many items were produced on a day in which the cost was:*

    a. 3(thousand)   **b.** 7(thousand)

(Answers on page 654)

Visit www.mhhe.com/bello to view helpful videos that provide step-by-step solutions to several of the problems below.

1. Find.

    a. $\sqrt{169}$ **b.** $-\sqrt{\dfrac{49}{81}}$

2. Find the square of each radical expression.

    a. $-\sqrt{121}$ **b.** $\sqrt{x^2 + 7}$

3. Classify each number as rational, irrational, or not a real number, and simplify if possible.

    a. $\sqrt{17}$ **b.** $-\sqrt{36}$

    c. $\sqrt{-100}$ **d.** $\sqrt{\dfrac{100}{49}}$

4. Find each root if possible.

    a. $\sqrt[4]{81}$ **b.** $-\sqrt[4]{625}$

    c. $\sqrt[3]{-8}$ **d.** $\sqrt[4]{-16}$

5. A diver jumps from a cliff 20 meters high. If the time t (in seconds) it takes an object dropped from a distance d (in meters) to reach the ground is given by the equation

$$t = \sqrt{\dfrac{d}{5}}$$

   how long does it take the diver to reach the water?

6. Simplify.

    a. $\sqrt{125}$ **b.** $\sqrt{54}$

7. Multiply.

    a. $\sqrt{3} \cdot \sqrt{11}$ **b.** $\sqrt{11} \cdot \sqrt{y}, y > 0$

8. Simplify.

    a. $\sqrt{\dfrac{7}{16}}$ **b.** $\dfrac{21\sqrt{50}}{7\sqrt{5}}$

9. Simplify.

    a. $\sqrt{144n^2}, n > 0$ **b.** $\sqrt{32y^7}, y > 0$

10. Simplify.

    a. $\sqrt[4]{96}$ **b.** $\sqrt[3]{\dfrac{-125}{8}}$

11. Simplify.

    a. $9\sqrt{13} + 7\sqrt{13}$ **b.** $14\sqrt{6} - 3\sqrt{6}$

12. Simplify.

    a. $\sqrt{28} + \sqrt{63}$ **b.** $\sqrt{40} + \sqrt{90} - \sqrt{160}$

13. Multiply and simplify.

    a. $\sqrt{3}\left(\sqrt{18} - \sqrt{5}\right)$ **b.** $\sqrt{5}\left(\sqrt{5} - \sqrt{7}\right)$

14. Write $\sqrt{\dfrac{3}{20}}$ with a rationalized denominator.

15. Write $\sqrt{\dfrac{y^2}{50}}, y > 0$ with a rationalized denominator.

16. Simplify.

    a. $8\sqrt{14} - \sqrt{7} \cdot \sqrt{2}$ **b.** $\dfrac{\sqrt{12x^3}}{\sqrt{4x^2}}, x > 0$

17. Simplify by rationalizing the denominator.

    a. $\dfrac{\sqrt[3]{500}}{\sqrt[3]{2}}$ **b.** $\dfrac{3}{\sqrt[3]{25}}$

18. Multiply and simplify.

    a. $\left(\sqrt{3} + 6\sqrt{2}\right)\left(\sqrt{3} - 2\sqrt{2}\right)$

    b. $\left(\sqrt{10} - 2\sqrt{20}\right)\left(\sqrt{10} + 2\sqrt{20}\right)$

19. Simplify by rationalizing the denominator.

    a. $\dfrac{11}{\sqrt{3} + 1}$ **b.** $\dfrac{2}{\sqrt{5} - \sqrt{2}}$

20. Simplify.

    a. $\dfrac{-6 + \sqrt{18}}{3}$ **b.** $\dfrac{-8 + \sqrt{8}}{4}$

21. Solve.

    a. $\sqrt{x + 1} = 2$ **b.** $\sqrt{x + 6} = x + 6$

22. Solve $\sqrt{x + 4} - x = 2$.

23. Solve $\sqrt{y + 3} = \sqrt{2y + 1}$.

24. Solve $\sqrt{y + 6} - 3\sqrt{2y - 5} = 0$.

25. The average length L of a long-distance call (in minutes) has been approximated by the equation $L = \sqrt{t} + 4$, where t is the number of years after 1995. In how many years would you expect the average length of a call to be 3 minutes?

⟩Answers to Practice Test **Chapter 8**

Answer		If You Missed	Review		
		Question	Section	Examples	Page
1. a. 13	**b.** $-\dfrac{7}{9}$	1	8.1	1	615
2. a. 121	**b.** $x^2 + 7$	2	8.1	2	616
3. a. Irrational	**b.** Rational; -6	3	8.1	3	616
c. Not a real number	**d.** Rational; $\dfrac{10}{7}$				
4. a. 3	**b.** -5	4	8.1	4	617
c. -2	**d.** Not a real number				
5. 2 sec		5	8.1	5	617
6. a. $5\sqrt{5}$	**b.** $3\sqrt{6}$	6	8.2	1	622
7. a. $\sqrt{33}$	**b.** $\sqrt{11y}$	7	8.2	2	622–623
8. a. $\dfrac{\sqrt{7}}{4}$	**b.** $3\sqrt{10}$	8	8.2	3	623
9. a. $12n$	**b.** $4y^3\sqrt{2y}$	9	8.2	4	624
10. a. $2\sqrt[4]{6}$	**b.** $\dfrac{-5}{2}$	10	8.2	5	624–625
11. a. $16\sqrt{13}$	**b.** $11\sqrt{6}$	11	8.3	1	629
12. a. $5\sqrt{7}$	**b.** $\sqrt{10}$	12	8.3	2	629
13. a. $3\sqrt{6} - \sqrt{15}$	**b.** $5 - \sqrt{35}$	13	8.3	3	630
14. $\dfrac{\sqrt{15}}{10}$		14	8.3	4	631
15. $\dfrac{y\sqrt{2}}{10}$		15	8.3	5	631
16. a. $7\sqrt{14}$	**b.** $\sqrt{3x}$	16	8.4	1	636
17. a. $5\sqrt[3]{2}$	**b.** $\dfrac{3\sqrt[3]{5}}{5}$	17	8.4	2	636
18. a. $-21 + 4\sqrt{6}$	**b.** -70	18	8.4	3	636–637
19. a. $\dfrac{11\sqrt{3} - 11}{2}$	**b.** $\dfrac{2\sqrt{5} + 2\sqrt{2}}{3}$	19	8.4	4	637–638
20. a. $-2 + \sqrt{2}$	**b.** $\dfrac{-4 + \sqrt{2}}{2}$	20	8.4	5	638
21. a. $x = 3$	**b.** $x = -5$ or $x = -6$	21	8.5	1	643–644
22. $x = 0$		22	8.5	2	644
23. $y = 2$		23	8.5	3	645–646
24. $y = 3$		24	8.5	3	645–646
25. $t = 5$ yr		25	8.5	4	646

> Cumulative Review **Chapters 1–8**

1. Add: $-\frac{2}{9} + \left(-\frac{1}{8}\right)$

2. Find: $(-3)^4$

3. Divide: $-\frac{1}{6} \div \left(-\frac{7}{12}\right)$

4. Evaluate $y \div 5 \cdot x - z$ for $x = 5$, $y = 50$, $z = 3$.

5. Simplify: $2x - (x + 4) - 2(x + 3)$

6. Write in symbols: The quotient of $(m + 3n)$ and p

7. Solve for x: $2 = 5(x - 3) + 1 - 4x$

8. Solve for x: $\frac{x}{2} - \frac{x}{3} = 1$

9. Graph: $-\frac{x}{6} + \frac{x}{2} \le \frac{x - 2}{2}$

10. Graph the point $C(-1, -3)$.

11. Determine whether the ordered pair $(-1, 3)$ is a solution of $4x - y = -1$.

12. Find x in the ordered pair $(x, 3)$ so that the ordered pair satisfies the equation $2x - 4y = -10$.

13. Graph: $5x + y = 5$

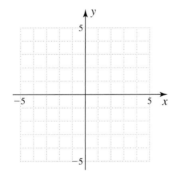

14. Graph: $4y - 8 = 0$

15. Find the slope of the line passing through the points $(0, 4)$ and $(6, 1)$.

16. What is the slope of the line $3x - 3y = -8$?

17. Find the pair of parallel lines.

(1) $15x - 12y = 4$

(2) $12y + 15x = 4$

(3) $-4y = -5x + 4$

18. Simplify: $(2x^4 y^{-3})^{-4}$

19. Write in scientific notation: 0.00000025

20. Divide and express the answer in scientific notation: $(2.72 \times 10^{-4}) \div (1.6 \times 10^4)$

21. Find (expand): $\left(4x^2 - \frac{1}{2}\right)^2$

22. Divide $(2x^3 - 7x^2 + x + 9)$ by $(x - 3)$.

23. Factor completely: $x^2 - 4x + 3$

24. Factor completely: $9x^2 - 27xy + 20y^2$

25. Factor completely: $4x^2 - 25y^2$

26. Factor completely: $-5x^4 + 5x^2$

27. Factor completely: $4x^3 - 8x^2 - 12x$

28. Factor completely: $3x^2 + 4x + 9x + 12$

29. Factor completely: $16kx^2 + 8kx + k$

30. Solve for x: $4x^2 + 17x = 15$

31. Write $\dfrac{2x}{5y}$ with a denominator of $15y^2$.

32. Reduce to lowest terms: $\dfrac{-16(x^2 - y^2)}{4(x - y)}$

33. Reduce to lowest terms: $\dfrac{x^2 + 4x - 21}{3 - x}$

34. Multiply: $(x - 8) \cdot \dfrac{x + 4}{x^2 - 64}$

35. Divide: $\dfrac{x + 5}{x - 5} \div \dfrac{x^2 - 25}{5 - x}$

36. Add: $\dfrac{7}{3(x + 5)} + \dfrac{5}{3(x + 5)}$

37. Subtract: $\dfrac{x + 4}{x^2 + x - 20} - \dfrac{x + 5}{x^2 - 16}$

38. Simplify: $\dfrac{\frac{2}{3x} + \frac{1}{2x}}{\frac{1}{x} + \frac{1}{4x}}$

39. Solve for x: $\dfrac{4x}{x - 4} + 1 = \dfrac{3x}{x - 4}$

40. Solve for x: $\dfrac{x}{x^2 - 4} + \dfrac{2}{x - 2} = \dfrac{1}{x + 2}$

41. Solve for x: $\dfrac{x}{x + 6} - \dfrac{1}{7} = \dfrac{-6}{x + 6}$

42. Solve for x: $1 + \dfrac{2}{x - 5} = \dfrac{20}{x^2 - 25}$

43. A van travels 100 miles on 4 gallons of gas. How many gallons will it need to travel 625 miles?

44. Solve for x: $\dfrac{x - 5}{5} = \dfrac{5}{4}$

45. Janet can paint a kitchen in 3 hours and James can paint the same kitchen in 4 hours. How long would it take to paint the kitchen if they worked together?

46. Find an equation of the line that passes through the points $(6, 4)$ and has slope $m = 5$.

47. Find an equation of the line having slope 4 and y-intercept 3.

48. Graph: $x - 5y < -5$

49. Graph:
$-y \geq -5x - 5$

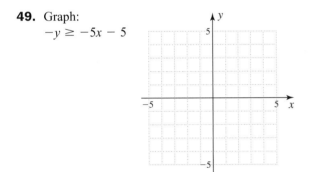

50. An enclosed gas exerts a pressure P on the walls of the container. This pressure is directly proportional to the temperature T of the gas. If the pressure is 3 lb/in.^{2} when the temperature is 240°F, find k.

51. If the temperature of a gas is held constant, the pressure P varies inversely as the volume V. A pressure of 1800 lb/in.^{2} is exerted by 6 ft^3 of air in a cylinder fitted with a piston. Find k.

52. Graph the system and find the solution (if possible):

$$x + 4y = 16$$
$$4y - x = 12$$

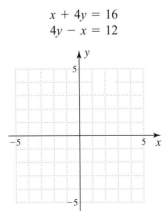

53. Graph the system and find the solution if possible:

$$y + 3x = -3$$
$$2y + 6x = -12$$

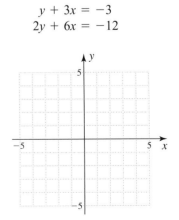

54. Solve by substitution (if possible):

$$x + 4y = -18$$
$$-2x - 8y = 32$$

55. Solve by substitution (if possible):

$$x + 3y = 10$$
$$-2x - 6y = -20$$

56. Solve the system (if possible):

$$x - 2y = 3$$
$$2x - y = 0$$

57. Solve the system (if possible):

$$5x + 4y = -18$$
$$-10x - 8y = 3$$

58. Solve the system (if possible):

$$4y + 3x = 11$$
$$6x + 8y = 22$$

59. Sara has $3.50 in nickels and dimes. She has three times as many dimes as nickels. How many nickels and how many dimes does she have?

60. The sum of two numbers is 215. Their difference is 85. What are the numbers?

61. Evaluate: $\sqrt[3]{-64}$

62. Simplify: $\sqrt[7]{(-3)^7}$

63. Simplify: $\sqrt{\dfrac{7}{243}}$

64. Simplify: $\sqrt[3]{128a^8b^9}$

65. Add and simplify: $\sqrt{48} + \sqrt{12}$

66. Perform the indicated operations:
$\sqrt[3]{2x}\left(\sqrt[3]{4x^2} - \sqrt[3]{81x}\right)$

67. Rationalize the denominator: $\dfrac{\sqrt{3}}{\sqrt{2p}}$

68. Find the product: $\left(\sqrt{125} + \sqrt{343}\right)\left(\sqrt{245} + \sqrt{175}\right)$

69. Rationalize the denominator: $\dfrac{\sqrt{x}}{\sqrt{x} - \sqrt{5}}$

70. Reduce: $\dfrac{6 + \sqrt{18}}{3}$

71. Solve: $\sqrt{x + 3} = -2$

72. Solve: $\sqrt{x - 6} - x = -6$

Chapter

9
nine

▶ **Quadratic Equations**

The Human Side of Algebra

The study of quadratic equations dates back to antiquity. Scores of clay tablets indicate that the Babylonians of 2000 B.C. were already familiar with the quadratic formula that you will study in this chapter. Their solutions using the formula are actually verbal instructions that amounted to using the formula

$$x = \sqrt{\left(\frac{a}{2}\right)^2 + b} - \frac{a}{2}$$

to solve the equation $x^2 + ax = b$.

By Euclid's time (circa 300 B.C.), Greek geometry had reached a stage of development where geometric algebra could be used to solve quadratic equations. This was done by reducing them to the geometric equivalent of one of the following forms:

$$x(x + a) = b^2$$
$$x(x - a) = b^2$$
$$x(a - x) = b^2$$

These equations were then solved by applying different theorems dealing with specific areas.

Later, the Arabian mathematician Muhammed ibn Musa al-Khowarizmi (read more about him in the Exercises for Section 9.2) (circa A.D. 820) divided quadratic equations into three types:

$$x^2 + ax = b$$
$$x^2 + b = ax$$
$$x^2 = ax + b$$

with only positive coefficients admitted. All of these developments form the basis for our study of quadratic equations.

9.1 Solving Quadratic Equations by the Square Root Property

▶ Objectives

Solve quadratic equations of the form

A ▷ $X^2 = A$

B ▷ $(AX \pm B)^2 = C$

▶ To Succeed, Review How To . . .

1. Find the square roots of a number (pp. 61, 614–615).
2. Simplify expressions involving radicals (pp. 623–624, 634–638).

▶ Getting Started
Square Chips

The Pentium chip is used in many computers. Because of space limitations, the chip is very small and covers an area of only 324 square millimeters. If the chip is square, how long is each side? As you recall, the area of a square is obtained by multiplying the length X of its side by itself. If we assume that the length of the side of the chip is X millimeters, then its area is X^2. The area is also 324 square millimeters; we thus have the equation $X^2 = 324$.

Suppose the side of a square is 5 units. Then the area is 5^2 or 25 square units. Now, suppose a side of the square is X units. What is the area? X^2. Now, let's go backwards. If we know that the area of a square is 100 square units, what is the length of a side? The corresponding equation is $X^2 = 100$ and the solution is $X = 10$.

In this section, we will learn how to solve quadratic equations that can be written in the form $X^2 = A$ or $(AX \pm B)^2 = C$ by introducing a new property called the *square root property of equations.*

5 units	X units
Area 25 square units	Area X^2 square units

A ▷ Solving Quadratic Equations of the Form $X^2 = A$

The equation $X^2 = 16$ is a *quadratic equation.* In general, a **quadratic equation** is an equation that can be written in the form $ax^2 + bx + c = 0$. How can we solve $X^2 = 16$? First, the equation tells us that a certain number X multiplied by itself gives 16 as a result. Obviously, one possible answer is $X = 4$ because $4^2 = 16$. But wait, what about $X = -4$? It's also true that $(-4)^2 = (-4)(-4) = 16$. Thus, the solutions of the equation $X^2 = 16$ are 4 and -4.

In mathematics, the number 4 is called the **positive square root** of 16, and -4 is called the **negative square root** of 16. These roots are usually denoted by

$$\sqrt{16} = 4 \qquad \text{Read "the positive square root of 16 is 4."}$$

and

$$-\sqrt{16} = -4 \qquad \text{Read "the negative square root of 16 is } -4\text{."}$$

Many of the equations we are about to study have irrational numbers for their solutions. For example, the equation $x^2 = 3$ has two *irrational* solutions. How do we obtain these solutions? By taking the square root of both sides of the equation:

$$x^2 = 3$$

Then,

$$x = \pm\sqrt{3} \qquad \text{Note that } (\sqrt{3})^2 = 3 \text{ and } (-\sqrt{3})^2 = 3.$$

The notation $\pm\sqrt{3}$ is a shortcut to indicate that x can be $\sqrt{3}$ or $-\sqrt{3}$.

> **NOTE**
>
> With a calculator, the answers to the equation $x^2 = 3$ can be approximated as $x \approx \pm 1.7320508$.

On the other hand, the equation $x^2 = 100$ has *rational* roots. To solve this equation, we proceed as follows:

$$x^2 = 100$$

Then,

$$x = \pm\sqrt{100}$$
$$x = \pm 10$$

Thus, the solutions are 10 and -10. Here is the property we just used.

SQUARE ROOT PROPERTY OF EQUATIONS	If A is a positive number and $X^2 = A$, then $X = \pm\sqrt{A}$; that is, $$X = \sqrt{A} \qquad \text{or} \qquad X = -\sqrt{A}$$

EXAMPLE 1   **Solving quadratic equations using the square root property**

Solve:

a. $x^2 = 36$ **b.** $x^2 - 49 = 0$ **c.** $x^2 = 10$

SOLUTION

a. Given: $x^2 = 36$. Then,

$$x = \pm\sqrt{36} \qquad \text{Use the square root property.}$$
$$x = \pm 6$$

Thus, the solutions of the equation $x^2 = 36$ are 6 and -6, since $6^2 = 36$ and $(-6)^2 = 36$. (Both solutions are rational numbers.)

b. Given: $x^2 - 49 = 0$. Unfortunately, this equation is not of the form $X^2 = A$. However, by adding 49 to both sides of the equation, we can remedy this situation.

$$x^2 - 49 = 0 \qquad \text{Given}$$
$$x^2 = 49 \qquad \text{Add 49.}$$
$$x = \pm\sqrt{49} \qquad \text{Use the square root property.}$$
$$x = \pm 7$$

PROBLEM 1

Solve:

a. $x^2 = 81$

b. $x^2 - 1 = 0$

c. $x^2 = 13$

(continued)

Answers to PROBLEMS

1. a. $x = \pm 9$ **b.** $x = \pm 1$

   c. $x = \pm\sqrt{13}$

Thus, the solutions of $x^2 - 49 = 0$ are 7 and -7. (Both solutions are rational numbers.)

Note that the equation $x^2 - 49 = 0$ can also be solved by factoring to write $x^2 - 49 = 0$ as $(x + 7)(x - 7) = 0$. Thus, $x + 7 = 0$ or $x - 7 = 0$; that is, $x = -7$ or $x = 7$, as before.

c. Given: $x^2 = 10$. Then, using the square root property,

$$x = \pm\sqrt{10}$$

Since 10 does not have a rational square root, the solutions of the equation $x^2 = 10$ are written as $\sqrt{10}$ and $-\sqrt{10}$. (Here, both solutions are irrational.)

In solving Example 1(b), we added 49 to both sides of the equation to obtain an equivalent equation of the form $X^2 = A$. Does this method work for more complicated examples? The answer is yes. In fact, when solving any quadratic equation in which the *only* exponent of the variable is 2, we can always transform the equation into an equivalent one of the form $X^2 = A$. Here is the idea.

PROCEDURE

Solving Quadratic Equations of the Form $AX^2 - B = 0$

To solve any equation of the form

$$AX^2 - B = 0$$

write

$$AX^2 = B \qquad \text{Add } B.$$

and

$$X^2 = \frac{B}{A} \qquad \text{Divide by } A.$$

so that, using the square root property,

$$X = \pm\sqrt{\frac{B}{A}}, \quad \frac{B}{A} \geq 0$$

Thus, to solve the equation

$$16x^2 - 81 = 0$$

we write it in the form $AX^2 = B$ or equivalently $X^2 = \frac{B}{A}$:

$$16x^2 - 81 = 0 \qquad \text{Given}$$
$$16x^2 = 81 \qquad \text{Add 81.}$$
$$x^2 = \frac{81}{16} \qquad \text{Divide by 16.}$$
$$x = \pm\sqrt{\frac{81}{16}} \qquad \text{Use the square root property.}$$
$$x = \pm\frac{9}{4}$$

The solutions are $\frac{9}{4}$ and $-\frac{9}{4}$. Since $16 \cdot \left(\frac{9}{4}\right)^2 - 81 = 16 \cdot \frac{81}{16} - 81 = 0$ and $16\left(-\frac{9}{4}\right)^2 - 81 = 16 \cdot \frac{81}{16} - 81 = 0$, our result is correct. Just remember to first rewrite the equation in the form $X^2 = \frac{B}{A}$.

EXAMPLE 2   **Solving a quadratic equation of the form** $AX^2 - B = 0$

Solve: $36x^2 - 25 = 0$

SOLUTION

$$36x^2 - 25 = 0 \qquad \text{Given}$$

$$36x^2 = 25 \qquad \text{Add 25.}$$

$$x^2 = \frac{25}{36} \qquad \text{Divide by 36.}$$

$$x = \pm\sqrt{\frac{25}{36}} \qquad \text{Use the square root property.}$$

$$x = \pm\frac{5}{6}$$

The solutions are $\frac{5}{6}$ and $-\frac{5}{6}$. (Here, both solutions are rational numbers.)

PROBLEM 2

Solve: $9x^2 - 16 = 0$

Of course, not all equations of the form $X^2 = A$ have solutions that are real numbers. For example, to solve the equation $x^2 + 64 = 0$, we write

$$x^2 + 64 = 0 \qquad \text{Given}$$

$$x^2 = -64 \qquad \text{Subtract 64.}$$

But there is no *real* number whose square is -64. If you square a nonzero real number, the answer is always positive. Thus, x^2 is positive and can never equal -64. The equation $x^2 + 64 = 0$ has *no real-number* solution.

As we have seen, not all equations have *rational-number* solutions—that is, solutions of the form $\frac{a}{b}$, where a and b are integers, $b \neq 0$. For example, the equation

$$16x^2 - 5 = 0$$

is solved as follows:

$$16x^2 - 5 = 0 \qquad \text{Given}$$

$$16x^2 = 5 \qquad \text{Add 5.}$$

$$x^2 = \frac{5}{16} \qquad \text{Divide by 16.}$$

$$x = \pm\sqrt{\frac{5}{16}} \qquad \text{Use the square root property.}$$

However, $\sqrt{\frac{5}{16}}$ is *not* a rational number even though the denominator, 16, is the square of 4. As you recall, using the quotient rule for radicals, we may write

$$\pm\sqrt{\frac{5}{16}} = \pm\frac{\sqrt{5}}{\sqrt{16}} = \pm\frac{\sqrt{5}}{4}$$

Thus, the solutions of the equation $16x^2 - 5 = 0$ are

$$\frac{\sqrt{5}}{4} \qquad \text{and} \qquad -\frac{\sqrt{5}}{4}$$

Both solutions are irrational numbers.

Answers to PROBLEMS

2. $\frac{4}{3}$ and $-\frac{4}{3}$

EXAMPLE 3 Solving a quadratic equation using the quotient rule

Solve:

a. $4x^2 - 7 = 0$ **b.** $8x^2 + 49 = 0$

SOLUTION

a.

$4x^2 - 7 = 0$	Given
$4x^2 = 7$	Add 7.
$x^2 = \dfrac{7}{4}$	Divide by 4.
$x = \pm\sqrt{\dfrac{7}{4}}$	Use the square root property.
$x = \pm\dfrac{\sqrt{7}}{\sqrt{4}}$	Use the quotient rule for radicals.
$x = \pm\dfrac{\sqrt{7}}{2}$	

Thus, the solutions of $4x^2 - 7 = 0$ are $\dfrac{\sqrt{7}}{2}$ and $-\dfrac{\sqrt{7}}{2}$. They are both irrational numbers.

b.

$8x^2 + 49 = 0$	Given
$8x^2 = -49$	Subtract 49.
$x^2 = -\dfrac{49}{8}$	Divide by 8.

Since the square of a real number x cannot be negative and $-\dfrac{49}{8}$ is negative, this equation has *no* real-number solution.

PROBLEM 3

Solve:

a. $49x^2 - 3 = 0$

b. $10x^2 + 9 = 0$

B ⟩ Solving Quadratic Equations of the Form $(AX \pm B)^2 = C$

We've already mentioned that to solve an equation in which the only exponent of the variable is 2, we must transform the equation into an equivalent one of the form $X^2 = A$. Now consider the equation

$$(x - 2)^2 = 9$$

If we think of $(x - 2)$ as X, we have an equation of the form $X^2 = 9$, which we just learned how to solve! Thus, we have the following.

$(x - 2)^2 = 9$	Given
$X^2 = 9$	Write $x - 2$ as X.
$X = \pm\sqrt{9}$	Use the square root property.
$X = \pm 3$	
$x - 2 = \pm 3$	Write X as $x - 2$.
$x = 2 \pm 3$	Add 2.

Hence,

$$x = 2 + 3 \quad \text{or} \quad x = 2 - 3$$

The solutions are $x = 5$ and $x = -1$.

Clearly, by thinking of $(x - 2)$ as X, we can solve a more complicated equation. In the same manner, we can solve $9(x - 2)^2 - 5 = 0$:

$$9(x - 2)^2 - 5 = 0 \qquad \text{Given}$$

$$9(x - 2)^2 = 5 \qquad \text{Add 5.}$$

$$(x - 2)^2 = \frac{5}{9} \qquad \text{Divide by 9.}$$

$$x - 2 = \pm\sqrt{\frac{5}{9}} \qquad \textit{Think of } x - 2 \textit{ as X and use the square root property.}$$

$$x - 2 = \pm\frac{\sqrt{5}}{\sqrt{9}} \qquad \text{Use the quotient rule for radicals.}$$

$$x - 2 = \pm\frac{\sqrt{5}}{3}$$

$$x = 2 \pm \frac{\sqrt{5}}{3} \qquad \text{Add 2.}$$

The solutions are

$$2 + \frac{\sqrt{5}}{3} = \frac{6 + \sqrt{5}}{3} \qquad \text{and} \qquad 2 - \frac{\sqrt{5}}{3} = \frac{6 - \sqrt{5}}{3}$$

EXAMPLE 4   Solving quadratic equations of the form $(AX + B)^2 = C$

Solve:

a. $(x + 3)^2 = 9$ **b.** $(x + 1)^2 - 4 = 0$ **c.** $25(x + 2)^2 - 3 = 0$

SOLUTION

a. $(x + 3)^2 = 9$ Given

$\quad x + 3 = \pm\sqrt{9}$ *Think of $(x + 3)$ as X.*

$\quad x + 3 = \pm 3$

$\qquad x = -3 \pm 3$ Subtract 3.

$x = -3 + 3 \qquad$ or $\qquad x = -3 - 3$

$x = 0 \qquad\qquad$ or $\qquad x = -6$

The solutions are 0 and -6.

b. $(x + 1)^2 - 4 = 0$ Given

$\quad (x + 1)^2 = 4$ Add 4 (to have an equation of the form $X^2 = A$).

$\qquad x + 1 = \pm\sqrt{4}$ *Think of $(x + 1)$ as X.*

$\qquad x + 1 = \pm 2$

$\qquad\qquad x = -1 \pm 2$ Subtract 1.

$x = -1 + 2 \qquad$ or $\qquad x = -1 - 2$

$x = 1 \qquad\qquad$ or $\qquad x = -3$

The solutions are 1 and -3.

c. $25(x + 2)^2 - 3 = 0$ Given

$\quad 25(x + 2)^2 = 3$ Add 3.

$\qquad (x + 2)^2 = \frac{3}{25}$ Divide by 25.

$\qquad x + 2 = \pm\sqrt{\frac{3}{25}}$ *Think of $(x + 2)$ as X.*

$\qquad x + 2 = \pm\frac{\sqrt{3}}{\sqrt{25}} = \pm\frac{\sqrt{3}}{5}$ Since $\sqrt{\frac{3}{25}} = \frac{\sqrt{3}}{\sqrt{25}}$

$\qquad\qquad x = -2 \pm \frac{\sqrt{3}}{5}$ Subtract 2.

The solutions are

$$-2 + \frac{\sqrt{3}}{5} = \frac{-10 + \sqrt{3}}{5} \qquad \text{and} \qquad -2 - \frac{\sqrt{3}}{5} = \frac{-10 - \sqrt{3}}{5}$$

PROBLEM 4

Solve:

a. $(x + 6)^2 = 36$

b. $(x + 2)^2 - 9 = 0$

c. $16(x + 1)^2 - 5 = 0$

Answers to PROBLEMS

4. a. The solutions are 0 and -12.

  b. The solutions are 1 and -5.

  c. The solutions are

$$-1 + \frac{\sqrt{5}}{4} = \frac{-4 + \sqrt{5}}{4}$$

and

$$-1 - \frac{\sqrt{5}}{4} = \frac{-4 - \sqrt{5}}{4}.$$

> **NOTE**
>
> It is easy to spot quadratic equations that have no real-number solution. Here is how: if the equation can be written in the form (Expression)^2 = Negative number [that is, (   )^2 = Negative number], the equation has no solution.

As we mentioned before, if we square any real number, the result is not negative. For this reason, an equation such as

$$(x - 4)^2 = -5$$ If $(x - 4)$ represents a real number, $(x - 4)^2$ *cannot* be negative. But -5 is negative, so $(x - 4)^2$ and -5 can *never* be equal.

has *no* real-number solution. Similarly,

$$(x - 3)^2 + 8 = 0$$

has no real-number solution, since

$$(x - 3)^2 + 8 = 0$$ is equivalent to $$(x - 3)^2 = -8$$

by subtracting 8. We use this idea in Example 5.

EXAMPLE 5   Solving a quadratic equation when there is no real-number solution

Solve: $9(x - 5)^2 + 1 = 0$

SOLUTION

$$9(x - 5)^2 + 1 = 0$$ Given

$$9(x - 5)^2 = -1$$ Subtract 1.

$$(x - 5)^2 = -\frac{1}{9}$$ Divide by 9.

Since $(x - 5)$ is to be a real number, $(x - 5)^2$ *can never* be negative. But $-\frac{1}{9}$ is negative. Thus, the equation $(x - 5)^2 = -\frac{1}{9}$ [which is equivalent to $9(x - 5)^2 + 1 = 0$] has *no* real-number solution.

PROBLEM 5
Solve: $16(x - 3)^2 + 7 = 0$

EXAMPLE 6   Solving a quadratic equation of the form $(AX - B)^2 = C$

Solve: $3(2x - 3)^2 = 54$

SOLUTION   We want to write the equation in the form $X^2 = A$.

$$3(2x - 3)^2 = 54$$ Given

$$(2x - 3)^2 = \frac{54}{3} = 18$$ Divide by 3.

$$X^2 = 18$$ Write $2x - 3$ as X.

$$X = \pm\sqrt{18}$$ Use the square root property.

$$X = \pm 3\sqrt{2}$$ Simplify the radical ($\sqrt{18} = \sqrt{9 \cdot 2} = 3\sqrt{2}$).

$$2x - 3 = \pm 3\sqrt{2}$$ Write X as $2x - 3$.

$$2x = 3 \pm 3\sqrt{2}$$ Add 3.

$$x = \frac{3 \pm 3\sqrt{2}}{2}$$ Divide by 2.

The solutions are

$$\frac{3 + 3\sqrt{2}}{2}   \text{ and }   \frac{3 - 3\sqrt{2}}{2}$$

PROBLEM 6
Solve: $2(3x - 2)^2 = 36$

Answers to PROBLEMS

5. No real-number solution

6. The solutions are $\frac{2 \pm 3\sqrt{2}}{3}$

EXAMPLE 7   **Application: BMI and weight**

Your body mass index (BMI) is a measurement of your healthy weight relative to your height. According to the National Heart and Lung Institute, the formula for your BMI is given by the equation BMI $= \frac{705W}{H^2}$, where W is your weight in pounds and H is your height in inches. If a person has a BMI of 20, which is in the "normal" range, and weighs 140 pounds, how tall is the person?

SOLUTION   Substituting 20 for BMI and 140 for W, we have

$$20 = \frac{705 \cdot 140}{H^2}$$

Multiplying both sides by H^2 $20H^2 = 705 \cdot 140$
Dividing both sides by 20 $H^2 = 705 \cdot 7$
Or $H^2 = 4935$
Thus, $H = \sqrt{4935} \approx 70$ inches

PROBLEM 7

What is the height of a person weighing 180 pounds with a BMI of 30, which is in the "overweight" range?

> **Exercises 9.1**

MathZone +x

Boost your grade at mathzone.com!
> Practice Problems   > Self-Tests
> NetTutor   > e-Professors
                                      > Videos

⟨ **A** ⟩   **Solving Quadratic Equations of the Form $X^2 = A$**   In Problems 1–20, solve the given equation.

1. $x^2 = 100$

2. $x^2 = 1$

3. $x^2 = 0$

4. $x^2 = 121$

5. $y^2 = -4$

6. $y^2 = -16$

7. $x^2 = 7$

8. $x^2 = 3$

9. $x^2 - 9 = 0$

10. $x^2 - 64 = 0$

11. $x^2 - 3 = 0$

12. $x^2 - 5 = 0$

13. $25x^2 - 1 = 0$

14. $36x^2 - 49 = 0$

15. $100x^2 - 49 = 0$

16. $81x^2 - 36 = 0$

17. $25y^2 - 17 = 0$

18. $9y^2 - 11 = 0$

19. $25x^2 + 3 = 0$

20. $49x^2 + 1 = 0$

⟨ **B** ⟩   **Solving Quadratic Equations of the Form $(AX \pm B)^2 = C$**   In Problems 21–60, solve the given equation.

21. $(x + 1)^2 = 81$

22. $(x + 3)^2 = 25$

23. $(x - 2)^2 = 36$

24. $(x - 3)^2 = 16$

25. $(z - 4)^2 = -25$

26. $(z + 2)^2 = -16$

27. $(x - 9)^2 = 81$

28. $(x - 6)^2 = 36$

29. $(x + 4)^2 = 16$

30. $(x + 7)^2 = 49$

31. $25(x + 1)^2 - 1 = 0$

32. $16(x + 2)^2 - 1 = 0$

33. $36(x - 3)^2 - 49 = 0$

34. $9(x - 1)^2 - 25 = 0$

35. $4(x + 1)^2 - 25 = 0$

Answers to PROBLEMS

7. $\sqrt{4230} \approx 65$ inches

36. $49(x + 2)^2 - 16 = 0$

37. $9(x - 1)^2 - 5 = 0$

38. $4(x - 2)^2 - 3 = 0$

39. $16(x + 1)^2 + 1 = 0$

40. $25(x + 2)^2 + 16 = 0$

41. $x^2 = \dfrac{1}{81}$

42. $x^2 = \dfrac{1}{9}$

43. $x^2 - \dfrac{1}{16} = 0$

44. $x^2 - \dfrac{1}{36} = 0$

45. $6x^2 - 24 = 0$

46. $3x^2 - 75 = 0$

47. $2(v + 1)^2 - 18 = 0$

48. $3(v - 2)^2 - 48 = 0$

49. $8(x - 1)^2 - 18 = 0$

50. $50(x + 3)^2 - 72 = 0$

51. $4(2y - 3)^2 = 32$

52. $2(3y - 1)^2 = 24$

53. $8(2x - 3)^2 - 64 = 0$

54. $5(3x - 1)^2 - 60 = 0$

55. $3\left(\dfrac{1}{2}x + 1\right)^2 = 54$

56. $4\left(\dfrac{1}{3}x - 1\right)^2 = 80$

57. $2\left(\dfrac{1}{3}x - 1\right)^2 - 40 = 0$

58. $3\left(\dfrac{1}{3}x - 2\right)^2 - 54 = 0$

59. $5\left(\dfrac{1}{2}y - 1\right)^2 + 60 = 0$

60. $6\left(\dfrac{1}{3}y - 2\right)^2 + 72 = 0$

〉〉〉 Applications

61. *Curves on roads* Naomi was traveling on a curved concrete highway of radius 64 feet when her car began to skid. The velocity v (in miles per hour) at which skidding starts to happen on a curve of radius r is given by the equation $v^2 = 9r$. How fast was Naomi traveling when she started to skid? If the speed limit was 25 miles per hour, was she speeding?

62. *Skid marks* Have you seen your local law enforcement person measuring skid marks at the scene of an accident? The speed s (in miles per hour) a car was traveling if it skidded d feet after the brakes were applied on a dry concrete road is given by the equation $s^2 = 24d$, where d is the length of the skid mark. Find the speed s of a car leaving a 50-foot skid mark after the brakes were applied. Give the final answer to the nearest mph.

63. *Terminal velocity* If air resistance is neglected, the terminal velocity v (in meters per second) of an object with initial velocity v_0 starting at a height of h feet is given by the equation $v^2 = 20h + v_0$. Find the terminal velocity of an object starting at a height of 10 meters with initial velocity of 25 meters per second.

64. *Investing in quadratics* Tran invested $100 at r percent compounded for 2 years. At the end of the 2 years, he received $121. Use the formula $(r + 1)^2 = \dfrac{A}{P}$, where A is the amount received at the end of the two years, and P is the principal (original amount), to find r.

Lenders Earn
8% to 29%

〉〉〉 *Using Your Knowledge*

Going in Circles   The area A of a circle of radius r is given by $A = \pi r^2$. Find the radius of a circle with the specified area.

65. 25π square inches

66. 12π square feet

The surface area A of a sphere of radius r is given by $A = 4\pi r^2$. Find the radius of a sphere with the specified surface area.

67. 49π square feet

68. 81π square inches

〉〉〉 *Write On*

69. Explain why the equation $x^2 + 6 = 0$ has no real-number solution.

70. Explain why the equation $(x + 1)^2 + 3 = 0$ has no real-number solution.

Consider the equation $X^2 = A$.

71. What can you say about A if the equation has no real-number solution?

72. What can you say about A if the equation has exactly one solution?

73. What can you say about A if the equation has two solutions?

74. What types of solution does the equation have if A is a prime number?

75. What types of solution does the equation have if A is a positive perfect square?

〉〉〉 *Concept Checker*

Fill in the blank(s) with the correct word(s), phrase, or mathematical statement.

76. A **quadratic equation** is an equation that **can be written** in the **form** _____.

77. The **positive square root** of **81** is _____.

78. The **negative square root** of **81** is _____.

79. If A is a **positive** number and $X^2 = A$, then $X =$ _____.

-9	$-A$
A	$ax^2 + bx + c = 0$
9	$\pm A$
$ax^2 + bx + c$	

〉〉〉 *Mastery Test*

Solve if possible:

80. $16(x - 3)^2 - 7 = 0$

81. $3(x - 1)^2 - 24 = 0$

82. $(x + 5)^2 = 25$

83. $(x + 2)^2 - 9 = 0$

84. $16(x + 1)^2 - 5 = 0$

85. $49x^2 - 3 = 0$

86. $10x^2 + 9 = 0$

87. $9x^2 - 16 = 0$

88. $x^2 = 121$

89. $x^2 - 1 = 0$

90. $x^2 = 13$

〉〉〉 *Skill Checker*

Expand:

91. $(x + 7)^2$

92. $(x + 5)^2$

93. $(x - 3)^2$

94. $(x - 5)^2$

9.2

Solving Quadratic Equations by Completing the Square

▶ **Objective**

A ⟩ Solve a quadratic equation by completing the square.

▶ **To Succeed, Review How To . . .**

1. Recognize a quadratic equation (p. 660).
2. Expand $(x \pm a)^2$ (pp. 369, 370).

▶ **Getting Started**

Completing the Square for Round Baseballs

The man has just batted the ball straight up at 96 feet per second. At the end of t seconds, the height h of the ball will be

$$h = -16t^2 + 96t$$

How long will it be before the ball reaches 44 feet? To solve this problem, we let $h = 44$ to obtain

$$-16t^2 + 96t = 44$$

$$t^2 - 6t = -\frac{44}{16} = -\frac{11}{4} \qquad \text{Divide by } -16 \text{ and simplify.}$$

This equation is a quadratic equation (an equation that can be written in the form $ax^2 + bx + c = 0$, $a \neq 0$). Can we use the techniques we studied in Section 9.1 to solve it? The answer is yes, if we can write the equation in the form

$$(t - N)^2 = A \qquad \begin{array}{l} N \text{ and } A \text{ are the} \\ \text{numbers we need to find} \\ \text{to solve the problem.} \end{array}$$

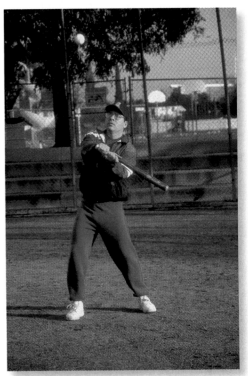

To do this, however, we should know a little bit more about a technique used in algebra called **completing the square**. We will present several examples and then generalize the results. Why? Because when we generalize the process and learn how to solve the quadratic equation $ax^2 + bx + c = 0$ by completing the square, we can use the formula obtained, called the *quadratic formula,* to solve *any* equation of the form $ax^2 + bx + c = 0$, by simply substituting the values of a, b, and c in the quadratic formula. We will do this in the next section, but right now we need to review how to expand binomials, because completing the square involves that process.

A › Solving Quadratic Equations by Completing the Square

Recall that

Thus,

$$(x + 7)^2 = x^2 + 14x + 7^2$$
$$(x + 2)^2 = x^2 + 4x + 2^2$$
$$(x + 5)^2 = x^2 + 10x + 5^2$$

Do you see any relationship between the coefficients of x (14, 4, and 10, respectively) and the last terms? Perhaps you will see it better if we write it in a table.

Coefficient of X	Last Term Squared
14	7^2
4	2^2
10	5^2

It seems that half the coefficient of x gives the number to be squared for the last term. Thus,

$$\frac{14}{2} = 7$$

$$\frac{4}{2} = 2$$

$$\frac{10}{2} = 5$$

Now, what numbers would you add to complete the given squares?

$$(x + 3)^2 = x^2 + 6x + \square$$
$$(x + 4)^2 = x^2 + 8x + \square$$
$$(x + 6)^2 = x^2 + 12x + \square$$

The correct answers are $\left(\frac{6}{2}\right)^2 = 3^2 = 9$, $\left(\frac{8}{2}\right)^2 = 4^2 = 16$, and $\left(\frac{12}{2}\right)^2 = 6^2 = 36$. We then have

$$(x + 3)^2 = x^2 + 6x + 3^2$$

The last term is always the square of half of the coefficient of the middle term.

$$6 \div 2 = 3$$

$$(x + 4)^2 = x^2 + 8x + 4^2$$

$$8 \div 2 = 4$$

$$(x + 6)^2 = x^2 + 12x + 6^2$$

$$12 \div 2 = 6$$

You can also find the missing number to complete the square by using a diagram. For example, to complete the square in $(x + 3)^2$ complete the diagram:

$$x \; + \; 3$$

	x^2	$3x$
	$3x$	?

(with row labels x + 3 on the left)

You need 3^2 or 9 in the lower right corner, so $(x + 3)^2 = x^2 + 6x + 9$, as before. You can see this technique used in Problems 41–46.

Here is the procedure we have just used to complete the square.

PROCEDURE

Completing the Square $x^2 + bx + \square$

1. Find the coefficient of the x term. (b)

2. Divide the coefficient by 2. $\left(\dfrac{b}{2}\right)$

3. Square this number to obtain the last term. $\left(\dfrac{b}{2}\right)^2 = \dfrac{b^2}{4}$

The idea is to add a term $\square$ to the binomial $x^2 + bx$ to change it into a perfect square trinomial.

Thus, to complete the square in

$$x^2 + 16x + \square$$

we proceed as follows:

1. Find the coefficient of the x term $\longrightarrow$ 16.
2. Divide the coefficient by 2 $\longrightarrow$ 8.
3. Square this number to obtain the last term $\longrightarrow$ 8^2.

Hence,

$$x^2 + 16x + \boxed{8^2} = (x + 8)^2$$

Now consider

$$x^2 - 18x + \square$$

Our steps to fill in the blank are as before:

1. Find the coefficient of the x term $\longrightarrow$ -18.
2. Divide the coefficient by 2 $\longrightarrow$ -9.
3. Square this number to obtain the last term $\longrightarrow$ $(-9)^2$.

Hence,

$$x^2 - 18x + (-9)^2$$
$$= x^2 - 18x + 9^2 \qquad \text{Recall that } (-9)^2 = (9)^2; \text{ they are both 81.}$$
$$= x^2 - 18x + 81 = (x - 9)^2$$

EXAMPLE 1   Completing the square

Find the missing term to complete the square:

a. $x^2 + 20x + \square$ **b.** $x^2 - x + \square$

SOLUTION

a. We use the three-step procedure:

   1. The coefficient of x is 20.

   2. $\dfrac{20}{2} = 10$

   3. The missing term is $\boxed{10^2} = \boxed{100}$. Hence, $x^2 + 20x + 100 = (x + 10)^2$.

b. Again we use the three-step procedure:

   1. The coefficient of x is -1.

   2. $\dfrac{-1}{2} = -\dfrac{1}{2}$

   3. The missing term is $\boxed{\left(-\dfrac{1}{2}\right)^2} = \boxed{\dfrac{1}{4}}$. Hence, $x^2 - x + \dfrac{1}{4} = \left(x - \dfrac{1}{2}\right)^2$.

PROBLEM 1

Find the missing terms:

a. $(x + 11)^2 = x^2 + 22x + \square$

b. $\left(x - \dfrac{1}{4}\right)^2 = x^2 - \dfrac{1}{2}x + \square$

Can we use the patterns we've just studied to look for further patterns? Of course! For example, how would you fill in the blanks in

$$x^2 + 16x + \square = (\quad)^2$$

Here the coefficient of x is 16, so $\left(\dfrac{16}{2}\right)^2 = 8^2$ goes in the box. Since

Same

$$X^2 + 2AX + A^2 = (X + A)^2$$

Same

$$x^2 + 16x + 8^2 = (x + 8)^2$$

Similarly,

$$x^2 - 6x + \square = (\quad)^2$$

is completed by reasoning that the coefficient of x is -6, so

$$\left(-\dfrac{6}{2}\right)^2 = (-3)^2 = (3)^2$$

goes in the box. Now

Same

$$X^2 - 2AX + A^2 = (X - A)^2$$

Same

Since the middle term on the left has a negative sign, the sign inside the parentheses must be negative.

Hence,

$$x^2 - 6x + \square = (\quad)^2$$

becomes

$$x^2 - 6x + 3^2 = (x - 3)^2$$

EXAMPLE 2   Completing the square

Find the missing terms:

a. $x^2 - 10x + \square = ($ $)^2$ **b.** $x^2 + 3x + \square = ($ $)^2$

SOLUTION

a. The coefficient of x is -10; thus, the number in the box should be $\left(-\frac{10}{2}\right)^2 = (-5)^2 = 5^2$, and we should have

Same

$$x^2 - 10x + 5^2 = (x - 5)^2$$

b. The coefficient of x is 3; thus, we have to add $\left(\frac{3}{2}\right)^2$. Then

$$x^2 + 3x + \left(\frac{3}{2}\right)^2 = \left(x + \frac{3}{2}\right)^2$$

PROBLEM 2

Find the missing terms:

a. $x^2 - 12x + \square = ($ $)^2$

b. $x^2 + 5x + \square = ($ $)^2$

We are finally ready to solve the equation from the *Getting Started* that involves the time it takes the ball to reach 44 feet, that is,

$$t^2 - 6t = -\frac{11}{4}$$ As you recall, t represents the time it takes the ball to reach 44 feet (p. 670).

Since the coefficient of t is -6, we must add $\left(-\frac{6}{2}\right)^2 = (-3)^2 = 3^2$ to both sides of the equation. We then have

$$t^2 - 6t + 3^2 = -\frac{11}{4} + 3^2$$

$$(t - 3)^2 = -\frac{11}{4} + \frac{36}{4}$$ Note that $3^2 = 9 = \frac{36}{4}$.

$$(t - 3)^2 = \frac{25}{4}$$

Then

$$(t - 3) = \pm\sqrt{\frac{25}{4}} = \pm\frac{5}{2}$$

$$t = 3 \pm \frac{5}{2}$$

$$t = 3 + \frac{5}{2} = \frac{11}{2} \quad \text{or} \quad t = 3 - \frac{5}{2} = \frac{1}{2}$$

This means that the ball reaches 44 feet after $\frac{1}{2}$ second (on the way up) and after $\frac{11}{2} = 5\frac{1}{2}$ seconds (on the way down).

Here is a summary of the steps needed to solve a quadratic equation by completing the square. The solution of our original equation follows so you can see how the steps are carried out.

PROCEDURE

Solving a Quadratic Equation by Completing the Square

1. Write the equation with the variables in descending order on the left and the constants on the right.
2. If the coefficient of the squared term is not 1, divide each term by this coefficient.
3. Add the square of one-half of the coefficient of the first-degree term to both sides.
4. Rewrite the left-hand side as the square of a binomial.
5. Use the square root property to solve the resulting equation.

Answers to PROBLEMS

2. a. $x^2 - 12x + 6^2 = (x - 6)^2$

b. $x^2 + 5x + \left(\frac{5}{2}\right)^2 = \left(x + \frac{5}{2}\right)^2$

For the equation from the *Getting Started,* we have:

1. $-16t^2 + 96t = 44$

2. $\dfrac{-16t^2}{-16} + \dfrac{96t}{-16} = \dfrac{44}{-16}$

$t^2 - 6t = -\dfrac{11}{4}$

3. $t^2 - 6t + \left(-\dfrac{6}{2}\right)^2 = -\dfrac{11}{4} + \left(-\dfrac{6}{2}\right)^2$

$t^2 - 6t + 3^2 = -\dfrac{11}{4} + 3^2$

4. $(t - 3)^2 = \dfrac{25}{4}$

5. $(t - 3) = \pm\sqrt{\dfrac{25}{4}}$

$t - 3 = \pm\dfrac{5}{2}$

$t = 3 \pm \dfrac{5}{2}$

Thus, $t = 3 + \dfrac{5}{2} = 5\dfrac{1}{2}$ or $t = 3 - \dfrac{5}{2} = \dfrac{1}{2}$.

We now use this procedure in another example.

EXAMPLE 3   Solving a quadratic equation by completing the square

Solve: $4x^2 - 16x + 7 = 0$

SOLUTION   We use the five-step procedure just given.

$4x^2 - 16x + 7 = 0$ Given

1. Subtract 7.

$4x^2 - 16x = -7$

2. Divide by 4 so the coefficient of x^2 is 1.

$x^2 - 4x = -\dfrac{7}{4}$

3. Add $\left(-\dfrac{4}{2}\right)^2 = (-2)^2 = 2^2$.

$x^2 - 4x + 2^2 = -\dfrac{7}{4} + 2^2$

4. Rewrite.

$(x - 2)^2 = -\dfrac{7}{4} + \dfrac{16}{4} = \dfrac{9}{4}$

$(x - 2)^2 = \dfrac{9}{4}$

5. Use the square root property to solve the resulting equation.

$(x - 2) = \pm\sqrt{\dfrac{9}{4}}$

$x - 2 = \pm\dfrac{3}{2}$

$x = 2 \pm \dfrac{3}{2}$

That is,

$x = \dfrac{4}{2} + \dfrac{3}{2} = \dfrac{7}{2}$ or $x = \dfrac{4}{2} - \dfrac{3}{2} = \dfrac{1}{2}$

Thus, the solutions of $4x^2 - 16x + 7 = 0$ are $\dfrac{7}{2}$ and $\dfrac{1}{2}$.

PROBLEM 3

Solve by completing the square:

$4x^2 - 24x + 27 = 0$

Answers to PROBLEMS

3. The solutions are $3 \pm \dfrac{3}{2}$, that is, $\dfrac{9}{2}$ and $\dfrac{3}{2}$.

Finally, we must point out that in many cases the answers you will obtain when you solve quadratic equations by completing the square are *not* rational numbers. (Remember? A rational number can be written as $\frac{a}{b}$, a and b integers, $b \neq 0$.) As a matter of fact, quadratic equations with rational solutions can be solved by factoring, often a simpler method.

EXAMPLE 4   Solving a quadratic equation by completing the square
Solve: $36x + 9x^2 + 31 = 0$

SOLUTION   We proceed using the five-step procedure.

$$36x + 9x^2 + 31 = 0 \quad \text{Given}$$

1. Subtract 31 and write in descending order.

$$9x^2 + 36x = -31$$

2. Divide by 9.

$$x^2 + 4x = -\frac{31}{9}$$

3. Add $\left(\frac{4}{2}\right)^2 = 2^2$.

$$x^2 + 4x + 2^2 = -\frac{31}{9} + 2^2 = -\frac{31}{9} + \frac{36}{9}$$

4. Rewrite as a perfect square.

$$(x + 2)^2 = \frac{5}{9}$$

5. Use the square root property.

$$(x + 2) = \pm\sqrt{\frac{5}{9}}$$

$$x + 2 = \pm\frac{\sqrt{5}}{3}$$

$$x = -2 \pm \frac{\sqrt{5}}{3} = \frac{-6 \pm \sqrt{5}}{3}$$

Thus, the solutions of $36x + 9x^2 + 31 = 0$ are $\frac{-6 + \sqrt{5}}{3}$ and $\frac{-6 - \sqrt{5}}{3}$.

PROBLEM 4
Solve:
$$4x^2 + 24x + 31 = 0$$

MathZone

⟩ Exercises 9.2

⟨ **A** ⟩ **Completing the square**   In Problems 1–20, find the missing term(s) to make the expression a perfect square.

1. $x^2 + 18x + \square$

2. $x^2 + 2x + \square$

3. $x^2 - 16x + \square$

4. $x^2 - 4x + \square$

5. $x^2 + 7x + \square$

6. $x^2 + 9x + \square$

7. $x^2 - 3x + \square$

8. $x^2 - 7x + \square$

9. $x^2 + x + \square$

10. $x^2 - x + \square$

11. $x^2 + 4x + \square = (\ \ )^2$

12. $x^2 + 6x + \square = (\ \ )^2$

13. $x^2 + 3x + \square = (\ \ )^2$

14. $x^2 + 9x + \square = (\ \ )^2$

15. $x^2 - 6x + \square = (\ \ )^2$

16. $x^2 - 24x + \square = (\ \ )^2$

17. $x^2 - 5x + \square = (\ \ )^2$

18. $x^2 - 11x + \square = (\ \ )^2$

19. $x^2 - \frac{3}{2}x + \square = (\ \ )^2$

20. $x^2 - \frac{5}{2}x + \square = (\ \ )^2$

⟩ Web IT go to **mhhe.com/bello** for more lessons

Answers to PROBLEMS

4. The solutions are

$-3 \pm \frac{\sqrt{5}}{2} = \frac{-6 \pm \sqrt{5}}{2}$

In Problems 21–40, solve the given equation if possible.

21. $x^2 + 4x + 1 = 0$

22. $x^2 + 2x + 7 = 0$

23. $x^2 + x - 1 = 0$

24. $x^2 + 2x - 1 = 0$

25. $x^2 + 3x - 1 = 0$

26. $x^2 - 3x - 4 = 0$

27. $x^2 - 3x - 3 = 0$

28. $x^2 - 3x - 1 = 0$

29. $4x^2 + 4x - 3 = 0$

30. $2x^2 + 10x - 1 = 0$

31. $4x^2 - 16x = 15$

32. $25x^2 - 25x = -6$

33. $4x^2 - 7 = 4x$

34. $2x^2 - 18 = -9x$

35. $2x^2 + 1 = 4x$

36. $2x^2 + 3 = 6x$

37. $(x + 3)(x - 2) = -4$

38. $(x + 4)(x - 1) = -6$

39. $2x(x + 5) - 1 = 0$

40. $2x(x - 4) = 2(9 - 8x) - x$

A visual method to complete the square   In the ninth century, an Arab mathematician named al-Khowarizmi (from which the word "algorithm" was derived) developed a method to complete the square in his book (*Hisab al-jabr w'al-muqabala, The Compendious Book on Calculation by Completion and Balancing*). To solve the equation

$$x^2 + 10x = 39$$

he used a square like the one on the left and then literally completed it by adding 25 to the lower right corner.

A page from al-Khowarizmi's book

The shaded area of the squares on the left is $x^2 + 10x$, so adding the 25 to the $x^2 + 10x = 39$ yields

$$x^2 + 10x + 25 = 39 + 25$$

or
$$(x + 5)^2 = 64$$

Taking roots
$$x + 5 = \pm 8$$

Subtracting 5
$$x = \pm 8 - 5$$

Which means
$$x = 3 \text{ and } x = -13.$$

In Problems 41–43, add a term to the binomial to change it into a trinomial square. Use the diagram to find the term that needs to be added to complete the square.

41. $x^2 + 6x$

42. $x^2 + 8x$

43. $25x^2 + 70x$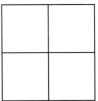

In Problems 44–46, use the results of Problems 41–43 to solve the equations.

44. $x^2 + 6x = 16$

45. $x^2 + 8x = 9$

46. $25x^2 + 70x = 32$

〉〉〉 *Using Your Knowledge*

Finding a Maximum or a Minimum   Many applications of mathematics require finding the maximum or the minimum of certain algebraic expressions. Thus, a certain business may wish to find the price at which a product will bring *maximum* profits, whereas engineers may be interested in *minimizing* the amount of carbon monoxide produced by automobiles.

Now suppose you are the manufacturer of a certain product whose average manufacturing cost $\overline{C}$ (in dollars), based on producing x (thousand) units, is given by the expression

$$\overline{C} = x^2 - 8x + 18$$

How many units should be produced to *minimize* the cost per unit? If we consider the right-hand side of the equation, we can complete the square and leave the equation unchanged by adding and subtracting the appropriate number. Thus,

$$\overline{C} = x^2 - 8x + 18$$
$$\overline{C} = (x^2 - 8x + \quad) + 18$$
$$\overline{C} = (x^2 - 8x + 4^2) + 18 - 4^2 \quad \text{Note that we have added and subtracted the square of one-half of the coefficient of } x, \left(\frac{-8}{2}\right)^2 = 4^2.$$

Then

$$\overline{C} = (x - 4)^2 + 2$$

Since the smallest possible value that $(x - 4)^2$ can have is 0 for $x = 4$, let $x = 4$ to give the minimum cost, $\overline{C} = 2$.

Use your knowledge about completing the square to solve the following problems.

47. A manufacturer's average cost $\overline{C}$ (in dollars), based on manufacturing x (thousand) items, is given by the equation

$$\overline{C} = x^2 - 4x + 6$$

   a. How many units should be produced to minimize the cost per unit?

   b. What is the minimum average cost per unit?

48. The demand D for a certain product depends on the number x (in thousands) of units produced and is given by the equation

$$D = x^2 - 2x + 3$$

What is the number of units that have to be produced so that the demand is at its lowest?

49. Have you seen people adding chlorine to their pools? This is done to reduce the number of bacteria present in the water. Suppose that after t days, the number of bacteria per cubic centimeter is given by the expression

$$B = 20t^2 - 120t + 200$$

In how many days will the number of bacteria be at its lowest?

》》》 *Write On*

50. What is the first step you would perform to solve the equation $6x^2 + 9x = 3$ by completing the square?

51. Can you solve the equation $x^3 + 2x^2 = 5$ by completing the square? Explain.

52. Describe the procedure you would use to find the number that has to be added to $x^2 + 5x$ to make the expression a perfect square trinomial.

53. Make a perfect square trinomial that has a term of $-6x$ and explain how you constructed your trinomial.

》》》 *Concept Checker*

Fill in the blank(s) with the correct word(s), phrase, or mathematical statement.

54. The **first** step to **complete the square** when solving an equation is to write the equation with the variables in **descending order** on the _____ side of the equation and the constants on the _____ side of the equation.

multiply	left
right	divide

55. When **completing the square**, if the **coefficient** of the **square** term is **not 1**, _____ each term by this coefficient.

》》》 *Mastery Test*

Find the missing term:

56. $(x + 11)^2 = x^2 + 22x + \square$

57. $\left(x - \dfrac{1}{4}\right)^2 = x^2 - \dfrac{1}{2}x + \square$

58. $x^2 - 12x + \square = (\quad)^2$

59. $x^2 + 5x + \square = (\quad)^2$

Solve:

60. $4x^2 - 24x + 27 = 0$

61. $4x^2 + 24x + 33 = 0$

》》》 *Skill Checker*

Write each equation in the standard form $ax^2 + bx + c = 0$, where a, b, and c are integers. (*Hint:* For equations involving fractions, first multiply each term by the LCD.)

62. $10x^2 + 5x = 12$

63. $x^2 = 2x + 2$

64. $9x = x^2$

65. $\dfrac{x^2}{2} + \dfrac{3}{5} = \dfrac{x}{4}$

66. $\dfrac{x}{4} = \dfrac{3}{5} - \dfrac{x^2}{2}$

67. $\dfrac{x}{3} = \dfrac{3}{4}x^2 - \dfrac{1}{6}$

9.3

Solving Quadratic Equations by the Quadratic Formula

▶ Objectives

A ⟩ Write a quadratic equation in the form $ax^2 + bx + c = 0$ and identify a, b, and c.

B ⟩ Solve a quadratic equation using the quadratic formula.

▶ To Succeed, Review How To . . .

1. Solve a quadratic equation by completing the square (pp. 670–676).
2. Write a quadratic equation in standard form (p. 442).

▶ Getting Started
Name That Formula!

As you were going through the preceding sections of this chapter, you probably wondered whether there is a surefire method for solving any quadratic equation. Fortunately, there is! As a matter of fact, this new technique incorporates the method of completing the square. As you recall, in each of the problems we followed the same procedure. Why not use the method of completing the square and solve the equation $ax^2 + bx + c = 0$ once and for all? We shall do that now. (You can refer to the procedure we gave to solve equations by completing the square on page 674.) Here is how the quadratic formula is derived:

"My final recommendation is that we reject all these proposals and continue to call it the quadratic formula."
Courtesy of Ralph Castellano

$$ax^2 + bx + c = 0, \quad a \neq 0 \qquad \text{Given}$$

Rewrite with the constant c on the right.
$$ax^2 + bx = -c$$

Divide each term by a.
$$x^2 + \frac{b}{a}x = -\frac{c}{a}$$

Add the square of one-half the coefficient of x.
$$x^2 + \frac{b}{a}x + \left(\frac{b}{2a}\right)^2 = \left(\frac{b}{2a}\right)^2 - \frac{c}{a}$$

Rewrite the left side as a perfect square.
$$\left(x + \frac{b}{2a}\right)^2 = \frac{b^2}{4a^2} - \frac{c}{a}$$

$$\left(x + \frac{b}{2a}\right)^2 = \frac{b^2}{4a^2} - \frac{4ac}{4a^2}$$

Write the right side as a single fraction.
$$\left(x + \frac{b}{2a}\right)^2 = \frac{b^2 - 4ac}{4a^2}$$

Now, take the square root of both sides.
$$x + \frac{b}{2a} = \frac{\pm\sqrt{b^2 - 4ac}}{2a}$$

Subtract $\frac{b}{2a}$ from both sides.
$$x = -\frac{b}{2a} \pm \frac{\sqrt{b^2 - 4ac}}{2a}$$

Combine fractions.
$$x = \frac{-b \pm \sqrt{b^2 - 4ac}}{2a}$$

There you have it, *the* surefire method for solving quadratic equations! We call this the *quadratic formula,* and we shall use it in this section to solve quadratic equations.

Do you realize what we've just done? We've given ourselves a powerful tool! *Any* time we have a quadratic equation in the standard form $ax^2 + bx + c = 0$, we can solve the equation by simply substituting a, b, and c in the formula.

THE QUADRATIC FORMULA

The solutions of $ax^2 + bx + c = 0$ $\quad(a \neq 0)$ are

$$x = \frac{-b \pm \sqrt{b^2 - 4ac}}{2a}$$

that is,

$$x = \frac{-b + \sqrt{b^2 - 4ac}}{2a} \quad \text{and} \quad x = \frac{-b - \sqrt{b^2 - 4ac}}{2a}$$

This formula is so important that you should memorize it right now. Before using it, however, you must remember to do two things:

1. First, write the given equation in the standard form $ax^2 + bx + c = 0$; then
2. Determine the values of a, b, and c.

A › Writing Quadratic Equations in Standard Form

In order to solve the equation $x^2 = 5x - 2$, we must first write the equation in standard form and then determine the values of a, b, and c as follows.

1. We must *first* write it in the standard form $ax^2 + bx + c = 0$ by subtracting $5x$ and adding 2; we then have

$$x^2 - 5x + 2 = 0 \quad \text{In standard form}$$

2. When the equation is in standard form, it's very easy to find the values of a, b, and c:

$$\underbrace{x^2}_{a\,=\,1} \quad - \quad \underbrace{5x}_{b\,=\,-5} \quad + \quad \underbrace{2}_{c\,=\,2} = 0 \qquad \text{Recall that } x^2 = 1x^2; \text{ thus, the coefficient of } x^2 \text{ is 1.}$$

If the equation contains fractions, multiply each term by the least common denominator (LCD) to clear the fractions. For example, consider the equation

$$\frac{x}{4} = \frac{3}{5} - \frac{x^2}{2}$$

Since the LCD of 4, 5, and 2 is 20, we multiply each term by 20:

$$20 \cdot \frac{x}{4} = \frac{3}{5} \cdot 20 - \frac{x^2}{2} \cdot 20$$

$$5x = 12 - 10x^2$$

$$10x^2 + 5x = 12 \qquad \text{Add } 10x^2.$$

$$10x^2 + 5x - 12 = 0 \qquad \text{Subtract 12.}$$

Now the equation is in standard form:

$$\underbrace{10x^2}_{a\,=\,10} \quad + \quad \underbrace{5x}_{b\,=\,5} \quad - \quad \underbrace{12}_{c\,=\,-12} = 0$$

You can get a lot of practice by completing this table:

	Given	Standard Form	a	b	c
1.	$2x^2 + 7x - 4 = 0$				
2.	$x^2 = 2x + 2$				
3.	$9x = x^2$				
4.	$\dfrac{x^2}{4} + \dfrac{2}{3}x = -\dfrac{1}{3}$				

STOP!   Do not proceed until you complete the table!

We will now use these four equations as examples.

B 〉 Solving Quadratic Equations with the Quadratic Formula

EXAMPLE 1   Solving a quadratic equation using the quadratic formula

Solve using the quadratic formula: $2x^2 + 7x - 4 = 0$

SOLUTION

1. The equation is already written in standard form:

$$\underbrace{2x^2}_{a\,=\,2} + \underbrace{7x}_{b\,=\,7} - \underbrace{4}_{c\,=\,-4} = 0$$

2. As Step 1 shows, it is clear that $a = 2$, $b = 7$, and $c = -4$.

3. Substituting the values of a, b, and c in the quadratic formula, we obtain

$$x = \frac{-7 \pm \sqrt{(7)^2 - 4(2)(-4)}}{2(2)}$$

$$x = \frac{-7 \pm \sqrt{49 + 32}}{4}$$

$$x = \frac{-7 \pm \sqrt{81}}{4}$$

$$x = \frac{-7 \pm 9}{4}$$

Thus,

$$x = \frac{-7 + 9}{4} = \frac{2}{4} = \frac{1}{2} \quad \text{or} \quad x = \frac{-7 - 9}{4} = \frac{-16}{4} = -4$$

The solutions are $\frac{1}{2}$ and -4.

PROBLEM 1

Solve using the quadratic formula:

$$3x^2 + 2x - 5 = 0$$

In Example 1, you could also solve $2x^2 + 7x - 4 = 0$ by factoring $2x^2 + 7x - 4$ and writing $(2x - 1)(x + 4) = 0$. You will get the same answers!

NOTE

Try factoring first unless otherwise directed.

EXAMPLE 2   Writing in standard form before using the quadratic formula

Solve using the quadratic formula: $x^2 = 2x + 2$

SOLUTION   We proceed by steps as in Example 1.

1. To write the equation in standard form, subtract $2x$ and then subtract 2 to obtain

PROBLEM 2

Solve using the quadratic formula:

$$x^2 = 4x + 4$$

Answers to PROBLEMS

1. The solutions are 1 and $-\frac{5}{3}$.   **2.** The solutions are $2 \pm 2\sqrt{2}$.

$$x^2 \underbrace{}_{a=1} - \underbrace{2x}_{b=-2} - \underbrace{2}_{c=-2} = 0$$

2. As Step 1 shows, $a = 1$, $b = -2$, and $c = -2$.

3. Substituting these values in the quadratic formula, we have

$$x = \frac{-(-2) \pm \sqrt{(-2)^2 - 4(1)(-2)}}{2(1)}$$

$$x = \frac{2 \pm \sqrt{4 + 8}}{2}$$

$$x = \frac{2 \pm \sqrt{12}}{2}$$

$$x = \frac{2 \pm \sqrt{4 \cdot 3}}{2}$$

$$x = \frac{2 \pm 2\sqrt{3}}{2}$$

Thus,

$$x = \frac{2 + 2\sqrt{3}}{2} = \frac{2(1 + \sqrt{3})}{2} = 1 + \sqrt{3}$$

or

$$x = \frac{2 - 2\sqrt{3}}{2} = \frac{2(1 - \sqrt{3})}{2} = 1 - \sqrt{3}$$

The solutions are $1 + \sqrt{3}$ and $1 - \sqrt{3}$.

  Note that $x^2 - 2x - 2$ is *not* factorable. You *must* know the quadratic formula or you must complete the square to solve this equation!

EXAMPLE 3   Rewriting a quadratic equation before using the quadratic formula

Solve using the quadratic formula: $9x = x^2$

SOLUTION   We proceed in steps.

1. Subtracting $9x$, we have

$$0 = x^2 - 9x$$

or

$$x^2 \underbrace{}_{a=1} - \underbrace{9x}_{b=-9} + \underbrace{0}_{c=0} = 0$$

2. As Step 1 shows, $a = 1$, $b = -9$, and $c = 0$ (because the c term is missing).

3. Substituting these values in the quadratic formula, we obtain

$$x = \frac{-(-9) \pm \sqrt{(-9)^2 - 4(1)(0)}}{2(1)}$$

$$x = \frac{9 \pm \sqrt{81 - 0}}{2}$$

$$x = \frac{9 \pm \sqrt{81}}{2}$$

$$x = \frac{9 \pm 9}{2}$$

Thus,

$$x = \frac{9 + 9}{2} = \frac{18}{2} = 9 \quad \text{or} \quad x = \frac{9 - 9}{2} = \frac{0}{2} = 0$$

The solutions are 9 and 0.

PROBLEM 3

Solve using the quadratic formula:

$$6x = x^2$$

Answers to PROBLEMS

3. The solutions are 0 and 6.

> **NOTE**
>
> $x^2 - 9x = x(x - 9) = 0$ could have been solved by factoring. Always try factoring first unless otherwise directed!

How do we know that the original equation could have been solved by factoring? We know because the answers are rational numbers (fractions or whole numbers). If your answer is a rational number, the equation could have been solved by factoring. Can we determine this in advance? Yes, and we will explain how to do so before the end of the section.

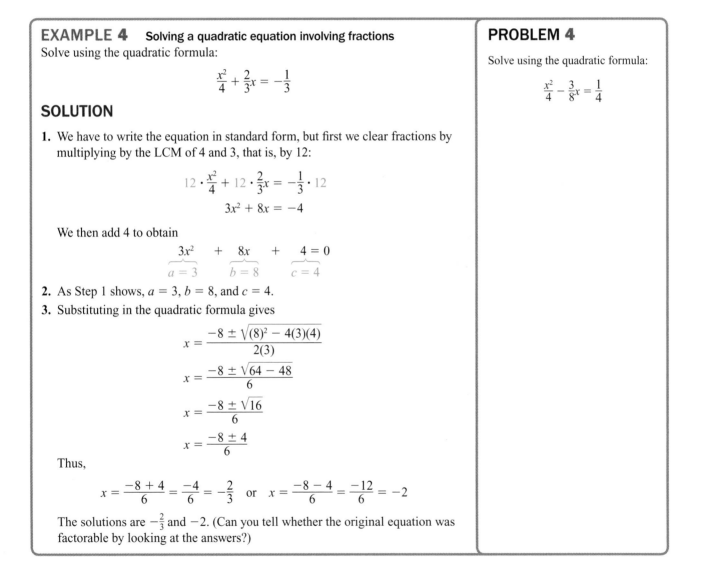

EXAMPLE 4   Solving a quadratic equation involving fractions

Solve using the quadratic formula:

$$\frac{x^2}{4} + \frac{2}{3}x = -\frac{1}{3}$$

SOLUTION

1. We have to write the equation in standard form, but first we clear fractions by multiplying by the LCM of 4 and 3, that is, by 12:

$$12 \cdot \frac{x^2}{4} + 12 \cdot \frac{2}{3}x = -\frac{1}{3} \cdot 12$$

$$3x^2 + 8x = -4$$

We then add 4 to obtain

$$\underbrace{3x^2}_{a=3} + \underbrace{8x}_{b=8} + \underbrace{4}_{c=4} = 0$$

2. As Step 1 shows, $a = 3$, $b = 8$, and $c = 4$.

3. Substituting in the quadratic formula gives

$$x = \frac{-8 \pm \sqrt{(8)^2 - 4(3)(4)}}{2(3)}$$

$$x = \frac{-8 \pm \sqrt{64 - 48}}{6}$$

$$x = \frac{-8 \pm \sqrt{16}}{6}$$

$$x = \frac{-8 \pm 4}{6}$$

Thus,

$$x = \frac{-8 + 4}{6} = \frac{-4}{6} = -\frac{2}{3} \quad \text{or} \quad x = \frac{-8 - 4}{6} = \frac{-12}{6} = -2$$

The solutions are $-\frac{2}{3}$ and -2. (Can you tell whether the original equation was factorable by looking at the answers?)

PROBLEM 4

Solve using the quadratic formula:

$$\frac{x^2}{4} - \frac{3}{8}x = \frac{1}{4}$$

Now, a final word of warning. As you recall, some quadratic equations do not have real-number solutions. This is still true, even when you use the quadratic formula. The next example shows how this happens.

Answers to PROBLEMS

4. The solutions are 2 and $-\frac{1}{2}$.

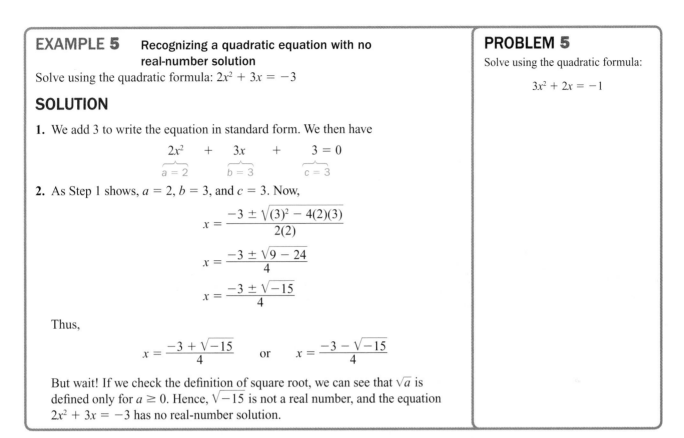

EXAMPLE 5   Recognizing a quadratic equation with no real-number solution

Solve using the quadratic formula: $2x^2 + 3x = -3$

SOLUTION

1. We add 3 to write the equation in standard form. We then have

$$\underbrace{2x^2}_{a\,=\,2} + \underbrace{3x}_{b\,=\,3} + \underbrace{3}_{c\,=\,3} = 0$$

2. As Step 1 shows, $a = 2$, $b = 3$, and $c = 3$. Now,

$$x = \frac{-3 \pm \sqrt{(3)^2 - 4(2)(3)}}{2(2)}$$

$$x = \frac{-3 \pm \sqrt{9 - 24}}{4}$$

$$x = \frac{-3 \pm \sqrt{-15}}{4}$$

Thus,

$$x = \frac{-3 + \sqrt{-15}}{4} \quad \text{or} \quad x = \frac{-3 - \sqrt{-15}}{4}$$

But wait! If we check the definition of square root, we can see that $\sqrt{a}$ is defined only for $a \geq 0$. Hence, $\sqrt{-15}$ is not a real number, and the equation $2x^2 + 3x = -3$ has no real-number solution.

PROBLEM 5

Solve using the quadratic formula:

$$3x^2 + 2x = -1$$

As you can see from Example 5, the number -15 under the radical determines the type of roots the equation has. The expression $b^2 - 4ac$ under the radical in the quadratic formula is called the **discriminant.**

THE DISCRIMINANT

If $b^2 - 4ac \geq 0$, the equation has real-number solutions.

If $b^2 - 4ac < 0$, the equation has no real-number solutions.

Note that in Example 5, $a = 2$, $b = 3$, and $c = 3$. Thus,

$$b^2 - 4ac = 3^2 - 4(2)(3) = 9 - 24 = -15 < 0$$

so the equation has no real-number solutions.

Is the discriminant good for anything else? Of course. The discriminant can also suggest the method you use to solve quadratic equations. Here are the suggestions:

Start by writing the equation in standard form. Then, if $b^2 - 4ac \geq 0$ is a perfect square, try factoring. (*Note:* Some perfect squares are 0, 1, 4, 9, 16, and so on.)

In Example 1, $2x^2 + 7x - 4 = 0$ and $b^2 - 4ac = 81$, so we could try to solve the equation by factoring.

In Example 2, $x^2 - 2x - 2 = 0$ and $b^2 - 4ac = 12$. The equation cannot be factored!

In Example 3, $x^2 - 9x = 0$ and $b^2 - 4ac = 81$. Try factoring!

In Example 4, $3x^2 + 8x + 4 = 0$ and $b^2 - 4ac = 16$. Try to solve by factoring.

In Example 5, $2x^2 + 3x + 3 = 0$ and $b^2 - 4ac = -15$, so you know there are no real-number solutions.

Answers to PROBLEMS

5. No real-number solution

We list the best method to solve each type of quadratic equation in the following chart.

Type of Equation	Method to Solve
$X^2 = A$	Square root property
$(AX \pm B)^2 = C$	Square root property
$ax^2 + bx + c = 0$	Factor if $b^2 - 4ac$ is a perfect square.
$ax^2 + bx + c = 0$	Quadratic formula if $b^2 - 4ac$ is not a perfect square

Note: If $b^2 - 4ac < 0$, there is no real-number solution. You do not even have to try a method!

› Exercises 9.3

‹ A › Writing Quadratic Equations in Standard Form
‹ B › Solving Quadratic Equations with the Quadratic Formula

In Problems 1–30, write the given equation in standard form, then solve the equation using the quadratic formula.

1. $x^2 + 3x + 2 = 0$

2. $x^2 + 4x + 3 = 0$

3. $x^2 + x - 2 = 0$

4. $x^2 + x - 6 = 0$

5. $2x^2 + x - 2 = 0$

6. $2x^2 + 7x + 3 = 0$

7. $3x^2 + x = 2$

8. $3x^2 + 2x = 5$

9. $2x^2 + 7x = -6$

10. $2x^2 - 7x = -6$

11. $7x^2 = 12x - 5$

12. $-5x^2 = 16x + 8$

13. $5x^2 = 11x - 4$

14. $7x^2 = 12x - 3$

15. $\dfrac{x^2}{5} - \dfrac{x}{2} = -\dfrac{3}{10}$

16. $\dfrac{x^2}{7} + \dfrac{x}{2} = -\dfrac{3}{4}$

17. $\dfrac{x^2}{2} = \dfrac{3x}{4} - \dfrac{1}{8}$

18. $\dfrac{x^2}{10} = \dfrac{x}{5} + \dfrac{3}{2}$

19. $\dfrac{x^2}{8} = -\dfrac{x}{4} - \dfrac{1}{8}$

20. $\dfrac{x^2}{3} = -\dfrac{x}{3} - \dfrac{1}{12}$

21. $6x = 4x^2 + 1$

22. $6x = 9x^2 - 4$

23. $3x = 1 - 3x^2$

24. $3x = 2x^2 - 5$

25. $x(x + 2) = 2x(x + 1) - 4$

26. $x(4x - 7) - 10 = 6x^2 - 7x$

27. $6x(x + 5) = (x + 15)^2$

28. $6x(x + 1) = (x + 3)^2$

29. $(x - 2)^2 = 4x(x - 1)$

30. $(x - 4)^2 = 4x(x - 2)$

› › ›   Applications

31. *Customer service cost*   The customer service department at a store has determined that the cost C of serving x customers is modeled by the equation $C = (0.1x^2 + x + 50)$ dollars. How many customers can be served if they are willing to spend $250?

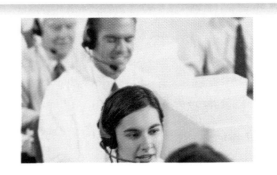

32. *Customer service cost*   The cost C of serving x customers at a boat dealership is given by the equation $C = (x^2 + 10x + 100)$ dollars. If the dealership spent \$1300 serving customers, how many customers were served?

33. *Computer manufacturing*   A computer manufacturer will produce x computers when the price P is given by the equation $P = (x^2 + 25x)$ dollars per unit. How many units will be produced when the price is \$350 per unit?

34. *Manufacturing*   When the price P of a ton of raw materials is modeled by the equation $P = (0.01x^2 + 5x)$ dollars, a supplier will produce x tons of it. How many tons will be produced when the price is \$5000 per ton?

❯❯❯   Using Your Knowledge

Deriving the Quadratic Formula and Deciding When to Use It   In this section, we derived the quadratic formula by completing the square. The procedure depends on making the x coefficient a equal to 1. But there's another way to derive the quadratic formula. See whether you can identify the change being made to $ax^2 + bx + c = 0$ in each step.

$$ax^2 + bx + c = 0 \qquad \text{Given}$$

35. $4a^2x^2 + 4abx + 4ac = 0$

36. $4a^2x^2 + 4abx = -4ac$

37. $4a^2x^2 + 4abx + b^2 = b^2 - 4ac$

38. $(2ax + b)^2 = b^2 - 4ac$

39. $2ax + b = \pm\sqrt{b^2 - 4ac}$

40. $2ax = -b \pm \sqrt{b^2 - 4ac}$

41. $x = \dfrac{-b \pm \sqrt{b^2 - 4ac}}{2a}$

The quadratic formula is not the only way to solve quadratic equations. We've actually studied four methods for solving them. The following table lists these methods and suggests the best use for each.

Method	When Used
1. The square root property	When you can write the equation in the form $X^2 = A$
2. Factoring	When the equation is factorable. Use the *ac* test (from Section 5.3) to find out!
3. Completing the square	You can always use this method, but it requires more steps than the other methods.
4. The quadratic formula	This formula can always be used, but you should try factoring first.

In Problems 42–56, solve the equation by the method of your choice.

42. $x^2 = 144$

43. $x^2 - 17 = 0$

44. $x^2 - 2x = -1$

45. $x^2 + 4x = -4$

46. $x^2 - x - 1 = 0$

47. $y^2 - y = 0$

48. $y^2 + 5y - 6 = 0$

49. $x^2 + 5x + 6 = 0$

50. $5y^2 = 6y - 1$

51. $(z + 2)(z + 4) = 8$

52. $(y - 3)(y + 4) = 18$

53. $y^2 = 1 - \dfrac{8}{3}y$

54. $x^2 = \dfrac{3}{2}x + 1$

55. $3z^2 + 1 = z$

56. $\dfrac{1}{2}x^2 - \dfrac{1}{2}x = \dfrac{1}{4}$

〉〉〉 *Write On*

In the quadratic formula, the expression $D = b^2 - 4ac$ appears under the radical, and it is called the **discriminant** of the equation $ax^2 + bx + c = 0$.

57. How many solutions does the equation have if $D = 0$? Is (are) the solution(s) rational or irrational?

58. How many solutions does the equation have if $D > 0$?

59. How many solutions does the equation have if $D < 0$?

60. Suppose that one solution of $ax^2 + bx + c = 0$ is

$$x = \frac{-b + \sqrt{D}}{2a}$$

What is the other solution?

〉〉〉 *Concept Checker*

Fill in the blank(s) with the correct word(s), phrase, or mathematical statement.

61. The **solutions** of the equation $ax^2 + bx + c = 0$ are _____.

62. When **solving** the equation $ax^2 + bx + c = 0$, if $b^2 - 4ac \geq 0$ the equation has _____ number solutions.

no

$$\dfrac{b \pm \sqrt{b^2 - 4ac}}{2a}$$

$$\dfrac{-b \pm \sqrt{b^2 - 4ac}}{2a}$$

real

〉〉〉 *Mastery Test*

Solve if possible:

63. $3x^2 + 2x = -1$

64. $6x = x^2$

65. $x^2 = 4x + 8$

66. $3x^2 + 2x - 5 = 0$

67. $\dfrac{x^2}{4} - \dfrac{3}{8}x = \dfrac{1}{4}$

68. $\dfrac{2}{3}x^2 + x = -1$

〉〉〉 *Skill Checker*

Find:

69. $(-2)^2$

70. -2^2

71. $(-1)^2$

72. -1^2

9.4   Graphing Quadratic Equations

▶ Objectives

A ⟩ Graph quadratic equations.

B ⟩ Find the intercepts, the vertex, and graph parabolas involving factorable quadratic expressions.

▶ To Succeed, Review How To . . .

1. Evaluate an expression (pp. 61–62, 69–72).
2. Graph points in the plane (p. 205).
3. Factor a quadratic (pp. 407–409, 419–421).

▶ Getting Started
Parabolic Water Streams

Look at the streams of water from the fountain. What shape do they have? This shape is called a **parabola.** The *graph* of a quadratic equation of the form $y = ax^2 + bx + c$ is a parabola. The simplest of these equations is $y = x^2$. This equation can be graphed in the same way as lines were graphed—that is, by selecting values for x and then finding the corresponding y-values as shown in the table on the left. The usual shortened version is shown in the table on the right.

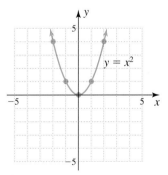

x-value	*y*-value
$x = -2$	$y = x^2 = (-2)^2 = 4$
$x = -1$	$y = x^2 = (-1)^2 = 1$
$x = 0$	$y = x^2 = (0)^2 = 0$
$x = 1$	$y = x^2 = (1)^2 = 1$
$x = 2$	$y = x^2 = (2)^2 = 4$

x	*y*
-2	4
-1	1
0	0
1	1
2	4

We graph these points on a coordinate system and draw a smooth curve through them. The result is the graph of the parabola $y = x^2$. Note that the arrows at the end indicate that the curve goes on indefinitely.

In this section, we shall learn how to graph parabolas.

A ⟩ Graphing Quadratic Equations

The graph of a quadratic equation of the form $y = ax^2 + bx + c$ is a **parabola.** Now that we know how to graph $y = x^2$, what would happen if we graph $y = -x^2$? To start, we could alter the table in the *Getting Started* by using the negative of the y-value on $y = x^2$, as shown next.

EXAMPLE 1   Graphing $y = ax^2$ when a is negative

Graph: $y = -x^2$

SOLUTION   We could always make a table of x- and y-values as before. However, note that for any x-value, the y-value will be the *negative* of the y-value on the parabola $y = x^2$. (If you don't believe this, go ahead and make the table and check it.) Thus, the parabola $y = -x^2$ has the same shape as $y = x^2$, but it's turned in the *opposite* direction (opens *downward*). The graph of $y = -x^2$ is shown in Figure 9.1.

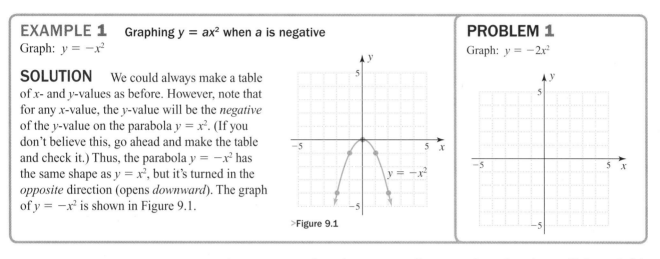

>Figure 9.1

PROBLEM 1

Graph: $y = -2x^2$

As you can see from the two preceding examples, when the coefficient of x^2 is positive (as in $y = x^2 = 1x^2$), the parabola opens *upward,* but when the coefficient of x^2 is negative (as in $y = -x^2 = -1x^2$), the parabola opens *downward*. In general, we have the following definition.

> **GRAPH OF A QUADRATIC EQUATION**
>
> The graph of a quadratic equation of the form $y = ax^2 + bx + c$ is a parabola that
> **1.** Opens upward if $a > 0$
> **2.** Opens downward if $a < 0$

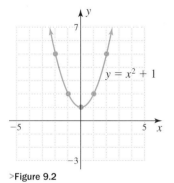

>Figure 9.2

Now, what do you think will happen if we graph the parabola $y = x^2 + 1$? Two things: First, the parabola opens upward, since the coefficient of x^2 is understood to be 1. Second, all of the points will be 1 unit higher than those for the same value of x on the parabola $y = x^2$. Thus, we can make the graph of $y = x^2 + 1$ by following the pattern of $y = x^2$. Similarly, the graph of $y = x^2 + 2$ is 2 units higher than for the parabola $y = x^2$. The graphs of $y = x^2 + 1$ and $y = x^2 + 2$ are shown in Figures 9.2 and 9.3, respectively.

You can verify that these graphs are correct by graphing several of the points for $y = x^2 + 1$ and $y = x^2 + 2$ that are shown in the following tables:

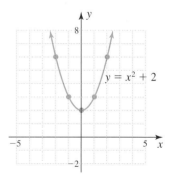

>Figure 9.3

Answers to PROBLEMS

1.

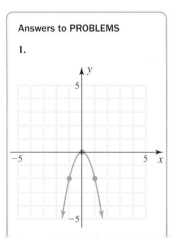

$y = x^2 + 1$	
x	**y**
0	1
1	2
−1	2

$y = x^2 + 2$	
x	**y**
0	2
1	3
−1	3

EXAMPLE 2   Graphing a parabola that opens downward

Graph: $y = -x^2 - 2$

SOLUTION   Since the coefficient of x^2 (which is understood to be -1) is *negative*, the parabola opens downward. It is also 2 units *lower* than the graph of $y = -x^2$. Thus, the graph of $y = -x^2 - 2$ is as shown in Figure 9.4.

You can verify the graph by checking the points in the table for $y = -x^2 - 2$:

$y = -x^2 - 2$	
x	**y**
0	-2
1	-3
-1	-3

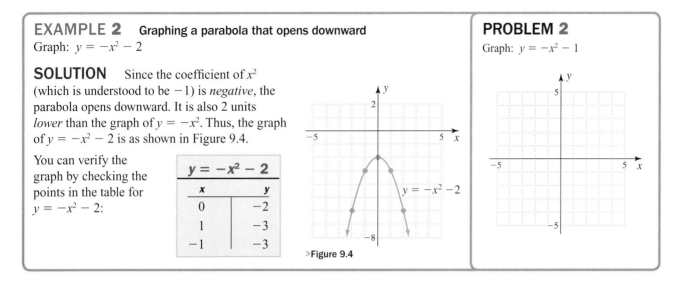

>Figure 9.4

PROBLEM 2

Graph: $y = -x^2 - 1$

So far, we've graphed parabolas only of the form $y = ax^2 + b$. How do you think the graph of $y = (x - 1)^2$ looks? As before, we make a table of values. For example,

For $x = -1$,	$y = (-1 - 1)^2 = (-2)^2 = 4$
For $x = 0$,	$y = (0 - 1)^2 = (-1)^2 = 1$
For $x = 1$,	$y = (1 - 1)^2 = (0)^2 = 0$
For $x = 2$,	$y = (2 - 1)^2 = 1^2 = 1$

$y = (x - 1)^2$	
x	**y**
-1	4
0	1
1	0
2	1

The completed graph is shown in Figure 9.5. Note that the shape of the graph is identical to that of $y = x^2$, but it is shifted 1 unit to the *right*.

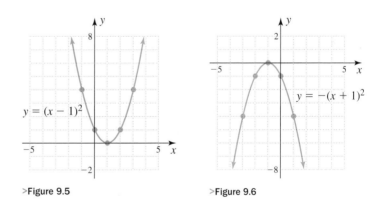

>Figure 9.5   >Figure 9.6

Similarly, the graph of $y = -(x + 1)^2$ is identical to that of $y = -x^2$ but shifted 1 unit to the *left,* as shown in Figure 9.6.

You can verify this using the table for $y = -(x + 1)^2$:

$y = -(x + 1)^2$	
x	**y**
0	-1
-1	0
-2	-1

Answers to PROBLEMS

2.

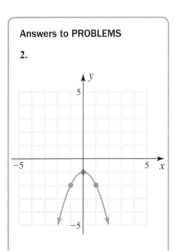

EXAMPLE 3   Graphing by vertical and horizontal shifts

Graph: $y = -(x - 1)^2 + 2$

SOLUTION   The graph of this equation is identical to the graph of $y = -x^2$ except for its position. The parabola $y = -(x - 1)^2 + 2$ opens downward (because of the first negative sign), and it's shifted 1 unit to the right (because of the -1) and 2 units up (because of the $+2$):

$$y = -(x - 1)^2 \quad + \quad 2$$

    ↑   ↑   ↑

Opens   Shifted   Shifted
downward 1 unit 2 units
(negative) right   up

>**Figure 9.7**

Figure 9.7 shows the finished graph of $y = -(x - 1)^2 + 2$.

As usual, you can verify this using the table for $y = -(x - 1)^2 + 2$:

$y = -(x - 1)^2 + 2$	
x	**y**
0	1
1	2
2	1

PROBLEM 3

Graph: $y = -(x - 2)^2 - 1$

In conclusion, we follow these directions for changing the graph of $y = ax^2$:

$$y = a(x \;\pm\; b)^2 \quad \pm \quad c \qquad \text{b and c positive}$$

   ↑   ↑   ↑

Opens upward   Shifts the   Moves the
for $a > 0$,   graph right   graph up
downward $(-b)$ or $(+c)$ or
for $a < 0$ left $(+b)$ down $(-c)$

Note that as a increases, the parabola "stretches." (We shall examine this fact in the *Calculator Corner*.)

B ❭ Graphing Parabolas Using Intercepts and the Vertex

You've probably noticed that the graph of a parabola is **symmetric**; that is, if you draw a vertical line through the **vertex** (the high or low point on the parabola) and fold the graph along this line, the two halves of the parabola coincide. If a parabola crosses the x-axis, we can use the x-intercepts to find the vertex. For example, to graph $y = x^2 + 2x - 8$, we start by finding the x- and y-intercepts:

For $x = 0$, $y = 0^2 + 2 \cdot 0 - 8 = -8$ The y-intercept

For $y = 0$, $0 = x^2 + 2x - 8$

$0 = (x + 4)(x - 2)$

Thus,

$$x = -4 \quad \text{or} \quad x = 2 \quad \text{The x-intercepts}$$

We enter the points $(0, -8)$, $(-4, 0)$, and $(2, 0)$ in a table like this:

x	y	
0	-8	y-intercept
-4	0	x-intercepts
2	0	
?	?	Vertex

Answers to PROBLEMS

3.

How can we find the vertex? Since the parabola is symmetric, the x-coordinate of the vertex is *exactly* halfway between the x-intercepts -4 and 2, so

$$x = \frac{-4 + 2}{2} = \frac{-2}{2} = -1$$

We then find the y-coordinate by letting $x = -1$ in $y = x^2 + 2x - 8$ to obtain

$$y = (-1)^2 + 2 \cdot (-1) - 8 = -9$$

Thus, the vertex is at $(-1, -9)$. The table now looks like this:

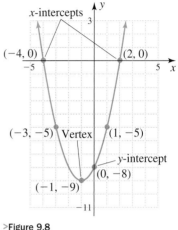

x	y	
0	-8	y-intercept
-4	0	$\left.\vphantom{\begin{matrix}a\\b\end{matrix}}\right\}$ x-intercepts
2	0	
-1	-9	Vertex

We can draw the graph using these four points or plot one or two more points. We plotted $(1, -5)$ and $(-3, -5)$. The graph is shown in Figure 9.8.

Here is the complete procedure.

>Figure 9.8

PROCEDURE

Graphing a Factorable Quadratic Equation

1. Find the y-intercept by letting $x = 0$ and then finding y.
2. Find the x-intercepts by letting $y = 0$, factoring the equation, and solving for x.
3. Find the vertex by averaging the solutions of the equation found in Step 2 (the average is the x-coordinate of the vertex) and substituting in the equation to find the y-coordinate of the vertex.
4. Plot the points found in Steps 1–3 and one or two more points, if desired. The curve drawn through the points found in Steps 1–4 is the graph.

Now let's use this procedure to graph a parabola.

EXAMPLE 4   Graphing parabolas using the intercepts and vertex
Graph: $y = -x^2 + 2x + 8$

SOLUTION   We use the four steps discussed.

1. Since $-x^2 = -1 \cdot x^2$ and -1 is negative, the parabola opens downward. We then let $x = 0$ to obtain $y = -(0)^2 + 2 \cdot 0 + 8 = 8$. Thus, $(0, 8)$ is the y-intercept.
2. We find the x-intercepts by letting $y = 0$ and solving for x. We have

$$0 = -x^2 + 2x + 8$$
$$0 = x^2 - 2x - 8 \quad \text{Multiply both sides by } -1 \text{ to}$$
$$0 = (x - 4)(x + 2) \quad \text{make the factorization easier.}$$
$$x = 4 \quad \text{or} \quad x = -2$$

We now have the x-intercepts: $(4, 0)$ and $(-2, 0)$.

PROBLEM 4

Graph: $y = -x^2 - 2x + 8$

Answer on Page 694

(continued)

3. The x-coordinate of the vertex is found by averaging 4 and -2 to obtain

$$x = \frac{4 + (-2)}{2} = \frac{2}{2} = 1$$

Letting $x = 1$ in $y = -x^2 + 2x + 8$, we find $y = -(1)^2 + 2 \cdot (1) + 8 = 9$, so the vertex is at $(1, 9)$.

4. We plot these points and the two additional points $(2, 8)$ and $(3, 5)$. The completed graph is shown in Figure 9.9.

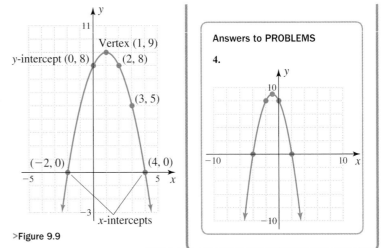

>**Figure 9.9**

Answers to PROBLEMS

4.

Actually, there is another way of finding the vertex of a parabola (without having to take averages). The technique involves completing the square! Suppose we wish to graph the equation

$$y = ax^2 + bx + c$$

Factor a from the first two terms
$$y = a\left(x^2 + \frac{b}{a}x\right) + c$$

Complete the square on $x^2 + \frac{b}{a}x$

by subtracting $\left(\frac{b}{2a}\right)^2$ and adding $\left(\frac{b}{2a}\right)^2$
$$y = a\left[x^2 + \frac{b}{a}x + \left(\frac{b}{2a}\right)^2\right] + c - a\left(\frac{b}{2a}\right)^2$$

Rewrite as a binomial square
$$y = a\left(x + \frac{b}{2a}\right)^2 + c - a\left(\frac{b}{2a}\right)^2$$

If $a > 0$ the parabola is $\cup$ shaped and y is a **minimum** when $x + \frac{b}{2a} = 0$, that is, when $x = -\frac{b}{2a}$. This means that the vertex is at $x = -\frac{b}{2a}$.

If $a < 0$ the parabola is $\cap$ shaped and y is a **maximum** when $x + \frac{b}{2a} = 0$, that is, when $x = -\frac{b}{2a}$. In either case, we have the following:

VERTEX OF A PARABOLA

The vertex of the parabola $y = ax^2 + bx + c$ has coordinate $x = -\frac{b}{2a}$. You **do not** have to memorize the y-coordinate of the vertex. After you find the x-coordinate, substitute $x = -\frac{b}{2a}$ in $y = ax^2 + bx + c$ and find y.

For example, to find the vertex of the parabola $y = -x^2 + 2x + 8$ of Example 4 simply let $x = -\frac{b}{2a} = -\frac{2}{2(-1)} = 1$. Substituting $x = 1$, we get $y = -(1)^2 + 2(1) + 8 = 9$. So the vertex is $(1, 9)$, as before.

EXAMPLE 5   Recording a maximum revenue

A company's revenue is modeled by the equation $R = 10x - 0.01x^2$, where R is revenue and x is the number of CDs made and sold per week. How many CDs does the company have to sell to obtain maximum revenue?

PROBLEM 5

A farmer has 200 feet of fencing he wants to use to enclose a rectangular plot of land for his wife's vegetable garden. What dimensions will give her the maximum area?

$100 - w$

Answers to PROBLEMS

5. 50 ft $\times$ 50 ft

SOLUTION   We first write the equation as $R = -0.01x^2 + 10x$. Since the coefficient of x^2 is negative, the parabola opens downward (is concave down), and the vertex is its highest point. Letting $a = -0.01$ and $b = 10$, we have

$$x = -\frac{b}{2a} = -\frac{10}{-0.02} = 500$$

$R = 10(500) - 0.01(500)^2 = 5000 - 2500 = 2500$. When the company sells $x = 500$ CDs a week, the revenue is a maximum: \$2500.

▢ ◈ ▥   Calculator Corner

Graphing Parabolas

If you have a graphing calculator, you can graph parabolas very easily. Let's explore the effect of different values of a, b, and c in the equation

$$y = a(x \pm b) \pm c$$

Using a standard window, graph $y = x^2$, $y = x^2 + 3$, and $y = x^2 - 3$. Clearly, adding the positive number k "moves" the graph of the parabola k units *up*. Similarly, subtracting the positive number k "moves" the graph of the parabola k units *down*, as shown in Window 1. Graph $y = -x^2$, $y = -x^2 + 3$, and $y = -x^2 - 3$ to check the effect of having a negative sign in front of x^2.

     Now, let's try some stretching exercises! Look at the graphs of $y = x^2$, $y = 2x^2$, and $y = 5x^2$ using a decimal window. As you can see in Window 2, as the coefficient of x^2 increases, the parabola "stretches"!

     Can you find the vertex and intercepts with your calculator? Yes, they're quite easy to find using a calculator. For example, let's use a decimal window to graph

$$y = x^2 - x - 2$$

which is shown in Window 3. You can use the [TRACE] key to find the vertex $(0.5, -2.25)$. Better yet, some calculators find the minimum of a function by pressing [2nd] [TRACE] [3] and following the prompts. The **minimum,** of course, occurs at the vertex, as shown in Window 4. On the other hand, the **maximum** of $y = -x^2 + x + 2$ occurs at the vertex. Can you find this maximum?

     Let's return to the parabola $y = x^2 - x - 2$. This parabola has two x-intercepts, as shown in Window 5. Using your [TRACE] key and a decimal window, you can find both of them! Window 5 shows that one of the intercepts occurs at $x = -1$. The other x-intercept (not shown) occurs when $x = 2$. Now solve the equation $0 = x^2 - x - 2$. What are the solutions of this equation? Do you see a relationship between the graph of $y = ax^2 + bx + c$ and the solutions of $0 = ax^2 + bx + c$? Can you devise a procedure so that you can solve quadratic equations using a calculator?

     If you're lucky, your calculator also has a "root" or "zero" feature. With this feature, you can graph $y = x^2 - x - 2$ and ask the calculator to find the root. The prompts ask "Left bound?" Use the ◆ key to move the cursor to a point left of the intercept (where this graph is above the x-axis). Find a point below the x-axis for the "Right bound" and press [ENTER] when asked for a "Guess." The root -1 is shown in Window 6.

Window 1

Window 2

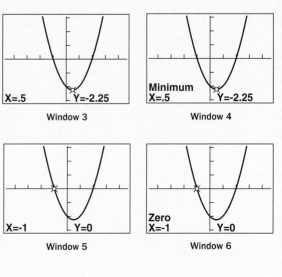

X=.5   Y=-2.25

Window 3

Minimum
X=.5   Y=-2.25

Window 4

X=-1   Y=0

Window 5

Zero
X=-1   Y=0

Window 6

> Exercises **9.4**

〈 **A** 〉 **Graphing Quadratic Equations**   In Problems 1–16, graph the given equation.

1. $y = 2x^2$

2. $y = 2x^2 + 1$

3. $y = 2x^2 - 1$

4. $y = 2x^2 - 2$

5. $y = -2x^2 + 2$

6. $y = -2x^2 - 1$

7. $y = -2x^2 - 2$

8. $y = -2x^2 + 1$

9. $y = (x - 2)^2$

10. $y = (x - 2)^2 + 2$

11. $y = (x - 2)^2 - 2$

12. $y = (x - 2)^2 - 1$

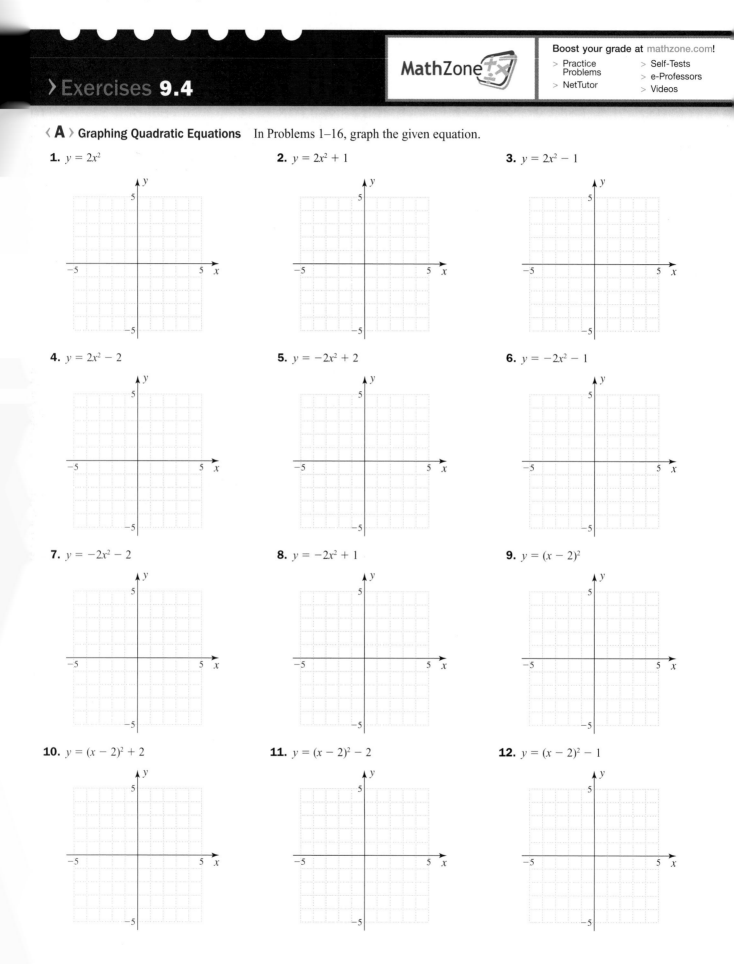

13. $y = -(x - 2)^2$

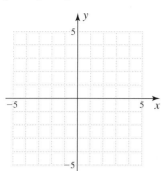

14. $y = -(x - 2)^2 + 2$

15. $y = -(x - 2)^2 - 2$

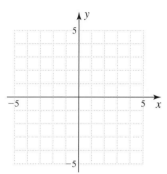

16. $y = -(x - 2)^2 - 3$

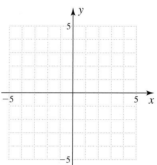

‹ **B** › **Graphing Parabolas Using Intercepts and the Vertex**   In Problems 17–22, find the x-intercepts, the y-intercept, the vertex, and then draw the graph.

17. $y = x^2 + 4x + 3$

18. $y = x^2 - 4x + 4$

19. $y = x^2 + 2x - 3$

20. $y = -x^2 - 4x - 3$

21. $y = -x^2 + 4x - 3$

22. $y = -x^2 - 2x + 3$

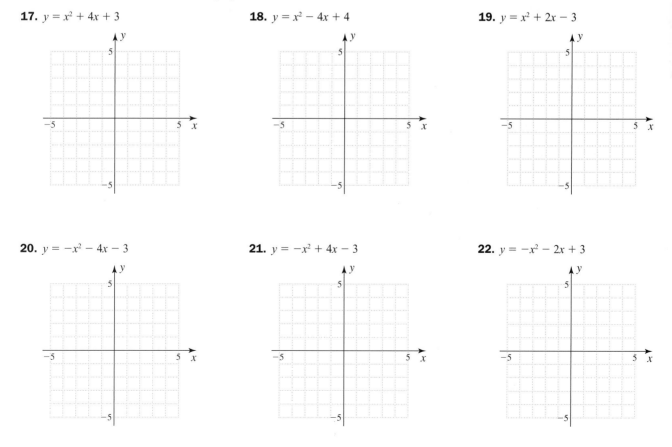

> > > **Applications**

Travel from Earth to Moon   The following graph looks somewhat like one-half of a parabola. It indicates the time it takes to travel from the earth to the moon at different speeds. Use it to solve Problems 23–26.

23. How long does the trip take if you travel at 10,000 kilometers per hour?

24. How long does the trip take if you travel at 25,000 kilometers per hour?

25. *Apollo 11* averaged about 6400 kilometers per hour when returning from the moon. How long was the trip?

26. What should the speed of the shuttle be if you wish to make the trip in 50 hours?

27. *Maximizing profit*   The profit P (in dollars) for a company is modeled by the equation $P = -5000 + 8x - 0.001x^2$, where x is the number of items produced each month. How many items does the company have to produce to obtain maximum profit? What is this profit?

28. *Maximizing revenue*   The revenue R for Shady Glasses is given by the equation $R = 1500p - 75p^2$, where p is the price of each pair of sunglasses (R and p in dollars). What should the price be to maximize revenue?

29. *Maximizing sales*   After spending x thousand dollars on an advertising campaign, the number of units N sold is given by the equation $N = 50x - x^2$. How much should be spent in the campaign to obtain maximum sales?

30. *Maximizing sales*   The number N of units of a product sold after a television commercial blitz is modeled by the equation $N = 40x - x^2$, where x is the amount spent in thousands of dollars. How much should be spent on television commercials to obtain maximum sales?

> > > **Using Your Knowledge**

Maximizing Profits   The ideas studied in this section are used in business to find ways to maximize profits. For example, if a manufacturer can produce a certain item for $10 each, and then sell the item for x dollars, the profit per item will be $x - 10$ dollars. If it is then estimated that consumers will buy $60 - x$ items per month, the total profit will be

$$\text{Total profit} = \begin{pmatrix}\text{Number of} \\ \text{items sold}\end{pmatrix}\begin{pmatrix}\text{Profit per} \\ \text{item}\end{pmatrix}$$
$$= (60 - x)(x - 10)$$

The graph for the total profit is this parabola:

When will the profits be at a maximum? When the manufacturer produces 35 items. Note that 35 is exactly halfway between

$$10 \quad \text{and} \quad 60 \quad \left(\frac{10 + 60}{2} = 35\right)$$

At this price, the total profits will be

$$P_T = (60 - 35)(35 - 10)$$
$$= (25)(25)$$
$$= \$625$$

Use your knowledge to answer the following questions.

31. What price would maximize the profits of a certain item costing $20 each if $60 - x$ items are sold each month (where x is the selling price of each item)?

32. Sketch the graph of the resulting parabola.

33. What will the maximum profits be?

⟩⟩⟩ *Write On*

34. What is the vertex of a parabola?

For Problems 35–40, consider the graph of the parabola $y = ax^2$.

35. What can you say if $a > 0$?

36. What can you say if $a < 0$?

37. If a is a positive integer, what happens to the graph of $y = ax^2$ as a increases?

38. What causes the graph of $y = ax^2$ to be wider or narrower than that of $y = x^2$?

39. What happens to the graph of $y = ax^2$ if you add the positive constant k? What happens if k is negative?

40. If a parabola has two x-intercepts and the vertex is at $(1, 1)$, does the parabola open upward or downward? Explain.

⟩⟩⟩ *Concept Checker*

Fill in the blank(s) with the correct word(s), phrase, or mathematical statement.

41. The graph of $y = ax^2 + bx + c$ is a **parabola** and opens **upward** (∪) when _____.

42. The graph of $y = ax^2 + bx + c$ is a **parabola** and opens **downward** (∩) when _____.

a < 0

a = 0

a > 0

⟩⟩⟩ *Mastery Test*

Graph:

43. $y = -(x - 2)^2 - 1$

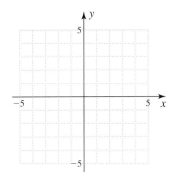

44. $y = -x^2 - 1$

45. $y = -2x^2$

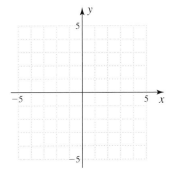

46. $y = -2x^2 + 3$

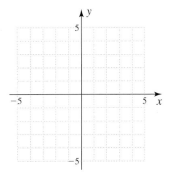

Graph and label the vertex and intercepts:

47. $y = -x^2 - 2x + 3$

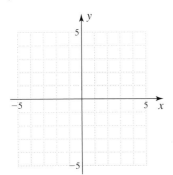

48. $y = x^2 + 3x + 2$

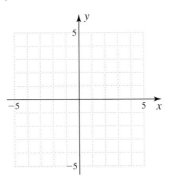

〉〉〉 *Skill Checker*

Find:

49. $3^2 + 4^2$

50. $5^2 + 12^2$

51. $\dfrac{120}{0.003}$

52. $\dfrac{150}{0.005}$

9.5

The Pythagorean Theorem and Other Applications

▶ Objectives

A ⟩ Use the Pythagorean theorem to solve right triangles.

B ⟩ Solve word problems involving quadratic equations.

▶ To Succeed, Review How To . . .

1. Find the square of a number (p. 61).
2. Divide by a decimal (pp. 22–23).
3. Solve a quadratic by factoring (pp. 440–446).

▶ Getting Started
A Theorem Not Dumb Enough!

In this section we discuss several applications involving quadratic equations. One of them pertains to the theorem mentioned in the cartoon. This theorem, first proved by Pythagoras, will be stated next.

By Permission of John L. Hart FLP, and Creators Syndicate, Inc.

A › Using the Pythagorean Theorem

Before using the Pythagorean theorem, you have to know what a right triangle is! A right triangle is a triangle containing one right angle (90°), and the hypotenuse is the side opposite the 90° angle. If that is the case, here is what the theorem says:

PYTHAGOREAN THEOREM	The square of the hypotenuse of a right triangle equals the sum of the squares of the other two sides; that is, $$c^2 = a^2 + b^2$$ where c is the hypotenuse of the triangle, and a and b are the two remaining sides (called the legs).

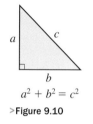

$$a^2 + b^2 = c^2$$

> Figure 9.10

Figure 9.10 shows a right triangle with sides a, b, and c. So if $a = 3$ units and $b = 4$ units, then the hypotenuse c is such that

$$c^2 = 3^2 + 4^2$$
$$c^2 = 9 + 16$$
$$c^2 = 25$$
$$c = \pm\sqrt{25} = \pm 5$$

Since c represents length, c cannot be negative, so the length of the hypotenuse is 5 units.

EXAMPLE 1   Using the Pythagorean theorem

The distance from the bottom of the wall to the base of the ladder in Figure 9.11 is 5 feet, and the distance from the floor to the top of the ladder is 12 feet. Can you find the length of the ladder?

SOLUTION   We use the RSTUV method given earlier.

1. Read the problem. We are asked to find the length of the ladder.

2. Select the unknown. Let L represent the length of the ladder.

3. Think of a plan. First we draw a diagram (see Figure 9.12) that gives the appropriate dimensions. Then we translate the problem using the Pythagorean theorem:

$$\underbrace{\left(\begin{array}{c}\text{Distance from}\\\text{wall to base of}\\\text{ladder}\end{array}\right)^2}_{5^2} + \underbrace{\left(\begin{array}{c}\text{Distance from}\\\text{floor to top of}\\\text{ladder}\end{array}\right)^2}_{12^2} = \underbrace{\left(\begin{array}{c}\text{Length of}\\\text{ladder}\end{array}\right)^2}_{L^2}$$

> Figure 9.11

12 ft

5 ft

PROBLEM 1

Find the length of the ladder if the distance from the wall to the base of the ladder is 5 feet but the height of the top of the ladder is 14 feet.

12 ft L (length of ladder)

5 ft

> Figure 9.12

(continued)

Answers to PROBLEMS

1. $\sqrt{221}$ ft

4. Use algebra to solve the problem. We use algebra to solve this equation:

$$5^2 + 12^2 = L^2$$
$$25 + 144 = L^2$$
$$L^2 = 169$$
$$L = \pm 13$$

Since L is the length of the ladder, L is positive, so $L = 13$.

5. Verify the solution. This solution is correct because $5^2 + 12^2 = 13^2$ since $25 + 144 = 169$.

B › Solving Word Problems Involving Quadratic Equations

Many applications of quadratic equations come from the field of engineering. For example, it is known that the pressure p (in pounds per square foot) exerted by a wind blowing at v miles per hour is given by the equation

$$p = 0.003v^2 \qquad \text{Since } v \text{ is raised to the second power, the equation is a quadratic.}$$

If the pressure p is known, the equation $p = 0.003v^2$ is an example of a quadratic equation in v. We show how this information is used in the next example.

EXAMPLE 2   Determining wind speed

A wind pressure gauge at Commonwealth Bay registered 120 pounds per square foot during a gale. If the pressure $p = 0.003\,v^2$, where v is the wind speed, what was the wind speed at that time?

SOLUTION   We will again use the RSTUV method.

1. Read the problem. We are asked to find the wind speed.

2. Select the unknown. Since we know that $p = 120$ and that $p = 0.003v^2$, we let v be the wind speed.

3. Think of a plan. We substitute for p to obtain

$$0.003v^2 = 120$$

4. Use algebra to solve the problem. The easiest way to solve this equation is to divide by 0.003 first. We then have

$$v^2 = \frac{120}{0.003}$$

$$v^2 = \frac{120 \cdot 1000}{0.003 \cdot 1000} = \frac{120{,}000}{3} = 40{,}000 \qquad \text{You can also divide 120 by 0.003 to obtain}$$

$$v = \pm\sqrt{40{,}000} = \pm 200$$

$$\begin{array}{r} 40{,}000 \\ 0.003\overline{)120{,}000} \\ 12 \end{array}$$

Thus, the wind speed v was 200 miles per hour. (We discard the negative answer as not suitable.)

5. Verify the solution. We leave the verification to you.

PROBLEM 2

Find the wind speed at the Golden Gate Bridge when a pressure gauge is registering 4.8 lb/ft^2.

Answers to PROBLEMS

2. 40 mi/hr

It can be shown in physics that when an object is dropped or thrown downward with an initial velocity v_0, the distance d (in meters) traveled by the object in t seconds is given by the formula

$$d = 5t^2 + v_0 t$$ This is an approximate formula. A more exact formula is $d = 4.9t^2 + v_0 t$.

Suppose an object is dropped from a height of 125 meters. How long would it be before the object hits the ground? The solution to this problem is given in Example 3.

EXAMPLE 3   Falling the distance

Use the formula $d = 5t^2 + v_0 t$ to find the time it takes an object dropped from a height of 125 meters to hit the ground.

SOLUTION   Since the object is dropped, the initial velocity is $v_0 = 0$, and we are given $d = 125$; we then have

$$125 = 5t^2$$
$$5t^2 = 125$$
$$t^2 = 25 \qquad \text{Divide by 5.}$$
$$t = \pm\sqrt{25} = \pm 5$$

Since the time must be positive, the correct answer is $t = 5$; that is, it takes 5 seconds for the object to hit the ground.

PROBLEM 3

Use the formula $d = 5t^2 + v_0 t$ to find the time it takes an object dropped from a height of 180 meters to hit the ground.

Finally, quadratic equations are also used in business. Perhaps you already know what it means to "break even." In business, the break-even point is the point at which the revenue R equals the cost C of the manufactured goods. In symbols,

$$R = C$$

Now suppose a company produces x (thousand) items. If each item sells for $3, the revenue R (in thousands of dollars) can be expressed by the equation

$$R = 3x$$

If the manufacturing cost C (in thousands of dollars) is given by the equation

$$C = x^2 - 3x + 5$$

and it is known that the company always produces more than 1000 items, the break-even point occurs when

$$R = C$$
$$3x = x^2 - 3x + 5$$
$$x^2 - 6x + 5 = 0$$
$$(x - 5)(x - 1) = 0 \qquad \text{Factor.}$$
$$x - 5 = 0 \quad \text{or} \quad x - 1 = 0 \qquad \text{Use the principle of zero product.}$$
$$x = 5 \qquad\qquad x = 1$$

But since it's known that the company produces more than 1 (thousand) item(s), the break-even point occurs when the company manufactures 5 (thousand) items.

EXAMPLE 4   Finding the break-even point

Find the number x of items (in thousands) that have to be produced to break even when the revenue R (in thousands of dollars) of a company is given by the equation $R = 3x$ and the cost (also in thousands of dollars) is given by the equation $C = 2x^2 - 3x + 4$.

SOLUTION   In order to break even,

$$R = C$$
$$3x = 2x^2 - 3x + 4$$
$$2x^2 - 6x + 4 = 0$$
$$x^2 - 3x + 2 = 0 \qquad \text{Divide by 2.}$$
$$(x - 1)(x - 2) = 0 \qquad \text{Factor.}$$
$$x - 1 = 0 \quad \text{or} \quad x - 2 = 0 \qquad \text{Use the property zero product.}$$
$$x = 1 \qquad\qquad x = 2$$

This means that the company breaks even when it produces either 1 (thousand) or 2 (thousand) items.

PROBLEM 4

Find the number x of items (in thousands) that have to be produced to break even when the revenue R of a company (in thousands of dollars) is given by the equation $R = 4x$ and the cost C (also in thousands of dollars) is given by the equation $C = x^2 - 3x + 6$.

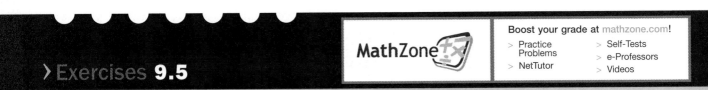

Boost your grade at mathzone.com!

> Practice Problems
> NetTutor

> Self-Tests
> e-Professors
> Videos

MathZone

> ## Exercises 9.5

⟨ **A** ⟩   **Using the Pythagorean Theorem**   In Problems 1–10, let a and b represent the lengths of the sides of a right triangle and c the length of the hypotenuse. Find the missing side.

1. $a = 12, c = 13$

2. $a = 4, c = 5$

3. $a = 5, c = 15$

4. $a = b, c = 4$

5. $b = 3, a = \sqrt{6}$

6. $b = 7, a = 9$

7. $a = \sqrt{5}, b = 2$

8. $a = 3, b = \sqrt{7}$

9. $c = \sqrt{13}, a = 3$

10. $c = \sqrt{52}, a = 4$

⟨ **B** ⟩   **Solving Word Problems Involving Quadratic Equations**   In Problems 11–24, solve the resulting quadratic equation.

11. How long is a wire extending from the top of a 40-foot telephone pole to a point on the ground 30 feet from the base of the pole?

12. Repeat Problem 11 where the point on the ground is 16 feet away from the pole.

13. The pressure p (in pounds per square foot) exerted on a surface by a wind blowing at v miles per hour is given by the equation

$$p = 0.003v^2$$

Find the wind speed when a wind pressure gauge recorded a pressure of 30 pounds per square foot.

14. Repeat Problem 13 where the pressure is 2.7 pounds per square foot.

15. An object is dropped from a height of 320 meters. How many seconds does it take for this object to hit the ground? (See Example 3.)

16. Repeat Problem 15 where the object is dropped from a height of 45 meters.

17. An object is thrown downward with an initial velocity of 3 meters per second. How long does it take for the object to travel 8 meters? (See Example 3.)

18. The revenue of a company is given by the equation $R = 2x$, where x is the number of units produced (in thousands). If the cost is given by the equation $C = 4x^2 - 2x + 1$, how many units have to be produced before the company breaks even?

Answers to PROBLEMS

4. 1 (thousand) or 6 (thousand) items

19. Repeat Problem 18 where the revenue is given by the equation $R = 5x$ and the cost is given by the equation $C = x^2 - x + 9$.

20. If P dollars are invested at r percent compounded annually, then at the end of 2 years the amount will have grown to $A = P(1 + r)^2$. At what rate of interest r will \$1000 grow to \$1210 in 2 years? (*Hint:* $A = 1210$ and $P = 1000$.)

21. A rectangle is 2 feet wide and 3 feet long. Each side is increased by the same amount to give a rectangle with twice the area of the original one. Find the dimensions of the new rectangle. (*Hint:* Let x feet be the amount by which each side is increased.)

22. The hypotenuse of a right triangle is 4 centimeters longer than the shortest side and 2 centimeters longer than the remaining side. Find the dimensions of the triangle.

23. Repeat Problem 22 where the hypotenuse is 16 centimeters longer than the shortest side and 2 centimeters longer than the remaining side.

24. The square of a certain positive number is 5 more than 4 times the number itself. Find this number.

〉〉〉 *Applications*

Economists and managers use quadratic equations to determine **market equilibrium,** the point at which the demand D for a quantity q of a product equals the supply S.

25. *Supply and demand* The supply S of custom vans produced is given by the equation $S = q^2 + 6q + 20$ and the demand is given by the equation $D = -4q + 220$.

    a. Find q the number of vans produced at which the supply S equals the demand D **(market equilibrium).**

    b. How many vans can be supplied at this point?

26. *Supply and demand* Windy manufacturing produces q sailboats a month. The supply S for the boats is given by the equation $S = -3q^2 - 6q + 86$ and the demand is given by the equation $D = \frac{1}{4}q^2 + 10$.

    a. How many boats are produced at market equilibrium?

    b. What is the demand for the boats at market equilibrium?

27. *Supply and demand* A jewelry manufacturing company produces q rings each day. The supply for the rings is given by the equation $S = -\frac{1}{2}q^2 - 2q + 30$ and the demand is given by the equation $D = q^2 + q - 6$.

    a. How many rings can they produce each day so that the supply is the same as the demand?

    b. What will be the supply of rings at this point?

28. *Supply and demand* The supply S of cars at a dealership is given by the equation $S = -\frac{3}{2}q^2 - 16q + 252$ and the demand is given by the equation $D = q^2 + 8q - 100$.

    a. How many cars does the dealer have to have in stock so that the supply is the same as the demand?

    b. How many cars are demanded at that point?

〉〉〉 *Using Your Knowledge*

Solving for Variables   Many formulas require the use of some of the techniques we've studied when solving for certain unknowns in these formulas. For example, consider the formula

$$v^2 = 2gh$$

How can we solve for v (the speed) in this equation? In the usual way, of course! Thus, we have

$$v^2 = 2gh$$

$$v = \pm\sqrt{2gh}$$

Since the speed v is positive, we simply write

$$v = \sqrt{2gh}$$

Use this idea to solve for the indicated variables in the given formulas.

29. Solve for c in the formula $E = mc^2$.

30. Solve for r in the formula $A = \pi r^2$.

31. Solve for r in the formula $F = \frac{GMm}{r^2}$.

32. Solve for v in the formula $KE = \frac{1}{2}mv^2$.

33. Solve for P in the formula $I = kP^2$.

〉〉〉 Write On

34. When solving for the hypotenuse c in a right triangle whose sides were 3 and 4 units, respectively, a student solved the equation $3^2 + 4^2 = c^2$ and obtained the answer 5. Later, the same student was given the equation $3^2 + 4^2 = x^2$ and gave the same answer. The instructor said that the answer was not complete. Explain.

35. In Example 2, one of the answers was discarded. Explain why.

36. Referring to Example 4, explain why you "break even" when $R = C$.

〉〉〉 Concept Checker

Fill in the blank(s) with the correct word(s), phrase, or mathematical statement.

37. A **right triangle** is a **triangle** containing one _____ **angle**.

38. According to **Pythagoras theorem,** in a **right** triangle with legs of length a and b and hypotenuse c, _____.

obtuse	$a^2 + c^2 = b^2$
right	$a^2 + b^2 = c^2$

〉〉〉 Mastery Test

39. Find the number x of hundreds of items that have to be produced to break even when the revenue R of a company (in thousands of dollars) is given by the equation $R = 4x$ and the cost C (also in thousands of dollars) is given by the equation $C = x^2 - 3x + 6$.

40. Find the time t (in seconds) it takes an object dropped from a height of 245 meters to hit the ground by using the formula $d = 5t^2 + v_0 t$ (meters).

41. The pressure p (in pounds per square foot) exerted by a wind blowing at v miles per hour is given by the equation

$$p = 0.003v^2$$

Find the wind speed at the Golden Gate Bridge when a pressure gauge is registering 7.5 pounds per square foot.

42. Find the length L of the ladder in the diagram if the distance from the wall to the base of the ladder is 5 feet and the height to the top of the ladder is 13 feet.

〉〉〉 Skill Checker

43. Evaluate $2x + 3$ when $x = -2$.

44. Evaluate $x^3 - 2x^2 + 3x - 4$ when $x = -3$.

9.6   Functions

▶ Objectives

A ▸ Find the domain and range of a relation.

B ▸ Determine whether a given relation is a function.

C ▸ Use function notation.

D ▸ Solving applications involving functions.

▶ To Succeed, Review How To . . .

1. Evaluate an expression (pp. 61–62, 69–72).
2. Use set notation (p. 45).

▶ Getting Started

Functions for Fashions

Did you know that women's clothing sizes are getting smaller? According to a recent J. C. Penney catalog, "Simply put, you will wear one size smaller than before." Is there a relationship between the new sizes and waist size? The table gives a $1\frac{1}{2}$-inch leeway for waist sizes. Let's use this table to consider the first number in the waist sizes corresponding to different dress sizes—that is, 30, 32, 34, 36. For Petite sizes 14–22, the waist size is 16 inches more than the dress size. If we wish to formalize this relationship, we can write

$$w(s) = s + 16 \qquad \text{Read "}w\text{ of }s\text{ equals }s\text{ plus 16."}$$

What would be the waist size of a woman who wears size 14? It would be

$$w(14) = 14 + 16 = 30$$

For size 16,

$$w(16) = 16 + 16 = 32$$

and so on. It works! Can you do the same for hip sizes? If you get

$$h(s) = s + 26.5$$

you are on the right track. As you can see, there is a *relationship* between hip size and dress size.

Women's Petite Size	14 WP	16 WP	18 WP	20 WP	22 WP	24 WP	26 WP	28 WP	30 WP
Women's Size	-	16 W	18 W	20 W	22 W	24 W	26 W	28 W	30 W
Bust	38-39½	40-41½	42-43½	44-45½	46-47½	48-49½	50-51½	52-53½	54-55½
Waist	30-31½	32-33½	34-35½	36-37½	38-40	40½-42½	43-45	45½-47½	48-50
Hips	40½-42	42½-44	44½-46	46½-48	48½-50	50½-52	52½-54	54½-56	56½-58

The word *relation* might remind you of members of your family—parents, brothers, sisters, and so on—but in mathematics, relations are expressed using ordered pairs. Thus, the relationship between Petite sizes 14–22 in the table and the waist size can be expressed by the set of ordered pairs:

$$w = \{(14, 30), (16, 32), (18, 34), (20, 36), (22, 38)\}$$

Note that we've specified that the waist size corresponding to the dress size is the first number in the row labeled "Waist." Thus, for every dress size, there is only *one* waist size. Relations that have this property are called *functions*. In this section, we shall learn about *relations* and *functions,* and how to evaluate them using notation such as $w(s) = s + 16$ and $h(s) = s + 26.5$.

A ⟩ Finding Domains and Ranges

In previous chapters, we studied ordered pairs that satisfy linear equations, graphed these ordered pairs, and then found the graph of the corresponding line. Thus, to find the graph of $y = 3x - 1$, we assigned the value 0 to x to obtain the corresponding y-value -1. Thus, a y-value of -1 is paired to an x-value of 0, which results in the ordered pair $(0, -1)$ as assigned by the rule $y = 3x - 1$.

RELATION, DOMAIN, AND RANGE	In mathematics, a **relation** is a set of ordered pairs. The **domain** of the relation is the set of all possible x-values, and the **range** of the relation is the set of all possible y-values.

EXAMPLE 1   Finding the domain and range of a relation given as a set of ordered pairs

Find the domain and range of the relation:

$$S = \{(4, -3), (2, -5), (-3, 4)\}$$

SOLUTION   The domain D is the set of all possible x-values. Thus, $D = \{4, 2, -3\}$.

The range R is the set of all possible y-values. Thus, $R = \{-3, -5, 4\}$.

PROBLEM 1

Find the domain and range of the relation:

$$T = \{(6, -4), (4, -6), (-1, 3)\}$$

In many cases, relations are defined by a **rule** for finding the y-value for a given x-value. Thus, the relation

$$\underbrace{S = \{}_{\text{"The set}} \quad \underbrace{(x, y)}_{\substack{\text{of all} \\ (x, y)}} \quad \underbrace{\mid}_{\substack{\text{such} \\ \text{that}}} \quad \underbrace{y = 3x - 1, x \text{ a natural number less than } 5\}}_{\substack{y = 3x - 1 \text{ and } x \text{ is a natural} \\ \text{number less than 5."}}}$$

defines—that is, gives a *rule* about—a set of ordered pairs obtained by assigning a natural number less than 5 to x and obtaining the corresponding y-value. Thus,

For $x = 1$, $y = 3(1) - 1 = 2$ and $(1, 2)$ is part of the relation.
For $x = 2$, $y = 3(2) - 1 = 5$ and $(2, 5)$ is part of the relation.
For $x = 3$, $y = 3(3) - 1 = 8$ and $(3, 8)$ is part of the relation.
For $x = 4$, $y = 3(4) - 1 = 11$ and $(4, 11)$ is part of the relation.

Can you see the pattern? For $S = \{(1, 2), (2, 5), (3, 8), (4, 11)\}$, the domain is $\{1, 2, 3, 4\}$, and the range is $\{2, 5, 8, 11\}$.

Answers to PROBLEMS

1. $D = \{6, 4, -1\}$; $R = \{-4, -6, 3\}$

> **NOTE**
>
> Unless otherwise specified, the *domain* of a relation is the largest set of real numbers that can be substituted for x that result in a real number for y. The *range* is then determined by the rule of the relation.

For example, for the relation

$$S = \left\{ (x, y) \,\middle|\, y = \frac{1}{x} \right\}$$

the domain consists of all real numbers *except zero* since you can substitute any real number for x in $y = \frac{1}{x}$ *except zero* because division by zero is not defined. The range is also the set of real numbers *except zero* because the rule $y = \frac{1}{x}$ will never yield a value of zero for any given value of x.

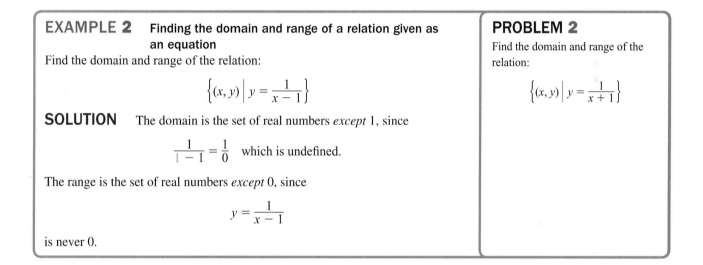

EXAMPLE 2   **Finding the domain and range of a relation given as an equation**

Find the domain and range of the relation:

$$\left\{ (x, y) \,\middle|\, y = \frac{1}{x - 1} \right\}$$

SOLUTION   The domain is the set of real numbers *except* 1, since

$$\frac{1}{1 - 1} = \frac{1}{0} \quad \text{which is undefined.}$$

The range is the set of real numbers *except* 0, since

$$y = \frac{1}{x - 1}$$

is never 0.

PROBLEM 2

Find the domain and range of the relation:

$$\left\{ (x, y) \,\middle|\, y = \frac{1}{x + 1} \right\}$$

B ⟩ Determining Whether a Relation Is a Function

In mathematics, an important type of relation is one where for each element in the domain, there corresponds one and only one element in the range. Such relations are called *functions*.

FUNCTION

A **function** is a set of ordered pairs in which each domain value has exactly one range value; that is, no two different ordered pairs have the same first coordinate.

Note that the graph of a function will never have two points stacked vertically, that is, two points that can be connected with a vertical line. (See the *Using Your Knowledge*.)

Answers to PROBLEMS

2. The domain is the set of all real numbers except −1. The range is the set of all real numbers except 0.

EXAMPLE 3   Determining whether a relation is a function

Determine whether the given relations are functions:

a. $\{(3, 4), (4, 3), (4, 4), (5, 4)\}$

b. $\{(x, y) \mid y = 5x + 1\}$

SOLUTION

a. The relation is not a function because the two ordered pairs $(4, 3)$ and $(4, 4)$ have the same first coordinate.

b. The relation $\{(x, y) \mid y = 5x + 1\}$ is a function because if x is any real number, the expression $y = 5x + 1$ yields one and only one value, so there will never be two ordered pairs with the same first coordinate.

PROBLEM 3

Determine whether the relations are functions:

a. $\{(5, 6), (6, 5), (6, 6), (7, 6)\}$

b. $\{(x, y) \mid y = 3x + 2\}$

C ⟩ Using Function Notation

We often use letters such as f, F, g, G, h, and H to designate functions. Thus, for the relation in Example 3(b), we use set notation to write

$$f = \{(x, y) \mid y = 5x + 1\}$$

because we know this relation to be a function. Another very commonly used notation for denoting the range value that corresponds to a given domain value x is $f(x)$. (This is usually read, "f of x.")

The $f(x)$ notation, called **function notation,** is quite convenient because it denotes the value of the function for the given value of x. For example, if

$$f(x) = 2x + 3$$

then

$$f(1) = 2(1) + 3 = 5$$
$$f(0) = 2(0) + 3 = 3$$
$$f(-6) = 2(-6) + 3 = -9$$
$$f(4) = 2(4) + 3 = 11$$
$$f(a) = 2(a) + 3 = 2a + 3$$
$$f(w + 2) = 2(w + 2) + 3 = 2w + 7$$

and so on. Whatever appears between the parentheses in $f(\ \ )$ is to be substituted for x in the rule that defines $f(x)$.

Instead of describing a function in set notation, we frequently say, "the function defined by $f(x) = \ldots$" where the three dots are replaced by the expression for the value of the function. For instance, "the function defined by $f(x) = 5x + 1$" has the same meaning as "the function $f = \{(x, y) \mid y = 5x + 1\}$."

EXAMPLE 4   Evaluating functions

Let $f(x) = 3x + 5$. Find:

a. $f(4)$ **b.** $f(2)$ **c.** $f(2) + f(4)$ **d.** $f(x + 1)$

SOLUTION   Since $f(x) = 3x + 5$,

a. $f(4) = 3 \cdot 4 + 5 = 12 + 5 = 17$

b. $f(2) = 3 \cdot 2 + 5 = 6 + 5 = 11$

c. Since $f(2) = 11$ and $f(4) = 17$,

$$f(2) + f(4) = 11 + 17 = 28$$

d. $f(x + 1) = 3(x + 1) + 5 = 3x + 8$

PROBLEM 4

Let $f(x) = 4x - 5$. Find:

a. $f(3)$ **b.** $f(4)$

c. $f(3) + f(4)$ **d.** $f(x - 1)$

EXAMPLE 5   Evaluating functions

A function g is defined by $g(x) = x^3 - 2x^2 + 3x - 4$. Find:

a. $g(2)$ **b.** $g(-3)$ **c.** $g(2) - g(-3)$

SOLUTION   Since $g(x) = x^3 - 2x^2 + 3x - 4$,

a. $g(2) = 2^3 - 2(2)^2 + 3(2) - 4$
$\quad\quad = 8 - 8 + 6 - 4 = 2$

b. $g(-3) = (-3)^3 - 2(-3)^2 + 3(-3) - 4$
$\quad\quad\quad = -27 - 18 - 9 - 4 = -58$

c. $g(2) - g(-3) = 2 - (-58) = 60$

PROBLEM 5

If $g(x) = x^3 - 3x^2 + 2x - 1$, find:

a. $g(3)$ **b.** $g(-2)$

c. $g(3) - g(-2)$

In Examples 4 and 5, we evaluated a specified function. Sometimes, as was the case in *Getting Started,* we must find the function, as shown in Example 6.

EXAMPLE 6   Finding the relationship between ordered pairs

Consider the ordered pairs $(2, 6)$, $(3, 9)$, $(1.2, 3.6)$, and $\left(\frac{2}{5}, \frac{6}{5}\right)$. There is a functional relationship $y = f(x)$ between the numbers in each pair. Find $f(x)$ and use it to fill in the missing numbers in the pairs $(\_\_, 12)$, $(\_\_, 3.3)$, and $(5, \_\_)$.

SOLUTION   The given pairs are of the form (x, y). A close examination reveals that each of the y's in the pairs is 3 times the corresponding x; that is, $y = 3x$ or $f(x) = 3x$. Now in each of the ordered pairs $(\_\_, 12)$, $(\_\_, 3.3)$, and $(5, \_\_)$, the y-value must be three times the x-value. Thus,

$$(\_\_, 12) = (4, 12)$$
$$(\_\_, 3.3) = (1.1, 3.3)$$
$$(5, \_\_) = (5, 15)$$

PROBLEM 6

Consider the ordered pairs $(3, 6)$, $(4, 8)$, $(1.1, 2.2)$, and $\left(\frac{3}{5}, \frac{6}{5}\right)$. There is a functional relationship $y = f(x)$ between the numbers in each pair. Find $f(x)$ and use it to fill in the missing numbers in the pairs $(\_\_, 10)$, $(\_\_, 4.4)$, and $(6, \_\_)$.

D › Applications Involving Functions

In recent years, aerobic exercises such as jogging, swimming, bicycling, and roller blading have been taken up by millions of Americans. To see whether you are exercising too hard (or not hard enough), you should stop from time to time and take your pulse to determine your heart rate. The idea is to keep your rate within a range known as the **target zone,** which is determined by your age. Example 7 explains how to find the **lower limit** of your target zone.

EXAMPLE 7   Exercises and functions

The lower limit L (heartbeats per minute) of your target zone is a function of your age a (in years) and is given by the equation

$$L(a) = -\frac{2}{3}a + 150$$

a. Find the value of L for a person who is 30 years old.

b. Find the value of L for a person who is 45 years old.

PROBLEM 7

a. Find L for a 21-year-old person.

b. Find L for a 33-year-old person.

(continued)

Answers to PROBLEMS

5. a. 5   **b.** -25 **c.** 30

6. $f(x) = 2x$; $(5, 10)$, $(2.2, 4.4)$, $(6, 12)$

7. a. $L = 136$ **b.** $L = 128$

SOLUTION

a. We need to find $L(30)$, and because

$$L(a) = -\frac{2}{3}a + 150,$$

$$L(30) = -\frac{2}{3}(30) + 150$$

$$= -20 + 150 = 130$$

This result means that a 30-year-old person should try to attain at least 130 heartbeats per minute while exercising.

b. Here, we want to find $L(45)$. Proceeding as before, we obtain

$$L(45) = -\frac{2}{3}(45) + 150$$

$$= -30 + 150 = 120$$

(Find the value of L for your own age.)

▣ ◈ ▦ Calculator Corner

Entering Functions

The idea of a function is so important in mathematics that even your calculator will only accept functions when you use the graphing feature. Thus, if you are given an equation with variables x and y and you can solve for y, you have a function. What does this mean?

Let's take $2x + 4y = 4$. Solving for y, we obtain $y = -\frac{1}{2}x + 1$, which can be entered and graphed. Similarly, $y = x^2 + 1$ can be entered and graphed (see Window 1). We cannot solve for y in the equation $x^2 + y^2 = 4$. The equation $x^2 + y^2 = 4$, whose graph happens to be a circle of radius 2, does not describe a function.

Your calculator can also evaluate functions. For example, to find the value $L(30)$ in Example 7, let's start by using x's instead of a's. Store the 30 in the calculator's memory by pressing 30 [STO▶] [X,T,θ,n] [ENTER], then enter the function $-\frac{2}{3}x + 150$ [ENTER]; the answer, 130, is then displayed as shown in Window 2.

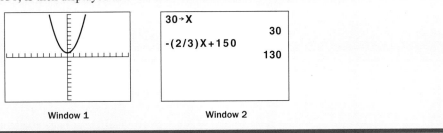

Window 1   Window 2

MathZone

Boost your grade at mathzone.com!
> Practice   > Self-Tests
  Problems   > e-Professors
> NetTutor   > Videos

❯ Exercises **9.6**

❮ **A** ❯ **Finding Domains and Ranges**   In Problems 1–14, find the domain and the range of the given relation.
[*Hint:* Remember that you cannot divide by zero.]

1. $\{(1, 2), (2, 3), (3, 4)\}$

2. $\{(3, 1), (2, 1), (1, 1)\}$

3. $\{(1, 1), (2, 2), (3, 3)\}$

4. $\{(4, 1), (5, 2), (6, 1)\}$

5. $\{(x, y) \mid y = 3x\}$

6. $\{(x, y) \mid y = 2x + 1\}$

7. $\{(x, y) \mid y = x + 1\}$

8. $\{(x, y) \mid y = 1 - 2x\}$

9. $\{(x, y) \mid y = x^2\}$

10. $\{(x, y) \mid y = 2 + x^2\}$

11. $\{(x, y) \mid y^2 = x\}$
(*Hint:* y^2 is never negative.)

12. $\{(x, y) \mid x = 1 + y^2\}$

13. $\left\{(x, y) \mid y = \dfrac{1}{x - 3}\right\}$

14. $\left\{(x, y) \mid y = \dfrac{1}{x - 2}\right\}$

In Problems 15–22, find the domain and the range of the relation. List the ordered pairs in each relation.

15. $\{(x, y) \mid y = 2x, x \text{ an integer between } -1 \text{ and } 2, \text{ inclusive}\}$

16. $\{(x, y) \mid y = 2x - 1, x \text{ a counting number not greater than } 5\}$

17. $\{(x, y) \mid y = 2x - 3, x \text{ an integer between } 0 \text{ and } 4, \text{ inclusive}\}$

18. $\{(x, y) \mid y = \frac{1}{x}, x \text{ an integer between } 1 \text{ and } 5, \text{ inclusive}\}$

19. $\{(x, y) \mid y = \sqrt{x}, x = 0, 1, 4, 9, 16, \text{ or } 25\}$

20. $\{(x, y) \mid y \le x + 1, x \text{ and } y \text{ positive integers less than } 4\}$

21. $\{(x, y) \mid y > x, x \text{ and } y \text{ positive integers less than } 5\}$

22. $\{(x, y) \mid 0 < x + y < 5, x \text{ and } y \text{ positive integers less than } 4\}$

‹ **B** ›   **Determining Whether a Relation Is a Function**   In Problems 23–30, decide whether the given relation is a function. State the reason for your answer in each case.

23. $\{(0, 1), (1, 2), (2, 3)\}$

24. $\{(1, 2), (2, 1), (1, 3), (3, 1)\}$

25. $\{(-1, 1), (-2, 2), (-3, 3)\}$

26. $\{(-1, 1), (-1, 2), (-1, 3)\}$

27. $\{(x, y) \mid y = 5x + 6\}$

28. $\{(x, y) \mid y = 3 - 2x\}$

29. $\{(x, y) \mid x = y^2\}$

30. $\{(x, y) \mid y = x^2\}$

‹ **C** ›   **Using Function Notation**   In Problems 31–35, find the indicated value of the function.

31. A function f is defined by $f(x) = 3x + 1$. Find:
   a. $f(0)$ **b.** $f(2)$ **c.** $f(-2)$

32. A function g is defined by $g(x) = -2x + 1$. Find:
   a. $g(0)$ **b.** $g(1)$ **c.** $g(-1)$

33. A function F is defined by $F(x) = \sqrt{x - 1}$. Find:
   a. $F(1)$ **b.** $F(5)$ **c.** $F(26)$

34. A function G is defined by $G(x) = x^2 + 2x - 1$. Find:
   a. $G(0)$ **b.** $G(2)$ **c.** $G(-2)$

35. A function f is defined by $f(x) = 3x + 1$. Find:
   a. $f(x + h)$
   b. $f(x + h) - f(x)$
   c. $\dfrac{f(x + h) - f(x)}{h}, \quad h \ne 0$

36. Given are the ordered pairs: (2, 1), (6, 3), (9, 4.5), and (1.6, 0.8). There is a simple functional relationship, $y = f(x)$, between the numbers in each pair. What is $f(x)$? Use this to fill in the missing number in the pairs (____, 7.5), (____, 2.4), and (____, $\frac{1}{7}$).

37. Given are the ordered pairs: $\left(\frac{1}{2}, \frac{1}{4}\right)$, (1.2, 1.44), (5, 25), and (7, 49). There is a simple functional relationship, $y = g(x)$, between the numbers in each pair. What is $g(x)$? Use this to fill in the missing number in the pairs $\left(\frac{1}{4}, \underline{\quad}\right)$, (2.1, $\underline{\quad}$), and ($\underline{\quad}$, 64).

39. If $g(x) = 2x^3 + x^2 - 3x + 1$, find:

    a. $g(0)$ **b.** $g(-2)$ **c.** $g(2)$

38. Given that $f(x) = x^3 - x^2 + 2x$, find:

    a. $f(-1)$ **b.** $f(-3)$ **c.** $f(2)$

⟨ D ⟩ Applications Involving Functions

40. *Fahrenheit and Celsius temperatures* The Fahrenheit temperature reading F is a function of the Celsius temperature reading C. This function is given by

$$F(C) = \frac{9}{5}C + 32$$

    a. If the temperature is 15°C, what is the Fahrenheit temperature?

    b. Water boils at 100°C. What is the corresponding Fahrenheit temperature?

    c. The freezing point of water is 0°C or 32°F. How many Fahrenheit degrees below freezing is a temperature of -10°C?

    d. The lowest temperature attainable is -273°C; this is the zero point on the absolute temperature scale. What is the corresponding Fahrenheit temperature?

42. *Target zone* Refer to Example 7 and Problem 41. The target zone for a person a years old consists of all the heart rates between $L(a)$ and $U(a)$, inclusive. Thus, if a person's heart rate is R, that person's target zone is described by $L(a) \leq R \leq U(a)$. Find the target zone for a person who is

    a. 30 years old   **b.** 45 years old

44. *Car rental costs* The cost C in dollars of renting a car for 1 day is a function of the number m of miles traveled. For a car renting for $20 per day and 20¢ per mile, this function is given by

$$C(m) = 0.20m + 20$$

    a. Find the cost of renting a car for 1 day and driving 290 miles.

    b. If an executive paid $60.60 after renting a car for 1 day, how many miles did she drive?

46. *Distance traveled by dropped ball* If a ball is dropped from a point above the surface of the earth, the distance s (in meters) that the ball falls in t seconds is a function of t. This function is given by

$$s(t) = 4.9t^2$$

Find the distance that the ball falls in:

    a. 2 seconds   **b.** 5 seconds

48. *Estimation of gravitational constant* An experiment, carefully carried out, showed that a ball dropped from rest fell 64.4 feet in 2 seconds. What is a more accurate value of g than that given in Problem 47?

41. *Upper limit of target zone* Refer to Example 7. The **upper limit** U of your target zone when exercising is also a function of your age a (in years), and is given by

$$U(a) = -a + 190$$

Find the highest safe heart rate for a person who is

    a. 50 years old

    b. 60 years old

43. *Ideal weight for men* The ideal weight w (in pounds) of a man is a function of his height h (in inches). This function is defined by

$$w(h) = 5h - 190$$

    a. If a man is 70 inches tall, what should his weight be?

    b. If a man weighs 200 pounds, what should his height be?

45. *Pressure below ocean surface* The pressure P (in pounds per square foot) at a depth of d feet below the surface of the ocean is a function of the depth. This function is given by

$$P(d) = 63.9d$$

Find the pressure on a submarine at a depth of:

    a. 10 feet   **b.** 100 feet

47. *Distance traveled by falling object* The function $S(t) = \frac{1}{2}gt^2$ gives the distance that an object falls from rest in t seconds. If S is measured in feet, then the gravitational constant, g, is approximately 32 feet/second per second. Find the distance that the object will fall in:

    a. 3 seconds

    b. 5 seconds

49. *Kinetic energy of moving object*   The kinetic energy (E) of a moving object (measured in joules) is given by the function $E(v) = \frac{1}{2}mv^2$, where m is the mass in kilograms and v is the velocity in meters per second. An automobile of mass 1200 kilograms is going at a speed of 36 kilometers per hour. What is the kinetic energy of the automobile?

50. *Kinetic energy of automobile*   If the kinetic energy of the automobile in Problem 49 is 93,750 joules, how fast (in kilometers per hour) is the automobile traveling?

⟩ ⟩ ⟩   *Using Your Knowledge*

The Vertical Line Test   The graph of an equation can be used to determine whether a relation is a function by applying the **vertical line test.** Since a function is a set of ordered pairs in which no two ordered pairs have the same first coordinate, the graph of a function will never have two points "stacked" vertically; that is, the graph of a function *never* contains two points that can be connected with a vertical line.

     Thus, the first two graphs shown here represent functions (no two points on the graph can be connected with a vertical line), but the last two do not.

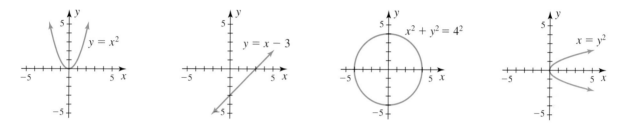

In Problems 51–54, use the vertical line test to determine whether the graph represents a function.

51.   **52.**   **53.**   **54.**

⟩ ⟩ ⟩   *Write On*

55. Is every relation a function? Explain.

56. Is every function a relation? Explain.

57. Explain how you determine the domain of a function defined by a set of ordered pairs.

58. Explain how you determine the domain of a function defined by a formula.

⟩ ⟩ ⟩   *Concept Checker*

Fill in the blank(s) with the correct word(s), phrase, or mathematical statement.

59. A **relation** is a set of _____ **pairs.**

60. The _____ of a **relation** is the set of all possible *x*-values.

61. The _____ of a **relation** is the set of all possible *y*-values.

62. A _____ is a **set of ordered** pairs in which each **domain value** has **exactly one range value.**

                                             range   **domain**

                                             relation   **function**

                                             ordered

> > >   *Mastery Test*

In Problems 63–66, find the domain and range of the relation.

63. $\{(-5, 5), (-6, 6), (-7, 7)\}$

64. $\{(5, -5), (-5, 5), (3, -3), (-3, 3)\}$

65. $\left\{(x, y) \,\middle|\, y = \dfrac{1}{x - 3}\right\}$

66. $\{(x, y) \mid y = -x + 2\}$

In Problems 67–71, determine whether the relation is a function, and explain why or why not.

67. $\{(-4, 5), (-5, 5), (-6, 5)\}$

68. $\{(5, -3), (5, -4), (5, -5)\}$

69. $\{(x, y) \mid y = 2x + 3\}$

70. $\{(x, y) \mid y = x^2\}$

71. $\{(x, y) \mid x = 3y^2\}$

72. If $h(x) = x^2 - 3$, find $h(-2)$.

73. If $g(x) = x^3 - 1$, find $g(-1)$.

74. The monthly cost for searches (questions) in an online service is \$9.95 with 100 free searches. After that, the cost is \$0.10 per search. The monthly cost for the service (in dollars) can be represented by the following function, where q is the number of searches:

For $q \le 100$: $f(q) = 9.95$

For $q > 100$: $f(q) = 9.95 + 0.10(q - 100)$

a. What would your monthly cost be if you made 99 searches during the month.

b. What would your monthly cost be if you made 130 searches during the month.

> **Collaborative Learning**

This exercise concerns an emergency ejection from a fighter jet and how it could relate to completing the square.

The height H (in feet relative to the ground) of a pilot after t seconds is given by the equation

$$H = -16t^2 + 608t + 4482$$

At about 10,000 feet a malfunction occurs. Eject! Eject!

We want to answer three questions:

1. What is the maximum height (relative to the ground) reached by the pilot after being ejected?

2. How many feet above the jet was the pilot ejected?

3. When did the pilot reach the ground with the aid of his or her parachute?

Form three teams to investigate the incident.

Team 1 is in charge of graphing the path of the pilot and the point of ejection using a graphing calculator. A possible viewing window to graph the path is shown. What expression Y_1 does the team have to graph? What is the equation of the line Y_2 shown in the screen and representing the point at which the pilot ejected?

Team 2 is in charge of answering the first two questions.

Start with: $H = -16(\quad)^2 + C$

Here is a hint: $H = -16(t^2 - 38t) + 4482$

Now, complete the square inside the parentheses to put the equation in the standard form for a parabola. What is the maximum for this parabola? This answers question 1. Since we know that the pilot ejected at 10,000 feet, how many feet above the jet was the pilot ejected? This answers question 2.

WINDOW
Xmin=0
Xmax=50
Xscl=10
Ymin=-2000
Ymax=12000
Yscl=2000
Xres=1

Team 3 has to answer the third question. When the pilot reaches the ground, what is H? The point (t, H) is a solution (zero) of the equation whose graph is shown and it occurs when $H = 0$. To find this point press ⟨2nd⟩ ⟨TRACE⟩ 2. The value of t is the time it took the pilot to land back on earth!

> **Research Questions**

1. Write a short paper detailing the work and techniques used by the Babylonians to solve quadratic equations.

2. Show that the solution of the equation $x^2 + ax = b$ obtained by the Babylonians and the solution obtained by using the quadratic formula developed in this chapter are identical.

3. Write a short paper detailing the methods used by Euler to solve quadratic equations.

4. Write a short paper detailing the methods used by the Arabian mathematician al-Khowarizmi to solve quadratic equations.

> **Summary Chapter 9**

Section	Item	Meaning	Example
9.1A	Quadratic equation	An equation that can be written in the form $ax^2 + bx + c = 0$	$x^2 = 16$ is a quadratic equation since it can be written as $x^2 - 16 = 0$.
	$\sqrt{a}$, the principal square root of a, $a \geq 0$	$\sqrt{a}$ is a nonnegative number b such that $b^2 = a$	$\sqrt{16} = 4$ since $4^2 = 16$
	Irrational number	A number that cannot be written in the form $\frac{a}{b}$, where a and b are integers and b is not 0	$\sqrt{2}$, $\sqrt{5}$, and $3\sqrt{7}$ are irrational.
	Real numbers	The rational and the irrational numbers	-7, $-\frac{3}{2}$, 0, $\sqrt{5}$, and 17 are real numbers.
	Square root property of equations	If A is a positive number and $X^2 = A$, then $X = \pm\sqrt{A}$.	If $X^2 = 7$, then $X = \pm\sqrt{7}$.
9.2A	Completing the square	A method used to solve quadratic equations	To solve $\quad 2x^2 + 4x - 1 = 0$ **1.** Add 1. $\quad 2x^2 + 4x = 1$ **2.** Divide by 2. $\quad x^2 + 2x = \frac{1}{2}$ **3.** Add 1^2. $\quad x^2 + 2x + 1 = \frac{3}{2}$ **4.** Rewrite. $\quad (x + 1)^2 = \frac{3}{2}$ **5.** Solve. $\quad x + 1 = \pm\sqrt{\frac{3}{2}}$ $\qquad\qquad x + 1 = \pm\frac{\sqrt{6}}{2}$ $\qquad x = -1 \pm \frac{\sqrt{6}}{2} = \frac{-2 \pm \sqrt{6}}{2}$

(continued)

Section	Item	Meaning	Example
9.3	Quadratic formula	The solutions of $ax^2 + bx + c = 0$ are $$x = \frac{-b \pm \sqrt{b^2 - 4ac}}{2a}$$	The solutions of $2x^2 + 3x + 1 = 0$ are $$\frac{-3 \pm \sqrt{3^2 - 4 \cdot 2 \cdot 1}}{2 \cdot 2}$$ that is, -1 and $-\frac{1}{2}$.
9.3A	Standard form of a quadratic equation	$ax^2 + bx + c = 0$ is the standard form of a quadratic equation.	The standard form of $x^2 + 2x = -11$ is $x^2 + 2x + 11 = 0$.
9.4	Graph of a quadratic equation	The graph of $y = ax^2 + bx + c$ $(a \neq 0)$ is a parabola.	
9.5	Pythagorean Theorem	The square of the hypotenuse of a right triangle equals the sum of the squares of the other two sides.	If a and b are the lengths of the sides and h is the length of the hypotenuse, $c^2 = a^2 + b^2$.
9.6A	Relation	A set of ordered pairs	$\{(2, 5), (2, 6), (3, 7)\}$ is a relation.
	Domain	The domain of a relation is the set of all possible x-values.	The domain of the relation $\{(2, 5), (2, 6), (3, 7)\}$ is $\{2, 3\}$.
	Range	The range of a relation is the set of all possible y-values.	The range of the relation $\{(2, 5), (2, 6), (3, 7)\}$ is $\{5, 6, 7\}$.
9.6B	Function	A set of ordered pairs in which no two ordered pairs have the same x-value	The relation $\{(1, 2), (3, 4)\}$ is a function.

❯ Review Exercises Chapter 9

(If you need help with these exercises, look at the section indicated in brackets.)

1. ❮ **9.1A** ❯ *Solve.*
  a. $x^2 = 1$
  b. $x^2 = 100$
  c. $x^2 = 81$

2. ❮ **9.1A** ❯ *Solve.*
  a. $16x^2 - 25 = 0$
  b. $25x^2 - 9 = 0$
  c. $64x^2 - 25 = 0$

3. ❮ **9.1A** ❯ *Solve.*
  a. $7x^2 + 36 = 0$
  b. $8x^2 + 49 = 0$
  c. $3x^2 + 81 = 0$

4. ❮ **9.1B** ❯ *Solve.*
  a. $49(x + 1)^2 - 3 = 0$
  b. $25(x + 2)^2 - 2 = 0$
  c. $16(x + 1)^2 - 5 = 0$

5. ⟨ **9.2A** ⟩ *Find the missing term in the given expression.*

   a. $(x + 3)^2 = x^2 + 6x + \square$

   b. $(x + 7)^2 = x^2 + 14x + \square$

   c. $(x + 6)^2 = x^2 + 12x + \square$

6. ⟨ **9.2A** ⟩ *Find the missing terms in the given expression.*

   a. $x^2 - 6x + \square = (\quad)^2$

   b. $x^2 - 10x + \square = (\quad)^2$

   c. $x^2 - 12x + \square = (\quad)^2$

7. ⟨ **9.2A** ⟩ *Find the number that must divide each term in the given equation so that the equation can be solved by the method of completing the square.*

   a. $7x^2 - 14x = -4$

   b. $6x^2 - 18x = -2$

   c. $5x^2 - 15x = -3$

8. ⟨ **9.2A** ⟩ *Find the term that must be added to both sides of the given equation so that the equation can be solved by the method of completing the square.*

   a. $x^2 - 4x = -4$

   b. $x^2 - 6x = -9$

   c. $x^2 - 12x = -3$

9. ⟨ **9.3B** ⟩ *Solve.*

   a. $dx^2 + ex + f = 0, d \neq 0$

   b. $gx^2 + hx + i = 0, g \neq 0$

   c. $jx^2 + kx + m = 0, j \neq 0$

10. ⟨ **9.3B** ⟩ *Solve.*

   a. $2x^2 - x - 1 = 0$

   b. $2x^2 - 2x - 5 = 0$

   c. $2x^2 - 3x - 3 = 0$

11. ⟨ **9.3B** ⟩ *Solve.*

   a. $3x^2 - x = 1$

   b. $3x^2 - 2x = 2$

   c. $3x^2 - 3x = 2$

12. ⟨ **9.3B** ⟩ *Solve.*

   a. $9x = x^2$

   b. $4x = x^2$

   c. $25x = x^2$

13. ⟨ **9.3B** ⟩ *Solve.*

   a. $\dfrac{x^2}{9} - x = -\dfrac{4}{9}$

   b. $\dfrac{x^2}{5} - \dfrac{x}{2} = \dfrac{3}{10}$

   c. $\dfrac{x^2}{3} + \dfrac{x}{6} = -\dfrac{1}{2}$

14. ⟨ **9.4A** ⟩ *Graph.*

   a. $y = x^2 + 1$

   b. $y = x^2 + 2$

   c. $y = x^2 + 3$

15. ⟨ **9.4A** ⟩ *Graph.*

   a. $y = -x^2 - 1$

   b. $y = -x^2 - 2$

   c. $y = -x^2 - 3$

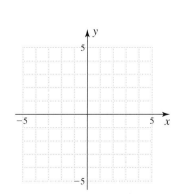

16. ⟨ **9.4A** ⟩ *Graph.*

   a. $y = -(x - 2)^2$

   b. $y = -(x - 3)^2$

   c. $y = -(x - 4)^2$

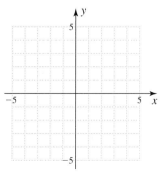

17. ⟨ **9.4 A** ⟩ *Graph.*

    a. $y = (x - 2)^2 + 1$

    b. $y = (x - 2)^2 + 2$

    c. $y = (x - 2)^2 + 3$

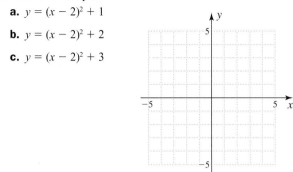

18. ⟨ **9.4 B** ⟩ *Graph and label the vertex and intercepts.*

    a. $y = -x^2 + 6x - 8$

    b. $y = -x^2 + 6x - 5$

    c. $y = -x^2 + 6x$

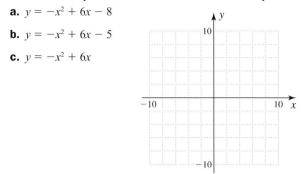

19. ⟨ **9.5 A** ⟩ *Find the length of the hypotenuse of a right triangle if the lengths of the two sides are:*

    a. 5 inches and 12 inches

    b. 2 inches and 3 inches

    c. 4 inches and 5 inches

20. ⟨ **9.5 B** ⟩ *After t seconds, the distance d (in meters) traveled by an object thrown downward with an initial velocity v_0 is given by the equation*

$$d = 5t^2 + v_0 t$$

Find the number of seconds it takes an object to hit the ground if the object is dropped $(v_0 = 0)$ from a height of:

    a. 125 meters   **b.** 245 meters

    c. 320 meters

21. ⟨ **9.6 A** ⟩ *Find the domain and range.*

    a. $\{(-3, 1), (-4, 1), (-5, 2)\}$

    b. $\{(2, -4), (-1, 3), (-1, 4)\}$

    c. $\{(-1, 2), (-1, 3), (-1, 4)\}$

22. ⟨ **9.6 A** ⟩ *Find the domain and range.*

    a. $\{(x, y) \mid y = 2x - 3\}$

    b. $\{(x, y) \mid y = -3x - 2\}$

    c. $\{(x, y) \mid y = x^2\}$

23. ⟨ **9.6 B** ⟩ *Which of the following are functions?*

    a. $\{(-3, 1), (-4, 1), (-5, 2)\}$

    b. $\{(2, -4), (-1, 3), (-1, 4)\}$

    c. $\{(-1, 2), (-1, 3), (-1, 4)\}$

24. ⟨ **9.6 C** ⟩ *If $f(x) = x^3 - 2x^2 + x - 1$, find:*

    a. $f(2)$

    b. $f(-2)$

    c. $f(1)$

25. ⟨ **9.6 D** ⟩ *The average price P(n) of books depends on the number n of millions of books sold and is given by the function*

$$P(n) = 25 - 0.3n \text{ (dollars)}$$

    a. Find the average price of a book when 10 million copies are sold.

    b. Find the average price of a book when 20 million copies are sold.

    c. Find the average price of a book when 30 million copies are sold.

(Answers on pages 723–724)

Visit www.mhhe.com/bello to view helpful videos that provide step-by-step solutions to several of the problems below.

1. Solve $x^2 = 64$.

2. Solve $49x^2 - 25 = 0$.

3. Solve $6x^2 + 49 = 0$.

4. Solve $36(x + 1)^2 - 7 = 0$.

5. Solve $9(x - 3)^2 + 1 = 0$.

6. Find the missing term in the expression
$(x + 4)^2 = x^2 + 8x + \square$.

7. The missing terms in the expressions
$x^2 - 8x + \square = (\quad)^2$ are _____ and _____,
respectively.

8. To solve the equation $8x^2 - 32x = -5$ by completing
the square, the first step will be to divide each term
by _____.

9. To solve the equation $x^2 - 8x = -15$ by completing
the square, one has to add _____ to both sides of the
equation.

10. The solution of $ax^2 + bx + c = 0$ is _____.

11. Solve $2x^2 - 3x - 2 = 0$.

12. Solve $x^2 = 3x - 2$.

13. Solve $16x = x^2$.

14. Solve $\dfrac{x^2}{2} + \dfrac{5}{4}x = -\dfrac{1}{2}$.

15. Graph $y = 3x^2$.

16. Graph $y = (x - 1)^2 + 1$.

17. Graph $y = -(x - 1)^2 + 1$.

18. Graph $y = -x^2 - 2x + 8$. Label the vertex and intercepts.

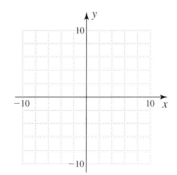

19. Find the length of the hypotenuse of a right triangle if the lengths of the two sides are 2 inches and 5 inches.

20. The formula $d = 5t^2 + v_0 t$ gives the distance d (in meters) an object thrown downward with an initial velocity v_0 will have gone after t seconds. How long would it take an object dropped ($v_0 = 0$) from a distance of 180 meters to hit the ground?

21. Find the domain and range of the relation

$$\{(-1, 1), (-2, 1), (-3, 2)\}$$

22. Find the domain and range of the relation

$$\left\{ (x, y) \,\middle|\, y = \frac{1}{x - 8} \right\}$$

23. State whether each of the following relations is a function.

a. $\{(2, -4), (-1, 3), (-1, 4)\}$
b. $\{(2, -4), (-1, 3), (1, -3)\}$

24. If $f(x) = x^3 + 2x^2 - x - 1$, find $f(-2)$.

25. The average price $P(n)$ of books depends on the number n of millions of books sold and is given by the function

$$P(n) = 25 - 0.4n \text{ (dollars)}$$

Find the average price of a book when 20 million copies are sold.

〉Answers to Practice Test **Chapter 9**

Answer	If You Missed	Review		
	Question	Section	Examples	Page
1. ± 8	1	9.1	1	661–662
2. $\pm\dfrac{5}{7}$	2	9.1	1, 2	661–663
3. No real-number solution	3	9.1	3	664
4. $-1 \pm \dfrac{\sqrt{7}}{6} = \dfrac{-6 \pm \sqrt{7}}{6}$	4	9.1	4	665
5. No real-number solution	5	9.1	5	666
6. 16	6	9.2	1	673
7. $16; x - 4$	7	9.2	2	674
8. 8	8	9.2	3, 4	675, 676
9. 16	9	9.2	3, 4	675, 676
10. $x = \dfrac{-b \pm \sqrt{b^2 - 4ac}}{2a}$	10	9.3	1	682
11. $2; -\dfrac{1}{2}$	11	9.3	1	682
12. $1; 2$	12	9.3	2	682–683
13. $0; 16$	13	9.3	3	683
14. $-\dfrac{1}{2}; -2$	14	9.3	4	684
15.	15	9.4	1	690
16.	16	9.4	3	692

Answer	If You Missed		Review	
	Question	Section	Examples	Page
17. $y = -(x - 1)^2 + 1$	17	9.4	3	692
18. $(-1, 9)$ $(0, 8)$ $(-4, 0)$ $(2, 0)$	18	9.4	4	693–694
19. $\sqrt{29}$ in.	19	9.5	1	701–702
20. 6 sec	20	9.5	2, 3	702, 703
21. $D = \{-1, -2, -3\}; R = \{1, 2\}$	21	9.6	1	708
22. Domain: All real numbers except 8; Range: All real numbers except zero.	22	9.6	2	709
23. a. Not a function **b.** A function	23	9.6	3	710
24. $f(-2) = 1$	24	9.6	4, 5	710, 711
25. $17	25	9.6	7	711, 712

> Cumulative Review **Chapters 1–9**

1. Add: $-\dfrac{3}{8} + \left(-\dfrac{1}{7}\right)$

2. Find: $(-4)^4$

3. Divide: $-\dfrac{3}{8} \div \left(-\dfrac{5}{24}\right)$

4. Evaluate $y \div 4 \cdot x - z$ for $x = 4$, $y = 16$, $z = 2$.

5. Simplify: $2x - (x + 4) - 2(x + 3)$

6. Write in symbols: The quotient of $(x - 5y)$ and z

7. Solve for x: $5 = 3(x - 1) + 5 - 2x$

8. Solve for x: $\dfrac{x}{8} - \dfrac{x}{9} = 1$

9. Graph: $-\dfrac{x}{4} + \dfrac{x}{8} \geq \dfrac{x - 8}{8}$

10. Graph the point $C(-1, -2)$.

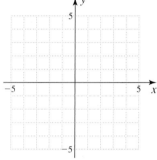

11. Determine whether the ordered pair $(-2, -3)$ is a solution of $2x - y = -1$.

12. Find x in the ordered pair $(x, 2)$ so that the ordered pair satisfies the equation $2x - 3y = -10$.

13. Graph: $2x + y = 4$

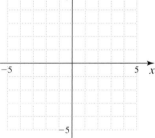

14. Graph: $2y - 8 = 0$

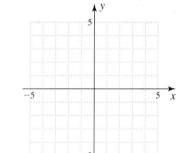

15. Find the slope of the line passing through the points $(-5, -8)$ and $(6, 6)$.

16. What is the slope of the line $8x - 4y = -13$?

17. Find the pair of parallel lines.

(1) $8y + 12x = 7$

(2) $12x - 8y = 7$

(3) $-2y = -3x + 7$

18. Simplify: $(3x^3y^{-5})^4$

19. Write in scientific notation: 0.0050

20. Divide and express the answer in scientific notation: $(1.65 \times 10^{-4}) \div (1.1 \times 10^3)$

21. Expand: $\left(2x^2 - \dfrac{1}{5}\right)^2$

22. Divide $(2x^3 + 5x^2 - 4x - 2)$ by $(x + 3)$.

23. Factor completely: $x^2 - 8x + 15$

24. Factor completely: $12x^2 - 25xy + 12y^2$

25. Factor completely: $25x^2 - 36y^2$

26. Factor completely: $-5x^4 + 80x^2$

27. Factor completely: $3x^3 - 3x^2 - 18x$

28. Factor completely: $4x^2 + 12x + 5x + 15$

29. Factor completely: $25kx^2 + 20kx + 4k$

30. Solve for x: $3x^2 + 13x = 10$

31. Write $\frac{6x}{5y}$ with a denominator of $10y^3$.

32. Reduce to lowest terms: $\dfrac{-4(x^2 - y^2)}{4(x - y)}$

33. Reduce to lowest terms: $\dfrac{x^2 - 4x - 21}{7 - x}$

34. Multiply: $(x - 4) \cdot \dfrac{x - 3}{x^2 - 16}$

35. Divide: $\dfrac{x + 2}{x - 2} \div \dfrac{x^2 - 4}{2 - x}$

36. Add: $\dfrac{5}{4(x + 1)} + \dfrac{7}{4(x + 1)}$

37. Subtract: $\dfrac{x + 7}{x^2 + x - 56} - \dfrac{x + 8}{x^2 - 49}$

38. Simplify: $\dfrac{\dfrac{1}{x} + \dfrac{3}{2x}}{\dfrac{2}{3x} - \dfrac{1}{4x}}$

39. Solve for x: $\dfrac{x}{x - 3} + 6 = \dfrac{5x}{x - 3}$

40. Solve for x: $\dfrac{x}{x^2 - 16} + \dfrac{4}{x - 4} = \dfrac{1}{x + 4}$

41. Solve for x: $\dfrac{x}{x + 4} - \dfrac{1}{5} = -\dfrac{4}{x + 4}$

42. Solve for x: $2 + \dfrac{8}{x - 2} = \dfrac{32}{x^2 - 4}$

43. A bus travels 250 miles on 10 gallons of gas. How many gallons will it need to travel 725 miles?

44. Solve for x: $\dfrac{x - 8}{4} = \dfrac{6}{5}$

45. Janet can paint a kitchen in 4 hours and James can paint the same kitchen in 5 hours. How long would it take for both working together to paint the kitchen?

46. Find an equation of the line that passes through the point $(-2, -4)$ and has slope $m = 6$.

47. Find an equation of the line having slope -2 and y-intercept -5.

48. Graph: $3x - 2y < -6$

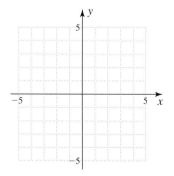

49. Graph: $-y \geq -5x + 5$

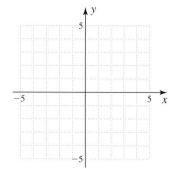

50. An enclosed gas exerts a pressure P on the walls of the container. This pressure is directly proportional to the temperature T of the gas. If the pressure is 7 pounds per square inch when the temperature is 490°F, find k.

51. If the temperature of a gas is held constant, the pressure P varies inversely as the volume V. A pressure of 1960 pounds per square inch is exerted by 7 cubic feet of air in a cylinder fitted with a piston. Find k.

52. Graph the system and find the solution if possible:

$$x + 3y = -12$$
$$2y - x = -2$$

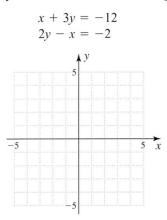

53. Graph the system and find the solution if possible:

$$y + 3x = 3$$
$$3y + 9x = 18$$

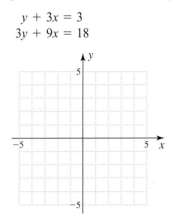

54. Solve (if possible) by substitution:

$$x + 3y = 10$$
$$4x + 12y = 42$$

55. Solve (if possible) by substitution:

$$x - 3y = 3$$
$$-3x + 9y = -9$$

56. Solve the system (if possible):

$$3x - 4y = 3$$
$$2x - 3y = 3$$

57. Solve the system (if possible):

$$4x + 3y = 18$$
$$8x + 6y = -3$$

58. Solve the system (if possible):

$$3y + x = -13$$
$$-4x - 12y = 52$$

59. Kaye has $2.50 in nickels and dimes. She has twice as many dimes as nickels. How many nickels and how many dimes does she have?

60. The sum of two numbers is 170. Their difference is 110. What are the numbers?

61. Evaluate: $\sqrt[5]{243}$

62. Simplify: $\sqrt[4]{(-5)^4}$

63. Simplify: $\sqrt{\dfrac{5}{243}}$

64. Rationalize the denominator: $\dfrac{\sqrt{2}}{\sqrt{7j}}$

65. Add: $\sqrt{32} + \sqrt{18}$

66. Perform the indicated operations:
$\sqrt[3]{3x}\left(\sqrt[3]{9x^2} - \sqrt[3]{16x}\right)$

67. Find the product: $\left(\sqrt{125} + \sqrt{63}\right)\left(\sqrt{245} + \sqrt{175}\right)$

68. Reduce: $\dfrac{20 - \sqrt{32}}{4}$

69. Solve: $\sqrt{x + 4} = -3$

70. Solve: $\sqrt{x - 4} - x = -4$

71. Solve for x: $9x^2 - 4 = 0$

72. Solve for x: $64x^2 + 9 = 0$

73. Solve for x: $36(x - 4)^2 - 7 = 0$

74. Find the missing term: $(x + 2)^2 = x^2 + 4x +$ ____

75. Find the number that must divide each term in the equation so that the equation can be solved by completing the square: $4x^2 + 8x = 14$

76. Find the number that must be added to both sides of the equation so that the equation can be solved by the method of completing the square: $x^2 + 4x = 5$

77. Solve for x: $jx^2 + kx + m = 0$

78. Solve for x: $2x^2 + 3x - 9 = 0$

79. Solve for x: $16x = x^2$

80. Graph: $y = -(x + 3)^2 + 2$

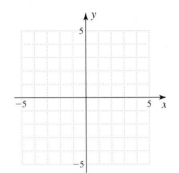

81. The distance d (in meters, m) traveled by an object thrown downward with an initial velocity of v_0 after t seconds is given by the formula $d = 5t^2 + v_0 t$. Find the number of seconds it takes an object to hit the ground if the object is dropped ($v_0 = 0$) from a height of 20 meters.

▶ Selected Answers

The brackets preceding answers for the Chapter Review Exercises indicate the Chapter, Section, and Objective for you to review for further study. For example, [3.4C] appearing before answers means those exercises correspond to Chapter 3, Section 4, Objective C.

Chapter R

Exercises R.1

1. $\frac{28}{1}$ **3.** $-\frac{42}{1}$ **5.** $\frac{0}{1}$ **7.** $-\frac{1}{1}$ **9.** 3 **11.** 42 **13.** 25 **15.** 21
17. 32 **19.** 6 **21.** 14 **23.** 30 **25.** 3 **27.** 1 **29.** 2 **31.** $\frac{5}{4}$
33. $\frac{1}{4}$ **35.** $\frac{7}{3}$ **37.** $\frac{2}{3}$ **39.** $\frac{4}{1} = 4$ **41. (a)** $\frac{1000}{45}$; **(b)** $\frac{200}{9}$; **(c)** $22\frac{2}{9}$;
(d) $\frac{1000}{24} = \frac{125}{3} = 41\frac{2}{3}$ **43.** $\frac{41}{100}$ **45. (a)** $\frac{2}{6}$; **(b)** 2 pieces;
(c) $\frac{2}{6} = \frac{1}{3}$; **(d)** They ate the same amount. **47.** $\frac{1}{8}$ **49.** $\frac{5}{8}$
51. (a) 2880 seconds; **(b)** 120 shots; **(c)** 24 **53.** $\frac{2}{15}$
55. Nat'l & Int'l News, $\frac{1}{2}$ **57.** $\frac{7}{30}$
59. Most: News & Beyond the Bay, $\frac{13}{30}$; Least: Traffic, $\frac{1}{10}$

Exercises R.2

1. $\frac{14}{9}$ **3.** $\frac{7}{5}$ **5.** $\frac{2}{1}$ or 2 **7.** $\frac{8}{1}$ or 8 **9.** $\frac{39}{7}$ **11.** $\frac{35}{3}$ **13.** $\frac{2}{3}$ **15.** $\frac{3}{2}$
17. $\frac{16}{5}$ **19.** $\frac{2}{5}$ **21.** $\frac{3}{5}$ **23.** $\frac{1}{1}$ or 1 **25.** $\frac{13}{8}$ **27.** $\frac{17}{15}$ **29.** $\frac{23}{6}$
31. $\frac{31}{10}$ **33.** $\frac{193}{28}$ **35.** $\frac{3}{8}$ **37.** $\frac{1}{6}$ **39.** $\frac{11}{20}$ **41.** $\frac{34}{75}$ **43.** $\frac{9}{20}$ **45.** $\frac{5}{4}$
47. $\frac{11}{18}$ **49.** $\frac{37}{15}$ **51.** 75 lb **53.** 5600 **55.** $16\frac{5}{8}$ **57.** $\frac{3}{10}$ **59.** $7\frac{7}{10}$
61. 4 pkg of hot dogs, 5 pkg of buns **67.** 8 **69.** $\frac{1}{5}$

Exercises R.3

1. $4 + \frac{7}{10}$ **3.** $5 + \frac{6}{10} + \frac{2}{100}$ **5.** $10 + 6 + \frac{1}{10} + \frac{1}{100} + \frac{3}{1000}$
7. $40 + 9 + \frac{1}{100} + \frac{2}{1000}$ **9.** $50 + 7 + \frac{1}{10} + \frac{4}{1000}$ **11.** $\frac{9}{10}$
13. $\frac{3}{50}$ **15.** $\frac{3}{25}$ **17.** $\frac{27}{500}$ **19.** $\frac{213}{100}$ **21.** \$989.07 **23.** 919.154
25. 182.103 **27.** 4.077 **29.** 26.85 **31.** \$2.38 **33.** 3.024
35. 6.844 **37.** 9.0946 **39.** 12.0735 **41.** 5.6396 **43.** 95.7
45. 0.024605 **47.** 12.90516 **49.** 0.002542 **51.** 0.6 **53.** 6.4
55. 1700 **57.** 80 **59.** 0.046 **61.** 0.2 **63.** 0.875 **65.** 0.1875
67. $0.\overline{2}$ **69.** $0.\overline{54}$ **71.** $0.\overline{27}$ **73.** $0.1\overline{6}$ **75.** $1.\overline{1}$ **77.** 0.33 **79.** 0.05
81. 3 **83.** 0.118 **85.** 0.005 **87.** 5% **89.** 39% **91.** 41.6%
93. 0.3% **95.** 100% **97.** $\frac{3}{10}$ **99.** $\frac{3}{50}$ **101.** $\frac{7}{100}$ **103.** $\frac{9}{200}$
105. $\frac{1}{75}$ **107.** 60% **109.** 50% **111.** $83\frac{1}{3}$% **113.** 50%
115. $133\frac{1}{3}$% **117.** 27.6 **119.** 26.7467 **121.** 35.250 **123.** 52.38
125. 74.84601 **127.** 0.955 **129.** $\frac{1}{2}$; 0.5; 50% **131.** 70%
133. (a) 40%; **(b)** $\frac{2}{5}$ **135. (a)** $\frac{49}{100}$; **(b)** 0.49 **137.** 5%
139. 1% **141.** 30

Chapter 1

Exercises 1.1

1. $a + c$ **3.** $3x + y$ **5.** $9x + 17y$ **7.** $3a - 2b$ **9.** $-2x - 5$
11. $7a$ **13.** $\frac{1}{7}a$ **15.** bd **17.** xyz **19.** $-b(c + d)$ **21.** $(a - b)x$
23. $(x - 3y)(x + 7y)$ **25.** $(c - 4d)(x + y)$ **27.** $\frac{y}{3x}$ **29.** $\frac{2b}{a}$
31. $\frac{a}{x + y}$ **33.** $\frac{a - b}{c}$ **35.** $\frac{y}{x}$ **37.** $\frac{p - q}{p + q}$ **39.** $\frac{x + 2y}{x - 2y}$ **41.** 16 **43.** 61
45. 9 **47.** 3 **49.** 56 **51.** 6 **53.** 50 **55.** 1 **57.** 4 **59.** $\frac{1}{4}$
61. $V = IR$ **63.** $P = P_A + P_B + P_C$ **65.** $D = RT$

67. $E = mc^2$ **69.** $c^2 = a^2 + b^2$ **71.** Multiplication **75.** $a + b$
77. ab or $a \cdot b$ **79.** evaluating **81.** $2x - y$ **83.** $7x + 4y$
85. 5 **87.** $\frac{1}{3}$ **89.** 0 **91.** 0

Exercises 1.2

1. -4 **3.** 49 **5.** $-\frac{7}{3}$ **7.** 6.4 **9.** $-3\frac{1}{7}$ **11.** -0.34 **13.** $0.\overline{5}$
15. $-\sqrt{7}$ **17.** $-\pi$ **19.** 2 **21.** 48 **23.** 3 **25.** $\frac{4}{5}$ **27.** 3.4
29. $1\frac{1}{2}$ **31.** $-\frac{3}{4}$ **33.** $-0.\overline{5}$ **35.** $-\sqrt{3}$ **37.** $-\pi$
39. Natural, whole, integer, rational, real **41.** Rational, real
43. Whole, integer, rational, real **45.** Rational, real **47.** Rational, real
49. Irrational, real **51.** Rational, real **53.** Rational, real **55.** 8
57. 8 **59.** 0, 8 **61.** $-5, \frac{1}{5}, 0, 8, 0.\overline{1}, 3.666...$ **63.** True
65. False; $|0| = 0$ **67.** True **69.** False; $\frac{3}{5}$ is not an integer.
71. False; $0.12345...$ is not rational. **73.** True **75.** $+20$ yards
77. -1312 feet from sea level **79.** $-4°$F to $+45°$F **81.** $+55$%
83. -3.1% **85.** $+22.50$% **87.** -30.78 million **89.** about $-43°$F
91. about $-40°$F **95.** all numbers with either terminating or repeating
decimal representations. **99.** 0 **101.** irrational **103.** $\sqrt{19}$ **105.** $8\frac{1}{4}$
107. $0.\overline{4}$ **109.** $-\frac{1}{2}$ **111.** Integer, rational, real **113.** Rational, real
115. Whole, integer, rational, real **117.** 0 **119.** 6.4 **121.** $\frac{31}{12}$

Exercises 1.3

1. 6 **3.** -4 **5.** 1 **7.** -7 **9.** 0 **11.** 3 **13.** 13 **15.** 2 **17.** -9
19. -12 **21.** 3.1 **23.** -4.7 **25.** -5.4 **27.** -8.6 **29.** $\frac{3}{7}$ **31.** $-\frac{1}{2}$
33. $-\frac{1}{12}$ **35.** $\frac{7}{12}$ **37.** $-\frac{13}{21}$ **39.** $-\frac{31}{18}$ **41.** -16 **43.** -20 **45.** -6
47. 16 **49.** -4 **51.** -2.6 **53.** 5.2 **55.** -3.7 **57.** $\frac{4}{7}$ **59.** $-\frac{29}{12}$
61. -7 **63.** 9 **65.** -4 **67.** 3500°C **69.** \$46 **71.** 14°C
73. -380 **75.** -355 **77.** 799 yr **79.** 284 yr **81.** 2262 yr
87. larger **89.** inverse **91.** -3.1 **93.** -7.7 **95.** -20
97. 1.2 **99.** -0.3 **101.** -5.7 **103.** $\frac{14}{5}$ **105.** $\frac{2}{5}$ **107.** $\frac{703}{20}$ or $35\frac{3}{20}$
109. $\frac{36}{121}$

Exercises 1.4

1. 36 **3.** -40 **5.** -81 **7.** -36 **9.** -54 **11.** -7.26 **13.** 2.86
15. $-\frac{25}{42}$ **17.** $\frac{1}{4}$ **19.** $-\frac{15}{4}$ **21.** -81 **23.** 25 **25.** -125 **27.** 1296
29. $-\frac{1}{32}$ **31.** 7 **33.** -5 **35.** -3 **37.** 0 **39.** Undefined **41.** 0
43. 5 **45.** 5 **47.** -2 **49.** -6 **51.** $-\frac{21}{20}$ **53.** $\frac{4}{7}$ **55.** $-\frac{5}{7}$
57. -0.5 or $-\frac{1}{2}$ **59.** $0.1\overline{6}$ or $\frac{1}{6}$ **61.** 15 calories; gain
63. 5 calories; loss **65.** 6 min **67.** 15.4 mi/hr each second
69. \$4.45 **71. (a)** $4\frac{2}{5}$; **(b)** 25 days **73. (a)** $4\frac{1}{5}$; **(b)** 50 days
75. -1.35 **77.** -3.045 **79.** 3.875 **85.** negative **87.** negative
89. 0.25 or $\frac{1}{4}$ **91.** -33 **93.** 55 **95.** 64 **97.** -7.04 **99.** $\frac{12}{35}$
101. $-\frac{7}{20}$ **103.** $-\frac{1}{2}$ **105.** -13 **107.** 14

Exercises 1.5

1. 26 **3.** 13 **5.** 53 **7.** 5 **9.** 3 **11.** 10 **13.** 7 **15.** 3 **17.** 8
19. 10 **21.** 10 **23.** 20 **25.** 27 **27.** -31 **29.** 36 **31.** -5
33. -5 **35.** 25 **37.** -15 **39.** -24 **41.** 87 **43. (a)** 144; **(b)** 126

45. (a) 115 lb; **(b)** 130 lb   **47. (a)** $99.99; **(b)** $77.99

49. (a) $5; **(b)** $18   **51.** 5 milligrams   **53.** $1\frac{1}{3}$ tablets every 12 hours

57. PEMDAS   **59.** exponents   **61.** division   **63.** subtraction

65. 0   **67.** 18   **69.** 38   **71.** 10   **73.** 119   **75.** 1   **77.** 0   **79.** 1

Exercises 1.6

1. Commutative property of addition

3. Commutative property of multiplication   **5.** Distributive property

7. Commutative property of multiplication

9. Associative property of addition

11. Associative property of addition; 5

13. Commutative property of multiplication; 7

15. Commutative property of addition; 6.5

17. Associative property of multiplication; 2   **19.** $13 + 2x$ **21.** 0

23. 1   **25.** Multiplicative inverse property   **27.** Identity element for addition   **29.** 3   **31.** 5   **33.** a **35.** -2 **37.** -2 **39.** $24 + 6x$

41. $8x + 8y + 8z$ **43.** $6x + 42$ **45.** $ab + 5b$ **47.** $30 + 6b$

49. $-4x - 4y$ **51.** $-9a - 9b$ **53.** $-12x - 6$ **55.** $-\frac{3a}{2} + \frac{6}{7}$

57. $-2x + 6y$ **59.** $-2.1 - 3y$ **61.** $-4a - 20$ **63.** $-6x - xy$

65. $-8x + 8y$ **67.** $-6a + 21b$ **69.** $0.5x + 0.5y - 1.0$

71. $\frac{6}{5}a - \frac{6}{5}b + 6$ **73.** $-2x + 2y - 8$ **75.** $-0.3x - 0.3y + 1.8$

77. $-\frac{5}{2}a + 5b - \frac{5}{2}c + \frac{5}{2}$ **79.** $(101 + 96) + 62$; 259 mi

81. The associative property of addition   **83. (a)** $A = a(b + c)$;

(b) $A_1 = ab$; **(c)** $A_2 = ac$; **(d)** The distributive property

85. (a) $5h - 195$; 145 lb; **(b)** $3h - 46$; 170 lb   **87.** 266   **89.** 276

93. No   **95.** Yes   **97.** $a \cdot (b \cdot c) = (a \cdot b) \cdot c$ **99.** $a \cdot b = b \cdot a$

101. $a \cdot 1 = 1 \cdot a = a$ **103.** $\frac{1}{a}$ **105.** $a(b + c) = ab + ac$

107. $-4a + 20$ **109.** $4a + 24$ **111.** 7   **113.** $16 + 4x$

115. 0   **117.** 1   **119.** 2   **121.** Identity element for multiplication

123. Identity element for addition

125. Associative property for multiplication

127. Associative property for addition

129. Associative property for multiplication; 2

131. Commutative property for multiplication; 1

133. Associative property for addition; 3   **135.** -7 **137.** $2x + 2$

Exercises 1.7

Translate This

1. H   **3.** M   **5.** N   **7.** L   **9.** G

1. $11a$ **3.** $-5c$ **5.** $12n^2$ **7.** $-7ab^2$ **9.** $3abc$ **11.** $1.3ab$

13. $0.2x^2y - 0.9xy^2$ **15.** $3ab^2c$ **17.** $-6ab + 7xy$

19. $\frac{3}{5}a + \frac{4}{7}a^2b$ **21.** $11x$ **23.** $8ab$ **25.** $-7a^2b$ **27.** 0   **29.** 0

31. $10xy - 4$ **33.** $-2R + 6$ **35.** $-L - 2W$ **37.** $-3x - 1$

39. $\frac{x}{9} + 2$ **41.** $6a + 2b$ **43.** $3x - 4y$ **45.** $x - 5y + 36$

47. $-2x^3 + 9x^2 - 3x + 12$ **49.** $x^2 - \frac{2}{5}x + \frac{1}{2}$

51. $10a - 18$ **53.** $-23a + 18b + 9$ **55.** $-4.8x + 3.4y + 5$

57. $\frac{1}{f} = \frac{1}{u} + \frac{1}{v}$; 2 terms   **59.** $F = \frac{n}{4} + 40$; 2 terms

61. $I = Prt$; 1 term   **63.** $P = a + b + c$; 3 terms

65. $A = \frac{bh}{2}$; 1 term   **67.** $K = Cmv^2$; 1 term

69. $d = 16t^2$; 1 term   **71.** $C = 12W - 500$; 2 terms

73. (a) $C = 55h$; **(b)** $C = 85h$; **(c)** $C = 160h$ **75.** $4S$; $P = 4S$

77. 18 ft, 6 in.   **79.** $2h + 2w + L$ **81.** $(7x + 2)$ in.

87. $a + b$ **89.** =   **91.** $A = LW$; 1 term

93. $14y - 11$ **95.** $-3ab^3$ **97.** $-3x^2 + 12x + 6$ **99.** 11

101. $-2x - 10$

Review Exercises

1. [1.1A] **(a)** $a + b$; **(b)** $a - b$; **(c)** $7a + 2b - 8$ **2.** [1.1A] **(a)** $3m$;

(b) $-mnr$; **(c)** $\frac{1}{7}m$; **(d)** $8m$ **3.** [1.1A] **(a)** $\frac{m}{9}$; **(b)** $\frac{9}{n}$ **4.** [1.1A] **(a)** $\frac{m + n}{r}$;

(b) $\frac{m + n}{m - n}$ **5.** [1.1B] **(a)** 12; **(b)** 6; **(c)** 36   **6.** [1.1B] **(a)** 3; **(b)** 9; **(c)** 7

7. [1.2A] **(a)** 5; **(b)** $-\frac{2}{3}$; **(c)** -0.37 **8.** [1.2B] **(a)** 8; **(b)** $3\frac{1}{2}$; **(c)** -0.76

9. [1.2C] **(a)** Integer, rational, real; **(b)** Whole, integer, rational, real;

(c) Rational, real; **(d)** Irrational, real; **(e)** Irrational, real; **(f)** Rational, real

10. [1.3A] **(a)** 2; **(b)** -0.8; **(c)** $-\frac{11}{20}$; **(d)** -2.2; **(e)** $\frac{1}{4}$

11. [1.3B] **(a)** -20; **(b)** -2.4; **(c)** $-\frac{17}{12}$ **12.** [1.3C] **(a)** 30; **(b)** -15

13. [1.4A] **(a)** -35; **(b)** -18.4; **(c)** -19.2; **(d)** $\frac{12}{35}$ **14.** [1.4B] **(a)** 16;

(b) -9; **(c)** $-\frac{1}{27}$; **(d)** $\frac{1}{27}$ **15.** [1.4C] **(a)** -4; **(b)** 2; **(c)** $-\frac{2}{3}$; **(d)** $\frac{3}{2}$

16. [1.5A] **(a)** 10; **(b)** 0   **17.** [1.5B] **(a)** 8; **(b)** -41 **18.** [1.5C] 152

19. [1.6A] **(a)** Commutative property of addition; **(b)** Associative

property of multiplication; **(c)** Associative property of addition

20. [1.6B] **(a)** $4x - 4$; **(b)** $-8x + 2$ **21.** [1.6C, E] **(a)** 1; **(b)** $-\frac{3}{4}$;

(c) -3.7; **(d)** $\frac{2}{3}$; **(e)** 0; **(f)** -5 **22.** [1.6E] **(a)** $-3a - 24$; **(b)** $-4x + 20$;

(c) $-x + 4$ **23.** [1.7A] **(a)** $-5x$; **(b)** $-10x$; **(c)** $-8x$; **(d)** $3a^2b$

24. [1.7B] **(a)** $12a - 5$; **(b)** $5x - 9$ **25.** [1.7C] **(a)** $m = \frac{p}{n}$ (2.75);

one term; **(b)** $W = \frac{11h}{2} - 220$; two terms

Chapter 2

Exercises 2.1

1. Yes   **3.** Yes   **5.** Yes   **7.** No   **9.** No   **11.** $x = 14$ **13.** $m = 19$

15. $y = \frac{10}{3}$ **17.** $k = 21$ **19.** $z = \frac{11}{12}$ **21.** $x = \frac{7}{2}$ **23.** $c = 0$

25. $x = -3$ **27.** $y = 0$ **29.** $c = -2.5$ **31.** $p = -9$ **33.** $x = -3$

35. $m = 6$ **37.** $y = 18$ **39.** $a = -7$ **41.** $c = -8$ **43.** $x = -2$

45. $g = -3$ **47.** No solution   **49.** All real numbers   **51.** $b = 6$

53. $p = \frac{20}{3}$ **55.** $r = \frac{9}{8}$ **57.** $9.41   **59.** 184.7   **61.** 88.1 million tons;

145.9 million tons   **63.** 21%   **65.** 27%   **67. (a)** No; **(b)** 110 lb

69. (a) No; **(b)** 2 kg   **71. (a)** No; **(b)** 1 kg   **73. (a)** yes;

(b) $W = 56.2 + 1.41(70 - 60) = 56.2 + 14.1 = 70.3$ **75. (a)** No;

(b) 100 over   **77.** No   **79.** No   **85.** $a + c = b + c$ **87.** No solution

89. $x = 9$ **91.** $x = 5.7$ **93.** $x = 4$ **95.** $x = \frac{7}{9}$ **97.** $z = 7$

99. No solution   **101.** No   **103.** -20 **105.** $-\frac{1}{2}$ **107.** $\frac{2}{3}$ **109.** 48

111. 40

Exercises 2.2

1. $x = 35$ **3.** $x = -8$ **5.** $b = -15$ **7.** $f = 6$ **9.** $v = \frac{4}{3}$

11. $x = -\frac{15}{4}$ **13.** $z = 11$ **15.** $x = -7$ **17.** $c = -7$ **19.** $x = 7$

21. $y = -\frac{11}{3}$ **23.** $a = -0.6$ **25.** $t = \frac{3}{2}$ **27.** $x = -2.25$

29. $C = -8$ **31.** $a = 12$ **33.** $y = -0.5$ **35.** $p = 0$ **37.** $t = -30$

39. $x = -0.02$ **41.** $y = 12$ **43.** $x = 21$ **45.** $x = 20$ **47.** $t = 24$

49. $x = 1$ **51.** $c = 15$ **53.** $W = 10$ **55.** $x = 4$ **57.** $x = 10$

59. $x = 3$ **61.** 12   **63.** 28   **65.** 50%   **67.** 150   **69.** 20   **71.** 7.6%

73. 9.5%   **75.** 3.8%   **77.** 5.3%   **79.** 3.8%   **81.** $85h = 3500$; $h = 41\frac{3}{17}$

83. $260h = 3500$; $h = 13\frac{6}{13}$ **85.** $450h = 3500$; $h = 7\frac{7}{9}$

87. $16,000   **91.** Multiply by the reciprocal of $-\frac{3}{4}$ **93.** $\frac{a}{c} = \frac{b}{c}$

95. LCM   **97.** 20%   **99.** $x = 6$ **101.** $y = 30$ **103.** $y = 10$

105. $y = 14$ **107.** $x = -14$

109. $4x - 24$ **111.** $40 - 5y$ **113.** $54 - 27y$ **115.** $-15x + 20$
117. 4   **119.** 3

Exercises 2.3

1. $x = 4$ **3.** $y = 1$ **5.** $z = 2$ **7.** $y = \frac{14}{5}$ **9.** $x = 3$ **11.** $x = -4$
13. $v = -8$ **15.** $m = -4$ **17.** $z = -\frac{2}{3}$ **19.** $x = 0$ **21.** $a = \frac{1}{13}$
23. $c = 35.6$ **25.** $x = -\frac{11}{80}$ **27.** $x = \frac{9}{2}$ **29.** $x = -20$ **31.** $x = -5$
33. $h = 0$ **35.** $w = -1$ **37.** $x = -3$ **39.** $x = \frac{5}{2}$ **41.** All real
numbers   **43.** $x = 2$ **45.** $x = 10$ **47.** $x = \frac{27}{20}$ **49.** $x = \frac{69}{7}$
51. $r = \frac{C}{2\pi}$ **53.** $y = \frac{6 - 3x}{2}$ **55.** $s = \frac{A - \pi r^2}{\pi r}$ **57.** $V_2 = \frac{P_1 V_1}{P_2}$
59. $H = \frac{f + Sh}{S}$ **61. (a)** $L = \frac{S + 22}{3}$; **(b)** $L = \frac{S + 21}{3}$
63. $H = \frac{W + 200}{5}$ **65.** 11 inches   **67.** Size 8   **69.** 15 hr
71. (a) 0.723; **(b)** 0.68   **73. (a)** 1.569; **(b.)** 1.61   **75. (a)** 1.992;
(b) 1.95; **(c)** Younger employees   **77.** $66\frac{2}{3}$ miles   **79. (a)** $S = 1.2C$;
(b) \$9.60   **83.** $x = 0$ and division by 0 is undefined.   **85.** literal
87. $m = 150$ **89.** $x = -1$ **91.** $x = 1$ **93.** $x = 1$ **95.** $\frac{a + b}{c}$
97. $a(b + c)$ **99.** $a - bc$

Exercises 2.4

Translate This

1. D   **3.** I   **5.** B   **7.** C   **9.** J

1. 44, 46, 48   **3.** $-10, -8, -6$ **5.** $-13, -12$ **7.** 7, 8, 9
9. Any consecutive integers   **11.** \$20, \$44   **13.** 39 to 94
15. \$104, \$43   **17.** 37, 111, 106   **19.** 21, 126   **21.** 38% from home,
13% from work   **23.** \$392 per week   **25.** 222 feet   **27.** 3 sec
29. 304 and 676   **31.** 10 minutes   **33. (a)** 22 mi;
(b) The limo is cheaper.   **35.** 20°   **37.** 45°   **39.** 42°
41. 157 million tons   **43.** 140 million tons   **45.** 5.6 million tons
47. 236 million tons   **49.** 12.1%   **51.** 84   **55.** RSTUV
57. complementary   **59.** 81, 83, 85   **61.** Pizza: 230 calories,
Shake: 300 calories   **63.** $T = \frac{20}{11}$ **65.** $T = 8$ **67.** $x = 8$
69. $P = 3500$ **71.** $x = 1000$

Exercises 2.5

Translate This

1. G   **3.** A   **5.** M   **7.** D   **9.** E

1. 50 mi/hr   **3.** 2 meters/min   **5.** $\frac{1}{2}$ hr   **7.** 4 hr   **9.** 10 mi   **11.** 6 hr
13. 1920 mi   **15.** 24 in./sec first 180 sec, 1.2 in./sec last 60 sec
17. 6 L   **19.** 30 lb copper, 40 lb zinc   **21.** 20 lb   **23.** 32 oz
of each   **25.** 12 quarts   **27.** Impossible; it would be 120% of the
concentrate.   **29.** \$5000 at 6%, \$15,000 at 8%   **31.** \$8000
33. \$6000 at 5%, \$4000 at 6%   **35.** \$25,000 at 6%, \$15,000 at 10%
41. $I = Pr$ **43.** 10 gal   **45.** 5 hr   **47.** 158   **49.** 100   **51.** $7\frac{2}{3}$

Exercises 2.6

Translate This

1. C   **3.** I   **5.** F   **7.** K   **9.** D

1. (a) 120; **(b)** 275 mi; **(c)** $R = \frac{D}{T}$; **(d)** 60 mi/hr   **3. (a)** 110 pounds;
(b) $H = \frac{W + 190}{5}$; **(c)** 78 inches tall (6 ft, 6 in.)   **5. (a)** 59°F;
(b) $C = \frac{5}{9}(F - 32)$; **(c)** 10°C   **7. (a)** EER $= \frac{\text{Btu}}{w}$; **(b)** 9;
(c) Btu $= (EER)(w)$; **(d)** 20,000   **9. (a)** $S = C + M$; **(b)** \$67;
(c) $M = S - C$; **(d)** \$8.25   **11. (a)** 60 cm; **(b)** 90 cm
13. (a) 62.8 in.; **(b)** $r = \frac{C}{2\pi}$; **(c)** 10 in.   **15. (a)** 13.02 m²; **(b)** $W = \frac{A}{L}$;
(c) 6 m   **17.** 35° each   **19.** 140° each   **21.** 175° each   **23.** 70°, 20°
25. 23°, 67°   **27.** 142°, 38°   **29.** 69°, 111°   **31.** 75 m   **33.** 4.5 in.
35. (a) $A = C - 10$; **(b)** 40   **37. (a)** 13.74 million; **(b)** $t = \frac{N - 9.74}{0.40}$ or

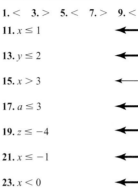

$t = \frac{5(N - 9.74)}{2}$; **(c)** 1995   **39. (a)** \$700;   **(b)** $x = \frac{A - 420}{28}$;
(c) 21 yr (2021)   **41. (a)** 67 in.; **(b)** $r = \frac{H - 34}{3.3}$; **(c)** 11 in.
43. (a) 149 cm; **(b)** $t = \frac{H - 62}{2.9}$; **(c)** 31 cm   **45.** 90 ft   **51.** 50%
53. $P = 2L + 2W$ **55.** $A = \pi r^2$ **57.** equal   **59.** 44°, 46°   **61.** 128°,
52°   **63.** 46°, 134°   **65. (a)** 126.48 cm; **(b)** $h = \frac{H - 71.48}{2.75}$; **(c)** 25 cm
67. (a) 75 in.²; **(b)** $b = \frac{2A}{h}$; **(c)** 4 in.   **69. (a)** 5°C; **(b)** $F = \frac{9}{5}C + 32$;
(c) 68°F   **71.** $x = 4$ **73.** $x = 3$ **75.** $x = -6$

Exercises 2.7

1. $<$ **3.** $>$ **5.** $<$ **7.** $>$ **9.** $<$

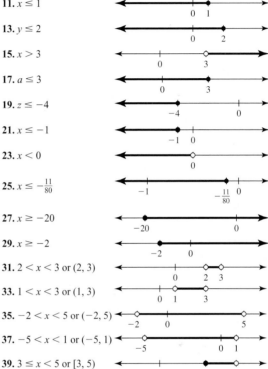

11. $x \le 1$
13. $y \le 2$
15. $x > 3$
17. $a \le 3$
19. $z \le -4$
21. $x \le -1$
23. $x < 0$
25. $x \le -\frac{11}{80}$
27. $x \ge -20$
29. $x \ge -2$
31. $2 < x < 3$ or $(2, 3)$
33. $1 < x < 3$ or $(1, 3)$
35. $-2 < x < 5$ or $(-2, 5)$
37. $-5 < x < 1$ or $(-5, 1)$
39. $3 \le x < 5$ or $[3, 5)$

41. $20°F < t < 40°F$ **43.** $\$12,000 < s < \$13,000$ **45.** $2 \le e \le 7$
47. $\$3.50 < c < \$4.00$ **49.** $a < 41$ ft   **51.** After 1996   **53.** 1991
55. $J = 5$ ft $= 60$ in.   **57.** $F = S - 3$ **59.** $S = 6$ ft 5 in. $= 77$ in.
61. $B > 74$ in. or $B > 6$ ft 2 in.   **63.** When multiplying or dividing
by a negative number.   **67.** $x - a < y - a$ **69.** $ax > ay$
71. $\frac{x}{a} > \frac{y}{a}$
73. $-3 \le x \le -1$
or $[-3, -1]$
75. $x > 3$
77. $x > 3$
79. $x \ge 2$
81. $x < 1$
83. $<$
85. $>$
87.
89.
91.

93.

$-5\ -4\ -3\ -2\ -1\ \ 0\ \ 1\ \ 2\ \ 3\ \ 4\ \ 5$

95.

$-5\ -4\ -3\ -2\ -1\ \ 0\ \ 1\ \ 2\ \ 3\ \ 4\ \ 5$

Review Exercises

1. [2.1A] **(a)** No; **(b)** Yes; **(c)** Yes   **2.** [2.1B] **(a)** $x = \frac{2}{3}$; **(b)** $x = 1$;
(c) $x = \frac{2}{3}$ **3.** [2.1B] **(a)** $x = \frac{2}{9}$; **(b)** $x = \frac{4}{7}$; **(c)** $x = \frac{1}{6}$
4. [2.1C] **(a)** $x = 5$; **(b)** $x = 0$; **(c)** $x = 3$ **5.** [2.1C] **(a)** No solution;
(b) $x = -\frac{1}{8}$; **(c)** No solution   **6.** [2.1C] **(a)** All real numbers;
(b) All real numbers; **(c)** All real numbers   **7.** [2.2A] **(a)** $x = -15$;
(b) $x = -14$; **(c)** $x = -2$ **8.** [2.2B] **(a)** $x = 12$; **(b)** $x = 15$; **(c)** $x = 9$
9. [2.2C] **(a)** $x = 6$; **(b)** $x = \frac{24}{7}$; **(c)** $x = 20$ **10.** [2.2C] **(a)** $x = 12$;
(b) $x = 28$; **(c)** $x = 40$ **11.** [2.2C] **(a)** $x = 17$; **(b)** $x = 7$; **(c)** $x = 9$
12. [2.2D] **(a)** 20%; **(b)** 10%; **(c)** 20%   **13.** [2.2D] **(a)** 50; **(b)** $33\frac{1}{3}$;
(c) $33\frac{1}{3}$ **14.** [2.3A] **(a)** $x = -3$; **(b)** $x = -4$; **(c)** $x = -5$
15. [2.3B] **(a)** $h = \frac{2A}{b}$; **(b)** $r = \frac{C}{2\pi}$; **(c)** $b = \frac{3V}{h}$ **16.** [2.4A] **(a)** 32, 52;
(b) 14, 33; **(c)** 29, 52   **17.** [2.4B] **(a)** Chicken: 278, pie: 300;
(b) Chicken: 291, pie: 329; **(c)** Chicken: 304, pie: 346
18. [2.4C] **(a)** 35°; **(b)** 30°; **(c)** 25°   **19.** [2.5A] **(a)** 4 hours;
(b) $1\frac{1}{2}$ hours; **(c)** 2 hours   **20.** [2.5B] **(a)** 10 pounds; **(b)** 15 pounds;
(c) 25 pounds   **21.** [2.5C] **(a)** $10,000 at 6%, $20,000 at 5%;
(b) $20,000 at 7%, $10,000 at 9%; **(c)** $25,000 at 6%, $5000 at 10%
22. [2.6A] **(a)** $m = \frac{C-3}{3.05}$; 8 minutes; **(b)** $m = \frac{C-3}{3.15}$; 10 minutes;
(c) $m = \frac{C-2}{3.25}$; 6 minutes   **23.** [2.6C] **(a)** 100°, 80°; **(b)** Both 60°;
(c) 65°, 25°   **24.** [2.7A] **(a)** $<$; **(b)** $>$; **(c)** $<$

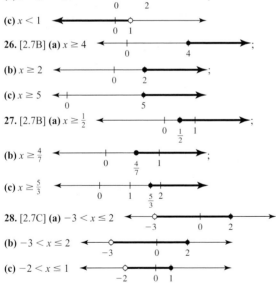

25. [2.7B] **(a)** $x < 3$;
(b) $x < 2$;
(c) $x < 1$
26. [2.7B] **(a)** $x \geq 4$;
(b) $x \geq 2$;
(c) $x \geq 5$
27. [2.7B] **(a)** $x \geq \frac{1}{2}$;
(b) $x \geq \frac{4}{7}$;
(c) $x \geq \frac{5}{3}$
28. [2.7C] **(a)** $-3 < x \leq 2$;
(b) $-3 < x \leq 2$
(c) $-2 < x \leq 1$

Cumulative Review Chapters 1–2

1. 7   **2.** $9\frac{9}{10}$ **3.** $-\frac{32}{63}$ **4.** 8.2   **5.** -8.64 **6.** -16 **7.** $\frac{21}{5}$
8. 69   **9.** Commutative property of multiplication   **10.** $30x + 42$
11. cd **12.** $-3x - 11$ **13.** $\frac{a-4b}{c}$ **14.** Yes   **15.** $x = 13$ **16.** $x = 9$
17. $x = 15$ **18.** $x = 8$ **19.** $b = \frac{S}{6a^2}$ **20.** 60 and 95
21. $350 from stocks and $245 from bonds   **22.** 24 hr
23. $2000 in bonds and $3000 in certificates of deposit
24. $x \geq 6$

Chapter 3

Exercises 3.1

1.

3.

5.

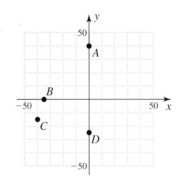

7. $A(0, 2\frac{1}{2})$; $B(-3, 1)$; $C(-2, -2)$; $D(0, -3\frac{1}{2})$; $E(1\frac{1}{2}, -2)$;
Quadrant II: B; Quadrant III: C; Quadrant IV; E; y-axis: A, D
9. $A(10, 10)$; $B(-30, 20)$; $C(-25, -20)$; $D(0, -40)$; $E(15, -15)$;
Quadrant I: A; Quadrant II: B; Quadrant III: C; Quadrant IV: E;
y-axis: D **11.** (20, 140)   **13.** (45, 147)   **15.** 2.68   **17.** 1.77
19. about 4.45   **21.** 236.2   **23.** 2.2   **25.** $950   **27.** $800   **29.** 70
31. 3 A.M.   **33.** 55   **35.** Before 3 A.M.: 195; after 3 A.M.: 288
37. Before 3 A.M.: 162; after 3 A.M.: 13   **39. (a)** Auto batteries;
(b) Tires   **41.** 26.5%; 1.43%   **43.** $500   **45.** $25

47. and **49.**

Equivalent Human Age

51. 75–100 lb   **53.** Less/younger, or more/older
55. Ages 6, 7, 9, 12, 13, and 15   **57. (a)** About $120,000;
(b) About $50,000; **(c)** About 23 months

59. (a)

Altitude (kilometers)	Temperature Change	Ordered Pair
1	$-7°C$	$(1, -7)$
2	$-14°C$	$(2, -14)$
3	$-21°C$	$(3, -21)$
4	$-28°C$	$(4, -28)$
5	$-35°C$	$(5, -35)$

(b)

61. 86°F   **63.** Less than 10%   **65.** 12°F   **67.** a is negative, b is positive; no   **69.** $a = c$ and $b = d$ **71.** abscissa
73. and **75.**

77.

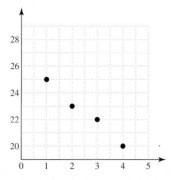

79. 11 human years   **81. (a)** Quadrant II; **(b)** Quadrant IV;
(c) Quadrant I; **(d)** Quadrant III   **83.** $x = 3$ **85.** $y = 9$ **87.** $y = 3$

Exercises 3.2

1. Yes   **3.** Yes   **5.** No   **7.** 0   **9.** -2 **11.** -3 **13.** 3   **15.** 2
17. $2x + y = 4$

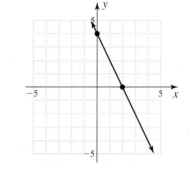

19. $-2x - 5y = -10$

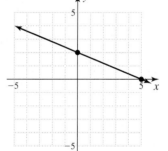

21. $y + 3 = 3x$

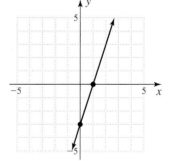

23. $6 = 3x - 6y$

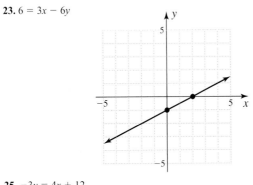

25. $-3y = 4x + 12$

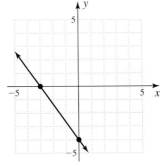

27. $-2y = -x + 4$

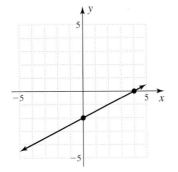

29. $-3y = -6x + 3$

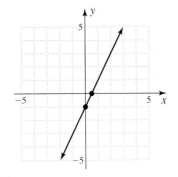

31. $y = 2x + 4$

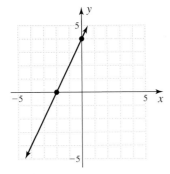

33. $y = -2x + 4$

35. $y = -3x - 6$

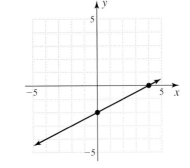

37. $y = \frac{1}{2}x - 2$

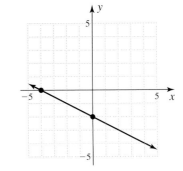

39. $y = -\frac{1}{2}x - 2$

41. (a) -9; **(b)** 2;
(c)

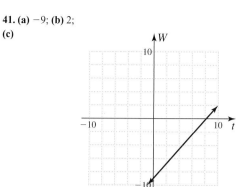

43. (a) $7000; **(b)** $8000; **(c)** $9000; $9800;
(d)

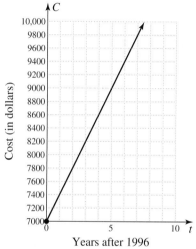

Years after 1996

45. (a) 56°F; **(b)** 20°F; **(c)** 60°F;
(d)

Temperature above Sea Level

Altitude (1000 ft)

(e) The temperature is negative.

47. (a) $1400; **(b)** $2200; **(c)** $200;
(d)

Catering Costs

Number of Guests

49. Yes **51.** No; 16 in. **53.** Yes **59.** straight **61. (a)** No; **(b)** No
63. $y = -3x + 6$

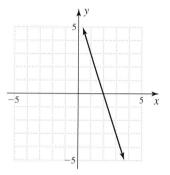

65. (a) -18; **(b)** -5;
(c)

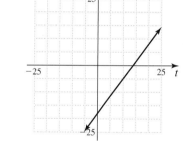

67. $x = -2$ **69.** $t = 0$

Exercises 3.3

1. $x + 2y = 4$

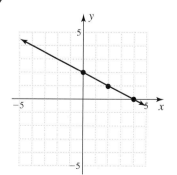

3. $-5x - 2y = -10$

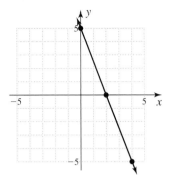

11. $3x + y = 0$

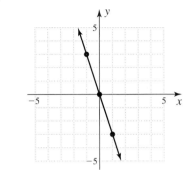

5. $y - 3x - 3 = 0$

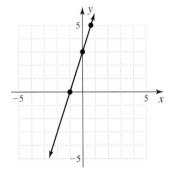

13. $2x + 3y = 0$

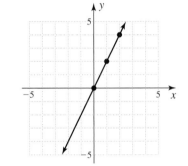

7. $6 = 6x - 3y$

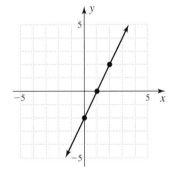

15. $-2x + y = 0$

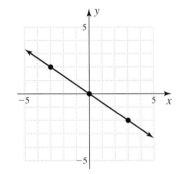

9. $3x + 4y + 12 = 0$

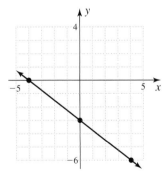

17. $2x - 3y = 0$

19. $-3x = -2y$

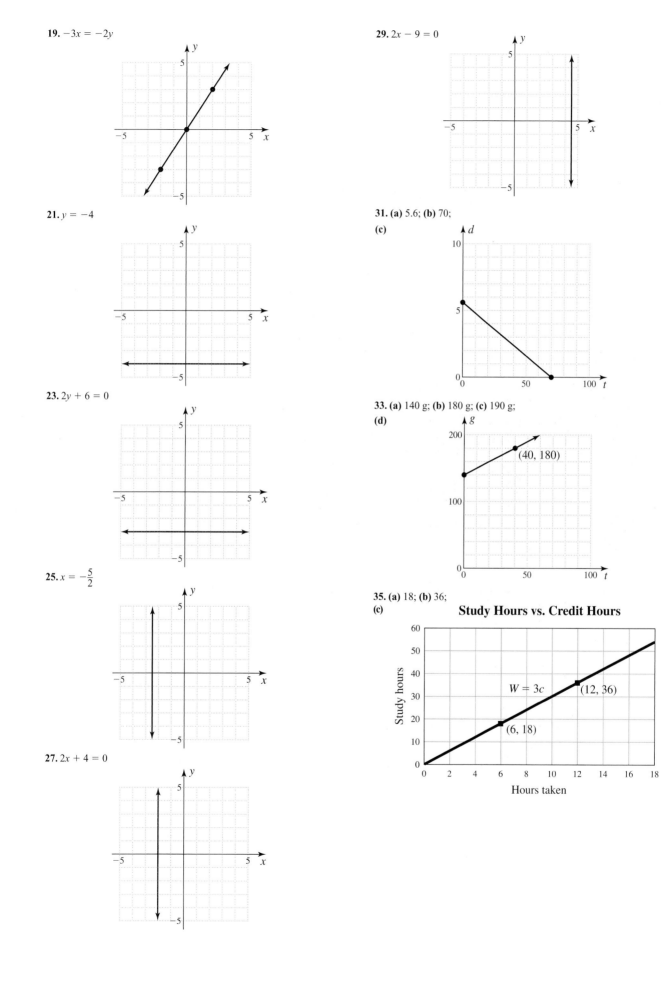

21. $y = -4$

23. $2y + 6 = 0$

25. $x = -\dfrac{5}{2}$

27. $2x + 4 = 0$

29. $2x - 9 = 0$

31. (a) 5.6; **(b)** 70;
(c)

33. (a) 140 g; **(b)** 180 g; **(c)** 190 g;
(d)

(40, 180)

35. (a) 18; **(b)** 36;
(c) **Study Hours vs. Credit Hours**

$W = 3c$ (12, 36)

(6, 18)

Study hours

Hours taken

37. (a) 0; **(b)** 624 lb/ft³; **(c)** 2496 lb/ft³;
(d)

Pressure at Depth of *f* Feet

39. 215
41.

43. A horizontal line   **45.** A line through the origin   **47.** Two
49. $(0, y)$ **51.** origin   **53.** vertical
55. $-2x + y = 4$

57. $4 + 2y = 0$

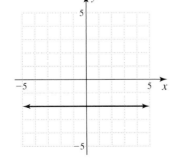

59. $-x + 4y = 0$

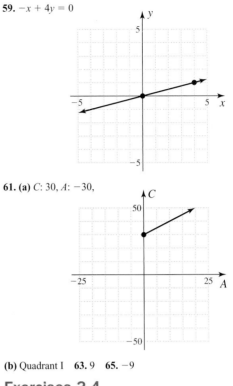

61. (a) C: 30, A: -30,

(b) Quadrant I   **63.** 9   **65.** -9

Exercises 3.4

1. $m = 1$ **3.** $m = -1$ **5.** $m = -\frac{1}{8}$ **7.** $m = \frac{1}{4}$ **9.** $m = 0$
11. $m = 0$ **13.** Undefined   **15.** $m = 3$ **17.** $m = -\frac{2}{3}$
19. $m = -\frac{1}{3}$ **21.** $m = \frac{2}{5}$ **23.** $m = 0$ **25.** Undefined
27. Parallel   **29.** Perpendicular   **31.** Parallel   **33.** Parallel
35. Neither; the lines coincide   **37.** Perpendicular
39. (a) $m = -0.1$; **(b)** Decreasing; **(c)** The rate at which the average
hospital stay is changing.   **41. (a)** 0.15; **(b)** Increasing;
(c) The slope 0.15 represents the annual increase in the life span
of American women.   **43. (a)** -32; **(b)** Decreasing;
(c) The slope -32 represents the decrease in the velocity of the ball.
45. (a) 0.4; **(b)** Increasing; **(c)** The slope 0.4 represents the annual
increase in the consumption of fat.   **47. (a)** From year 0 to year 1;
(b) \$160; **(c)** $m = -160$; **(d)** The decrease from year 0 to year 1
49. (a) From year 3 to year 4; **(b)** $m = 52$; **(c)** $m = 43$; **(d)** The slope
for year 3 to 4
51.   **53.**

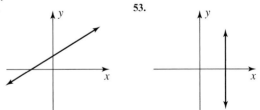

57. $\frac{y_2 - y_1}{x_2 - x_1}$ **59.** parallel   **61.** $m = -1$ **63.** Undefined   **65.** $m = -\frac{3}{2}$
67. Parallel   **69.** Perpendicular   **71.** $2x + 8$ **73.** $-2x - 2$

Exercises 3.5

1. $y = \frac{1}{2}x + \frac{3}{2}$

3. $y = -x + 6$

5. $y = 5$

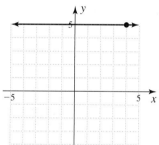

7. $y = 2x - 3$ **9.** $y = -4x + 6$ **11.** $y = \frac{3}{4}x + \frac{7}{8}$ **13.** $y = 2.5x - 4.7$
15. $y = -3.5x + 5.9$
17. $y = \frac{1}{4}x + 3$

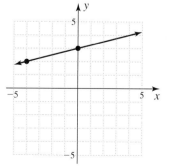

19. $y = -\frac{3}{4}x - 2$

21. $x - y = -1$

23. $3x - y = 4$

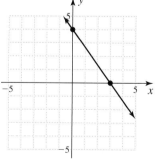

25. $4x + 3y = 12$

27. $x = 3$

29. $y = -3$

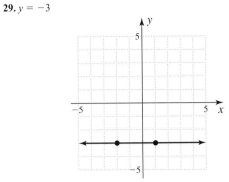

31. $y = x + 3$ **33.** $y = 2x$ **35.** $y = -2x - 3$ **37.** $y = 2x + 50$
39. (a) $2; (b) $75   **41.** The y-axis   **43.** $Ax + 0y = C$
47. $y = mx + b$ **49.** $Ax + By = C$
51. $y = -\frac{3}{4}x + 2$

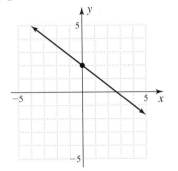

53. $2x - 3y = 9$

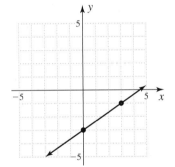

55. $5x + y = -11$

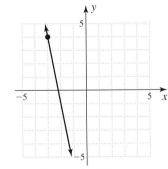

57. $y = 0.35x + 50$ **59.** $4x - y = 12$

Exercises 3.6

1. $C = 1.7m + 0.3$; $51.30   **3.** 6 mi   **5.** (a) $2; (b) $m - 1$; (c) $1.70;
(d) $1.70(m - 1)$; **(e)** $C = 2 + 1.70(m - 1)$ or $C = 1.7m + 0.3$; yes
7. (a) $C = 175$, if $m \le 60$; (b) $C = 175$, if $m \le 60$;
(c) $C = \frac{175}{m}$, if $m \le 60$ **9.** (a) $m = \frac{6}{5}$; (b) $m = \frac{6}{5}$; (c) Yes;
(d) $y = \frac{6}{5}x - 11$; **(e)** $y = \frac{6}{5}x - 11$; yes; **(f)** -5
11. (a) $C = 37.5h + 100$; (b) 3 hr   **13.** (a) $C = 2.3m + 40$;
(b) 70 min; **(c)** $C = 7.2m$; **(d)** About 8 min

15. (a) $C = 0.35m + 50$; (b) 110 min   **17.** 100 min; after 100 min
19. (a) $y = 1.95x + 29$; (b) $m = 1.95$; (c) $b = 29$
21. $y = -0.1x + 8$; 5 days; 4 days
23. (a) 3 human yr corresponds to 30 dog yr; 9 human yr corresponds
to 60 dog yr; **(b)** $m = 5$; **(c)** $d = 5h + 15$; **(d)** 35 dog yr; **(e)** 10 yr;
(f) 1.2 yr   **25.** (a) $m = 0.0255$; (b) $c = 0.0255b - 0.0245$;
(c) 0.0775; 0.1285; **(d)** 4   **27.** (a) $W = 2A + 8$; (b) $m = 2$; (c) $b = 8$;
(d) $W = 10$; **(e)** $W = 20$;
(f)

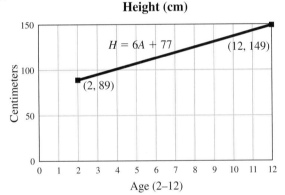

29. (a) $H = 6A + 77$; (b) $m = 6$; (c) $b = 77$; (d) $H = 89$; (e) $H = 149$;
(f)

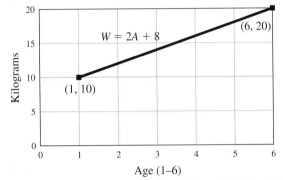

31. (a) $H = 62.92 + 2.39 f$; (b) $m = 2.39$; (c) $b = 62.92$
37.

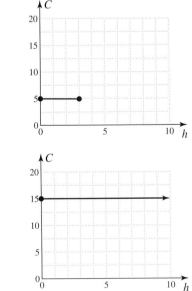

39.

41. $C = 0.15m + 40$; $107.50   **43.** (a) 30 hr cost $24;
(b) 33 hr cost $26.70; **(c)** $C = 0.9x - 3$ **45.** >   **47.** >   **49.** <

Exercises 3.7

1. $2x + y > 4$

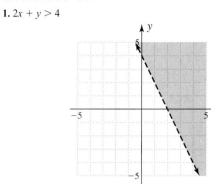

3. $-2x - 5y \leq 10$

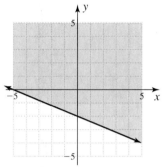

5. $y \geq 3x - 3$

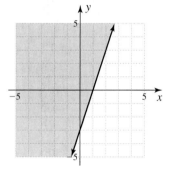

7. $6 < 3x - 6y$

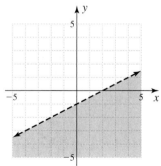

9. $3x + 4y \geq 12$

11. $10 < -2x + 5y$

13. $x \geq 2y - 4$

15. $y < -x + 5$

17. $2y < 4x + 5$

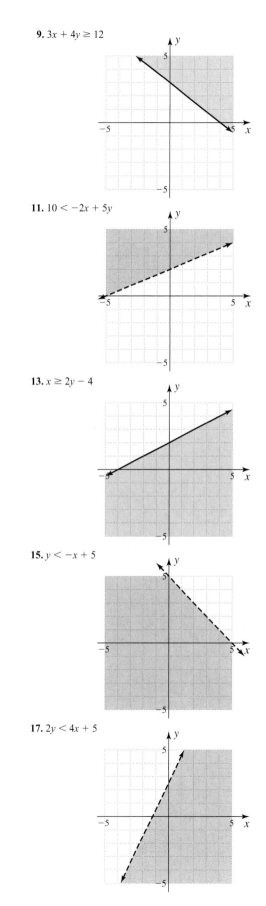

19. $x > 1$

21. $x \le \frac{5}{2}$

23. $y \le -\frac{3}{2}$

25. $x - \frac{2}{3} > 0$

27. $y + \frac{1}{3} \ge \frac{2}{3}$

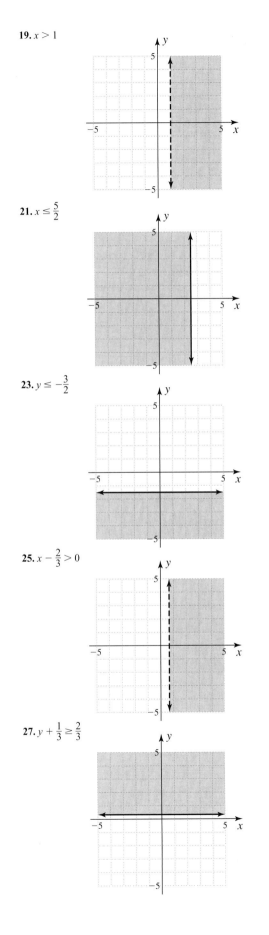

29. $2x + y < 0$

31. $y - 3x > 0$

33. rectangle

35. <18; 17.6%   **37.** 1966

39.

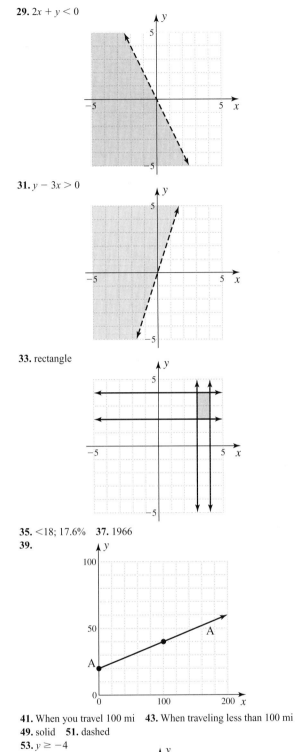

41. When you travel 100 mi   **43.** When traveling less than 100 mi
49. solid   **51.** dashed
53. $y \ge -4$

55. $y - 3 > 0$

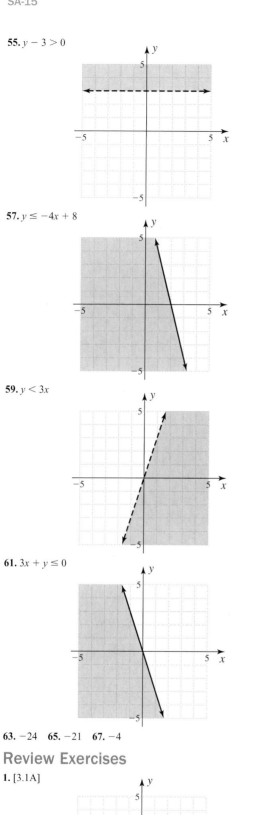

57. $y \le -4x + 8$

59. $y < 3x$

61. $3x + y \le 0$

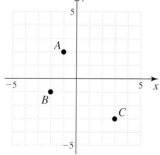

63. -24 **65.** -21 **67.** -4

Review Exercises

1. [3.1A]

2. [3.1B] A: (1, 1); B: $(-1, -2)$; C: $(3, -3)$
3. [3.1C] **(a)** For a wind speed of 10 mi/hr, the wind chill temperature is $-10°$F; **(b)** $-22°$F; **(c)** $-45°$F
4. [3.1D] **(a)** $13; **(b)** $15; **(c)** $25
5. [3.1E] **(a)** Quadrant II; **(b)** Quadrant III; **(c)** Quadrant IV
6. [3.1F] **(a)**

(b) For a wind speed of 10 mi/hr, the wind chill temperature is $-10°$F;
(c) $s = 20$ **7.** [3.2A] **(a)** No; **(b)** No; **(c)** Yes
8. [3.2B] **(a)** $x = 3$; **(b)** $x = 4$; **(c)** $x = 2$
9. [3.2C] **a.** $x + y = 4$
           b. $x + y = 2$
           c. $x + 2y = 2$

10. [3.2C]

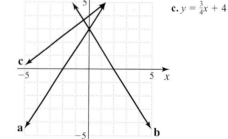

    a. $y = \frac{3}{2}x + 3$
    b. $y = -\frac{3}{2}x + 3$
    c. $y = \frac{3}{4}x + 4$

11. [3.2D]

    a. $g = 30 - 0.2t$
    b. $g = 20 - 0.2t$
    c. $g = 10 - 0.2t$

12. [3.3A]

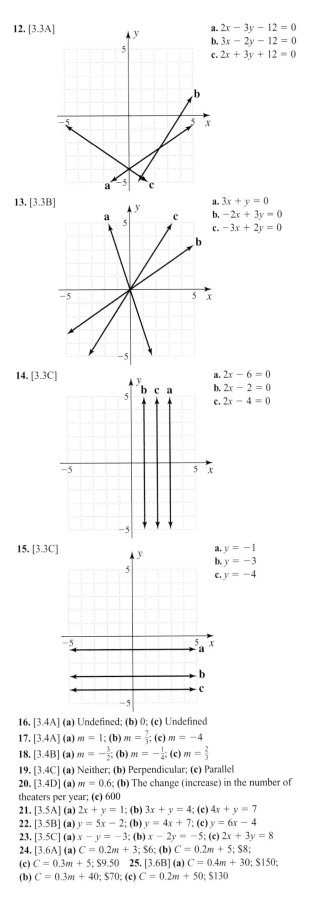

a. $2x - 3y - 12 = 0$
b. $3x - 2y - 12 = 0$
c. $2x + 3y + 12 = 0$

13. [3.3B]

a. $3x + y = 0$
b. $-2x + 3y = 0$
c. $-3x + 2y = 0$

14. [3.3C]

a. $2x - 6 = 0$
b. $2x - 2 = 0$
c. $2x - 4 = 0$

15. [3.3C]

a. $y = -1$
b. $y = -3$
c. $y = -4$

16. [3.4A] **(a)** Undefined; **(b)** 0; **(c)** Undefined
17. [3.4A] **(a)** $m = 1$; **(b)** $m = \frac{7}{3}$; **(c)** $m = -4$
18. [3.4B] **(a)** $m = -\frac{3}{2}$; **(b)** $m = -\frac{1}{4}$; **(c)** $m = \frac{2}{3}$
19. [3.4C] **(a)** Neither; **(b)** Perpendicular; **(c)** Parallel
20. [3.4D] **(a)** $m = 0.6$; **(b)** The change (increase) in the number of theaters per year; **(c)** 600
21. [3.5A] **(a)** $2x + y = 1$; **(b)** $3x + y = 4$; **(c)** $4x + y = 7$
22. [3.5B] **(a)** $y = 5x - 2$; **(b)** $y = 4x + 7$; **(c)** $y = 6x - 4$
23. [3.5C] **(a)** $x - y = -3$; **(b)** $x - 2y = -5$; **(c)** $2x + 3y = 8$
24. [3.6A] **(a)** $C = 0.2m + 3$; \$6; **(b)** $C = 0.2m + 5$; \$8;
(c) $C = 0.3m + 5$; \$9.50   **25.** [3.6B] **(a)** $C = 0.4m + 30$; \$150;
(b) $C = 0.3m + 40$; \$70; **(c)** $C = 0.2m + 50$; \$130

26. [3.6C] **(a)** $C = 0.40m + 1$; **(b)** $C = 0.40m + 2$; **(c)** $C = 0.40m + 3$
27. [3.7A] **(a)** $2x - 4y < -8$

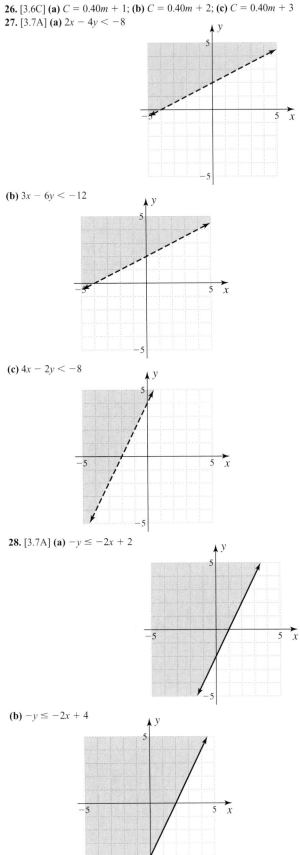

(b) $3x - 6y < -12$

(c) $4x - 2y < -8$

28. [3.7A] **(a)** $-y \leq -2x + 2$

(b) $-y \leq -2x + 4$

(c) $-y \le -x + 3$

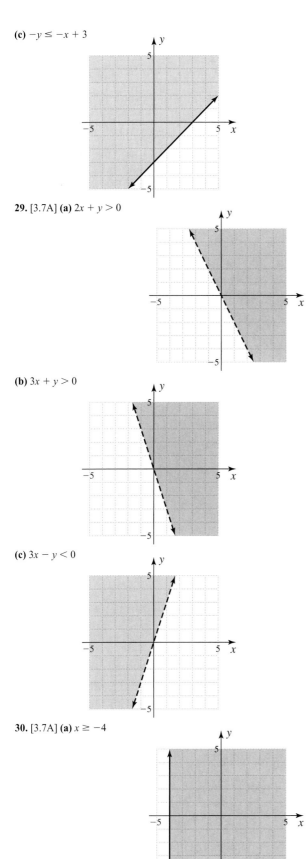

29. [3.7A] **(a)** $2x + y > 0$

(b) $3x + y > 0$

(c) $3x - y < 0$

30. [3.7A] **(a)** $x \ge -4$

(b) $y - 4 < 0$

(c) $2y - 4 \ge 0$

Cumulative Review Chapters 1–3

1. 1   **2.** $3\frac{1}{7}$ **3.** $-\frac{13}{24}$ **4.** 13.0   **5.** -19.76 **6.** 1296   **7.** 12   **8.** 12
9. Commutative property of addition   **10.** $18x - 21$ **11.** cd^2
12. $-x - 14$ **13.** $\frac{m+n}{p}$ **14.** No   **15.** $x = 6$ **16.** $x = 9$
17. $x = 54$ **18.** $x = 10$ **19.** $d = \frac{S}{7c^2}$ **20.** 65 and 105
21. \$435 from stocks and \$190 from bonds   **22.** 2 hr
23. \$12,000 in bonds, \$13,000 in certificates of deposit
24.

25.

26. $(2, -4)$ **27.** No   **28.** $x = -2$
29. $2x + y = 8$

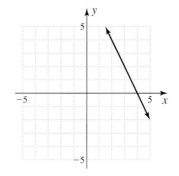

30. $4x - 8 = 0$

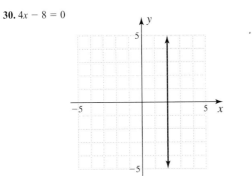

31. 9   **32.** 2   **33.** (1) and (3)   **34. (a)** Undefined; **(b)** 0
35. $-\frac{25}{72}$ **36.** 8.2   **37.** 16   **38.** 2   **39.** 5   **40.** $6x - 11$ **41.** $\frac{d + 4e}{f}$
42. $x = 3$ **43.** $x = 63$ **44.** 35 and 75   **45.** \$4000 in bonds; \$5000 in certificates of deposit
46.

47.

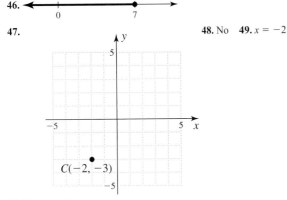

48. No   **49.** $x = -2$

50. $3x + y = 3$

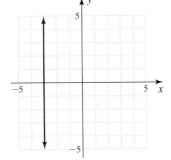

51. $3x + 9 = 0$

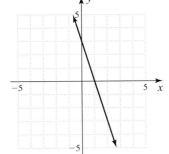

52. $-\frac{7}{3}$ **53.** 3   **54.** (1) and (3)   **55. (a)** 0; **(b)** Undefined
56. $y = -3x + 6$ **57.** $y = 3x - 1$

58. $6x - y < -6$

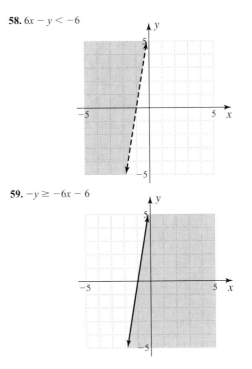

59. $-y \geq -6x - 6$

Chapter 4

Exercises 4.1

1. $24x^3$ **3.** $30a^4b^3$ **5.** $3x^3y^3$ **7.** $-\frac{b^5c}{5}$ **9.** $\frac{3x^3y^3z^6}{2}$ **11.** $-30x^5y^3z^8$
13. $15a^4b^3c^4$ **15.** $24a^3b^5c^5$ **17.** x^4 **19.** $-\frac{a^2}{2}$ **21.** $2x^3y^2$ **23.** $-\frac{x^3y^2}{2}$
25. $\frac{2a^3y^4}{3}$ **27.** $\frac{3ab^3c}{4}$ **29.** $\frac{3a^4}{2}$ **31.** $-x^6y$ **33.** $-x^2y^2$ **35.** $2^6 = 64$
37. $3^2 = 9$ **39.** x^9 **41.** y^6 **43.** $-a^6$ **45.** $8x^9y^6$ **47.** $4x^4y^6$
49. $-27x^9y^6$ **51.** $9x^{12}y^6$ **53.** $-8x^{12}y^{12}$ **55.** $\frac{16}{81}$ **57.** $\frac{27x^6}{8y^9}$ **59.** $\frac{16x^8}{81y^{12}}$
61. $12x^6y^3$ **63.** $-576a^2b^3$ **65.** $-144a^4b^6$ **67.** 96   **69.** x^5y^2
71. \$1259.71   **73.** \$1464.10   **75.** $V = \frac{2}{3}x^3$ **77.** 144 sandwiches
79. \$0.0075 or 0.75¢   **81.** $V = x^3$ **83.** 16 in.³   **85.** ≈ \$23.63
87. $V = \frac{9}{16}x^3$ in.³   **89.** 37.5 servings   **91.** $V = \pi r^2 h$ **93.** $V = \frac{1}{3}\pi r^2 h$
95. (a) 123,454,321; **(b)** 12,345,654,321   **97. (a)** $5^2 = 25$; **(b)** $7^2 = 49$
99. The bases are different.   **103.** x^{m+n} **105.** negative   **107.** negative
109. x^{mn} **111.** $\frac{x^m}{y^n}$ **113.** $30a^{11}$ **115.** $-12x^4yz^5$ **117.** $-5x^2y^6$
119. $\frac{ab^3c}{4}$ **121.** y^{20} **123.** $\frac{27y^6}{x^{15}}$ **125.** 4   **127.** 64   **129.** $\frac{1}{8}$ **131.** $\frac{81}{16}$

Exercises 4.2

1. $\frac{1}{4^2} = \frac{1}{16}$ **3.** $\frac{1}{5^3} = \frac{1}{125}$ **5.** $8^2 = 64$ **7.** x^7 **9.** $\frac{3^3}{4^2} = \frac{27}{16}$
11. $\frac{3^4}{5^2} = \frac{81}{25}$ **13.** $\frac{b^6}{a^5}$ **15.** $\frac{y^9}{x^9}$ **17.** 2^{-3} **19.** y^{-5} **21.** q^{-5} **23.** 3
25. 4   **27.** $\frac{1}{16}$ **29.** $\frac{1}{216}$ **31.** $\frac{1}{64}$ **33.** x^2 **35.** y^2 **37.** $\frac{1}{a^5}$ **39.** $\frac{1}{x^2}$
41. $\frac{1}{x^2}$ **43.** $\frac{1}{a^5}$ **45.** 1   **47.** $3^5 = 243$ **49.** $\frac{1}{4^3} = \frac{1}{64}$ **51.** $\frac{1}{y^2}$ **53.** x^3
55. $\frac{1}{x^2}$ **57.** $\frac{1}{x^7}$ **59.** x^3 **61.** $\frac{a^2}{b^6}$ **63.** $-\frac{8b^6}{27a^3}$ **65.** a^8b^4 **67.** $\frac{1}{x^{15}y^6}$
69. $x^{27}y^6$ **71.** $\frac{1}{2^4}, \frac{1}{16}$ **73. (a)** $\frac{1}{1000}$; **(b)** $\frac{1}{10^5}$; **(c)** 10^{-3}; **(d)** 10^{-3}
75. 6.242 billion   **77.** 4.922 billion   **79. (a)** \$412; **(b)** \$463.71
81. \$61.68   **83.** \$180,405.36   **85.** \$106.14   **87. (a)** 50% per day;
(b) $A = 40(1 + 0.50)^n$; **(c)** 203; **(d)** 11,677; **(e)** Answers may vary.
91. No   **93.** 1   **95.** x^n **97.** x^{m+n} **99.** $\frac{6^2}{7^2} = \frac{36}{49}$ **101.** $9^2 = 81$ **103.** 7^{-6}
105. $5^2 = 25$ **107.** z^2 **109.** $\frac{81x^8}{16y^{24}}$ **111. (a)** \$3645.00; **(b)** \$6172.84
113. 839   **115.** 0.0816

Exercises 4.3

1. 5.5×10^7 **3.** 3×10^8 **5.** 1.9×10^9 **7.** 2.4×10^{-4}
9. 2×10^{-9} **11.** 153 **13.** $171{,}000{,}000$ **15.** $6{,}850{,}000{,}000$ **17.** 0.23
19. 0.00025 **21.** 1.5×10^{10} **23.** 3.06×10^4 **25.** 1.24×10^{-4}
27. 2×10^3 **29.** 2.5×10^{-3} **31. (a)** 5.1×10^{10}; **(b)** $51{,}000{,}000{,}000$
33. 4.374074074 **35.** 2×10^7 **37.** Now: $2{,}000{,}000$; later: $12{,}000{,}000$
39. 365 **41. (a)** Mercury; **(b)** Mars; **(c)** Mercury, Venus, Earth, Mars
43. (a) Jupiter; **(b)** Uranus; **(c)** Uranus, Neptune, Saturn, Jupiter
45. 2.99792458×10^8 **47.** 3.09×10^{13} **49.** 3.27 **53.** 10, integer
55. powers   **57.** Distance: 2.89×10^5; mass: 1.2456×10^{-1}
59. $6 \times 10^0 = 6$ **61.** $1.32 \times 10^8 = 132{,}000{,}000$ **63.** 102 **65.** 8
67. 4

Exercises 4.4

1. Binomial; 1   **3.** Monomial; 1   **5.** Trinomial; 2   **7.** Monomial; 0
9. Binomial; 3   **11.** $8x^3 - 3x$; 3   **13.** $8x^2 + 4x - 7$; 2   **15.** $x^2 + 5x$; 2
17. $x^3 - x^2 + 3$; 3   **19.** $4x^5 - 3x^3 + 2x^2$; 5   **21. (a)** 4; **(b)** -8
23. (a) 7; **(b)** 7   **25. (a)** 9; **(b)** 13   **27. (a)** 9; **(b)** -3
29. (a) 3; **(b)** -1 **31. (a)** $(-16t^2 + 150)$ ft; **(b)** 134 ft; **(c)** 86 ft
33. (a) $(-4.9t^2 + 200)$ m; **(b)** 195.1 m; **(c)** 180.4 m
35. (a) 251; **(b)** About 610; about 1318
37. (a) 30%; **(b)** 25%; **(c)** Decreasing; **(d)** $2.71 billion;
(e) $3.25 billion   **39. (a)** 28°; **(b)** $-122°$; 122° below 0 is
unreasonable   **41. (a)** 34 million; **(b)** $N(5) = 34.6$; the result is close
43. (a) $21,236; **(b)** $C(5) = $21,237$; close! **(c)** $26,402
45. (a) $2191; **(b)** $C(5) = 2192; close! **(c)** $2742   **47. (a)** $12,127;
(b) $C(5) = $12,129$; close! **(c)** $15,819   **49.** $(-32t - 10)$ ft/sec
51. (a) -11.8 m/sec; **(b)** -21.6 m/sec   **55.** No; negative exponent on x.
57. 0; $7^4 = 2401x^0$ **59.** polynomial   **61.** binomial   **63.** 12   **65.** 18
67. 8   **69.** No degree   **71.** $5x^2 - 3x - 8$ **73.** Monomial
75. Polynomial   **77. (a)** 0.115; **(b)** 0.105; **(c)** 0.1149; very close;
(d) 0.1057; very close   **79.** $-7ab$ **81.** $3x^2y$ **83.** $8xy^2$

Exercises 4.5

1. $12x^2 + 5x + 6$ **3.** $-2x^2 - x - 8$ **5.** $-3x^2 + 7x - 5$
7. $-x^2 - 5$ **9.** $x^3 - 2x^2 - x - 2$ **11.** $2x^4 - 8x^3 + 2x^2 + x + 5$
13. $-\frac{3}{5}x^2 + x + 1$ **15.** $0.4x^2 + 0.3x - 0.4$ **17.** $-2x^2 - 2x + 3$
19. $3x^4 + x^3 - 5x - 1$ **21.** $-5x^4 + 5x^3 - 2x^2 - 2$
23. $5x^3 - 4x^2 + 5x - 1$ **25.** $-\frac{1}{7}x^3 - \frac{2}{3}x^2 + 5x$
27. $-\frac{3}{7}x^3 + \frac{1}{9}x^2 + x - 6$ **29.** $-4x^4 - 4x^3 - 3x^2 - x + 2$
31. $4x^2 + 7$ **33.** $-x^2 - 4x - 6$ **35.** $4x^2 - 7x + 6$
37. $-3x^3 + 6x^2 - 2x$ **39.** $7x^3 - x^2 + 2x - 6$ **41.** $3x^2 - 7x + 7$
43. 0 **45.** $4x^3 - 3x^2 - 7x + 6$ **47.** $3x^3 - 2x^2 + x - 8$
49. $-5x^3 - 5x^2 + 4x - 9$ **51.** $2x^2 + 6x$ **53.** $3x^2 + 4x$ **55.** $27x^2 + 10x$
57. (a) $P(t) = 0.05t^3 - 0.85t^2 + 0.6t + 45$; **(b)** $P(10) = 16$ lb;
$P(20) = 117$ lb   **59. (a)** $T(t) + B(t) + R(t) = 77t^2 + 337.5t + 8766$;
(b) $8766; **(c)** $12,378.50
61. $T(t) + B(t) + R(t) - S(t) = -485.5t^2 + 25t + 6141$
63. (a) $-14t^2 + 73t + 1078$; **(b)** $1078; **(c)** $408
65. (a) $0.3t^3 - 3t^2 + 6t + 40$; **(b)** $40,000; $100,000
67. $R = x - 50$ **69.** $R = 2x$ **71.** 30   **77.** $a - b$
79. $5x^2 + 3x + 1$ **81.** $4x^3 + 10x^2 + 1$ **83.** $4x^2 + 8x$ **85.** $-6x^5$
87. $-6x^9$ **89.** $6y - 24$

Exercises 4.6

1. $45x^5$ **3.** $-10x^3$ **5.** $6y^3$ **7.** $3x + 3y$ **9.** $10x - 5y$
11. $-8x^2 + 12x$ **13.** $x^5 + 4x^4$ **15.** $4x^2 - 4x^3$ **17.** $3x^2 + 3xy$
19. $-8xy^2 + 12y^3$ **21.** $x^2 + 3x + 2$ **23.** $y^2 - 5y - 36$
25. $x^2 - 5x - 14$ **27.** $x^2 - 12x + 27$ **29.** $y^2 - 6y + 9$
31. $6x^2 + 7x + 2$ **33.** $6y^2 + y - 15$ **35.** $10z^2 + 43z - 9$
37. $6x^2 - 34x + 44$ **39.** $16z^2 + 8z + 1$ **41.** $6x^2 + 11xy + 3y^2$
43. $2x^2 + xy - 3y^2$ **45.** $10z^2 + 13yz - 3y^2$ **47.** $12x^2 - 11xz + 2z^2$
49. $4x^2 - 12xy + 9y^2$ **51.** $6 + 17x + 12x^2$ **53.** $6 - 7x - 3x^2$
55. $8 - 16x - 10x^2$ **57.** $(x^2 + 7x + 10)$ square units
59. $(96t - 16t^2)$ ft   **61.** $V_2CP + V_2PR - V_1CP - V_1PR$ **63.** 1200 ft²

65. $800 + 160 + 40 + 200 = 1200$ ft²   **67. (a)** $S_1 = 20x$ ft²;
(b) $S_2 = xy$ ft²; **(c)** $S_3 = 40y$ ft²   **69.** $800 + 20x + xy + 40y$; yes
71. $Y = 1000x - x^2$ **73. (a)** $R = 1000p - 30p^2$; **(b)** $8000
75. $P = -3p^2 + 65p - 100$ **77.** $18   **79.** Yes   **81.** No; consider the
results in Problem 82.   **83.** first   **85.** inner   **87.** $-35x^6$
89. $x^2 + 4x - 21$ **91.** $9x^2 + 9x - 4$ **93.** $10x^2 - 19xy + 6y^2$
95. $6x - 18y$ **97.** $64x^2$ **99.** $9x^2$ **101.** $9A^2$

Exercises 4.7

1. $x^2 + 2x + 1$ **3.** $4x^2 + 4x + 1$ **5.** $9x^2 + 12xy + 4y^2$
7. $x^2 - 2x + 1$ **9.** $4x^2 - 4x + 1$ **11.** $9x^2 - 6xy + y^2$
13. $36x^2 - 60xy + 25y^2$ **15.** $4x^2 - 28xy + 49y^2$ **17.** $x^2 - 4$
19. $x^2 - 16$ **21.** $9x^2 - 4y^2$ **23.** $x^2 - 36$ **25.** $x^2 - 144$ **27.** $9x^2 - y^2$
29. $4x^2 - 49y^2$ **31.** $x^4 + 7x^2 + 10$ **33.** $x^4 + 2x^2y + y^2$
35. $9x^4 - 12x^2y^2 + 4y^4$ **37.** $x^4 - 4y^4$ **39.** $4x^2 - 16y^4$
41. $x^3 + 4x^2 + 8x + 15$ **43.** $x^3 + 3x^2 - x + 12$
45. $x^3 + 2x^2 - 5x - 6$ **47.** $x^3 - 8$ **49.** $-x^3 + 2x^2 - 3x + 2$
51. $-x^3 + 8x^2 - 15x - 4$ **53.** $2x^3 + 6x^2 + 4x$ **55.** $3x^3 + 3x^2 - 6x$
57. $4x^3 - 12x^2 + 8x$ **59.** $5x^3 - 20x^2 - 25x$
61. $x^3 + 15x^2 + 75x + 125$ **63.** $8x^3 + 36x^2 + 54x + 27$
65. $8x^3 + 36x^2y + 54xy^2 + 27y^3$ **67.** $16t^4 + 24t^2 + 9$
69. $16t^4 + 24t^2u + 9u^2$ **71.** $9t^4 - 2t^2 + \frac{1}{9}$ **73.** $9t^4 - 2t^2u + \frac{1}{9}u^2$
75. $9x^4 - 25$ **77.** $9x^4 - 25y^4$ **79.** $16x^6 - 25y^6$
81. $\frac{(x^3 + 8x^2 + 20x + 16)\pi}{3} = \frac{1}{3}\pi x^3 + \frac{8}{3}\pi x^2 + \frac{20}{3}\pi x + \frac{16}{3}\pi$
83. $\frac{4(x^3 + 3x^2 + 3x + 1)\pi}{3} = \frac{4}{3}\pi x^3 + 4\pi x^2 + 4\pi x + \frac{4}{3}\pi$
85. $(x^3 + 4x^2 + 5x + 2)\pi = \pi x^3 + 4\pi x^2 + 5\pi x + 2\pi$
87. $I_D = 0.004 + 0.004V + 0.001V^2$ **89.** $T_1^4 - T_2^4$
91. $Kt_n^2 - 2Kt_nt_a + Kt_a^2$ **93.** $v_i^2 - v_0^2$ **95. (a)** 9; **(b)** 5; **(c)** No
97. (a) x^2; **(b)** xy; **(c)** y^2; **(d)** xy **99.** They are equal.   **103.** When the
binomials are the sum and difference of the same two terms.
105. $X^2 + XB + AX + AB$ **107.** $X^2 - 2AX + A^2$
109. $x^2 + 14x + 49$ **111.** $4x^2 + 20xy + 25y^2$
113. $x^4 + 4x^2y + 4y^2$ **115.** $9x^2 - 4y^2$
117. $15x^2 + 2x - 24$ **119.** $21x^2 - 34xy + 8y^2$
121. $x^3 + 3x^2 + 3x + 2$ **123.** $x^4 + x^3 - x^2 - 2x - 2$
125. $x^3 + 6x^2y + 12xy^2 + 8y^3$ **127.** $3x^5 - 3x$ **129.** $2x$ **131.** $-\frac{1}{2}$

Exercises 4.8

1. $x + 3y$ **3.** $2x - y$ **5.** $-2y + 8 - \frac{4}{y}$ **7.** $10x + 8$ **9.** $3x - 2$
11. $x + 2$ **13.** $(y - 2)\,R -1$ **15.** $x + 6$ **17.** $3x - 4$
19. $(2y - 5)\,R -1$ **21.** $x^2 - x - 1$ **23.** $y^2 - y - 1$
25. $(4x^2 + 3x + 7)\,R\ 12$ **27.** $x^2 + 2x + 4$ **29.** $4y^2 + 8y + 16$
31. $(x^2 + x + 1)\,R\ 3$ **33.** $x^3 + x^2 - x + 1$ **35.** $(m^2 - 8)\,R\ 10$
37. $x^2 + xy + y^2$ **39.** $(x^2 + 2x + 4)\,R\ 16$ **41. (a)** $2 + \frac{20}{x}$; **(b)** $22;
(c) $4   **43. (a)** $x + 2$ lb; **(b)** 7 lb   **45.** $R_b = \frac{R_1R_2}{R_3} + R_2 + R_1$
47. $\overline{C(x)} = 3x + 5$ **49.** $\overline{P(x)} = 50 + x - \frac{7000}{x}$ **55.** divisor, quotient,
remainder   **57.** $2x^2 + 2x + 3$ **59.** $4x^2 - 3x$ **61.** $2y - 4 - \frac{1}{y^2}$
63. 180   **65.** 120   **67.** 180

Review Exercises

1. [4.1A] **(a) (i)** $-15a^3b^4$; **(ii)** $-24a^3b^5$; **(iii)** $-35a^3b^4$; **(b) (i)** $6x^3y^3z^5$;
(ii) $12x^3y^4z^3$; **(iii)** $20x^2y^3z^4$ **2.** [4.1B] **(a) (i)** $-2x^5y^4$; **(ii)** $-6x^6y^3$;
(iii) $-2x^7y^5$; **(b) (i)** $\frac{x^3y^6}{8}$; **(ii)** $\frac{xy^7}{2}$; **(iii)** $\frac{xy^6}{3}$ **3.** [4.1C] **(a)** $2^6 = 64$;
(b) $2^4 = 16$; **(c)** $3^4 = 81$ **4.** [4.1C] **(a)** y^6; **(b)** x^6; **(c)** a^{20}
5. [4.1C] **(a)** $16x^2y^6$; **(b)** $8x^6y^3$; **(c)** $27x^6y^6$ **6.** [4.1C] **(a)** $-8x^3y^9$;
(b) $9x^4y^6$; **(c)** $-8x^6y^6$ **7.** [4.1C] **(a)** $\frac{4y^4}{x^3}$; **(b)** $\frac{27x^3}{y^9}$; **(c)** $\frac{16x^8}{y^{16}}$
8. [4.1C] **(a)** $32x^{12}y^4$; **(b)** $-72x^4y^9$; **(c)** $256x^6y^{16}$
9. [4.2A] **(a)** $\frac{1}{2^3} = \frac{1}{8}$; **(b)** $\frac{1}{3^4} = \frac{1}{81}$; **(c)** $\frac{1}{5^2} = \frac{1}{25}$
10. [4.2A] **(a)** x^4; **(b)** y^3; **(c)** z^5 **11.** [4.2B] **(a)** x^{-5}; **(b)** y^{-7}; **(c)** z^{-8}
12. [4.2C] **(a) (i)** $2^3 = 8$; **(ii)** $2^2 = 4$; **(iii)** $3^3 = 27$; **(b) (i)** $\frac{1}{y^3}$; **(ii)** $\frac{1}{y^5}$; **(iii)** $\frac{1}{y^6}$
13. [4.2C] **(a) (i)** $\frac{1}{x^4}$; **(ii)** $\frac{1}{x^5}$; **(iii)** $\frac{1}{x^6}$; **(b) (i–iii)** 1; **(c) (i)** x; **(ii)** x^3;
(iii) x^2 **14.** [4.2C] **(a)** $\frac{9x^2}{4y^{14}}$; **(b)** $\frac{8x^6y^{21}}{27}$; **(c)** $\frac{4}{9x^3y^8}$
15. [4.3A] **(a) (i)** 4.4×10^7; **(ii)** 4.5×10^6; **(iii)** 4.6×10^5;
(b) (i) 1.4×10^{-3}; **(ii)** 1.5×10^{-4}; **(iii)** 1.6×10^{-5}

16. [4.3B] **(a)** **(i)** 2.2×10^5; **(ii)** 9.3×10^6; **(iii)** 1.24×10^8; **(b)** **(i)** 5; **(ii)** 6; **(iii)** 7   **17.** [4.4A] **(a)** Trinomial; **(b)** Monomial; **(c)** Binomial   **18.** [4.4B] **(a)** 4; **(b)** 2; **(c)** 2
19. [4.4C] **(a)** $9x^4 + 4x^2 - 8x$; **(b)** $4x^2 - 3x - 3$; **(c)** $-4x^2 + 3x + 8$
20. [4.4D] **(a)** 284; **(b)** 156; **(c)** -100 **21.** [4.5A] **(a)** $5x^2 - x - 10$;
(b) $-5x^2 + 15x + 2$; **(c)** $9x^2 - 7x + 1$ **22.** [4.5B] **(a)** $-x^2 - 7x + 4$;
(b) $7x^2 - 7x + 3$; **(c)** $-5x^2 + 4x - 11$ **23.** [4.6A] **(a)** $-18x^7$;
(b) $-40x^9$; **(c)** $-27x^{11}$ **24.** [4.6B] **(a)** $-2x^3 - 4x^2y$;
(b) $-6x^4 - 9x^3y$; **(c)** $-20x^4 - 28x^3y$ **25.** [4.6C] **(a)** $x^2 + 15x + 54$;
(b) $x^2 + 5x + 6$; **(c)** $x^2 + 16x + 63$ **26.** [4.6C] **(a)** $x^2 + 4x - 21$;
(b) $x^2 + 4x - 12$; **(c)** $x^2 + 4x - 5$ **27.** [4.6C] **(a)** $x^2 - 4x - 21$;
(b) $x^2 - 4x - 12$; **(c)** $x^2 - 4x - 5$ **28.** [4.6C] **(a)** $6x^2 - 13xy + 6y^2$;
(b) $20x^2 - 27xy + 9y^2$; **(c)** $8x^2 - 26xy + 15y^2$
29. [4.7A] **(a)** $4x^2 + 12xy + 9y^2$; **(b)** $9x^2 + 24xy + 16y^2$;
(c) $16x^2 + 40xy + 25y^2$ **30.** [4.7B] **(a)** $4x^2 - 12xy + 9y^2$;
(b) $9x^2 - 12xy + 4y^2$; **(c)** $25x^2 - 20xy + 4y^2$
31. [4.7C] **(a)** $9x^2 - 25y^2$; **(b)** $9x^2 - 4y^2$; **(c)** $9x^2 - 16y^2$
32. [4.7D] **(a)** $x^3 + 4x^2 + 5x + 2$; **(b)** $x^3 + 5x^2 + 8x + 4$;
(c) $x^3 + 6x^2 + 11x + 6$ **33.** [4.7E] **(a)** $3x^3 + 9x^2 + 6x$;
(b) $4x^3 + 12x^2 + 8x$; **(c)** $5x^3 + 15x^2 + 10x$
34. [4.7E] **(a)** $x^3 + 6x^2 + 12x + 8$; **(b)** $x^3 + 9x^2 + 27x + 27$;
(c) $x^3 + 12x^2 + 48x + 64$ **35.** [4.7E] **(a)** $25x^4 - 5x^2 + \frac{1}{4}$;
(b) $49x^4 - 7x^2 + \frac{1}{4}$; **(c)** $81x^4 - 9x^2 + \frac{1}{4}$ **36.** [4.7E] **(a)** $9x^4 - 4$;
(b) $9x^4 - 16$; **(c)** $4x^4 - 25$ **37.** [4.8A] **(a)** $2x^2 - x$; **(b)** $4x^2 - 2x$;
(c) $4x^2 - 2x$ **38.** [4.8B] **(a)** $x + 6$; **(b)** $x + 7$; **(c)** $x + 8$
39. [4.8B] **(a)** $4x^2 - 4x - 4$; **(b)** $6x^2 - 6x - 6$; **(c)** $2x^2 - 2x - 2$
40. [4.8B] **(a)** $(2x^2 + 6x - 2)$ R 2; **(b)** $(2x^2 + 6x - 3)$ R 3;
(c) $(3x^2 + 3x - 1)$ R 4   **41.** [4.8B] **(a)** $(x^2 + x)$ R $(4x + 1)$;
(b) $(x^2 + x)$ R $(5x + 1)$; **(c)** $(x^2 + x)$ R $(6x + 1)$

Cumulative Review Chapters 1–4

1. 7   **2.** $9\frac{9}{10}$ **3.** $-\frac{5}{18}$ **4.** 13.6   **5.** -6.24 **6.** -16 **7.** $\frac{6}{7}$ **8.** 47
9. Associative property of multiplication   **10.** $6x - 24$ **11.** $-7xy^3$
12. $4x + 10$ **13.** $\frac{m + 3n}{p}$ **14.** No   **15.** $x = 15$ **16.** $x = 49$
17. $x = 36$ **18.** $x = 6$ **19.** $b = \frac{S}{6a^2}$ **20.** 40 and 60
21. \$430 from stocks and \$185 from bonds   **22.** 18 hr
23. \$8000 in bonds; \$9000 in certificates of deposit

24.

25.

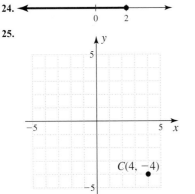

$C(4, -4)$

26. $(3, -2)$ **27.** Yes   **28.** $x = 1$
29. $5x + y = 5$

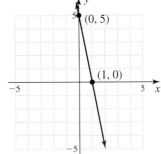

$(0, 5)$

$(1, 0)$

30. $3y - 15 = 0$

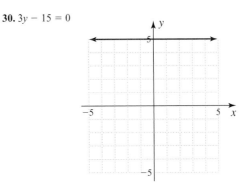

31. $\frac{7}{5}$ **32.** 2   **33.** (2) and (3)   **34.** $18x^7y^3$ **35.** $-\frac{5x}{y^2}$ **36.** x^3
37. x^5 **38.** $\frac{64x^{12}}{y^{12}}$ **39.** 8.0×10^6 **40.** 4.2×10^{-8} **41.** Monomial
42. 2   **43.** $-3x^3 - 3x^2 - x + 4$ **44.** 32   **45.** $4x^3 + 9x - 3$
46. $-9x^3 - 12xy$ **47.** $9x^2 + 12xy + 4y^2$ **48.** $16x^2 - 9y^2$
49. $25x^4 - 5x^2 + \frac{1}{4}$ **50.** $16x^4 - 81$ **51.** $(3x^2 - 5x + 4)$ R 3
52. $y - 6 = 4(x + 2)$ or $y = 4x + 14$
53. $y = -2x + 5$
54. $2x - 3y < -6$

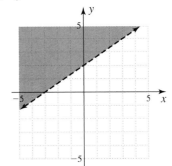

55. $3x - y \geq 0$

Chapter 5

Exercises 5.1

1. 4   **3.** 8   **5.** 1   **7.** a^3 **9.** x^3 **11.** $5y^6$ **13.** $2x^3$ **15.** $3bc$
17. 3   **19.** $9ab^3z$ **21.** $3(x + 5)$ **23.** $9(y - 2)$ **25.** $-5(y - 4)$
27. $-3(x + 9)$ **29.** $4x(x + 8)$ **31.** $6x(1 - 7x)$
33. $-5x^2(1 + 5x^2)$ **35.** $3x(x^2 + 2x + 3)$ **37.** $9y(y^2 - 2y + 3)$
39. $6x^2(x^4 + 2x^3 - 3x^2 + 5)$ **41.** $8y^3(y^5 + 2y^2 - 3y + 1)$
43. $\frac{1}{7}(4x^3 + 3x^2 - 9x + 3)$ **45.** $\frac{1}{8}y^2(7y^7 + 3y^4 - 5y^2 + 5)$
47. $(x + 4)(3 - y)$ **49.** $(y - 2)(x - 1)$ **51.** $(t + s)(c - 1)$
53. $4x^3(1 + x - 3x^2)$ **55.** $6y^7(1 - 2y^2 - y^4)$ **57.** $(x + 2)(x^2 + 1)$
59. $(y - 3)(y^2 + 1)$ **61.** $(2x + 3)(2x^2 + 1)$ **63.** $(3x - 1)(2x^2 + 1)$
65. $(y + 2)(4y^2 + 1)$ **67.** $(2a + 3)(a^2 + 1)$ **69.** $(x^2 + 4)(3x^2 + 1)$
71. $(2y^2 + 3)(3y^2 + 1)$ **73.** $(y^2 + 3)(4y^2 + 1)$ **75.** $(a^2 - 2)(3a^2 - 2)$
77. $(6a - 5b)(1 + 2d)$ **79.** $(x^2 - y)(1 - 3z)$ **81.** $\alpha L(t_2 - t_1)$

83. $(R - 1)(R - 1)$ **85.** $2\pi r(h + r)$ **87.** $P(Q_2 - Q_1)$ **89.** $V(k - PV)$
91. $f(u + v_s)$ **93.** $-w(\ell - z)$ **95.** $a(a + 2s)$ **97.** $-16(t^2 - 5t - 15)$
101. reverse   **103.** smallest   **105.** $(3x^2 + 1)(2x^2 - 3)$
107. $(2x - 3)(3x^2 - 1)$ **109.** $3x^2(x^4 - 2x^3 + 4x^2 + 9)$
111. $(2x + y)(6x - 5y)$ **113.** $-3(y - 7)$ **115.** $(x + b)(5x + 6y)$
117. 2   **119.** $x^2 + 8x + 15$ **121.** $x^2 - 5x + 4$ **123.** $2x^2 - x - 3$

Exercises 5.2

1. $(y + 2)(y + 4)$ **3.** $(x + 2)(x + 5)$ **5.** $(y + 5)(y - 2)$
7. $(x + 7)(x - 2)$ **9.** $(x + 1)(x - 7)$ **11.** $(y + 2)(y - 7)$
13. $(y - 2)(y - 1)$ **15.** $(x - 4)(x - 1)$ **17.** Not factorable (prime)
19. Not factorable (prime)   **21.** $(x + a)(x + 2a)$
23. $(z + 3b)(z + 3b) = (z + 3b)^2$ **25.** $(r + 4a)(r - 3a)$
27. $(x + 10a)(x - a)$ **29.** $-(b - y)(b + 3y)$ **31.** $(m - 2a)(m + a)$
33. $2t(t + 1)(t + 4)$ **35.** $ax^2(ax + 3a + 2)$ **37.** $b^3x^5(x - 3)(x + 4)$
39. $2c^5z^4(z - 3)(z + 5)$ **41. (a)** $-5t^2 + 5t + 10$; **(b)** $-5(t - 2)(t + 1)$;
(c) $t = 2$ and $t = -1$; **(d)** 2 seconds   **43. (a)** $-16(t - 5)(t + 3)$;
(b) $t = 5$ and $t = -3$; **(c)** 5 seconds   **45. (a)** $-16(t + 1)(t - 2)$;
(b) 32 ft; **(c)** 0 ft   **47. (a)** $(x + 4)(x - 5)$; **(b)** 20   **49.** $(D + 6)(D + 2)$
51. $-(x - 2L)(x - 2L) = -(x - 2L)^2$ **53.** $-(x - 20)(x - 5)$
63. positive   **65.** positive, negative   **67.** $(y - 2)(3y^2 - 1)$
69. $(y - 3)(2y^2 - 1)$ **71.** $(x + 4y)(x + 3y)$ **73.** $(x - 2y)(x - 5y)$
75. not factorable (prime)   **77.** $2y^3(1 + 3y^2 + 5y^4)$
79. (a) -12; **(b)** -4 **81. (a)** 4; **(b)** -5

Exercises 5.3

1. $(2x + 3)(x + 1)$ **3.** $(2x + 3)(3x + 1)$ **5.** $(2x + 1)(3x + 4)$
7. $(x + 2)(2x - 1)$ **9.** $(x + 6)(3x - 2)$ **11.** $(4y - 3)(y - 2)$
13. $2(2y^2 - 4y + 3)$ **15.** $2(3y + 1)(y - 2)$ **17.** $(3y + 2)(4y - 3)$
19. $3(3y + 1)(2y - 3)$ **21.** $(3x + 1)(x + 2)$ **23.** $(5x + 1)(x + 2)$
25. Not factorable (prime)   **27.** $(3x + 1)(x - 2)$ **29.** $(5x + 2)(3x - 1)$
31. $4(2x + y)(x + 2y)$ **33.** $(2x + 3y)(3x - y)$ **35.** $(7x - 3y)(x - y)$
37. $(3x + y)(5x - 2y)$ **39.** Not factorable (prime)   **41.** $(3r + 5)(4r - 1)$
43. $(11t + 2)(2t - 3)$ **45.** $3(3x - 2)(2x - 1)$ **47.** $a(3b + 1)(2b + 1)$
49. $x^3y(6x + y)(x + 4y)$ **51.** $-(3x + 2)(2x + 1)$
53. $-(3x + 2)(3x - 1)$ **55.** $-(4m + n)(2m - 3n)$
57. $-(8x - y)(x - y)$ **59.** $-x(x + 3)(x + 2)$ **61.** $(2g + 9)(g - 4)$
63. $(2R - 1)(R - 1)$ **65. (a)** $(0.02t + 1)(0.2t + 2.72)$; **(b)** \$2.72;
(c) \$3.72   **67.** $5(m + 49)(m - 5)$ **69.** $-5(t - 1)(t + 2)$
71. $(3x - 5)(x - 4)$ **73.** $(2L - 3)(L - 3)$ **75.** $(2L - x)(L - x)$
77. (a) $(2x + 3)(3x + 1)$; **(b)** $(3x + 1)(2x + 3)$; **(c)** Both
79. ac, b **81.** $(3x + 2)(x - 2)$ **83.** $(2x + 3y)(x - 2y)$
85. $2(4x - 1)(2x + 1)$ **87.** $x^2(3x^2 + 5x - 3)$ **89.** $x^2 + 16x + 64$
91. $9x^2 - 12x + 4$ **93.** $4x^2 + 12xy + 9y^2$ **95.** $9x^2 - 25y^2$
97. $x^4 - 16$

Exercises 5.4

1. Yes   **3.** No   **5.** Yes   **7.** No   **9.** Yes   **11.** $(x + 1)^2$ **13.** $3(x + 5)^2$
15. $(3x + 1)^2$ **17.** $(3x + 2)^2$ **19.** $(4x + 5y)^2$ **21.** $(5x + 2y)^2$
23. $(y - 1)^2$ **25.** $3(y - 4)^2$ **27.** $(3x - 1)^2$ **29.** $(4x - 7)^2$
31. $(3x - 2y)^2$ **33.** $(5x - y)^2$ **35.** $(x + 7)(x - 7)$
37. $(3x + 7)(3x - 7)$ **39.** $(5x + 9y)(5x - 9y)$
41. $(x^2 + 1)(x + 1)(x - 1)$ **43.** $(4x^2 + 1)(2x + 1)(2x - 1)$
45. $\left(\frac{1}{3}x + \frac{1}{4}\right)\left(\frac{1}{3}x - \frac{1}{4}\right)$ **47.** $\left(\frac{1}{2}z + 1\right)\left(\frac{1}{2}z - 1\right)$ **49.** $\left(1 + \frac{1}{2}s\right)\left(1 - \frac{1}{2}s\right)$
51. $\left(\frac{1}{2} + \frac{1}{3}y\right)\left(\frac{1}{2} - \frac{1}{3}y\right)$ **53.** Not factorable (prime)
55. $3x(x + 2)(x - 2)$ **57.** $5t(t + 2)(t - 2)$ **59.** $5t(1 + 2t)(1 - 2t)$
61. $(7x + 2)^2$ **63.** $(x + 10)(x - 10)$ **65.** $(x + 10)^2$
67. $(3 + 4m)(3 - 4m)$ **69.** $(3x - 5y)^2$ **71.** $(z^2 + 4)(z + 2)(z - 2)$
73. $3x(x + 5)(x - 5)$ **75.** $(9 + x)(9 - x)$ **77.** $(C + kp)(C - kp)$
79. $\frac{3S}{2bd^3}(d + 2z)(d - 2z)$ **81.** $D(x) = (x - 7)^2$ **83.** $C(x) = (x + 6)^2$
87. $(x + 2)$ **89.** $(X + A)^2$ **91.** $(X + A)(X - A)$ **93.** $(x + 1)(x - 1)$
95. $(3x + 5y)(3x - 5y)$ **97.** $(3x - 4y)^2$ **99.** $(4x + 3y)^2$
101. $(3x + 5)^2$ **103.** Not factorable (prime)   **105.** $\left(\frac{1}{9} + \frac{1}{2}x\right)\left(\frac{1}{9} - \frac{1}{2}x\right)$
107. $2x(3x + 5y)(3x - 5y)$ **109.** Yes; $(x + 3)^2$ **111.** No
113. Yes; $(2x - 5y)^2$ **115.** $R^2 - r^2$ **117.** $6(x + 1)(x - 4)$
119. $2(x + 3)(x - 3)$

Exercises 5.5

1. $(x + 2)(x^2 - 2x + 4)$ **3.** $(2m - 3)(4m^2 + 6m + 9)$
5. $(3m - 2n)(9m^2 + 6mn + 4n^2)$ **7.** $s^3(4 - s)(16 + 4s + s^2)$
9. $x^4(3 + 2x)(9 - 6x + 4x^2)$ **11.** $3(x - 3)(x + 2)$ **13.** $(5x + 1)(x + 2)$
15. $3x(x^2 + 2x + 7)$ **17.** $2x^2(x^2 - 2x - 5)$ **19.** $2x^2(2x^2 + 6x + 9)$
21. $(x + 2)(3x^2 + 1)$ **23.** $(x + 1)(3x^2 + 2)$ **25.** $(x + 1)(2x^2 - 1)$
27. $3(x + 4)^2$ **29.** $k(x + 2)^2$ **31.** $4(x - 3)^2$ **33.** $k(x - 6)^2$
35. $3x(x + 2)^2$ **37.** $2x(3x + 1)^2$ **39.** $3x^2(2x - 3)^2$
41. $(x^2 + 1)(x + 1)(x - 1)$ **43.** $(x^2 + y^2)(x + y)(x - y)$
45. $(x^2 + 4y^2)(x + 2y)(x - 2y)$ **47.** $-(x + 3)^2$ **49.** $-(x + 2)^2$
51. $-(2x + y)^2$ **53.** $-(3x + 2y)^2$ **55.** $-(2z - 3y)^2$
57. $-2x(3x + 2y)^2$ **59.** $-2x(3x + 5y)^2$ **61.** $-x(x + 1)(x - 1)$
63. $-x^2(x + 2)(x - 2)$ **65.** $-x^2(2x + 3)(2x - 3)$
67. $-2x(x - 2)(x^2 + 2x + 4)$ **69.** $-2x^2(2x + 1)(4x^2 - 2x + 1)$
71. $(x + 6)(x^2 - 6x + 36)$ **73.** $(2x - 1)(4x^2 + 2x + 1)$
75. $\frac{2\pi A}{360}(R_1 + Kt)$ **77.** $\frac{3S}{2bd^3}(d + 2z)(d - 2z)$ **81.** $x^2 - 9$ can be
factored as $(x + 3)(x - 3)$ **83.** $(X - A)(X^2 + AX + A^2)$ **85.** look
87. $5x^2(x^2 - 2x + 4)$ **89.** $(3x + 7)(2x - 5)$
91. $(3t - 4)(9t^2 + 12t + 16)$ **93.** $(x^2 + 9)(x + 3)(x - 3)$
95. $-(3x + 5y)^2$ **97.** $(4y + 3x)(16y^2 - 12xy + 9x^2)$
99. $-x^2(x + y)(x^2 - xy + y^2)$ **101.** $(2x + 3)(5x - 1)$
103. $(2x + 1)(x - 3)$

Exercises 5.6

1. $x = -3$ or $x = -\frac{1}{2}$ **3.** $x = 1$ or $x = -\frac{3}{2}$ **5.** $y = 3$ or $y = \frac{2}{3}$
7. $y = 1$ or $y = -\frac{1}{3}$ **9.** $x = -\frac{4}{3}$ or $x = -\frac{1}{2}$ **11.** $x = 2$ or $x = -\frac{1}{3}$
13. $x = \frac{4}{5}$ or $x = -2$ **15.** $x = 1$ or $x = \frac{8}{5}$ **17.** $y = 5$ or $y = \frac{2}{3}$
19. $y = -2$ or $y = -\frac{1}{2}$ **21.** $x = -\frac{1}{3}$ **23.** $y = 4$ **25.** $x = -\frac{2}{3}$
27. $y = \frac{5}{2}$ **29.** $x = -5$ **31.** $x = 4$ or $x = 1$ **33.** $x = -2$ or $x = -\frac{5}{2}$
35. $x = 1$ **37.** $x = \frac{1}{2}$ or $x = -\frac{1}{3}$ **39.** $y = \frac{5}{2}$ or $y = -\frac{5}{2}$
41. $z = 3$ or $z = -3$ **43.** $x = \frac{7}{5}$ or $x = -\frac{7}{5}$ **45.** $m = 0$ or $m = 5$
47. $n = 0$ or $n = 5$ **49.** $y = -3$ or $y = -8$ **51.** $y = 9$ or $y = 7$
53. $v = 2$ or $v = -1$ or $v = -2$ **55.** $m = 2$ or $m = 1$ or $m = 4$
57. $n = 4$ or $n = -1$ or $n = -2$ **59.** $x = 1$ or $x = -3$
61. (a) $m = -49$ or $m = 5$; **(b)** No; $m = 5$ **63. (a)** 1 sec;
(b) $t = 1$ or $t = -2$; No; $t = 1$ **65.** $x = 7$ **69.** One
71. $A = 0, B = 0$ **73.** $x = 3$ or $x = -\frac{2}{3}$ **75.** $x = -2$ or $x = \frac{1}{5}$
77. $x = 0$ or $x = 1$ **79.** $x = 4$ or $x = -5$ or $x = 1$
81. $m = -\frac{2}{3}$ or $m = -1$ **83.** $2H^2 + 6H + 9$ **85.** $H^2 - 6H - 27$

Exercises 5.7

Translate This

1. K   **3.** B   **5.** A   **7.** G   **9.** J

1. 8, 9 or 1, 2   **3.** 3, 5 or $-3, -1$ **5.** 4, 6 or 2, 4
7. $b = 10$ in., $h = 8$ in.   **9.** $L = 15$ in., $h = 10$ in.
11. $x = 4$; $r = 7$ units   **13.** $L = 50$ ft, $W = 5$ ft   **15.** About 247 ft
17. 6 in., 8 in., 10 in.   **19.** 9 in., 12 in., 15 in.   **21.** 30 mi/hr
23. (a) 35 mi/hr; **(b)** Yes   **25.** 40 mi/hr   **27.** 5 teams   **29.** 8 delegates
31. 30 students   **33.** 1 sec   **35.** 1 sec   **37.** The first formula
41. $a^2 + b^2 = c^2$ **43.** 3, 5 or $-7, -5$ **45.** 5 in., 12 in., 13 in.
47. 50 mi/hr   **49.** $\frac{15}{18}$ **51.** $18(x - 2y)$ **53.** $(x + 6)(x - 1)$

Review Exercises

1. [5.1A] **(a)** 30; **(b)** 6; **(c)** 1   **2.** [5.1B] **(a)** $6x^5$; **(b)** $2x^8$; **(c)** x^3
3. [5.1C] **(a)** $5x^3(4 - 11x^2)$; **(b)** $7x^4(2 - 5x^2)$; **(c)** $8x^7(2 - 5x^2)$
4. [5.1C] **(a)** $\frac{1}{7}x^2(3x^4 - 5x^3 + 2x^2 - 1)$; **(b)** $\frac{1}{9}x^3(4x^4 - 2x^3 + 2x^2 - 1)$;
(c) $\frac{1}{8}x^5(3x^4 - 7x^3 + 3x^2 - 1)$ **5.** [5.1D] **(a)** $(x - 7)(3x^2 - 1)$;
(b) $(x + 6)(3x^2 + 1)$; **(c)** $(x - 2y)(4x^2 + 1)$
6. [5.2A] **(a)** $(x + 7)(x + 1)$; **(b)** $(x - 9)(x + 1)$; **(c)** $(x + 5)(x + 1)$
7. [5.2A] **(a)** $(x - 5)(x - 2)$; **(b)** $(x - 7)(x - 2)$; **(c)** $(x + 4)(x - 2)$

8. [5.3B] **(a)** $(2x + 3)(3x - 2)$; **(b)** $(2x + 1)(3x - 1)$;
(c) $(2x + 5)(3x - 1)$ **9.** [5.3B] **(a)** $(3x - y)(2x - 5y)$;
(b) $(3x - 2y)(2x - y)$; **(c)** $(3x - 4y)(2x - y)$
10. [5.4B] **(a)** $(x + 2)^2$; **(b)** $(x + 5)^2$; **(c)** $(x + 4)^2$
11. [5.4B] **(a)** $(3x + 2y)^2$; **(b)** $(3x + 5y)^2$; **(c)** $(3x + 4y)^2$
12. [5.4B] **(a)** $(x - 2)^2$; **(b)** $(x - 3)^2$; **(c)** $(x - 6)^2$
13. [5.4B] **(a)** $(2x - 3y)^2$; **(b)** $(2x - 5y)^2$; **(c)** $(2x - 7y)^2$
14. [5.4C] **(a)** $(x + 6)(x - 6)$; **(b)** $(x + 7)(x - 7)$; **(c)** $(x + 9)(x - 9)$
15. [5.4C] **(a)** $(4x + 9y)(4x - 9y)$; **(b)** $(5x + 8y)(5x - 8y)$;
(c) $(3x + 10y)(3x - 10y)$ **16.** [5.5A] **(a)** $(m + 5)(m^2 - 5m + 25)$;
(b) $(n + 4)(n^2 - 4n + 16)$; **(c)** $(y + 2)(y^2 - 2y + 4)$
17. [5.5A] **(a)** $(2y - 3x)(4y^2 + 6xy + 9x^2)$;
(b) $(4y - 5x)(16y^2 + 20xy + 25x^2)$;
(c) $(2m - 5n)(4m^2 + 10mn + 25n^2)$
18. [5.5B] **(a)** $3x(x^2 - 2x + 9)$; **(b)** $3x(x^2 - 2x + 10)$;
(c) $4x(x^2 - 2x + 8)$ **19.** [5.5B] **(a)** $2x(x - 2)(x + 1)$;
(b) $3x(x - 3)(x + 1)$; **(c)** $4x(x - 4)(x + 1)$
20. [5.5B] **(a)** $(x + 4)(2x^2 + 1)$; **(b)** $(x + 5)(2x^2 + 1)$;
(c) $(x + 6)(2x^2 + 1)$ **21.** [5.5B] **(a)** $k(3x + 2)^2$; **(b)** $k(3x + 5)^2$;
(c) $k(2x + 5)^2$ **22.** [5.5C] **(a)** $-3x^2(x + 3)(x - 3)$;
(b) $-4x^2(x + 4)(x - 4)$; **(c)** $-5x^2(x + 2)(x - 2)$
23. [5.5C] **(a)** $-(x + y)(x^2 - xy + y^2)$;
(b) $-(2m + 3n)(4m^2 - 6mn + 9n^2)$;
(c) $-(4n + m)(16n^2 - 4mn + m^2)$
24. [5.5C] **(a)** $-(y - x)(y^2 + xy + x^2)$;
(b) $-(2m - 3n)(4m^2 + 6mn + 9n^2)$;
(c) $-(4t - 5s)(16t^2 + 20st + 25s^2)$
25. [5.5C] **(a)** $-(4x^2 + 12xy - 9y^2)$; **(b)** $-(25x^2 + 30xy - 9y^2)$;
(c) $-(16x^2 + 24xy - 9y^2)$ **26.** [5.6A] **(a)** $x = 5$ or $x = -1$;
(b) $x = 6$ or $x = -1$; **(c)** $x = 7$ or $x = -1$
27. [5.6A] **(a)** $x = -\frac{5}{2}$ or $x = 2$; **(b)** $x = -\frac{5}{2}$ or $x = 1$;
(c) $x = -\frac{3}{2}$ or $x = 1$ **28.** [5.6A] **(a)** $x = 1$ or $x = \frac{4}{3}$;
(b) $x = 3$ or $x = \frac{7}{2}$; **(c)** $x = 3$ or $x = -1$ **29.** [5.7A] **(a)** 4, 6;
(b) 2, 4 or -2, 0; **(c)** 10, 12 **30.** [5.7C] 18 in., 24 in., 30 in.

Cumulative Review Chapters 1–5

1. $-\frac{13}{24}$ **2.** 16.4 **3.** -31.35 **4.** -25 **5.** 3 **6.** 69

7. Associative property of multiplication **8.** xy^4 **9.** $-x - 11$
10. $\frac{d + 5e}{f}$ **11.** $x = 3$ **12.** $x = 63$ **13.** $x = 6$ **14.** 40 and 65
15. 10 hr **16.** \$4000 in bonds; \$2000 in certificates of deposit

17.

18.

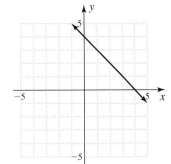

$C(3, -3)$

19. No **20.** $x = 2$
21. $x + y = 4$

22. $4y - 20 = 0$

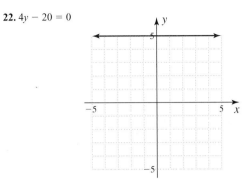

23. $\frac{2}{11}$ **24.** 3 **25.** (1) and (2) **26.** $y - 3 = 6(x + 5)$ or $y = 6x + 33$
27. $y = -3x + 4$
28.

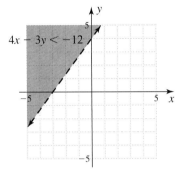

$4x - 3y < -12$

29.

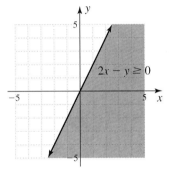

$2x - y \geq 0$

30. $-\frac{4x^2}{y^4}$ **31.** x^2 **32.** $\frac{1}{x^4}$ **33.** $\frac{y^9}{27x^9}$ **34.** 4.8×10^{-5} **35.** 4.60×10^{-5}
36. 2 **37.** 1 **38.** $3x^3 - 7x^2 - 13$ **39.** $-32x^6 - 16x^4y$
40. $9x^2 + 12xy + 4y^2$ **41.** $25x^2 - 49y^2$ **42.** $4x^4 - \frac{4}{5}x^2 + \frac{1}{25}$
43. $25x^4 - 81$ **44.** $(2x^2 + 5x + 5)$ R 1 **45.** $2x^6(6 - 7x^3)$
46. $\frac{1}{5}x^3(4x^4 - 3x^3 + 4x^2 - 2)$ **47.** $(x - 3)(x - 9)$
|48. $(5x - 2y)(4x - 3y)$ **49.** $(5x + 7y)(5x - 7y)$
50. $-5x^2(x + 4)(x - 4)$ **51.** $3x(x + 1)(x - 3)$ **52.** $(2x + 5)(x + 3)$
53. $k(3x + 1)^2$ **54.** $x = -5$ or $x = \frac{3}{4}$

Chapter 6

Exercises 6.1

1. $x = 7$ **3.** $y = -4$ **5.** $x = 3$ or $x = -3$ **7.** None **9.** $y = 2$ or $y = 4$
11. $x = -2$ **13.** $x = 3$ **15.** $x = -3$ **17.** $y = -2$ or $y = 8$
19. $x = 0$ or $x = -1$ **21.** $\frac{9}{21}$ **23.** $-\frac{16}{22}$ **25.** $\frac{20xy}{24y^3}$ **27.** $-\frac{9xy^3}{21y^4}$
29. $\frac{4x(x - 2)}{x^2 - x - 2}$ or $\frac{4x^2 - 8x}{x^2 - x - 2}$ **31.** $\frac{-5x(x - 2)}{x^2 + x - 6}$ or $\frac{-5x^2 + 10x}{x^2 + x - 6}$
33. $\frac{x^2}{2y^3}$ **35.** $\frac{-3y^4}{x}$ **37.** $\frac{1}{2xy^3}$ **39.** $\frac{5x}{y^2}$ **41.** $\frac{x - y}{3}$
43. $-3(x - y)$ or $3y - 3x$ **45.** $\frac{-1}{4(x - y)}$ or $\frac{1}{4y - 4x}$
47. $\frac{1}{2(x + 2)}$ or $\frac{1}{2x + 4}$ **49.** $\frac{1}{x + y}$ **51.** $\frac{1}{2}$ **53.** $\frac{1}{3}$ **55.** $\frac{-1}{1 + 2y}$

57. $\frac{-1}{1+2y}$ **59.** $\frac{-1}{x+2}$ **61.** -1 **63.** $-(x+5)$ or $-x-5$
65. $-(x-2)$ or $2-x$ **67.** $\frac{-1}{x+6}$ **69.** $\frac{1}{x-2}$

71. (a) \$50 million; \$130 million; \$80 million; **(b)** \$30 million;
\$78 million; \$63 million; **(c)** \$20 million; \$52 million; \$17 million;
(d) The fraction of the total advertising that is spent on national
advertising   **73. (a)** 10%; **(b)** 21.25%; **(c)** The fraction of the total
advertising that is spent on TV advertising
75. (a) \$137.91; \$185.12; \$241.25; **(b)** Values are close;
(c) \$830 million; **(d)** $\frac{N(t)}{T(t)} = \frac{0.83t^3 - 5.5t^2 + 55t}{13.45t^3 - 96.4t^2 + 262t}$; **(e)** $\frac{830}{6430} = \frac{83}{643} \approx 13\%$
77. (a) \$61.5; \$48; \$102.5; **(b)** 2003 and 2005 are close, 2004
is quite different; **(c)** \$3180 million; **(d)** $\frac{C(t)}{T(t)} = \frac{8.5t^3 - 68t^2 + 148t}{13.45t^3 - 96.4t^2 + 262t}$;
(e) $\frac{3180}{6430} = \frac{318}{643} \approx 49\%$ **79. (a)** $\frac{1.1t + 33}{2.8t + 281}$;
(b) In 2000: $\frac{33}{281} \approx 12\%$; In 2050: $\frac{88}{421} \approx 21\%$; **(c)** Increasing   **81.** $\frac{2}{3}$
83. (a) 4 to 1; **(b)** 2500 rpm   **85. (a)** Yes; **(b)** No   **87.** $P(x) = -Q(x)$
89. $\frac{A \cdot C}{B \cdot C}$ **91.** $\frac{-a}{b}$ **93.** -1 **95.** $x - 3$ **97.** $\frac{5}{2}$ **99.** $\frac{-1}{x-4}$ or $\frac{1}{4-x}$
101. $2(x+y)$ or $2x + 2y$ **103.** $\frac{9xy}{24y^3}$ **105.** $x = 2$ or $x = -2$
107. $\frac{2}{3}$ **109.** $(x+3)(x-1)$ **111.** $(x-5)(x-2)$

Exercises 6.2

1. $\frac{8x}{3y}$ **3.** $\frac{-4xy}{3}$ **5.** $\frac{3y}{2}$ **7.** $-8xy$ **9.** $\frac{x+1}{x+7}$ **11.** $\frac{2-2x}{3-2x}$
13. $\frac{3x+15}{x+1}$ **15.** $-(x+2)$ or $-x-2$ **17.** $\frac{x+3}{x-3}$ **19.** $\frac{1}{4}$
21. $\frac{2x^2 - 2x}{x+4}$ **23.** $\frac{y^2 + 5y + 6}{y+6}$ **25.** $\frac{y^2}{2-3y}$ **27.** $\frac{5x-2}{x-2}$ **29.** $2y - 5$
31. $\frac{x-1}{x+2}$ **33.** $\frac{x-5}{5x-15}$ **35.** $\frac{x+4}{x-3}$ **37.** $\frac{-5x-20}{2x+6}$ **39.** $\frac{x-2}{x-1}$
41. $\frac{x+3}{1-x^2}$ **43.** $\frac{2x-4}{x-3}$ **45.** $\frac{x^2 + 11x + 30}{4}$ **47.** $\frac{x^2 - 4x + 4}{x^2 + 2x - 3}$
49. $\frac{x+6}{5}$ **51.** $x - 1$ **53.** $\frac{x^2 - 6x + 5}{x^2 - 6x + 8}$ **55.** 1   **57.** $\frac{x+y}{x}$
59. $\frac{xy + 2x + 6y + 3y^2}{xy - 2x - 12y + 6y^2}$ **61.** $\frac{x^2 - x - 12}{2x^2 - x - 1}$ **63.** $\frac{a}{a+3}$ **65.** $\frac{450}{x}$ **67.** $\frac{1000}{x}$
69. $\frac{5t(t+2)}{t^2 + 5}$ or $\frac{5t^2 + 10t}{t^2 + 5}$ **71.** $\frac{RR_T}{R - R_T}$ **73.** $C_R = \frac{60,000 + 9000x}{x}$ **79.** $\frac{AD}{BC}$
81. $\frac{x^2 + 4x + 3}{x^2 - 4}$ **83.** $\frac{1}{6-x}$ **85.** $x^2 - 6x + 9$ **87.** $2 - x$
89. $12x$ **91.** $\frac{x^2 - x - 12}{2x^2 - x - 1}$ **93.** $\frac{51}{40}$ **95.** $\frac{19}{40}$ **97.** $\frac{7}{3}$

Exercises 6.3

1. (a) $\frac{5}{7}$; **(b)** $\frac{11}{x}$ **3. (a)** $\frac{2}{3}$; **(b)** $\frac{4}{x}$ **5. (a)** 2; **(b)** $\frac{5}{x}$
7. (a) 2; **(b)** $\frac{2}{3(x+1)}$ **9. (a)** $\frac{4}{3}$; **(b)** $\frac{5x}{2(x+1)}$
11. (a) $\frac{5}{12}$; **(b)** $\frac{56-3x}{8x}$ **13. (a)** $\frac{12}{35}$; **(b)** $\frac{x^2 + 36}{9x}$
15. (a) $\frac{2}{15}$; **(b)** $\frac{5}{14(x-1)}$ **17. (a)** $\frac{53}{56}$; **(b)** $\frac{8x-1}{(x+1)(x-2)}$
19. (a) $\frac{13}{24}$; **(b)** $\frac{3x+12}{(x-2)(x+1)}$ **21.** $\frac{2x^2 - 2x - 6}{(x+4)(x-4)(x-1)}$
23. $\frac{5x^2 + 19x}{(x+5)(x-2)(x+3)}$ **25.** $\frac{6x-4y}{(x+y)^2(x-y)}$ **27.** $\frac{10-x}{(x+5)(x-5)}$
29. $\frac{-6x-10}{(x+2)(x+1)(x+3)}$ **31.** $\frac{y^2}{(y+1)(y-1)}$ **33.** $\frac{5-4y-2y^2}{(y+4)(y-4)}$ or $\frac{-2y^2 - 4y + 5}{(y+4)(y-4)}$
35. $\frac{2x^2 - x + 3}{(x-2)(x+1)^2}$ **37.** 0   **39.** $\frac{a^2 + 3a + 3}{(a+2)(a^2 - 2a + 4)}$ **41. (a)** $\frac{f+u}{f+u}$; **(b)** 1
43. $\frac{g_m + 8}{g_m^2}$ **45.** $\frac{nR_2 - nR_1 + (n-1)d}{nR_1R_2}$ **47.** $\frac{-w_0x^3 + 3w_0L^2x - 2w_0L^3}{6L}$
49. $\frac{p^2 - 2gm^2rM}{2mr^2}$ **51.** 1.5   **53.** $1.41\overline{6}$ **55.** $0.0024\overline{6}$ or 0.0025   **59.** LCD
61. $\frac{R(t) - P(t)}{G(t)} = \frac{0.02t^2 - 0.34t + 1.42}{0.04t^2 + 2.34t + 90}$ **63.** $\frac{2x+18}{(x-1)(x+3)}$ **65.** $\frac{2}{x-2}$
67. $\frac{7x+9}{(x+2)(x-2)(x+1)}$ **69.** $\frac{9}{20}$ **71.** 42   **73.** $x^2 - 1$

Exercises 6.4

1. $\frac{1}{5}$ **3.** $\frac{1}{3}$ **5.** $\frac{ab-a}{b+a}$ **7.** $\frac{b+a}{b-a}$ **9.** $\frac{6b+4a}{48b - 9a}$ **11.** $\frac{26}{5}$
13. $\frac{x}{2x-1}$ **15.** 2   **17.** $\frac{1}{x-y}$ **19.** $\frac{x+2}{x+1}$ **21.** $\frac{x-5}{4}$
23. $\frac{1}{2(x-4)}$ **25.** $\frac{6}{25}$ yr   **27.** $11\frac{43}{56}$ yr   **29.** $84\frac{11}{1000}$ yr   **31.** $4\frac{3}{5}$ yr
33. $248\frac{27}{50}$ yr   **35.** a fraction that has one or more fractions in its
numerator, denominator or both.   **41.** LCD **43.** $\frac{11}{6}$ **45.** $\frac{w+4}{w+1}$
47. $\frac{m+5}{m-4}$ **49.** $w = 124$ **51.** $x = -3$ **53.** $x = 3$

Exercises 6.5

1. $x = 6$ **3.** $x = 4$ **5.** $x = -6$ **7.** $x = 12$ **9.** $x = 4$
11. $x = 5$ **13.** $x = 2$ **15.** $x = 2$ **17.** $x = -11$ **19.** $x = -2$
21. No solution   **23.** $x = -\frac{3}{2}$ **25.** $x = -6$ **27.** $x = -8$
29. No solution   **31.** No solution   **33.** $x = -3$ **35.** $x = -\frac{5}{2}$
37. $x = \frac{1}{7}$ **39.** $x = -\frac{3}{5}$ **41.** $z = \frac{45}{2}$ **43.** $y = \frac{5}{7}$ **45.** $v = \frac{2}{5}$
47. $z = 3$ **49.** $x = 4$ **51.** $V_2 = \frac{P_1V_1}{P_2}$ **53.** $V_2 = 4$ L
55. $V_2 = \frac{299 \cdot 21}{273} = 23$ L   **57.** $h = \frac{2A}{b_1 + b_2}$ **59.** $Q_1 = \frac{PQ_2}{P+1}$
61. $f = \frac{ab}{a+b}$ **65.** term   **67.** $x = -3$ **69.** $x = -4$
71. $x = -2$ **73.** $F = \frac{9}{5}C + 32$ **75.** $x = \frac{33}{4}$ **77.** $h = \frac{12}{5}$ **79.** $R = 15$

Exercises 6.6

1. 9250   **3.** 100   **5.** 6   **7.** 3.6   **9. (a)** 2250; **(b)** 36
11. (a) 3 to 1; **(b)** 35 to 1; **(c)** Minneapolis; climate   **13.** 37 million
15. $5\frac{1}{7}$ hr   **17.** $1\frac{7}{8}$ min   **19.** 6 hr   **21.** 10 mi/hr   **23.** 100 mi/hr
25. (a) 58; **(b)** No   **27.** 189; no   **29.** $x = 5\frac{1}{3}$; $y = 6\frac{2}{3}$
31. $DE = 13\frac{5}{7}$ in.   **33.** $x = 16$; $y = 8$ **35.** $x = 12$; $y = 12$
37. $a = 24$; $b = 30$ **39.** $x = 8$; $y = 8\frac{4}{5}$ **41.** $x = 10$; $y = 25$
43. 1.5 in. $\times$ 2 in.   **45. (a)** Approximately 32 hr;
(b) Approximately $6\frac{1}{2}$ days   **47.** $18\frac{2}{3}$ lb   **49.** 6 ft   **51.** 2650
57. $ad = bc$ **59.** similar   **61.** $3\frac{1}{13}$ hr   **63.** 200   **65.** $k = 5$
67. $k = 240$

Exercises 6.7

1. (a) $S = \frac{W}{16}$; **(b)** $k = \frac{1}{16}$; **(c)** 10 lb   **3. (a)** $R = \frac{W}{10}$; **(b)** $k = \frac{1}{10}$;
(c) 16   **5. (a)** $R = kt$; **(b)** $k = 45$; **(c)** 2.4 min   **7. (a)** $T = kh^3$;
(b) $k \approx 0.00057$; **(c)** 240 lb   **9. (a)** $f = \frac{k}{d}$; **(b)** $k = 4$; **(c)** 16
11. 10.8 in.^{3} **13. (a)** $w = \frac{k}{s}$; **(b)** $k = 7200$; **(c)** 720 words
15. (a) $b = \frac{k}{a}$; **(b)** $k = 2970$; **(c)** 90 (per 1000 women)   **17. (a)** $d = ks$;
(b) $k \approx 17.63$; **(c)** The time it took to drive d mi at s mph
19. (a) $C = 4(F - 37)$ or $C = 4F - 148$; **(b)** 212 chirps per minute
21. (a) $I = kn^2$; **(b)** $k = 0.05$; **(c)** 245%   **23. (a)** $BAC = k(N-1)$;
(b) $k = 0.026$; **(c)** 0.104; **(d)** 4   **25. (a)** $BAC = \frac{7.8}{W}$; **(b)** 0.03
(c) 97.5 lb; **(d)** Less than 0.08   **27. (a)** $T = 15S$; **(b)** 4
33. directly   **35. (a)** $C = kn$; **(b)** $k = 100$; **(c)** 2 strips
37. (a) $v = \frac{k}{t}$; **(b)** $k = 120$; **(c)** The distance traveled
39. $x + 2y = 4$

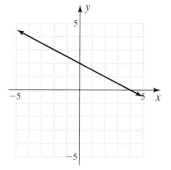

41. $y - 2x = 4$

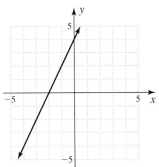

43. Not parallel

Review Exercises

1. [6.1B] **(a)** $\frac{10xy}{16y^2}$; **(b)** $\frac{12xy}{16y^3}$; **(c)** $\frac{10x^2y}{15x^5}$

2. [6.1C] **(a)** $-3(x - y)$ or $3y - 3x$;
(b) $-2(x - y)$ or $2y - 2x$; **(c)** $4(x - y)$ or $4x - 4y$

3. [6.1A, C] **(a)** $\frac{-1}{x+1}$; $0, -1$; **(b)** $\frac{-1}{x-1}$ or $\frac{1}{1-x}$; $0, 1$;
(c) $\frac{-1}{1-x}$ or $\frac{1}{x-1}$; $0, 1$ **4.** [6.1C] **(a)** $-(x + 3)$ or $-x - 3$;
(b) $-(x + 2)$ or $-x - 2$; **(c)** $-(x + 3)$ or $-x - 3$

5. [6.2A] **(a)** $\frac{2xy}{3}$; **(b)** $\frac{3xy}{2}$; **(c)** $8xy^2$

6. [6.2A] **(a)** $\frac{x+2}{x+3}$; **(b)** $\frac{x+1}{x+5}$; **(c)** $\frac{x+5}{x+4}$

7. [6.2B] **(a)** $\frac{x-3}{x+2}$; **(b)** $\frac{x-4}{x+1}$; **(c)** $\frac{x-5}{x+4}$

8. [6.2B] **(a)** $\frac{-1}{x-5}$ or $\frac{1}{5-x}$; **(b)** $\frac{-1}{x-1}$ or $\frac{1}{1-x}$; **(c)** $\frac{-1}{x-2}$ or $\frac{1}{2-x}$

9. [6.3A] **(a)** $\frac{2}{x-1}$; **(b)** $\frac{2}{x-2}$; **(c)** $\frac{1}{x+1}$

10. [6.3A] **(a)** $\frac{2}{x+1}$; **(b)** $\frac{2}{x+2}$; **(c)** $\frac{2}{x+3}$

11. [6.3B] **(a)** $\frac{3x-2}{(x+2)(x-2)}$; **(b)** $\frac{4x-2}{(x+1)(x-1)}$; **(c)** $\frac{5x-9}{(x+3)(x-3)}$

12. [6.3B] **(a)** $\frac{-6x-10}{(x+1)(x+2)(x+3)}$; **(b)** $\frac{7x+1}{(x-2)(x+1)^2}$;
(c) $\frac{-4x}{(x+2)(x+1)(x-1)}$ **13.** [6.4A] **(a)** $\frac{6}{17}$; **(b)** $\frac{14}{27}$; **(c)** $\frac{6}{25}$

14. [6.5A] **(a)** $x = 3$; **(b)** $x = 7$; **(c)** $x = 6$

15. [6.5A] **(a)** $x = -3$; **(b)** $x = -5$; **(c)** $x = -6$

16. [6.5A] **(a)** No solution; **(b)** No solution; **(c)** No solution

17. [6.5A] **(a)** $x = -6$ or $x = \frac{13}{3}$; **(b)** $x = -7$ or $x = \frac{11}{2}$;
(c) $x = -8$ or $x = \frac{33}{5}$ **18.** [6.5B] **(a)** $a_1 = \frac{A(1 - b^n)}{1 - b}$; **(b)** $b_1 = \frac{B(1 - c)}{1 - c^n}$;
(c) $c_1 = \frac{C(d + 1)}{(1 - d^n)}$ **19.** [6.6A] **(a)** $10\frac{1}{2}$ gal; **(b)** $13\frac{1}{2}$ gal; **(c)** 18 gal

20. [6.6A] **(a)** $x = 18$; **(b)** $x = 10\frac{2}{5}$; **(c)** $x = \frac{-13}{5}$

21. [6.6B] **(a)** $3\frac{3}{7}$ hr; **(b)** $4\frac{4}{9}$ hr; **(c)** $3\frac{3}{5}$ hr

22. [6.6B] **(a)** 4 mi/hr; **(b)** 8 mi/hr; **(c)** 12 mi/hr

23. [6.6B] **(a)** 32; **(b)** 35; **(c)** 40

24. [6.6B] **(a)** 1125; **(b)** 1250; **(c)** 1375

25. [6.6B] **(a)** $10\frac{1}{2}$; **(b)** $10\frac{2}{3}$; **(c)** $2\frac{2}{3}$ **26.** [6.7A] **(a)** $C = km$; 150 calories;
(b) $C = km$; 420 calories; **(c)** $C = km$; 45 calories;

27. [6.7B] **(a)** $F = \frac{30}{L}$; $F = 5$; **(b)** $F = 3$; **(c)** $F = 2$

Cumulative Review Chapters 1–6

1. $-\frac{17}{30}$ **2.** 2.7   **3.** 25   **4.** 2   **5.** 69   **6.** $6x - 14$

7. $\frac{a - b}{c}$ **8.** $x = 4$ **9.** $x = 63$ **10.** 30 and 65

11. $9000 in bonds; $3000 in certificates of deposit

12.

13.

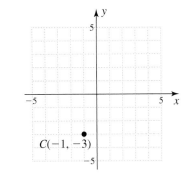

14. Yes   **15.** $x = 3$

16.

17.

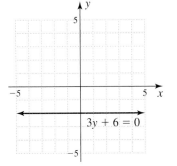

18. $\frac{7}{16}$ **19.** 2   **20.** (2) and (3)   **21.** $y - 3 = 6(x + 4)$ or $y = 6x + 27$

22. $y = -2x + 5$

23. $3x - 4y < -12$

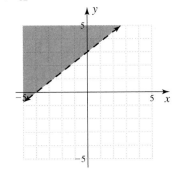

24. $3x - y \geqslant 0$

25. x **26.** $\frac{1}{x^2}$

27. $\frac{y^8}{81x^{12}}$ **28.** 3.6×10^{-4} **29.** 1.80×10^{-5}

30. -4 **31.** $-12x^6 + 9x^4 - 4$ **32.** $25x^4 - 2x^2 + \frac{1}{25}$

33. $81x^4 - 4$ **34.** $(3x^2 - x + 4)$ R 2   **35.** $3x^2(2 - 3x^2)$

36. $\frac{1}{5}x^3(2x^4 - 4x^3 + 4x^2 - 1)$ **37.** $(x - 7)(x - 8)$

38. $(5x - 4y)(3x - 5y)$ **39.** $(4x + 3y)(4x - 3y)$

40. $-4x^2(x + 1)(x - 1)$ **41.** $3x(x + 1)(x - 3)$ **42.** $(x + 1)(4x + 3)$

43. $k(4x - 1)^2$ **44.** $x = -3, x = \frac{5}{2}$ **45.** $\frac{14xy^2}{12y^3}$

46. $-2(x + y)$ or $-2x - 2y$ **47.** $-(x + 5)$ or $-x - 5$ **48.** $\frac{x + 3}{x + 7}$

49. $\frac{-1}{x - 3}$ or $\frac{1}{3 - x}$ **50.** $\frac{6}{x + 8}$ **51.** $\frac{-2x - 11}{(x + 6)(x + 5)(x - 5)}$ **52.** 42

53. $x = 3$ **54.** $x = -8$ **55.** No solution   **56.** $x = -5$ **57.** 23 gal

58. $\frac{-70}{11}$ **59.** $3\frac{3}{13}$ hr   **60.** $k = \frac{1}{50}$ **61.** $k = 13{,}720$

Chapter 7

Exercises 7.1

1. Consistent; $(1, 3)$
  ① $x + y = 4$; ② $x - y = -2$

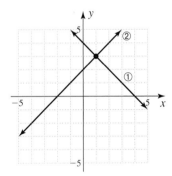

3. Consistent; $(-2, 1)$
  ① $x + 2y = 0$; ② $x - y = -3$

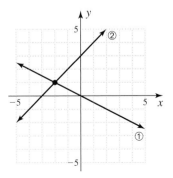

5. Dependent
  $3x - 2y = 6$; $6x - 4y = 12$

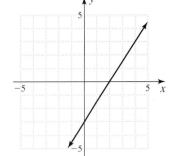

7. Dependent
  $3x - y = -3$; $y - 3x = 3$

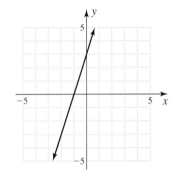

9. Inconsistent
  ① $2x - y = -2$; ② $y = 2x + 4$

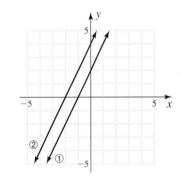

11. Consistent; $(-2, -2)$
  ① $y = -2$; ② $2y = x - 2$

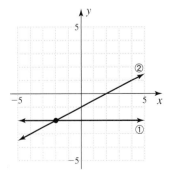

13. Consistent; (3, 2)
  ① $x = 3$; ② $y = 2x - 4$

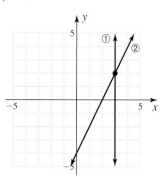

15. Consistent; (1, 2)
  ① $x + y = 3$; ② $2x - y = 0$

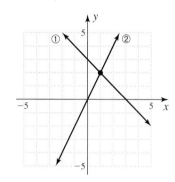

17. Consistent; (0, 5)
  ① $5x + y = 5$; ② $5x = 15 - 3y$

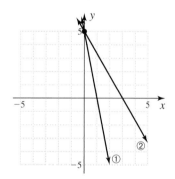

19. Dependent
  $3x + 4y = 12$; $8y = 24 - 6x$

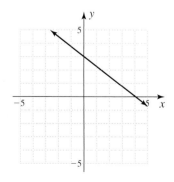

21. Consistent; (0, 3)
  ① $y = x + 3$; ② $y = -x + 3$

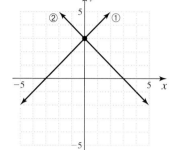

23. Consistent; (1, 0)
  ① $y = 2x - 2$; ② $y = -3x + 3$

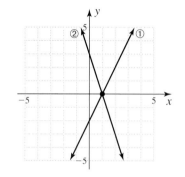

25. Consistent; (−2, −3)
  ① $-2x = 4$; ② $y = -3$

27. Consistent; (3, −3)
  ① $y = -3$; ② $y = -3x + 6$

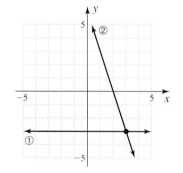

29. Inconsistent

① $x + 4y = 4$; ② $y = -\frac{1}{4}x + 2$

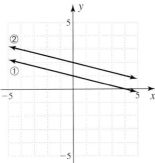

31. (a) $C = 20 + 35m$; **(b)**

m	C
6	230
12	440
18	650

(c)

33.

35. DVD player and rental option is cheaper if used more than 18 months. (The options are equal if used for 18 months.)

37. (a) $W = 100 + 3t$; **(b)**

t	W
5	115
10	130
15	145
20	160

(c)

39. (a) $C = 0.60m$; **(b)** $C = 0.45m + 45$;

(c)

41. (a) $C_B = 5x$; **(b)** $C_H = 6y$; **(c)** $x + y$; $x + y = 50$; **(d)** $5x + 6y = 270$; **(e)** 30 baskets, 20 buffets

43. (a) $C_B = 6x$; **(b)** $C_C = 7y$; **(c)** $x + y$; $x + y = 20$; **(d)** $6x + 7y = 126$; **(e)** 14 croissants, 6 omelets

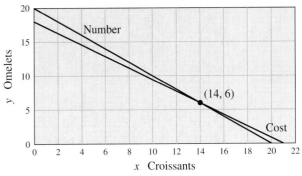

45. (a) $x + y = 16$; (b) $P_Q = 2x$; (c) $P_W = y$;
(d) $2x + y = 24$; (e) Both 8 grams

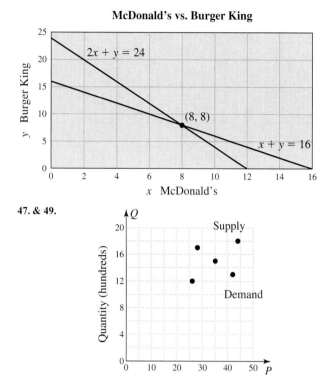

McDonald's vs. Burger King

47. & 49.

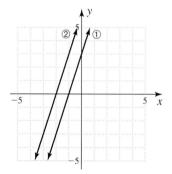

51. (35, 15)   **53.** 1500   **57.** No solution   **59.** One solution
61. one   **63.** no   **65.** infinitely
67. No solution; ① $y - 3x = 3$; ② $2y = 6x + 12$

69. Consistent; (2, 2)
   ① $x + y = 4$; ② $2x - y = 2$

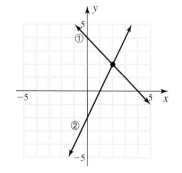

71. Inconsistent; no solution
   ① $2x + y = 4$; ② $2y + 4x = 6$

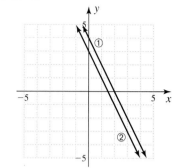

73. $x = 4$ **75.** Yes   **77.** No

Exercises 7.2

1. Consistent; (2, 0)   **3.** Consistent; (2, 3)   **5.** Inconsistent; no solution
7. Inconsistent; no solution   **9.** Dependent; infinitely many solutions
11. Consistent; $(-1, -1)$ **13.** Inconsistent; no solution
15. Consistent; (4, 1)   **17.** Consistent; (5, 3)   **19.** Consistent; (0, 2)
21. Inconsistent; no solution   **23.** Consistent; (2, 1)
25. Consistent; (3, 1)   **27.** Dependent; infinitely many solutions
29. Inconsistent; no solution   **31.** Consistent; $(2, -3)$
33. (a) $p = 20 + 3(h - 15)$ when $h > 15$; (b) $p = 20 + 2(h - 15)$
when $h > 15$; (c) When $h \le 15$ hours, $p = \$20$.   **35.** When
150 minutes are used   **37.** 10 tables   **39.** 1st   **41.** At $-40°$, $F = C$.
43. 4 days   **45.** 5 days   **47.** (a) $x + y = \$15,127$; (b) $x = y + 9127$;
(c) $x = \$12,127$, $y = \$3000$ **49.** (a) $T + PM = \$3222$;
(b) $PM - T = \$1350$; (c) $PM = \$2286$, $T = \$936$
51. $L_1 = 80$; $L_2 = 320$ **53.** (a) $5x = 4x + 500$; (b) 500 units
57. one   **59.** no   **61.** Inconsistent   **63.** Consistent; (2, 3)
65. Dependent   **67.** Inconsistent   **69.** 10 months   **71.** $x = 8$
73. $x = -1$

Exercises 7.3

1. (1, 2)   **3.** (0, 2)   **5.** Inconsistent   **7.** Inconsistent   **9.** $(10, -1)$
11. $(-26, 14)$ **13.** (6, 2)   **15.** $(2, -1)$ **17.** (3, 5)   **19.** $(-3, -2)$
21. (8, 6)   **23.** (1, 2)   **25.** (5, 3)   **27.** (4, 3)   **29.** Dependent
31. 0.8 lb Costa Rican; 0.2 lb Indian Mysore   **33.** 10 lb Oolong;
40 lb regular tea   **35.** $c = 1770$, $m = 665$ **37.** $c = 7$, $t = 31$
39. $c = 190$, $a = 340$ **41.** Tweedledee $120\frac{2}{3}$ lb; Tweedledum $119\frac{2}{3}$ lb
45. equivalent   **47.** $\left(\frac{41}{13}, \frac{7}{13}\right)$ **49.** (5, 1)   **51.** Inconsistent
53. $n + d = 300$ **55.** $4(x - y) = 48$ **57.** $m = n - 3$

Exercises 7.4

Translate This

1. K   **3.** A   **5.** B   **7.** F   **9.** J

1. 5 nickels; 20 dimes   **3.** 15 nickels; 5 dimes   **5.** 4 fives; 6 ones
7. 59; 43   **9.** 105; 21   **11.** Impossible   **13.** Pikes Peak = 14,110 ft;
Longs Peak = 14,255 ft   **15.** 1050 mi   **17.** Boat speed = 15 mi/hr;
current speed = 3 mi/hr   **19.** 20 mi   **21.** $5000 at 6%; $15,000 at 8%
23. $8000   **25.** Public: 11.7 million; private: 3.3 million
27. HBO: 14,353,650; Showtime: 14,346,350   **29.** $K = 96$, $A = 26.5$
31. $K = 140$, $A = 165$ **37.** $10d + 25q$ **39.** RT **41.** 20 nickels;
10 dimes   **43.** Wind speed = 50 mi/hr; plane speed = 350 mi/hr
45. Democratic: 59 million; Republican: 62 million

47. $x - 2y > 6$

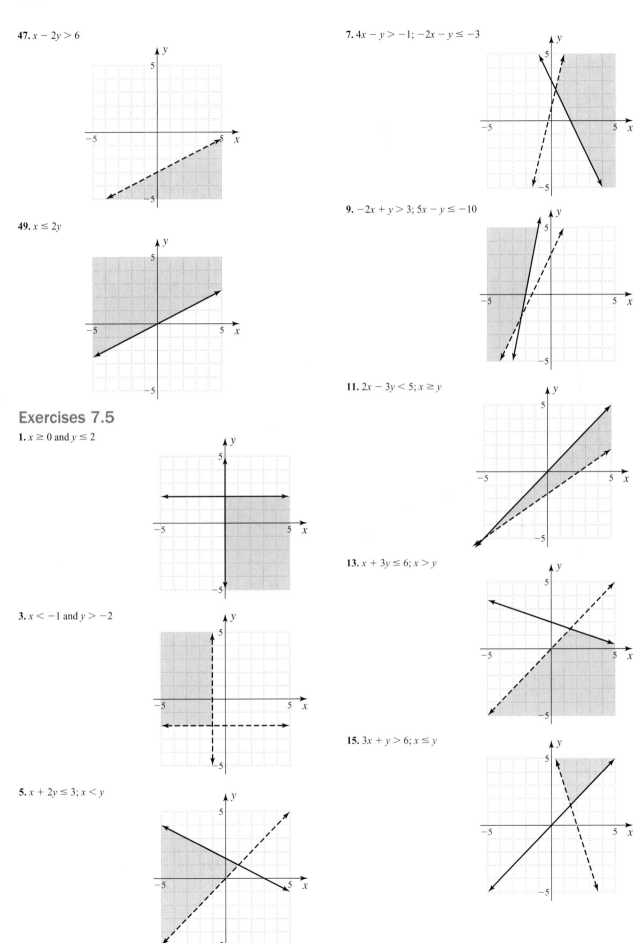

49. $x \leq 2y$

Exercises 7.5

1. $x \geq 0$ and $y \leq 2$

3. $x < -1$ and $y > -2$

5. $x + 2y \leq 3$; $x < y$

7. $4x - y > -1$; $-2x - y \leq -3$

9. $-2x + y > 3$; $5x - y \leq -10$

11. $2x - 3y < 5$; $x \geq y$

13. $x + 3y \leq 6$; $x > y$

15. $3x + y > 6$; $x \leq y$

17. For example, (1, 3), (2, 3)

19. & 21.

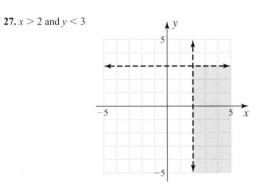

25. solid

27. $x > 2$ and $y < 3$

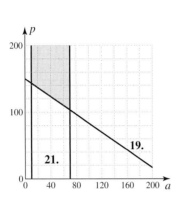

29. $3x - y < -1; x + 2y \le 2$

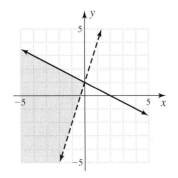

31. 16   **33.** 8

Review Exercises

1. [7.1A, B]
(a) Solution: (1, 2)
$2x + y = 4; y - 2x = 0$

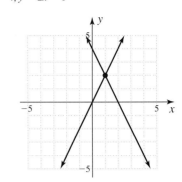

b. Solution: (2, 2)
$x + y = 4; y - x = 0$

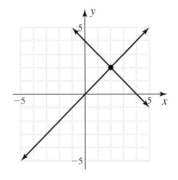

c. Solution: (1, 3)
$x + y = 4; y - 3x = 0$

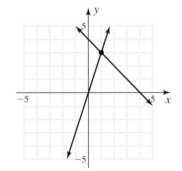

2. [7.1A, B]
(a) Inconsistent; no solution
$y - 3x = 3; 2y - 6x = 12$

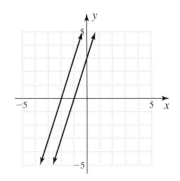

(b) Inconsistent; no solution
  $y - 2x = 2; 2y - 4x = 8$

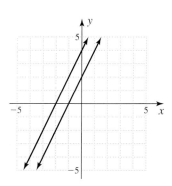

(c) Inconsistent; no solution
  $y - 3x = 6; 2y - 6x = 6$

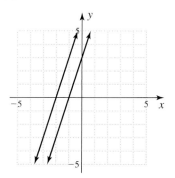

3. [7.2A, B] **(a)** Inconsistent; no solution; **(b)** Inconsistent; no solution; **(c)** $(1, 1)$ **4.** [7.2A, B] **(a)** Dependent; infinitely many solutions; **(b)** Dependent; infinitely many solutions; **(c)** Dependent; infinitely many solutions   **5.** [7.3A] **(a)** $(-1, 2)$; **(b)** $(2, -1)$; **(c)** $(-1, -2)$ **6.** [7.3] **(a)** Inconsistent; no solution; **(b)** Inconsistent; no solution; **(c)** Inconsistent; no solution
7. [7.3B] **(a)** Dependent; infinitely many solutions; **(b)** Dependent; infinitely many solutions; **(c)** Dependent; infinitely many solutions
8. [7.4A] **(a)** 20 nickels; 20 dimes; **(b)** 40 nickels; 10 dimes; **(c)** 50 nickels; 5 dimes   **9.** [7.4B] **(a)** 70; 110; **(b)** 60; 120; **(c)** 50; 130
10. [7.4C] **(a)** 550 mi/hr; **(b)** 520 mi/hr; **(c)** 500 mi/hr
11. [7.4D] **(a)** Bonds: $5000; CDs: $15,000; **(b)** Bonds: $17,000; CDs: $3000; **(c)** Bonds: $10,000; CDs: $10,000
12. [7.5] **(a)** $x > 4; y < -1$

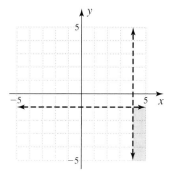

(b) $x + y > 3; x - y < 4$

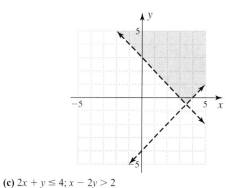

(c) $2x + y \le 4; x - 2y > 2$

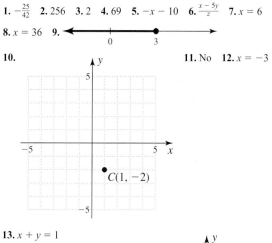

Cumulative Review Chapters 1–7

1. $-\frac{25}{42}$ **2.** 256   **3.** 2   **4.** 69   **5.** $-x - 10$ **6.** $\frac{x - 5y}{z}$ **7.** $x = 6$

8. $x = 36$ **9.**

10.   **11.** No   **12.** $x = -3$

13. $x + y = 1$

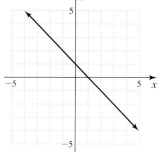

14. $2y + 8 = 0$

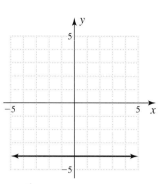

15. $-\frac{1}{7}$ **16.** 3   **17.** (1) and (2)   **18.** $\frac{y^9}{8x^{12}}$ **19.** 3.5×10^{-5}

20. 2.10×10^{-7} **21.** $36x^4 - 3x^2 + \frac{1}{16}$ **22.** $(2x^2 + x - 1)$ R 3

23. $(x - 6)(x - 5)$ **24.** $(3x - 4y)(5x - 3y)$

25. $(9x + 8y)(9x - 8y)$ **26.** $-3x^2(x + 2)(x - 2)$

27. $4x(x + 1)(x - 2)$ **28.** $(4x + 3)(x + 1)$ **29.** $k(5x - 3)^2$

30. $x = -3; x = \frac{5}{3}$ **31.** $\frac{20xy}{15y^2}$ **32.** $-3(x + y)$ **33.** $-(x + 2)$

34. $\frac{x - 2}{x + 5}$ **35.** $\frac{1}{4 - x}$ **36.** $\frac{6}{x - 9}$ **37.** $\frac{-2x - 5}{(x + 3)(x + 2)(x - 2)}$

38. $\frac{39}{2}$ **39.** $x = -5$ **40.** $x = -3$ **41.** No solution   **42.** $x = -6$

43. 13 gal   **44.** $x = \frac{31}{5}$ **45.** $2\frac{2}{9}$ hr   **46.** $y + 2 = -4(x + 6)$ or
$y = -4x - 26$ **47.** $y = 5x + 2$

48. $4x - 3y > -12$

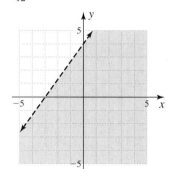

49. $-y \le -3x + 6$

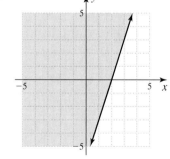

50. $\frac{1}{50}$ **51.** 6240

52. $x + 2y = 6; 2y - x = -2$

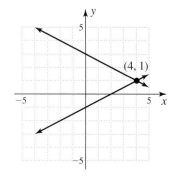

(4, 1)

53. $y - x = -1; 2y - 2x = -4$. No solution (inconsistent)

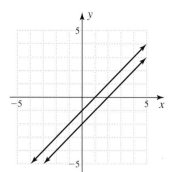

54. Inconsistent; no solution
55. Dependent; infinitely many solutions   **56.** (1, 1)
57. Inconsistent; no solution
58. Dependent; infinitely many solutions
59. 12 nickels and 36 dimes
60. 35 and 130

Chapter 8

Exercises 8.1

1. 5   **3.** -3 **5.** $\frac{4}{3}$ **7.** $-\frac{2}{9}$ **9.** $\frac{5}{9}$ and $-\frac{5}{9}$ **11.** $\frac{7}{10}$ and $-\frac{7}{10}$ **13.** 5
15. 11   **17.** $x^2 + 1$ **19.** $3y^2 + 7$ **21.** 6; rational   **23.** Not a real
number   **25.** -8; rational   **27.** $\frac{4}{3}$; rational   **29.** -2.449489743;
irrational   **31.** -1.414213562; irrational   **33.** 3   **35.** -3 **37.** -4
39. 5   **41.** 10 sec   **43.** $\sqrt{20}$ sec ≈ 4.5 sec   **45.** 5 in.   **47.** 244 mi
49. 11 ft $\times$ 11 ft   **51.** 30 mi/hr   **53.** 72 ft/sec   **55. (a)** $5\frac{1}{11}$; **(b)** $5\frac{3}{11}$;
(c) $5\frac{5}{11}$ **57.** One   **59.** One   **61.** One   **63.** $a \ge 0$ **65.** $b^2 = a$
67. 0   **69.** a **71.** $\frac{4}{7}$ **73.** 12   **75.** 17   **77.** $-\frac{7}{11}$; rational
79. 3.872983346; irrational   **81.** -5 **83.** 3   **85.** 2   **87.** 7

Exercises 8.2

1. $3\sqrt{5}$ **3.** $5\sqrt{5}$ **5.** $6\sqrt{5}$ **7.** $10\sqrt{2}$ **9.** $8\sqrt{6}$ **11.** $5\sqrt{5}$ **13.** $10\sqrt{6}$
15. 19   **17.** $10\sqrt{7}$ **19.** $12\sqrt{3}$ **21.** $\sqrt{15}$ **23.** 9   **25.** 7   **27.** $\sqrt{3x}$
29. $6a$ **31.** $\frac{\sqrt{2}}{5}$ **33.** 2   **35.** 3   **37.** $5\sqrt{3}$ **39.** 12   **41.** $10a$ **43.** $7a^2$
45. $-4a^3\sqrt{2}$ **47.** $m^6\sqrt{m}$ **49.** $-3m^5\sqrt[3]{3m}$ **51.** $2\sqrt[3]{5}$ **53.** $-2\sqrt[3]{2}$
55. $\frac{2}{3}$ **57.** $2\sqrt[4]{3}$ **59.** $\frac{4}{3}$ **61.** 9 mi   **63.** 50 mi   **67. (a)** Irrational;
(b) Yes   **69.** $\frac{\sqrt{a}}{\sqrt{b}}$ **71.** $\sqrt[n]{a} \cdot \sqrt[n]{b}$ **73.** $3\sqrt[3]{5}$ **75.** $\frac{3}{4}$ **77.** $10x^3$ **79.** $\sqrt{2}$
81. $\frac{2}{3}$ **83.** 12   **85.** $12x$ **87.** $14x^3$

Exercises 8.3

1. $10\sqrt{7}$ **3.** $5\sqrt{13}$ **5.** $3\sqrt{2}$ **7.** $4\sqrt{2}$ **9.** $27\sqrt{3}$ **11.** $-7\sqrt{7}$
13. $29\sqrt{3}$ **15.** $-8\sqrt{5}$ **17.** $10\sqrt{2} - \sqrt{30}$ **19.** $2\sqrt{21} + \sqrt{30}$
21. $3 - \sqrt{6}$ **23.** $\sqrt{10} + 5$ **25.** $2\sqrt{3} - 3\sqrt{2}$ **27.** $2\sqrt{2} - 10$
29. $2\sqrt{3} - 3\sqrt{2}$ **31.** $\frac{\sqrt{6}}{2}$ **33.** $-2\sqrt{5}$ **35.** 2   **37.** $\frac{-\sqrt{10}}{5}$
39. $\frac{1}{2}$ **41.** $\frac{x\sqrt{2}}{2}$ **43.** $\frac{a\sqrt{b}}{b}$ **45.** $\frac{\sqrt{30}}{8}$ **47.** $\frac{x\sqrt{2}}{2}$ **49.** $\frac{x\sqrt{5}}{10}$ **51.** 1 hr
53. $r = \frac{\sqrt{\pi S}}{2\pi}$ **59.** exactly   **61.** $13\sqrt{3}$ **63.** $6\sqrt{2}$ **65.** $12\sqrt{2}$
67. $5\sqrt{3} - 3$ **69.** $\frac{3\sqrt{7}}{7}$ **71.** $\frac{\sqrt{2x}}{2}$ **73.** $x^2 - 9$ **75.** $2x + 4$

Exercises 8.4

1. 16   **3.** 13   **5.** -4 **7.** 5   **9.** $18\sqrt{10}$ **11.** $12\sqrt{11}$ **13.** $3\sqrt[3]{2} - 2$
15. $\sqrt[3]{2}$ **17.** $3\sqrt{x}$ **19.** $\frac{9y}{4}$ **21.** $\frac{8ab\sqrt{3a}}{3}$ **23.** $\frac{2bc^3\sqrt[3]{b^2}}{3}$ **25.** $5\sqrt[3]{2}$
27. $2\sqrt[3]{3}$ **29.** $2\sqrt{15} - 117$ **31.** $6\sqrt{6} + 29$ **33.** $59 - 20\sqrt{6}$ **35.** 5
37. $3\sqrt{2} - 3$ **39.** $\frac{2\sqrt{7} + 2}{3}$ **41.** $2\sqrt{2} - \sqrt{6}$ **43.** $2\sqrt{5} + \sqrt{15}$
45. $-\sqrt{10} + \sqrt{15}$ **47.** $\frac{\sqrt{15} + \sqrt{6}}{3}$ **49.** $5 + 2\sqrt{6}$ **51.** -2 **53.** $\frac{-4}{3}$
55. $\frac{1 + \sqrt{3}}{3}$ **57.** $\frac{-1 + \sqrt{23}}{2}$ **59.** $\frac{-2 + \sqrt{10}}{3}$ **61.** $\frac{-4 + \sqrt{7}}{3}$ **63.** $\frac{-3 + 3\sqrt{3}}{2}$

65. $\frac{2}{\sqrt{5}+1}$ **67.** $\frac{\sqrt{5}+1}{2}$ **69.** The golden mean equals its reciprocal
plus 1; yes.   **71.** $\frac{x+1}{x}$ **73. (a)** $x+1=x^2$; **(b)** $x^2-x-1=0$
75. $\sqrt{\frac{2(P_2-P_1)}{\gamma P_1}}$ **79.** No; $\sqrt{2}-1$ **81.** a^2b-c^2d **83.** $-2+\sqrt{7}$
85. $\frac{6-3\sqrt{2}}{2}$ **87.** $\frac{3\sqrt[3]{4}}{2}$ **89.** $4\sqrt[3]{2}$ **91.** $2\sqrt[3]{a^2}$ **93.** $-13-\sqrt{15}$
95. $x-1$ **97.** $4x$ **99.** x^2+2x+1 **101.** $x(x-3)$
103. $(x-2)(x-1)$

Exercises 8.5

1. $x=16$ **3.** No solution   **5.** $y=4$ **7.** $y=8$ **9.** $x=8$ **11.** $x=0$
13. $x=5$ **15.** $y=20$ **17.** $y=25$ **19.** $y=9$ or $y=1$ **21.** $y=6$
23. $x=5$ **25.** $x=4$ **27.** $x=\frac{11}{3}$ **29.** $y=5$ **31.** $S=50.24$ ft^2
33. 144 ft   **35.** 3.24 ft   **37.** 2000   **39.** $r=100$ ft   **41.** 31 m
43. $x=27$ **45.** $x=7$ **47.** $x=16$ **49.** No solution   **55.** among
57. 10 thousand   **59.** $y=2$ **61.** $x=2$ **63.** $x=7$ **65.** No solution
67. $x=2$ or $x=3$ **69.** 7   **71.** $\frac{\sqrt{5}}{4}$

Review Exercises

1. [8.1A, C] **(a)** 9; **(b)** Not a real number; **(c)** $\frac{6}{5}$ **2.** [8.1A, C] **(a)** -6;
(b) $-\frac{8}{5}$; **(c)** Not a real number   **3.** [8.1B] **(a)** 8; **(b)** 25; **(c)** 17
4. [8.1B] **(a)** 36; **(b)** 17; **(c)** 64   **5.** [8.1B] **(a)** x^2+1; **(b)** x^2+4;
(c) x^2+5 **6.** [8.1C] **(a)** Irrational; 3.3166; **(b)** Rational; -5;
(c) Not a real number   **7.** [8.1C] **(a)** Rational; **(b)** Rational;
(c) Not a real number   **8.** [8.1D] **(a)** 4; **(b)** -2; **(c)** -3
9. [8.1D] **(a)** 2; **(b)** -2; **(c)** Not a real number   **10.** [8.1E] **(a)** $2\frac{3}{4}$ sec;
(b) 3 sec; **(c)** $3\frac{1}{4}$ sec   **11.** [8.2A] **(a)** $4\sqrt{2}$; **(b)** $4\sqrt{3}$; **(c)** 14
12. [8.2A] **(a)** $\sqrt{21}$; **(b)** 6; **(c)** $\sqrt{5y}$ **13.** [8.2B] **(a)** $\frac{\sqrt{3}}{4}$; **(b)** $\frac{\sqrt{5}}{6}$; **(c)** $\frac{3}{2}$
14. [8.2B] **(a)** 2; **(b)** $\sqrt{7}$; **(c)** $3\sqrt{5}$ **15.** [8.2C] **(a)** $6x$; **(b)** $10y^2$; **(c)** $9n^4$
16. [8.2C] **(a)** $6y^5\sqrt{2}$; **(b)** $7z^4\sqrt{3}$; **(c)** $4x^6\sqrt{3}$ **17.** [8.2C] **(a)** $y^7\sqrt{y}$;
(b) $y^6\sqrt{y}$; **(c)** $5n^3\sqrt{2n}$ **18.** [8.2D] **(a)** $2\sqrt[3]{3}$; **(b)** $\frac{2}{3}$; **(c)** $-\frac{5}{4}$
19. [8.2D] **(a)** 3; **(b)** $2\sqrt[4]{3}$; **(c)** $2\sqrt[4]{5}$ **20.** [8.3A] **(a)** $15\sqrt{3}$; **(b)** $9\sqrt{2}$;
(c) $6\sqrt{3}$ **21.** [8.3A] **(a)** $3\sqrt{11}$; **(b)** $\sqrt{2}$; **(c)** $\sqrt{3}$
22. [8.3B] **(a)** $2\sqrt{15}-\sqrt{6}$; **(b)** $5-\sqrt{15}$; **(c)** $7-7\sqrt{14}$
23. [8.3C] **(a)** $\frac{\sqrt{10}}{4}$; **(b)** $\frac{x\sqrt{2}}{10}$; **(c)** $\frac{y\sqrt{3}}{9}$ **24.** [8.4A] **(a)** 3; **(b)** 3; **(c)** 4
25. [8.4A] **(a)** $7\sqrt{15}$; **(b)** $6\sqrt{6}$; **(c)** $7\sqrt{14}$
26. [8.4A] **(a)** $3\sqrt[3]{3}$; **(b)** 3; **(c)** 2   **27.** [8.4A] **(a)** $\frac{7\sqrt[3]{2}}{2}$; **(b)** $\frac{5\sqrt[3]{3}}{3}$; **(c)** $\frac{9\sqrt[3]{5}}{5}$
28. [8.4A] **(a)** $-27-2\sqrt{6}$; **(b)** $-23+\sqrt{35}$ **29.** [8.4A] **(a)** -5;
(b) -34 **30.** [8.4B] **(a)** $\frac{3\sqrt{3}-3}{2}$; **(b)** $5\sqrt{2}+5$
31. [8.4B] **(a)** $7\sqrt{3}+7\sqrt{2}$; **(b)** $\frac{2\sqrt{5}+2\sqrt{2}}{3}$ **32.** [8.4C] **(a)** $-4+\sqrt{2}$;
(b) $\frac{-8+\sqrt{3}}{2}$ **33.** [8.5A] **(a)** $x=7$; **(b)** No solution   **34.** [8.5A] **(a)** $x=4$;
(b) $x=6$ **35.** [8.5A] **(a)** $x=5$; **(b)** $x=7$ **36.** [8.5B] **(a)** $y=4$;
(b) $y=0$ **37.** [8.5B] **(a)** $y=1$; **(b)** $y=0$
38. [8.5C] **(a)** 40 thousand or 40,000; **(b)** 240 thousand or 240,000

Cumulative Review Chapters 1–8

1. $-\frac{25}{72}$ **2.** 81   **3.** $\frac{2}{7}$ **4.** 47   **5.** $-x-10$ **6.** $\frac{m+3n}{p}$ **7.** $x=16$
8. $x=6$ **9.**

10.
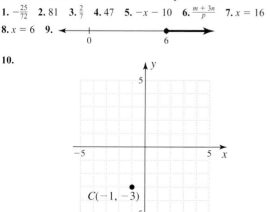

11. No   **12.** $x=1$
13. $5x+y=5$

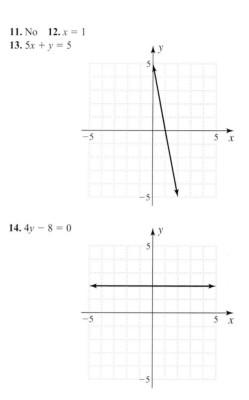

14. $4y-8=0$

15. $-\frac{1}{2}$ **16.** 1   **17.** (1) and (3)   **18.** $\frac{y^{12}}{16x^{16}}$ **19.** 2.5×10^{-7}
20. 1.7×10^{-8} **21.** $16x^4-4x^2+\frac{1}{4}$ **22.** $(2x^2-x-2)$ R 3
23. $(x-1)(x-3)$ **24.** $(3x-5y)(3x-4y)$ **25.** $(2x+5y)(2x-5y)$
26. $-5x^2(x+1)(x-1)$ **27.** $4x(x+1)(x-3)$ **28.** $(3x+4)(x+3)$
29. $k(4x+1)^2$ **30.** $x=-5$; $x=\frac{3}{4}$ **31.** $\frac{6xy}{15y^2}$ **32.** $-4(x+y)$
33. $-(x+7)$ **34.** $\frac{x+4}{x+8}$ **35.** $\frac{1}{5-x}$ **36.** $\frac{4}{x+5}$ **37.** $\frac{-2x-9}{(x+5)(x+4)(x-4)}$
38. $\frac{14}{15}$ **39.** $x=2$ **40.** $x=-3$ **41.** No solution   **42.** $x=-7$
43. 25 gal   **44.** $x=\frac{45}{4}$ **45.** $1\frac{5}{7}$ hr   **46.** $y-4=5(x-6)$ or
$y=5x-26$ **47.** $y=4x+3$
48. $x-5y<-5$

49. $-y\geq-5x-5$

50. $\frac{1}{80}$ **51.** 10,800
52. $x + 4y = 16$; $4y - x = 12$

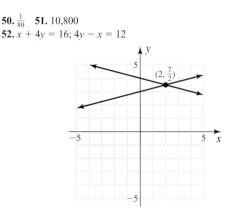

53. $y + 3x = -3$; $2y + 6x = -12$; No solution

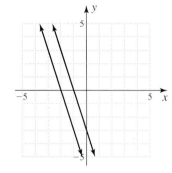

54. Inconsistent; no solution **55.** Dependent; infinitely many solutions
56. $(-1, -2)$ **57.** Inconsistent; no solution
58. Dependent; infinitely many solutions **59.** 10 nickels and 30 dimes
60. 65 and 150 **61.** -4 **62.** -3 **63.** $\frac{\sqrt{21}}{27}$ **64.** $4a^2b^3\sqrt[3]{2a^2}$
65. $6\sqrt{3}$ **66.** $2x - 3\sqrt[3]{6x^2}$ **67.** $\frac{\sqrt{6p}}{2p}$ **68.** $420 + 74\sqrt{35}$
69. $\frac{x + \sqrt{5x}}{x - 5}$ **70.** $2 + \sqrt{2}$ **71.** No real-number solution
72. $x = 6$; $x = 7$

Chapter 9

Exercises 9.1

1. $x = \pm 10$ **3.** $x = 0$ **5.** No real-number solution **7.** $x = \pm\sqrt{7}$
9. $x = \pm 3$ **11.** $x = \pm\sqrt{3}$ **13.** $x = \pm\frac{1}{5}$ **15.** $x = \pm\frac{7}{10}$
17. $y = \pm\frac{\sqrt{17}}{5}$ **19.** No real-number solution **21.** $x = 8$ or $x = -10$
23. $x = 8$ or $x = -4$ **25.** No real-number solution **27.** $x = 18$ or
$x = 0$ **29.** $x = 0$ or $x = -8$ **31.** $x = -\frac{4}{5}$ or $x = -\frac{6}{5}$ **33.** $x = \frac{25}{6}$
or $x = \frac{11}{6}$ **35.** $x = \frac{3}{2}$ or $x = -\frac{7}{2}$ **37.** $x = 1 \pm \frac{\sqrt{5}}{3} = \frac{3 \pm \sqrt{5}}{3}$
39. No real-number solution **41.** $x = \pm\frac{1}{9}$ **43.** $x = \pm\frac{1}{4}$
45. $x = \pm 2$ **47.** $v = 2$ or $v = -4$ **49.** $x = \frac{5}{2}$ or $x = -\frac{1}{2}$
51. $y = \frac{3}{2} \pm \sqrt{2} = \frac{3 \pm 2\sqrt{2}}{2}$ **53.** $x = \frac{3}{2} \pm \sqrt{2} = \frac{3 \pm 2\sqrt{2}}{2}$
55. $x = -2 \pm 6\sqrt{2}$ **57.** $x = 3 \pm 6\sqrt{5}$ **59.** No real-number solution
61. 24 mph; no **63.** 15 m/sec **65.** 5 in. **67.** $3\frac{1}{2}$ ft **71.** $A < 0$
73. $A > 0$ **75.** Rational **77.** 9 **79.** $\pm A$ **81.** $x = 1 \pm 2\sqrt{2}$
83. $x = 1$ or $x = -5$ **85.** $x = \pm\frac{\sqrt{3}}{7}$ **87.** $x = \pm\frac{4}{3}$ **89.** $x = \pm 1$
91. $x^2 + 14x + 49$ **93.** $x^2 - 6x + 9$

Exercises 9.2

1. 81 **3.** 64 **5.** $\frac{49}{4}$ **7.** $\frac{9}{4}$ **9.** $\frac{1}{4}$ **11.** 4; $x + 2$ **13.** $\frac{9}{4}$; $x + \frac{3}{2}$
15. 9; $x - 3$ **17.** $\frac{25}{4}$; $x - \frac{5}{2}$ **19.** $\frac{9}{16}$; $x - \frac{3}{4}$ **21.** $x = -2 \pm \sqrt{3}$

23. $x = -\frac{1}{2} \pm \frac{\sqrt{5}}{2} = \frac{-1 \pm \sqrt{5}}{2}$ **25.** $x = -\frac{3}{2} \pm \frac{\sqrt{13}}{2} = \frac{-3 \pm \sqrt{13}}{2}$
27. $x = \frac{3}{2} \pm \frac{\sqrt{21}}{2} = \frac{3 \pm \sqrt{21}}{2}$ **29.** $x = \frac{1}{2}$ or $x = -\frac{3}{2}$
31. $x = 2 \pm \frac{\sqrt{31}}{2} = \frac{4 \pm \sqrt{31}}{2}$ **33.** $x = \frac{1}{2} \pm \sqrt{2} = \frac{1 \pm 2\sqrt{2}}{2}$
35. $x = 1 \pm \frac{\sqrt{2}}{2} = \frac{2 \pm \sqrt{2}}{2}$ **37.** $x = 1$ or $x = -2$
39. $x = -\frac{5}{2} \pm \frac{3\sqrt{3}}{2} = \frac{-5 \pm 3\sqrt{3}}{2}$ **41.** $x^2 + 6x + 9 = (x + 3)^2$
43. $25x^2 + 70x + 49 = (5x + 7)^2$ **45.** $x = 1$ and $x = -9$
47. (a) 2 (thousand) = 2000; **(b)** $2 **49.** 3 days **51.** No; x is cubed
in the equation. **53.** $x^2 - 6x + 9$ **55.** divide **57.** $\frac{1}{16}$
59. $\frac{25}{4}$; $x + \frac{5}{2}$ **61.** $x = -3 \pm \frac{\sqrt{3}}{2} = \frac{-6 \pm \sqrt{3}}{2}$ **63.** $x^2 - 2x - 2 = 0$
65. $10x^2 - 5x + 12 = 0$ **67.** $9x^2 - 4x - 2 = 0$

Exercises 9.3

1. $x = -2$ or $x = -1$ **3.** $x = -2$ or $x = 1$ **5.** $x = \frac{-1 \pm \sqrt{17}}{4}$
7. $x = \frac{2}{3}$ or $x = -1$ **9.** $x = -\frac{3}{2}$ or $x = -2$ **11.** $x = \frac{5}{7}$ or $x = 1$
13. $x = \frac{11 \pm \sqrt{41}}{10}$ **15.** $x = \frac{3}{2}$ or $x = 1$ **17.** $x = \frac{3 \pm \sqrt{5}}{4}$ **19.** $x = -1$
21. $x = \frac{3 \pm \sqrt{5}}{4}$ **23.** $x = \frac{-3 \pm \sqrt{21}}{6}$ **25.** $x = 2$ or $x = -2$
27. $x = \pm 3\sqrt{5}$ **29.** $x = \pm\frac{2\sqrt{3}}{3}$ **31.** 40 **33.** 10 **35.** Multiply each
term by $4a$. **37.** Add b^2 on both sides. **39.** Take the square root of
each side. **41.** Divide each side by $2a$. **43.** $x = \pm\sqrt{17}$
45. $x = -2$ **47.** $y = 0$ or $y = 1$ **49.** $x = -3$ or $x = -2$ **51.** $z = 0$
or $z = -6$ **53.** $y = \frac{1}{3}$ or $y = -3$ **55.** No real-number solution
57. One; rational **59.** None **61.** $\frac{-b \pm \sqrt{b^2 - 4ac}}{2a}$
63. No real-number solution **65.** $x = 2 \pm 2\sqrt{3}$
67. $x = 2$ or $x = -\frac{1}{2}$ **69.** 4 **71.** 1

Exercises 9.4

1. $y = 2x^2$

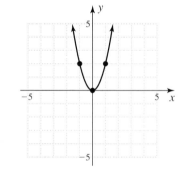

3. $y = 2x^2 - 1$

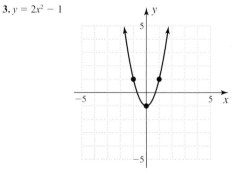

5. $y = -2x^2 + 2$

7. $y = -2x^2 - 2$

9. $y = (x - 2)^2$

11. $y = (x - 2)^2 - 2$

13. $y = -(x - 2)^2$

15. $y = -(x - 2)^2 - 2$

17. $y = x^2 + 4x + 3$ x-int: $(-3, 0)$, $(-1, 0)$; y-int: $(0, 3)$; $V(-2, -1)$

19. $y = x^2 + 2x - 3$ x-int: $(-3, 0)$, $(1, 0)$; y-int: $(0, -3)$; $V(-1, -4)$

21. $y = -x^2 + 4x - 3$ x-int: $(3, 0)$, $(1, 0)$; y-int: $(0, -3)$; $V(2, 1)$

23. About 40 hr   **25.** About 60 hr   **27.** $x = 4000$; $P = \$11,000$
29. \$25 (thousand) $= \$25,000$ **31.** \$40   **33.** \$400
35. The parabola opens upward.   **37.** The graph "Stretches" upward.
39. The graph is shifted upward; the graph is shifted downward.
41. $a > 0$

43. $y = -(x - 2)^2 - 1$

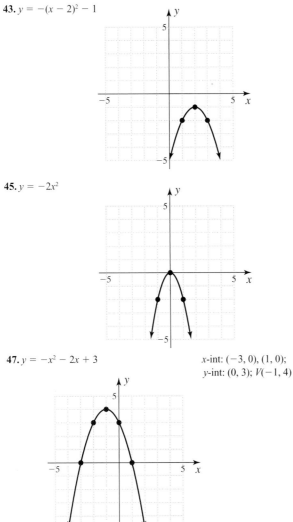

45. $y = -2x^2$

47. $y = -x^2 - 2x + 3$

x-int: $(-3, 0)$, $(1, 0)$;
y-int: $(0, 3)$; $V(-1, 4)$

49. 25 **51.** 40,000

Exercises 9.5

1. $b = 5$ **3.** $b = 10\sqrt{2}$ **5.** $c = \sqrt{15}$ **7.** $c = 3$ **9.** $b = 2$ **11.** 50 ft
13. 100 mi/hr **15.** 8 sec **17.** 1 sec **19.** 3000 units **21.** 3 ft by 4 ft
23. 10 cm; 24 cm; 26 cm **25.** (a) $q = 10$; (b) 180
27. (a) 4 rings (b) 14 **29.** $c = \sqrt{\frac{E}{m}} = \frac{\sqrt{Em}}{m}$ **31.** $r = \sqrt{\frac{GMm}{F}} = \frac{\sqrt{FGMm}}{F}$
33. $p = \sqrt{\frac{I}{k}} = \frac{\sqrt{Ik}}{k}$ **37.** right **39.** 1 or 6 (hundred) **41.** 50 mi/hr
43. -1

Exercises 9.6

1. $D = \{1, 2, 3\}$; $R = \{2, 3, 4\}$ **3.** $D = \{1, 2, 3\}$; $R = \{1, 2, 3\}$
5. $D =$ All real numbers; $R =$ All real numbers
7. $D =$ All real numbers; $R =$ All real numbers
9. $D =$ All real numbers; $R =$ All nonnegative real numbers
11. $D =$ All nonnegative real numbers; $R =$ All real numbers
13. $D =$ All real numbers except 3; $R =$ All real numbers except 0
15. $D = \{-1, 0, 1, 2\}$; $R = \{-2, 0, 2, 4\}$; $(-1, -2), (0, 0), (1, 2), (2, 4)$
17. $D = \{0, 1, 2, 3, 4\}$; $R = \{-3, -1, 1, 3, 5\}$; $(0, -3), (1, -1),$
$(2, 1), (3, 3), (4, 5)$ **19.** $D = \{0, 1, 4, 9, 16, 25\}$; $R = \{0, 1, 2, 3, 4, 5\}$;
$(0, 0), (1, 1), (4, 2), (9, 3), (16, 4), (25, 5)$ **21.** $D = \{1, 2, 3\}$;
$R = \{2, 3, 4\}$; $(1, 2), (1, 3), (1, 4), (2, 3), (2, 4), (3, 4)$

23. A function; one y-value for each x-value **25.** A function; one
y-value for each x-value **27.** A function; one y-value for each
x-value **29.** Not a function; two y-values for each positive
x-value **31.** (a) $f(0) = 1$; (b) $f(2) = 7$; (c) $f(-2) = -5$
33. (a) $F(1) = 0$; (b) $F(5) = 2$; (c) $F(26) = 5$
35. (a) $f(x + h) = 3x + 3h + 1$; (b) $f(x + h) - f(x) = 3h$;
(c) $\frac{f(x + h) - f(x)}{h} = 3$, $h \neq 0$ **37.** $y = g(x) = x^2$; $\left(\frac{1}{4}, \frac{1}{16}\right)$, $(2.1, 4.41)$,
$(\pm 8, 64)$ **39.** (a) $g(0) = 1$; (b) $g(-2) = -5$; (c) $g(2) = 15$
41. (a) 140; (b) 130 **43.** (a) 160 lb; (b) 78 in. **45.** (a) 639 lb/ft²;
(b) 6390 lb/ft² **47.** (a) 144 ft; (b) 400 ft **49.** 60,000 joules
51. A function **53.** A function **55.** No **59.** ordered **61.** range
63. $D = \{-5, -6, -7\}$; $R = \{5, 6, 7\}$
65. $D =$ All real numbers except 3; $R =$ All real numbers except 0
67. A function; one y-value for each x-value **69.** A function; one
y-value for each x-value **71.** Not a function; two y-values for each
positive x-value **73.** $g(-1) = -2$

Review Exercises

1. [9.1A] (a) $x = \pm 1$; (b) $x = \pm 10$; (c) $x = \pm 9$
2. [9.1A] (a) $x = \pm\frac{5}{4}$; (b) $x = \pm\frac{3}{5}$; (c) $x = \pm\frac{5}{8}$
3. [9.1A] (a) No real-number solution; (b) No real-number solution;
(c) No real-number solution
4. [9.1B] (a) $x = -1 \pm \frac{\sqrt{3}}{7} = \frac{-7 \pm \sqrt{3}}{7}$;
(b) $x = -2 \pm \frac{\sqrt{2}}{5} = \frac{-10 \pm \sqrt{2}}{5}$; (c) $x = -1 \pm \frac{\sqrt{5}}{4} = \frac{-4 \pm \sqrt{5}}{4}$
5. [9.2A] (a) 9; (b) 49; (c) 36 **6.** [9.2A] (a) 9; $x - 3$; (b) 25; $x - 5$;
(c) 36; $x - 6$ **7.** [9.2A] (a) 7; (b) 6; (c) 5 **8.** [9.2A] (a) 4; (b) 9;
(c) 36 **9.** [9.3B] (a) $x = \frac{-e \pm \sqrt{e^2 - 4df}}{2d}$; (b) $x = \frac{-h \pm \sqrt{h^2 - 4gi}}{2g}$;
(c) $x = \frac{-k \pm \sqrt{k^2 - 4jm}}{2j}$
10. [9.3B] (a) $x = 1$ or $x = -\frac{1}{2}$; (b) $x = \frac{1 \pm \sqrt{11}}{2}$; (c) $x = \frac{3 \pm \sqrt{33}}{4}$
11. [9.3B] (a) $x = \frac{1 \pm \sqrt{13}}{6}$; (b) $x = \frac{1 \pm \sqrt{7}}{3}$; (c) $x = \frac{3 \pm \sqrt{33}}{6}$
12. [9.3B] (a) $x = 0$ or $x = 9$; (b) $x = 0$ or $x = 4$; (c) $x = 0$ or $x = 25$
13. [9.3B] (a) $x = \frac{9 \pm \sqrt{65}}{2}$; (b) $x = -\frac{1}{2}$ or $x = 3$;
(c) No real-number solution
14. [9.4A]

(a) $y = x^2 + 1$;
(b) $y = x^2 + 2$;
(c) $y = x^2 + 3$

15. [9.4A]

(a) $y = -x^2 - 1$;
(b) $y = -x^2 - 2$;
(c) $y = -x^2 - 3$

16. [9.4A]

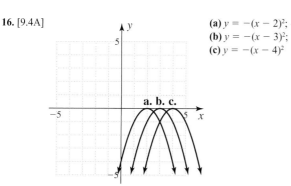

(a) $y = -(x - 2)^2$;
(b) $y = -(x - 3)^2$;
(c) $y = -(x - 4)^2$

13. $2x + y = 4$

17. [9.4A]

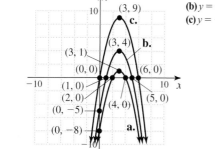

(a) $y = (x - 2)^2 + 1$;
(b) $y = (x - 2)^2 + 2$;
(c) $y = (x - 2)^2 + 3$

14. $2y - 8 = 0$

18. [9.4B]

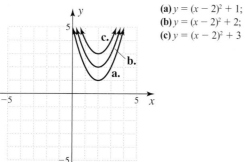

(a) $y = -x^2 + 6x - 8$
(b) $y = -x^2 + 6x - 5$
(c) $y = -x^2 + 6x$

15. $\frac{14}{11}$ **16.** 2   **17.** (2) and (3)   **18.** $\frac{81x^{12}}{y^{20}}$ **19.** 5.0×10^{-3}

20. 1.5×10^{-7} **21.** $4x^4 - \frac{4}{5}x^2 + \frac{1}{25}$ **22.** $(2x^2 - x - 1)$ R 1

23. $(x - 5)(x - 3)$ **24.** $(4x - 3y)(3x - 4y)$ **25.** $(5x + 6y)(5x - 6y)$

26. $-5x^2(x + 4)(x - 4)$ **27.** $3x(x + 2)(x - 3)$ **28.** $(x + 3)(4x + 5)$

29. $k(5x + 2)^2$ **30.** $x = -5; x = \frac{2}{3}$ **31.** $\frac{12xy^2}{10y^3}$ **32.** $-(x + y)$

33. $-(x + 3)$ **34.** $\frac{x-3}{x+4}$ **35.** $\frac{1}{2-x}$ **36.** $\frac{3}{x+1}$

37. $\frac{-2x - 15}{(x + 8)(x + 7)(x - 7)}$ **38.** 6   **39.** $x = 9$ **40.** $x = -5$

41. No solution   **42.** $x = -6$ **43.** 29 gal   **44.** $x = \frac{64}{5}$ **45.** $2\frac{2}{9}$ hr

46. $y + 4 = 6(x + 2)$ or $y = 6x + 8$ **47.** $y = -2x - 5$

48. $3x - 2y < -6$

19. [9.5A] **(a)** 13 in.; **(b)** $\sqrt{13}$ in.; **(c)** $\sqrt{41}$ in.   **20.** [9.5B] **(a)** 5 sec;
(b) 7 sec; **(c)** 8 sec   **21.** [9.6A] **(a)** $D = \{-3, -4, -5\}$; $R = \{1, 2\}$;
(b) $D = \{2, -1\}$; $R = \{-4, 3, 4\}$; **(c)** $D = \{-1\}$; $R = \{2, 3, 4\}$
22. [9.6A] **(a)** D = All real numbers; R = All real numbers;
(b) D = All real numbers; R = All real numbers; **(c)** D = All real
numbers; R = All nonnegative real numbers
23. [9.6B] **(a)** A function; **(b)** Not a function; **(c)** Not a function
24. [9.6C] **(a)** $f(2) = 1$; **(b)** $f(-2) = -19$; **(c)** $f(1) = -1$
25. [9.6D] **(a)** $22; **(b)** $19; **(c)** $16

Cumulative Review Chapters 1–9

1. $-\frac{29}{56}$ **2.** 256   **3.** $\frac{9}{5}$ **4.** 14   **5.** $-x - 10$ **6.** $\frac{x - 5y}{z}$ **7.** $x = 3$

8. $x = 72$ **9.**

10.

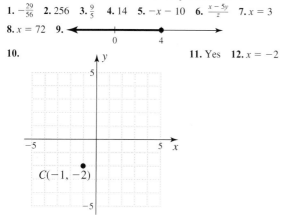

11. Yes   **12.** $x = -2$

49. $-y \geq -5x + 5$

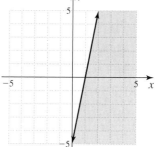

50. $\frac{1}{70}$ **51.** 13,720

52. $x + 3y = -12; 2y - x = -2$

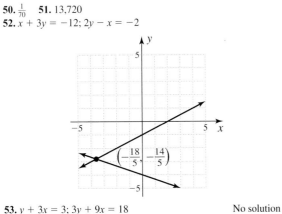

$\left(-\frac{18}{5}, -\frac{14}{5}\right)$

53. $y + 3x = 3; 3y + 9x = 18$ No solution

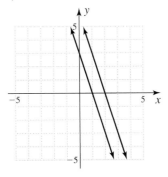

54. Inconsistent; no solution   **55.** Dependent; infinitely many solutions
56. $(-3, -3)$ **57.** Inconsistent; no solution   **58.** Dependent; infinitely
many solutions   **59.** 10 nickels and 20 dimes   **60.** 30 and 140
61. 3   **62.** 5   **63.** $\frac{\sqrt{15}}{27}$ **64.** $\frac{\sqrt{14j}}{7j}$ **65.** $7\sqrt{2}$ **66.** $3x - 2\sqrt[3]{6x^2}$
67. $280 + 46\sqrt{35}$ **68.** $5 - \sqrt{2}$ **69.** No real-number solution
70. $x = 4; x = 5$ **71.** $x = \frac{2}{3}; x = -\frac{2}{3}$ **72.** No real-number solution
73. $x = \frac{24 \pm \sqrt{7}}{6}$ or $x = 4 + \frac{\sqrt{7}}{6}; x = 4 - \frac{\sqrt{7}}{6}$ **74.** 4   **75.** 4   **76.** 4
77. $x = \frac{-k + \sqrt{k^2 - 4jm}}{2j}; x = \frac{-k - \sqrt{k^2 - 4jm}}{2j}$
78. $x = \frac{3}{2}; x = -3$ **79.** $x = 0; x = 16$
80. $y = -(x + 3)^2 + 2$

81. 2 sec

Chapter R

Opener R: 2: left © Thinkstock/Getty Images RF; **Opener R: 2:** right © Royalty Free/Corbis; **8:** left © Brand X Pictures/Getty Images RF; **8:** right © Chris Speedie/Getty Images; **9:** Courtesy Ignacio Bello; **12:** Courtesy Ignacio Bello; **20:** © McGraw-Hill Higher Education, Dr. Thomas D. Porter, photographer.

Chapter 1

Opener 35: Public Domain; **60:** © Royalty Free/Corbis; **66:** © The McGraw Hill Companies, Inc., Dr. Thomas Porter, photographer.

Chapter 2

Opener 105: © The British Museum; **106:** © Royalty Free/Corbis; **132:** © Jose Carrilo/Photo Edit; **143:** © Mary Kate Denny/PhotoEdit; **172:** © Fred Bellet/Tampa Tribune.

Chapter 3

Opener 203: © Stapleton Collection/Corbis; **267:** © Eunice Harris/Index Stock.

Chapter 4

Opener 307: © Bibliotheque national de France; **315:** © PhotoDisc Vol. OS23/Getty Images RF; **317:** top Courtesy Ignacio Bello; **317:** middle Courtesy Ignacio Bello; **317:** lower Courtesy Ignacio Bello; **320:** © Doug Struthers/Stone/Getty Images; **326:** © Steve Gschmeissner/SPL/Photo Researchers, Inc.; **329:** Courtesy Ignacio Bello; **333:** NASA/JPL-Caltech/R. Hurt (SSC/Caltech); **335:** top Neptune: NASA, L. Sromovsky, and P. Fry (University of Wisconsin-Madison); **335:** top Saturn: NASA, ESA, J. Clarke (Boston University), and Z. Levay (STScI); **335:** top Uranus: Credit: NASA, ESA, L. Sromovsky and P. Fry (University of Wisconsin), H. Hammel (Space Science Institute), and K. Rages (SETI Institute); **335:** lower Jupiter: STScI/NASA; **335:** lower Mercury: NASA; **335:** lower Venus: USGS MAGELLAN IMAGING RADAR; **335:** lower Earth: NASA; **335:** lower Mars: David Crisp and the WFPC2 Science Team (Jet Propulsion Laboratory/California Institute of Technology), and NASA; **337:** © Dallas and John Heaton/PictureQuest; **358:** Courtesy Ignacio Bello; **378:** © Royalty Free/Corbis.

Chapter 5

Opener 395: © Archivo Iconografico, S.A./Corbis; **414:** Courtesy Ignacio Bello; **425:** © Morton Beebe/Corbis; **432:** © Martin Rotker/Photo Researchers, Inc.; **439:** Courtesy Ignacio Bello.

Chapter 6

Opener 469: © The British Museum/Topham-HIP/The Image Works; **473:** Courtesy Ignacio Bello; **496:** © Tony Freeman/Photo Edit; **507:** Courtesy of Adler Planetarium and Astronomy Museum, Chicago, IL; **511:** © 2006 Calvin J. Hamilton; **512:** Courtesy Ignacio Bello; **519:** © Bloomimage/Corbis RF; **520:** Courtesy NASA/JSC; **533:** Courtesy Ignacio Bello.

Chapter 7

Opener 551: Courtesy National Library; **564:** © John A. Rizzo/PhotoDisc Vol. 30/Getty Images; **565:** left © William McKellar/Brand X/Corbis RF; **565:** right © Foodcollection/Getty Images RF; **578:** Courtesy Ignacio Bello; **586:** © Foodcollection/Getty Images RF; **586:** right © Foodcollection/Getty Images RF; **595:** Courtesy NASA.

Chapter 8

Opener 613: © Stock Montage/Hulton Archive/Getty Images; **614:** © AP Photo/Steve Miller; **618:** © Hal Whipple; **619:** top © Bloomimage/Corbis RF; **619:** bottom © AP Photo/Mike Derer; **621:** Courtesy Ignacio Bello; **626:** © PhotoDisc V245/Getty Images; **632:** © Jeremy Woodhouse/PhotoDisc V44/Getty Images; **633:** © Brian Bahr/Getty Images; **634:** Courtesy Ignacio Bello; **642:** © Tony Freeman/Photo Edit; **648:** © The McGraw-Hill Companies, Inc./Connie Mueller Photo.

Chapter 9

Opener 659: © Bettmann/Corbis; **660:** © Tony Freeman/Photo Edit; **668:** left © L. Clarke/Corbis RF; **668:** right © David R. Frazier/Photo Researchers, Inc.; **670:** © Tony Freeman/Photo Edit; **677:** Public Domain; **686:** © PhotoDisc/Getty Images RF; **687:** top © Digital Vision/Getty RF; **687:** bottom © Royalty Free/Corbis; **689:** Courtesy Ignacio Bello; **705:** left © Photolink/PhotoDisc/GettyRF; **705:** right Courtesy Ignacio Bello; **716:** Courtesy NASA Dryden Flight Research Center, Tony Landis, photographer.

▶ Index

⊙ Index